Zoology

For Degree Students

Molecular Biology | Principles of Genetics

About the Author

Dr V K Agarwal, MSc, PhD, has rich teaching experience of more than four decades at Meerut College, Meerut. He has taught subjects such as Cell Biology, Genetics, Molecular Biology, Chordate Embryology (Developmental Biology), Invertebrate Zoology, Chordate Zoology, Evolution, Ecology, Animal Behaviour (Ethology) to undergraduate and postgraduate students.

Dr Agarwal has to his credit the authorship of several widely acclaimed books such as *Cell Biology*, *Genetics*, *Chordate Embryology* (*Developmental Biology*), *Ecology* (*Environmental Biology*), *Ethology* (*Animal Behaviour*), *Zoology for Degree Students–BSc First Year* (includes *Invertebrate Zoology*), *Zoology for Degree Students–Semester I, II* and *III*, *Molecular Biology, Genetic Engineering,* and a multi-coloured edition of *Cell Biology, Genetics, Molecular Biology, Evolution and Ecology.* The author has also co-authored *CBSE – IX Biology.* He has too published few research papers on *Insect Physiology* and *Toxicology*.

Zoology

For Degree Students

Molecular Biology | Principles of Genetics

B.Sc. (Honours) Semester - V

As per UGC Choice Based Credit System (CBCS)

V K Agarwal
MSc, PhD
Former Head of the Zoology Department
Meerut College, Meerut, Uttar Pradesh

S. CHAND
PUBLISHING
S Chand And Company Limited
(ISO 9001 Certified Company)

S Chand And Company Limited

(ISO 9001 Certified Company)

Head Office: Block B-1, House No. D-1, Ground Floor, Mohan Co-operative Industrial Estate, New Delhi – 110 044 | Phone: 011-66672000

Registered Office: A-27, 2nd Floor, Mohan Co-operative Industrial Estate, New Delhi – 110 044 Phone: 011-49731800

www.**schandpublishing.com**; e-mail: **info@schandpublishing.com**

Branches

Chennai : Ph: 23632120; chennai@schandpublishing.com

Guwahati : Ph: 2738811, 2735640; guwahati@schandpublishing.com

Hyderabad : Ph: 40186018; hyderabad@schandpublishing.com

Jalandhar : Ph: 4645630; jalandhar@schandpublishing.com

Kolkata : Ph: 23357458, 23353914; kolkata@schandpublishing.com

Lucknow : Ph: 4003633; lucknow@schandpublishing.com

Mumbai : Ph: 25000297; mumbai@schandpublishing.com

Patna : Ph: 2260011; patna@schandpublishing.com

First Edition 2018

Reprint 2020 (Twice), 2021

Reprint 2022

ISBN: 978-93-525-3525-5 **Product Code:** H6ZOO68ZOOL10ENAA18O

PRINTED IN INDIA

By Vikas Publishing House Private Limited, Plot 20/4, Site-IV, Industrial Area Sahibabad, Ghaziabad – 201 010 and Published by S Chand And Company Limited, A-27, 2nd Floor, Mohan Co-operative Industrial Estate, New Delhi – 110 044.

Preface

It gives me immense pleasure and satisfaction to present this textbook, **Zoology for Degree Students - Semester V**, covering subjects **Molecular Biology** and **Principles of Genetics.** This comprehensively written textbook provides up-to-date information on the subject in an informative and systematic fashion.

This book has been designed as per the UGC Choice Based Credit System (CBCS) syllabus of Zoology, B.Sc. (Honours) - Semester V. It includes the following two core courses (CC):

1. **CC XI Molecular Biology**
2. **CC XII Principles of Genetics**

This book of molecular biology and genetics has been written for University students who have had prior exposure to the subjects of biology and biochemistry.

The molecular biology is perhaps the fastest progressing of all our intellectual endeavours (Malacisky 2005). Initially, most efforts in the early days of molecular biology were devoted to characterizing the physical and chemical properties of biological macromolecules such as DNA, RNA, and proteins. Once those parameters were generally understood, attention was given to elucidating the manner in which those macromolecules function in the living cell. Efforts of molecular biologists have contributed to an ever increasing understanding of the origin of human species; to the creation of new, genetically engineered drugs; and to the sequencing of the entire genome of a variety of species, including disease-causing (pathogenic) microorganisms, agriculturally important plants and of course, the human organism itself.

The scope of molecular biology is now so broad as to have an impact on virtually all areas of biology, from more traditional disciplines of taxonomy and systematic, to the more contemporary courses of study such as gene function. Currently a broad extent of molecular concepts have been integrated into most biology disciplines such as cell biology, developmental biology, genetics and even evolutionary biology.

The science of genetics includes the rules of inheritance in cells, individuals and populations, and molecular mechanisms by which genes control the growth, development and appearance of an organism. No area of biology can be truly appreciated or understood without the knowledge of genetics because genes not only control cellular processes, they also determine the course of evolution. Genetics is an exciting basic science, concepts of which, provide the framework for the study of modern biology (R.H. Tamarin, 1996).

To maintain the balance of text of genetics, equal importance has been given as far as possible, to discoveries of the Mendelian era, genetic mechanisms in microorganisms and recent advances in gene structure and function.

Some effort has been made to provide the earliest as well as the latest information in the field. Each chapter first introduces the reader to elementary principles. After the basic knowledge, advanced information is given in increasing depth so that the reader should get an insight into the recent developments in the field.

While rapid advances are being made, and new information is flowing in, we face some challenges. We are still groping in the dark for regulators of gene expression in eukaryotes. What are the controlling elements that turn genes on and off so that exact quantities of required proteins are synthesized at any moment is not known. Factors causing cancers in humans are still quite vague. However, in Chapter 18 of Molecular Biology (*i.e.,* Regulatory RNAs), it is described how small synthetic siRNAs (= small interfering RNAs) may prove to be an important new tool in the treatment of cancer and viral diseases, including AIDS.

Some of the special features of this textbook are:

1. It includes molecular biology having current topics such as: regulatory RNAs, DNA repair mechanisms and Transcriptional regulation of eukaryotic genes.
2. Genetics includes current topics such as, microbial genetics, plasmids, IS elements, transposons and Retroelements, Phenotypic plasticity, and Genomic imprinting due to DNA methylation.

Degree students will find this book extremely useful as it lucidly explains the concepts with the help of well-illustrated and adequately labelled diagrams.

To friends and family, I remain grateful for timely support during development of this textbook. My special thanks to Usha, Surabhi, Shubhra and Anubha, and also to my F3 generation of Praful, Muskan and Khushi. I also thank Trilok Jain who has illustrated the figures of this textbook.

Kudos to the Editorial Team of S Chand, especially, Surabhi Khare, Shalini Negi, Priyanka Sharan, Sanjeev Arora, Arun Kumar and Ketan Mithrani for timely release of the book. Their zeal to excel is indeed infectious.

Suggestions and feedback from the readers will be gratefully acknowledged.

V K Agarwal
Villa-20, Green Wood City
Godwin Colony
Haridwar Bypass Road
Meerut-250005
Uttar Pradesh (India)

Syllabus

BSc (Hons) Zoology – Semester V

CORE COURSE XI

MOLECULAR BIOLOGY

Unit 1: Nucleic Acids 4

Salient features of DNA and RNA
Watson and Crick model of DNA

Unit 2: DNA Replication 12

DNA Replication in prokaryotes and eukaryotes, Mechanism of DNA replication, Semi-conservative, bidirectional and semi-discontinuous replication, RNA priming, Replication of circular and linear *ds*-DNA, replication of telomeres

Unit 3: Transcription 10

RNA polymerase and transcription unit, Mechanism of transcription in prokayotes and eukaryotes, Synthesis of rRNA and mRNA, Transcription factors

Unit 4: Translation 12

Genetic code, Degeneracy of the genetic code and Wobble Hypothesis: Process of protein synthesis in prokaryotes, Ribosome strucutre and assembly in prokaryotes, Fidelity of protein synthesis, Aminoacyl tRNA synthetases and charging of tRNA; Proteins involved in initiation, Elongation and termination of polypeptide chain, Inhibitors of protein synthesis; Difference between prokaryotic and eukaryotic translation

Unit 5: Post Transcriptional Modifications and Processing of Eukaryotic RNA 6

Structure of globin mRNA; Split genes: Concept of introns and exons, Splicing mechanism, Alternative splicing, Exon shuffling, and RNA editing, Processing of tRNA

Unit 6: Gene Regualtion 10

Transcription regulation in prokaryotes: Principles of transcriptional regulation with examples from *lac* operon and *trp* operon; Transcription regulation in eukaryotes: Activators, Repressors, Enhancers, Silencer elements, Gene silencing, Genetic imprinting

Unit 7: DNA Repair Mechanims 3

Pyrimidine dimerization and mismatch repair

Unit 8: Regulatory RNAs 3

Ribo-switches, RNA interference, miRNA, siRNA

CORE COURSE XII

PRINCIPLES OF GENETICS

Unit 1: Mendelian Genetics and its Extension **8**

Principles of inheritance, Incomplete dominance and co-dominance, Multiple alleles, Lethal alleles, Epistasis, Pleiotropy, Sex-linked, Sex-influenced and sex-limited characters inhertiance

Unit 2: Linkage, Crossing Over and Chromosomal Mapping **12**

Linkage and crossing over, Cytological basis of crossing over, Molecular mechanisms of crossing over including models of recombination, Recombination frequency as a measure of linkage intensity, Two factor and three factor crosses, Interference and coincidence, Somatic cell hybridization

Unit 3: Mutations **10**

Types of gene mutations (classification), Types of chromosomal aberrations (Classification, figures and with one suitable example of each), Molecular basis of mutations in relation to UV light and chemical mutagens; Detection of mutations: CLB method, attached *X* method

Unit 4: Sex Determination **4**

Chromosomal mechanisms of sex determination in Dropsophila and Man

Unit 5: Extra-chromosomal Inhertiance **6**

Criteria for extra-chromosomal inheritance, Antibiotic resistance in *Chlamydomonas*, Mitochondrial mutations in *Saccharomyces*, Infective heredity in *Paramecium* and Maternal effects

Unit 6: Polygenic Inheritance **3**

Polygenic inheritance with suitable examples; simple numericals based on it

Unit 7: Recombination in Bacteria and Viruses **9**

Conjugation, Transformation, Transduction, Complementation test in Bacteriophage

Unit 8: Transposable Genetic Elements **8**

Transposons in bacteria, Ac-Ds elements in maize and P elements in *Drosophila*, Transposons in humans

Contents

MOLECULAR BIOLOGY (CC–XI)

PRINCIPLES OF GENETICS (CC–XII)

(CC–XI)
MOLECULAR BIOLOGY

1

Definition, History and Techniques of Molecular Biology

1.1 INTRODUCTION

The term molecular biology was first used in 1945 by **William Asbury** who was referring to the study of the chemical and physical structure of biological macromolecules. By that time, biochemists had discovered many fundamental intracellular chemical reactions. The importance of specific reactions and of protein structure in defining the numerous properties of cells was also appreciated. However, the development of molecular biology had to await the understanding that the most advantageous aproaches would be made by studying "simple" systems such as bacteria and bacteriophages which yield information about the basic biological processes more readily than animal cells. In fact, the faith in the basic uniformity of life processes was an important factor in rapid growth of molecular biology. That is, it was believed that fundamental biological principles that govern the activity of simple organisms, such as bacteria and viruses, must apply to more complex cells; only the details should vary. This faith has been amply justified by experimental results.

The roots of molecular biology were established in 1953 when an Englishman, **Francis Crick** and a young American, **James Watson** working at Medical Research Council Unit, Cavendish Laboratory, Cambridge, proposed a double helical model for the structure of DNA (deoxyribonucleic acid) molecule which was well known as the chemical bearer of genetic informations of certain microorganisms (bacteria, bacteriophages, etc.) due to pioneer discoveries made by **Grifith** (1928), **Avery**, **Macleod** and **McCarthy** (1944) and **Hershey** and **Chase** (1952). This discovery was followed by a thorough search of occurrence of DNA as the genetic material in other microorganisms, plants and animals and also by investigations of the molecular and atomic nature of different reactions of living cells. From all these studies has emerged the realization that the basic chemical organization and the metabolic processes of all living things are remarkably similar despite their morphological diversity and that the physical and chemical principles governing living systems are similar to those governing non-living systems.

The present understanding of molecular biology is that in most organisms the phenotype or the body structure and function ultimately depend for their determination on the structural and functional (*i.e.,* enzymatic) proteins or polypeptides. The synthesis of polypeptides is specified, directed and regulated by self-duplicating genes which are borne within molecules of DNA which is the universally accepted chemical bearer of genetic informations of most living organisms except certain viruses in which this function is carried by RNA, another nucleic acid. The genetic informations for polypeptide synthesis are initially dictated by the disposition of nitrogen bases in DNA molecule and are copied down by the process of **transcription**. During transcription stage copies (that is, transcripts) of an individual gene or genes are synthesized. These copies are molecules of RNA that include such familiar classes as ribosomal RNA, messenger RNA and transfer RNA. The biochemical interplay of these RNA copies which leads to the synthesis of a polypeptide chain, is called **translation**, meaning, literally, that the genetic message encoded in a messenger RNA molecule is translated into the linear sequence of amino acids in a polypeptide. The polypeptide in its turn determines the phenotype of the organism.

1.2 HISTORICAL BACKGROUND

The molecular biology is a very young biological discipline and has a very short history. Certain notable accomplishments of molecular biologists can be summarized as follows:

1928 **F. Grifth** discovered the phenomenon of transformation in bacteria.

1934 **M. Schlesinger** demonstrated that the bacteriophages are composed of DNA and protein.

1941 **G.W. Beadle** and **E.L.Tatum** published their classical study on the biochemical genetics of *Neurospora*.

1944 **O.T. Avery, CM. MacLeod** and **M. McCarthy** recognized the DNA nature of transforming principle of pneumococcus bacteria. The fact suggested that it is DNA and not protein which is the hereditary chemical.

1948 **A. Boivin**, **R.Vendrely** and **C. Vendrely** showed that in the different cells of an organism the quantity of DNA for each haploid set of chromosome is constant.

1950 **E. Chargaff** demonstrated that in DNA the numbers of adenine and thymine groups are always equal and so are the numbers of guanine and cytosine groups.

1952 **A.D. Hershey** and **M. Chase** demonstrated that only the DNA of T_2 bacteriophage enters the host, the bacterium *Escherichia coli*, whereas the protein remains behind.

1953 **J. D. Watson** and **F.H.C. Crick** proposed a model for DNA comprising of two helically intertwined chains tied together by hydrogen bonds between the purines and pyrimidines.

1956 **A. Gierer** and **G. Schramm** demonstrated that RNA is the genetic material of tobacco mosaic virus (TMV).

1957 **H. Fraenkel-Conrat** and **B. Singer** separated RNA from the protein of TMV viruses, produced hybrid RNA viruses and confirmed the view that RNA is the genetic material of some viruses. **Mathew Meselson** and **Franklin W. Stahl** performed a density-gradient experiment (using heavy isotope of nitrogen, ^{15}N) in bacteria to confirm the Watson and Crick's **semiconservative** theory of DNA replication.

Arthur Kornberg isolated DNA polymerase I enzyme from living *E.coli* and used it for *in vitro* DNA replication.

1958 **G. Beadle** and **E.Tatum** received Nobel Prize for their contribution in biochemical genetics of fungus.

J. Laderberg got Nobel Prize for the discovery of bacterial recombination.

1959 **R.L. Sinsheimer** isolated single-stranded DNA from a small virus ϕ-X174 which attacks *Escherichia coli.*

S. Ochoa; **A. Kornberg** received Nobel Prize for artificial synthesis of nucleic acids.

1961 **M.W. Nirenberg** and **J.H. Matthaei** cracked the messenger RNA code.

F.H.C. Crick and his colleagues showed that the genetic language is made up of three-letter words (*i.e.*, triplet codons).

F. Jacob and **J. Monod** put forward the operon concept.

1962 **J. Watson** and **F. Crick**; **M. Wilkens** got Nobel Prize for the discovery of molecular nature of DNA.

1963 **J.P. Waller** reported that nearly one-half of all proteins in *E.coli* cells have the amino acid methionine in the N-terminal position.

Sydney Brenner selected the nematode *Caenorhabditis elegans* as model organism for genetics; ultrastructural and molecular analysis of animal development.

1964 **K.A. Marcker** and **F. Sanger** discovered a peculiar aminoacyl- tRNA in *E.coli*, called **N-formyl- methionyl - tRNA** and suggested that this molecule may play a role in the special mechanism of chain elongation.

R.W. Holley and his colleagues gave detailed structure of alanyl tRNA ($tRNA^{aia}$) from yeast. Holley died in 1993.

1965 **F.H.C. Crick** proposed the **wobble hypothesis** for anticodons of tRNA and explained how several codons meant for same amino acid are recognized by same tRNA.

H. Wallace and **M.L. Birnstiel** isolated ribosomal RNA genes in *Xenopus*.

F. Jacob., **A. Lwoff**, and **J. Monod** received Nobel Prize for the discovery of protein synthesis mechanism in virus.

1967 **J. Cairns** established that DNA polymerase I enzyme is a DNA repair enzyme of *E.coli* (*i.e.*, it is not involved in DNA replication).

1968 **R.W. Holley**; **H.G. Khorana** and **M.W. Nirenberg** got Nobel Prize for deciphering the genetic code.

1969 **A.D. Hershey**, **M. Delbruck** and **S.E. Luria** shared Nobel Prize in medicine for their contribution to replication and recombination in viruses (bacteriophages).

Britten and **Davidson** proposed the **gene-battery** model for regulation of protein synthesis in eukaryotes.

1970 **Howard Temin** and **David Baltimore** demonstrated the synthesis of DNA on RNA template of tumour viruses. Both were awarded Nobel Prize in 1975 for the discovery of an enzyme called **RNA directed DNA polymerase** (or **reverse transcriptase**) which is present in the core of virus particle (rous sarcoma virus).

Biotechnology emerged as a new discipline due to marriage of biological science with technology (see **Dubey**, 1995).

Knippers; **Moses** and **Richardson** isolated DNA-polymerase-II enzyme.

R.D. Kornberg and **Gefter** discoverd DNA polymerase III enzyme.

1972 **Mertz** and **Davis** in 1972 demonstrated that cohesive termini of cleaved DNA molecule could be covalently sealed with *E.coli* DNA ligase and were able to produce recombinant DNA molecules.

Cohen *et al.*, for the first time reported the cloning of DNA by using plasmid as vector.

R. Porter; **G.M. Edelman** received Nobel Prize (physiology and medicine) for the discovery of chemical structure of antibodies.

C.B. Anfinsen; **S. Moore** and **W.H. Stein** got Nobel Prize (chemistry) for the discovery of chemical structure and activity of the enzyme ribonuclease.

1973 **S.H. Kim** suggested three dimensional structure, *i.e.*, L-shaped model, of tRNA.

1975 **E.M. Southern** developed a method, called **Southern blotting technique** for analysing the related genes in a DNA restriction fragment.

D. Pribnow discovered **Pribnow box** or **minus ten sequence** in *E.coli* genome.

1977 **P.A. Sharp** and **R.J. Roberts** discovered split genes of adenovirus.

D.S. Hogness, **I.B. David** and **N. Davidson** studied split genes for 28S rRNA in *Drosophila.*

P. Chambon, **P. Leder** and **R.A. Flavell** studied split genes of B'globin, ovalbumin and tRNA.

Itakura *et al.,* first of all produced human insulin (humulin) by means of recombinant technology.

1978-79 **W. Gilbert** first of all used the terms **exon** and **intron** (for split genes).

1978 **Hinnen** *et al.,* first of all described the transformation of yeast (*Saccharomyces cervisae*) by the help of plasmid of *E.coli.*

1979 **H.G. Khorana** reported completion of the total synthesis of a biologically functional gene.

Alwine *et al.,* developed **northern blotting technique** in which mRNA bands are blot transferred from the gel onto chemically reactive paper.

Towbin *et al.,* developed the **western blotting technique** to find out the newly encoded protein by a transformed cell.

1980 **Fredrick Sanger** got the second Nobel Prize for discovering complete sequence of 5400 nucleotides of single stranded DNA of $\phi \times 174$ bacteriophage.

1982 **A. Klug** was awarded Nobel Prize in chemistry for providing three-dimensional structure of tRNAs.

Rubin and **Spradling** for the first time introduced *Drosophila* gene of xanthine dehydrogenase into a P-element (= parental element) which then was microinjected into embryo deficient for this gene.

R.D. Palmiter and **R.L. Brinter** produced transgenic mice by genetic engineering.

1983 **Marilyn Kozak** proposed the scanning hypothesis for initiation of translation by eukaryotic ribosomes.

1984 **Robert Tijan** identified a DNA-binding protein called **SP1** which is involved in eukaryotic gene regulation.

1985 **Karry Mullis** discovered **polymerase chain reaction** (**PCR**) which is widely exploited in gene cloning for genetic engineering. He made use of a thermostable enzyme (acts best on 72° C temperature), called **Taq DNA polymerase**, isolated from *Thermus aquaticus.*

1984,86 **Alec Jeffreys** discovered the technique of **DNA fingerprinting**.

1987 **S. Tonegawa** was awarded Nobel Prize for discovering the mode of rearrangements of DNA sequences of mammalian immunoglobulin genes to produce a large variety of antibodies.

Stanford and coworkers developed the **particle bombardment gun** which shooted foreign DNA into plant cells or tissues at a very high speed.

1988 **J.W. Black**, **G.B. Elion** and **G.H. Hitchings** were awarded Nobel Prize for formulating drugs such as 6- mercaptopurine and thioguanine, which lead to inhibition of DNA synthesis and of cell division. This proved effective in cancer chemotherapy. They also designed drugs for treating gout, malaria and viral infections such as herpes.

1989 **T. Cech** and **S. Altman** awarded Nobel Prize for showing enzymatic role of some RNA molecules, called **ribozymes**.

1991 **Dr. Lalji Singh** at CCMB, Hyderabad has developed a new technique of DNA fingerprinting by using **BKM-DNA** probe (BKM = banded krait minor satellite). He discovered this probe while he was working on sex determination in snake, the banded krait (*Bungarus fasciatus*) for his Ph.D. work.

1992 **Edwin G. Krebs** and **Edmond H. Fisher** were awarded Nobel Prize for the pioneering work on "reversible protein phosphorylation as a biological regulator mechanism." Phosphorylation of proteins is shown to affect transcription, translation, cell division and many other cellular processes.

Prof. Asis Dutta of JNU, New Delhi, was selected for the **Birla Award** for **Science and Technology** for cloning and characterization of two novel genes-gene for oxalate decarboxylase from *Lathyrus sativus* (in 1991) and gene for a **seed specific nutritionally balanced protein** from *Amaranthus* (in 1992).

1993 **M.J. Chamberlain** proposed the **inchworm model** for elongation of transcript of DNA template.

This year's Nobel Prize in chemistry was shared by **Kary Mullis** (for the discovery of PCR) with **Michael Smith** (for site directed mutagenesis).

1994 **Alfred Gilman** and **Martin Rodbell**, got Nobel Prize (M & P) for research on structure and function of GTP-binding (G) proteins.

1995 **Edward B. Lewis**, **Christiane Nusslein-Volhard** and **Eric Wieschaus** got Nobel Prize (M & P) for research in Genetic control of embryonic develpoment.

1996 **Rolf M. Zinkernagel** and **Peter C. Doherty** got Nobel Prize (M & P) for research on Recognition of virus-infected cells by the immune system.

1997 **Jens C. Skou**, **Paul Boyer** and **John Walker** got Nobel Prize (chemistry) for discovery of Na^+/K^+ -ATPase (*i.e.,* sodium-potassium pump) and for mechanism of ATP synthesis.

Stanley B. Prusiner got Nobel Prize (chemistry) for research on protein nature of prions.

1998 **Robert Furchgott**, **Louis Ignarro** and **Ferid Murad** got Nobel Prize (M & P) for recognizing NO as intercellular messenger.

1999 **Gunter Blobal** got Nobel Prize (M & P) for research on protein trafficking.

2000 **Arvid Carlsson**, **Paul Greengard** and **Eric Kandel** got Nobel Prize (M & P) for research on synaptic transmission and signal transduction.

2001 **Leland H. Hartwell**, **Tim Hunt** and **Paul Nurse** got Nobel Prize (M & P) on research in control of the cell cycle.

2002 **Sydney Brenner**, **John Sulston** and **H. Robert Horvitz** got Nobel Prize (M & P) for introduction of *Caenorhabditis elegans* as a model organism for research in molecular biology.

2003 **Peter Agre** got Nobel Prize (chemistry) for research in structure of membrane channels.

2004 **Richard Axel** and **Linda B. Buck** got Nobel Prize (M & P) for research on olfactory receptor.

Aaron Ciechanover, **Avram Hershko** and **Irwin Rose** got Nobel Prize (chemistry) for discovery of ubiquitin and proteasomes.

2006 **Andrew Z. Fire** and **Craig C. Mello** got Nobel Prize (M & P) for the discovery of RNA interference (RNAi).

Roger D. Kornberg got Nobel Prize (chemistry) for research on transcription in eukaryotes (*i.e.,* discovery of DNA polymerase III).

2007 **Mario R. Capecchi**, **Martin J. Evans** and **Oliver Smithies** got Nobel Prize (M & P) for research on development of techniques for knockout mice.

2008 **Francoise Barre-Sinoussi**, **Luc Montagnier Herald jur Hausen** got Nobel Prize (M & P) for the Discovery of HIV.

Martin Chalfie, **Osamu Shimomura** and **Roger Tsien** got Nobel Prize (chemistry) for discovery and development of GFP (= Green fluorescent protein).

2016 Nobel Prize (M & P) was awarded to professor **Yoshinori Ohsumi**, a Japanese cell biologist of Tokya Institute of Technology in Yokohama, Japan, for his discoveries on mechanisms of autophagy (self eating). Major breakthrough of Ohsumi's researches are: 1. Autophagy exists in yeast cells. 2. 15 genes are essential for autophagy and functional characterisation of proteins encoded by these genes.

2017 Nobel Prize (M & P) has been awarded to a trio of American scientists **Jeffery C. Hall**, **Michael Rosbash** and **Michael W. Young** for their discoveries of the molecular mechanisms controlling circadian rhythms (the biological clocks, Box 1.1). In their study, they explained how plants, animals and humans adapt their biological rhythms ot synchromize it with Earth's revolutions.

Box 1.1

1. During 1970s, a gene called **period** was identified which controls circadian rhythms in fruit flies.
2. In 1984, **Hall** and **Rosbash** isolated the **period** gene. They discovered that **PER**, the protein encoded by period gene accumulates during the night and is degraded during the day. Thus, PER protein levels oscillate over a 24-hour cycle, in synchrony with the circadian rhythm. They also hypothesized that PER protein could prevent its own synthesis by an inhibitory feedback loop, thereby regulating its own level in a continuous cyclic rhythm.
3. In 1994, **Young** discovered a second clock gene called **"timeless"**, encoding **TIM** protein that was required for normal circadian rhythm. He also observed that only when PER combines TIM, the two proteins were able to enter the cell's nucleus and block **period** gene activity to close the inhibitory feedback loop. Young also identified a gene called, **doubletime**, encoding the DBT protein that delays the accumulation of the PER protein. This provides insight into how an oscillation is adjusted to more closely match a 24-hour cycle.

1.3 PROLOGUE

Viruses are noncellular biological entities. Only two basic types of cells are found in the cellular organisms: **Prokaryotic cells** (in Eubacteria and Archae bacteria) and **Eukaryotic cells** (in organisms of four kingdoms – animals, plants, fungi, and protists – the unicellular algae and protozoa).

Eubacteria (*e.g., Escherichia coli*) are structurally defined as prokaryotes. These cells have a plasma membrane (a fluid-mosaic membrane of molecules of lipids and proteins), usually enclosed in a rigid cell wall, but no intracellular compartments. They have a single major circular chromosome. They may be unicellular or multicellular.

Archaea (*e.g., Methanococcus jannaschii*) are structurally defined as prokaryotes but probably branched off from the eukaryotes after their common ancestor diverged from the eubacteria (**Cooper** and **Hausman**, 2007). They tend to inhabit extreme environments. They are biochemically closer to eubacteria in some ways but to eukaryotes in others. For example, the 1740 kb genome of the archaeon *M. jannaschii* encodes a maximum of 1738 proteins. Comparisons revealed that those (proteins) involved in energy production and metabolism are most like those of eubacteria, while those involved in replication, transcription and translation are more similar to those of eukaryotes.

Cells of Eukaryota possess well-defined subcellular compartment bounded by lipid-protein membrane (*e.g.,* nuclei, mitochondria, chloroplasts, endoplasmic reticulum, Golgi apparatus, *etc.*). These organelles are sites of distinct biochemical processes and define the eukaryotes.

In most multicellular eukaryotes, groups of cells **differentiate** during development of the organism to provide specialized functions (*e.g.*, as in liver, brain, kidney). In most cases, they contain the same DNA but transcribe different genes. Like all other cellular processes, differentiation is controlled by genes. Coordination of the activities of different cell types requires communication between them. This involves **signaling molecules** such as neurotransmitters, hormones, and growth factors which are secreted by one tissue and act upon another through cell-specific receptors.

Molecular Biology seeks to explain the relationships between the structure and function of biological molecules and how these relationships contribute to the operation and control of biochemical processes. Of principal interest are the processes of replication, transcription and translation. The new experimental technologies involved in manipulating these molecules are central to modern molecular biology. The **central dogma** is the original proposal that 'DNA makes RNA and RNA makes protein', which happen via the processes of transcription and translation respectively (Fig. 1.1). This is broadly correct, although a number of examples are known which contradict part of it. Retroviruses reverse transcribe RNA into DNA, a number of viruses are able to replicate RNA directly into an RNA copy, and a number of organisms can edit a messenger RNA sequence so that the protein coding sequence is not directly specified by DNA sequence.

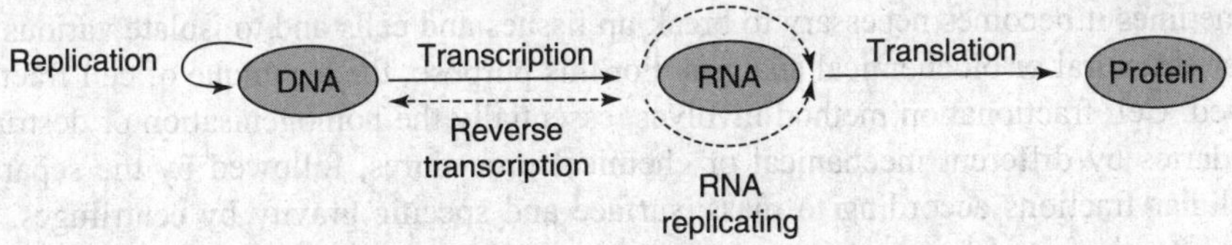

Fig. 1.1. Flow of genetic information.

In case of **prokaryotic gene expression** transcription of a single gene of an operon starts at the promoter, ends at the terminator and produces monocistronic or polycistronic messenger RNA. The coding regions of the message are translated by the ribosomes from the start codon (close to the ribosome binding site) to the stop codon. Transfer RNAs deliver the proper amino acid, according to the genetic code, to the growing protein chain.

In case of **eukaroyotic gene expression**, mostly monocistronic messenger RNA's are transcribed from a gene, initiated at a promoter. The resulting pre-messenger RNA is capped at the 5'-end and has a poly (A) tail added to the 3'end. Introns are removed from exons by splicing before the mature mRNA (formed by joining of exons) is exported from the nucleus to be translated by ribosomes in the cytoplasm.

Mutations are heritable permanent changes in the base sequence of DNA. Point mutations may be transitions (*e.g.*, G·C → A·T) or transversions (*e.g.*, G·C → T·A). Deletions and insertions involve the loss or addition of bases and can cause frame-shifts in reading the genetic code. Silent mutations have no phenotypic effects, while missense and nonsense mutations change the amino acid sequence effects, while missense and nonsense mutations change the amino acid sequence of the encoded protein.

DNA repair involves photoreactivation and excision repair (Chapter 17). In **photoreactivation**, cleavage of the cyclobutane ring of pyrimidine dimers by DNA photolyases restores the original DNA structure. Photolyases (the photoreactivating enzymes) have chromophores which absorb blue light to provide energy for the reaction. Photoreactivation is an example of direct reversal of a lesion of DNA and is error-free (**Turner** *et at.*, 2000). In **nucleotide excision repair**, an endonuclease makes nicks on either side of the lesion in DNA, which is then removed to leave a gap. This gap is filled by a DNA polymerase, and DNA ligase makes the final phosphodiester bond. In **base excision repair** the lesion is removed by a specific DNA glycosylase. The resulting **AP site** (AP = a purinic or a pyrimidinic) is cleaved and expanded by a gap by an AP endonuclease plus exonuclease. Thereafter the process is like nucleotide excision repair.

Cancer results from mutation that disrupt the controls regulating normal cell growth. The growth of normal cells in subject to many different types of control which may be lost independently of each other, during the multistep progression to malignancy.

1.4 MATERIAL AND METHODS IN MOLECULAR BIOLOGY

Different molecular biologists have made an intensive use of a variety of the microorganisms such as bacteriophages and other viruses; *Escherichia coli* and other bacteria; unicellular green algae, yeast, *Neurospora* and other fungi; protozoans, etc., in their investigations of varied nature (see Chapter 2). However, many significant discoveries of molecular biology have been made by working on larval salivary gland chromosomes of Dipteran insect (*Drosophila*) and lampbrush chromosomes of oocytes of different amphibians and other higher animals. To understand different biochemical events of prokaryotic and eukaryotic cells at molecular level, a wide array of bio-physico-chemical techniques such as electron microscopy, ultracentrifugation, colorimetry, spectrophotometry, chromatography, isotopic tracers, X-ray crystallography, electrophoresis, etc., are used in molecular biology.

1. Cell Fractionation and Centrifugation

Sometimes it becomes necessary to break up tissues and cells and to isolate various parts of the cell for structural or biochemical analysis. For this purpose, the technique of cell fractionation is employed. Cell fractionation method involves, essentially the homogenisation or destruction of cell boundaries by different mechanical or chemical procedures, followed by the separation of the subcellular fractions according to mass, surface and specific gravity by centrifuges.

The centrifuge is the most important tool available to cell biologists for the separation and analysis of subcellular organelles and their constituent macromolecules. Centrifugal force is generated by placing test tubes containing samples to be fractionated in a holder called a **rotor** and then spinning the rotor at high speeds by the help of a **motor** in a centrifuge. The ultra-centrifuge, developed by the Swedish chemist, **Theodor Svedberg** between 1920 to 1940, is a special high-speed instrument capable of spinning rotors at speeds of 100,000 revolutions per minute or more. The large forces generated by these high speeds make it possible to isolate not just organelles but macromolecules such as proteins and nucleic acids as well.

Stokes formula. The rate of movement of any particle subjected to a centrifugal force is given by **Stokes formula**:

$$\frac{dx}{dt} = \frac{2r^2(\rho_p - \rho_M)}{9\eta}.g$$

where dx/dt is the velocity with which the particle moves towards the bottom of the tube, r is the particle radius, ρ_p and ρ_M are the respective densities of the particle and the suspending medium and g is the centrifugal force exerted on the particle. This formula tells us that when the density of a particle exceeds that of the medium, its velocity will be positive and the particle will migrate in the direction of the force field (towards the bottom of tube). If the particle is less dense than the medium, its velocity will be negative, meaning that it will move towards the top of the tube. Finally, if the density of particle is exactly equal to the density of the suspending medium, the particle will not move because the term, $\rho_p - \rho_M = 0$.

Types of Centrifugation

(*a*) **Velocity centrifugation** (It separates organelles and molecules based on differences in size)

(*i*) **Differential centrifugation**

(*ii*) **Moving zone centrifugation**

(*b*) **Isodensity centrifugation** (It separates organelles and molecules on the basis of differences in density).

For these centrifugations, the following three centrifugation processes are employed:

1. **High speed centrifugation.** In high speed centrifugation rotor runs at a speed of about 20,000 rpm and generates a gravitational pull of about 50,000 g. At this speed most microorganisms, cell debris and cell organelles form sediments. It is used in **differential centrifugation**.
2. **Ultracentrifugation.** Ultracentrifuge (Fig. 1.2) of **Svedburg** operates at extremely high speed (about 75,000 rpm) and produces a gravitational pull of about 500,000 g. It is used for **differential centrifugation, zonal centrifugation, density gradient centrifugation** and **isopycnic centrifugation**.
3. **Analytical centrifugation.** This is the latest and most sophisticated instrument. Its sample container has a quartz window and the rotor has an optical system. It is possible to observe the material sedimenting and measure the sedimentation rate during the run.

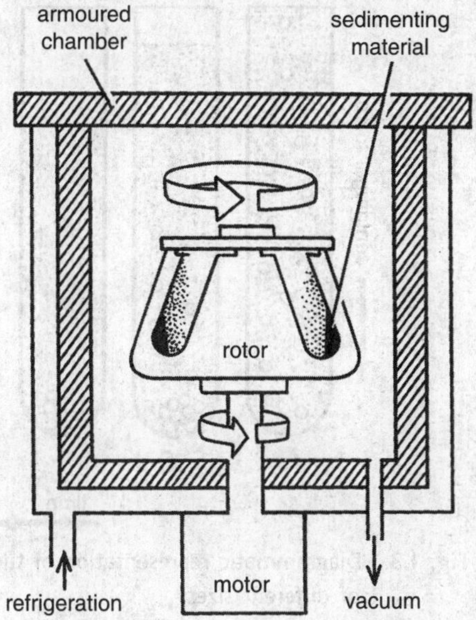

Fig. 1.2. The ultracentrifuge. The sample (*i.e.*, homogenate) is contained in tubes that are inserted into a ring of cylindrical holes in a metal rotor. Rapid rotation of the rotor generates immense centrifugal forces, which cause particles in sample to sediment. The vacuum tends to reduce friction, preventing heating of the rotor and allowing the refrigeration system to maintain the sample at 4°C.

Mechanism of Centrifugation (Cell Fractionation)

1. In the cell fractionation, the cells are gently broken by grinding a small piece of tissue in a homogeniser having a moving close-fitting glass or plastic pestle within a tube that contains a medium such as sucrose solution to preserve the cellular organelles (Fig. 1.4). The solution containing homogenised or disrupted cells, is called **homogenate**. In a homogenate the particles of different sizes are distributed homogeneously. The homogenate is subjected differential centrifugation of increasing velocity.
2. The tissue/liver homogenate is subjected first to low speed centrifugation at 700X g for 10 minutes. This sediments cell nuclei to the bottom of centrifuge tube (Fig 1.3). This **sediment-I** is called **nuclear fraction**. Nuclear fraction includes nucleus, intact cells, tissue debris and chloroplasts (in case of plant cells).
3. The overlying **supernatant-I** is transferred to another tube for second centrifugation at 10,000Xg for 20 minutes. The **sediment-II** is called **mitochondrial fraction**. It contains mitochondria and also lysosomes and peroxisomes.
4. Again the **supernatant-II** is removed and subjected to third centrifugation at (~ 100,000Xg) for 60 minutes. This causes the sedimentation of **microsomal fraction** which includes free ribosomes, fragments of smooth ER, rough ER, Golgi apparatus and plasma membrane.

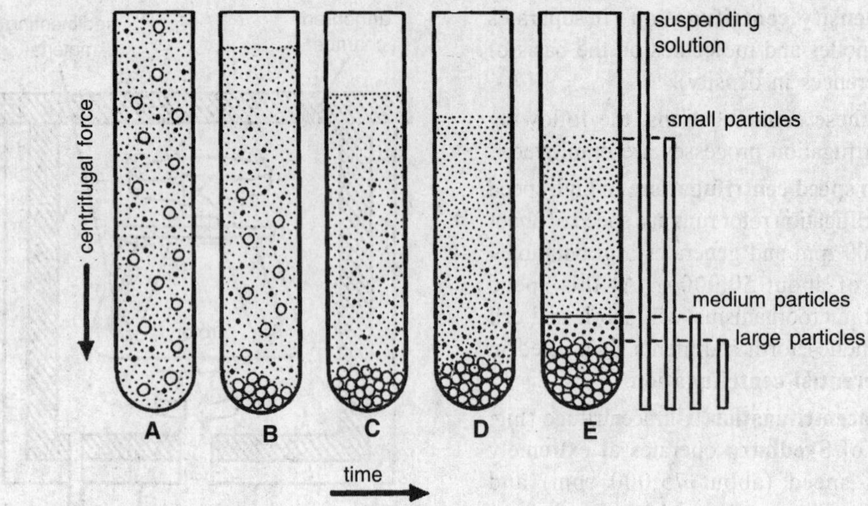

Fig. 1.3. Diagrammatic representation of the process of centrifugation and sedimentation of particles of different sizes.

5. The **final supernatant** contains soluble molecules of cytoplasm and is called **cytosol fractions**. It contains proteins, soluble nucleic acids, soluble polysaccharides, lipid droplets and other particles. These are separated by **chromatography**, **dialysis** and **electrophoresis**.

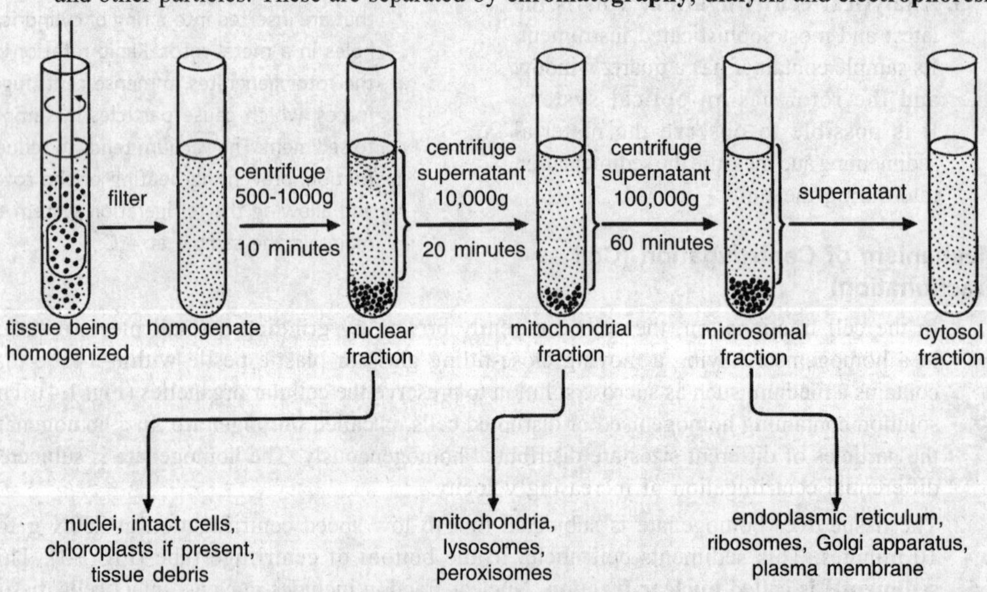

Fig. 1.4. Subcellular fractionation by differential centrifugation.

The technique of cell fractionation has been improved through the use of **density gradient centrifugation**. In this method, centrifuge tube is loaded with layers of solution of varying densities of either sucrose, heavy water, cesium chloride or albumin, in a gradient from top to bottom. Once the gradient is formed, the homogenate is layered on the top and centrifuged until the particles reach equilibrium with the gradient. For this reason, this type of separation is called **equilibrium density** or **isopycnic centrifugation**.

Marker enzymes. The **purity** (contamination) of the subcellular fraction can be checked by the use of **marker enzymes**. DNA polymerase is the marker enzyme for nucleus, while glutamate dehydrogenase and glucose 6-phosphatase are the markers for mitochondria and ribosomes, respectively. Hexokinase is the marker enzyme for cytosol.

2. Autoradiography

Autoradiography is a technique which is used to locate radioactive isotopes in cells, tissues, organs and whole organisms. A specimen is exposed to a solution containing molecules that have been made radioactive by the incorporation of radioactive isotopes, such as tritium (^3H), carbon 14 (^{14}C), phosphorus (^{32}P) and sulphur (^{35}S). The tagged molecules are often precursor molecules used by the cell in the synthesis of other needed molecules. At intervals, samples are removed from the solution; in case of smaller tissues, the samples are sectioned and mounted on glass slides or grids. The sections are then coated with a photographic emulsion and stored in the dark for periods ranging up to several months.

When a radioactive atom emits a beta particle (*i.e.*, electron) the photographic emulsion is affected in a manner similar to the exposure of a photographic emulsion to light. Over a period of time sufficient radioactive emissions occur to affect the silver grains of the emulsion. Black spots will appear at those sites when the emulsion is developed. Such spots will mark sites in the tissues where the radioactive atoms have accumulated. These sites can be identified by examining the stained tissue sections under the light microscope.

In the technique of autoradiography, for the study of DNA metabolism of cell ^3H-thymidine is used; for RNA metabolism ^3H-uridine is used; for protein synthesis various tritiated (^3H-tagged) amino acids are used; and for polysaccharides and glycoproteins tritiated monosaccharides such as ^3H-mannose and ^3H-fucose are employed.

Pulse-labelling technique. This technique is used for those cases where biological molecules undergo considerable modifications after their synthesis (*e.g.*, ribosomal RNA). Here, actively growing cells are exposed to a radioactive precursor for a short period. The labelled precursor is then removed and replaced by 'cold' (unlabelled) precursor molecules. The unlabelled precursors are incorporated into the newly synthesized molecules and have the effect of 'chasing' the previously synthesized molecules containing the radioactive precursor through any maturation process. If the molecular species under investigation is sampled shortly after the start of the experiment, only the primary synthetic product will contain radioactivity. After longer time intervals, the original radioactive molecules will have been replaced with non-radioactive molecules. In this way the flow of radioactivity through a maturation process can be followed, together with any movement of molecular species within the cell.

3. Chromatography

The separation of molecules from biological materials is an important part of biochemical work. Various methods are used for separating molecules. **Chromatography** is a method of separation which is utilised for separation and purification of both organic and inorganic molecules. The word 'chromatography' has its origin in Greek words '*chromo*' meaning colour and '*graphy*' meaning to measure. The Russian botatinst **Mikhail Tswett** is credited with the original development of this technique. In 1903, he reported the successful separation of a mixture of plant pigments using a column of calcium carbonate (Fig. 1.5). The early methods of isolation and purification of compounds of mixtures were empirical, slow and laborious. But with the advancements in separation procedures over the years, chromatography has become a highly efficient technique for separation and purification of compounds.

Chromatography is based on the phenomenon of adsorption and differential migration of components of a mixture or a compound. This needs the following two basic things:

1. A force for propelling the molecules of any solution, mixture or cytosol. It is called **mobile phase** and it involves moving solvent.

2. A selective impedance causing differential movement of various molecules (*i.e.*, called **impeding medium** or **adsorbent**). It is called **immobile phase** and it involves the matrix (filter paper, $CaCO_3$, etc.) through which the solvent is moving.

During chromatography the mixture or solution is allowed to percolate in an insoluble medium or adsorbent or matrix which has different affinity for molecules of different substances. The molecules migrate through the medium (matrix) at different rates and are separated. For example, if a solution of leaf pigment, chlorophyll in petroleum ether is allowed to percolate through a column of powdered calcium carbonate, the pigments are initially absorbed by pure solvent (*i.e.*, petroleum ether), the individual pigments—carotene, chlorophyll and xanthophyll separate and appear as the coloured bands or zones on the calcium carbonate column like the light rays in spectrum. Such a preparation is called **chromatogram** (Fig. 1.5 B).

The impeding medium (or immobile phase) is called **adsorbent**. The liquid that passes through adsorbent is **filterate** and the solution obtained after eluting adsorbed substances is **eluate**. The substances that are separated are **samples** or **solutes** and preparation having these solutes is called chromatogram.

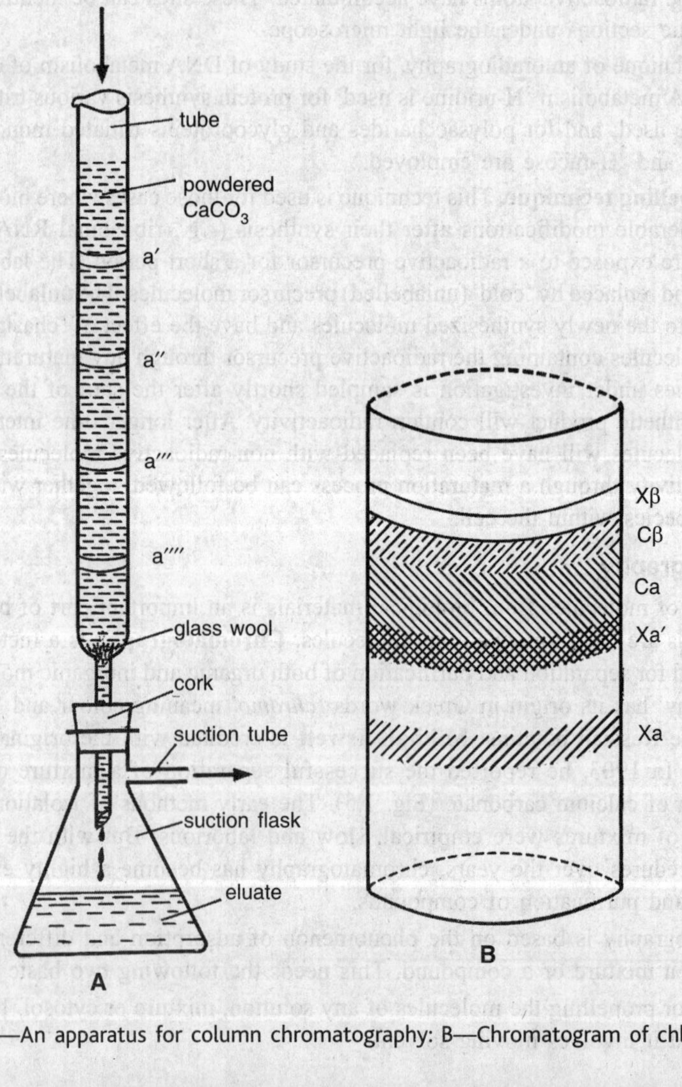

Fig. 1.5. A—An apparatus for column chromatography: B—Chromatogram of chlorophyll.

Principle of Chromatography

Chromatography is based on **partition** or **distribution coefficient (Kd)**. For two such immiscible (*i.e.*, liquids which are not able to be mixed together) phases A and B, the value for this coefficient is constant at a given temperature and is given as:

$$\text{Kd} = \frac{\text{Concentration in phase A}}{\text{Concentration in phase B}}$$

The term **effective distribution coefficient** is defined as the total amount of substance in one phase divided by the total amount present in the other phase.

$$\text{Effective distribution coefficient} = \frac{\text{Total amount in phase A}}{\text{Total amount in phase B}}$$

Basically all chromatographic systems consist of the **stationary phase** which may be a solid, gel, liquid or solid/liquid mixture that is immobilised, and the **mobile phase** which may be liquid or gas (Table 1.1). The mobile phase flows over or through the stationary phase. The choice of stationary and mobile phase is made that the compounds to be separated have different distribution coefficient.

Types of Chromatography

Chromatography can be classified into various types depending upon the type of solid support, stationary phase and the mobile phase (Fig. 1.6). Depending upon the physical state of the stationary phase, whether the stationary phase is solid or liquid, chromatography can be divided into two types:

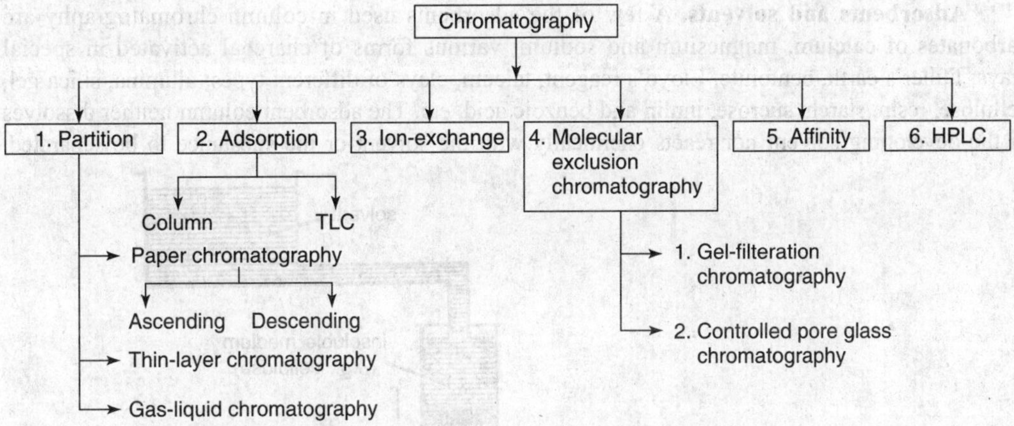

Fig. 1.6. Important types of chromatography. HPLC–High performance liquid chromatography; TLC–Thin layer chromatography.

1. **Adsorption chromatography.** The chromatography in which stationary phase is solid, *e.g.*, thin-layer chromatography and ion exchange chromatography are adsorption chromatography.

Table 1.1 Phase of various types of chromatographic techniques.

S. No.	*Technique*	*Stationary phase*	*Mobile phase*
1.	Adsorption chromatography	Solid	Liquid
	(*i*) Thin-layer chromatography	Solid	Liquid
	(*ii*) Ion-exchange chromatography	Solid	Liquid
	(*iii*) Affinity chromatography	Solid	Liquid

2.	Partition chromatography	Liquid	Liquid
	(i) Paper chromatography	Liquid (supported by cellulose fibers)	Liquid
	(ii) Gas-liquid chromatography	Liquid or solid	Gas

2. Partition chromatography. When the stationary phase is liquid, it is called partition chromatography, *e.g.*, paper chromatography.

In an alternative method, chromatography is divided into following two main types:

I. Column Chromatography

In column chromatography, the stationary phase is attached to a suitable matrix and it is packed into a glass or metal column (Fig. 1.5; Table 1.2). The mobile phase is passed through the column either by gravity or by use of a pumping system. Adsorption, ion exchange, molecular exclusion, affinity and gas-liquid chromatographies are examples of column chromatography.

In column chromatography an insoluble medium is packed into a glass tube; the length and width of this tube influence the separation of the molecules. The molecules to be separated are applied to the top of the column and their migration is started by adding a solvent. The characteristic separation which results depends on the choice of solvent and carrier material. A positively charged carrier binds negatively charged molecules; other carriers contain pores which are penetrated by the smaller molecules, which are, therefore, slowed down. The solvent flows through the column, the *eluate* (the liquid emerging from the bottom of the column) is collected in many fractions (Fig. 1.7). The eluate fractions contain the separated classes of molecules. Column chromatography is important for the separations of mixtures of proteins, that is, for the isolation of enzymes such as cytochrome c or RNA polymerase.

Adsorbents and solvents. A few of the adsorbents used in column chromatography are carbonates of calcium, magnesium and sodium; various forms of charcoal activated in special ways; Fuller's earth, bentonite, Lloyd's reagent, talcum, clays of different types; alumina, silica gel, cellulose, resin, starch, sucrose, inulin and benzoic acid, etc. The adsorbent column neither dissolves in the developing solvent nor reacts chemically with the solvent or the substance to be separated.

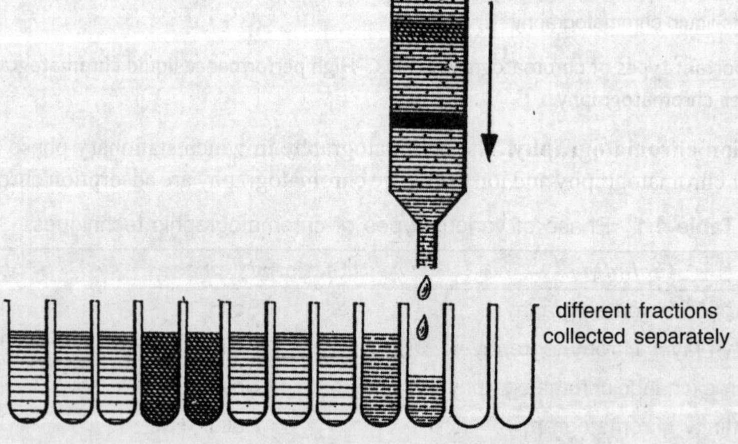

Fig. 1.7. Column chromatography.

The **solvents** commonly used are water, aqueous solutions of various acids, alkalis and salts; alcohols, ketones and cyclohexanone, hydrocarbons such as benzene, chloroform and carbon tetrachloride, etc.

Table 1.2 Differences between column and partition chromatography.

S. No.	*Column chromatography*	*S. No.*	*Partition chromatography*
1.	It was discovered by **Tswett**.	1.	It was discovered by **Martin Synge**.
2.	The adsorbent column may be of carbonates of calcium, magnesium or sodium; activated charcoal; Fuller's earth, resin, starch, etc.	2.	The adsorbent column may be of silica gel, cellulose or starch which are capable of holding water.
3.	Solvent used may be water, aqueous solution of inorganic acids or alkalies or salts, alcohol, ketones, hydrocarbons, benzene, chloroform or carbon tetrachloride.	3.	More than one solvents are used at a time. The substance to be separated are dissolved in alcohol and are then passed through the adsorbent column, already moistened with water.
4.	Column chromatography is based on the principle of differential rate of diffusion of the substances to be separated.	4.	Partition chromatography Is based on the principle of differential solubility of substances to be separated in two or more immiscible solvents.
5.	This technique is used to separate any type of substances from any solvent.	5.	This technique is used to separate water soluble compounds from a mixture of two or more immiscible solvents.

II. Thin Layer Chromatography or Planer Chromatography

In this case, the stationary phase, attached to a suitable matrix is coated thinly on to a glass, plastic or metal foil plate. The mobile liquid phase passes across the thin-layer plate by capillary action.

Paper chromatography is a planer chromatography in which the stationary phase is supported by cellulose fibers of a paper sheet. The mobile phase passes over by capillary action over the stationary phase.

A. Adsorption chromatography. In includes the following five types of chromatographies:

1. Adsorption chromatography. This is most commonly used to separate non-ionic and water-insoluble compounds such as triglycerides, vitamins and some drugs.

In this technique, column is packed with an adsorbent (Box 1.2). When the sample is passed through the column, the molecules with specific groups bind with adsorbent. The strength of binding of a particular anylate depends upon the functional groups present in its structure. Hydroxyl groups and aromatic groups tend to increase interaction depends upon the relative strength of the interaction of various molecules present in the eluent to the specific adsorption sites. As eluent is constantly passed down the column, differences in the binding strengths eventually lead to the separation of the analytes. The adsorption chromatography is influenced more by the presence of specific groups than by simple molecular size because only a specific group rather than the whole molecule can interact with the adsorption site.

Box 1.2

Certain solid materials have the ability to hold molecules at their surfaces. Such molecules are known as **adsorbents** and the process is known as **adsorption**. This involves weak, non-ionic attractive forces of van der Walls and hydrogen bonding type. The binding of the molecules occur at specific adsorption sites on the adsorption. Various materials such as silica, alumina and activated charcol are used as adsorbents.

2. Ion-exchange chromatography. In ion exchange chromatography molecules are separated by their difference in charge. The ion-exchange resin is used as chromatographic adsorbent which consists of porous polystyrene beads (polymers) to which ionizable groups are added chemically. Ion-exchange chromatography depends on the attraction between oppositely charged particles. The technique is used for the separation of many biological materials having ionizable groups such as amino acids and proteins.

Column is packed with an ion-exchanger which may be either cation or anion exchanger. **Cation exchangers** (*e.g.*, **DEAE-** or **diethylaminoethyl-cellulose**) possess negatively charged groups and attract positively charged anions. These are also known as **acidic ion-exchanger** as their negatively charges result from the ionization of acidic groups. **Anion-exchangers** (***e.g.*, CM or carboxy methyl cellulose**) have positively charged groups that attract negative charged cations. These are also called **basic-ion exchange materials** since positively charges generally result from the association of protons with basic groups.

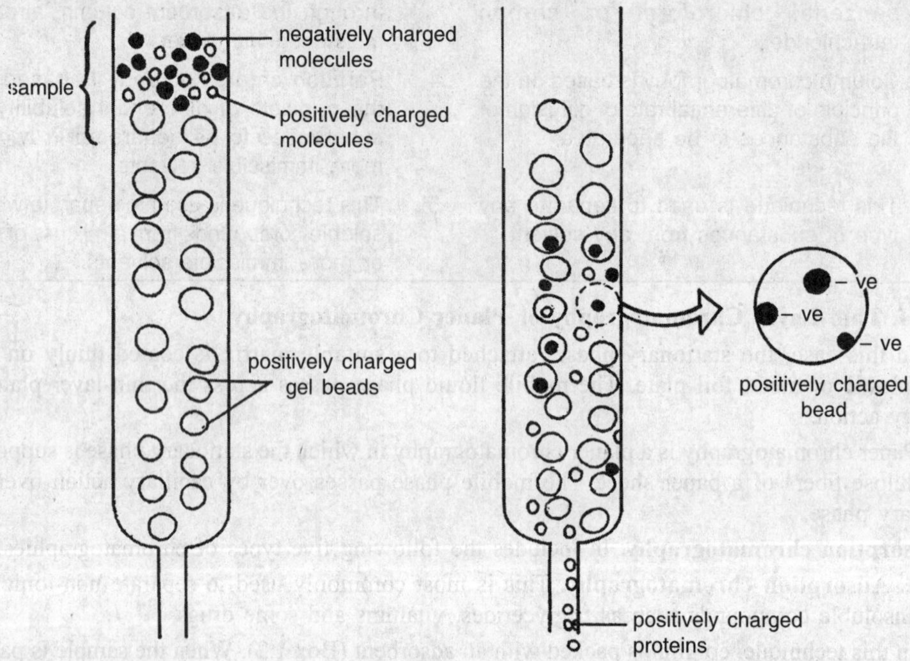

Fig. 1.8. Ion-exchange chromatography.

When a sample (a mixture of charged molecules such as proteins) is passed through the column carrying the ion-exchange material (*e.g.*, cation exchanger), the molecules with greater positive charges interact with the ion-exchange material and bind tighly to the molecules carrying lesser number of positive charges. Molecules with negative charge may just be washed out without any binding. Then, the bound molecules are eluted out one by one with an eluting buffer of a suitable pH. A number of ion-exchanger depends upon the stability of the sample components, their relative molecular mass and the specific requirements of the separation.

Amino acid analyzer (Fig. 1.9). A mixture of amino acids (protein hydrolysate) or proteins can be conveniently separated by ion-exchange chromatography. The amino acid mixture (at pH around 3) is passed through cation exchange and the individual amino acids can be eluted (*elute* means to remove the adsorbed material from an adsorbent by means of a solvent) by using buffers of different pH. The various fractions eluted, containing individual amino acids, are allowed to react with ninhydrin reagent to form coloured complex. This is continuously monitored for qualitative and quantitative identification of amino acids. The **amino acid analyzer**, first developed by **Moore** and **Stein**, is based on this principle (Fig. 1.9).

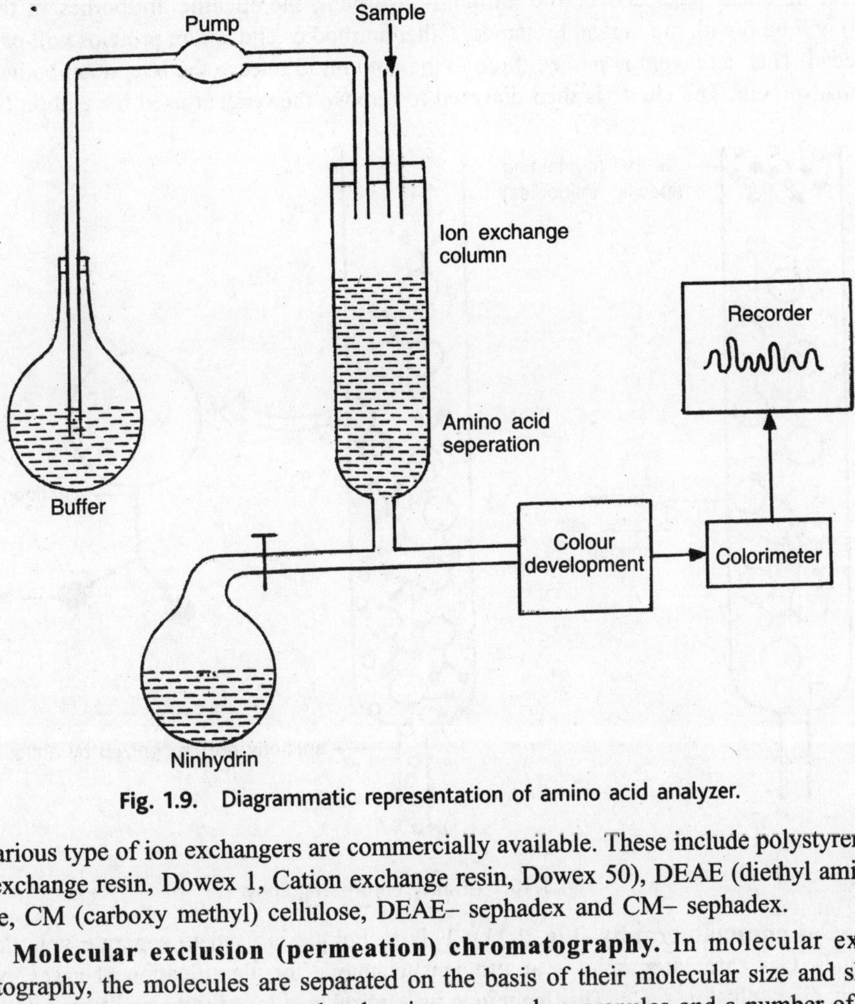

Fig. 1.9. Diagrammatic representation of amino acid analyzer.

Various type of ion exchangers are commercially available. These include polystyrene resins (anion exchange resin, Dowex 1, Cation exchange resin, Dowex 50), DEAE (diethyl aminoethyl) cellulose, CM (carboxy methyl) cellulose, DEAE– sephadex and CM– sephadex.

3. Molecular exclusion (permeation) chromatography. In molecular exclusion chromatography, the molecules are separated on the basis of their molecular size and shapes. A variety of porous materials act as molecular sieve, *e.g.*, glass granules and a number of the gels are used as molecular sieves. When the separation is carried out by using the gels as the molecular sieve, the technique is also called **gel filteration chromatography**. If porous glass granules are utilised as molecular sieve, the technique is called **controlled pore glass chromatography**. The term exclusion (permeation) chromatography describes all molecular separation processes using molecular sieves. When the mixture is poured over a column filled with gel particles or porous glass granules, large molecules in the mixture are completely excluded from the pores and pass through the interstitial spaces and appear in the eluate first. The smaller molecules are distributed between the mobile phase inside and outside the molecular sieve and then pass through the column at a slower rate, hence appear last in the eluate (Fig. 1.8).

4. Affinity chromatography. Affinity chromatography is the method to separate all the molecules of a particular specificity from a mixture. It is basically a type of adsorption chromatography in which the molecules to be purified are specifically and reversibly adsorbed by a complimentary binding substance or ligand immobilised on insoluble support matrix. The technique is used for the separation of specific antibodies from the blood serum. For the purification of antibodies an immunoadsorbent is prepared which consists of a solid matrix to which antigen 'ag' has been coupled. The matrix used must be stable during binding of macromolecules and its subsequent elution. Agarose, sephadex and some derivatives of cellulose can be used as matrix. When the serum is passed over the immuno-adsorbent, the specific antibodies in the mixture (serum) will bind with the 'ag' and retained. Other antibodies and serum proteins will pass through unimpeded. Then a reagent is passed through the column to release the bound antibodies from the immunoadsorbent. The eluate is then dialyzed to remove the reagent used for elution (Fig. 1.10).

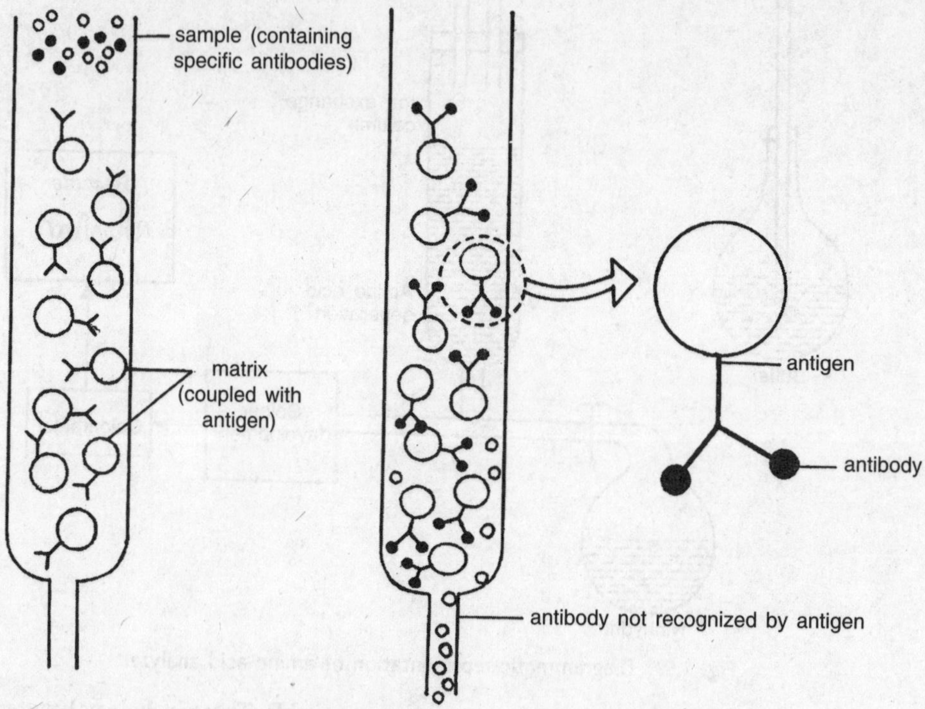

Fig. 1.10. Affinity chromatography.

5. Gas chromatography (Fig. 1.11). This technique is used to separate volatile organic compounds. Gas chromatography is an efficient technique for identification of gases, pollutants, drugs, vitamins, alkaloids, etc. This technique was introduced by **Martin** and **Synge** (1952). This type of chromatography involves injecting the volatile materials into a column contaning a liquid or solid stationary phase. A gas chromatograph consists of a flowing mobile phase, an injection port, a separation column containing the stationary phase and a detector. The stationary phase may be either a liquid or solid. In gas-liquid chromatography (GLC), the mobile phase is a gas and the stationary phase is a thin layer of non-volatile liquid bound to a solid support. In gas-solid chromatography a solid adsorbent acts as the stationary phase and an adsorption process occurs. The organic compounds are separated due to differences in their partitioning behaviour between the mobile gas phase and stationary phase in the column.

In gas chromatography, the sample is injected into the injection port from where it is carried into the column. The eluate released from the column pass through a detector which is linked via an amplifier to a chart recorder which records a peak as a particular compound passes through a detector.

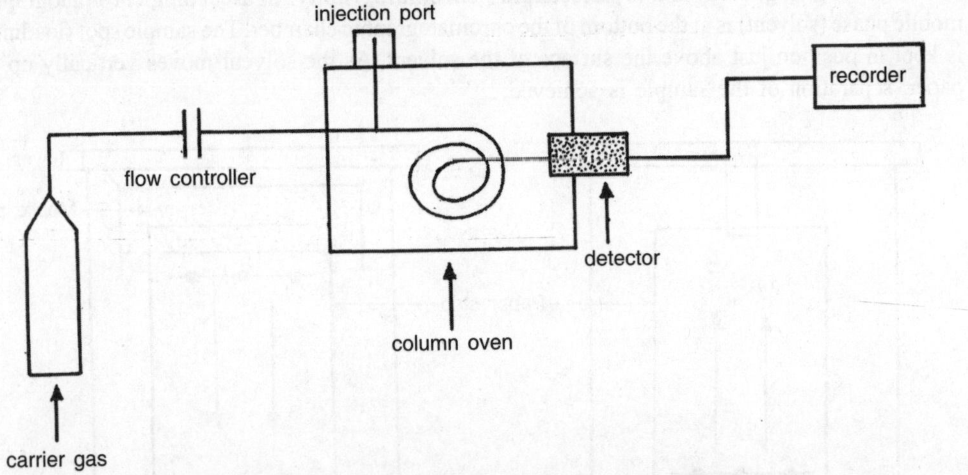

Fig. 1.11. Schematic representation of a gas chromatography.

B. Partition chromatography. It includes following two types of chromatographies:

1. Thin layer chromatography (TLC). This type of chromatography is extensively used in quantitative determination of high molecular weight compounds. Carbohydrates, dyes, pigments, vitamins, steroids, etc., can be separated and characterized by this method. In TLC, a thin layer of stationary phase is formed on a suitable flat surface such as glass, plastic or a metal foil. The movement of the mobile phase across the layer takes place rapidly by simple capillary action (Table 1.3). As the mobile phase moves across the layer from one edge to the opposite, it transfers analytes placed on the thin layer. The movement of the analyte is expressed by its **retention factor (Rf)** which is expressed as follows:

$$\text{Rf} = \frac{\text{Distance moved by the analyte from origin}}{\text{Distance moved by solvent from origin}}$$

For detection of the analyte, several methods are available. A general method is to spray the plate with 50% (v/v) sulphuric acid or 25% (v/v) sulphuric acid in ethanol and heating at 110°C. This results in most of the compounds showing up as brown spots.

Table 1.3 Some differences between TLC and GLC.

S. No.	*TLC or Thin-layer chromatography*	*S. No.*	*GLC or Gas-liquid chromatography*
1.	Adsorbent may be cellulose powder, alumina or cellulose.	1.	Adsorbent or the immobile phase is represented by liquid.
2.	The mobile phase is the solvent (liquid).	2.	The mobile phase is inert gas such as nitrogen, carbon dioxide, helium or argon.

2. Paper chromatography. Paper chromatography is a widely used technique in the separation of organic and biochemical products in laboratories. In this technique, stationary liquid phase is supported by cellulose fibers of a paper sheet. The mobile phase is developing solvent. The mobile phase passes over the stationary phase by capillary action. In paper chromatography the test solution is applied as a small spot on the chromatography paper (Whatman's No. 1 paper) and dried. The solvent is allowed to travel on the paper. As the solvent moves it carries the mixture components along with it. There are two types of paper chromatography:

(*a*) **Ascending** (Fig. 1.12) and (*b*) **descending chromatography**. In ascending chromatography, the mobile phase (solvent) is at the bottom of the chromatographic chamber. The sample spot (loading spot) is kept in position just above the surface of the solvent. As the solvent moves vertically up on the paper, separation of the sample is achieved.

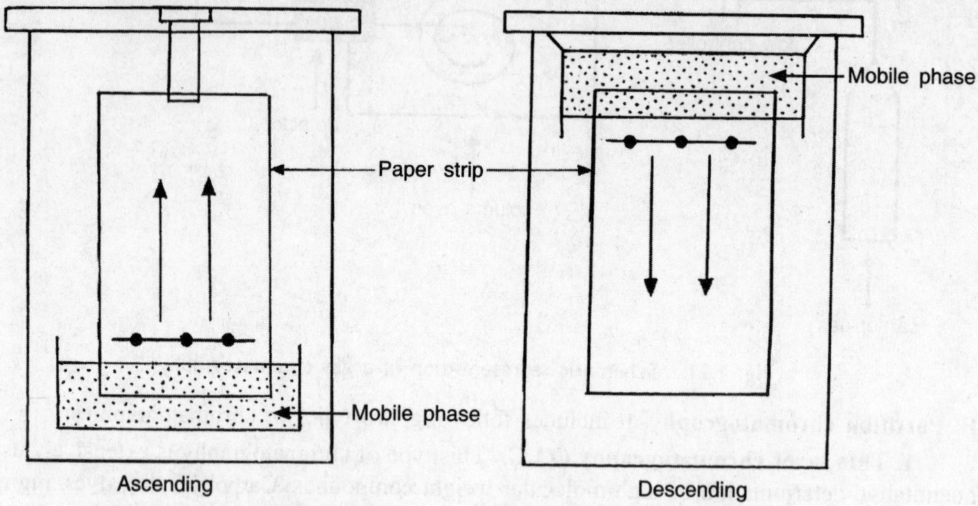

Fig. 1.12. Paper chromatography– ascending and descending types.

In descending chromatography, the mobile phase (solvent) is kept in the upper position of the chromatography chamber. The paper is inserted with the upper end with sample spot close to the mobile phase kept at the upper position of chromatographic chamber. Separation of the sample occurs as the solvent moves downwards due to capillary action and gravity.

A number of methods are available for the detection of eluate. Spraying with aqueous solution of 0.2% ninhydrin saturated with butyl alcohol is commonly employed for identification (Box 1.3). The identification of a given compound is made on the basis of its Rf value.

Box 1.3

Ninhydrin, which forms purple complex with α-amino acids, is frequently used as a colouring reagent.

HPLC or High Performance Liquid Chromatography. In general, the chromatographic techniques are slow and time consuming. The separation can be greatly improved by applying high pressure in the range of 5,000-10,000 psi (pounds per squase inch), hence this technique is also referred to as high pressuse liquid chromatography. HPLC requires the use of non-compressible resin materials and strong metal columns. The eluants of the column are detected by methods such as UV absorption and fluorescence.

4. Electrophoresis

Electrophoresis is a biochemical technique used to separate charged particles from a mixture (Fig. 1.13). The term 'electrophoresis' describes the migration of a charged particle under the influence of an electric field. This is a simple, rapid, sensitive and versatile analytical tool used to study and purify the charged molecules such as amino acids, peptides, proteins, nucleotides and nucleic acids. These molecules exist in solution as electrically charged species either as cations (–) and anions (+). Under the influence of an electric current these particles will migrate either to the cathode or anode, depending on the nature of their net charge.

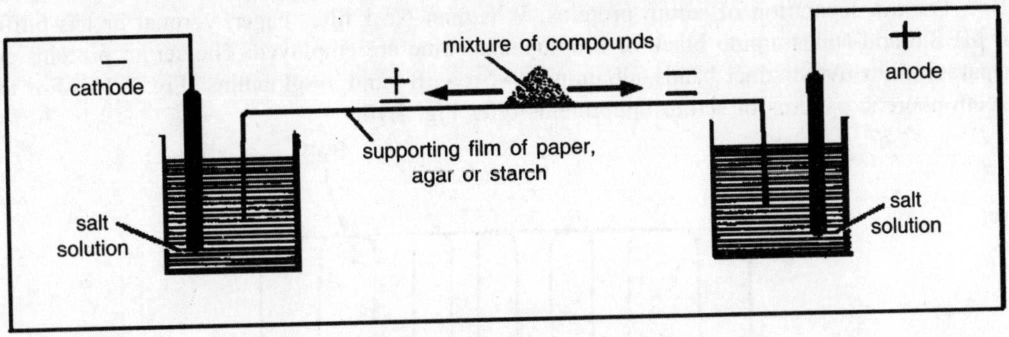

Fig. 1.13. Electrophoresis.

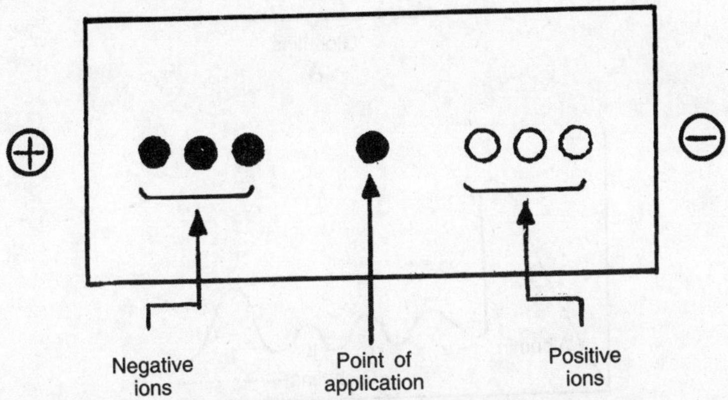

Fig. 1.14. Result of paper electrophosesis.

Electrophoresis is a technique, developed by **Tiselius** in 1937, for the analysis and separation of colloids and is based on the movement of charged colloidal particles in an electric field. **Arne Wilhelm Kaurin Tiselius** (1902–1971), Swedish Chemist awarded the 1948 Nobel Prize for Chemistry for his development in order to apply electrophoresis to the study of proteins, notably those of the blood.

Types of Electrophoresis. Among the electrophoretic techniques, zone electrophoresis (paper, gel), isoelectric focussing and immunoelectrophoresis are important and commonly employed in the laboratory. The original **moving boundary electrophoresis**, developed by **Tiselius**, is less freuently used these days. In this technique, the U-tube is filled with protein solution overlaid by a buffer solution. As the proteins move in solution during electrophoresis, they form boundaries which can be identified by refractive index.

1. Zone electrophoresis. This is a simple and modified method of moving boundary electrophoresis in which an inert supporting material such as paper or gel are used.

(*i*) Paper electrophoresis. Paper electrophoresis is most common type of electrophoresis which is run in clinical laboratories. This technique is used for separation of small molecules such as amino acids, peptides and carbohydrates. In this technique, the sample is applied on a strip of filter paper wetted with desired buffer solution. The ends of the strip are dipped into the buffer reservoirs in which electrodes are placed (Fig. 1.13). The electric current is applied allowing the molecules to migrate for significant time. The paper is removed, dried and stained with a dye that specifically colours the substances to be detected (Fig. 1.4). The coloured spots can be identified by comparing with a set of standards run simultaneously.

For the separation of serum proteins, Whatman No.1 filter paper, veronal or tris buffer at pH 8.6 and stains amido black or bromophenol blue are employed. The serum proteins are separated into five distinct bands–albumin, α_1–, α_2–, β– and γ– globulins (Fig. 1.15). For the electrophoretic patterns of serum lipoproteins refer Fig. 1.16.

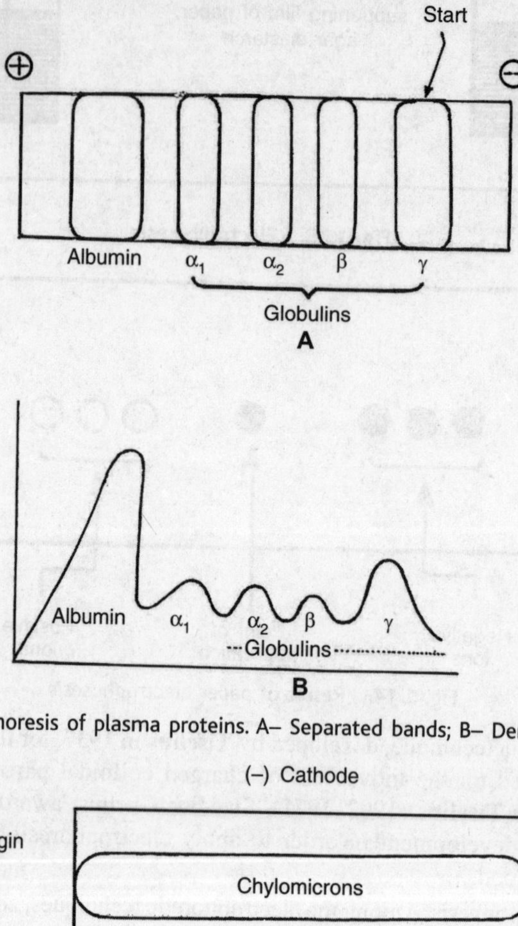

Fig. 1.15. Electrophoresis of plasma proteins. A– Separated bands; B– Densitometer scanning.

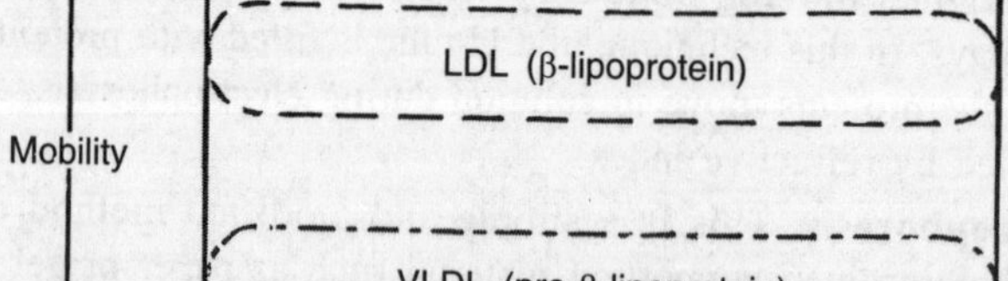

Fig. 1.16. Electrophoresis of plasma (serum) lipoproteins.

***(ii)* Gel electrophoresis.** The gel electrophoresis involves the separation of molecules based on their size, in addition to the electrical charge. The movement of large molecules is slow in gel electrophoresis. The resolution is much higher in this technique. Thus, serum proteins can be separated to about 15 bands, instead of 5 bands on paper electrophoresis.

The gels commonly used in gel electrophoresis are agarose and polyacrylamide, sodium dodecyl sulphate (SDS).

***(a)* Agarose gel electrophoresis.** Starch, agarose or polysaccharide gels are used for electrophoretic techniques. Agarose is a linear polysaccharide made up of the basic repeat unit agarobiose. Agarose gel is formed by boiling dry agarose in aqueous buffer until a clean solution forms. This is allowed to cool at room temperature to form a rigid gel. The pore size in the gel depends upon the initial concentration of agarose. If the concentration is low large pore sizes are formed. Agarose is usually used at concentration between 1% and 3%. Agarose gels are used for electrophoresis of both proteins and nucleic acids. Agarose gel electrophoresis run by placing the gel under the level of buffer is most commonly employed in all the nucleic acid research applications such as DNA fingerprinting, sequencing and recombinant DNA technology. Agarose gel is also used in techniques such as immuno-electrophoresis (used for antigens and antibodies).

***(b)* SDS-Page.** SDS-polyacrylamide gel electrophoresis is a widely used technique for the qualitative analysis of protein mixtures. The relative molecular mass of proteins can also be determined by this method. Sodium dodecyl sulphate (SDS) is an anionic detergent. In this method the protein mixture (sample) is first boiled in a buffer containing β-merceptoethanol and SDS. This denatures the proteins and SDS binds with the proteins. The sample buffer also contains an ionizable tracking dye which allows the electrophoretic run to be monitored. It also contains sucrose and glycerol. These allow the sample to settle easily through the electrophoresis buffer to the bottom when injected into the loading wells. Once the samples are all loaded, a current is passed through the gel. Normally 15% polyacrylamide gel is used as the separating gel for separating proteins in the range of molecular mass of 100,000 to 10,000.

Gel Electrophoresis Apparatuses. The electrophoresis equipment consists of two units—a power pack and an electrophoresis unit. Electrophoresis units are of two types—**vertical** (Fig. 1.17) and **horizontal** (Fig. 1.18). Vertical slab gel units are commercially available and routinely used to separate proteins in acrylamide gel. The gel is formed between two glass plates that are clamped together.

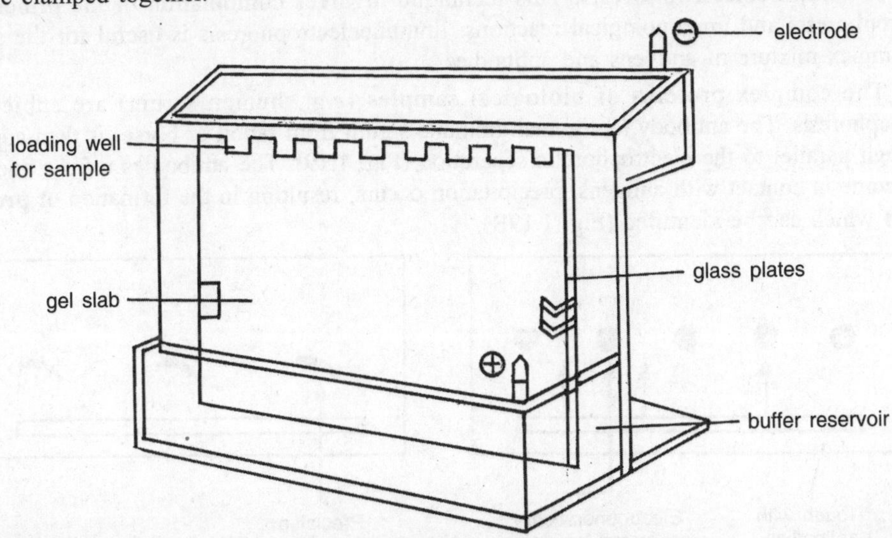

Fig. 1.17. A vertical gel electrophoresis apparatus.

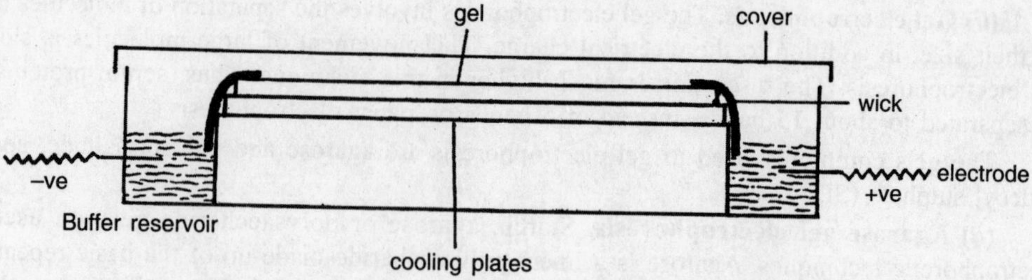

Fig. 1.18. A typical horizontal electrophoresis apparatus.

A **plastic comb** is placed in the gel solution and is removed after polymerization to provide **loading wells** for samples. When the apparatus is assembled, the electrophoresis tank buffer surrounds the gel plates and gives some cooling of the gel plates. In horizontal electrophoresis appartus, the gel is cast on a glass or plastic sheet and placed on a cooling plate. The power pack supplies a direct current between the electrodes in electrophoresis unit. All electrophoresis is carried out in an appropriate buffer. The buffer is essential to maintain a constant state of ionization of the molecules being separated. The sample moves under the influence of current. At the end of run, the position of the molecules on the gel is fixed with a fixative to prevent simple diffusion. Then the separated components are stained to visualize them. The **bands** can be quantified (by elution or by scanning with densitometer) as the uptake of the dye is directly proportional to the concentration of the molecules in each band.

The rate of electrophoretic migration (or 'mobility') depends on the pH of the medium, strength of electric field, magnitude of the net charge on the molecule, and the size and shape of the molecule.

2. Isoelectric focussing. The technique is primarily based on the **immobilization of the molecules of isoelectric pH** during electrophoresis. Stable pH gradients are set up (usually in a gel) covering the pH range to include the isoelectric points of the components in a mixture. As the electrophoresis occurs, the moleculer (*e.g.,* proteins) migrate to positions corresponding to their isoelectric points, get immobilized and form sharp stationary bonds. The gel block can be stained and identified. By isoelectric focussing, serum proteins can be separated to as many as 40 bands. Isoelectric focussing can be conveniently used for the purification of proteins.

3. Immunoelectrophoresis. This technique involves combination of the principles of electrophoresis and immunological reactions. Immunoelectrophoresis is useful for the analysis of complex mixture of antigens and antibodies.

The complex proteins of biological samples (*e.g.,* human serum) are subjected to electrophoresis. The antibody (antihuman immune serum from rabbit or horse) is then applied in a trough parallel to the electrophoretic separation (Fig. 1.19). The antibodies diffuse and, when they come in contact with antigens, precipitation occurs, resulting in the formation of **precipitin bands** which can be identified (Fig. 1.19B).

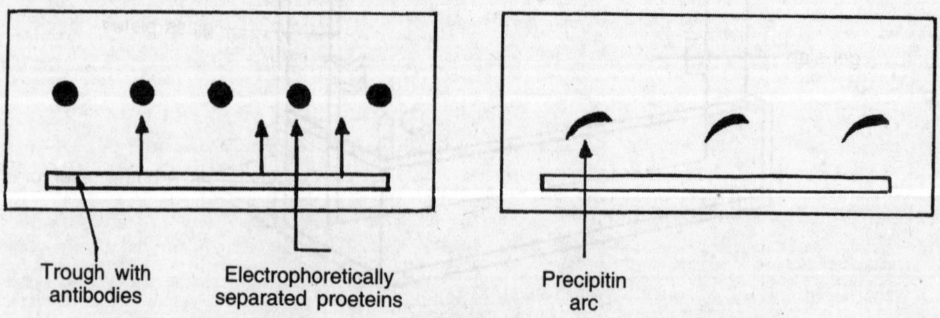

Fig. 1.19. Diagrammatic representation of immunoelectrophoresis.

5. Dialysis

It is a sensitive method for separating lower molecular-weight component from macromolecules. A thin membrane in the form of a tube is filled with the solution containing the molecules to be separated. The pore size of the membrane allows the diffusion of small molecules such as salt or amino acids; larger molecules such as proteins or nucleic acids cannot pass through the pores and so remain inside the dialysis tube.

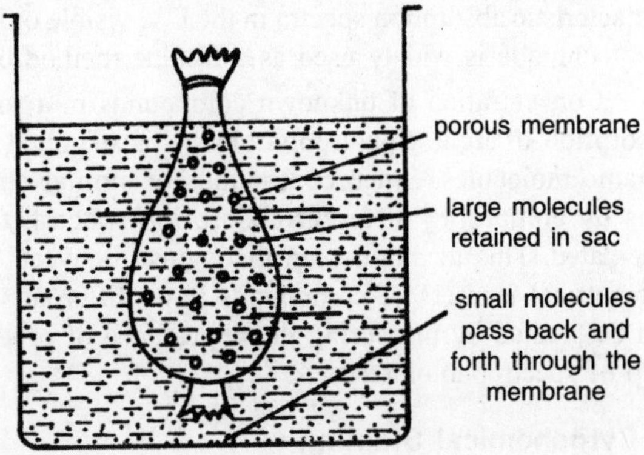

Fig. 1.20. Dialysis.

In dialysis, a protein solution is placed in a cylinder of cellophane tubing sealed at one end (Fig. 1.20). The tubing is then sealed at the other end and suspended in a large volume of buffer. The cellophane membrane is semipermeable: high molecular weight proteins are too large to pass through the pores of the membrane so proteins remain inside the tubing while low molecular weight solutes including, in this case, ammonium and sulphate ions diffuse out and are replaced by solutes in the buffer.

6. Spectrophotometry

Spectrophotometry is the measurement of the intensity of different colours (wavelengths) present in a beam of light, normally using a prism or diffraction grating and a photoelectrical cell. **Spectrophotometer** is an instrument for producing a spectrum and measuring the relation between absorption of electromagnetic radiation and frequency (or wavelength) of that radiation. It is a device used by researchers to determine how much radiation at various wavelengths a sample can absorb. The amount of absorption can be used to determine the concentration of particular molecules within a sample, because each type of molecule or macromolecule has its own characteristic wavelength(s) of absorption, called **absorption spectrum**. This instrument is used for colour comparison and in chemical analysis. Spectrophotometer is based on the principle of Beer–Lambert law.

A spectrophotometer typically has two light sources, which can emit ultraviolet or visible light. The light source is passed through a monochrometer, which emits light at a desired wavelength (Fig. 1.21). This incident light then strikes a sample contained within a cuvette. Some of the incident light will be absorbed, and some will not. The unabsorbed light passes through the sample and is detected by the spectrophotometer. The amount of light which strikes the detector is substracted from the amount of incident light, yielding the measure of absorption. In this way, the spectrophotometer provides an absorption reading for the sample. This reading can be used to calculate the concentration of particular molecules or macromolecules in a sample.

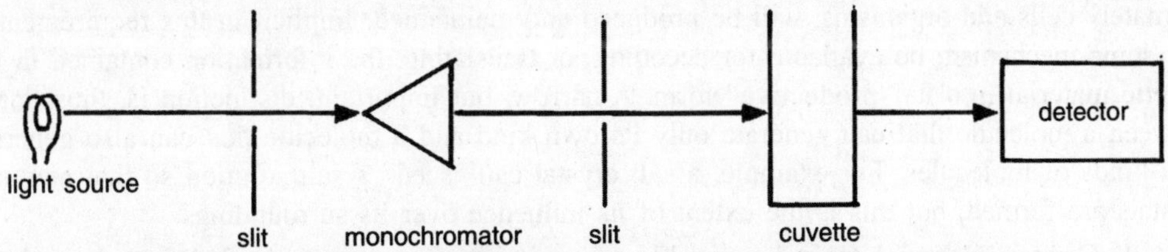

Fig. 1.21. Schematic representation of single beam spectrophotometer.

In detector, various types of photosensitive devices such as photovoltaic cells, phototubes or photoamultiplier tubes are used.

Applications of spectrometry. Spectrometry is one of most valuable analytical techniques of cytology. It has a wide variety of applications or uses. Compounds may be identified by their characteristic absorption spectra in the UV, visible or infrared regions of electromagnetic spectrum. The technique is widely used as a routine method of quantitative analysis of compounds.

Concentration of unknown compounds in a solution can be determined by measuring the absorption of light. Spectrophotometry is of much use in elucidating the structure of various organic molecules. Since compounds of similar structure have analogous absorption spectra, thus by comparing different absorption spectra, the structure of organic molecules can be elucidated. This technique can distinguish between 'cis' and 'trans' isomers of a complex. The geometrical isomers of compounds have different visible spectra. Enzyme catalyzed reactions can be studied by measuring the appearance of produces or disappearance of substrate with the help of spectrophotometry.

7. Cytochemical Staining

In addition, there are certain specific stains, called **cytochemical** stains that bind selectively to some specific groups of cellular macromolecules such as proteins, nucleic acids, polysaccharides and lipids. For example, Millon reaction, diazonium reaction and Napthol Yellow 5 stain are used for the proteins; alkaline fast green is used for histone (basic protein); Feulgen reaction (using Schiff's reagent) is used for DNA; methyl green-pyronine stain (Unna-Pappenheim stain) is used in distinguishing between DNA and RNA and it stains DNA green and RNA red; acetocarmine and acetoorcein stains are used to stain chromosomes of dividing cells; periodic acid-Schiff (PAS) reaction is used for the demonstration of polysaccharide materials such as starch, cellulose, hemicellulose, and pectin in the plant cells and mucoproteins (glycoproteins), hyaluronic acid and chitin in animal cells; and fat soluble dyes such as Sudan Red and Sudan Black B are used for the lipids. The Sudan Black B is a specific stain for phospholipids and is used to stain Golgi apparatus.

1.5 BASIC REQUIREMENTS TO BE MET BY GENETIC MATERIAL

According to the molecular biologists certain requirements must be met by any molecule if it is to be qualified as the substance that transmits genetic information from one generation to next. These requirements extend directly from what is known about the continuity of species and the process of evolutionary change.

1. The genetic material must contain biologically useful information that is maintained in a stable form.

2. The genetic information must be reproduced and transmitted faithfully from cell to cell or from generation to generation.

3. The genetic material must be able to express itself so that other biological molecules, and ultimately cells and organisms, will be produced and maintained. Implicit in this requirement is that some mechanism be available for decoding, or translating, the information contained in the genetic material into its "productive" form. A narrow, but important, distinction is, thus, made between a molecule that can generate only its own kind and a molecule that can also generate new kinds of molecules. For example, a salt crystal can "seed" a salt solution so that new salt crystals are formed, but this is the extent of its influence over its surroundings.

4. Genetic material must be capable of variation. Two sources of change have been recognized in present day genetic systems: mutation and recombination. A mutation changes the

nature of information transmitted from parent to offspring, and, thus, it represents a relatively drastic way of bringing about variation. Recombination is a more moderate way of producing variation. It occurs during the course of some sort of sexual process, and it involves the precise shuffling of parental genetic information so that new combinations of genes are produced. These are then inherited by the offspring.

With these four requirements in mind we can study in the forthcoming chapters — How the genetic material is discovered?, What is the physical and chemical properties of it?, How it carries genetic information from one generation to next?, How a genetic material determines the phenotype of an organism? and many problems of similar nature.

QUESTIONS

Long Answer Questions

1. What is molecular biology? Discuss its importance for the humans.
2. Describe the historical growth of molecular biology.
3. Describe the method of cell fractionation. What is centrifugation? Describe in brief the mechanism of centrifugation.
4. Write down the different types of centrifuges.
5. What is chromatography? Describe various types of chromatography. What is TLC? Describe the process of TLC.
6. Name various kinds of chromatographic procedures. Descibe the procedure of paper chromatography and column chromatography.
7. What is electrophoresis? Explain the principle of electrophoresis. Describe different types of apparatuses of electrophoresis.
8. What is spectrophotometry? Describe the structure of a spectrophotometer.

Short Answer Questions

1. Describe the contribution of the following instrumentation and techniques in the growth of molecular biology:

   (*i*) Electrophoresis;

   (*ii*) Chromatography;

   (*iii*) Radioactive labelling;

   (*iv*) Autoradiography.
2. Discuss the principle of autoradiography and its application.
3. What is chromatography? Write down the principle.
4. Give differences between column chromatography and partition chromatography.
5. What is the difference between TLC and GLC?
6. Write short notes on the following:

   (*i*) Gas chromatography (*ii*) Paper chromatography

   (*iii*) Chromatography (*iv*) Ion exchange chromatography

   (*v*) SDS–PAGE (*vi*) Electrophoresis

   (*vii*) Dialysis (*viii*) Application of spectrophotometry

   (*ix*) Radioisotopes

Very Short Answer Questions

1. What is autoradiography?
2. Define chromatography.
3. What is electrophoresis?
4. Define centrifugation.

Fill in the Blanks

1. DNA transcribes into .... which .... into proteins.
2. Mutations are heritable .... in the base sequence of DNA.
3. DNA repair involves .... and .... repair.
4. .... demonstrated that in DNA the numbers of adenine and thymine groups are always equal and so are .... and cytosine groups.
5. Watson and .... ; M. Wilkens got .... Nobel Prize for the discovery of .... of DNA.
6. .... proposed the wobble hypothesis for anticodons of tRNA.
7. R.W. Holley, ... and M.W. Nirenberg got 1968 Nobel Prize for deciphering the genetic code.
8. .... first of all used the terms exons and introns for the split genes.
9. .... got second Nobel Prize in 1980 for discovering complete sequence of single stranded DNA of .... bacteriophage.
10. .... discovered the technique of DNA .... by using BKM-DNA probe.
11. Roger D. Kornberg got 2006 Nobel Prize for the discovery of eukaryotic ....

Multiple Choice Questions

1. The physical measurement of a molecule in chromatography is
   (*a*) Rg value (*b*) Rf value
   (*c*) half life (*d*) Rs value
2. In electrophoresis, the buffer establishes contact between
   (*a*) two electrodes (*b*) the sample and electrodes
   (*c*) the sample and supporting medium (*d*) the supporting medium and electrode
3. Separation by electrophoresis in based on
   (*a*) charge (*b*) density
   (*c*) solubility (*d*) refraction

ANSWERS

Very Short Answer Questions

1. Autoradiography is a procedure for detecting the location of radioactive molecules by overlaying a sample with photographic film, which becomes darkened upon exposure to radioactivity.
2. It is a group of related techniques that utilize the flow of a fluid phase over a non-mobile absorbing phase to separate molecules based on their relative affinities for the two phases, which in turn reflect differences in size, charge, hydrophobicity or affinity for a particular chemical group.
3. It is a group of related techniques that utilize an electrical field to separate electrically charged molecules.
4. It is a process of rapidly spinning a tube containing a fluid to subject its contents to a centrifugal force.

Fill in the Blanks

1. RNA, translates.

2. Permanent changes.

3. Photoreactivation, excision.

4. E. Chargaff, guanine.

5. Crick, 1962, molecular nature.

6. F. H. C. Crick

7. H. G. Khorana

8. W. Gilbert.

9. Fredrick Sanger, $\phi \times 174$.

10. Lalji Singh, fingerprinting.

11. DNA polymerase III.

Multiple Choice Questions

1. (*b*) **2.** (*d*) **3.** (*a*).

2

Organisms Used in Molecular Biology

2.1 MOLECULAR BIOLOGICAL IMPORTANCE OF VIRUSES

Viruses are very small sized, infectious, host-specific and nucleoproteinous particles. They are submicroscopic biological entities which though lack cellular organization, possess their own genetic material, genetically determined macromolecular organization and characteristic mode of inheritance. Viruses are **obligate cellular parasites** of either bacteria, plants or animals (**Verma** and **Agarwal 2016**). Viruses are called **Border line organisms** or **organisms at the edge of life**, as these are metabolically inert outside their host cell but are living and capable of multiplying inside their host cell by controlling the cellular machinery of the host cell. Size of viruses varies widely. Most viruses have a diameter of 10 to 30 nm. Smallest sized virus is Foot and Mouth virus (of about 10 nm) while largest sized virus is Citrus Tristera Virus (about 2000 nm).

Shape of viruses also varies and may be simple helical, icosahedral or complex forms. TMV (tobacco mosaic virus) is long, cylindrical rod-like helical ribovirus while T_4 bacteriophage (Fig. 2.1) has a complex symmetry. For example, T-even phage looks like a tadpole (T_2 phage has an icosahedral head, helical tail sheath, hexagonal end plate and rod-shaped tail fibers).

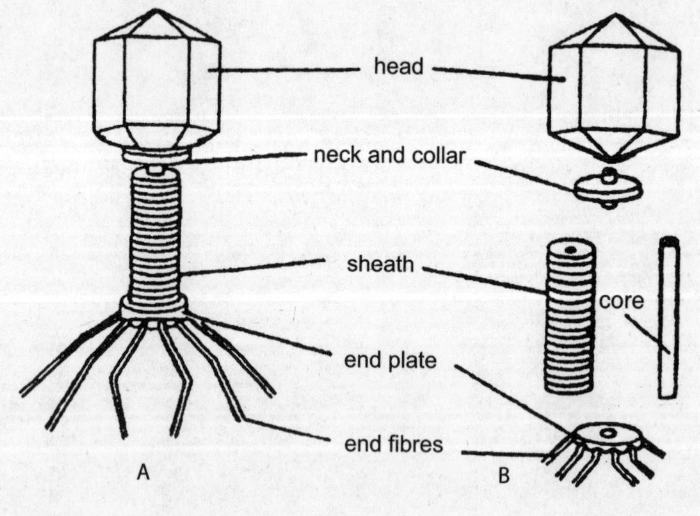

Fig. 2.1. A T_4 bacteriophage. A—External structure. B—Parts of T_4 phage.

Chemically, viruses are **nucleoproteinous** in nature and have nucleic acid (DNA or RNA acting as genetic material) at center which is surrounded by a protein coat, called **capsid**, formed of subunits called **capsomeres**. Depending upon the nucleic acid present, the viruses are either **riboviruses** (with RNA) or **deoxyriboviruses** (with DNA). Majority of plant viruses and some animal viruses (*e.g.*, Retro, polio and rabies viruses) are riboviruses, while most of animal viruses and cauliflower mosaic virus are deoxyriboviruses. Viruses can have a "**lipid envelope**" derived from the plasma membrane of the host (*e.g.*, Mumps virus, herpes virus and influenza virus).

Life Cycle of T_4 Bacteriophage

The T_4 bacteriophage is an obligatory, intracellular parasite of the bacterium *Escherichia coli*. When a T_4 bacteriophage infects a bacterial cell it gets attached to the cell wall of the host's cell by its tail which secretes certain proteolytic enzymes, for making perforation in the bacterial cell wall. This is followed by injection of DNA core (phage chromosome) into the bacterial cell. The capsid of T_4 phage remains outside the cell and called **ghost**. Inside the bacterial cell, the phage DNA either starts to replicate by taking the precursor substances from the host cytoplasm to form its own progeny (such a phage is called **virulent phage**), or becomes integrated into the host chromosome as **prophage** to replicate in synchrony with the bacterial chromosome (such a phage is called, **non-virulent** or **temperate phage**). A host cell having a virulent phage soon bursts (lyses) and releases several hundred new viral particles which can then repeat the same cycle (Fig. 2.2). A virulent phage, thus, destroys the host cell.

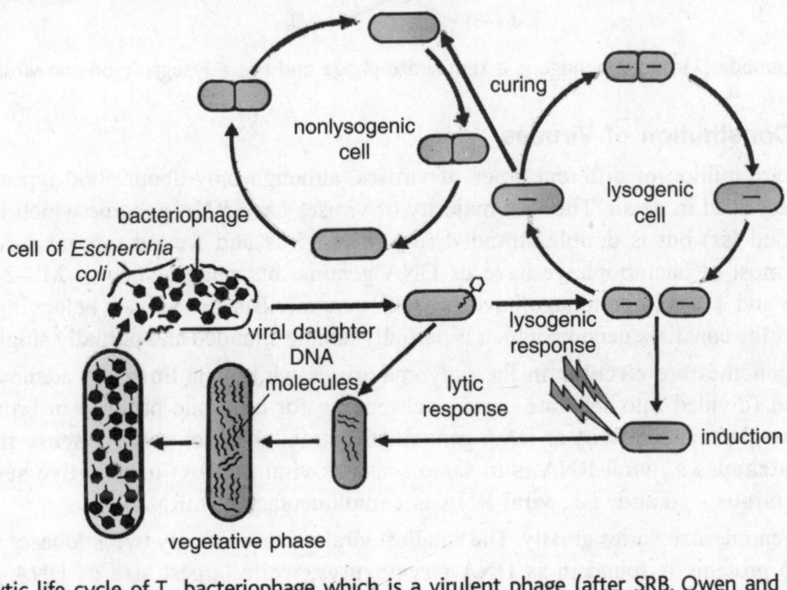

Fig. 2.2. Lytic life cycle of T_4 bacteriophage which is a virulent phage (after SRB, Owen and Edgar, 1965).

The temperate phage (Fig. 2.3) does not destroy bacterial cell. Very occasionally, however, a bacterium containing temperate phages will lyse and liberate several hundred infectious progeny of viral particles. Because such hosts are potentially subject to lysis they are termed **lysogenic**. A temperate phage, thus, remains quiescent for long time, but may convert into reproductive virulent phage by some radiation source such as ultraviolet light or X-rays and disrupts the bacterial cell.

Genetic applications of T_4 bacteriophage. T_4 bacteriophage has following genetical applications: (1) It helps in the identification of the nature of genetic material. (2) Its various mutant forms are easily detectable. (3) The recombination of it is also easily detectable.

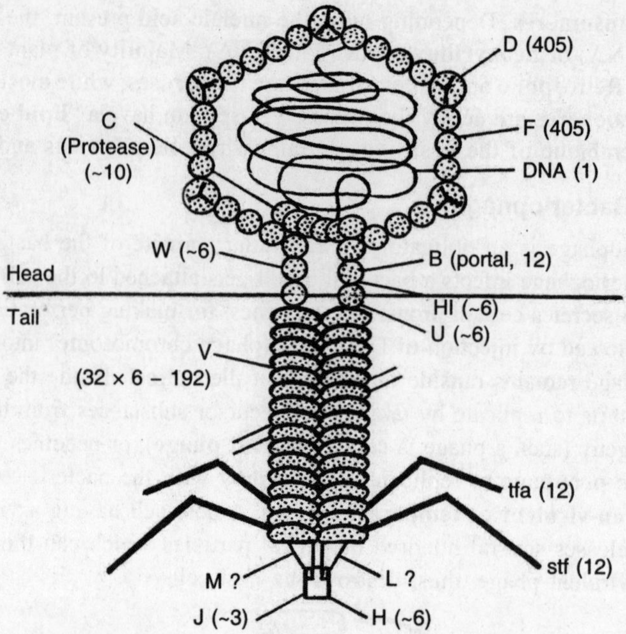

Fig. 2.3. Lambda (λ) bacteriophage is a temperate phage and has a lysogenic or non-virulent cycle.

Genomic Constitution of Viruses

There are million of different types of viruses, although only about 5000 types of viruses have been described in detail. The vast majority of viruses have RNA genome which is generally single stranded (ss) but is double stranded (ds) in reovirus and wound tumour virus. Animal viruses and most of bacteriophage have ds DNA genome but phages such as MS-2 coliphage, fd coliphage and $\phi \times 174$ coliphage have ss DNA genome. But the viruses belonging to family **Hepadnaviridae** contain a genome which is partially double stranded and partially single stranded.

Viral genomes are **circular** in the polyomaviruses while it is **linear** in adenoviruses and is **segmented** (divided into separate parts, each coding for only one protein) in brome mosaic virus. With single stranded RNA or DNA genome, the strand is either **positive sense strand** (also called **plus-strand**, *i.e.*, viral RNA is in same sense as viral mRNA) or **negative sense strand** (also called **minus - strand**; *i.e.*, viral RNA is complimentary to mRNA).

Viral genome size varies greatly. The smallest viral genome of only two kilobases and coding for only two proteins is found in **ss DNA circoviruses**, while largest size ds DNA genome of 2 mega-bases and coding for about 2500 proteins is found in **panderoviruses**. In general, RNA viruses have smaller genome size than DNA viruses because of a higher error rate when replicating.

Molecular Biological Importance of Viruses

Viruses provide simple biological systems so can be used to manipulate and investigate a number of aspects of cell biology, genetics and molecular biology.

(*i*) Genetic importance. Study of viruses has helped to understand the basic mechanisms of moleuclar genetics such as **DNA replication, transcription, RNA processing, translation, protein transport** and **immunology**.

(*a*) Genetists (and biotechnologists) often use viruses as **vectors** to introduce desired genes in the host cells. This is useful for mass production of a specific product or to study the effect of introducing a new gene into the genome.

(*ii*) **Virotherapy** uses the viruses to treat various diseases, as they can target specific cell types and DNA. It has shown promising result in the treatment of **cancer** and in **gene therapy**. **Phage therapy** is gaining increasing interest because of high level of antibiotic resistance found in many types of pathogenic bacteria.

Microbiological importance. Expression of heterologous proteins by viruses is the basis of several manufacturing processes that are currently being used for the production of various proteins such as **vaccines, antigens** and **antibodies**. Industiral processes have been developed using viral vectors.

Evolutionary importance. Viruses are an important natural means of transferring genes between different species which increases **genetic diversity** followed by natural selection and evolution. It is thought that viruses played a central role in the early evolution before the diversification of bacteria, archaea and eukaryotes at the time of the last universal common ancestor of life on earth (**Forterre** and **Phillipe** 1999).

2.2 MOLECULAR BIOLOGICAL IMPORTANCE OF *ESCHERICHIA COLI*

E.coli includes both asexual and sexual reproductions in its life cycle. Its asexual reproduction is achieved by simple **binary fission**. The sexual reproduction of *E.coli* takes place by **conjugation** which occurs only between genetically differentiated strains (Fig. 2.4). Some mating types of *E.coli* have

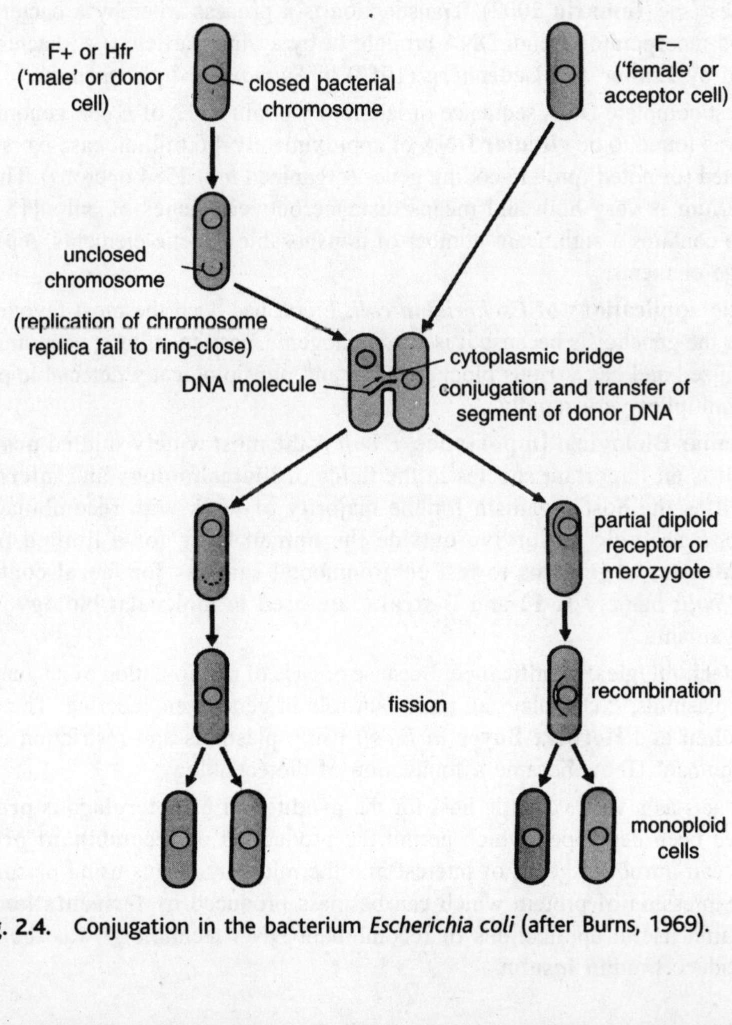

Fig. 2.4. Conjugation in the bacterium *Escherichia coli* (after Burns, 1969).

a small extra-chromosomal DNA particle called **sex factor** or **F**, while other mating types may lack the sex factor. The bacterial strains with sex factors are called **F^+** or **Hfr (male** or **donor)** strains, while bacterial strains lacking the sex factor are called **F^-** or **female** or **acceptor** strains. When a chance collision of F^+ and F^- bacteria takes place, both types of cells attach with each other and the walls between the two bacterial cells break open to form a small cytoplasmic bridge between the two. Through this cytoplasmic bridge the transfer of sex factor (DNA segment) of F^+ cell to F^- or female bacterial cell, takes place and thus conjugation completes. The conjugating bacterial cells now separate and are called **exconjugants**. The receptor F^- cell, thus, assumes a partial diploidy and is called **"merozygote"**. In the merozygote the sex factor remains incorporated with the circular DNA molecule of F^- cells. The F^- merozygote reproduces by cell division and ultimately regains monoploidy.

There are three methods of gene transfer in *E.coli* for causing recombination of genes: **conjugation**, **transformation** and **transduction**.

Conjugation was reported by **J. Lederberg** and **E. L. Tatum** in 1946 while its detail was given by **Jacob** and **Wollman** in 1956. In 1928, **F. Griffith** demonstrated **transformation** in *Pneumococcus* bacteria from capsulated and pathogenic S-type bacteria to non-capsulated and non-pathogenic R-type bacteria and change R-type into S-type bacteria. Transformation is a process whereby prokaryotes (*e.g.*, bacteria) takes up DNA from the environment and incorporates it into their genomes (see **Tamarin** 2002). Transduction is a process whereby a bacterial cell can gain access to and incorporate foreign DNA brought in by a viral particle (*i.e.*, bacteriophage). It was first reported by **Zinder** and **Lederberg** (1952) in *Salmonella typhimurium*.

The first complete DNA sequence of laboratory strain K-12 of *E.coli* genome was published in 1991. It was found to be **circular DNA** of approximately 4.6 million base pairs (bp), containing 4288 annotated (or noted) protein coding genes (organized into 2584 operons). The coding density of the bacterium is very high and means distance between genes of only 118 base pairs. The genome also contains a significant number of transposable genetic elements, repeat elements and bacteriophage elements.

Genetic applications of *Escherichia coli.* *E.coli* has been the most favourite experimental organism for the geneticists because it is non-pathogenic, easy to culture, economical, small-sized, simply organized and has various biochemically and physiologically detectable phenotypic traits. Further, it multiplies very rapidly.

Molecular Biological Importance. *E.coli* is the most widely studied **prokaryotic model organism**. It is an important species in the fields of **biotechnology** and **microbiology**, where it has served as the host organism for the majority of work with recombinant DNA. As the cells of *E.coli* are able to survive outside the human body for a limited period so act as potential **indicator organisms** to test environmental samples for faecal contamination. Two strains of *E.coli*; namely **K-12** and **B-strain**, are used in molecular biology as both tool and a model organism.

1. Biotechnological significance. Because of ease of manipulation of its genetic material and presence of plasmids, *E.coli* plays an important role in genetic engineering. The work of **Stanley Norman Cohen** and **Herbert Boyer** in *E.coli* using plasmids and restriction endonucleases to create recombinant DNA, became a foundation of biotechnology.

E.coli acts as a very versatile host for the **production of heterologous protein expression systems** have been developed which permit the production of **recombinant proteins** in *E.coli*. Researchers can introduce genes of interest into the microorganisms using plasmids which allow high level expression of protein which can be mass produced by **fermentation biotechnology**. One of the first useful applications of recombinant DNA technology was the manipulation of *E.coli* to produce **human insulin**.

Similarly proteins requiring post-transcripitional modification such as **glycosylation** for stability or functioning have been expressed using the N-linked glycosylation system of *Campylobacter jejuni* engineered into *E.coli* (**Huang** *et al.*, 2012).

Genetically modified *E.coli* cells have also been used in **vaccine development**, **bioremediation, production of biofuels** (**Summers**, 2013) and production of **immobilized enzymes** (**Halverson**, 2013).

2. Microbiological significance of *E.coli.* The bacterium, *E.coli* is frequently used as a **model organism** in microbiological studies. Its laboratory strains such as *E.coli K-12* have lost their capacity to form biofilm which enables them to protect from antibiotics and other harmful chemicals.

E.coli is a good model organism to study conjugation. **Seymour Benzer** (1961) used *E.coli* to understand the topography of gene strucutre, *i.e.*, linear arrangement of genes.

Bacillus subtilus is one of the best studied prokaryotes in terms of molecular biology and cell biology. It acts as a model for **differentiation**, **gene/protein regulation** and **cell cycle regulation**.

3. Therapeutic significance. Non-pathogenic strains of *E.coli* such as **mutafor** and **Colinfant**, are used as **probiotic agents** in medicines, mainly for the treatment of various gastrointestinal diseases including **inflammatory bowel disease (IBD)** (**Kamada** *et al.*, 2005).

4. Actinobacter, a gram-negative and non-motile bacterium, is a soil-bacterium and helps in mineralization of aromatic organic compounds.

5. Corynebacterium is found in a range of different ecological niches such as soil, vegetables, sewage, skin and cheese smear. So this bacterium is highly suitable for **adaptive physiological** studies.

2.3 MOLECULAR BIOLOGICAL IMPORTANCE OF *CAENORHABDITIS ELEGANS*

After searching for the ideal animal system to use as a model in dissecting development, **S. Brenner** (1965) chose the small free-living soil nematode (worm), *Caenorhabditis elegans* (Fig. 2.5) as the species of choice. Progress in dissecting and understanding the morphogenesis of *C.elegans* has been extremely rapid since that time. *C.elegans* is particularly suited to correlate genetic and ultrastructural dissections.

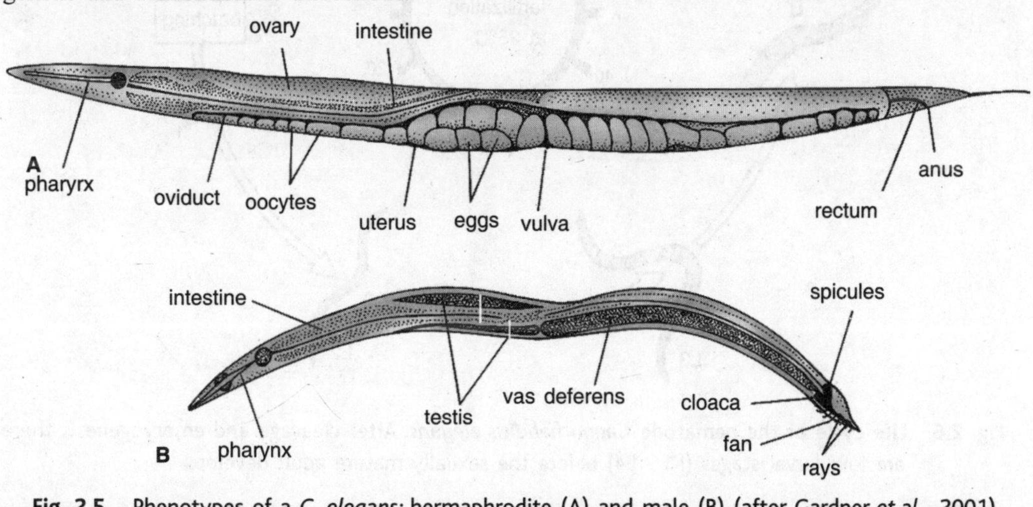

Fig. 2.5. Phenotypes of a *C. elegans:* hermaphrodite (A) and male (B) (after Gardner *et al.*, 2001).

As a suitable genetic research material, *C.elegans* has the following advantages:

1. Small size. The adult worms are about 1 mm in length and just 70 mm in diameter.

2. Short generation time. It reproduces with a 3 day life cycle under optimal growth conditions.

3. Simple growth conditions. It grows on agar plates with *E.coli* as food. Early larval stages can be stored frozen and later resuscitated (*i.e.,* revived from unconsciousness).

4. Two sexes: hermaphrodites and males. There are two sexual phenotypes in *C.elegans*: males which have only testes and hermaphrodites that contain both testes and ovaries. Hermaphrodites have two X chromosomes while males have only one X chromosome. Y chromosome is absent in *C.elegans*.

The existence of hermaphrodites (having male and female sex organs in the same worm) allows geneticists to produce worms homozygous for newly induced mutations very easily by simply allowing self-fertilization, just as Mendel allowed his pea plants to self-pollinate. In hermaphrodites there are about 300 amoeboid sperms and almost all will fertilize the eggs (**Wolpert** *et al.,* 2002).

5. Large number of progeny. A single hermaphrodite will produce about 300 eggs during its life (Fig. 2.6).

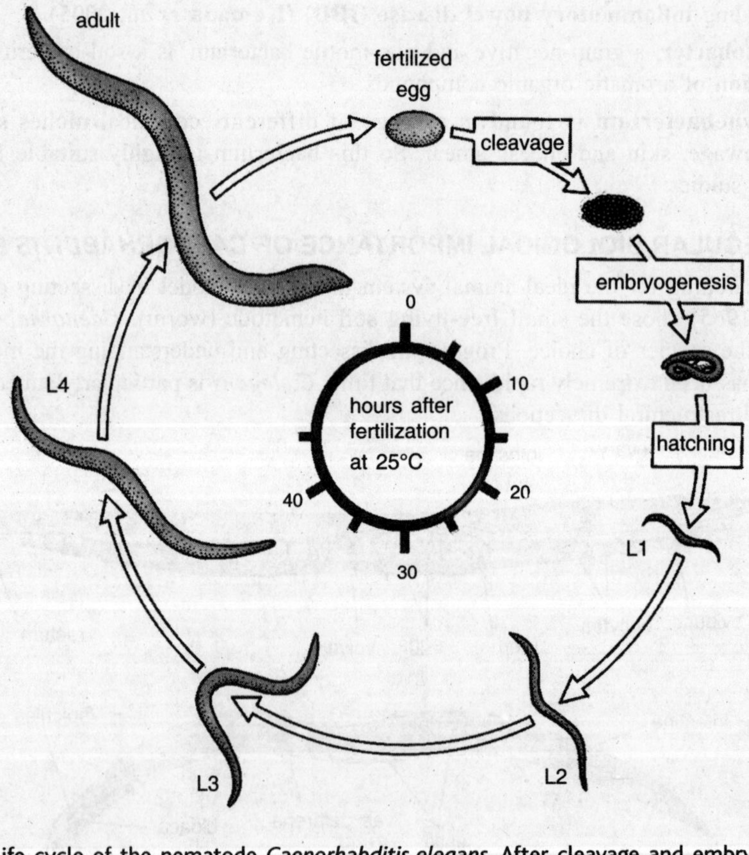

Fig. 2.6. Life cycle of the nematode *Caenorhabditis elegans*. After cleavage and embryogenesis there are four larval stages (L1 - L4) before the sexually mature adult develops.

6. Small genome. *C.elegans* is diploid (2N) and its genome has 6 chromosomes: 5 autosomes and a sex chromosome (X). Hermaphrodites are XX while males are XO. Sex is determined by X chromosome to autosome ratio. Its genome contains about 8×10^7 nucleotide pairs which greatly facilitates molecular analyses such as cloning, sequencing and physical mapping of genome. There are about 20,000 genes on 6 chromosomes, of which only 19,099 are protein-coding genes.

7. No exoskeleton, transparent bodies. All cells of the body can be observed directly with the light microscope at all stages of development.

8. Precise and invariant developmental programme. Each adult hermaphrodite is composed of exactly 945 cells and these are produced from the zygote by precise pathways of cell division, cell migration, cell growth and cell death (apoptosis); these are called **invariant cell lineages** (Box 2.1).

Box 2.1

A constancy in the number of somatic cells (in an organ, or in the entire body of an animal) is called **eutely** (see **Brusca** and **Brusca** 2002). An adult hermaphrodite form of *C.elegans* has 959 somatic cells in its body. An adult male of this nematode has 1031 somatic cells in its body.

The transparent body of *C. elegans* has permitted **J. E. Sulston** and coworkers (1977) to establish the entire cell lineage for the adult worm from the single-celled zygote, an accomplishment that would be impossible in almost any other higher animal. Now, a small army of developmental geneticists are dissecting this pathway at the molecular, cellular and morphological levels. The dominant approach is **induce mutation**, select those of the desire type and use biochemical and microscopic tools to define the resulting blocks in the pathway. The final result of their efforts will be a molecular map of the morphogenesis of the worm just like that worked out for phage T4.

Molecular Biological Importance

1. Embryological studies. In *C.elegans*, the process of embryogenesis by a stereotyped series of cell divisions and transparency of the embryo enabled **John Sulston** *et al.* (1977), to trace the entire cell lineage from fertilization to hatching so that ancestory of most cells is predetermined (Box 2.1). This facilitates light microscopic and ultra-structural analysis of developing animal at higher magnification especially using differential interference microscopy. The first cell divisions of early embryogenesis in this ogranism are among the best understood examples of **asymmetric cell divisions**. Genetic and molecular biological analysis of the cleavage state has revealed that the fates of the cells in the early embryo are determined by a combination of maternally-derived factors, which segregate asymmetrically to specific cells during cleavage and certain cell-signalling mechanisms of the neighbouring cells.

Elaboration of larval structures involves **cell interactions** through evolutionary conserved **signalling pathways**. Those involving **epidermal growth factors (EGF), transforming growth factors (TGF)**, etc. These cellular interactions include guidance clues by which several cells, including the growing axonal processes of motor neurons, which reach their final positions along the basement membrane of the pseudocoelom, guide these cells in the nematode are similar to the molecules which guide axonal growth from the floor of neural plate to the spinal cord in developing avian and mammalian embryos. So, *C.elegans* acts as a *useful model system for the study of nerve regeneration*.

2. Genetic significance. Small sized genome of *C.elegans* makes the nematode highly suitable for the genetic studies. DNA sequencing of its genome has helped in identifying genes required for a biological process of interest; determining the molecular organization of these genes after their isolation. Genetical studies of *C.elegans* has helped in understanding mechanisms of pattern formation in development, origin of left-right asymmetry, cell fate determination, programmed cellular death (apoptosis), cell migration, developmental timing mechanisms, etc. Limited life span (*i.e.*, less than 3 weeks) is considered to be due to genetic control of aging in which certain genes are switched off by development of dauer larval stage. *C.elegans* can take up RNA from the bacteria that they eat for RNAi (**RNA interference**). This ability has been mapped down to a single gene **scd-2**, which when inserted as a **transgene** in other species allows them to take up RNA for RNAi (employed for silencing the function of a gene).

3. Cancer studies. The nematode *C.elegans* has become an *important model system for cancer research* as the molecular components which regulate vulval development in the nematode are similar to *ras* pathways found in mammalian cells, in which conversion of *ras* to *oncogenes* causes a variety of cancers.

4. Strains of the nematode are cheap to breed and can be frozen. When subsequently thawed, they remain viable, allowing long-term storage.

5. *C.elegans* acts as a model organism for research into **ageing**, *e.g.*, inhibition of an **insulin-like growth factor** signalling pathway has been shown to increase adult life-span three fold (**Wilkow** *et al.*, 2000).

6. It is notable in **animal sleep studies** as **lethargus phase** occurs shortly before each moult.

7. It was found to have survived the **Space Shuttle Columbia** in February, 2003 so is being used to explore the effects of zero gravity on genetic basis of muscle atrophy.

2.4 MOLECULAR BIOLOGICAL IMPORTANCE OF *SACCHAROMYCES CEREVISIAE* (YEAST)

The yeast is a one-celled ascomycete fungus which includes in its life cycle two morphologically identical diploid and haploid generations that alternate with each other (Fig. 2.7). The diploid adult usually multiplies by budding in which the nuclei divide by mitosis, but under certain environmental conditions they undergo meiosis to form four monoploid **meiospores** or **ascospores** within the old cell wall which now called **ascus**. The ascospores are liberated from the ascus and each ascospore germinates into a monoploid adult cell. Out of four ascospores two develop into adults of + mating type, while rest two develop into adults of - mating type. Sometime two yeast cells of opposite strains contact and an inter-cellular cytoplasmic bridge is formed between the two cells. Both cells ultimately fuse (plasmogamy) and their nuclei also fuse to form a diploid nucleus (karyogamy). The resultant vegetative cell repeats the same life cycle.

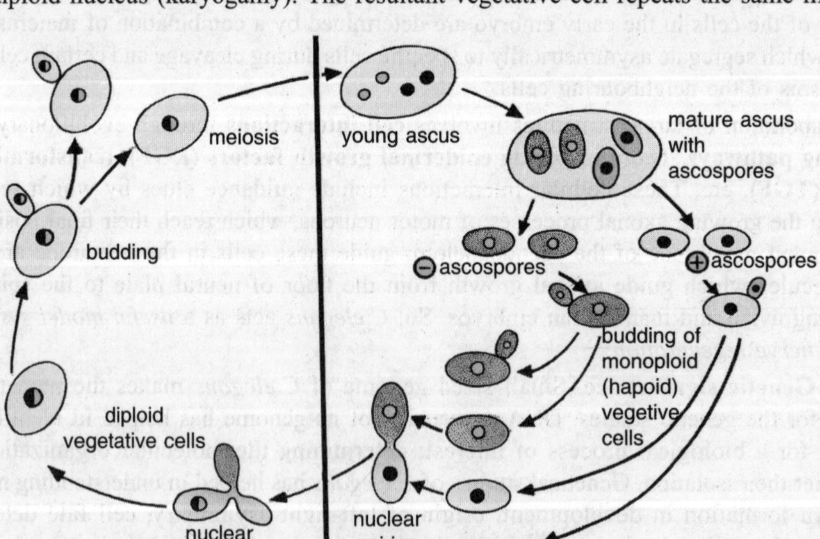

Fig. 2.7. Life cycle of baker's yeast, *Saccharomyces cerevisiae* (after Burns, 1969).

Genetic applications of *Saccharomyces cerevisiae*. The yeast is a favourite genetic material because it is easy to culture, economical, small-sized, easily available and has short and simple life cycle. Further, all the meiotic products (ascospores) can be analysed easily and quickly from the cytogenetical point of view.

Genomic Constitution

The genome of *S.cerevisiae* is composed of about 200 to 2,220 kbp and 6,275 genes, compactly arranged on 16 chromosomes. It was first eukaryotic genome to be sequenced (**Goffeau** *et al.*, 1996) and has been maintained in the **Saccharomyces Genome Data base**.

Molecular Biological Importance

The yeast *S. cerevisiae* is one of the most intensively studied **eukaryotic model organism** in molecular biology and cell biology due to a number of reasons, such as single-cellular nature, short generation time (doubling time is of only 1.5 to 2 hours); easy culturability; well studied genome, having many protein homologs to those of humans; genetic transformation ability, etc. It is estimated that at lest 31% of yeast genes have homologs in the human genome (**Botstein** *et al.*, 1997).

1. Ageing stuides. Studies on *S.cerevisiae* have helped in identification of mammalian genes affecting ageing. The common methods of measuring ageing in yeast are **Replicative Life Span** (**RLS**) (*i.e.*, number of times a cell divides) and **Chronological Life Span (CLS)** (*i.e.*, how long a cell survives in a nondividing state) through dietary restriction (*i.e.*, limiting the amount of glucose or amino acids in growth medium). Studies on yeast have revealed that the mother cell undergoing budding by mitosis undergoes ageing over successive generations but the age-related damage is lost when it undergoes meiosis and gametogenesis. So the meiosis rejuvenates the mother cell.

2. Genetic importance. Studies on *S.cerevisiae* have proved the **adaptive function of meiosis and recombination**. It was found that mutations in the genes (*e.g.*, genes *rad52*) essential for mitotic and meiotic recombination in *S.cerevisiae* cause increased sensititvity to radiations or DNA damaging chemicals such as **methyl methane sulfonate** (mutagens) and show reduced recombination. The studies have shown that the main selective force maintaining meiosis is enchanced recombination repair of DNA damage.

3. Cell biological importance. The availability of the genome sequence of *S.cerevisiae* and a set of deletion mutants covering 90% of the yeast genome has further enhanced the power of the fungus to act as a model for understanding the regulation of eukaryotic cells.

Meiotic analysis is the traditional method for genetically determining the sequence and distance between the linked genes. The yeast, *S.cerevisiae* is suited for meiotic mapping as the four spores in the ascus are arranged in linear manner so is suitable for **tetrad analysis**.

4. Biotechnological importance. In yeast, mating occurs between two haploid cells of opposite mating types. It has been exploited by molecular biologists to combine genes, plasmids or proteins at will. As the mating pathway employs a **G-protein-coupled receptor**, **G protein** and **RGS protein signalling cascade** that is homologous to those found in humans so has been exploited to investigate basic mechanisms of **signal transduction** (Box 2.2) and **desensitization**.

Box 2.2

Signal transduction pathway is a pathway in which action of kinase enzymes that free transcription factors, or some other action, translates an environmental signal into some form of gene action (**Tamarin**, 2002).

5. Commercial importance. *Saccharomyces cerevisiae* is a **top-fermenting yeast** so is most commonly used yeast in brewing beer. In May, 2013 **Oregon legislature** (USA) declared it as **official state microbe** for its use in brewing industry. This fungus is also used in baking. The CO_2 generated by the yeast fermentation is used as a **leavening agent** in bread and other baked goods.

2.5 MOLECULAR BIOLOGICAL IMPORTANCE OF *NEUROSPORA CRASSA*

The mycelium of red bread mold, *Neurospora crassa* (an ascomycete fungus) is composed of intertwined, segmented filaments called **hyphae**. Each hyphal segment contains a haploid nucleus with seven chromosomes. *N.crassa* reproduces asexually by haploid spores called **conidia**, and sexually by the contact of hyphae of two different mating types or strains, as A (+) and a (–) or B strains (Fig. 2.8). The plasmogamy is followed by pairing of A or + nuclei in a single hyphal cell. Both contacting hyphae develop into a fruiting body called **perithecium**, in which many elongated sac-like sporangia called **asci** (singular **ascus**) are developed from diploid or A (+) and a (–) nuclei containing hyphae. Eventually, both nuclei of opposite strain unite forming a diploid zygote nucleus which soon undergoes meiosis to form four haploid **meiospores**. Each meiospore then divides mitotically and thus, eight monoploid **ascospores** are formed. Four of these ascospores give rise to A or + mating strain, while the other four to a or–strain plant. The ascospores are liberated from the asci and germinate into new adults.

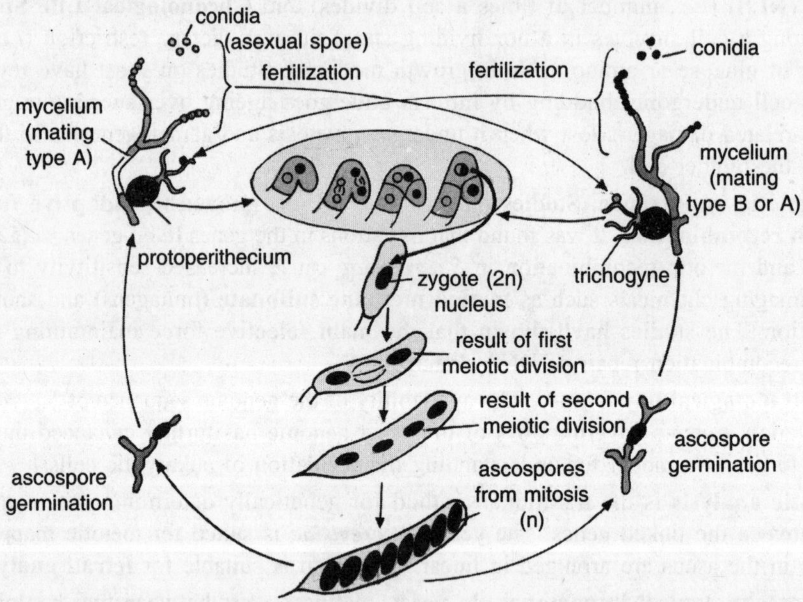

Fig. 2.8. Life cycle of the red bread mold, *Neurospora crassa* (modified after Winchester, 1967).

Genomic Consitution of *N.crassa*

N.crassa is a haploid fungus. Its genome contains 7 chromosomes and has been sequenced completely in 2003 (**Galagan** *et al.*, 2003). The genome of red bread mold (*N.crassa*) is about 43 megabases long and includes about 10,000 genes.

Genetic application of *Neurospora crassa*. *N.crassa* is a most favourite experimental material for researches of physiological genetics and cytogenetics due to its following peculiarities: (1) It has short life cycle and readily propagates asexually into unlimited population of given genotype. (2) It can be kept in pure cultures on chemically defined media. (3) Its eight linearly arranged ascospores in an ascus provide an opportunity to see that what has happened at meiosis because (*i*) all the products of a single meiosis are preserved within one structure and cannot be confused with product of other meiotic division, (*ii*) the positions of a particular ascopore within its ascus can be referred to the actual position of a nucleus in meiosis, as determined by the orientation of separating chromosomes on a division spindle.

Molecular Biological Importance

N.crassa acts as a model organism in the studies of molecular events, involved in **mutation studies, epigenetics, gene silencing, cell fusion, cell biology, development**, etc.

***(i)* Cell biological studies**. The fungus *N.crassa* was used by **Edward Tatum** and **C. W. Beadle** for the study of stage of meiosis. They found that in a mature ascus, pairs of adjacent spores are identical in genetic constitution and the ascospores are arranged in 2 by 2 manner and not 4 by 4 manner. This arrangement is possible only when the crossing over had occurred between non-sister chromatids at 4-strand stage (*i.e.*, post-replication phases).

***(ii)* Genetical studies**. *N.crassa* is highly suitable for genetic studies, especially mutational studies as even recessive traits will express themselves. **Tatum** and **Beadle** exposed the ascospores to X-rays causing mutations. They then observed failures in metabolic pathways caused by non-synthesis of certain enzymes. From their studies, they concluded "**one gene-one enzyme**" **hypothesis** which states that specific gene codes for specific enzymatic protein. This hypothesis laid the foundation of **biochemical genetics**.

Genetic studies of ascospores of *Neurospora* also established the phenomenon of "**gene conversion**". In *N.crassa,* the *gene pan-2* is necessary for the synthesis of vitamin pantothenic acid. *N.crassa* mutant strain defective in pan-2 gene grows in only that growth medium in which the vitamin pantothenic acid is added from outside. Two pan-2 mutations B5 and B3, are located at different sites in the pan-2 gene, so that a cross between B5 and B3 mutants yield wild-type recombinants at low frequency. It was suggested that gene conversion events arise when a molecular recombination event happens to occur near the genetic markers under study. Studies of gene conversion provide insight into the details of molecular mechanism of recombination.

In *N.crassa*, mating occurs only between different strains which suggests some degree of outcrossing is favoured by natural selection. This outcrossing provides the benefit of masking recessive mutations that otherwise be deleterious to sexual spore formation.

2.6 MOLECUAR BIOLOGICAL IMPORTANCE OF *ARABIDOPSIS THALIANA*

Arabidopsis thaliana (Fig. 2.9) is a small economically unimportant member of the mustard family. This annual crucifer grows in wild as a weed and commonly called **thale cress** or vetch. The plant develops as a small ground-hugging rosette of leaves, from which a branched flowering stem is produced with an inflorescence at the end of each branch. This little flowering weed has been used for genetic analysis for over four decades. Its advantages for genetic studies have been extensively documented by **G. Redi, M. Koorneef, A. R. Kranz** and colleagues (Box 2.3).

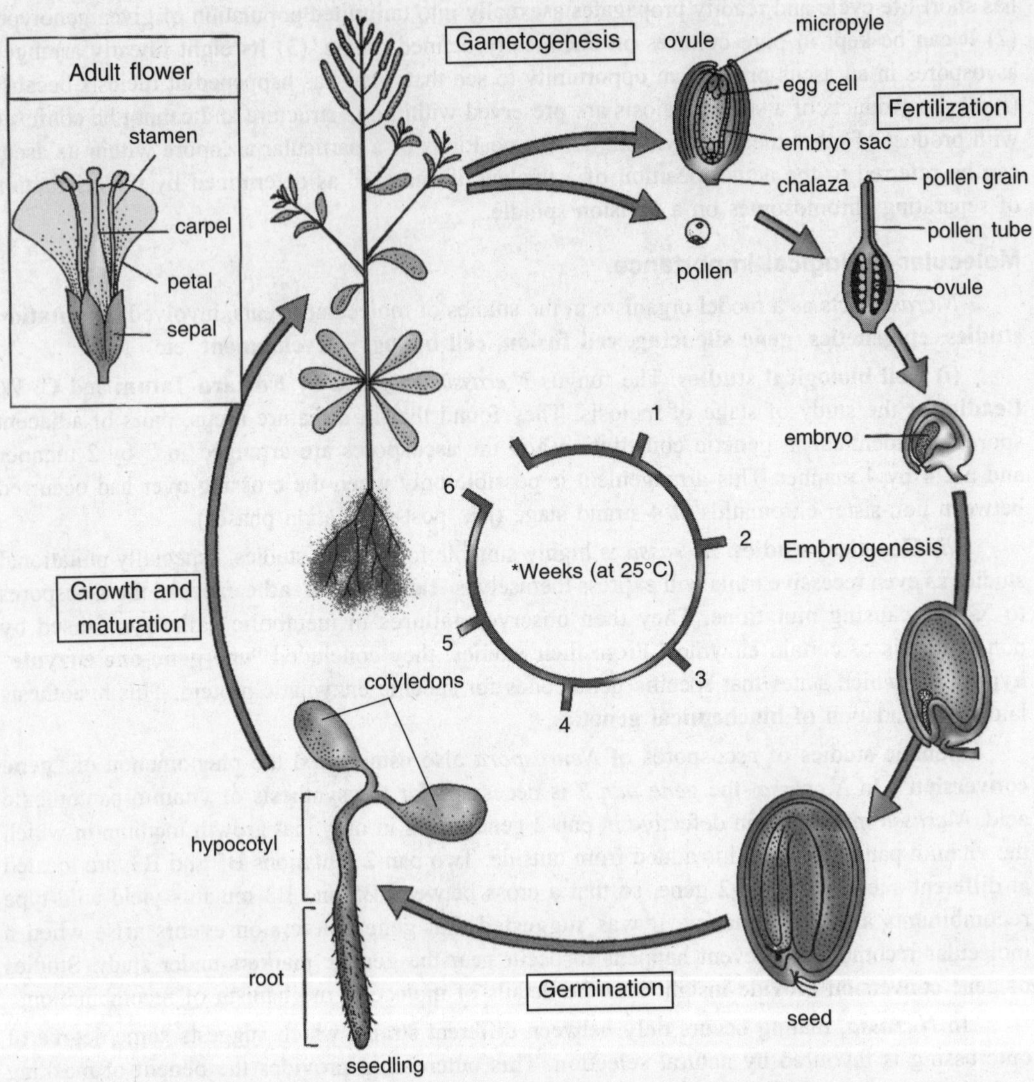

Fig. 2.9. Life cycle of *Arobidopsis thaliana* (* in continuous illumination).

Box 2.3

A. thaliana did not become a major genetic research subject until the mid 1980s when **E. Meyerowitz** and **R. Pruitt** demonstrated that the *A. thaliana* genome is very small and contains almost no interspersed repetitive DNA. **C. Somerville** and coworkers demonstrated the possibility of mutational dissections in *A. thaliana* by mutagenesis and "brute-force" (no genetic selection) screening. In 1988, *A. thaliana* has emerged on the plant model system of choice, the *Drosophila melanogaster* or *C. elegans* of the plant kingdom. *A. thaliana* is now considered to be the best plant system for genetic dissections of plant development and other biological process unique to plants (*e.g.*, photosynthesis).

As the model genetic plant, *A.thaliana* has the following advantages:

1. Small size. Very large numbers of plants can be grown in relatively small growth chambers.

2. Short generation time. The plant completes its life cycle within 5 weeks.

3. High seed production. Each plant can produce up to 40,000 seeds.

4. Very small genome. Its genome is 20 times larger than *E.coli* genome and less than 1/100 the size of the maize genome.

5. Very little interspersed repitive DNA. Unique sequences of DNA are on average 1,20,000 nucleotide pairs in length (as compared to about 1000 nucleotide pairs in length in maize).

6. Natural self-pollination. Very low levels of outcrossing (as opposed to natural cross pollination in maize).

The small size of *A. thaliana* permits one to grow large populations in growth chambers under controlled environmental conditions and greatly facilitates mutational dissections. Large populations of plants can be screened to permit the detection of rare mutant phenotypes, and these screenings can be done at much lower expense than for large plants such as maize. The short generation time allows one to collect eight to nine generations of genetic data per year.

A. thaliana acts as a **popular model organism** in plant biology, genetics, population interactions and molecular biology.

1. Genetic studies. *A. thaliana* is a diploid organism having very small sized genome formed of 5 chromosomes (Haploid number) of approximately 157 megabase pairs (Mbp) (**Bennett** *et al.*, 2003). So it is highly suitable for **genetic mapping and sequencing**. It was the first plant to be sequenced in 2000 by *Arabidopsis* **Genome Initiative**. It contains 27000 genes encoding for about 35,000 proteins. In 1976, **Bennett** and **Smith** showed that *Arabidopsis* has the smallest nuclear DNA content.

2. Metabolic studies. Biochemical studies on this plant have provided useful information about the effects of changes of environmental factors on metabolic processes.

A. thaliana is extensively used in the study of genetic basis of **phototropism, chloroplast alignment, stomata apperture size, circadian rhythms** and **blue-light influenced processes**. It was proposed that the plant has certain light receptors such as **phototropin** and **cryptochrome**.

3. *A. thaliana* is highly suitable plant for **Agro-bacterium-mediated transformation** to transfer foreign DNA into it by simply dipping its flower into a solution containing *Agrobacterium*, DNA of interest and a detergent (**Clough** *et al.*, 1998).

4. Mutation studies. Insertional mutants are available for most of genes of the plant *A. thaliana*, which make it unique for plant biology. The site of T-DNA insertions has been determined for over 300,000 transgenic lines. First mutant of *Arabidopsis thaliana* having a double flower phenotype was documented in 1873 by **Alexander Braun**. In 1945, **Erna Reinholz** described the first collection of its mutants generated by X-ray mutagenesis.

5. Development studies. As the young embryo of *A. thaliana* and its roots are translucent, so is highly suitable for developmental studies by using **fluorescent protein markers** followed by fluorescent or confocal laser screening microscope. So the developmental stage of each cell can be inferred. This avoids the need of fixation and section cutting.

A. thaliana has been extensively studied as a **model for flower development**. **Homeotic mutations** in this plant result in interchange of floral leaves. These studies led to the formulation of **ABC model of flower development** (**Coen** and **Meyerowitz**, 1991) according to which class A genes affect the development of sepals and petals, class B genes affecting the development of petals and stamens, and class C genes affecting the development of stamens and carpels.

6. Plant-patnogen interactions. The plant *A. thaliana* has been successfully used in the study of **plant pathology**, *i.e.*, interactions with plants and disease-causing pathogens. It was concluded that when a plant is exposed to a microbe, the plant detects **pathogen associated**

molecular patterns (PAMPs) with specialized **pattern recognition receptors (PRRs)** on the plant cell surface. The best characterized PRR in *A. thaliana* is **FLS-2 (Flagellin sensing-2** which recognizes bacterial flagellin (**Gomez** and **Boller**, 1999) which has a ligand **flg-22** (formed of 22 amino acids). Pathogenic organisms employed for studying such interactions are bacteria (*Pseudomonas syringe*), fungi (*Colletotrichum destructivum*), viruses (TMV) and nematode (*Heterodera*).

7. Ecological studies. The genus *Arabidopsis* consists of 27 species and approximately 200 ecotypes (wild populations). These ecotypes represent a natural source of heritable variations. Differences are observed in flowering time, response to cold, fresh weight production and pathogen resistance. Among ecotypes, a **DNA sequence polymorphism** of upto 1.4% in low-copy DNA has been measured. DNA polymorphism are exploited in F_2 populations for **mapping and quantitative trait locus (QTL) analysis.**

2.7 MOLECULAR BIOLOGICAL IMPORTANCE OF FRUITFLY (*DROSOPHILA MELANOGASTER*)

The fruit fly *Drosophila melanogaster* is a dipteran insect, which has attracted the attention of many geneticists (Box 2.4). The adult flies are sexually dimorphic and both sexes display various morphological differences in primary and secondary sexual characters. For example, male fly has 740 ommatidia in its sessile compound eyes, while female fly has 780 ommatidia in its eye. Further, the male *Drosophila* has **sex combs** on the fore legs, while female *Drosophila* lacks them. The life cycle of *Drosophila* includes the stages such as **egg, embryonic stage, first instar larva, second instar larva, third instar larva, pupa** and **adult** (see Fig. 2.10).

Box 2.4

Geneticists have employed a variety of animals in their experiments. However, their choice usually fell upon *Drosophila melanogaster*, guinea pig, fowl, rabbit, cattle and humans. All these animals are fundamentally multicellular and sexually reproducing with well marked sexual dimorphism between male and female sexes. Their life cycles include an alternation of generation between a haploid and diploid generation. The haploid generation includes male (sperms) and female (ova) sex cells, which unite to produce a diploid zygote (fertilization). The diploid generation is started from the unicellular zygote which by undergoing mitosis of cleavage and various complicated embryological events, forms a multicellular adult animal having **somatic cells** (form body) and **germinal cells** (form sex cells). The primary function of somatic cells is to provide nourishment and sufficient protection from various ecological hazards such as heat, desiccation and radiations to the germinal cells. The germinal cells produce sperms and ova by undergoing gametogenesis and thus, the life cycle of a sexually reproducing animal continues.

The *Drosophila* egg (a centrolecithal egg) is sausage-shaped and the future anterior end is easily recognizable by the micropyle, a nipple-shaped structure in the tough external coat surrounding the egg. During fertilization, sperm enter the anterior end of the egg through the micropyle. The zygote nucleus undergoes a series of rapid mitotic divisions, one about every 9 minutes but, unlike most animal embryos, there is no cleavage of the cytoplasm. The result is a **syncytium** in which many nuclei are present in a common cytoplasm. After nine divisions the nuclei move to the periphery of egg to form **syncytial blastoderm**. Shortly afterwards, membranes grow in from the surface to enclose a nucleus and form cells and the blastoderm becomes truely cellular after about 13 mitoses. Not all the nuclei give rise to the cells of the cellular blastoderm; some 15 or so end up at the posterior end of the embryo and develop into **pole cells**, which later give rise to germ cells, that is, sperm or eggs.

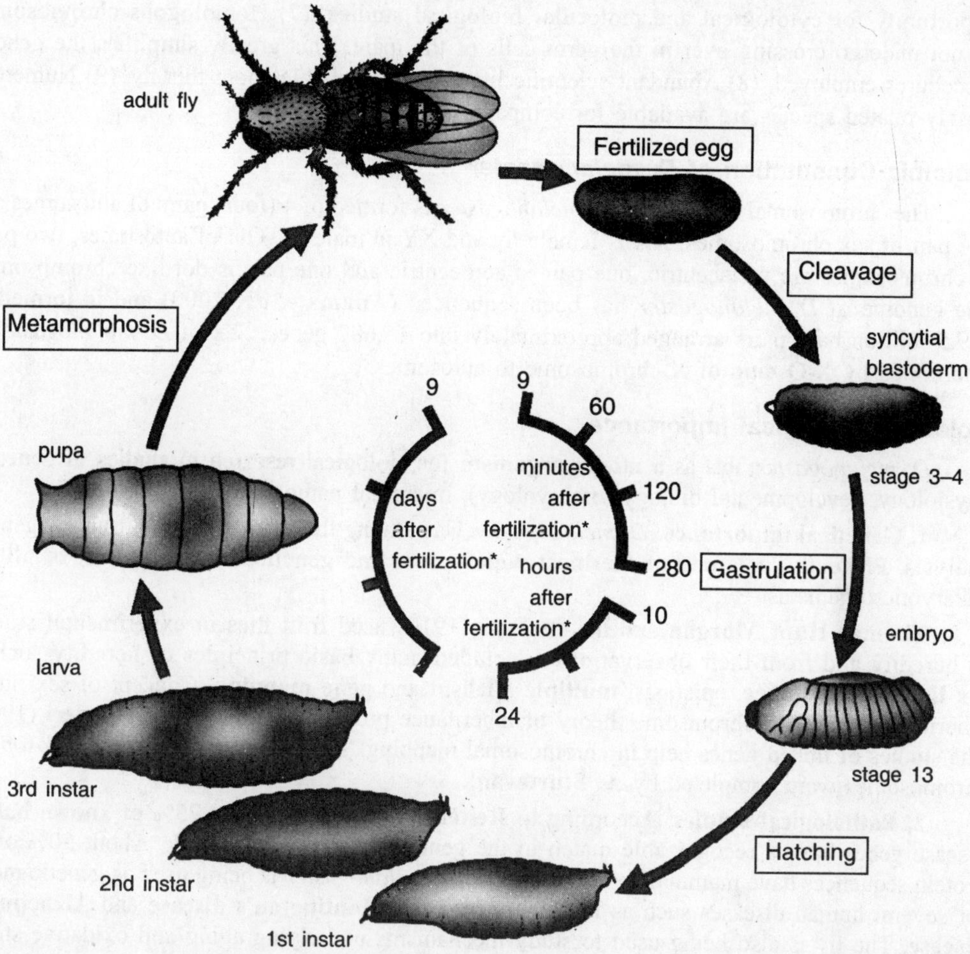

Fig. 2.10. Life cycle of *Drosophila melanogaster* ; * = at 25°C incubation (after Wolpert, 2002).

Gastrulation starts about 3 hours after fertilization. The larva hatches about 24 hours after fertilization. The head of larva is a complex structure and has **acorn**. At the posterior end, the most terminal structures are called the **telson**. Between these extremities, three thoracic segments and eight abdominal segments can be distinguished by specializations in the cuticle secreted by the epidermis. As the larva feeds and grows, it molts, shedding its cuticle. Such ecdysis occurs twice, each stage being called an **istar**.

When the larva undergoes hormone-induced **metamorphosis** after the third instar larva, adult fly is formed.

Genetic Applications of *Drosophila melanogaster*

From the standpoint of genetics, *D. melanogaster* is the best known of all multicellular organisms and the one which is mostly used in an elementary genetics laboratory. This is because of its following advantages—(1) The fly is small, anatomically complex, readily handled, and breeds prolifically in the laboratory. (2) Conditions for culturing of *Drosophila* are simple, cheap and readily controlled. (3) Its life cycle is short, takes only two weeks for its completion. (4) Many mutants of this fly are already known. (5) Its number of chromosomes is small (*i.e.*, its diploid number is 8). (6) The polytene chromosomes of salivary glands of its larva provide best opportunity for cytological and molecular biological studies. (7) Homologous chromosomes do not undergo crossing over in the germ cells of the male. This greatly simplifies the genetic procedures employed. (8) Abundant scientific literature is available about this fly. (9) Numerous closely related species are available for comparative studies.

Genomic Constitution of *D. melanogaster*

The chromosomal make up of *D. melanogaster* is formed of 4 (four) pairs of autosomes and one pair of sex chromosomes: XX is female fly and XY in male fly. Out of autosomes, two pairs of chromosomes are metacentric, one pair is acrocentric and one pair is dot-like chromosomes. The genome of *D. melanogaster* has been sequenced (**Adams** *et al.*, 2000) and is formed of 139.5 million base pairs arranged approximately into 15,682 genes. Sex of *D. melanogaster* is determined by X-O ratio of X chromosome to autosomes.

Molecular Biological Importance

D. melanogaster acts as a **model organism** for biological research in studies of genetics, physiology, developmental biology (embryology), microbial pathogenesis, evolution, etc.

1. Genetical importance. *D. melanogaster* was among the first organisms used for genetic analysis. Presently, it is one of the most widely used and genetically best-known of all the eukaryotic organisms.

Thomas Hunt Morgan and his students (1910) used fruit flies in experimental studies of heredity and from their observations concluded many basic principles of heredity such as **sex linked inheritance**, **epistasis**, **multiple allelism** and **gene mapping**. Concept of sex linked inheritance supported chromsome theory of inheritance proposed by **Sutton** and **Boveri** (1903). The studies of linked genes help in chromosomal mapping. The first genetic maps of *Drosophila* chromosomes were completed by **A. Sturtevant**.

2. Pathological studies. According to **Reiter** *et al.*, (2001), about 75% of known human disease genes have a recognizable match in the genome of *D. melanogaster*. About 50% of fly protein sequences have mammalian homologs. So *D. melanogaster* is being used as genetic model for several human diseases such as **Parkinson disease, Huntington's disease** and **Alzheimer's disease**. The fly is also being used to study mechanisms underlying aging and oxidative stress, diabetes and cancer, as well as drug abuse.

Embryogenesis is *Drosophila* is peculiar that the **cleavage occurs in the syncitium** so the early embryo contains approximately 5000 to 6000 nuclei in an unseparated cytoplasm of the embryo. During gastrulation and early development, embryo undergoes well characterized **morphogenetic movements** including germ band extension, formation of endoderm (gut), etc.

4. Immunological studies. Humoral immune response in *Drosophila* is a systemic response (systemic means affecting the body generally). It is mediated through the **Toll and imd pathways** which help in detecting the microbes. The Toll pathway in *Drosophila* has homologues of Toll-like pathways in mammals. **Spatzle**, a known ligand for Toll pathways in flies, is produced in response to **Gram positive bacteria**, parasitic and fungal infections (Box 2.5). On the other hand, the imd pathway is triggered by **Gram negative bacteria**.

Box 2.5

During the embryogenesis of *Drosophila,* **Toll protein** and **Spatzle protein** are involved in generating dorsal-ventral polarity. The proteolytic cleavage of Easter and Spatzle is strictly limited to the area around the most ventral embryonic cells. The cleaved spatzle protein is now able to bound to its receptor in the oocyte cell (=plasma) membrane, the product of the *toll gene*. Toll protein is a maternal product that is evenly distributed throughout the cell membrane of the egg (**Hashimoto** *et al.*, 1988,1991), but it becomes activated only by binding spatzle protein, which is produced only on the ventral side of egg. Therefore, the Toll receptors on the ventral side of the egg are transducing a signal into the egg, while Toll receptors on the dorsal side of egg are not. This *localized activation* establishes the dorsal-ventral polarity of the oocyte (see **Gilbert**, 2010); (also see Chapter 15 of Molecular Biology).

5. Behavioural studies. Ron Konopka and **Seymour Benzer** (1971) described mutation-caused behaviour pattern changes in *Drosophila*. They reported that the mutants have faster or slower or broken rhythms. They proposed that these mutations affect a group of genes and their products which comprise **biological clock**.

2.8 MOLECULAR BIOLOGICAL IMPORTANCE OF ZEBRA FISH (*DANIO RERIO*)

The zebra fish, *Danio rerio*, is a bony fish belonging to family **Cyprinidae** (order **Cypriniformes**) of class **Osteichthyes** (Teleostomi). It is a tropical fresh-water fish. It is also a popular aquarium fish. Zebra fish is a native of rivers of south eastern Himalaya region and is found in parts of India, Pakistan, Nepal, Bangla-desh and Burma.

Body of zebra fish is elongated and laterally flattened having dark blue of silver-gold stripes along the longitudinal axis which extend to the end of caudal fin. This fish becomes sexually mature at 3 months of age and its mating behaviour is photoperiodic. Spawning takes place at dawn under natural conditions due to their circadian rhythms while in the laboratory conditions spawning occurs at the beginning of the light cycle. Under the natural conditions, it has external fertilization. However, its eggs and sperms can be harvested from anaesthesized fishes and fertilized *in vitro*. Its sperms can be deep freezed for the later use. The life span of zebra fish is 2 to 3 years. In captive breeding, it produces hundreds of eggs on a weekly basis.

Zebra fish can tolerate a wide range of water quality, so large number of fishes can be raised and maintained at low cost and labour.

Molecular Biological Importance

Due to a number of peculiar features such as fully sequenced genome, rapid embryonic development, relatively robust and transparent embryos, etc., zebra fish serves as a model organism for the embryological (especially regenerative studies), genetic and toxicological studies. Its use as a laboratory animal was pioneered by **George Streisinger** *et al.*

(*i*) Importance in Zebra fish in embryological studies. In zebra fish, the fertilized egg, embryo and the surrounding chorion are translucent which allow the detailed microscopic examination of development. The larger size of embryo (0.7 mm in diameter) of zebra fish facilitates micro-injection, dissection, transplantation and other embryological manipulations. These features also help in the identification and subsequent analysis of mutant phenotypes.

The embryological development studies of zebra fish have revealed the following phenomena:

1. In zebra fish, the dorso-ventral symmetry is established due to transport of substances from the vegetal hemisphere into marginal blastomeres via cortical microtubules.

2. Dorsal axis is established due to accumulation of β-carotene in the nuclei of blastomeres of dorsal side.

3. In zebra fish, embryonic body is shaped by three morphogenetic movements of gastrulation: epiboly, involution and convergence.

(*ii*) Importance of zebra fish in genomic studies. The haploid genome of zebra fish is formed of 25 metacentric chromosomes of similar size and is formed of a DNA of about 1.7×10^9 bp. Genetic linkage map of zebra fish has been developed by employing the biotechnological tools and techniques such as PCR, RAPD (Random Amplified Polymorphic DNA). CA-dinucleotide repeats and SSLP (Simple Sequence Length Polymorphism). Genetic map of zebra fish has about 1200 markers and average interval between the markers is about 1500 bp.

By using chemical mutagens (*e.g.*, N-ethyl nitrosourea), physical mutagens (*e.g.*, γ-rays irradiation) and retro-virus mediated gene transfer techniques, it has been identified that there are mutations in over 400 genes of zebra fish. Insertional mutations tend to alter the embryonic morphology and gene expression pattern. These methods have identified genes involved in early embryonic and larval development in the zebra fish. By the analysis of normal and mutant embryos, zebra fish will contribute significantly to our understanding of molecular mechanisms governing early embryonic development.

***(iii)* Decorative importance of zebra fish**. In late 2003, **transgenic zebra fish** that express green, red and yellow fluorescent proteins became commerically available in USA. The fluorescent strains are trade named **Glofish**. Other cultivated varieties of zebra fish are **golden**, **sandy**, **longfish**, and **leopard**. Leopard danio is a spotted colour morph of zebra fish which arose due to a pigment modulation (**Watanable** *et al.*, 2006).

***(iv)* Significance of zebra fish in regeneration studies**. Zebra fish has the ability to regenerate its fins, skin, heart, lateral line hair cells and even photoreceptors and neurons of retina. It has been shown to be mediated by dedifferentiation and proliferation of neuroglia (**Bemardos** *et al.*, 2007). Study of gene expression during regeneration has shown the involvement of several important **signalling pathways** such as Wnt signalling and fibrobalst growth factor (FGF).

The zebra fish has an online database of genetic, genomic and developmental information, the **Zebra fish Information Network (ZIN)**.

2.9 MOLECULAR BIOLOGICAL IMPORTANCE OF MOUSE (*MUS MUSCULUS*)

Mouse (*Mus musculus*) belongs to order **Rodentia**, class **Mammalia**, phylum **Chordata**. This small-sized common mammal is small plant-eater having chisel-like incisor teeth.

Haploid genome of mouse consists of 20 chromosomes, *i.e.*, 19 autosomes and 1 sex chromosome (either X or Y chromosome). However, the normal somatic cells of mouse are diploid and haploid genome is about 3×10^9 bp. Number of genes in mouse genome is estimated to be about 100,000 genes. To assess exact number of murine genes, a large **expressed sequence tag (EST)**, project is being carried out. As mice are genetically so well characterized through numerous mutation studies, they have become important tool in studying vertebrate biology including genetics, evolution, ecology, development (embryology), physiology and pathology.

Molecular Biological Importance of Mouse

1. Medical (clinical) research. The hose mouse complex consists of four commensal species: (*i*) *Mus muscularis domesticus*; (*ii*) *M. musculus musculus*; (*iii*) *M. musculus castaneus* and (*iv*) *M. musculus bacterianus*. In addition there are three aboriginal species: *M. spicilegus*, *M. macedonicus* and *M. spretus*. Laboratory strains of mouse have been established from diverse wild-caught animals by brother-sister matings (inbreeding) and back-crossing over many generations. Most of them are predominantly of *M. musculus domesticus* origin. Approximately 430 different mouse inbred strains are presently known. From the inbred strains, many substrains have been developed.

These mouse strains and their substrains are used for different medical research purposes, *e.g.*, (*i*) **IQ9/Re substrain** is used to study high incidence of spontaneous testicular teratomas.

(*ii*) **AKR/J substrain** is employed for the study of high incidence of lymphatic leukemia, and presence of 10 T-cell antigen.

(*iii*) **BALB/cHE**. A substrain is employed to study low incidence of mammary tumours.

***(iv)* C3H/B1 substrain** is used for study of high incidence of mammary tumours.

***(v)* CBA/Ca** substrain is used for study of low incidence of mammary tumours.

2. Genetic importance of mouse. In mouse genome contains several different types of **repetitive sequences**. Though some simple-sequence repetitive elements such as **microsatellite** and **minisatellites** are present but mostly these are of high-sequences, *e.g.*, **SINE** and **LINE** families of retroposons.

Mice are also used as models for all types of genetic variability. Mutations in the mouse genome can arise spontaneously or can be induced. Mutants can also be created by introducing additional genetic informations into the mouse genome so producing **transgenic mice** which can be used to study the effect of introduced gene construct *in vivo*. Transgenic mice are produced by any one methods such as microinjection of a DNA construct into the fertilized oocyte; by treating the mouse embryos with recombinant viruses, introducing DNA of interest into embryonic stem cells. By introducing desired DNA segments, the specific sites in the genome can be modified.

By **target mutations**, it has become possible to introduce virtually any change into the mouse genome. Some of the common mutants of mouse are:

***(i)* Agouti mouse** in which agouti (*a*) gene is altered which in homozygous condition causes black coat colour and immunity defects leading to lethality.

***(ii)* SCID-mutant** in which protein-kinase gene is mutated which causes non-functioning of B- and T-cells (lymphocytes) of immune system.

***(iii)* Nude-mutant** in which Hfh-II (H-forked homologue) gene is mutated and causes hairless condition and reduced fertility.

***(iv)* Mdx-mutant** in which dystrophin gene is mutated and causes muscular dystrophy.

3. Mutagenesis model. Two mouse models such as **Big Blue mouse** and **Mutamouse** have been developed which are being used for accurate analysis of mutagenesis in mammals. The Big Blue mouse is a transgenic mouse and carries a DNA construct from λ-phage containing **lac 1 gene** which codes for the lac repressor which prevents the expression of lac operon. This mouse is being used to investigate exact nucleotide change of the mutation. Mutamouse is also a transgenic mouse carrying a λ-phase derived DNA construct having an intact *lac 1 gene.*

Comparative Genomics

In 2002, a draft sequence of the mouse (*Mus musculus*) genome was completed by an international consortium of investigators, allowing for the first time a comparison of two mammalian genomes (*i.e.*, mice and human beings). The human genome has about 400 million more nucleotides than that of mouse. A comparison of the two genomes reveals that both have about 30,000 genes, and they share the bulk of them; in fact, the human genome shares 99% of its genes with mice. Humans and mice diverged about 75 million years ago, too little time for many evolutionary differences to accumulate. There are only 300 genes unique to either organism, comprising about 1% of the genome. Most of the nearly 150 genes unique to mice are linked with either the sense of smell, which is highly developed in rodents, or with reproduction. Comparing these similar genomes is already yielding information on the *functions of genes*. It is must easier to design experiments to identify gene function in an experimental system like the mouse than in humans. Once the mouse sequence was known, the function of 1000 previously unidentified human genes was understood.

The genomes of humans and mice are so similar that one wonders why mice and humans are so different. The best explanation for why a mouse develops into a mouse and not a human is that the genes are expressed at different times and possibly in different tissues. This may be the case for **cystic fibrosis gene**, which has been identified in both species. Defects in the human cystic fibrosis gene cause especially devastating effects in the lungs, but mice with cystic fibrosis do not have lung symptoms.

Comparison of the mouse and human genomes reveals that since mice and humans last shared a common ancestor about 75 million years ago (MYA), mouse DNA has **mutated** about twice as fast as human DNA. This is a fascinating observation in search of an explanation. The difference in generation time between mice and humans could account for some of this distinct mutant rate, because mice would have had more opportunities to mix and match genomic components during **meiosis**.

Perhaps the most unexpected finding in comparing the mouse and human genomes lies in the similarities between the **"junk" DNA**, mostly **retrotransposons**, in the two species. This DNA does not code for proteins. A survey of the location of retrotransposon DNA in both species shows that it has independently ended up in comparable regions of the genome. Recently, it has become clear that this "junk" DNA has more of a function than was previously assumed. It is found rich in **regulatory RNA sequences** (*e.g.,* small interfering RNA or siRNA, microRNAs or miRNAs, etc.). While most of the transcripts code for mouse proteins, as many as 4280 could not be matched to any known mouse protein. This suggests that a large part of the transcribed genome consists of nonprotein-encoding genes-that is, transcripts that a single retrotransposon can cuase heritable differences in coat colour in mice.

One of the most exciting aspects of comparing rat and mice genomes is the potential to capitalise on the extensive research on rat physiology, especially **heart disease** and long history of genetics in mice.

Variation in the organization of genomes is as intriguing as gene sequence differences. Over long segments of chromosomes, the linear order of mouse and human genes is the same–the common ancestral sequence has been preserved in both species. This **conservation of syntheny** was anticipated from earlier gene mapping studies and provides strong evidence that evolution actively shapes the organization of the mammalian genome (**Raven** *et al.*, 2005).

QUESTIONS

Long Answer Questions

1. Write short notes on the following

(*i*) Modes of reproduction in viruses;

(*ii*) Genetical and microbiological importance of viruses.

2. Write short notes on the following

(*i*) Modes of reproduction in bacteria;

(*ii*) Molecular biological importance of bacteria.

3. Write short notes on following topics of *Caenorhabditis*

(*a*) As a model organism; (*b*) Developmental biological importance.

4. Write short notes on the following

(*i*) Microbiological importance of yeast; (*ii*) Genetical importance of *Neurospora.*

5. Write short notes on the following

(*i*) Developmental studies of *Arabidopsis;*

(*ii*) Genomic constitution of *Drosophila.*

6. Write short notes on the following

(*i*) Importance of genetic studies of mouse;

(*ii*) Importance of developmental studies of zebra fish.

Short Answer Questions

1. Write about genomic constitution and life cycle of viruses.
2. Discuss the molecular biological importance of viruses.
3. Write a note on molecular biological importance of the bacteria.
4. How is the nematode *Caenorhabditis elegans* highly suitable as a model organism for molecular biological studies?
5. Describe the significance of genomic constitution and developmental biological studies of *Caenorhabditis elegans.*
6. Discuss mode of reproduction and genome of *Saccharomyces.*
7. Write about molecular biological significance of yeast (*Saccharomyces*).
8. How is *Neurospora* useful in genetical studies?
9. Discuss sexual reproduction and genomic constitution of *Neurospora.*
10. Discuss the genomic constitution and genetic significance of *Arabidopsis thaliana.*
11. Write a note on sex-determination and genetic and developmental biological importance of *Drosophila.*
12. Discuss the importance of zebra fish in development biological and regeneration studies.
13. Write a note on genetic importance of studies of mouse.

Very Short Answer Questions

1. Why viruses are called "border line organisms".
2. State biochemical nature of viruses.
3. State differences between lytic cycle and lysogenic cycle of viruses.
4. What is the difference between positive sense strand and negative sense strand of virus?
5. Write about one genetic significance of viruses.
6. How is *Escherichia coli* highly suitable as a model organism?
7. Name various modes of gene transfer in *E.coli* (or bacteria).
8. State the difference between transformation and transduction.
9. How many genes for proteins are found in the genome of strain K-12 of *E. coli*?
10. Enlist significance of genetically modified *E.coli.*
11. What do you mean by eutely in *Caenorhabditis*?
12. Discuss the genetic constitution of *Caenorhabditis.*
13. Enlist two uses of *Caenorhabditis* in biological research.
14. Discuss RNA interference in *Caenorhabditis.*
15. Write about genomic constitution of yeast.
16. Discuss cell biological significance of yeast.
17. What is genomic constitution of *Neurospora.*
18. What is molecular biological importance of *Neurospora crassa*?
19. State the significance of outcrossing in *Neurospora.*
20. What is genetical importance of *Arabidopsis thaliana*?
21. Why *Arabidopsis* is highly suitable for developmental studies.
22. What is ABC model of flower development?
23. What is genomic constitution of *Drosophila melanogaster*?
24. What is peculiar about the embryological studies of *Drosophila*?

25. What is molecular biological importance of zebra fish (*Danio rerio*)?

26. What are the important genetic features of *Mus musculus* ?

Fill in the Blanks

1. Majority of plant viruses and some animal viruses (*e.g.*, polio virus) are ....
2. Most animal viruses and cauliflower mosaic virus are ....
3. Lambda (λ) phage is a temperate phage and has a .... life cycle.
4. T_4 phage is a .... bacteriophage and it destroys the .... cell.
5. Laboratory strains of *Escherichia coli* are .... and ....
6. .... in *Neurospora* helps in masking the deleterious recessive mutations.
7. Yeast genome has .... chromosomes and .... genes.
8. *Caenorhabditis elegans* show constancy in the number of somatic cells, it is called ....
9. Tetrad analysis can be done in one-celled ascomycete fungus, called ....
10. Common name of *Arabidopsis thaliana* is ....
11. During embryonic development of *Drosophila melanogaster*, ... protein and .... protein are involved in generating dorsal-ventral polarity.
12. Bacteria tend to do gene transfer by three methods: .... , transformation and ....

Multiple Choice Questions

1. Which of the following are called "Organisms at the edge of life"?

   (*a*) bacteria (*b*) viruses

   (*c*) cyanobacteria (*d*) archaebacteria

2. Transfer of genetic material from one bacterium to another bacterium by a bacteriophage is called

   (*a*) transduction (*b*) transformation

   (*c*) conjugation (*d*) translation

3. Which type of sex determination is found in the nematode *Caenorhabditis*?

   (*a*) XX(♀) – XY(♂) (*b*) XX(♂) – XY(♀)

   (*c*) XO(♀) – XX(♂) (*d*) XX(♀) – XO(♂)

4. *Saccharomyces cerevisiae* is characterized by all except

   (*a*) heterothalism

   (*b*) can grow on sugar such as lactose and cellobiose

   (*c*) suitable for meiotic analysis

   (*d*) commercially used in brewing and baking industries

5. Fruiting body of *Neurospora crassa* is called

   (*a*) ascocarp (*b*) ascogonium

   (*c*) apothecium (*d*) perithecium

6. *Arabidopsis* is highly suitable for genetical mapping because it has

   (*a*) small sized DNA content

   (*b*) easy for *Agrobacterium*-mediated gene transfer

   (*c*) pattern recognition receptors

   (*d*) transparent young embryo and roots

7. In *Drosophila melanogaster*, the sex is determined by

(*a*) X-Y ratio (*b*) X-X ratio

(*c*) Y-autosome ratio (*d*) X-autosome ratio

8. Mutamouse is a transgenic mouse for analysis of mutagenesis. It has one of the following genes

(*a*) lac Z gene (*b*) lac A gene

(*c*) lac Y gene (*d*) lac I gene

9. Use of zebra fish as a laboratory animal was pioneered by

(*a*) S. Benzer (*b*) F. Laibach

(*c*) G. Streisinger (*d*) Thomas Hunt

10. Which fungus was of great use in finding out genetic principles?

(*a*) *Mucor* (*b*) *Neurospora*

(*c*) Yeast (*d*) *Agaricus*

11. A lot of information about linkage in genetics has been obtained by researches on the organism [*CSIR-UGC (NET)*]

(*a*) *Oenothera* (*b*) *Esherichia coli*

(*c*) *Pisum* (*d*) *Drosophila*

12. *Neurospora crassa* [*CSIR-UGC (NET)*]

(*a*) was the organism on which Beadle and Tatum conducted experimental work to elucidate biochemical genetics

(*b*) was the fungus in which Pontecorvo demonstrated mitotic crossing over

(*c*) was the organism used by T.H. Morgan to illustrate linkage and crossing over

(*d*) none of the above was known from *N.crassa*

ANSWERS

Very Short Answer Questions

1. Viruses are cellular parasites. They are metabolically inert outside their host cell but are living and capable of multiplying inside their host cell by controlling the cellular machinery of the host cell.

2. Chemically viruses are nucleoproteinous in nature and have nucleic acid (DNA or RNA acting as genetic material) at center which is surrounded by a protein coat, called **capsid**, formed of subunits called **capsomeres.**

3. *Lytic cycle* is followed by **virulent pahges** such as T_2 and T_4 phages of *Escherichia coli* (bacterium). In this, the bacteriophage adsorbs at the specific receptor site of cell wall of host cell and injects its nucleic acid inside the host cell by making a hole in its wall with the help of a lysozyme. Inside the host cell, nucleic acid undergoes **eclipse stage**, lyses the host's DNA, uses host's machinery and synthesizes large number of copies of its nucleic acid molecule (by replicaton) and capsid proteins. Nucleic acid and capsid assemble to produce phage progeny. Finally, the host's cell wall is lysed and progeny viruses are released which repeat the process. *Lysogenic cycle* is followed by temperate phages or prophages such as λ-bacteriophage, in which viral DNA become associated with host's DNA and replicate with host's DNA.

4. Virus with single stranded RNA or DNA genome, the strand is either **positive sense** (called **plus strand,** *i.e.*, viral RNA is in same sense as viral mRNA or **negative sense** called **minus strand** *i.e.*, viral RNA is complimentary to mRNA).

5. Study of viruses has helped to understand the basic mechanisms of molecular biology such as DNA replication, transcription, RNA processing, translation, protein transport and immunology.

6. The bacterium *E.coli* has been the most favourite experimental organism for genetics, molecular biology and microbiology because it is non-pathogenic, easy to culture, economical, small-sized, simply organized and has various biochemically and physiologically detectable phenotypic traits. This bacterium multiplies very rapidly also.

7. In *E. coli* and other bacteria, there are three methods of gene transfer from one bacterium to another bacterium so producing recombination of genes: conjugation, transformation and transduction.

8. **Transformation** is a process whereby prokaryotes (*e.g.*, bacteria) takes up DNA from the environment and incorporate it into their genomes (*e.g.*, *Pneumococcus*). **Transduction** is a process whereby a bacterial cell can gain access to and incorporate foreign DNA brought in by a viral particle (*e.g.*, bacteriophage, *e.g.*, *Salmonella typhimurium*).

9. 4288 genes.

10. Genetically-modified (GM) *E.coli* cells have been used in the production of human insulin, vaccine development, bioremediation, production of biofuels and immobilized enzymes.

11. A constancy in the number of somatic cells (in a given organ, or in the entire body of an animal) is called **eutely**. It is common in nematodes. For example, each adult hermaphrodite of *C.elegans* has a fixed number of 959 somatic cells in its body. Each adult male of *C.elegans* has 1031 somatic cells in its body.

12. Six chromosomes of diploid *Caenorhabditis elegans* contain about 13,000 functional genes. It was first multicellular organism whose whole genome was sequenced.

13. 1. *Caenorhadbitis* is useful model organism for study of nerve regeneration.

    2. This nematode is also an important model for cancer research.

14. The nematode *Caenorhabditis elegans* can take up RNA from the bacteria that they eat for **RNAi** (RNA interference). This ability has been mapped down to a single gene **scd-2** which when inserted as a **transgene** in other species allows them to take up RNA for RNAi (employed for silencing the function of a gene).

15. The genome of Baker's yeast, *Saccharomyces cerevisiae* is about 2,200 kbp long and is composed of 6,275 genes, compactly arranged on 16 chromosomes.

16. Yeast is commonly used in tetrad analysis.

17. The genome of *Neurospora crassa* contains 7 chromosomes and has been sequenced completely in 2003. Its genome is about 43 megabases long and includes about 10,000 genes.

18. The fungus red bread mould, *Neurospora crassa* acts as a model organism because it is easy to grow, has a haploid life cycle, ascospores arranged linearly in the asci and simple genetic constitution. It is important in studies of molecular events involved in mutation studies, epigenesis, gene silencing, cell fusion, cell biology and development.

19. In *N. crassa*, mating occurs only between different strains which suggest some degree of outcrossing which is favoured by natural selection. The outcrossing provides the benefit of masking recessive mutations that would otherwise be deleterious to sexual reproduction.

20. The angiospermic plant, *Arabidopsis thaliana* acts as a popular model organism in genetics. It is a diploid organism having very small sized genome formed of 5 chromosomes of approximately 157 megabase pairs (Mbp). So is highly suitable for genetic mapping and sequencing. It contains 27,000 genes encoding for about 35,000 proteins. Its small size, short life span, self-pollinating nature and production of many seeds (*e.g.*, about 10,000 seeds per plant) make it suitable for cultivation and genetic experiments.

21. *Arabidopsis thaliana* has been extensively studied as a model for flower development. Homeotic mutation in this plant result in interchange of floral leaves. These studies led to the formation of ABC model of flower development.

22. The ABC model was proposed by **Coen** and **Meyerowitz** (1991) in context with *Arabidopsis*. According to it, class A genes affect development of sepals and petals, class B genes affect development of petals and stamens and class C genes affect the development of stamens and carpels.

23. The chromosome make up of *D. melanogaster* is formed of 4 pairs of autosomes and one pair of sex chromosomes : XX in female and XY in male fly. The genome of *D. melanogaster* has been sequenced by **Adams** *et al.,* (2000) and is formed of 139.5 million base pairs arranged in approximately 15,682 genes.

24. Embryogenesis in *Drosophila* is peculiar that the cleavage occurs in a syncitium. During the gastrulation, embryo undergoes well characterized morphogenetic movemetns including germ band extension, formation of several furrows, ventral immigration of mesoderm, posterior and anterior invagination of endoderm (gut), etc.

25. Zebra fish (*Danio rerio*) is a good model animal for research in molecular biology for its following peculiar features: fully sequenced genome, rapid embryonic development, testable developmental behaviour, relatively robust and transparent embryos, etc.

26. The physical size of a haploid genome (20 chromosomes : 19 autosomes and 1 sex chromosome) of mouse, *Mus musculus* is about 3×10^9 bp. Number of genes in mouse genome is estimated about 100,000 genes. The mouse genome contains several different types of **repetitive sequences**, *e.g.*, microsatellites, ministatellites, SINE and LINE families of retroposons.

Fill in the Blanks

1. Riboviruses.
2. Deoxyriboviruses.
3. Lysogenic.
4. Virulent, host.
5. K-12, B-strain.
6. Outcrossing.
7. 16,6275.
8. Eutely.
9. Yeast.
10. Thale cress.
11. Toll, Spatzle.
12. Conjugation, transduction.

Multiple Choice Questions

1. (*b*) **2.** (*a*) **3.** (*d*) **4.** (*b*) **5.** (*d*) **6.** (*a*)
7. (*b*) **8.** (*a*) **9.** (*c*) **10.** (*b*) **11.** (*d*) **12.** (*a*).

3

Identification of the Genetic Materials

When it became evident that the chromosomes were the organs of heredity, because (*i*) they formed the only link between two generations, (*ii*) they carried linearly arranged genes, (*iii*) they occurred in every organism in specific number and had specific morphology for a particular species, and (*iv*) any variation in their number of morphology affected the phenotype of the species, then, various attempts were made by early molecular geniticists to identify the physical and chemical nature of genes. But the genes were found so minute structures that their physical identity remained almost impossible. However, the extensive chemical analysis of chromosomes of different organisms have revealed that chromosomes contain proteins and nucleic acids (DNA and RNA), and it was thought that genes might have either proteins or nucleic acids as their component molecules. Early molecular geneticists have assigned the informational roles of genes to the chromosomal proteins because, they found nucleic acids too simple to carry genetic informations. (This is called **tetranucleotide hypothesis).**

The controversy about the assignment of genetic role either to chromosomal proteins or to chromosomal DNA, existed up to 1949 when **A. Mirsky** and **H. Ris** had found that all cells of an organism appeared to contain the same amount of DNA, whereas different cell types contained quite different amounts and kinds of protein. Its constancy, therefore, favoured DNA as the genetic material. Around 1953, it was universally accepted that DNA is the genetic substance (*i.e.*, chemical of which genes are composed) of most microorganisms and higher organisms. Later on, RNA was found to be the genetic material of some viruses. The concept that DNA or RNA is the genetic material of most organisms has been developed and supported by the following evidences.

3.1 DIRECT EVIDENCES FOR DNA AS THE GENETIC MATERIAL

The most conclusive evidences in support of DNA as the genetic material came from the following three avenues of approach on microorganism: transformation of bacteria, mode of infection of bacteriophages and conjugation of bacteria.

1. The Transformation Experiments

In 1928, **Frederick Griffith** encountered a phenomenon now known as genetic transformation. Colonies of virulent strain (pathogenic) of pneumonia causing bacterium, *Diplococcus pneumoniae* grown on nutrient agar, have a **smooth (S)** glistening appearance owing to the presence of a type specific, polysaccharide (a polymer of glucose and glucuronic acid) capsule. The avirulent (non-pathogenic) strains, on the other hand, lack this capsule and they produce dull, **rough (R)** colonies. Smooth (S) and rough (R) characters are directly related to the presence or absence of the capsule and this trait is known to be genetically determined. Both S and R forms occur in several types and are designated as S-I, S-II, S-III, etc., and R-I, R-II, R-III, etc., respectively. All these subtypes of S and R bacteria differ with each other in the type of antigens, they produce. The kind of antigen produced is likewise genetically determined. Smooth (S) forms sometimes mutate to rough (R) forms, but this change has not been found reversible.

In the course of his work, **Griffith** injected laboratory mice with live R-II pneumococci; the mice suffered no illness because R-II pneumococci was avirulent (Fig. 3.1). But, when the mice were injected with virulent S-III pneumococci, the mice suffered from pneumonia and died. However, when he injected the heat killed S-III bacteria in mice, they did not suffer from pneumonia. But, when the mice were injected with the mixture of living avirulent R-II and heat killed S-III virulent, the unexpected symptoms of pneumonia appeared and high mortality resulting in them (Fig. 3.2). By postmorteming the dead mice, it was found that their heart blood had both R-II and S-III pneumococci. From these results, Griffith concluded that the presence of the heat-killed S-III bacteria must have caused a **transformation** of the living R-II bacteria, so as to restore to them the capacity for capsule formation they had earlier lost by gene mutation. This was called "**Griffith effect**" or more popularly "**bacterial transformation**".

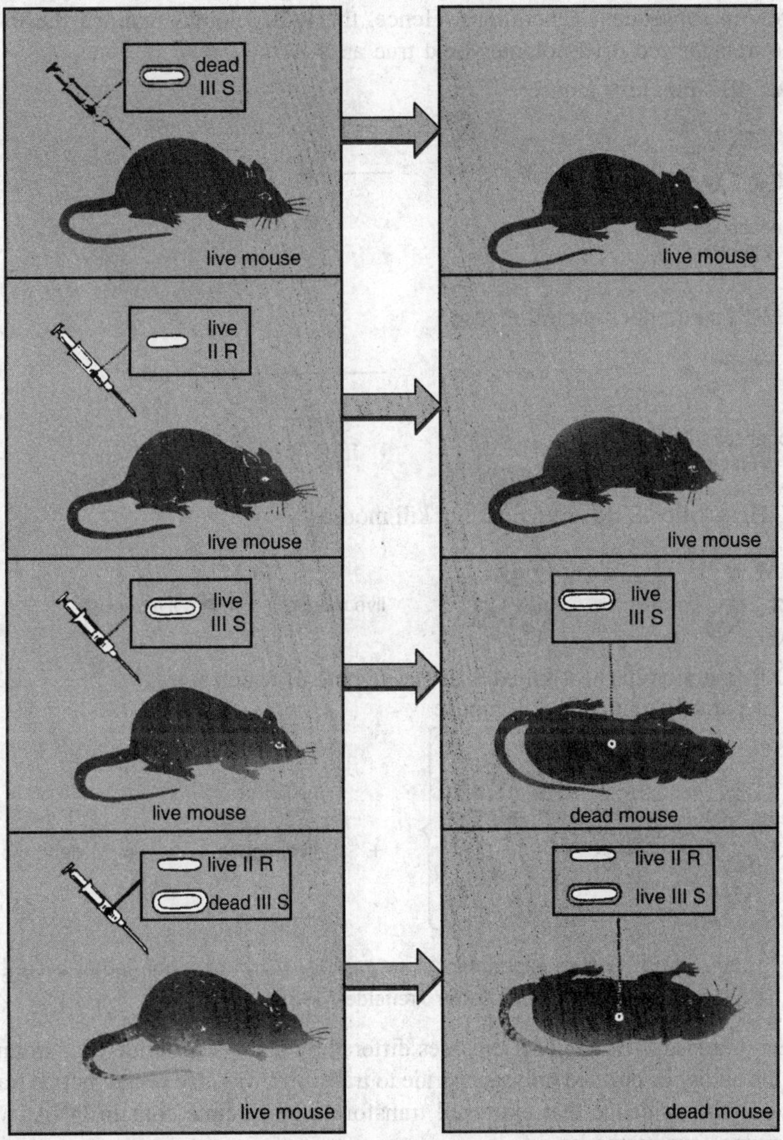

Fig. 3.1. Griffith experiment that demonstrated the principle of transformation.

Identification of the "transforming principle or substance". Griffith could not understand the cause of bacterial transformation and that is first of all identified by **Oswald Avery, Colin MacLeod** and **Maclyn McCarty** (1944). They partially purified the **transforming principle** from the cell extract (*i.e.,* cell free extract of S-III bacteria) and demonstrated that it was DNA. These workers modified the known schemes for isolating DNA and prepared samples of DNA from S-III bacteria. They added this DNA to a live R bacterial culture; after a period of time they placed a sample of S-III containing R-II bacterial culture on an agar surface and allowed it to grow to form colonies. Some of the colonies (about 1 in 10^4) that grew were S-III type (Fig. 3.3). To show that this was a permanent genetic change, they dispersed many of the newly formed S-III colonies and placed them on a second agar surface. The resulting colonies were again S-III type. If an R-II colony arising from the original mixture was dispersed, only R-II bacteria grew in subsequent generations. Hence, the R-II colonies retained the R-II character, whereas the transformed S-III colonies bred true as S-III.

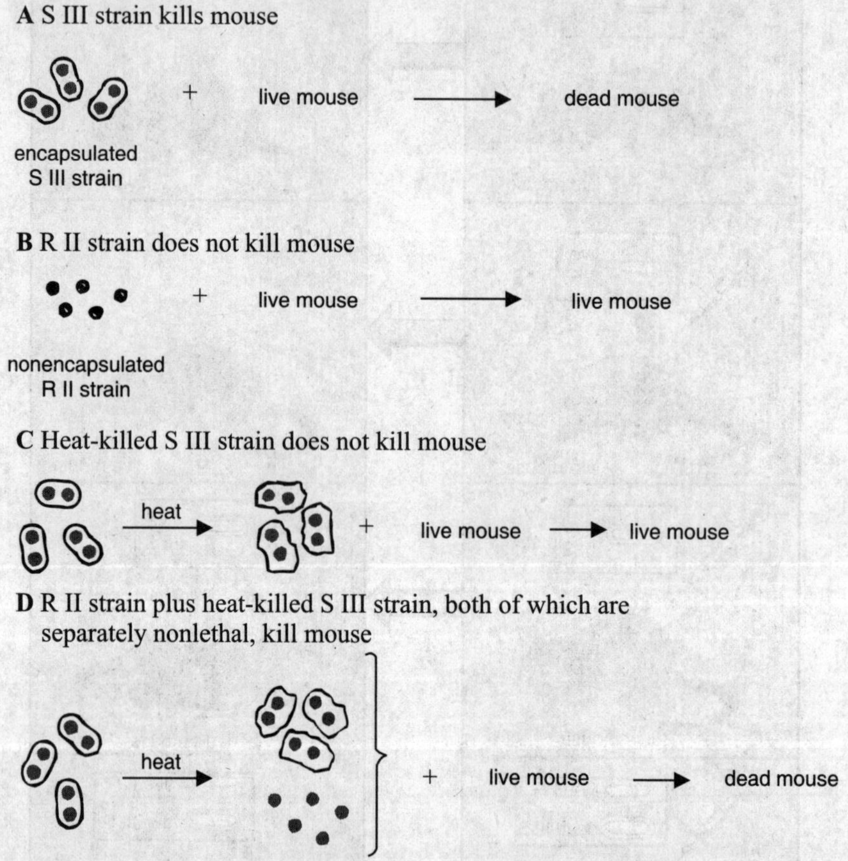

Fig. 3.2. Explanation of Griffith experiment showing conversion of a non-lethal bacterial strain to a lethal form by a cell extract (after Freifelder, 1985).

Further, because S-III and R-II colonies differed by a polysaccharide coat around each S-III bacterium, the ability of purified polysaccharide to transform was also tested, but no transformation was observed. The evidence that extracted transforming principle contain DNA, was provided by the following four methods:

1. Chemical analysis showed that the major component was deoxyribose containing nucleic acid.

2. Physical measurement showed that the sample contained a highly viscous substance having the properties of DNA.

3. Experiments demonstrated that transforming activity is not lost by reaction with either (*a*) purified proteolytic (protein - hydrolyzing) enzymes trypsin, chymotrypsin, or a mixture of both or (*b*) ribonuclease (an enzyme that depolymerizes RNA). The lipase too has no effect.

4. It was demonstrated that the treatment with materials, known to contain DNA depolymerizing activity (DNAase enzyme), inactivated the transforming principle. Later on, such a transforming substance, the DNA, was found in a variety of bacteria (*e.g., Hemophilus influenzae, Bacillus subtilus, Shigella paradysenteriae,* etc.) and several other organisms.

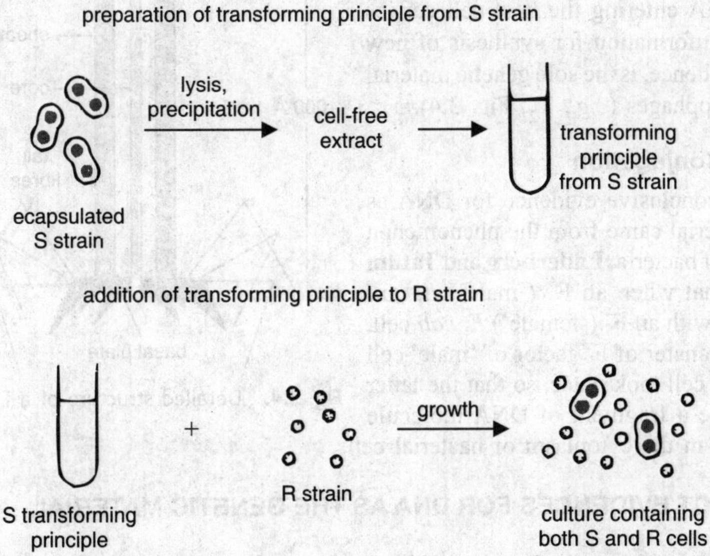

Fig. 3.3. Transformation experiment of **Avery, MacLeod** and **McCarthy** (after Freifelder, 1985).

2. The Blender Experiment

An elegant confirmation of the genetic nature of DNA came from an experiment with *E. coli* phage T_2. This experiment known as the **blender experiment** because a kitchen waring blender was used as a major piece of apparatus, was performed by **Alfred Hershey** and **Martha Chase** in 1952. They demonstrated that the DNA injected by a phage particle into a bacterium contains all the information required to synthesize progeny phage particles.

A single particle of phage T_2 consists of DNA (now known to be a single molecule) encased in a protein shell (Fig. 3.4). The DNA is the only phosphorus containing substance in the phage particle; the proteins of the shell, which contain the amino acids methionine and cysteine, have the only sulphur atoms. In these experiments, phage DNA was made radioactive by growing infected bacteria on a medium containing radioactive phosphate ($^{32}PO_4$). Since phage proteins do not contain phosphorus, only DNA would be labelled. Similarly, phage proteins were labelled with the help of $^{35}SO_4$. Since DNA does not contain sulphur, only protein would be labelled with ^{35}S. Such differential labelling would enable one to distinguish between DNA and proteins of the phage without performing any chemical tests. **Hershey** and **Chase** then allowed both kinds of labelled phage particles to infect *E. coli* bacteria (Fig. 3.5). The infected bacteria were immediately agitated in a waring blender. After shaking, only radioactive ^{32}P was found associated with bacterial cells and ^{35}S was found only in surrounding medium and not in bacterial cells. When phage progeny was studied for radioactivity in this experiment, it was found that the phage progeny carried labelled only ^{32}P. The progeny was not labelled with ^{35}S. This clearly indicates that only DNA and not protein is injected into bacterial cells. The empty protein coat (ghost) left outside. Thus, **Hershey** and **Chase** experiment proved that DNA entering the host cell carries all the genetic information for synthesis of new phage particles, hence, is the sole genetic material in DNA bacteriophages (*e.g.,* T_2; Fig. 3.4).

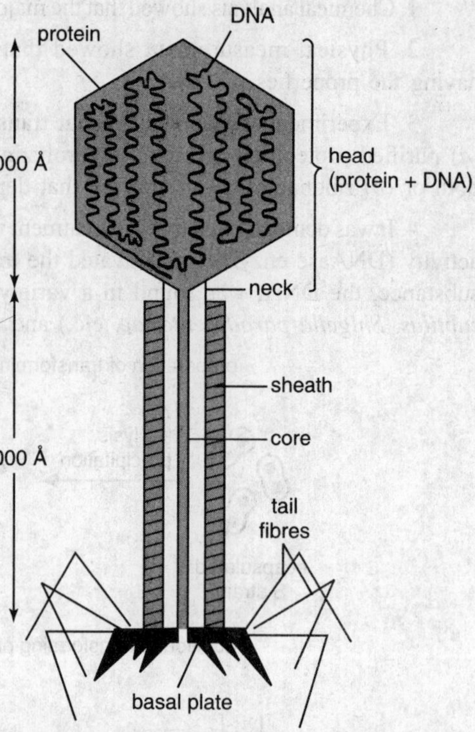

Fig. 3.4. Detailed structure of a T_2 bacteriophage.

3. Bacterial Conjugation

Another conclusive evidence for DNA as the genetic material came from the phenomenon of conjugation of bacteria. **Laderberg** and **Tatum** (1946) found that when an F^+ ('male') *E. coli* cell conjugated with an F^- ('female') *E. coli* cell, unidirectional transfer of F^+ factor of 'male' cell to F or 'female' cell took place, so that the latter was found to be a fragment of DNA molecule which occurred in the cytoplasm of bacterial cell.

3.2 INDIRECT EVIDENCES FOR DNA AS THE GENETIC MATERIAL

The indirect evidences in favour of DNA as the genetic meterial came from higher organisms which are not as easy to manipulate as bacteria and viruses. The fact that DNA is the genetic material of higher organisms has been supported by the following facts:

1. The Feulgen techniques have shown that DNA entirely remains restricted to the chromosomes and it forms one of the major components of chromosomes.

Table 3.1 DNA content and ploidy (**Pollister** *et al.,* 1951).

Cells	*Mean DNA Feulgen content (Picograms)*	*Presumed chromosome set (Ploidy)*
Spermatids	1.68	Haploid (2n)
Liver	3.16	Diploid (2n)
Liver	6.30	Tetraploid (4n)
Liver	12.80	Octaploid (8n)

2. Various quantitative measurements of amount of DNA in different cells have shown that there is a correlation between the amount of DNA and the number of chromosomes sets (ploidy). Diploid cells contain twice as much DNA as do haploid cells of the same species as shown in the Table 3.1.

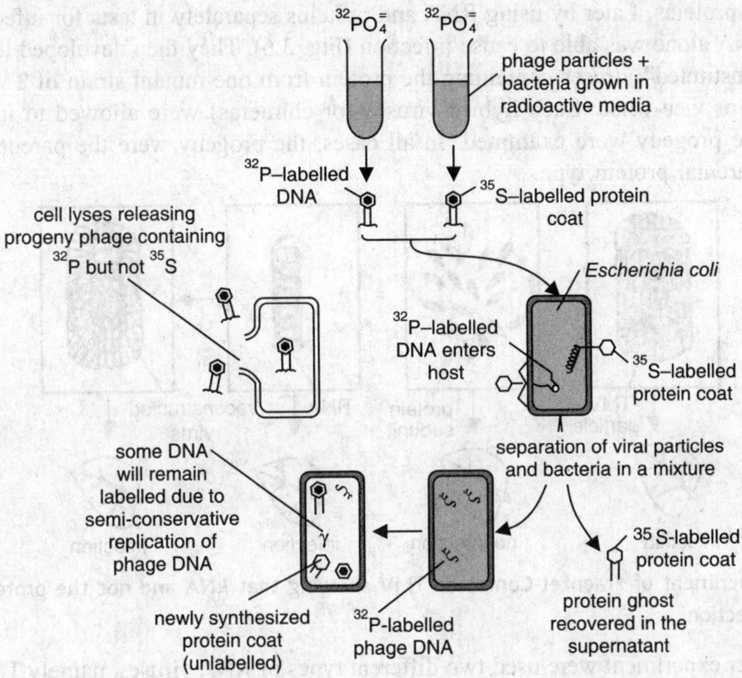

Fig. 3.5. A classical experiment of Hershey and Chase, where DNA of T_2 phage is labelled with ^{32}P and the protein is labelled with ^{35}S.

Table 3.2 DNA contents in picograms (gram × 10^{-12}) of various cells.

S.No.	*Organism*	*Kidney*	*Liver*	*Erythrocytes*	*Sperm*
1.	Chicken	2.4	2.5	2.5	1.3
2.	Bovine	6.4	6.4	—	3.3
3.	Carp	—	3.0	3.3	1.6
4.	Human	5.6	5.6	—	2.5

3. The diploid amount of DNA is constant within the species but varies from one species to another, as shown in the Table 3.2.

4. In 1950 **Swift** found the amount of DNA per resting nucleus (interphase nucleus) in mitotically active tissues of animals ranged from twice (2n) the value found in the gametes (1n) to four times (4n) this value. He concluded that DNA synthesis occurred during the interphase.

This parallelism of behaviour in DNA and chromosomes clearly indicates that DNA is the genetic material of higher animals.

3.3 EVIDENCES FOR RNA AS THE GENETIC MATERIAL OF SOME VIRUSES

The demonstration that RNA is the genetic material in RNA-containing viruses came in 1956, when **A. Gierer** and **G. Schramm** showed that tobacco plants could be inoculated with purified RNA from the tobacco mosaic virus (TMV), and TMV-like lesions could later be identified on the tobacco leaves. A different approach was taken by **H. Fraenkel-Conrat** and **B. Singer** in experiments published in 1957. They first developed the techniques for separating TMV particles into RNA and proteins. Later by using RNA and proteins separately in tests for infectivity, it was shown that RNA alone was able to cause infection (Fig. 3.6). They then developed techniques for forming "reconstituted" viruses containing the protein from one mutant strain of TMV and RNA from another, or vice-versa. Such hybrid viruses (or chimeras) were allowed to infect tobacco leaves, and the progeny were examined. In all cases, the progeny were the parental RNA type and not the parental protein type.

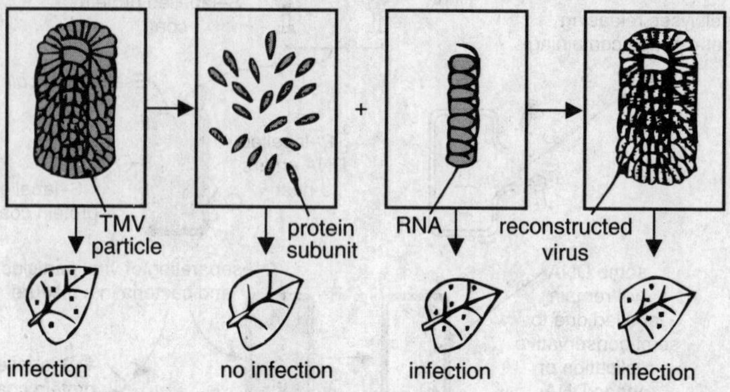

Fig. 3.6. Experiment of Fraenkel-Conrat on TMV showing that RNA and not the protein can cause infection.

In another experiment were used two different types of RNA viruses, namely TMV (tobacco mosaic virus) and HRV (Holmes rib-grass virus). The latter is also called **Plantago strain,** since this virus was isolated from *Plantago lanceolata.* Proteins of these two viruses differ in having different frequencies and sequences of amino acids. Also these viruses give different symptoms on leaves of a particular variety of tobacco, while TMV produces mottling of leaves, HRV produces distinct ring patterns. Reciprocal chimeras using RNA of one strain and protein of other strain were obtained. It was found that when these chimeras were used for infection, the progeny had proteins which correspond to the virus from which RNA of the infecting virus particle was derived (Fig. 3.7). This proved that specificity of virus proteins was determined by RNA alone and that proteins carried no genetic information.

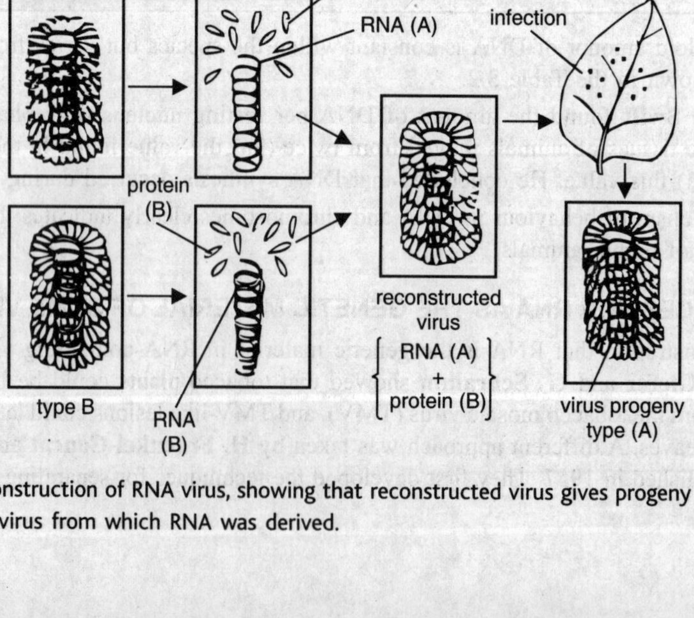

Fig. 3.7. Reconstruction of RNA virus, showing that reconstructed virus gives progeny which resembles the virus from which RNA was derived.

In a much improved technique **N. Pace** and **S. Spiegelman** in 1966 purified RNA from two different mutant strains of the RNA phage QB which had quite distinct base compositions. The isolated RNAs were then incubated separately in the presence of an *E. coli* cell extract containing an enzyme capable of RNA replication. The new RNA synthesized was in each case identical in base composition to the particular phage RNA presented to the *in vitro* system, thus, indicating that the phage RNA can serve as a template for its self-replication.

The genetic RNA is found to be either single-stranded or double-stranded (see Table 3.3).

Table 3.3 Different RNA viruses and the nature of genetic RNA associated with them.

S. No.	*Virus*	*Types of RNA*
1.	**Plant viruses**	
	TMV	Single stranded
	Wound tumour	Double stranded
2.	**Animal viruses**	
	Influenza virus	Single stranded
	Rous sarcoma	Single stranded
	Poliomyelitis	Single stranded
	Reovirus	Double stranded
3.	**Bacteriophages**	
	MS2, F2, r17	Single stranded

QUESTIONS

Long Answer Questions

1. DNA is the genetic material–furnish experimental evidence.

2. Describe two classical experiments which demonstrated that DNA is the genetic material.

3. Describe an experiment which established the genetic nature of RNA in certain viruses.

Short Answer Questions

1. Write about the contribution of the following molecular biologists regarding the identification of the genetic material:

(*i*) F. Griffith;

(*ii*) O.T. Avery, C.M. Macleod and M. McCarthy;

(*iii*) A.D. Hershey and M.J. Chase;

(*iv*) Fraenkel–Conrat.

Very Short Answer Questions

1. Differentiate between transformation and transduction.

2. (*a*) How did the transformation experiments of Griffith differ from those of Avery and his associates?

(*b*) What was the significant contribution of each?

(*c*) Why was Griffith's work not evidence for DNA as the genetic material whereas the experiments of Avery *et al.*, provided direct proof that DNA carried the genetic information?

3. How could it be demonstrated that the mixing of heat-killed Type III *pneumococcus* with live Type II resulted in a transfer of genetic material from Type III to Type II rather than a restoration of viability to Type III by Type II?

4. What is the macromolecular composition of a bacteriophage such as phage T_2?

5. What chemical properties do DNA and proteins possess that allow researchers to specifically label one or the other of these macromolecules with a radioactive isotope?

6. (*a*) What was the objective of the experiment carried out by Hershey and Chase?

(*b*) How was the objective accomplished?

(*c*) What is the significance of this experiment?

Fill in the Blanks

1. .... and H. Ris had reported that all cells of an organism contain .... amount of DNA.

2. In 1928, Frederick Griffith discovered the phenomenon of .... in *D. melanogaster*.

3. .... and Chase in 1952 demonstrated that the DNA injected by a phage particle into a bacterium carries all .... for synthesis of new phage particle.

4. Experiment of .... and .... in 1957 showed that TMV has .... as the genetic material.

Multiple Choice Questions

1. Who performed transformation experiments in bacteria?

(*a*) Louis Pasteur (*b*) Griffith

(*c*) Macleod (*d*) Meselson and Stahl

2. Bacterial transformation was discovered by

(*a*) Avery *et al.* (*b*) Beadle and Tatum

(*c*) Robert Brown (*d*) Har Govind Khurana

3. The dye involved in Feulgen reaction to stain DNA is

(*a*) Janus Green B (*b*) Neutral Red

(*c*) Basic Fuchsin (*d*) Haematoxylin

4. *Pneumococcus* experiment proved that

(*a*) bacteria undergo binary fission

(*b*) bacteria do not reproduce asexually

(*c*) RNA may sometimes control production of DNA and protein

(*d*) DNA is the genetic material

5. The process in which DNA of a bacterial cell is transferred into another bacterial cell through a virus is

(*a*) transduction (*b*) transformation

(*c*) conjugation (*d*) reproduction

6. Transformation experiment was performed on which bacteria

(*a*) *E. coli* (*b*) *Diplococcus pneumoniae*

(*c*) *Salmonella* (*d*) *Pasteurella pestis*

7. Who proved that DNA is the basic genetic material?

(*a*) Griffith (*b*) Boveri and Sutton

(*c*) Hershey and Chase (*d*) Watson

8. Who established that RNA is the genetic material?

(*a*) Nirenberg and Holley (*b*) Lederberg

(*c*) Fraenkel-Conrat (*d*) Griffith

9. A bacteriophage with radioactive DNA and protein when infects a bacterium the radioactivity inside the bacterium will be located

(*a*) in DNA (*b*) in protein

(*c*) both in DNA and protein (*d*) in all parts of bacterial cell

ANSWERS

Very Short Answer Questions

1. Transformation involves naked DNA from one cell becoming incorporated into another's DNA, whereas in transduction, a virus serves as the vector transferring DNA derived from one cell into another.
2. (*a*) Griffith's *in vivo* experiments demonstrated the occurrence of transformation in *Pneumococcus*. They provided no indication as to the molecular basis of the transformation phenomenon. Avery and colleagues carried out *in vitro* experiments, employing biochemical analyses to demonstrate that transformation was mediated by DNA.

   (*b*) Griffith showed that a transforming substance existed; Avery *et al.* defined it as DNA.

   (*c*) Griffith's experiments detected the substance responsible for transformation. Avery *et al.* isolated DNA in "pure" form and demonstrated that it could mediate transformation.
3. Purified DNA from Type III cells was shown to be sufficient to transform Type II cells. This occurred in the absence of any dead Type III cells.
4. About 1/2 protein, 1/2 DNA. A single long molecule of DNA is enclosed within a complex "coat" that is composed of many proteins.
5. DNA contains phosphorus (normally ^{31}P), but no sulphur, it can be labelled with radioactive phosphate ^{32}P. Proteins contain sulphur (normally ^{32}S), but usually no phosphorus; they can be labelled with radioactive sulphur ^{35}S.
6. (*a*) The objective was to determine whether the genetic material was DNA or protein.

   (*b*) By labelling phosphorus, a constituent of DNA, and sulphur, a constituent of protein, in a virus, it was possible to demonstrate that only the labelled phosphorus was introduced into the host cell during the viral reproductive cycle. This was enough to reproduce new phages.

   (*c*) Therefore DNA, not protein, is the genetic material.

Fill in the Blanks

1. A. Mirsky, same.
2. Transformation.
3. Hershey, genetic information.
4. Fraenkel-Conrat, Singer, RNA.

Multiple Choice Questions

1. (*b*) **2.** (*a*) **3.** (*c*) **4.** (*d*) **5.** (*a*) **6.** (*b*)

7. (*c*) **8.** (*c*) **9.** (*a*).

4

Chemical Nature of Genetic Materials (DNA and RNA) (Watson and Crick Model of DNA)

DNA (deoxyribonucleic acid) and RNA (ribonucleic acid), the principal genetic materials of living organisms, are chemically called **nucleic acids** and are complex molecules larger than most proteins and contain carbon, oxygen, hydrogen, nitrogen and phosphorus.

4.1 HISTORICAL

In 1869, **Friedrich Miescher**, 22-year-old Swiss physician (working in the laboratory of Felix Hoppe-Seyler in Tubingen) had isolated from pus cells obtained from discarded bandages in the Franko-Prussian War, and from salmon sperm, a previously identified macromolecular substance, to which he gave the name **nuclein**. Although he was unaware of the structure and function of nuclein, he submitted his findings for publication. The editor who received the paper was doubtful about some aspects of the Miescher's report and delayed publication for two years while he tried repeating some of the more questionable aspects of Miescher's work. Finally, in 1871, Miescher's report was published, but it made little immediate impact. He continued his careful work up to his death in 1895, recognizing (with the help of his student **Richard Altmann** in 1889) that nuclein was of higher molecular weight and was associated in some way with a basic protein, to which he gave the name **protamine**.

In 1880, **Emil Fischer** identified pyrimidines and purines. The biochemist, **Albrecht Kossel** identified the constituent nitrogenous bases of nuclein as well as its 5-carbon sugar and phosphoric acid. It was **Altmann** who first suggested, in 1899, the use of the term **nucleic acid** to describe phosphorus-containing nuclein. **Kossel** was awarded the 1910 Nobel Prize for demonstrating the presence of two pyrimidines (cytosine and thymine) and two purines (adenine and guanine) in nucleic acids. Kossel's work and later investigations of **Ascoli**, **Levine** and **Jones** during the first quarter of the 1900s disclosed the two kinds of nucleic acids, deoxyribonucleic acid (DNA) and ribonucleic acid (RNA). Development of DNA-specific staining techniques by **Feulgen** and **Rossenbeck** in 1924 enabled Feulgen to demonstrate in 1937 that most of the DNA content of the cell is located in the nucleus. It was not until the 1950s that the inter-nucleotide bond was established by **A.R. Todd** (**Judson**,1979). Because DNA is the only genetic material of most living organisms, so deserves priority in discussion.

4.2 DEOXYRIBONUCLEIC ACID OR DNA

Highly purified DNA, extracted from a wide variety of plants, animals, bacteria and viruses, has been found to be complex macro-molecular or polymeric chemical compound which contains four kinds of smaller building blocks (monomers) called **deoxyribotids** or **deoxyribonucleotides**. Each deoxyribonucleotide is made up of three moieties: a **phosphoric acid molecule** (biologically called phosphate); a pentose sugar called **2 deoxyribose**; and **pyrimidine** and **purine** nitrogenous bases. Four major kinds of nitrogenous bases have been found in four kinds of deoxyribonucleotides of DNA: two are heterocyclic and two-ringed purines, **adenine (A)** and **guanine (G)**, and two are one ringed pyrimidines, **cytosine (C)** and **thymine (T)**. The pyrimidine ring can be numbered in two different ways (Fig. 4.1 to 4.6). In this book, we will follow the old numbering system.

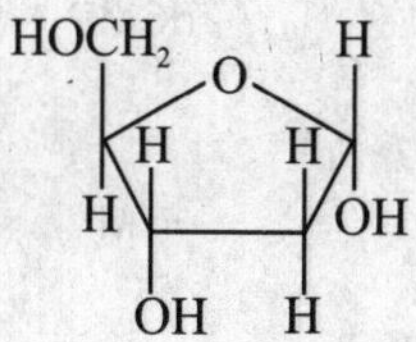

Fig. 4.1. Chemical formula of deoxyribose.

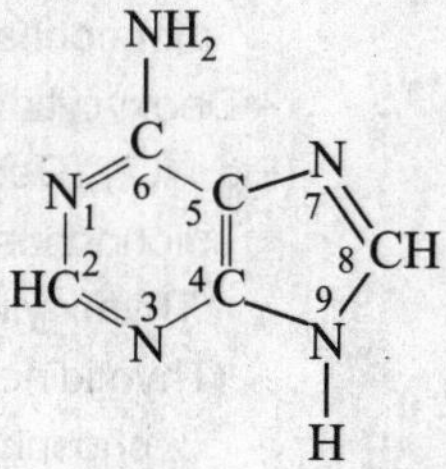

Fig. 4.2. Chemical formula of adenine (6-amino-purine, mw = 151.13 daltons).

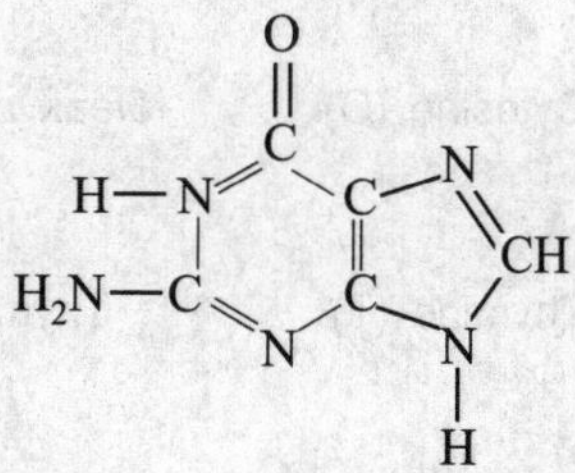

Fig. 4.3. Chemical formula of guanine (2-amino-6-hydroxipurine, purine, mw = 135.13 daltons).

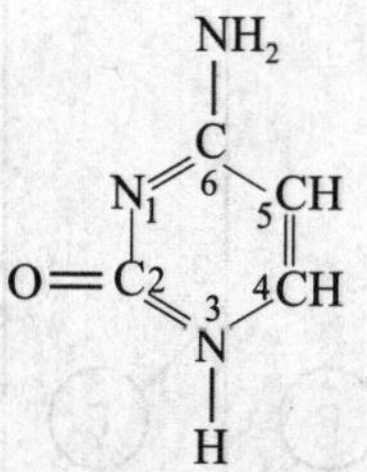

Fig. 4.4. Chemical formula of cytosine (2-hydroxy-6-amino pyrimidine, mw = 111.10 daltons).

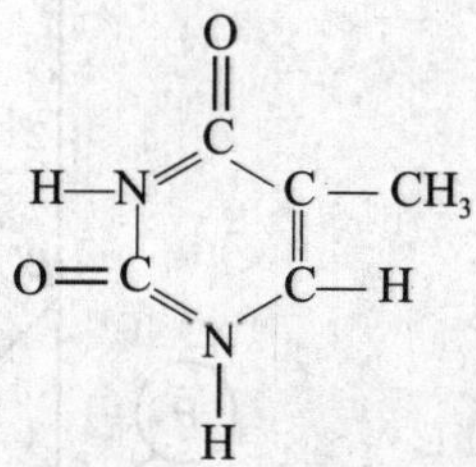

Fig. 4.5. Chemical formula of thymine (2,6-dihydroxy -5-methylpyrimidine, mw = 126.12 daltons.

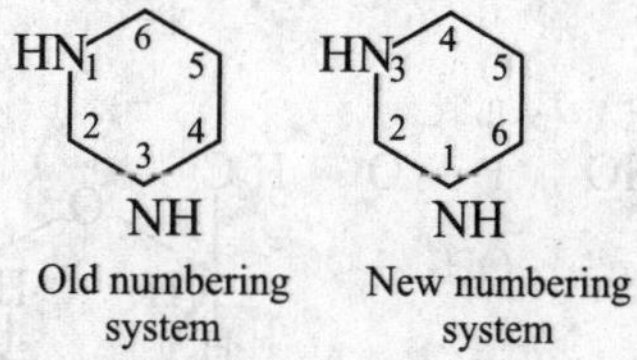

Fig. 4.6. Numbering of carbon atoms in a pyrimidine ring.

That part of each nucleotide which contains a nitrogenous base and deoxyribose is called **deoxyribonucleoside**. The four kinds of deoxyribonucleosides and deoxyribonucleotides can be tabulated as follows (see Table 4.1):

The four deoxyribonucleotides besides occurring in DNA molecule, occur also in nucleoplasm and cytoplasm, but in their triphosphate forms such as deoxyadenosine triphosphate (dATP), deoxyguanosine triphosphate (dGTP), deoxycytidine triphosphate (dCTP) and thymidine triphosphate (TTP). The significance of occurrence of deoxyribonucleotides in triphosphate forms lies in the fact that during DNA replication, the DNA polymerase enzyme can act only on triphosphate of deoxyribonucleotides.

Table 4.1 Four nitrogen bases, nucleosides and nucleotides of DNA molecule.

S. No.	*Nitrogen base*	*Base + deoxyribose = deoxyribonucleoside*	*Deoxyribonucleoside + phosphoric acid = deoxyribonucleotide*	*Abbreviation for nucleotide*
1.	Adenine (A)	Deoxyadenosine	Deoxyadenylic acid (Deoxyadenosine monophosphate)	dAMP
2.	Guanine (G)	Deoxyguanosine	Deoxyguanylic acid (Deoxyguanosine monophosphate)	dGMP
3.	Cytosine (C)	Deoxycytidine	Deoxycytidylic acid (Fig. 4.7) (Deoxycytidine monophosphate)	dCMP
4.	Thymine (T)	Thymidine	Thymidylic acid (Thymidine monophosphate)	TM

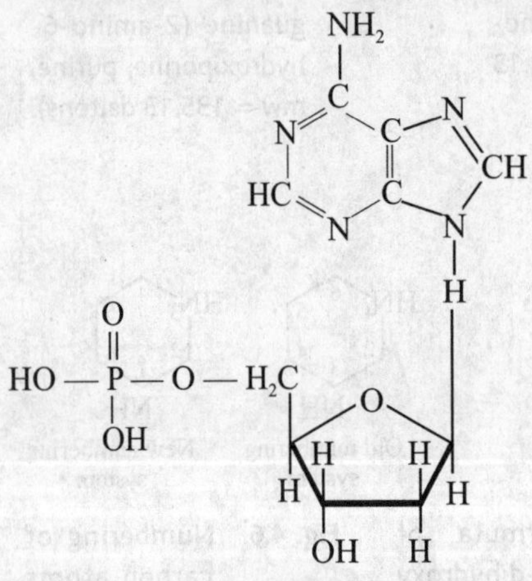

Fig. 4.7. Chemical formula of deoxycytidylic acid.

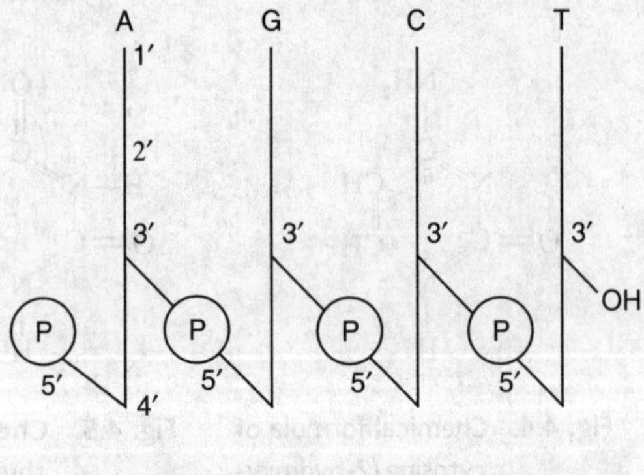

Fig. 4.8. A short hand system of representation of a polynucleotide with four nucleotides.

Molar Ratios of Nitrogen Bases in DNA Molecule

When many samples of DNA were isolated, purified and analysed by various techniques, such as paper chromatography, etc., it was found by **Hotchkiss** and **Chargaff** in 1948 that contrary to Levene's (1920's) tetranucleotide theory which considered DNA as a monotonous polymer having four DNA bases in approximately equal molar proportions, the four nucleotide bases are not necessarily present in DNA in exactly equal proportions. Thus, **Chargaff** reported that the DNA extracted from calf thymus nuclei contains the four bases in the following molar proportions: 28% adenine, 24% guanine, 20% cytosine and 28% thymine. When he analyzed the DNA samples of varied animals, he found that the exact base composition of DNA differs according to its biological source. The relative amounts of purines and pyrimidines in samples of DNA of different living organisms are tabulated as follows (Table 4.2):

Table 4.2 Relative amounts of nitrogen bases in different samples of DNA (**Chargaff** and **Davidson**,1955).

S. No.	*Source*	*Adenine*	*Guanine*	*Cytosine*	*Thymine*	$\frac{A+T}{G+C}$
1.	Beef sperm	28.7	22.2	22.0	27.2	1.26
2.	Human thymus	30.9	19.9	19.8	29.4	1.52
3.	Human liver	30.3	19.5	19.9	30.3	1.53
4.	Human sperm	30.7	19.3	18.8	31.2	1.62
5.	Hen red blood cells	28.8	20.5	21.5	29.2	1.38
6.	Rat bone marrow	28.6	21.4	21.5	28.4	1.33
7.	Herring sperm	27.8	22.2	22.6	27.5	1.23
8.	*Paracentrotus lividus* (sea urchin) sperm	32.8	17.7	18.4	32.1	1.85
9.	Salmon	29.7	20.8	20.4	29.1	1.43
10.	Wheat germ	26.5	23.5	23.0	27.0	1.19
11.	Yeast	31.3	18.7	17.1	32.9	1.79
12.	*Diplococcus pneumoniae*	29.8	20.5	18.0	31.6	1.59
13.	*K-12 Escherichia coli*	26.0	24.9	25.2	23.9	1.00
14.	*Mycobacterium tuberculosis*	15.1	34.9	35.4	14.6	0.42
15.	Bacteriophage T_2	32.5	18.2	16.7	32.6	1.86

These different values of A, G, C, and T in different samples of DNA are suggesting that DNA rather than being a monotonous polymer, carries genetic information in the form of specific nucleotide base sequences.

The equivalence rule. Erwin Chargaff, in 1950, formulated the equivalence rule which suggested that despite wide compositional variations exhibited by different types of DNA, the total amount of purines equaled the total amount of pyrimidines (A + G = T + C); the amount of adenine equaled the amount of thymine (A = T) and the amount of guanine equaled the amount of cytosine (G = C). **Chargaff's equivalence rule** has been found to apply almost universally in different organisms (viruses, bacteria, plants and animals). However, DNA isolated from higher plants and animals was in general rich in adenine and thymine (A : T) and relatively poor in guanine and cytosine (G : C) (*e.g.*, AT/GC ratio of DNA of man was 1.40 : 1); whereas DNA isolated from microorganisms (viruses, bacteria and lower plants and animals) was in general rich in guanine and cytosine and relatively poor in adenine and thymine (*e.g.*, AT/GC ratio of DNA of *Mycobacterium tuberculosis* was 0.60 : 1). These differences in AT/GC ratios of microorganism and higher organisms undoubtedly reflect the difference in genetic information carried by these hereditary molecules and also phylogenetic, evolutionary and taxonomical significance of them.

Physical, Molecular or Geometrical Organization of DNA

The first person to give any thought to the three dimensional structure of DNA was **W.T. Astbury** who by his X-ray crystallographic studies of DNA molecule concluded in 1938 that because DNA has high density, so, its polynucleotide was a stack of flat nucleotides, each of which was oriented perpendicularly to the long axis of the molecule and was situated every 3.4 Å along the stack. The X-ray crystallographic studies of Astbury were continued by **Wilkins** and his associates who managed to prepare highly oriented DNA fibres that allowed them to obtain an X-ray diffraction photograph (Fig. 4.9) One of his female associates **Rosalind Franklin** obtained a superior X-ray diffraction photograph of DNA which confirmed Astbury's earlier inference of 3.4 internucleotide distance. Such studies demonstrated that DNA was a helical structure with a diameter of 20 Å and a pitch (one round) of about 34 Å. **Watson** and **Crick** who were already engaged in constructing some suitable model for DNA structure, when observed Franklin's picture of DNA molecule, they immediately utilized that information in constructing a molecular model for DNA. In April 1953, **Watson** and **Crick** published their conclusions about the structure of the DNA in the same issue of '**Nature**', in which **Wilkins** and his colleagues presented the X-ray evidence for that structure.

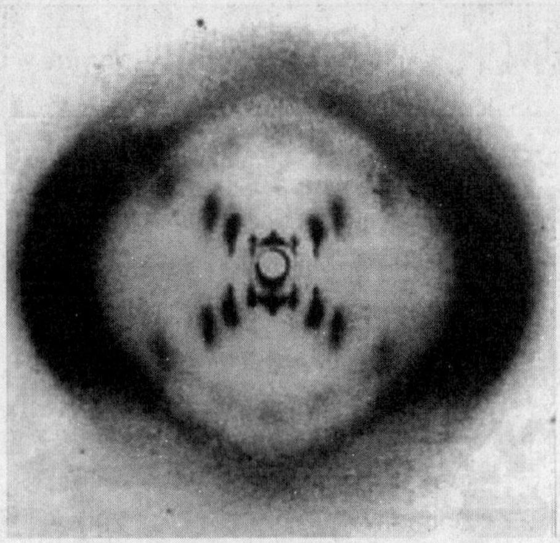

Fig. 4.9. Franklin's original X-ray diffraction photograph.

Considerations of Watson and Crick in the construction of double helical structure of DNA molecule. **Watson** and **Crick** concluded directly from the X-ray diffraction photograph of DNA taken by **Franklin** that (1) the DNA polynucleotide chain has the form of a regular helix, (2) the helix has a diameter of about 20Å and, (3) the helix makes one complete turn every 34Å along its length, and hence, since the internucleotide distance is 3.4Å, consists a stack of ten nucleotides per turn. Considering the known density of the DNA molecule, **Watson** and **Crick** next concluded that the helix must contain two polynucleotide chains, or two stacks of ten nucleotides each per turn, since the density of a cylinder 20Å in diameter and 34Å long would be too low if it contained but a single stack of ten, and too high if it contained three or more stacks of ten nucleotides each. Before trying to arrange these two polynucleotide chains into a regular helix of the required dimensions, however, **Watson** and **Crick** placed a further restriction on their model–a restriction that derived from their knowledge that DNA is, after all, the genetic material. If DNA is to contain hereditary information as they reasoned, and if that information is inscribed as a specific sequence of the four bases along the polynucleotide chain, then the molecular structure of DNA must be able to accommodate any arbitrary sequence of bases along its polynucleotide chains. Otherwise, the capacity of DNA as an information carrier would be too severely limited. Hence, they felt the need of construction of such a regular helix that though is composed of two polynucleotide chains containing an arbitrary sequence of nucleotide bases every 3.4Å along their length, it would, nevertheless, have a constant diameter of 20Å. Since the dimension of the purine ring is greater than that of the pyrimidine ring, **Watson** and **Crick** hit upon the idea that the two-chain helix could have a constant diameter if there existed a complementary relation between the two nucleotide stacks, so that at every level one stack harbours a purine base and the other a pyrimidine base. Finally, to endow the helix with thermodynamics stability, the structure would have ample opportunities for the formation of hydrogen bonds between amino or hydroxyl-hydrogens and keto-oxygens or immino-nitrogens of the purines and pyrimidine bases. These considerations led them to construct a double helix model for the molecular structure of DNA molecule.

4.3 WATSON AND CRICK'S STRUCTURAL MODEL OF DNA

In DNA molecule the adjacent deoxyribonucleotides are joined in a chain by phosphodiester bridges or bonds which link the 5′ carbon of the deoxyribose of one mononucleotide unit with the 3′ carbon of the deoxyribose of the next mononucleotide unit (Fig. 4.8). According to **Watson** and **Crick** DNA molecule consists of two such polynucleotide chains wrapped helically around each other, with the sugar-phosphate chain on the outside (forming ribbon-like backbone of double helix) and purines and pyrimidines on the inside of the helix (projecting between two sugar phosphate backbones as transverse bars). The two polynucleotide strands are held together by hydrogen bonds between specific pairs of purines and pyrimidines (Box 4.1).

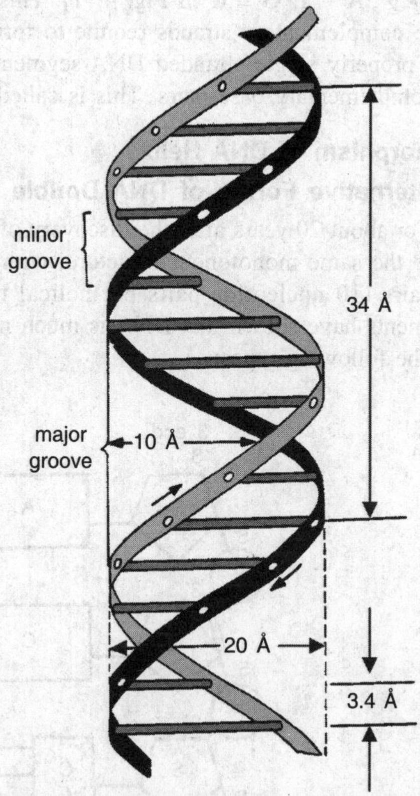

Fig. 4.10. Watson-Crick double helical DNA molecule.

The hydrogen bonds between purines and pyrimidines are such that adenine can bond only to thymine by two hydrogen bonds, and guanine can bond only to cytosine by three hydrogen bonds and no other alternative is possible between them. The specificity of the kind of hydrogen bonds that can be formed assures that for every adenine in one chain there will be thymine in the other. For every guanine in first chain there will be a cytosine in the other and so on. Thus, the two chains are complementary to each other; that is, the sequence of nucleotides in one chain dictates the sequence of nucleotides in the other. The two strands run anti-parallely–that is, have opposite directions. One strand has phosphodiester linkage in 3′ → 5′ direction, while other strand has phosphodiester linkage in just reverse or 5′ → 3′ direction. Further, both polynucleotides strands remain separated by 20Å distance. The coiling of double helix is right handed and a complete turn occurs every 34 Å. Since each nucleotide occupies 3.4 Å distance along the length of a polynucleotide strand, ten mononucleotides occur per complete turn. (The base pairs are rotated 36° with respect to each adjacent pair). The helix has two external grooves, a deep wide one, called **major groove** and a shallow narrow one, called **minor groove** (Fig. 4.10): both of these grooves are large enough to allow protein molecules to come in contact with the bases.

Box 4.1

1. The width (or diameter) of a DNA double helix is 20 Å (2 nm). Each turn (pitch) of the helix is 34 Å (3.4 nm) with 10 pairs of nucleotides, each pair placed at a distance of about 3.4 Å (0.34 nm).
2. Each strand of DNA has a **hydrophilic** deoxyribose phosphate backbone (3'-5' phosphodiester bonds) on the outside (periphery) of the molecule while **hydrophobic** bases are stacked inside (core).
3. In the DNA molecule, the genetic information resides on one of the two strands known as **template strand** or **anti-sense** strand. The complementary strand is the **sense strand.**

Denaturation and Renaturation of DNA

The two complementary polynucleotide strands of DNA easily separate on heating (or by titration with alkali or acid) by the breakdown of hydrogen bonds between their nitrogenous bases (*e.g.*, A = T, G ≡ C in Fig. 4.11). This is called **melting** or **denaturation**. On cooling, the two complementary strands reunite to form double stranded DNA (**renaturation**). Because of this property single stranded DNA segments from different sources can join, provided these have complementary base pairs. This is called **annealing**.

Polymorphism of DNA Helix
(Or Alternative Forms of DNA Double Helix)

For about 20 years after the discovery of the DNA double helix in 1953, DNA was thought to have the same monotonous structure, with exactly 36° of helical twist between its adjacent base pairs (10 nucleotide pairs per helical turn) and a uniform helix geometry. Subsequent experiments have shown that DNA is much more **polymorphic** than expected. Thus, DNA can be of the following types:

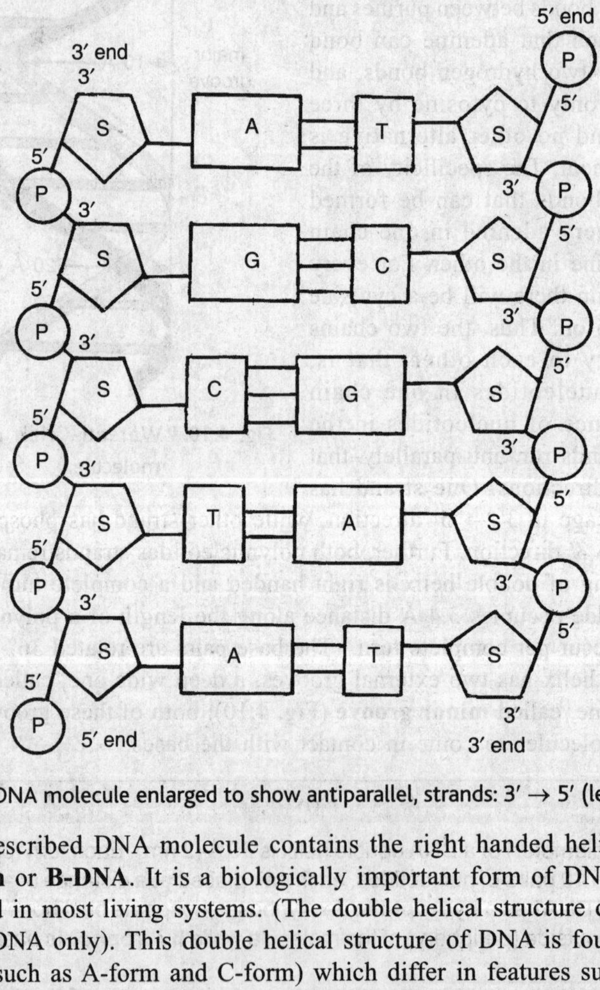

Fig. 4.11. A part of DNA molecule enlarged to show antiparallel, strands: 3′ → 5′ (left) and 5′ → 3′ (right).

The above described DNA molecule contains the right handed helical coiling and has been called **B-form** or **B-DNA**. It is a biologically important form of DNA that is commonly and naturally found in most living systems. (The double helical structure described by Watson and Crick was B–DNA only). This double helical structure of DNA is found to exist in other alternative forms (such as A-form and C-form) which differ in features such as (*i*) number of residues (monomers) per turn ("n") or (*ii*) the spacing of residues along the helical axis ("h")

(Table 4.3). For example, **A - form DNA** (A-DNA) is right handed but less hydrated than the B-form DNA. A-DNA is more compact with 11 base pairs per turn of the helix and it is 23 angstroms in diameter. The bases are tilted more in relation to the axis of the helix than in the B-DNA. The A-DNA may occur under experimental conditions.

Table 4.3 Important features of different forms of DNA double helical structures.

Helix type	*Conditions*	*Base pair per turn*	*Rotation per bp*	*Vertical rise per bp*	*Helical diameter*
A	75 per cent relative humidity; Na^+, K^+, Cs^+, ions	11	+ 32.7° (right handed)	2.56Å	23Å
B	92 per cent relative humidity, low ionic strength	10	+ 36.0° (right handed)	3.38Å	19Å
C	66 per cent relative humidity; Li^+ ions	9.33	+ 38.6° (right handed)	3.32Å	19Å
Z	Very light salt concentration	12	– 30.0° (left handed)	3.71Å	18Å

The B-DNA is found in fibres of living cells at a very high 92 per cent relative humidity and low ionic strength. Likewise, A-form of DNA is found at 75 per cent humidity in the presence of high ionic strength of NA^+, Ka^+ or Cs^+ ions. C-form of DNA is found at 66 per cent relative humidity in the presence of lithium (Li^+) ions. These three forms are assumed to be found in all DNAs.

There are certain other forms of DNA such as **D-form** and **E-form**, both of which are found as rare extreme variants and contain only 8 and 7½ base pairs per turn respectively. These rare DNA variants are found only in some DNA molecules which lack guanine.

Z-DNA (or Left-handed DNA). Crystallographic studies on synthetic nucleotides consisting of alternating purines and pyrimidines such as GCGCGCGC have shown that left-handed DNA can also exist. This DNA is called **Z-DNA** because of its zigzag structure. Z-DNA can also be found in solutions of high-ionic strength, such as 2M NaCI. In Z-form of DNA, the molecule still consists of two anti-parallel chains, but, otherwise, it is quite different from the A or B form (Fig. 4.12). The helix of Z-DNA is 18 angstrom in diameter, containing 12 base pairs per turn. The differences and similarities of Z-DNA and B-DNA have been summarized in Table 4.3; some other features are the following :

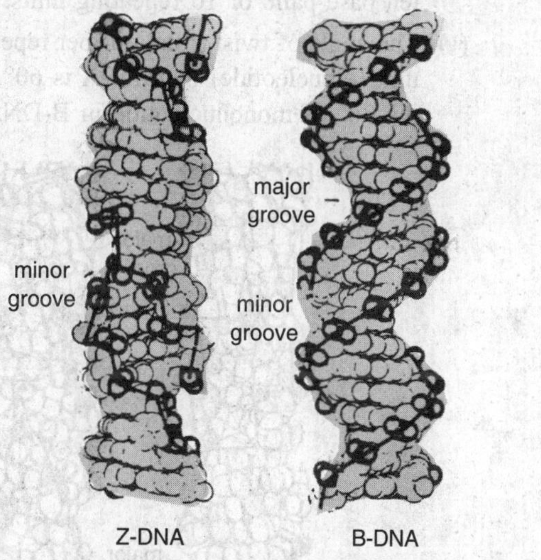

Fig. 4.12. Two molecular forms of DNA double helix, right-handed B-DNA and left-handed Z-DNA. The heavy black line in each molecule goes from phosphate group to phosphate group, indicating a smooth right-handed helix in B-DNA and irregular or zigzag left-handed helix in Z-DNA (after De Robertis and De Robertis Jr., 1987).

A. Similarities between Z-DNA and B-DNA

(1) Both are double helical.

(2) In both DNAs, two polynucleotide strands of double helix are antiparallel.

(3) Both forms exhibit G ≡ C pairing.

B. Differences between Z-DNA and B-DNA

(*i*) Z-DNA has left-handed helical sense, while B-DNA has right-handed helical sense (Fig. 4.13).

(*ii*) The phosphate backbone of Z-DNA follows a zigzag course, while in B-DNA this backbone is regular.

(*iii*) In Z-DNA, the adjacent sugar residues have opposite orientation, while in B-DNA they have same orientation. Due to this, repeating unit is a dinucleotide in Z-DNA as against a mononucleotide unit in B-DNA (Fig. 4.14).

(*iv*) In Z-DNA, one complete helix (*i.e.*, a twist through 360°) has twelve base pairs or six repeating dinucleotide units, while in B-DNA one complete helix has only ten base pairs or 10 repeating units.

(*v*) The angle of twist (rotation) per repeating unit (dinucleotide) in Z-DNA is 60° than the 36° of mononucleotide in B-DNA.

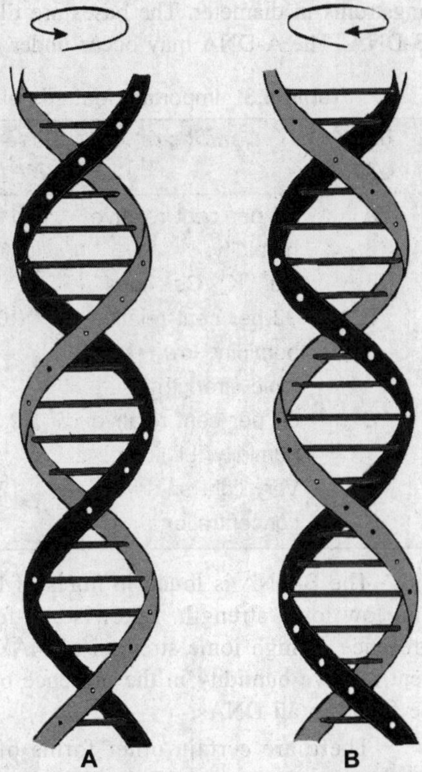

Fig. 4.13. The normal right handed helical sense of B-DNA (A) and rare left handed helical sense of Z-DNA(B).

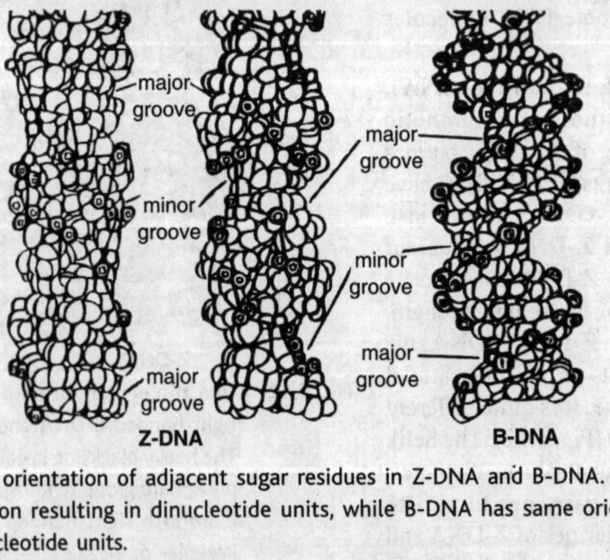

Fig. 4.14. Mode of orientation of adjacent sugar residues in Z-DNA and B-DNA. Z-DNA has opposite orientation resulting in dinucleotide units, while B-DNA has same orientation resulting in mononucleotide units.

(*vi*) In Z-DNA, one complete turn of helix is 45Å long, while in B-DNA it is 34Å long.

(*vii*) Since bases get more length spread out in Z-DNA and since the angle of tilt is 60°, they are more closer to the axis. Due to this fact, the diameter of Z-DNA molecule is 18 Å than the 20 Å diameter of B-DNA.

Regions of Z-DNA may be involved in gene regulation in the cells of "higher" organisms. Short regions of such drastically altered helical geometry could be specifically recognized by gene regulatory proteins, and thereby have important biological roles (see **Alberts** *et al.*, 1989). Lastly, the presence of any alternate configuration (*e.g.*, Z-DNA) suggests that DNA is a more flexible molecule than was previously thought and that it can adopt in the genome of a variety of forms (**Rich**, 1980).

Biological Significance of Watson and Crick's Model of DNA

DNA as the hereditary or genetic material of cell is expected to possess following three separate special features:

1. Self duplication. The Watson and Crick's model of DNA provides the only possible device whereby a particular molecule with specific sequence of base pairs could be duplicated easily. Each strand of double helix of DNA can construct its complementary strand by acting as a template. The sequence of nitrogenous bases determines the laying down of bases in the second or complementary strand.

2. Transmitting information. The DNA structure (*i.e.*, double helix) explains transmitting of information for the synthesis of proteins. Since the information are coded in the sequence of nitrogenous bases, the same sequence if coded in some other molecule (messenger RNA or mRNA) can be carried out to the site of protein synthesis. One of the two strands of DNA molecule can synthesize a new strand of mRNA, which in turn controls the proper alignment of amino acids in the protein chain to be formed.

Thus, DNA stores all the information to control growth, development, differentiation, reproduction, etc., of most living organisms. These genetic informations of DNA are in the form of base triplet called **codes.** Each code is formed of sequence of three nucleotides of DNA. Thus informations of DNA are carried by a single stranded mRNA formed by the process of **transcription.** Genetic message or mRNA is also present in the form of base triplets called **codons.** The sequence of codons on mRNA is called **genetic code.** The genetic code is decoded on the ribosomes (*i.e.*, protein factories which contain another kind of RNA transcribed by DNA) to form proteins by linking specific number and types of amino acids in a specific sequence (of polypeptides and proteins). This process is called **translation.** Some proteins act as **structural proteins** (*e.g.*, cytoskeleton, collagen, muscles, etc.) while some proteins act as **enzymes** and regulate the functioning (metabolism) of the cell. As already has been mentioned earlier, DNA is also capable of its self-duplication or replication so that daughter cells have same blue print or genetic material as in the parent cell. This unidirectional flow of genetic information is called **central dogma** (Fig. 4.15).

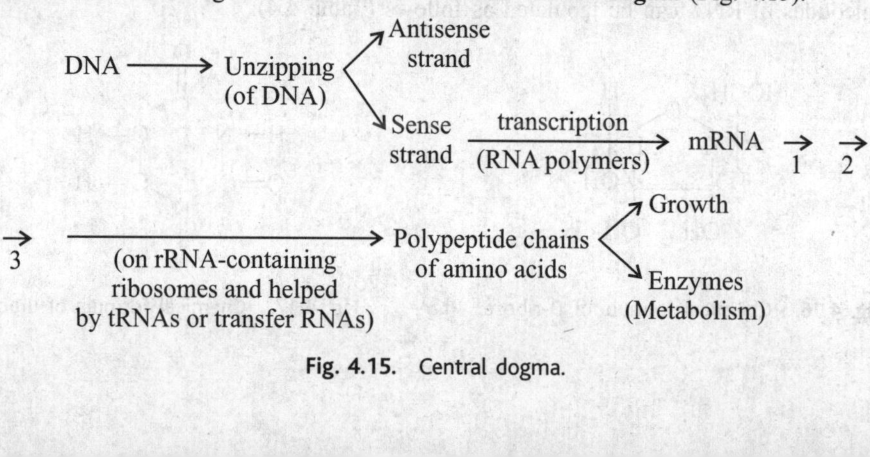

Fig. 4.15. Central dogma.

3. DNA repair. DNA molecule is relatively fragile and may get damaged from shear bending forces and by local changes in pH. It base sequence changes by chemical agents from environment and also by X-ray and ultraviolet radiations. Actually the double helical structure of DNA presents the most stable structure. It is most suited for the long term preservation of the coded informations. Because the base sequence in the two polynucleotide chains is complementary, the coded informations in a DNA molecule are preserved in duplicate. If sequence of bases on one chain is disturbed, the correct sequence of the other chain may help in correcting the errors that occur in the complementary strand and sends the correct message.

Single Stranded DNA

Sinsheimer (1959) isolated an unusual type of DNA from a small bacteriophage, $\phi \times 174$ coliphage (ϕ = phi, a Greek alphabet in lower case), which attacks *Eshcerichia coli* bacterium. This DNA is formed of about 5,400 nucleotides (5,386 nucleotides according to **Malacinski** 2005) and its is circular and single stranded. Unlike double stranded DNA, it is not resistant to formaldehyde. Its X-ray crystallographic studies showed :

1. It is **circular** in shape and does not show helical configuration.

2. It does not follow equivalence rule of nitrogen bases. The proportion of A : T : G : C is 1 : 1 : 33 : 0.98 : 0.75. So A ≠ T and G ≠ C.

3. It is not a molecule of uniform diameter. EM studies showed it to be more kinky and less thicker than double-stranded DNA.

4. The amino groups of the bases reacted rapidly with formaldehyde which indicated that its bases are exposed.

On entering inside the host (*i.e.*, bacteria), its DNA polymerase enzyme synthesizes a "**minus strand**" which is complimentary to the "**plus strand**" of parental virus. Two strands together form a **plus-minus DNA-helix** which is also called the **replication form (RF)**.

DNA of $\phi \times 174$ virus (one of the smallest known viruses) is circular and single stranded with a molecular weight of about 1.7 million and a contour length of only 1.77 μm.

4.4 RIBONUCLEIC ACID (RNA)

Some plant viruses (*e.g.*, TMV, turnip yellow mosaic viruses, wound tumour viruses, etc.,), animal viruses (*e.g.*, influenza viruses, foot and mouth viruses; rous sarcoma viruses, poliomyelitis viruses, reoviruses, etc.) and bacteriophages (*e.g.*, MS_2, etc.) contain ribonucleic acid (RNA) as their genetic material. Like DNA, RNA is polymeric nucleic acid of four monomeric **ribotids** or **ribonucleotides**.

Each ribonucleotide contains a pentose sugar (**D-ribose**; Fig. 4.16); a molecule of phosphate group and nitrogen base. The nitrogen bases of RNA are two purines, **adenine** and **guanine** and two pyrimidines, **cytosine** and **uracil** (Fig. 4.17). The four bases, ribonucleosides and ribonucleotides of RNA can be tabulated as follows (Table 4.4):

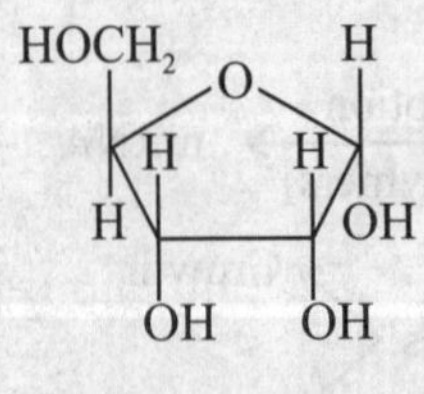

Fig. 4.16. Chemical formula of D-ribose.

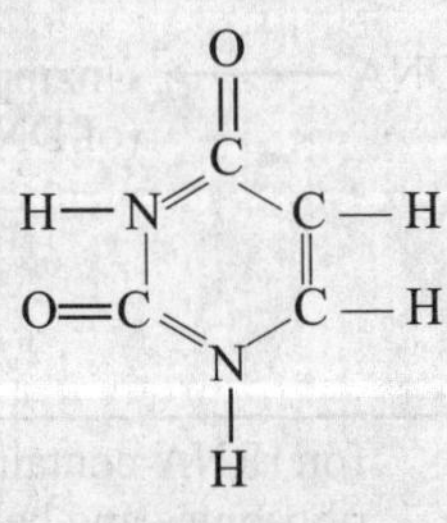

Fig. 4.17. Chemical formula of uracil.

Table 4.4 Four components of RNA.

S. No.	*Base*	*Ribonucleoside*	*Ribonucleotide*	*Abbreviation for ribonucleotide*
1.	Adenine (A)	Adenosine	Adenylic acid (Adenosine monophosphate)	AMP
2.	Guanine (G)	Guanosine	Guanylic acid (Guanosine monophosphate)	GMP
3.	Cytosine (C)	Cytidine	Cytidylic acid (Cytidine monophosphate)	CMP
4.	Uracil (U)	Uridine	Uridylic acid (Uridine monophosphate)	UMP

The four ribonucleotides also occur freely in nucleoplasm but in the form of triphosphates of ribonucleosides such as adenosine triphosphate (ATP) and uridine triphosphate (UTP).

Molecular Structure of RNA

RNA molecule may be either single stranded or double stranded but not helical like DNA molecule (Table 4.5). Single stranded RNA occurs as genetic material in plant viruses (*e.g.*, TMV, TYM), animal viruses (*e.g.*, influenza viruses, foot and mouth viruses, rous sarcoma viruses, poliomyelitis viruses) and bacteriophages (*e.g.*, MS_2). The non-genetic RNAs except tRNA of prokaryotes and eukaryotes, also have single stranded RNA molecules. The double stranded but non-helical RNA occurs as the genetic material in some plant viruses (*e.g.*, reoviruses). The transfer or soluble RNA (tRNA or sRNA) which is non-genetic RNA of prokaryotes and eukaryotes, is double stranded but non-helical structure.

Each strand of RNA is polynucleotidic, that is, made up of many ribonucleotides. In the polynucleotide strand of RNA, the ribose and phosphoric acids of nucleotides remain linked by phosphodiester bonds. The organisms which have only RNA, is called **genetic RNA**. While the organisms which have DNA along with RNA, they use the RNA in carrying the orders of DNA and in them because RNA has no genetic role, so called **non-genetic RNA**. The non-genetic RNA is heterogeneous and includes the following three genera: ribosomal RNA (rRNA), transfer RNA (tRNA) and messenger RNA (mRNA). Each genera of non-genetic RNA has **DNA-dependent replication** (transcription) of itself, that is, it is not self-replicating like DNA and is transcribed by DNA.

Replication of Genetic RNA

The genetic RNA of viruses is self-replicating, that is, it can produce its own replica by itself. So, its mode of replication is called **RNA-dependent RNA synthesis**. The genetical research on genetic RNA has revealed the following facts about it.

(1) The viral RNA functions directly as a messenger RNA which, in association with the ribosomal apparatus of the host, directs the synthesis of both the **RNA polymerase enzyme** (required for RNA replication) and the proteins of the viral coat.

(2) With the mediation of RNA polymerase and on the standard base-pairing principles, the viral RNA serves as a template in the synthesis of a complementary RNA chain, and thus a double stranded structure is produced.

Table 4.5 DNA - RNA similarities and differences. (*Source:* **Sylvia S. Mader,** 1998).

A. DNA - RNA similarities

1. Both are nucleic acids.
2. Both are composed of nucleotides.
3. Both have a sugar-phosphate backbone.
4. Both have four different types of bases.

B. DNA - RNA differences

S. No.	*DNA*	*S. No.*	*RNA*
1.	Found in the nucleus	1.	Found in nucleus and cytoplasm
2.	The genetic material	2.	Helper to DNA
3.	Sugar is deoxyribose	3.	Sugar is ribose
4.	Bases are A,T,C,G	4.	Bases are A,U,C,G
5.	Double stranded	5.	Single stranded
6.	Is transcribed (to give mRNA)	6.	Is translated (to give proteins)

QUESTIONS

Long Answer Questions

1. Discuss, in brief, the structure of deoxyribonucleic acid and compare it with that of ribonucleic acid.

2. Give an account of Watson and Crick's double-stranded model of DNA molecule.

3. Do the two strands of DNA helix carry the same genetic information? Explain.

4. Give an account of single stranded DNA of coliphage $\phi \times 174$.

Short Answer Questions

1. What kind of evidence indicates (1) that DNA can reproduce itself and (2) that nucleotides occur in matched pairs in DNA molecule?

2. Write short notes on the following:

(*i*) Z-DNA, (*ii*) B-DNA,

(*iii*) Tetranucleotide hypothesis, (*iv*) Left-handed DNA versus right-handed DNA.

3. Explain denaturation and renaturation of DNA.

4. Write about biological significance of Watson and Crick's model of DNA molecule.

Very Short Answer Questions

1. Why is orientation of complementary strands of DNA duplex called anti-parallel.

2. Assume the following base sequence was found in a 20 base DNA strand:

3′ ATT CGA CCT TAT TAC TGC AC 5′

(*a*) What would be the first 5 bases in the 3′ end of the complementary strand?

(*b*) What would be the 10 bases of the 5′ end of the complementary strand?

(*c*) Assuming the presence of complementary strands, what is the per cent composition of the polymer with respect to AT base pairs and with respect to GC base pairs.

3. Analysis of four double-stranded DNA samples yielded the following information:

1.15% cytosine; 2.12% guanine; 3.35% thymine; 4.28% adenine.

(*a*) What would be the percentage of the other bases in each sample?

(*b*) Could any of these samples have been obtained from the same organism? If so, which ones?

4. Ratio of the bases present in different samples of nucleic acid yielded the following results:

(a) (A + C)/(T + G) = 1, (b) (A + C)/(U + G) = 0.8,

(c) (A + G)/(T + C) = 1.5.

Which were RNA and which were DNA? Which were single and double-stranded?

5. If one DNA sample had a melting temperature of 85.5°C and another showed a melting temperature of 88°C. What might you conclude concerning the base composition of the two samples?

6. Assume an average-sized gene consisting of a linear sequence of 1000 bases and there were 1000 genes in a bacterial chromosome.

(a) How many bases would such a chromosome contain in each strand of the double helix?

(b) If 10 nucleotides = 34Å, how long would this chromosome would be in millimeters?

7. (a) What background material did Watson and Crick's have available for developing a model of DNA?

(b) What was their contribution to the building of the model?

8. (a) Why was a double helix chosen for the basic pattern of the molecule?

(b) Why were hydrogen bonds placed in model to connect the base?

9. What are differences between DNA and RNA?

10. RNA was extracted from TMV (tobacco mosaic virus) particles and found to contain 20 per cent cytosine. Using this information, is it possible to predict what percentage of the bases in TMV are adenine? If so, what percentage? If not, why not?

Fill in the Blanks

1. The equivalence rule for DNA molecule was formulated by .... in 1950.
2. Double helical model for DNA molecule was postulated by .... and .... in ....
3. B-DNA molecule contains .... helical coiling.
4. Z-DNA molecule contains .... helical coiling.
5. Both polynucleotide strand of DNA are separated by .... distance.
6. In a polynucleotide strand of DNA deoxyribonucleotides are joined by ....
7. DNA molecule makes a complete turn after every .... base pairs.
8. .... is the technique to determine the 3-dimensional structure of DNA.
9. Viruses such as TMV and TYMV contain .... as their genetic material.

Multiple Choice Questions

1. In a sample, DNA is found to have the base composition (mole ratio) of adenine = 40, T = 22, G = 21 and cytosine = 17. It shows that

(a) DNA is circular duplex (b) DNA is linear duplex

(c) DNA is single stranded (d) DNA has melting point

2. In the double helix model of DNA, how far is each base pair from the next base pair?

(a) 3.4 nm (b) 0.34 nm

(c) 2.0 nm (d) 34 nm

3. The scientist who developed cytochemical technique for identification DNA was

(a) McClintok (b) Abbe

(c) Fuelgen and Rossenbeck (d) Waldeyer

4. What is common point of similarity between DNA and RNA?

(*a*) both are double stranded (*b*) both have identical sugar molecules

(*c*) both have identical pyrimidine bases (*d*) both are polymers of nucleotides

5. DNA strands are antiparallel because of

(*a*) hydrogen bonds (*b*) phosphodiester bonds

(*c*) disulphide bonds (*d*) glycosidic bonds

6. Chargaff's rule is applicable to

(*a*) single stranded RNA (*b*) single stranded DNA and RNA

(*c*) single stranded DNA (*d*) double stranded DNA

7. The diameter of Z-DNA molecule is

(*a*) 18Å (*b*) 22Å

(*c*) 45Å (*d*) 34Å

8. What is the type of coiling in DNA?

(*a*) right handed (*b*) left handed

(*c*) opposite (*d*) zig-zag

9. Diameter of DNA helix is

(*a*) 20Å (*b*) 10Å

(*c*) 30Å (*d*) 40Å

10. Which ratio is constant for DNA?

(*a*) A + G/T + C (*b*) A + T/G + C

(*c*) A + C/U + G (*d*) A + U/C + G

ANSWERS

Very Short Answer Questions

1. Two chains of DNA molecule run anti-parallely, *i.e.*, both have opposite directions. One strand has phosphodiester bonds in 3′ → 5′ direction, while other strand has phosphodiester bonds in just reverse or 5′ → 3′ direction.

2. (*a*) Bases at the 3′ end of the opposite strand will be complementary to those at the end 5' end of the polymer given. Therefore, the first 5 bases will be 3′ GTGCA.

(*b*) 5′ TAA GCT GGAA

(*c*) A-T = 60 per cent, G-C = 40 per cent.

3. (*a*) 1.15% C, 15% G, 35% T, 35% A.

2.12% G, 12% C, 38% T, 38% A.

3.35% T, 35% A, 15% C, 15% G.

4.28% A, 28% T, 22% G, 15% C.

(*b*) Yes, samples 1 and 3.

4. (*a*) Double-stranded DNA, (*b*) Single-stranded RNA, (*c*) Single-stranded DNA.

5. The second sample has a higher G-C content.

6. (*a*) 1 million, (*b*) 3.4 mm.

7. (*a*) The ladder-like pattern was known from X-ray diffraction studies. Chemical analyses had shown that 1:1 relationship existed between the organic bases adenine and thymine and between cytosine and guanine. Physical data concerning the length of each spiral and the staking of bases were also available.

   (*b*) Watson and Crick developed the model of a double helix, with the rigid strands of sugar and phosphorus forming spirals around an axis, and hydrogen bonds connecting the complementary bases in base-pairs.

8. (*a*) A multistranded, spiral structure was suggested by the X-ray diffraction patterns. A double-stranded helix with specific base-pairing fits the 1:1 stoichiometry observed for A : T and G : C in DNA (*stoichiometry* is a branch of chemistry that treats of the proportions of elements or compounds involved in reactions, and the methods of calculating them).

   (*b*) Use of the known hydrogen-bonding potential of the bases provided a means of holding the two complementary strands in a stable configuration in such a double helix.

9. DNA has one atom less of oxygen than RNA in the sugar part of the molecule. In DNA, thymine replaces the uracil that is present in RNA. (In certain bacteriophages, DNA containing a uracil is present). DNA is most frequently double-stranded, but bacteriophages such as $\phi \times 174$ contain single-stranded DNA. RNA is most frequently single-stranded. Some viruses, such as reoviruses, however, contain double-stranded RNA chromosomes.

10. No. TMV RNA is single stranded. Thus, the base-pair stoichiometry of DNA does not apply.

Fill in the Blanks

1. Chargaff.
2. Watson, Crick, 1953.
3. Right-handed.
4. Left-handed.
5. 20 Å
6. Phosphodiester bonds.
7. 10.
8. X-ray diffraction.
9. RNA.

Multiple Choice Questions

1. (*c*) 2. (*b*) 3. (*c*) 4. (*d*) 5. (*b*) 6. (*d*)
7. (*a*) 8. (*a*) 9. (*a*) 10. (*a*).

5

Organization of Genetic Material (DNA of Nucleosomes, Centromeres and Telomers)

5.1 CHROMATIN

When chromatin is isolated from interphase nuclei, the individual chromosomes are not recognizeable. Instead one observes an irregular aggregate of nucleoprotein. Chemical analysis of isolated chromatin shows that it consists primarily of DNA and proteins and lesser amounts of RNA (see Table 5.1). These proteins are of two major classes:

(1) **Basic proteins** (positively charged at neutral pH) called **histones;** and (2) a heterogenous, largely acidic (negatively charged at neutral pH) group of proteins collectively referred to as **non-histone chromosomal proteins**.

Histones

Histones play a major structural role in chromatin. They are present in the chromatin of all higher eukaryotes in amounts equivalent to the amounts of DNA (weight to weight). Indeed histones are present in such enormous quantities in the cell (about 60 million molecules of each type per human cell) that their total mass in chromatin is about equal to that of DNA (see **Albert** *et al.,* 2002). The histones of all the higher plants and animals consists of five different major proteins. These five major histones, called **H1, H2a, H2b, H3** and **H4** are present in almost all cell types. The five histones are present in molar ratios of approximately 1H1:2H2a:2H2b:2H3:2H4. They are specifically complexed with DNA to produce the basic structural subunits of chromatin,

Table 5.1 The constituents of calf thymus chromatin (after **Tamarin,** 2002).

S. No.	*Constituent*	*Relative weight**
1.	DNA	100
2.	Histone proteins	114
3.	Nonhistone proteins	33
4.	RNA	1

* Weight relative to 100 units of DNA.

small (approximately 110 Å in diameter by 60 Å high) ellipsoidal "beads" called **nucleosomes** (Fig. 5.1) by **P.Oudet** *et al.* (1975).

The histones have been highly conserved during evolution, four of the five types histones being very similar in all higher eukaryotes. Indeed, the amino acid sequence of histone 4 from a pea and a cow differ at only at 2 of the 102 positions. This strong *evolutionary conservation* suggests that a change in any position is deleterious to the cell. This suggestion has been tested directly in yeast cells, in which it is possible to mutate a given histone gene *in vitro* and introduce it into the yeast genome in place of the normal gene. As might be expected, most changes in histone sequences are lethal; the few that are not lethal cause changes in the normal pattern of gene expression, as well as, other abnormalities.

The histones are basic because they contain 20–30 percent **arginine** and **lysine**, two positively charged amino acids (Table 5.2). The exposed – NH_3^+ group of arginine and lysine allow histones to act as **polycations**. This is important in their interaction with DNA, which is **polyanionic**, because of the negatively charged phosphate groups.

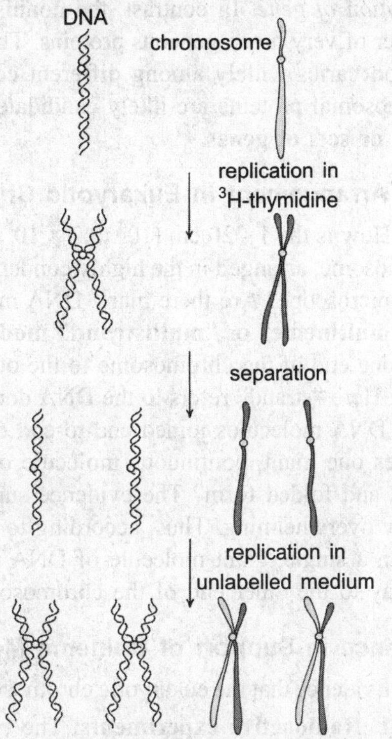

Fig. 5.1. Radioactive labeling of a uninemic eukaryotic chromosome following semiconservative replication. Replication occurs first in the presence of 3H – thymidine and then in its absence. After the second round of replication, one chromatid of each chromosome is labeled, whereas the other is not, confirming that there is only one DNA molecule per chromatid and that the chromosome is thus uninemic (after Tamarin, 2002).

Table 5.2 Composition of histones (*Source:* **Tamarin,** 2002).

S. No.	*Fraction*	*Class*	*Number of amino acids*	*Percentage of basic amino acids*
1.	H1	Very lysine rich	213	30
2.	H2A	Lysine, arginine rich	129	23
3.	H2B	Moderately lysine rich	125	24
4.	H3	Arginine rich	135	24
5.	H4	Arginine, glycine rich	102	27

The remarkable constancy of histones H2A, H2B, H3 and H4 in all cell types of an organism and even between widely divergent species is consistent with the idea that they are important in chromatin structure ("DNA packaging" and are only *nonspecifically involved in the regulation of gene.* In contrast, the nonhistone protein fraction of chromatin consists of a large number of very heterogeneous proteins. The composition of the nonhistone chromosomal protein fraction varies widely among different cell types of the same organism. Thus, the nonhistone chromosomal proteins are likely candidate for roles in **the regulation of expression of specific genes or sets of genes.**

DNA Arrangement in Eukaryotic Chromosome: the Unineme Model

How is the 1 - 20 cm (10^4 to $2 \times 10^5 \mu m$) of DNA, which is present in an average eukaryotic chromosome, arranged in the highly condensed mitotic and meiotic structures that are seen with the light microscope? Are there many DNA molecules that run parallel through out the chromosome (the "**multineme**" or "**multistrand**" **model**), or is there just one DNA double helix extending from one end of the chromosome to the other end (the "**unineme**" or "**single – strand**" **model**)? (*Note*: Here "strand" refers to the DNA double helix, not the single-strand of the DNA). Are there many DNA molecules joined end-to-end or arranged in some other fashion in the chromosome, or does one giant, continuous molecule of DNA extend from one end to the other in a highly coiled and folded form? The evidence supporting the unineme model of chromosome structure is now overwhelming. Thus, according to current view, each eukaryotic chromosome appears to contain a single, giant molecule of DNA that extends from one end through the centromere all the way to the other end of the chromosome.

Evidence in Support of Unineme Model

Evidence that the eukaryotic chromosome is **uninemic** comes from several sources as follows:

1. Radioactive experiments. The best data are provided by radioactive labeling studies, first done by **J.Taylor** and his colleagues in 1957. If a eukaryote is allowed to undergo one DNA replication in the presence of **tritiated (3H-1) thymidine**, each of the daughter chromatids would be expected to contain a double helix with one unlabeled DNA template strand and one labeled strand of newly synthesized bases (Fig. 5.1). This configuration is expected on the basis of semi-conservative replication, with each chromatid containing one double helix. A second round of DNA replication, in the absence of 3H-thymidine, should produce chromosomes in which one chromatid would have unlabelled DNA and one would have labeled DNA. Thus, one chromatid of every pair is labeled and one is not.

2. Lampbrush chromosomes in vertebrate oocytes. Another strong evidence supporting unineme model of chromosome structure has come from the studies on the lampbrush chromosomes. Such chromosomes are present during prophase I of oogenesis in many vertebrates, particularly amphibian. Lambrush chromosomes are upto 800 μm long; they thus provide very favourable material for cytological studies. The homologous chromosomes are paired, and each has duplicated to produce two chromatids at the lampbrush stage. Each lambrush chromosome contains a central axial region, where the two chromatids are highly condensed and numerous pairs of lateral loops (Fig. 5.2). The loops are transcriptionally active regions of single chromatids. The integrity of both the central axis and the lateral loops is dependent on DNA. Since treatment of both of these with **DNase enzyme** fragments them. In contrast, treatment with **RNase enzyme** or **protease enzyme** removes surrounding matrix material, but does not destroy the continuity of either the axis or the loops. Electron microscopy of RNase – and protease – treated lampbrush chromosomes reveals a central filament of just over 20Å in diameter in the lateral loops. Since each loop is a segment of one chromatid, and since the diameter of a DNA double helix is 20Å, these lampbrush chromosomes must be unineme structures.

Above conclusion is also supported by studies on the **kinetics of nuclear digestion** of lampbrush chromosomes. That is, the kinetics observed are those expected if a single double helix of DNA is the central filament in the loops. The axial region then contains two DNA molecules, one from each of the two tightly paired chromatids.

3. Viscoelasotometric evidence. In 1974, biophysicists such as **R.Kavenoff, L.Klotf** and **B. Zimm,** have used a technique called **viscoelastometry** (a procedure for analyzing viscosity parameters of molecules in solution) to estimate the sizes of DNA molecules from eukaryotic chromosomes. When DNA molecules in solution are exposed to a driving force (*e.g.,* by rotating a cylinder in the solution) they are stretched into an extended conformation. When the driving force is removed (*e.g.,* by stopping the rotation of the cylinder), the molecules will return to their lowest energy or relaxed state (Fig. 5.3).

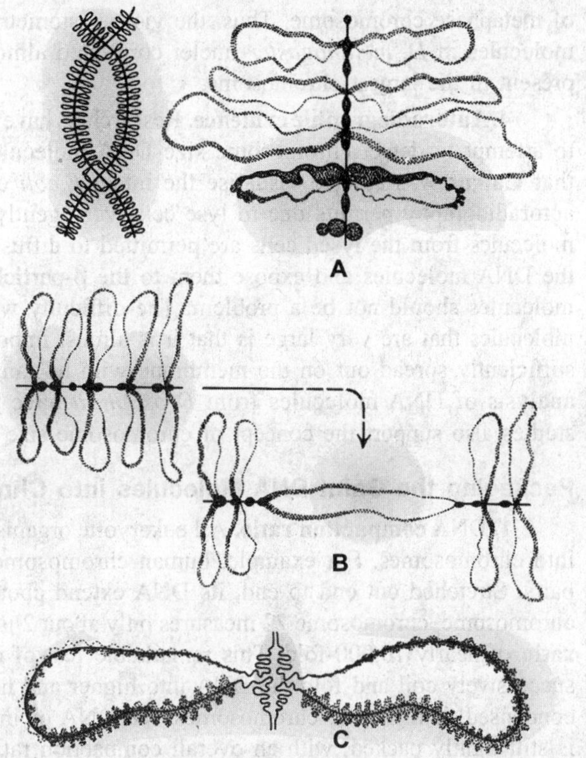

Fig. 5.2. Schematic representation of the structure of the giant lampbrush chromosomes of amphibian oocytes. A – semidiagrammatic view of the central chromosome axis with pairs of morphologically identical loops. Each chromosome has already duplicated in interphase. Each homologous chromosome thus contains two chromatids. B – A portion of a lampbrush chromosome that has been stretched during preparation, revealing continuity of the loops and the central axis. C – A single pair of loops displaying the continuity of the single DNA molecule that extends through the loop and axial regions of each chromatid. The matrix is primarily nascent RNA that is synthesized on the extended DNA in the loop region (after Gardner *et al.,* 2001).

In the solution being analysed, the *recoil time* is a function of the size of DNA molecule. The recoil decays exponentially over long time intervals, and the time constant, called the **retardation time (T)**, is a sensitive function of the size of the *largest DNA molecules* in the solution. Thus, even if some of the DNA molecules are broken by shearing forces, the viscoelastometric procedure will permit one to estimate the size of the largest DNA molecules present. For minimizing the chance of shearing of the molecules when they are released from the cells, the cells can be lysed right in the chamber of the viscoelastometer.

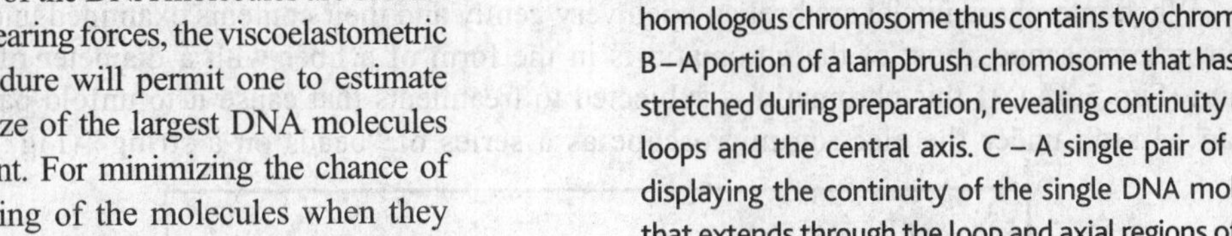

Fig. 5.3. Relaxed and extended state of a DNA molecule.

The largest chromosome in *Drosophila melanogaster* cells were estimated to have a mass of 4.1×10^{10} daltons by viscoelastometry. The largest chromosomes of *Drosophila* have been shown to contain about 4.3×10^{10} daltons of DNA by direct **microspectrophotometric** analysis of metaphase chromosome. Thus, the viscoelastometric estimate of the size of the largest DNA molecules in *D. melanogaster* nuclei correlated almost exactly with the total amount of DNA present in the largest chromosome.

4. Autoradiographic evidence. Researchers have also used the technique of **autoradiography** to attempt to detect chromosome size DNA molecules from eukaryotic cells in the same way that **Cairns** was able to visualise the intact *E.coli* chromosome. Similar to viscoelastometry, autoradiography permits one to lyse cells very gently so as to minimize shearing forces. DNA molecules from the lysed cells are permitted to diffuse onto a membrane that is used to pick up the DNA molecules and expose them to the β-particle – sentitive emulsion. Thus, breakage of molecules should not be a problem. The difficulty with autoradiographic examination of DNA molecules that are very large is that it is almost impossible to get all segments of the molecule sufficiently spread out on the membrane with no tangles or overlaps so that the entire graphic analysis of DNA molecules from *Drosophila* have been successful, and the results of these studies also support the concept of chromosome size DNA molecules (**Kavenoff** *et al.,* 1973).

Packaging the Gaint DNA Molecules into Chromosomes

1. DNA compaction ratio. All eukaryotic organisms have elaborate ways of packaging DNA into chromosomes. For example, human chromosome 22 contains about 48 million nucleotide pairs. Stretched out end to end, its DNA extend about 1.5 cm. Yet, when it exists as a mitotic chromosome, chromosome 22 measures only about 2 μm in length, giving end-to-end **compaction ratio** of nearly 10,000-fold. This remarkable feat of compression is performed by proteins that successively coil and fold the DNA into higher and higher levels of organization. Although less condensed than mitotic chromosomes, the DNA of interphase chromosomes, called **chromatin,** is still tightly packed, with an overall compaction ratio of approximately 1000 – fold.

2. Nucleosomes. Chromosome structure is **dynamic**: not only do chromosomes globally condense in accord with the cell cycle, but different regions of the *interphase chromosomes condense and decondense* as the cells gain access to specific DNA sequences for gene expression, DNA repair and DNA replication. The packaging of chromosomes must therefore be accomplished in a way that allows rapid localized, on demand access to the DNA.

When interphase nuclei are broken open very gently and their contents examined under the electron microscope, most of the chromatin is in the form of a fiber with a diameter of about 30 nm (Fig. 5.4A). If this chromatin is subjected to treatments that cause it to unfold partially, it can be seen under the electron microscope as a series of "beads on a string" (Fig. 5.4B).

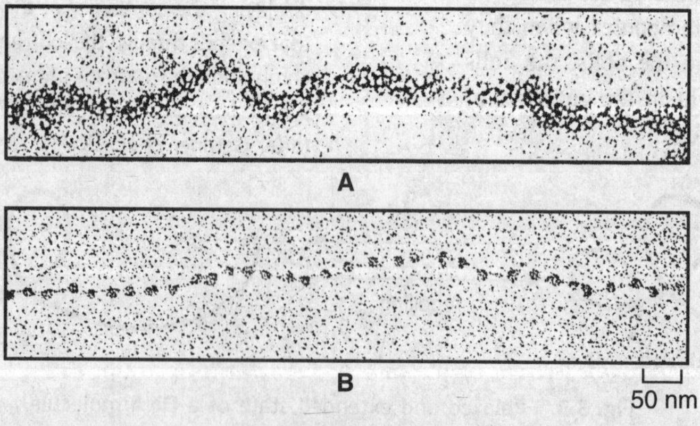

Fig. 5.4. Nucleosomes as seen in the electron microscope. A – Chromatin isolated directly from an interphase nucleus appears in the electron microscope as a thread 30 nm thick. B – This electron micrograph shows a length of chromatin that has been experimentally unpacked or decondensed, after isolation to show the nucleosomes.

The string is DNA, and each bead is a "nucleosome core particle" that consists of DNA wound around a protein core formed from histones. The beads on a string represent the *first level of chromosomal DNA packing*.

The structural organization of nucleosomes was determined after first isolating them from unfolded chromatin by digestion with particular enzymes (called **micrococcal nuclease**) that break down DNA by cutting between the nucleosome core particles, the **linker DNA**, is degraded. Each individual nucleosome core particle consists of a octamer or complex of eight histone proteins – two molecules each of histones H2A, H2B, H3 and H4 and double-stranded DNA (called **core DNA**) that is 146 nucleotide pair long. The histone octamer forms a protein core (**spool** according to **Tamarin,** 2002) around which the double-stranded DNA is bound (Fig. 5.5).

Each nucleosome core particle is separated next by a region of linker DNA, which can vary in length from a few nucleotide pairs upto about 80 (the term **nucleosome** technically refers to a nucleosome core particle plus one of its adjacent DNA linker, but it is often used synonymously with nucleosome core particle; see **Alberts** *et al.*, 2002).

On an average, therefore, nucleosome repeats at intervals of about 200 nucleotide pairs. For example, a diploid human cell with 6.4×10^9 nucleotide pairs contains approximately 30 million nucleosomes. The formation of nucleosomes converts a DNA molecule into a **chromatin thread** about one-third of its initial length, and this provides the first level of DNA packing.

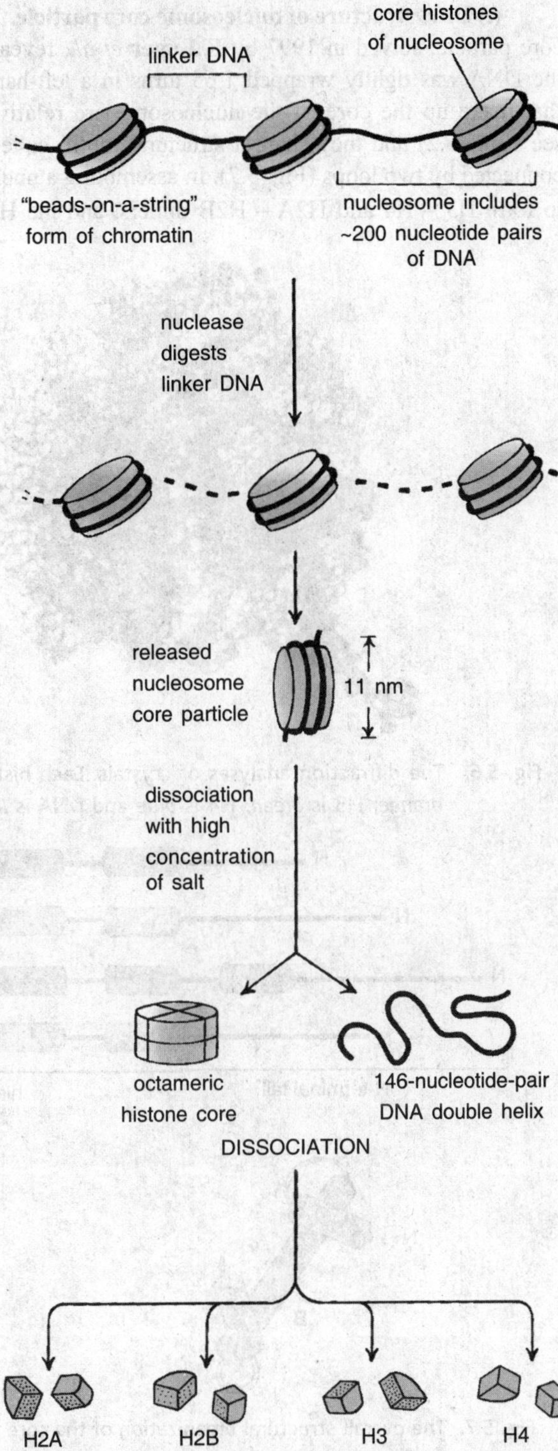

Fig. 5.5. Structural organization of the nucleosome (after Alberts *et al.*, 2002).

***(i)* Ultrastructure of nucleosome core particle.** The high–resolution structure of a nucleosome core particle, solved in 1997 by **K.Luger** *et al.*, revealed a disc-shaped histone core around which the DNA was tightly wrapped 1.65 turns in a left-handed coil (Fig. 5.6). All four of the histones that make up the core of the nucleosome are relatively small proteins (102 – 135 aminoacids; see Table 5.2) and they share a structural motif, called **histone fold**, formed from three α helices connected by two loops (Fig. 5.7). In assembling a nucleosome, histone folds first bind to each other to form H3 – H4 and H2A – H2B dimers, and the H3 – H4 dimers combine to form tetramers.

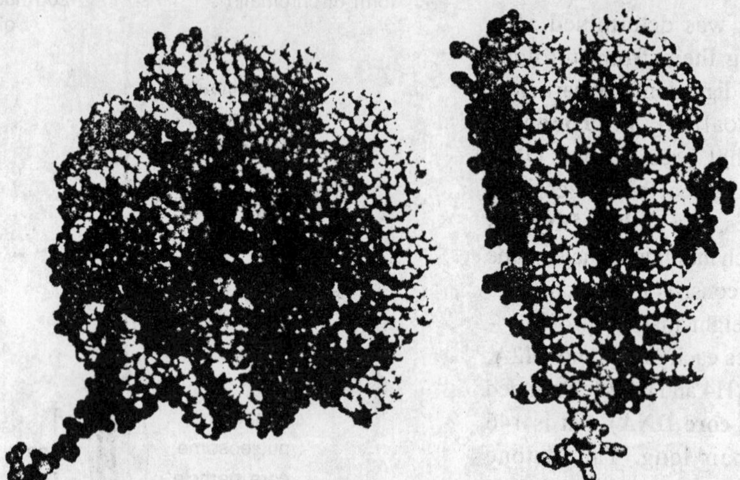

Fig. 5.6. The diffraction analyses of crystals. Each histone is coloured as follows: H2A is *red*, H2B is *orange*, H3 is *green*, H4 is *blue* and DNA is *light gray* (Colours are not shown in the figure).

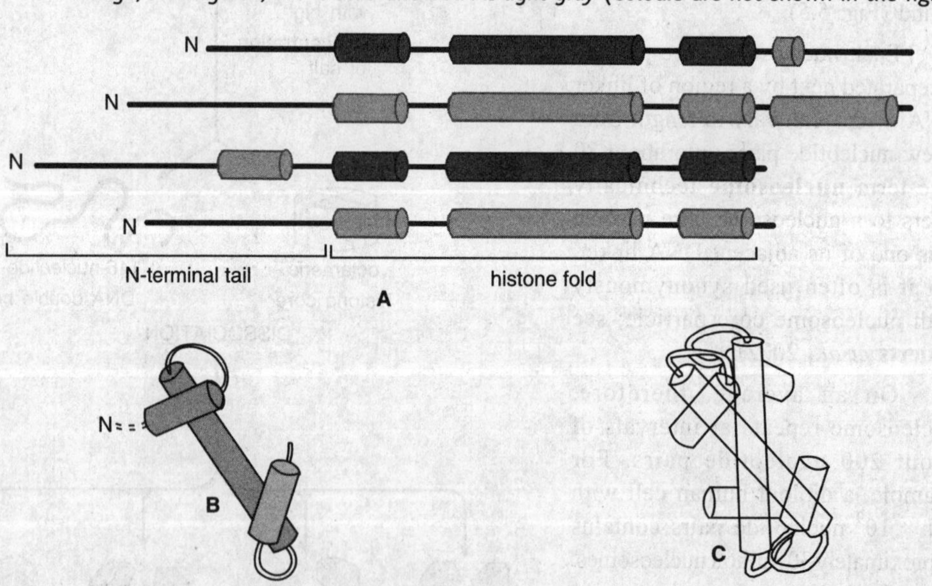

Fig. 5.7. The overall structural organization of the core histones. A – Each of the core histones contains an N – terminal tail, which is subject to several forms of covalent modification, and a histone fold region, as indicated. B – The structure of the histone fold, which is formed by all four of the core histones. C – Histones 2A and 2B form a dimer through an interaction known as the "hand shake" (after Alberts *et al.*, 2002).

An H3 – H4 tetramer then further combines with two H2A – H2B dimers to form the compact octamer core, around which the DNA is wound (Fig. 5.8 and Fig. 5.9).

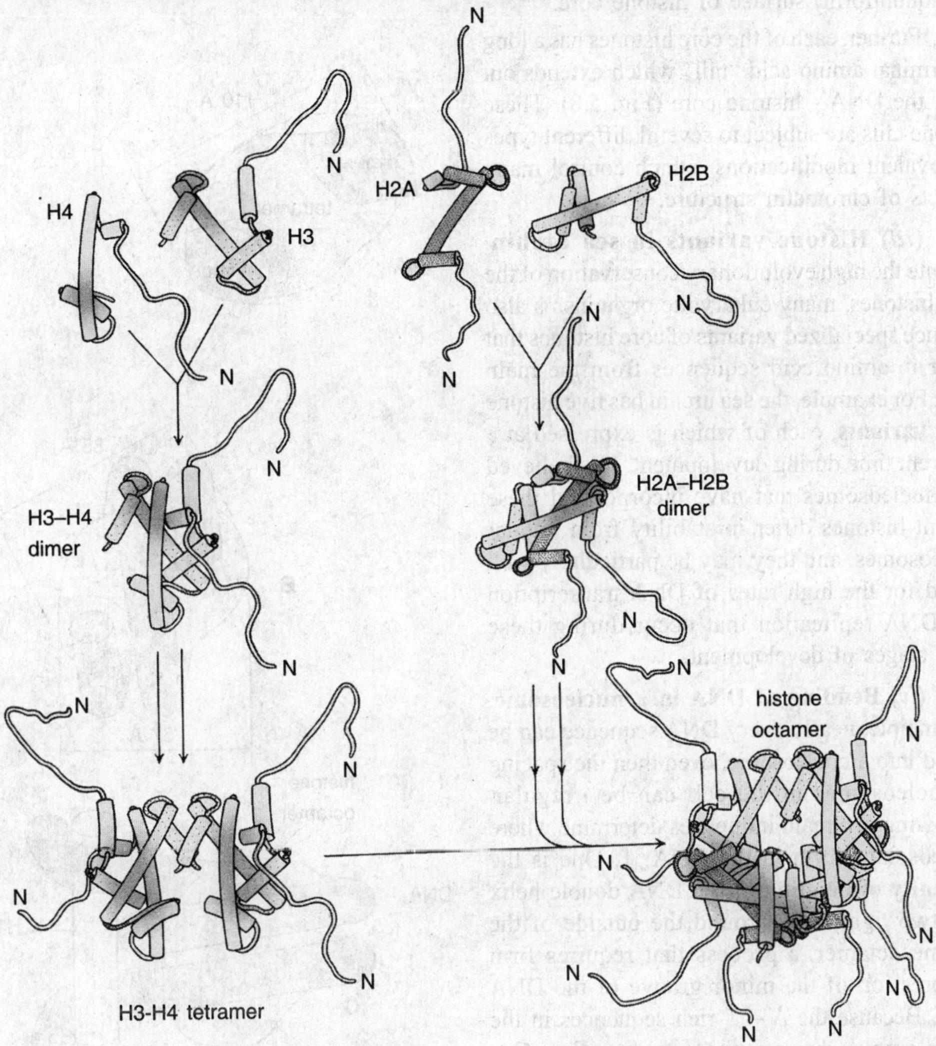

Fig. 5.8. The assembly of a histone octamer. The histone H3 – H4 dimer and the H2A – H2B dimer are formed from the hand shake interaction. An H3 – H4 tetramer forms the scaffold of the octamer onto which two H2A – H2B dimers are added, to complete the assembly. Note that all eight N – terminal tails of the histone protrude from the disc – shaped core structure (after Alberts *et al.*, 2002).

***(ii)* Bonding between histone core and DNA.** About 142 hydrogen bonds are formed between DNA and the histone core in each nucleosome. Nearly half of these bonds form between the amino acid backbone of the histones and the phosphodiester backbone of DNA. Numerous hydrophobic interactions and salt linkages also hold DNA and protein together in the nucleosome. For example, all the core histones are rich in lysine and arginine (two amino acids with basic side chains), and their positive charges can effectively neutralize the negatively charged DNA backbone. These numerous interactions explain in part why DNA of virtually any sequence can be bound on a histone octamer core. The path of the DNA around the histone core is not smooth; rather, several kinks (or bends) are seen in the DNA; this is expected from the rough (or nonuniform) surface of histone core.

Further, each of the core histones has a long N-terminal amino acid "tail" which extends out from the DNA - histone core (Fig. 5.8). These histone tails are subject to several different types of covalent modifications, which control many aspects of chromatin structure.

***(iii)* Histone variants in sea urchin.** Despite the high evolutionary conservation of the core histones, many eukaryotic organinsms also produce specialized variants of core histones that differ in amino acid sequences from the main ones. For example, the sea urchin has five histone **H2A variants**, each of which is expressed at a different time during development. It is believed that nucleosomes that have incorporated these variant histones differ in stability from regular nucleosomes, and they may be particularly well suited for the high rates of DNA transcription and DNA replication that occur during these early stages of development.

***(iv)* Bending of DNA in a nucleosome.** In principle, nearly every DNA sequence can be folded into a nucleosome, even then the spacing of nucleosomes in the cell can be irregular. Following *two* main influences determine where nucleosomes form in the DNA: 1. One is the **difficulty of bending** of the DNA double helix into two tight turns around the outside of the histone octamer, a process that requires firm compression of the minor groove of the DNA helix. Because the A – T rich sequences in the minor groove are easier to compress than G – C rich sequences; each histone octamer tend to "position itself on the DNA so as to maximize A – T rich minor grooves on the inside of the DNA coil (Fig. 5.10). Thus, a segment of DNA that contain short A – T rich sequences spaced by an integral number of DNA turns is easier to bend around the nucleosome than a segment of DNA lacking this feature.

2. The presence of certain other tightly bound proteins on the DNA also influences the position of nucleosomes on DNA molecule.

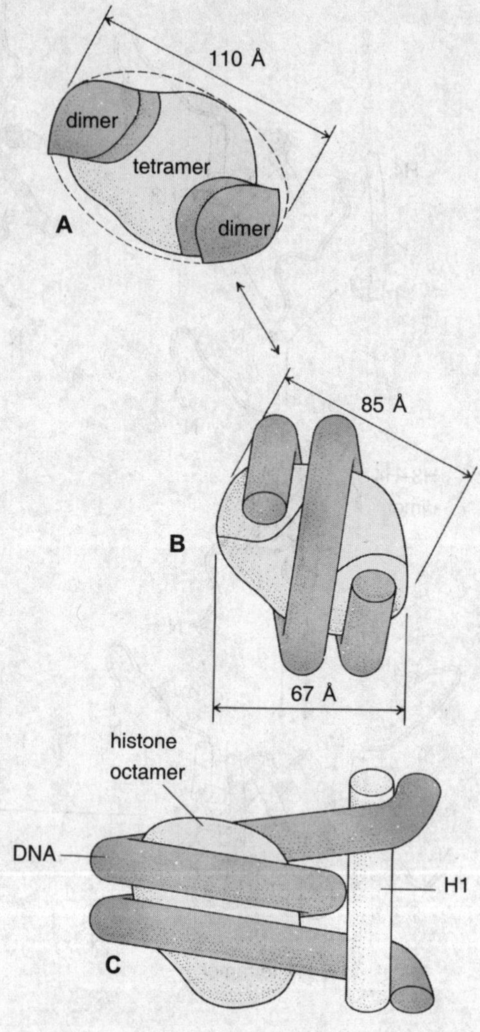

Fig. 5.9. Nucleosome structure. A – schematic comparison of the eight histones comprising the nucleosome in salt solution. A dimer consists of one H2A and one H2B histone molecule: a tetramer consists of two H3 and two H4 histones. B – DNA fits in surface grooves on the more compacted structure found in physiological conditions. C – The diagram shows the presumed position of the histone (after Tamarin, 2002).

3. Packaging of nucleosomes into compact chromatin fiber. The nucleosomes are packed on top of one another generating regular arrays in which the DNA is even more highly condensed. Thus, when nuclei are very gently lysed on to an electron microscope grid, most of the chromatin is seen to be in the form of a fiber with a diameter of about 30 nm, which is considerably wider then chromatin in the "beads on a string" form (Fig. 5.4).

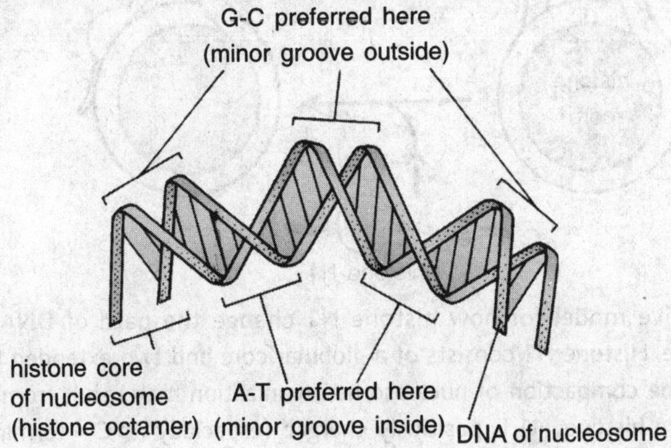

Fig. 5.10. Mode of bending of DNA in a nucleosome. The DNA helix makes 1.65 tight turns around the histone octamer. This diagram is showing how the minor groove is compressed on the inside of the turn. Owing to certain structural features of the DNA molecule, A – T base pairs are preferentially accommodated in such a narrow minor groove (after Alberts *et al.*, 2002).

The 30 nm chromatin fiber is found to have **zig–zag** structure (Fig. 5.11). The 30 nm structure found in chromosomes is probably a fluid mosaic of the different zig-zag variations. Differences, in linker lengths probably introduce further local perturbations into the zig-zag structure. Finally, the presence of other DNA-binding proteins and DNA sequences that are difficult to fold into nucleosome punctuate the 30 nm fiber with irregular features (Fig. 5.11).

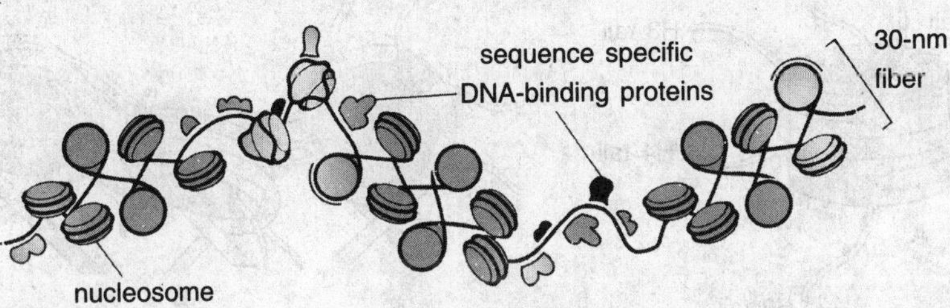

Fig. 5.11. Irregularities in the 30 nm fiber. This schematic view of the 30 nm fiber illustrates its interruption by sequence-specific DNA-binding proteins. The interruptions in the 30 nm fiber may be due to regions of DNA that lack nucleosomes altogether or more probably, to regions that contain altered or remodeled nucleosomes (after Alberts *et al.*, 2002).

Several mechanisms probably act together to form the 30 nm fiber from a linear string of nucleosomes. A additional histone, called **histone H1**, is involved in this process. H1 is larger than the core histones and is considerably less well conserved. In fact, the cells of most eukaryotic organisms make several histone H1 proteins of related but quite distinct amino acid sequences. A single histone H1 molecule binds to each nucleosome, contacting both DNA and protein, and changing the path of the DNA as it exits from the nucleosome. Histone H1 causes a change in the exit path in DNA which is crucial for compacting nucleosomal DNA and ultimately it interlocks to form the 30 nm fiber (Fig. 5.12).

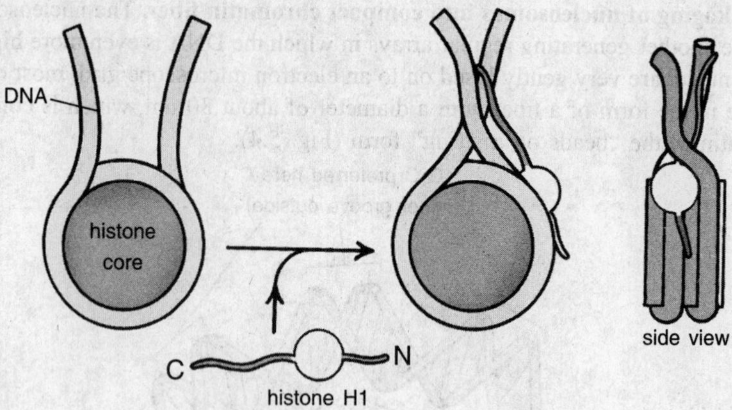

Fig. 5.12. A speculative model for how histone H1 change the path of DNA as it exits from the nucleosome. Histone H1 consists of a globular core and two extended tails. Part of the effect of H1 on the compaction of nucleosome organization may result from charge neutralization: like the core histone, H1 is positively charged (especially its C – terminal tail) and this helps to compact the negatively charged DNA. Unlike the core histones, H1 does not seem to be essential for cell viability; in one ciliated protozoan the nucleus expands nearly twofold in the absence of histone H1, but the cells otherwise appear normal (after Alberts *et al.,* 2002).

The second mechanism for forming the 30 nm fiber probably involves the highly flexible tails of the core histones, which, we saw above, extend from the nucleosome. It is thought that these tails may help attach one nucleosome to another – thereby allowing a string of them, with the aid of histone H1 to condense into the 30 nm fiber (Fig. 5.13). This model has been supported by X-ray diffraction analyses and evidence that the histone tails interact with DNA.

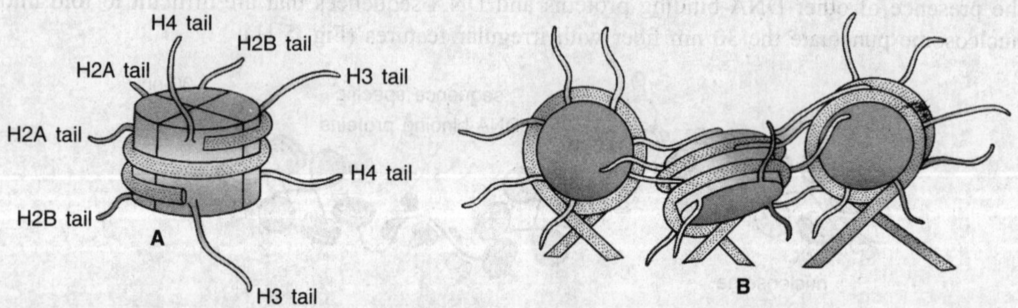

Fig. 5.13. A speculative model for histone tails in the formation of the 30 nm fiber. A – The approximate exit points of the eight histone tails (one from each histone subunit) from histone octamer. B – A speculative model showing how the histone tails may help to pack nucleosomes together (after Alberts *et al.,* 2002).

Roles of histone tails in gene regulation. The N – terminal tails of each of the four core histones are highly conserved in their sequence, and perform crucial functions in regulating chromatin structure. Each tail is subject to several types of covalent modifications, including **acetylation** of lysines, **methylation** of lysines, and **phosphorylation** of serines (Fig. 5.14). Histones are synthesized in the cytosol and then assembled into nucleosomes. Some of the modifications of histone tails occur just after their synthesis, but before their assembly. For us those modifications are significant however, which take place once the nucleosome has been assembled. These nucleosome modifications are added and removed by enzymes that reside in the nucleus; for example, acetyl groups are added to the histone tails by histone acetyl transferases (HATs) and taken off by histone deacetylase (HDACs).

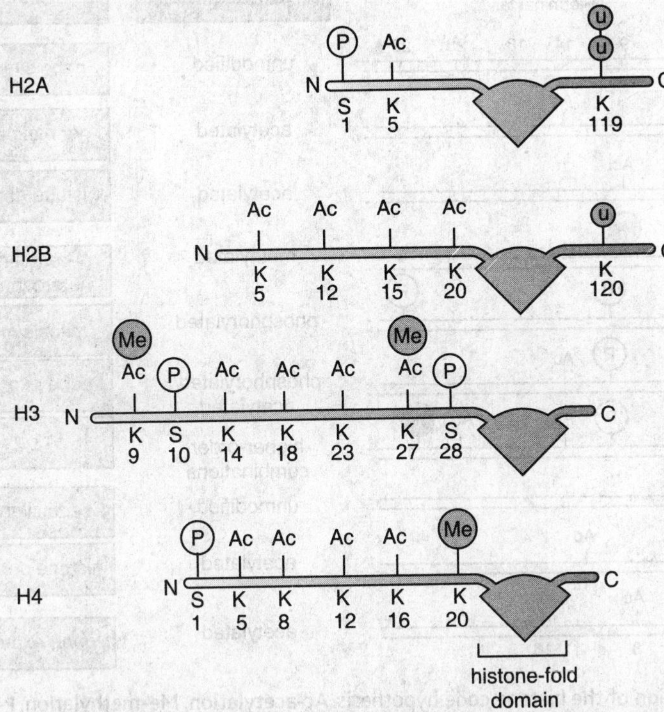

Fig. 5.14. Covalent modification of core histone tails. Known modifications of the four histone core proteins are indicated: K = lysine, S = serine, Me = methyl group, Ac = acetyl group, P = phosphate, u = ubiquitin. Note that some positions (*e.g.*, lysine of H3) can be modified in more than one way. Histone H2B can be modified by a single ubiquitin molecule; H2A can be modified by the addition of several ubiquitins (after Alberts *et al.*, 2002).

The various modifications of the histone tails have many important outcomes. Although modifications of the tail have little direct effect on the stability of an individual nucleosome, they seem to affect the stability of the 30 nm chromatin fiber and of the higher order structures discussed below. For example, **histone acetylation** tends to destabilize the chromatin structure, perhaps in part because adding an acetyl group removes the positive charge from the lysine, thereby making it more difficult for histones to neutralize the charges on DNA as chromatin is compacted. However, the most acute effect of modified histone tails is their ability to attract specific proteins to a stretch of chromatin that has been appropriately modified. Depending on the precise tail modifications, these additional proteins can either cause further compaction of the chromatin or can facilitate access to the DNA.

Histone code hypothesis. This hypothesis was proposed by **C.D. Allis** and coworkers in 2000 and 2001. According this hypothesis, histone tails can be marked by different combinations of modifications. Each such marking conveys a specific meaning to the stretch of chromatin on which it occurs. Only a few of the meanings of the modifications are known. For example, a doublyacetylated H4 tail is "read" by a protein required for gene expression. H3 tail methylated at lysine 9 is recognized by a set of proteins that creates an especially compact form of chromatin, which silences gene expression (Fig. 5.15).

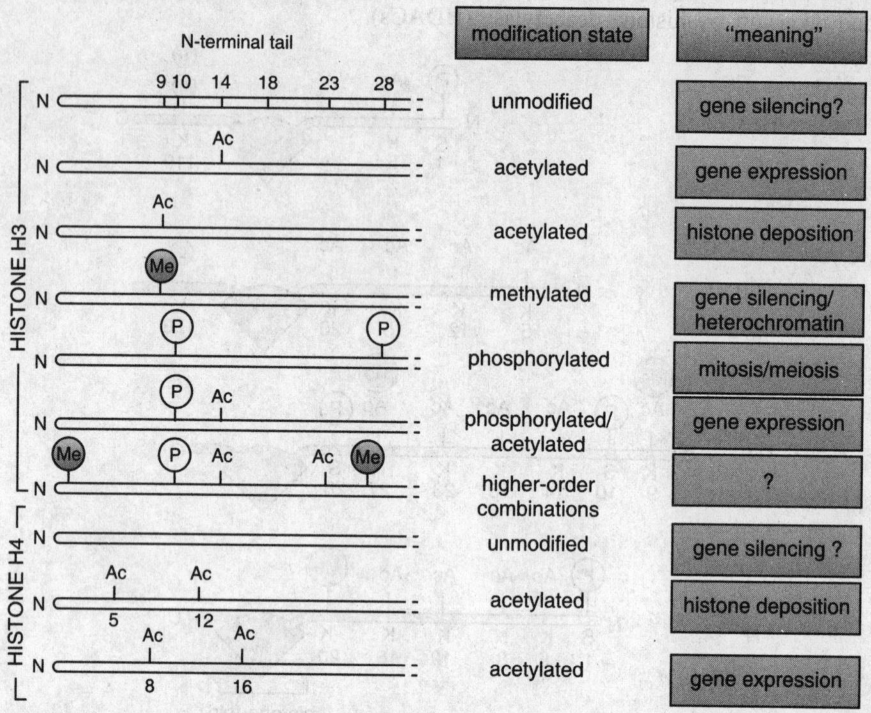

Fig. 5.15. Explanation of the histone code hypothesis. Ac-acetylation, Me-methylation, P-phosphorylation (after Alberts *et al.*, 2002).

Likewise, the acetylation of lysine 14 of histone H3 and lysine 8 and 16 of histone H4 – usually associated with gene expression – is performed by the *type A histone acetylases* (HATs) in the nucleus. In contrast, the acetylation of lysins 5 and 12 of histone H4 and a lysine of histone H3 takes place in the cytosol, after the histones have been synthesized but before they have been incorporated into nucleosomes; these modifications are catalyzed by the *type B histone acetylases*. These modified histones are deposited onto DNA after DNA replication, and their acetyl groups are taken off shortly afterwards by histone deacylases (HDACs). Thus, the acetylation at these positions signals newly replicated chromatin.

Further, modification of a particular position in a histone tail can take on different meanings depending on other features of the local chromatin structure. For example, the phosphorylation of position 10 of histone H3 is associated not only with the condensation of chromosomes that takes place in mitosis and meiosis but also with the expression of certain genes. Lastly, some histone tail modifications are interdependent. For example, methylation of H3 position 9 blocks the phosphorylation of H3 position 10 and vice versa.

Behavior of Nucleosomes During Chromosome Duplication

When DNA is replicated, twice as many nucleosomes are needed since one double helix becomes two. Recent studies indicate that a parental nucleosome is partly disassembled during DNA replication and reassembled on one or the other daughter strand, apparently randomly. The other DNA strand has a new nucleosome constructed of proteins called **chromatin assembly factors** which are of following *three* types: In fruit flies, a protein complex, called the **replication –coupling assembly factor** assembles new nucleosomes. In addition, a protein complex, called **condensin** is needed for the condensation of interphase chromosomes to mitotic chromosomes. This complex includes two **SMC proteins** (for structural maintenance of chromosomes) and two non-SMC proteins. SMC proteins also help other chromosomal activities, such as mitotic segregation, sister-chromatid adhesion, dosage compensation and recombination. Thus, a diverge array of proteins is involved in creating nucleosomes condensing interphase chromosomes and performing many activities of the chromosome.

Dosage compensation has recently been associated with a change in nucleosome structure. The activated X chromosome (sex chromatin or Barr body) appears to have a different type of histone present. Histone H2A is replaced by a variant, called **mH2A.**

Chromatin Remodeling

Nucleosomes apparently play a major role in controlling gene expression; DNA with nucleosomes has a much lower rate of transcription than DNA without nucleosomes. It makes sense that the positions of nucleosomes can provide or prevent access to promoters. There are regions of the DNA, known as **nuclease hypersensitive sites**, that appear to be nucleosome free. These sites, usually multiples of a region of about 200 base pairs, are particularly sensitive to digestion by different nuclease enzymes. When these regions are isolated they usually have sequences that control functions in *replication*, *transcription* and other activities of DNA. For example, numerous promoter regions in *Drosophila,* mouse and human DNA are in nuclease – hypersensitive sites. Hence, some specific DNA sequences are kept free of nucleosomes, and these sequences appear to be recognized by various enzymes such as RNA polymerase. In many other cases, however, nucleosomes do appear to cover promoters and repress transcription. For transcription to occur in these cases, some form of **chromatin remodeling** must take place.

Two general classcs of proteins are involved in chromatin remodeling. *First* are proteins that **acetylate** the N – terminal tails of the histones, a process that may cause the nucleosomes to bind less tightly to the core and thus make it available for attachment of **transcription factors**. These enzymes are called **histone acetyl transferases (HATs)**. Deacetylaling enzymes have the reverse effect. They act to repress transcription. *Second*, a class of ATP dependent proteins such as **SWI/SNF complex** in yeast also affect chromatin remodeling. (**Note:** some workers called the proteins SWI because they are involved in mating type *switching,* and called them SNF for sucrose nonfermenting). The SWI/SNF complex is a group of eleven proteins involved in transcription activation in many genes, presumably allowing transcription factors to access the promoters by remodeling chromatin. These proteins are able to reposition a nucleosome on DNA by sliding the nucleosome down the DNA.

From above findings, we can draw the conclusion that although nucleosomes serve as a general first order packing mechanism in eukaryotic DNA, they can be positioned precisely and can diminish transcription. However, once transcription begins, RNA polymerase enzyme apparently moves along nucleosomed DNA by translocation of the histone by 75 to 80 base pairs without the disrupting the nucleosome itself (**Studitsky** *et al.,* 1994). This seems to be accomplished by the RNA polymerase moving the DNA and then reforming the nucleosome in its wake (Fig. 5.16).

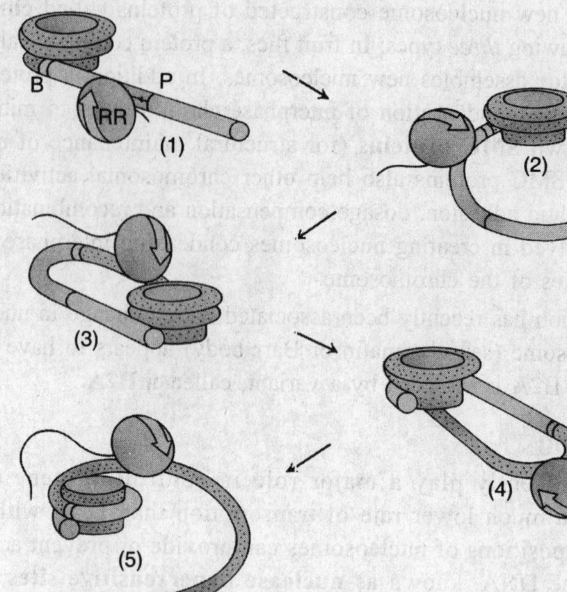

Fig. 5.16. RNA polymerase enzyme steps around a nucleosome without disrupting it. A – The RNA polymerase begins at a promoter (P) and heads for the nucleosomed DNA (*shaded*), whose border is noted with a line and letter B. As the polymerase encounters the nucleosome, it begins to unwrap the DNA from the histones. (2). The displaced DNA then re–encounters the histones (3), about 75 to 80 base pairs upstream from the original point of nucleosome formation. The polymerase continues on its way (4 and 5), and the nucleosome reforms in its displaced position without disrupting the histones or ever fully losing contact with the core DNA (after Tamarin, 2002).

Higher Order Structure of Chromatin

Since the nucleosome has a width of only 110Å and metaphase chromosomes appear to be constructed of a fibre having a diameter about 2400 Å, several additional levels of chromatin compaction lead to the metaphase chromosome. Various experiments, which change the ionic strength the chromatin is subject to, indicate that the 110 Å DNA spontaneously forms a 300 Å, solenoid-like fiber with increased ionic strength. It seems that this fiber results from the coiling of the nucleosomal DNA (Fig. 5.17). This 300 Å (30 nm) fiber is not, however, the final form of the DNA. We can account for the contraction of the 300 Å fiber to the 2,400 Å fiber found in metaphase

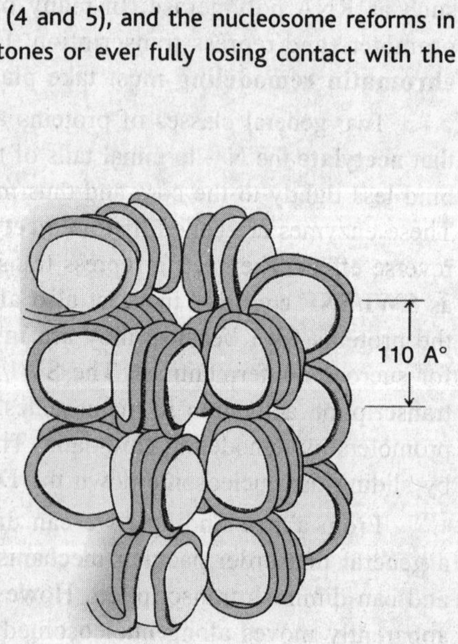

Fig. 5.17. Solenoid model for the formation of the 300 Å chromatin fiber. Nucleosomal DNA wraps in a helical fashion, forming a hollow core. Although histone H1 is not shown, it is known to be on the inside of the solenoid (after Tamarin, 2002).

chromosomes by the formation of a second solenoid-like structure from the winding of the 300 Å fiber (Fig. 5.18).

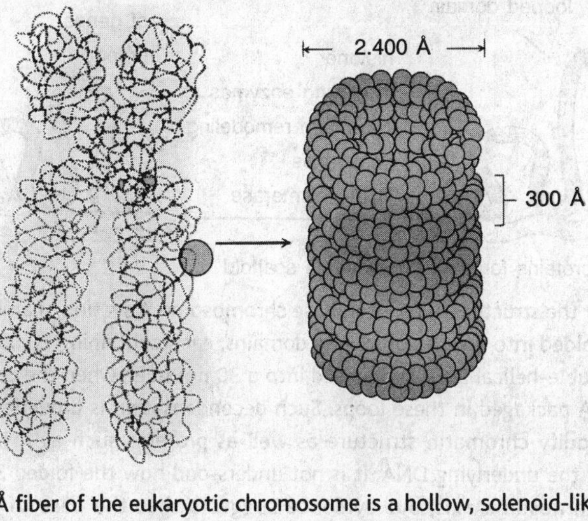

Fig. 5.18. **The 2,400 Å fiber of the eukaryotic chromosome is a hollow, solenoid-like structure. It is formed by the coiling of the 300 Å fiber, which itself is solenoid (after Tamarin, 2002).**

When the histones are removed from a chromosome, the DNA billows out, leaving a proteinaceous structure termed a **scaffold** (Fig. 5.19). The scaffold is involved in condensing the

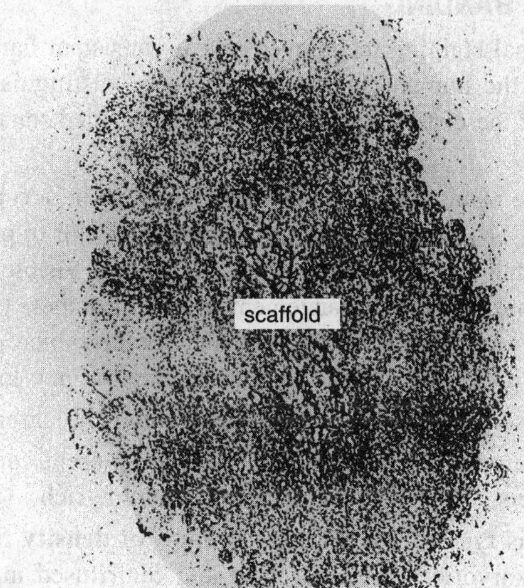

Fig. 5.19. Electron micrograph of a human metaphase chromosome to show the scaffold protein (**J. Paulson** and **U. Laemmli**, 1977). When the histones are removed from a eukaryotic chromosome, a fibrous scaffold remains. The DNA loops out from this scaffold.

300 Å chromatin fiber into the tightly packed metaphase chromosome (Fig. 5.20). The scaffold structure is formed from **nonhistone proteins**; two of them predominate, namely SC1 and SC2. **SC1** has been identified as topoisomerase II. Several hundred different proteins, many in minute quantities, are believed to be involved in replication, repair and transcription.

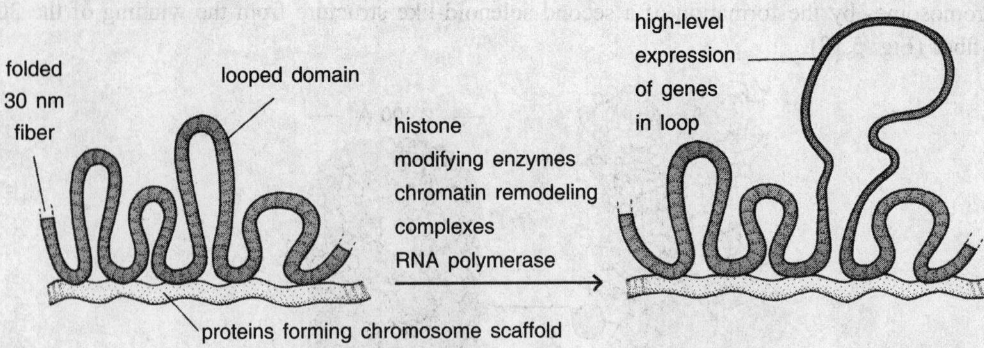

Fig. 5.20. A model for the structure of an interphase chromosome. A section of an interphase chromosome is shown folded into a series of looped domains, each containing 20,000 – 100,000 nucleotide pairs of double-helical DNA condensed into a 30 nm fiber, when the cell requires direct access to the DNA packaged in these loops. Such decondensation is brought about by enzymes that directly modify chromatin structure as well as proteins such as RNA polymerase, that act directly on the underlying DNA. It is not understood how the folded 30 nm fiber is anchored to the chromosome axis, but evidence suggests that the base of chromosomal loops is rich in DNA topoisomerases, which are enzymes that allow DNA to swivel when anchored (after Alberts *et al.*, 2002).

5.2 CHROMOSOME BANDING

Various chromosomal staining techniques reveal consistent banding patterns. By means of these patterns, all of the human chromosomes can be distinguished. Of possibly greater importance is the fact that these staining techniques have provided some insight into the structure of the chromosome.

1. G – bands. These bands are obtained with **Giemsa stain**, a complex of stain specific for the phosphate groups of DNA. Treatment of fixed chromatin with trypsin or hot salts brings out the G – bands. Giemsa stain enhances banding that is already visible in mitotic chromosomes. The banding pattern is caused by the arrangement of chromosomes. Under careful observation, the major G – bands prove to consists of many smaller chromosomeres. G – banding appearance has led **Comings** (1978) to suggest the mechanism of chromosome folding (Fig. 5.21).

2. C – bands. These are giemsa – stained bands after the chromosomes are treated with NaOH. The C is for "centromere", because these bands represent constitutive heterochromatin surrounding the centromers. The DNA is also usually satellite rich.

Satellite DNA. This type of DNA differs in **buoyan density** from the major portion of cellular DNA. When eukaryotic DNA is isolated and centrifused in CsCl, forming a density gradient, the majority of the DNA forms one band in the gradient at a single buoyant density. The buoyancy is determined by the G – C content of the DNA. However, smaller secondary bands are also usually present, indicating regions of DNA having sequences different from the majority of the cell's DNA. DNA isolated by this way is referred to as **satellite DNA** because of the secondary, or satellite, bands formed in the density gradient. In a chromosome, the satellite DNA is found primarily around the centromeres and consists of numerous repetitions of a short sequence.

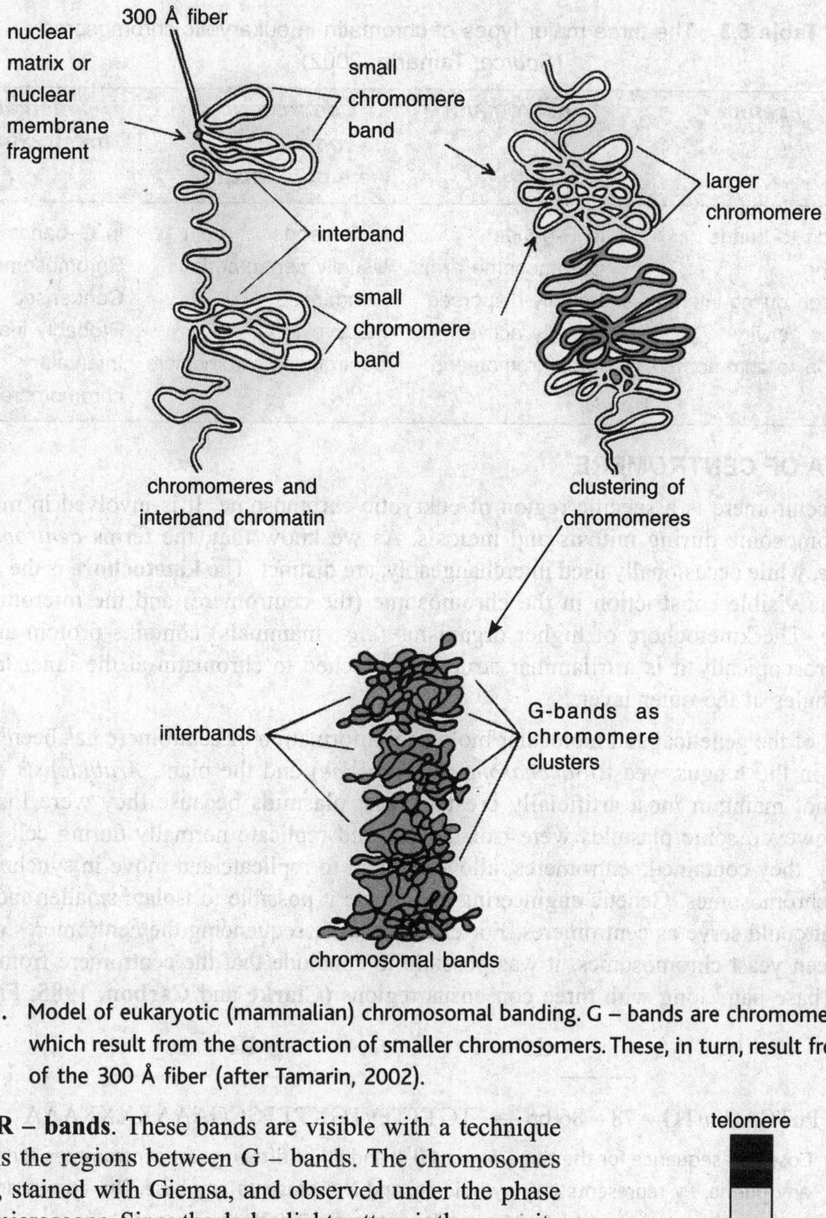

Fig. 5.21. Model of eukaryotic (mammalian) chromosomal banding. G – bands are chromomere clusters, which result from the contraction of smaller chromosomers. These, in turn, result from looping of the 300 Å fiber (after Tamarin, 2002).

3. R – bands. These bands are visible with a technique that stains the regions between G – bands. The chromosomes are fixed, stained with Giemsa, and observed under the phase contrast microscope. Since the dark – light pattern is the opposite of the G – band pattern, these bands are called **reverse bands.**

Based on the information obtained from these staining techniques, **Comings** (1978), distinguished between three basic chromatin types: euchromatin, constitutive heterochromatin and intercalary heterochromatin (Table 5.3). The only chromatin involved in transcription is **euchromatin. Constitutive heterochromatin surrounds** the centromere and is rich in satellite DNA (Fig. 5.22). **Intercalary heterochromatin** is found throughout the chromosome. Thus, it becomes apparent that the eukaryotic chromosome is a relatively complex structure.

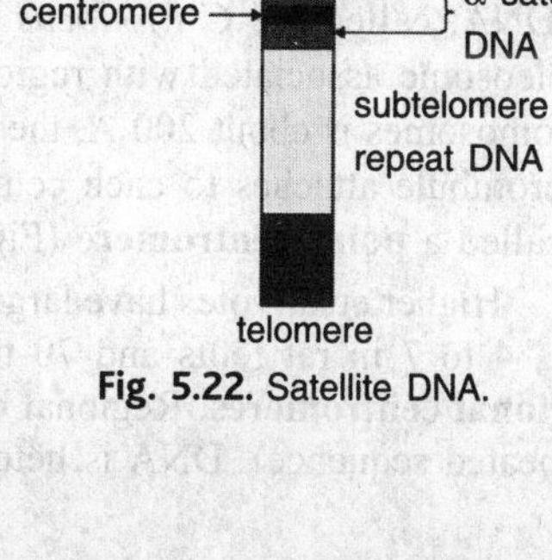

Fig. 5.22. Satellite DNA.

Table 5.3 The three major types of chromatin in eukaryotic chromosome (*Source:* **Tamarin,** 2002).

S. No.	*Feature*	*Euchromatin*	*Centromeric constitutive heterochromatin*	*Intercalary heterochromatin*
1.	Relation to bands	In R-bands	In C–bands	In G–bands
2.	Location	Chromosome arms	Usually centromeric	Chromosome arms
3.	Condition during interphase	Usually dispersed	Condensed	Condensed
4.	Genetic activity	Usually active	Inactive	Probably inactive
5.	Relation to chromomeres	Interchromomeric	Centromeric chromomere	Intercalary chromomeres

5.3 DNA OF CENTROMERE

The centromere is a specific region of eukryotic chromosome. It is involved in movement of the chromosome during mitosis and meiosis. As we know that, the terms *centromere* and *kinetochore*, while occasionally used interchangeably, are distinct. The **kinetochore** is the interface between the visible constriction in the chromosome (the centromere) and the microtubules of the spindle. The kinetochore of higher organisms (*e.g.,* mammals) contains protein and some RNA. Microscopically it is a trilaminar structure, attached to chromatin at the inner layer and to microtubules at the outer layer.

Most of the genetical and molecular biological information of centromere has been obtained by studies in the fungus, yeast (*Saccharomyces cervisiae*) and the plant, *Arabidopsis thaliana.* Cells did not maintain most artificially created yeast plasmids because they were lost during mitosis. However, some plasmids were isolated that did replicate normally during cell division. Presumably, they contained centromeres, allowing them to replicate and move in synchrony with the host's chromosomes. Genetic engineering also made it possible to isolate smaller and smaller regions that could serve as centromeres. For example, after sequencing the centromeres of fifteen or the sixteen yeast chromosomes, it was possible to conclude that the centromere from yeast is about 250 base pairs long with three consensus regions (**Clarke** and **Carbon,** 1985; Fig. 5.23).

I II III

$\longleftarrow$ $\longleftarrow$ $\longrightarrow$

PuTCACPuTG – 78 – 86 bp —— TGTTTPyTGXTTTCCGAAAXXXXAAA

Fig. 5.23. Cosensus sequence for the three regions (I, II and III) of fifteen yeast centromeres. *Pu* represents any purine, *Py* represents any pyrimidine, and *X* represents any base. The arrows are located over the inverted repeat sequences (after Tamarin, 2002).

In the light of current molecular biological data, the centromere can be defined as a sequence of DNA, called the *C EN locus* or C E N region. This region may contain a single modified nucleosome associated with region II. The 250 base-pair length of the C E N regions of yeast chromosomes is about 200 Å, the same as the diameter of a microtubule, indicating that only one microtubule attaches to each centromere during mitosis or meiosis in a yeast cell. This region is called a **point centromere** (Fig. 5.24).

Higher eukaryotes have large centromeric regions each of which attaches more microtubules, *e.g.,* 4 to 7 in rat fetus and 70 to 150 in Plant *Haemanthus.* These regions are referred to as **regional centromeres.** Regional centromeres range from 19 to 100 kb, with unique and satellite (repeated sequence). DNA is heterochromatic and may include expressed genes.

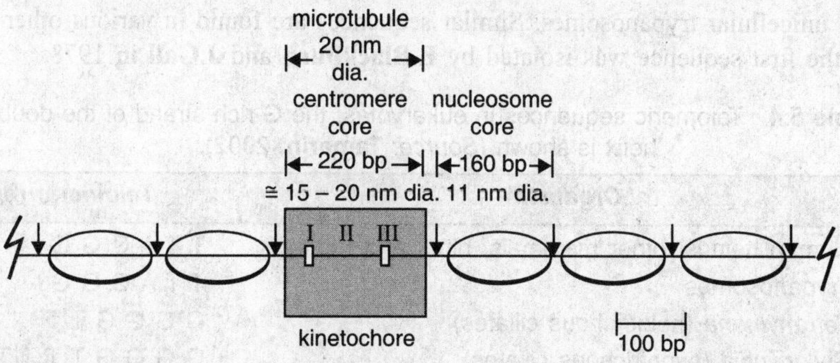

Fig. 5.24. Schematic view of a yeast centromeric region. The *arrows* indicate nuclease-hypersensitive sites. A microtubule is about the same width as the centromeric region (after Tamarin, 2002).

Arabidopsis centromeres span 0.9 – 1.2 Mb of DNA (Box 5.1) and each one is made up largely of 180 bp repeat sequences. In humans, the equivalent sequences are 171 bp and are called **alphoid DNA.** DNA of centromere of *Arabidopsis* contains multiple copies of genome-wide repeats, along with a few genes, the latter at a density of 7 – 9 per 100 kb compared with 25 genes per 100 bp for the noncentromeric regions of *Arabidopsis* chromosomes (**Copenhaver** *et al.*, 1999). The discovery that centromere DNA contains genes was a big surprise because it was thought that these regions were genetically inactive.

Box 5.1 Unit of length for DNA molecules

Because DNA is a double–stranded, the lengths of molecules are described as so many base pairs (bp). A **kilobase pair (kb)** is 10^3 bp and a megabase pair (**Mb**) is 10^6 bp. A **giga base pair (GP)** is 10^9 MB. In summary:

1 kb = 1000 bp
1 Mb = 1000 kb = 1000 000 bp
1 Gb = 1000 Mb = 1000 000 000 bp

Length of RNA molecules cannot be expressed in bp because most RNAs are single – stranded: their lengths is described as so many nucleotides (see **Brown,** 2002).

The **special centromeric proteins** in humans include at least seven that are not found else where in the chromosome (**Warburton,** 2001). One of these proteins, **CENP-A** is very similar to histone H3 and is thought to replace this histone in the centromeric nucleosomes. Part of the human kinetochore is made up of alphoid DNA plus CENP_A and other proteins, but its structure has not been described in detail (**Vafa and Sullivan,** 1997).

5.4 DNA OF TELOMERES

It is already known that eukaryotic chromosomes are linear, each has two ends, referred to as **telomeres**. The telomeres not only mark the termination of the linear chromosome but also have several specific functions. For example, telomeres must prevent the chromosomal ends from acting in a "sticky" fashion the way the broken chromosomal ends act. In other words, chromosomal ends must also prevent the ends of chromosomes from being degraded by **exonucleases** and must allow chromosomal ends to be properly replicated.

Most telomeres isolated so far are repetitions of sequences of five to eight bases. In human beings, the telomeric sequence is T T A G G G, repeated 300 to 5000 times at the end of each chromosome. The human telomere was discovered by **Robert K. Moyzis** and his colleagues (1988) when they probed *the highly repetitive segment of human DNA.* When a probe for this sequence was applied to human chromosomes, the sequence was found at the tip of each chromosome in roughly the same quantity. This is a highly conserved sequence, found in all vertebrates studied as well as in unicellular trypanosomes. Similar sequences are found in various other eukaryotes (Table 5.4); the first sequence was isolated by **E.Blackburn** and **J.Gall** in 1978.

Table 5.4 Telomeric sequences in eukaryotes; the G-rich strand of the double helix is shown (*Source:* **Tamarin,** 2002).

S. No.	*Organism*	*Telomeric repeat*
1.	Human beings, other mammals, birds and reptiles	T T A G G G
2.	Trypanosomes	T T A G G G
3.	*Tetrahymena* (holotrichous ciliates)	G G G G T T
4.	*Stylonychia* (hypotrichous ciliates)	G G G G T T T T
5.	Yeast	G T, G G T and G G G T
6.	Plants	T T T A G G G

When a linear DNA molecule is replicated, the 3′ – 5′ strand can be replicated up to the end. The 5′ – 3′ strand, however, is replicated with DNA primers that are then degraded, leaving a short gap on the progeny strand (Fig. 5.25). It is always the G-rich strand of telomeric DNA that ends up single stranded, forming a 3′ overhang of twelve to sixteen nucleotides. Thus, *the normal replication process of a linear DNA molecule leaves an incomplete terminus.* Hence, molecular biologists suspected that there would be a unique mechanism for the replication of telomeres.

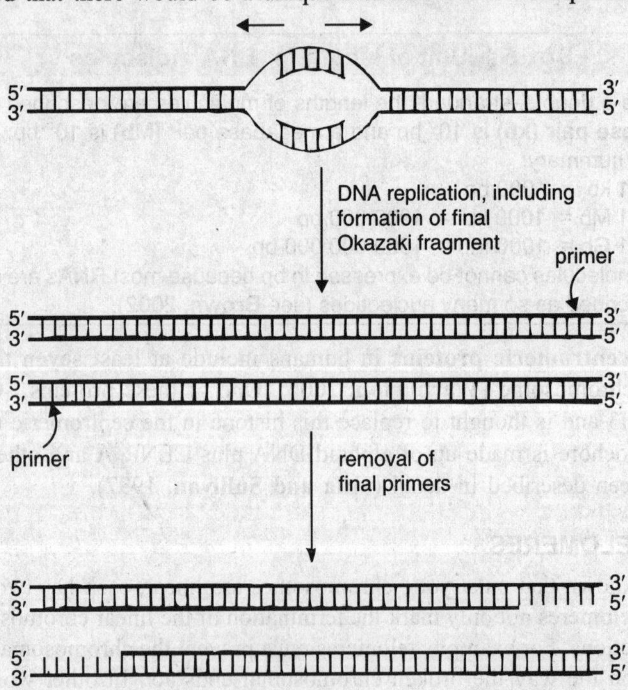

Fig. 5.25. Removal of final primers after the replication of linear DNA creates single stranded ends (after Tamarin, 2002).

Telomerase Enzyme

Telomeric sequences appear to be added *de novo* without DNA template assistance by an enzyme called **telomerase**. This enzyme was discovered by **Blackburn** and her colleagues (1990).

This was seen when telomeres from another species were engineered into yeast cells. After a cell cycle, the yeast telomeric sequence had been added on at the ends of the foreign chromosome, the result presumably of the telomerase enzyme.

Further, when **Blackburn** and her colleagues (1992) isolated telomerase, they discovered that a segment of RNA, about 160 nucleotides in length is an integral part of the enzyme. Telomerase RNA has a region that is complementary of the G – rich repeat of the telomeric DNA sequence of the species. After careful experimentation, including the modifying the gene for the telomerase RNA, **Blackburn** and her colleagues conducted that telomerase enzyme uses its RNA as a template for adding telomeric repeats to the ends of chromosomes. Telomerase enzyme is thus a **reverse transcriptase**, using RNA nucleotides as a template to polymerize DNA nucleotides (Box 5.2).

Box 5.2

Telomerase is an unusual enzyme in that it consists of both protein and RNA. In the human enzyme, the RNA component is 450 nucleotides in length and contains near its 5' end the sequence 5' – C U A A C C C U A A C – 3', whose central region is the reverse complement of the human telomere repeat sequence 5' – T T A G G G – 3' (**Feng** *et al.,* 1995). This enables telomerase to extend the telomeric DNA at the 3' end of a polynucleotide by the copying mechanism, in which the telomerase RNA is used as a template for each extension step, the DNA synthesis being carried out by the protein component of enzyme which is a reverse transcriptase (**Lingner** *et al.,* 1997).

Mode of working of telomerase. The first step in, telomerase extension is **hybridization** of the 3' end of the telomere with the RNA component of telomerase (Fig. 5.26A). Then, with the telomerase RNA as a template, the 3' end of the telomere is extended (Fig. 5.26B). Finally, a translocation step takes place that displaces the telomere in respect to the RNA, returning to the configuration at the beginning of the process (Fig. 5.26C). The single stranded C-rich strand is then synthesized with **DNA polymerase** and **DNA ligase**.

Protection of telomeres. Once the telomeres have been added to the ends of eukaryotic chromosomes, different organisms use any of *three* different method known to protect the ends of the chromosomes.

1. The guanine rich DNA can form complex structures. Biochemists have discovered that four guanines can form a planar **G-tetraplex,** with the four bases hydrogen bonded to each other (**Wang** and **Patel,** 1993).

2. Proteins have been discovered that bind to the 3' ends of telomeres. In the ciliate, *Oxytricha nova*, a protein called the **telomerase end binding protein (TEBP)** attaches to the 3' ends of telomeres and protects them (**Horvath,** 1998).

3. A novel structure called the **t – loop** has been discovered at the ends of mammalian telomeres (**Griffith** *et al.,*1999). The loop forms at the end of chromosomes under the direction of a protein called **TRF2** (Telomere Repeat Binding Factor) which causes the 3' end of the chromosome to loop around and interdigitate into the double helix, forming the loop (Fig. 5.27). A t-loop from mouse liver cell is about 10,000 bases around (**Griffith** *et al.*, 1999).

There are some telomeric proteins that are thought to form a linkage between the telomere and the periphery of the nucleus, the area in which the chromosome ends are localized (**Tham** and **Zakian,** 2000).

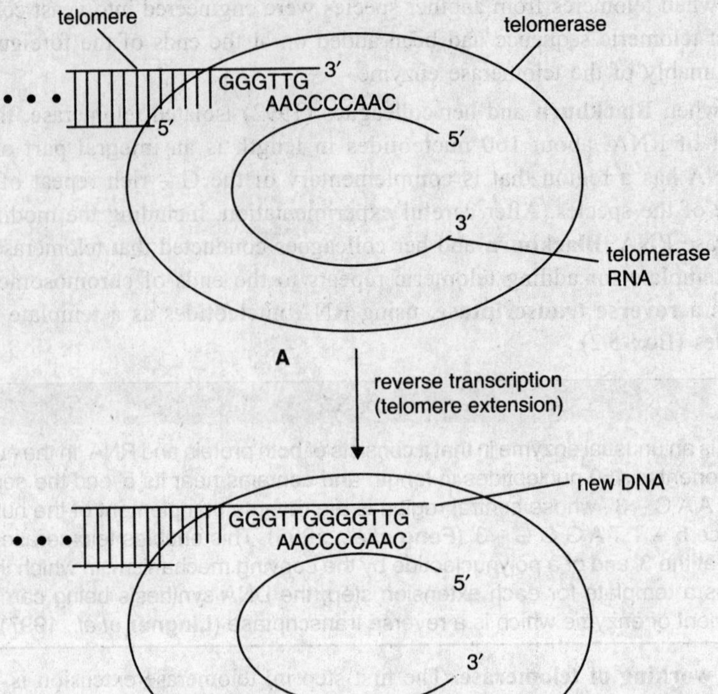

Fig. 5.26. Telomerase enzyme extends telomeres using telomerase RNA as a template. Gap filling by DNA polymerase and ligase complete the double helix (after Tamarin, 2002).

How do cells keep track of the number of their telomeric repeats? Proteins have isolated that bind to telomeres (Rap1 in yeast *Saccharomyces cerevisiae*, **Marcand** *et al.,* 1997; TRF1 in human beings. **Van Steensel** *et al*., 1997). By mutating these proteins or the telomeric sequences, molecular biologists have changed the equilibrium number of telomeric repeats. This led to the current model that the cell counts the number of these proteins bound to the telomeres, not the number of telomeres directly, to know whether telomeres should be added.

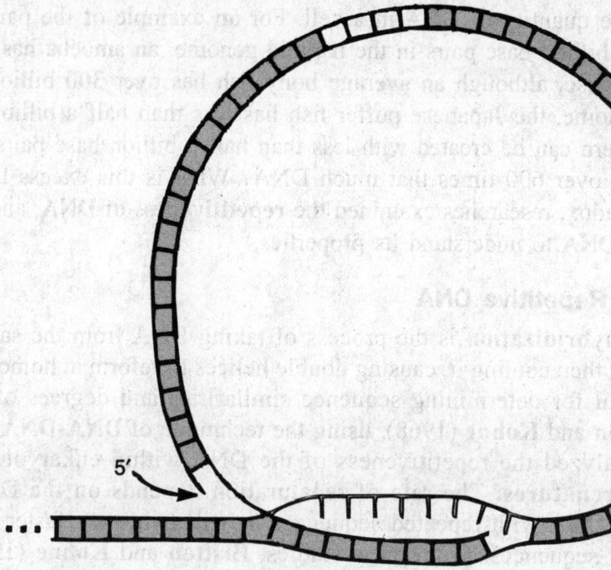

Fig. 5.27. The t-loop at the end of the mammalian telomere. The t-loop is formed by the interdigitation of the 3′ end of the telomere into the double helix (after Tamarin, 2002).

Telomerase and Aging

In yeast, protozoa and other single-celled organisms, telomerase enzyme is active, keeping the ends of the chromosomes at the appropriate length. These cells can divide potentially forever. In contrast, in higher organisms, telomere shortening leads to the termination of a cell lineage. For several years, the biologists have attempted to link this process with **cell senescense,** a phenomenon originally observed in cell cultures. All normal cell cultures have a limited lifetime: after a certain number of divisions the cells enter a senescent state in which they remain alive but cannot divide. With some mammalian cell lines notably fibroblast cultures (connective tissue cells), senescence can be delayed by engineering the cells so that they synthesize active telomerase (**Reddel** 1998). In mammals active telomerase is found essential in germ line cells (see **Klug** and **Cummings**, 2003). These experiments suggest a clear relationship between telomere shortening and senescence but the exactness of the link has been questioned (**Blackburn,** 2000).

Telomerase and Cancer

Not all cell lines display senescene. **Cancerous cells** are able to divide continuously in culture, their immortality being looked upon as analogus to tumour growth in an intact organism with several types of cancer, the absence of senescene is associated with activation of telomerase, sometimes to the extent that telomere length is maintained through multiple cell divisions, but often in such a way that the telomerase is overactive. In fact, human telomerase was isolated from an immortal cell line (HeLa) derived from cervical cancer cells.

This current information may be exploited for the clinical purpose: if, telomerase can be **deactivated** in tumour cells, the cells may stop dividing or die, thereby eliminating the cancer. One type of human cancer, **dyskeratosis cogenita**, appears to result from a mutation in the gene specifying the RNA component of human telomerase (**Marciniak** and **Curaente,** 2001).

5.5 THE C-VALUE PARADOX

Why do eukaryotes have so much DNA, and why is there huge variation in the DNA content between species of comparable complexity? These questions define the **C-value paradox**, in which C refers to the quantity of DNA in a cell. For an example of the paradox, although human beings have 3.3 billion base pairs in the haploid genome, an amoeba has more than 200 billion base pairs. Likewise, although an average bony fish has over 300 billion base pairs of DNA in its haploid genome, the Japanese puffer fish has less than half a billion base pairs. If the basic bony fish pattern can be created with less than half a billion base pairs, why does the average bony fish have over 600 times that much DNA? What is this excess DNA doing? To explain the C-value paradox, researches examined the **repetitiveness of DNA**, and more recently probed and sequenced DNA to understand its properties.

1. Unique DNA and Repetitive DNA

The **DNA-DNA hybridization** is the process of taking DNA from the same or different sources and heating and then cooling it, causing double helices to reform at homologous regions. This technique is useful for determining sequence similarities and degrees of repetitiveness among DNAs. **R.Britten** and **Kohne** (1968), using the technique of **DNA-DNA hybridization,** first systematically analyzed the repetitiveness of the DNA within eukaryotes. When DNA is heated, it cools, it **renatures.** The rate of renaturation depends on the DNA sequences. If the sample contains DNA with repeated sequences, it will hybridize faster than DNA that does not have repeated sequences. From these studies, **Britten** and **Kohne** (1968) found that eukaryotic chromosomes contain regions of unique, moderately repetitive and highly repetitive DNA. **Unique DNA** is, as its name implies, DNA with unrepeated sequences. **Repetitive DNA** is DNA whose sequences are repeated in the genome. During the last 39 years, hundreds of plants and animals have been analyzed for the relative proportion of repetitive DNA in their genome (Table 5.5).

Satellite DNA, found around centromeres is highly repetitive DNA with a unique repeat length of about two hundred base pairs. Unique DNA makes up most of the transcribed genes of an organism. The rest of the DNA is repetitive DNA in a few to several hundred thousand copies.

Table 5.5 Relative proportions (per cent of haploid genome) of repetitive sequences.

			Organism	***Proportion of repetitive sequences (%)***
I.	Animals	(i)	Sea urchin	38
		(ii)	Snail	60
		(iii)	Toad	45
		(iv)	Mouse	40
		(v)	Calf	45
		(vi)	Human	30
II.	Plants	(i)	Bread wheat	80
		(ii)	Oats	83
		(iii)	Maize	78
		(iv)	Garden pea	75
		(v)	Barley	76
		(vi)	Rye	92

This repetitive DNA comprises three categories. One is "junk" DNA, DNA that is not useful to the organism, made up of untranscribed and parasitic sequences (selfish DNA). Another category is transcribed genes in many copies that have diverged from each other, such as antibody, collagen, and globin genes. The term diver **gene family** has been given to the genes that have arisen by duplication, with or without divergence, from an ancestral gene. Finally, the transcribed genes in many copies that are virtually identical such as ribosomal RNA genes and histone genes, make up a third category of repetitive DNA.

2. Junk DNA

Transposons in prokaryotes are generally viewed as selfish or parasitic: they serve no purpose to the cell. The transposons replicate on their own, increasing in number. Eukaryotic transposons are mostly **retrotransposons**, *i.e.,* transposable elements that move by way of an RNA intermediate. In other words, the retrotransposon is transcribed into RNA and then, by reverse transcription, converted to a cDNA and then inserted into the genome. These elements can make up 50 per cent of the eukaryotic genome, existing in hundreds of thousands of copies. They generally fall into two categories: LINES and SINES.

Box 5.3

The term **'junk DNA'** is falling out of favour of molecular biologists. This is partly because the number of surprises resulting from genome research over the last few years has meant that molecular biologists have become less confident in asserting that any part of the genome is unimportant simply because we do not currently know what its function might be. One thing that is clear that the bulk of the intergenic DNA is made up of repeated sequences of one type or another (see **Brown**, 2002).

1. LINES or Long Interspersed Elements. They contain upto seven thousand pairs each and contain genes for **reverse transcription, RNA binding** and **endonuclease activity**. They thus have the ability to jump by way of an RNA intermediate. Human DNA is believed to be composed of about 15 per cent LINES.

2. SINES or Short Interspersed Elements. They are generally derivatives of transfer RNA (tRNA) genes and do not have the ability to retrotranspose on their own. That is, in the past, their transcripts were modified, converted to cDNA by reverse transcription and then reinserted into the host's genome. They depend on the reverse transcriptase enzyme provided by the genes of LINES or retroviruses. One group of SINES not derived from tRNA is derived from RNA of the signal recognition particle; members of this group occur in human beings in about five hundred thousand copies of a three hundred base pair sequence. Because these sequences are cleaved by the restriction endonuclease *alu I,* they are called the *Alu family*. The human genome is also pervaded by remnants of at least a dozen distinct families of ancient retroviruses scattered throughout our chromosomes.

At this juncture, some explanation can be forwarded for the C-value paradox. Much eukaryotic DNA is junk (Box 5.3). Apparently doing no harm. In some cases, 97% of the host genome is composed of junk DNA. Recent work has explained that gross differences in DNA content between higher organisms may be due to the differing abilities of different species to rid themselves of this parasitic DNA. If it builds up without being removed, the DNA content of the species can rise. Thus, wide differences in DNA content among higher eukaryotes mentioned at the beginning of this section have little to do with the complexity of the organism, but rather with the ability of the organism to remove the junk DNA as it forms (see **Tamarin**, 2002).

5.6 OVERLAPPING GENES

Bacteriophage $\phi \times 174$ contains a single stranded DNA approximately 5,400 nucleotides in length. Complete sequence of these 5,400 nucleotides was worked out by **Fredrick Sanger** and his colleagues. The genome of $\phi \times 174$ phage consists of nine cistrons. Amino acid sequences of major proteins coded by the genome of $\phi \times 174$ phage have also been worked out and even their molecular weights are known. From the information about protein coded, an estimate could be made of the number of nucleotides required. This estimate of number of nucleotides exceeds 6,000 which is much higher than the actual number of nucleotides present, *i.e.,* 5,400. Thus, it was difficult to explain how these proteins could be synthesized from a DNA segment which is not long enough to code for the required number of amino acids. When detailed study of system was done then it was discovered that sequences in the same segment could be utilized by two different cistrons coding for different proteins. Such **overlapping of genes** (cistrons) will be theoretically possible if the two cistron have two functions at different times and their nucleotide sequences are translated in two different reading frames.

In 1976 **Barrell** and his coworkers discovered that in $\phi \times 174$ phage which have nine cistrons (*A, B, C, D, E, J, F, G, H*), cistron *E* is present between *D* and *J* and that the cistron *E* overlap cistron *D*. This could be done by working out the nucleotide sequences of DNA segment representing gene *D, E* and *J*. It could be shown that amber (codon for chain termination) mutations in cistron *E* lie within cistron *D* and these amber mutations do not influence the translation of cistron *D* into its protein. Similarly, some nonsense mutations for cistron *D* lie in cistron *E* suggesting that cistrons *D* and *E* overlap in the DNA sequences and that the cistrons *D* and *E* are translated in two different reading frames so that amber codon in mRNA of one cistron will be read as termination of the other cistron.

Among genes *D, E,* and *J* in $\phi \times 174$ phage, cistron *D* codes for a simple protein, cistron *E* codes for a protein for lysis and cistron *J* codes for a particle component. It was shown that *enough space is not available where these three genes are located,* thus, conforming that there was overlapping between *D* and *E* cistrons (Fig. 5.28). Overlapping has also been suggested to be present in regions of genes *A* and *B*.

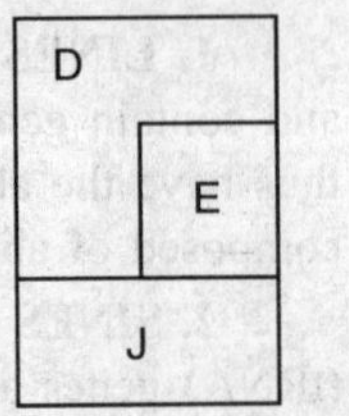

Fig. 5.28. Overlapping of *E* and *D* genes in phage $\phi \times 174$.

Overlapping sequences (=genes) have also been discovered in tryptophan mRNA of *E.coli* and in many other cases. In view of wide occurence of overlapping genes, it seems that phenomenon of overlapping genes, is an economic device to make better use of genetic information is less DNA.

5.7 PSEUDOGENES

In multicellular organisms, a wide variety of DNA sequences are found, which are of no apparent use for the organism. Some of these sequences are defective copies of functional gene and are therefore called **pseudogenes**. Such pseudogenes are found to occur in human beings, mouse (*e.g.,* α globin pseudogenes), *Drophilla* (*e.g.,* histone pseudogenes), etc. For example, human **α–globin** and **â-globin pseudogenes** (ψ), found in each of the two globin gene clusters. Complete nucleotide sequence of alpha globin pseudogene is now known. It has been also reported that both of these genes (α-globin and â-globin) are non-translatable, since they may have mutations in initiation codons and also frame-shift mutations along their length.

5.8 CRYPTIC GENES

Cryptic genes are phenotypically, silent DNA sequences that are not expressed during the normal life cycle of an organism. They may be activated in a few individuals of a population, due to events such as mutation, recombination, transposition and other genetic mechanisms due to specific altered environmental conditions and this leads to their detection. For example, genes for α-glucoside utilization in *E.coli* and *Salmonella.* Four cryptic genetic systems (*bgl, cel, ase, arbT*) involved in the α–glucoside utilization have been characterized in *E.coli.*

QUESTIONS

Long Answer Questions

1. Summerise the evidence that the eukaryotic chromosome is uninemic.
2. What are the major protein components of the eukaryotic chromosome? What are their functions?
3. Write short notes on:

   (*i*) Nucleosomal concept;

   (*ii*) Chromatin condensation.

Short Answer Questions

1. What is the protein composition of a nucleosome? What function does histone H1 have?
2. What do we know about the sequence within a yeast centromere?
3. Describe three ways in which cells protect their telomere.
4. Can nucleosome contain the DNA for one gene? Explain.

Very Short Answer Questions

1. Give the significance of lampbrush chromosomes.
2. How many DNA double helical fibers are present in an eukaryotic chromosome?
3. 'Beads in the string' appearance of chromatin represents what?
4. What is the importance of telomere in the chromosomes?
5. What evidence is used to determine the length of DNA associated with a nucleosome? What is a nuclease-hypersensitive site? What functions are associated with these sites?
6. What is the relationship among the 110Å, 300Å and 2,400Å fibers of the eukaryotic chromosome?
7. What is a telomere? What are its functions? What is its structure?
8. What functions exist in unique, repetitive and highly repetitive DNAs?
9. What is the C-value paradox, and how is it explained?
10. What is satellite DNA? What does it signify?
11. If chromatin is digested with an endonuclease enzyme to produce two hundred base-pair fragments, and these fragments are then used for transcription experiments, very little RNA is made. Provide an explanation for this observation.
12. If radioactive probes are made from highly repetitive DNA, these probes hybridize *in situ* mainly to centromeric and telomeric regions. What does this result suggest about the organization of chromosomes?
13. Why is higher-order chromosomal structure expected in eukaryotes but not in prokaryotes?
14. How could comparative DNA studies aid us in understanding the roles of the different kinds of DNA present in the eukaryotic chromosomes?

Fill in the Blanks

1. Lampbrush chromosomes are involved in ....
2. In .... chromosome, centromere is located near the center of the chromosome.
3. SAT-chromosomes always have ....
4. Folded fiber model of chromosome was proposed by ....
5. The number of base pairs in DNA of a nucleosome is ....
6. Histone HI causes a change in the exit path in the DNA which is important for .... nucleosomal DNA and ultimately HI histone .... to form the 30 nm fiber.

7. .... are synthesized in the .... and then assembled into nucleosomes (in nucleus).
8. Acetyl groups are added to tails of histones by .... (HATs) and taken off by .... (HDACs).
9. The DNA present in centromeres of human chromosomes is called ....
10. When the histones are removed from a eukaryotic chromosome, the DNA billows out leaving a proteinaceous structure called ....
11. Telomerase enzyme acts as a .... , using RNA nucleotides as a .... to polymerise DNA nucleotides of telomeres.
12. Overlapping genes are found in the coliphage, called ....

Multiple Choice Questions

1. Chromatin consists of
   (*a*) RNA (*b*) DNA
   (*c*) RNA and histones (*d*) DNA and histones
2. DNA is associated with highly basic proteins called
   (*a*) histones (*b*) non-histones
   (*c*) albumins (*d*) all of these
3. A nucleosome is made of
   (*a*) DNA
   (*b*) histone
   (*c*) histone wrapped over octameric core of nucleic acid
   (*d*) DNA wrapped over octameric core of histone
4. Solenoid is a structure of
   (*a*) nucleosomal organisation with 10 nm diameter.
   (*b*) condensed chromatin fiber with 30 nm diameter
   (*c*) highly condensed form of chromatid with 300 nm diameter
   (*d*) well-organized chromosome with 1400 nm thickness
5. The terminal end of a chromosome is called
   (*a*) metamere (*b*) telomere
   (*c*) centromere (*d*) basal granule
6. A gene that specifies the amino acid sequence of a polypeptide chain is termed as
   (*a*) structural gene (*b*) regulator gene
   (*c*) operator gene (*d*) split gene
7. The gene not expressing any protein is known as
   (*a*) epistatic gene (*b*) hypostatic gene
   (*c*) pseudogene (*d*) none of these
8. The intervening sequence of 'gene' are known as
   (*a*) introns (*b*) exons
   (*c*) cistron (*d*) codons
9. In split genes, the coding sequences are called
   (*a*) introns (*b*) operons
   (*c*) exons (*d*) cistrons

10. Exon part of mRNA have codes for

(*a*) protein (*b*) lipid

(*c*) carbohydrate (*d*) phospholipid

11. Complete the given flow chart about DNA compaction and packaging by selecting the correct option.

DNA ⟶ A ⟶ B ⟶ Chromatin fiber ⟶ C ⟶ Chromosome

	A	B	C
(*a*)	solenoid	nucleosome	chromatid
(*b*)	solenoid	chromatid	nucleosome
(*c*)	nucleosome	solenoid	chromatid
(*d*)	nucleosome	chromatid	solenoid

12. The term nucleosome was coined by

(*a*) Dupraw (*b*) Oudet

(*c*) Flemming (*d*) Emil Heitz

ANSWERS

Very Short Answer Questions

1. Lampbrush chromosomes are involved in rapid yolk synthesis during vitellogenesis period of growth phase of oogenesis in insects, fishes, amphibians, reptiles, birds, *etc.*

2. One DNA in each eukaryotic chromosome.

3. The 10 nm chromatin fiber (*i.e.,* nucleosomes in close apposition).

4. In eukaryotes, the chromosomal ends are called telomeres. Each telomere gives polarity to each chromosome and it prevents other chromosomes to fuse with it.

5. The length of DNA associated with nucleosomes was determined by footprinting, in which free DNA was digested, leaving only those segments protected by nucleosomes. Nucleosome hypersensitive sites are sites not in a nucleosomal state; they seems to be sites involved in initiation, transcription and other DNA activities.

6. See Fig. 5.17 and Fig. 5.18.

7. Telomeres are repetitive DNA sequences at chromosomal end. They are repetition of five or eight base sequence. Most telomeres are G-rich. Telomeres protect the ends of chromosomes and probably provide signals on senescence of cells.

8. Highly repetitive DNA usually makes up the centromeric and telomeric regions of the chromosome. Unique DNA, making up the bulk of structural genes, has a large component that is transcribed. Repetitive DNA is composed of dispersed DNA (*e.g.,* short and long interspersed elements – SINES and LINES), multiple copies of transcribed DNA (*e.g.,* ribosomal RNA, histones) and diverged copies of ancestral genes (*e.g.,* globin family genes).

9. The C-value paradox involves the issues of the excessive amounts of DNA in eukaryotic species that seem to have similar complexity. It is explained by the large amount of structural DNA in chromosomes as well as the large amounts of short and long interpersed elements (SINES and LINES).

10. Satellite DNA differs in its base sequence from the main quantity of DNA and thus forms a satellite band during buoyant density analysis. It is usually centromeric heterochromatin, composed of a highly repetitive DNA.

11. Spaces between the nucleosomes must contain many promoter sequences. DNA to be digested, it must be unprotected. Since we see little transcription, the promoters must be missing and must have been destroyed.

12. Highly repetitive DNA must be located in these chromosomal regions. Since most highly repetitive DNA is not transcribed, the results suggest that centromeric and telomeric regions are not transcribed.

13. The simplest explaination is the difference in amount of the genetic material in prokaryotes and eukaryotes. Since the average human chromosome has several centimeters of DNA, that DNA must be contracted to a size in which it can be moved during mitosis and meiosis, without tangling and breaking. Nucleosomes form the first order of coiling of the nucleosomed DNA bring it down to a manageable size for nuclear divisional processes.

14. Comparative DNA studies can be helpful in understanding the roles of the various types of DNA in the eukaryotic chromosomes if there are cases in which there are remarkably large differences in the amount of DNA in similar species. It can be then inferred that the basic developmental plan of an organism is contained in the one with the lower amount of DNA, and the extra DNA in the species with more DNA may be superfluous. We do have cases in which amphibian differ by as much as one hundred time the amount of DNA found in similar species. The puffer fish has only one-sixth amount of DNA as other higher eukaryotes.

Fill in the Blanks

1. Rapid yolk synthesis.

2. Metacentric.

3. Satellite.

4. Du Praw.

5. 200.

6. Compacting, interlocks.

7. Histones, cytosol.

8. Histone acetyl transferases, histone deacetylases.

9. Alphoid DNA.

10. Scaffold.

11. Reverse transcriptase, template.

12. $\phi \times 174$.

Multiple Choice Questions

1. (*d*)	**2.** (*a*)	**3.** (*d*)	**4.** (*b*)	**5.** (*b*)	**6.** (*a*)
7. (*c*)	**8.** (*a*)	**9.** (*c*)	**10.** (*a*)	**11.** (*c*)	**12.** (*b*).

6

Split Genes

Split genes means that the (DNA) sequences containing actual information of the gene (called **exons**) are interrupted by other sequences (called **introns**) which are spliced out after transcription.

6.1 DISCOVERY OF SPLIT GENES

With the discovery of DNA as the genetic material, a gene was regarded as a continuous segment of DNA. In 1964, it was also proved that there was co-linearity between sequence of nucleotides on DNA and sequence of amino acids in corresponding protein. Therefore, there was absolutely no doubt about the continuity of nucleotides in a gene represented by a DNA segment. This was proved to be true both for prokaryotes and eukaryotes. However, a big surprise came in 1977, when geneticists came to know that in some mammals, birds and amphibians, a gene may not be represented by a continuous sequence of nucleotides but may be interrupted by some intervening sequences which are not represented in mRNA transcribed from the gene and utilized for the synthesis of proteins. Such genes with intervening sequences were called **split genes** or **interrupted genes** (Fig. 6.1). Split genes are also called **discontinuous** or **mosaic genes** (see **Brown**, 1998).

The discovery of split genes was made in 1977 by various groups of biologists in a variety of materials: (*i*) Two research groups separately headed by **Philip A. Sharp** and **Richard J. Roberts** studied genes of adenovirus 2. (*ii*) Research groups of **P.Chambon**, **P.Leader** and **R.A. Flavell** studied globin genes, ovalbummin genes and tRNA genes. In all these cases, the genes were found to be interrupted by intervening sequences. The credit for discovery of split genes, however, goes to **Philip Sharp** and **Richard Roberts**, who won in 1993 Nobel for Medicine or Physiology for their independent work on split genes. They analyzed the hybrids of *late mRNA* of adenovirus 2 with the adenovirus genomic DNA. When these mRNA-DNA hybrids were examined under electron microscope, the adjoining sequences of mRNA were found to be hybridized with discontinuous stretches of genomic DNA of adenovirus. The intervening DNA sequences were observed as loops and the phenomenon was later described as **R-looping**.

Chambon's group compiled sequences of the boundaries of introns from a large number of protein coding eukaryotic genes (not ribosomal RNA or tRNA genes), which revealed the presence of *consensus sequences* at the intron-exon junctions. Of these GT (guanine-thymine) was always found at the 5' side of the intron (**left splice junction**) and AG (adenine-guanine) at the 3' side (**right splice junction**). This became popularly known as **GT-AG rule** or **Chambon's rule** (*Note:* Rarely, AT-AC occur at intron-exon junctions). Some of the diseases (*e.g.,* thalessemia) are caused by mutations, which created or abolished these splice junctions.

Methods of Investigations of Split Genes

A detailed study was conducted on **ovalbumin gene** found in chicken. This ovalbumin gene is responsible for synthesis of a protein, called **ovalbumin** consisting of 386 amino acids and synthesized only by highly specialized tubular gland cells of the oviduct at the time when the hen is laying eggs. The expression of this ovalbumin gene is controlled by some female sex hormones. Chambon and his collegues synthesized artificial ovalbumin gene in order to study its regulation. Such an artificial gene of ovalbumin could be synthesized by using ovalbumin mRNA which could give rise to cDNA (complementary DNA) with the help of enzyme **reverse transcriptase**. This cDNA was inserted into a plasmid and cloned in *E.coli* for its multiplication. When this cDNA was compared with corresponding genomic DNA, it was discovered (through DNA hybridization) that the genomic DNA had additional intervening sequences.

Another ingenious technique used for the study of split gene was the use of **restriction enzymes** which have the property of cleaving DNA at unique sites. More than 100 such restriction endonuclease enzyme are now available. In view of this, restriction endonucleases could be used to find out the presence or absence of a certain sequence in a particular gene. For example, when *Eco*R1 and *Hind III* enzymes were used with cDNA for ovalbumin, it was found that no cleavage occurred suggesting that the sequences of six base pairs each recognized by these two restriction enzymes were absent. On the basis of this, it was expected that if the DNA extracted from oviduct was cleaved by utilizing these two enzymes, the ovalbumin gene will not be broken and the cleavage will occur at other places thus making it possible to isolate ovalbumin gene from the living cells. This DNA segment representing ovalbumin gene was expected to be separated with the help of hybridization with cDNA artificially synthesized. When such hybridization was done with cDNA hybridized with different fragments of DNA rather than with a single fragment. Hybridization between the single stranded DNA having the gene for ovalbumin and its mRNA also showed the formation of distinct loops at specific sites as observed in electron microscope (Fig. 6.1). Such DNA which exists in the loops is obviously missing in mRNA. This kind of conclusion eventually led to the discovery of split genes in 1977.

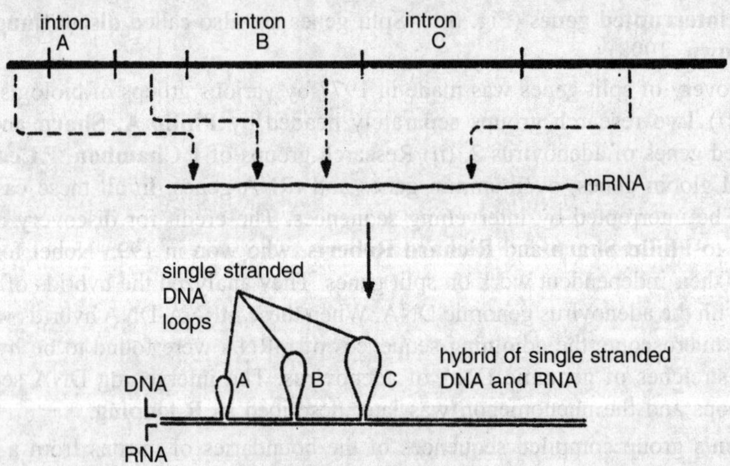

Fig. 6.1. The upper figure shows a DNA sequence representing an split or interrupted gene with their introns (A, B, C) and the synthesis of its mRNA; the bottom figure shows the result of hybridization of mRNA with single stranded DNA obtained after denaturation of native DNA (note the loops formed by three intron genes).

In subsequent years, it could be proved that split genes are present at least in two more cases, *i.e.,* gene for β-**globin** (a component of haemoglobin molecule) in rabbit and mouse and **immunoglobin gene** (antibody gene). Later on, split genes were reported to occur in higher organisms as a common phenomenon. For understanding the structure of a split gene, we can consider an example of a hypothetical split gene which has a DNA sequence including three pieces, called **exon1, exon 2** and **exon 3** (Fig. 6.2). These three exons are separated by two long intervening sequences, called **intron A** and **intron B**. The terms exon and intron were used by **Walter Gilbert** (1985) for the first time and are being followed ever since.

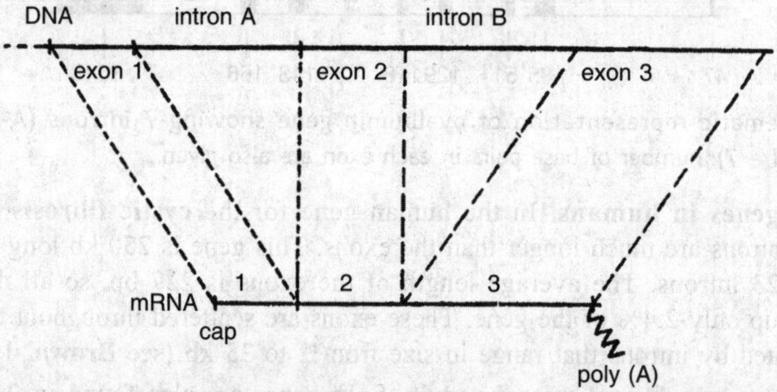

Fig. 6.2. Diagrammatic representation of a hypothetical split gene having three exons (exon1, exon2 and exon3) and two introns (intron A and intron B) and its relationship with mRNA. There are no sequences in mRNA corresponding to those in two introns.

Figure 6.2 shows that in the transcription of DNA, the ultimate product, *i.e.,* mRNA had only those sequences which correspond to exons, the sequences representing introns being entirely absent. In subsequent years, it is discovered that both exons and introns are first transcribed and this primary transcript in then modified. The sequences corresponding to introns are removed from the transcript and the sequences corresponding to exons are joined together, in correct order to give rise to mRNA. A generalization has been made that the order of exons on DNA is the same as the order in which they are found in the processed mRNA.

Examples of split genes or interrupted genes occur in a variety of organisms: (*i*) nuclear genes for proteins; (*ii*) nuclear genes for rRNAs; (*iii*) nuclear genes for tRNAs; (*iv*) mitochondrial genes in yeast; (*v*) chloroplast genes in a wide variety of plants; (*vi*) genes in archaebaceria and (*vii*) genes in bacteriophages of *E.coli.* They were initially believed to be entirely absent in eubacterial genomes but in 1990s introns were discovered even in eubacteria.

6.2 CERTAIN EXAMPLES OF SPLIT GENES

1. Structure of ovalbumin split gene of hen (chicken). In the split gene of ovalbumin, there are eight exons and seven intron (Fig. 6.3 and Fig. 6.4). The size of ovalbumin gene is 7,700 base pairs long although the mature mRNA for ovalbumin is only 1,872 nucleotides. The entire ovalbumin gene with its 7700 base pairs is first transcribed to a precursor RNA to which a cap consisting of 7 mG is added at the 5′ end and poly A tail is added at the 3′ end. After the addition of cap and tail, the introns are excised in the first step and the remaining two in second step (Fig. 6.5). The exons produced due to splicing are then joined with the help of enzyme ligase to produce the mature mRNA of ovalbumin gene.

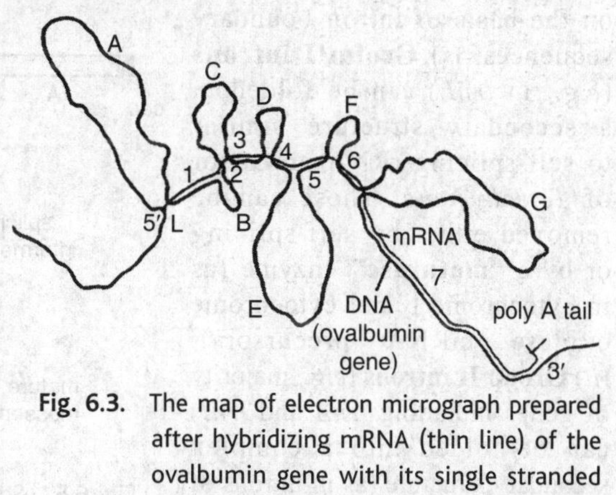

Fig. 6.3. The map of electron micrograph prepared after hybridizing mRNA (thin line) of the ovalbumin gene with its single stranded DNA (thick line). The loops A-G represent 7 different introns and segments L and 1-7 represent 8 exons.

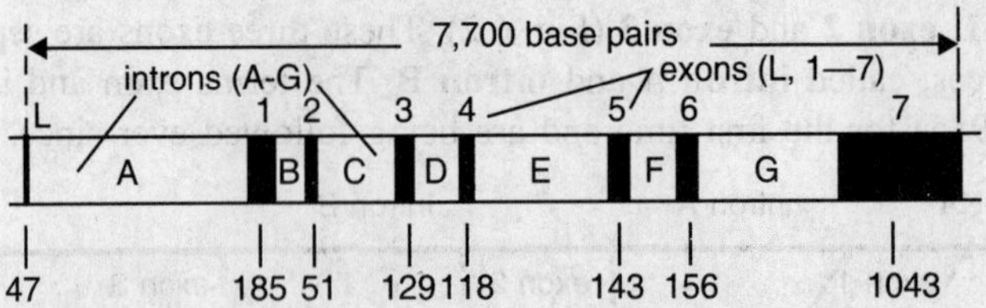

Fig. 6.4. Schematic representation of ovalbumin gene showing 7 introns (A-G) and 8 exons (L, 1 – 7); number of base pairs in each exon are also given.

2. Split genes in humans. In the human gene for the **cystic fibrosis** transmembrane regulator, the introns are much longer than the exons. This gene is 250 kb long and is split into 24 exons and 23 introns. The average length of the exons is 227 bp, so all the exons added together make up only 2.4% of the gene. These exons are scattered throughout the entire length of gene, separated by introns that range in size from 2 to 35 kb (see **Brown,** 1998).

3. Split genes in fungal mitochondria. Split genes are also found in the mitochondria. Introns of these split genes in fungal mitochondria are of two type: (*i*) **Group I introns** which are found in majority of the fungal mitochondria split genes, do not carry any conserved sequence, called **internal guide sequence**. By internal pairing, this guide sequence brings the two intron-exon junctions together and help in splicing out the introns. These group I introns are also found in the nuclear gene coding for rRNA in *Tetrahymena* (a ciliate) and *Physarum* (a slime mold). Features similar to those of group I introns are also reported from introns of phage T4 genes. (*ii*) **Group II introns** resemble nuclear genes and have consensus sequences (GT and APy) and a branch sequence that resembles the TACTAAC box. These introns are excised as lariats.

4. Split Genes in Chloroplasts

Split genes for ribosomal RNA (rRNA), transfer RNAs (tRNA) and some proteins have also been reported in the chloroplast genomes of several plants including *Chlamydomonas* and *Nicotiana*. Introns found in chloroplast genes can be classified into three groups on the basis of intron boundary sequences. (*i*) **Group I introns** (*e.g.,* in *trnL*) can be folded in a secondary structure similar to self-splicing rRNA precursor of *Tetrahymena*. These can be removed either by self-splicing or by a **"maturase"** enzyme (as in cytochrome b and cytochrome oxidase mRNA precursors). (*ii*) **Group II introns** (*e.g.,* majority of genes including *trnA* and *trnI*) can be folded into a complex secondary structure (as in introns of mitochondrial genes for cytochrome oxidase in maize and yeast).

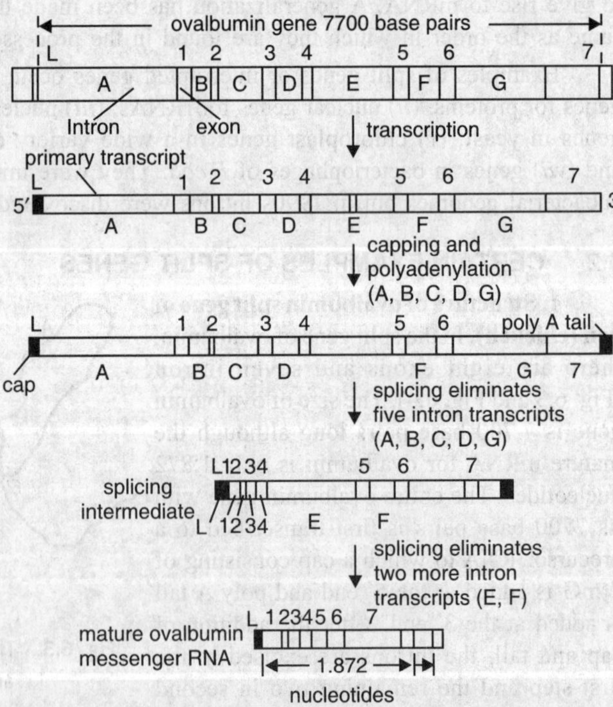

Fig. 6.5. Steps involved in the production of ovalbumin mRNA (1872 nucleotides long) from ovalbumin gene (7700 base pairs long); introns (A-G) and exons (L and 1 -7).

(*iii*) **Group III introns** (*e.g., trnG, trnK, trnV, rpl2, rps12, rps16, etc.*) have conserved sequences at their borders (GTGCGNY at 5' end, and ATCNRYY(N)YYAY at 3' end), similar to those in the eukaryotic nuclear genes (R = purine; Y = pyrimidine; N = any nucleotide).

Although introns are generally absent in protein-coding genes of bacteria, archaebacteria and some lower eukaryotes, class II type of self-splicing introns have recently been reported even in bacteria. Such a distribution of introns has been used for a study of the role of introns in the evolution of genes or alternatively for a study of the origin of genes or alternatively for a study of the origin of introns themselves.

6.3 EVOLUTIONARY ORIGIN OF SPLIT GENES

Following three hypothesis/theories have been proposed for explaining origin of split genes: 1. Exon (domain) shuffling hypothesis; 2. Exon theory; and 3. Intron early hypothesis.

Exon shuffling hypothesis is an interesting idea which holds that each exon of split gene contains a different subcomponent of the biological information carried by the gene as a whole. Individual subcomponents are not sufficient on their own to code for a complete protein, but the information they contain is meaningful in that it specifies a recognizable portion of the proteins function, for example, enabling the protein to bind a particular substrate or attach to a specific site in the cell. The hypothesis is that during evolution, exons from different discontinuous genes can be "shuffled", creating new combinations of biological information (Fig. 6.6). **Exon shuffling** would be more likely to produce new functional proteins than would be entirely random rearrangement of existing genes.

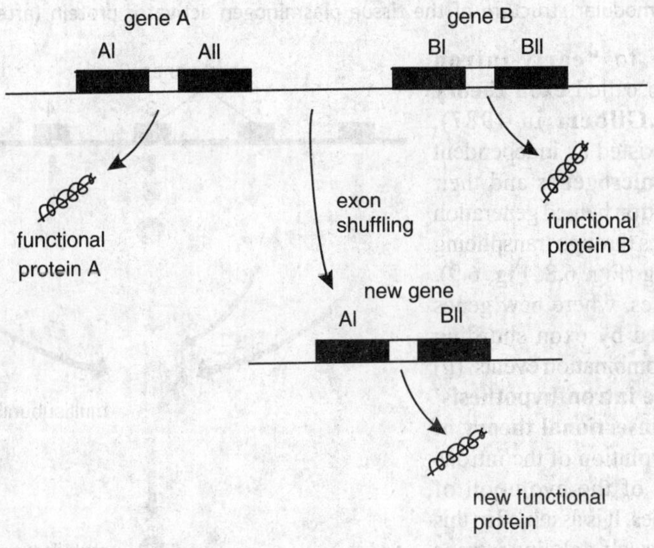

Fig. 6.6. The hypothesis of 'exon shuffling' suggests that new functional genes could be formed by rearranging exons of existing genes (after Brown, 1998).

Domain or exon shuffling is illustrated by **tissue plasminogen activator (TPA)**, a protein in the blood of vertebrates and which is involved in the blood clotting response. The TPA gene has four exons, each coding for a different structural domain (Fig. 6.7). The upstream exon coding for a 'finger' module that enables the TPA protein to bind to **fibrin**, a fibrous protein found in blood clots which activates TPA. This exon appears to be derived from a second fibrin-binding protein, **fibronectin,** and is absent from the gene for a related protein, **urokinase**, which is not activated by fibrin. The second TPA exon specifies a **growth-factor** domain which has apparently been obtained from the gene for **epidermal growth factor** and which may enable TPA to stimulate cell proliferation. The last two exons code for "Kringle" structures which TPA uses to bind to fibrin clots; these kringle exons come from the plasminogen gene (**Li** and **Graur,** 1991).

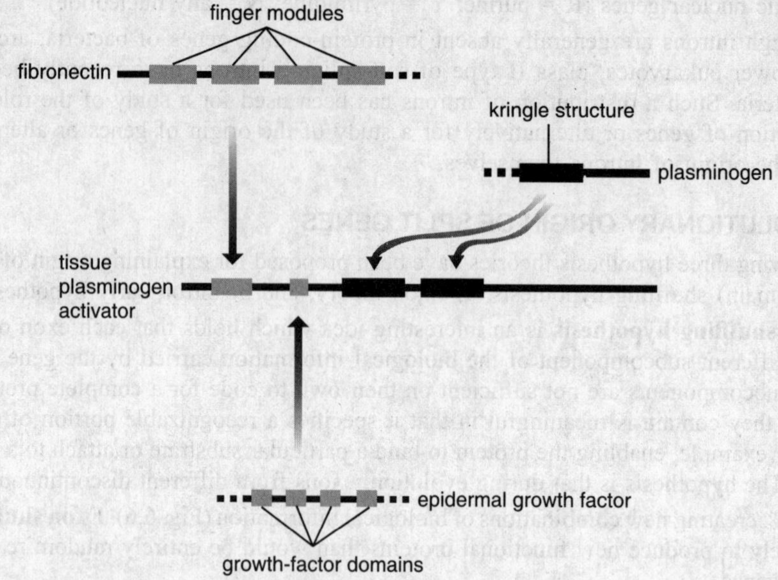

Fig. 6.7. The modular structure of the tissue plasminogen activator protein (after Brown, 2002).

According to **"early-intron hypothesis"** (also called **exon theory** of genes by **W.Gilbert** in 1987), ancestral exons existed as independent genetic units of **microgenes** and their association with intron helped generation of diversity in genes through transplicing and exon shuffling (Fig. 6.8; Fig. 6.9). There are examples, where new genes could be generated by exon shuffling through intron recombination events. (*ii*) According to **'late intron hypothesis'** (also called the **"insertional theory of intron orgin"**) evolution of the introns was independent of the evolution of protein-coding genes. It is assumed in this hypothesis that the self-splicing introns (class I introns) invaded the genome of eukaryotic ancestors, as endosymbionts, and revolved into mitochondria and chloroplasts containing class II introns (**Eickbush,** 2000). It is postulated

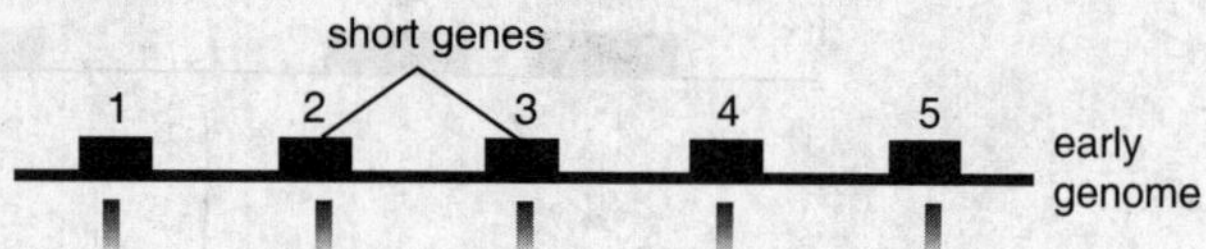

Fig. 6.8. The 'exon theory of genes'. The short genes of the first genomes probably coded for single domain polypeptides that would have had to associate together to form a multisubunit protein to produce an effective enzyme. Later the synthesis of this enzyme could have made more efficient by linking the short genes together into one discontinuous or split gene coding for a multidomain subunit protein (after Brown, 2002).

that invasion took place at the mRNA level, which was followed by reverse transcription and recombination with DNA, producing split genes. These introns gave other types of introns later. This explains, why introns are absent in bacteria, archaebacteria and lower eukaryotes. The recent report of introns in bacteria may be the initial stage of this process of evolution of introns.

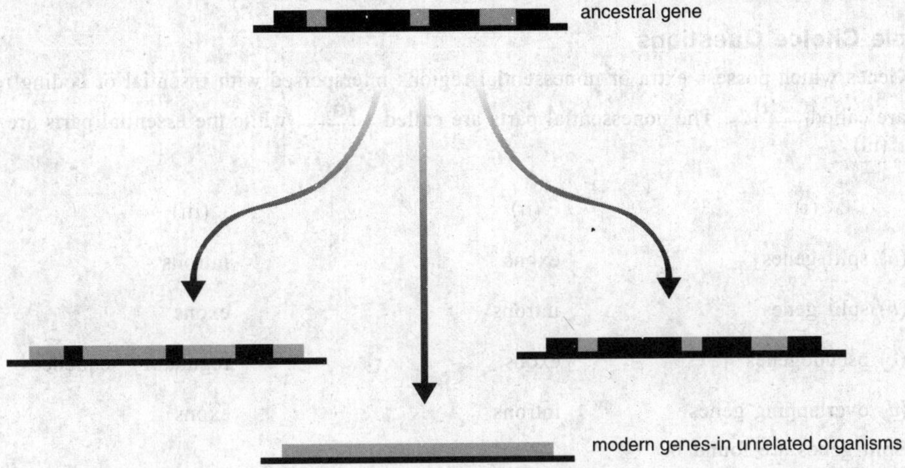

Fig. 6.9. One prediction of the "early intron" hypothesis is that the positions of introns in homologous genes should be similar in unrelated organisms, because all these genes are descended from an ancestral intron-containing gene (after Brown, 2002).

QUESTIONS

Long Answer Questions

1. Give an account of split genes.
2. Describe certain examples of split genes.

Short Answer Questions

1. Describe exon shuffling hypothesis.
2. What is early-intro hypothesis or exon theory.
3. Describe late-nitron hypothesis.

Very Short Answer Questions

1. What are split genes?
2. Define the exon.
3. What is exon shuffling?
4. What are nitrons?
5. Define splicing.

Fill in the Blanks

1. Split genes means that DNA sequences containing actual information of the gene (called ....) are interrupted by other sequences (called ....) which are spliced out after transcription.
2. The 1993 Nobel Prize for Medicine or Physiology was awarded to .... and .... for independent work on the .... genes.

3. .... enzymes have the property of cleaving DNA at unique sites.
4. P. Chambon and his colleagues synthesized artificial .... gene for investigating split genes of hen.
5. Exon theory of gene was formulated by .... in 1987.

Multiple Choice Questions

1. Genes which possess extra or nonessential regions interspersed with essential or coding regions are called ...(i)... The nonessential parts are called ...(ii)..., while the essential parts are called ...(iii)...

	(i)	(ii)	(iii)
(*a*)	split genes	exons	introns
(*b*)	split genes	introns	exons
(*c*)	pseudogenes	exons	regulatory sequences
(*d*)	overlapping genes	introns	exons

2. Split genes are found in
   (*a*) histone genes of eukaryotes
   (*b*) thymidylate synthase gene of prokaryotes
   (*c*) interferon gene of eukaryotes
   (*d*) all of these
3. Processed genes are those genes that
   (*a*) lack introns
   (*b*) show homology to functional genes
   (*c*) lack promoters
   (*d*) both (*a*) and (*c*)
4. What are nonconstitutive genes?
   (*a*) genes whose products are required for normal cellular activities
   (*b*) genes which are switched on or off according to the requirement of cellular activities
   (*c*) genes which express constantly in a cell
   (*d*) genes which lack introns

ANSWERS

Very Short Answer Questions

1. Genes having introns and exons.
2. In a split gene that has intervening sequences (introns), the exon is a region that is actually exported from the nucleus to be expressed or become part of a transfer RNA or ribosomal RNA.

3. The hypothesis put forward by Walter Gilbert that exons code for the functional units of a protein, and that the evolution of new genes proceeds by recombination or exclusion of exons.
4. DNA sequences within a gene (called split gene) that are transcribed but removed prior to translation.
5. For introns to be removed, the ends of the exons must be brought together and connected in a process called splicing (see chapter 9).

Fill in the Blanks

1. Exons, introns.
2. Philip Sharp, Richard Roberts, split.
3. Restriction.
4. Ovalbumin.
5. W. Gilbert.

Multiple Choice Questions

1. (*b*) 2. (*b*) 3. (*d*) 4. (*b*)

7

DNA Replication

As a carrier of genetic information DNA has the following two important functions: **1. Heterocatalytic function** – When DNA directs the synthesis of chemical molecules other than itself (*e.g.,* synthesis of RNA, proteins, etc.), then such functions of DNA are called heterocatalytic functions. **2. Autocatalytic functions** – The functions of DNA which directs the synthesis of DNA itself, are called autocatalytic functions. Here, we are concerned only with autocatalytic function of DNA.

7.1 WATSON AND CRICK'S MODEL FOR DNA REPLICATION

The double helix model of DNA molecule of Watson and Crick is beautifully embodied a built-in template system for self-replication or autocatalytic function. Because of the specificity of base pairing, the sequence of bases along one chain automatically determines the base sequence along the other. Thus, each chain of the double helix can serve as **template** for the synthesis of the other. For the replication of **DNA** molecule, **Watson** and **Crick** proposed that replication involved the disruption of hydrogen bonds followed by a rotation and separation of the two polynucleotide strands. Each purine and pyrimidine base of each polynucleotide strand is thought to attract a complementary free nucleotide available for polymerization in cell and to hold it in place by means of the specific hydrogen bonds.

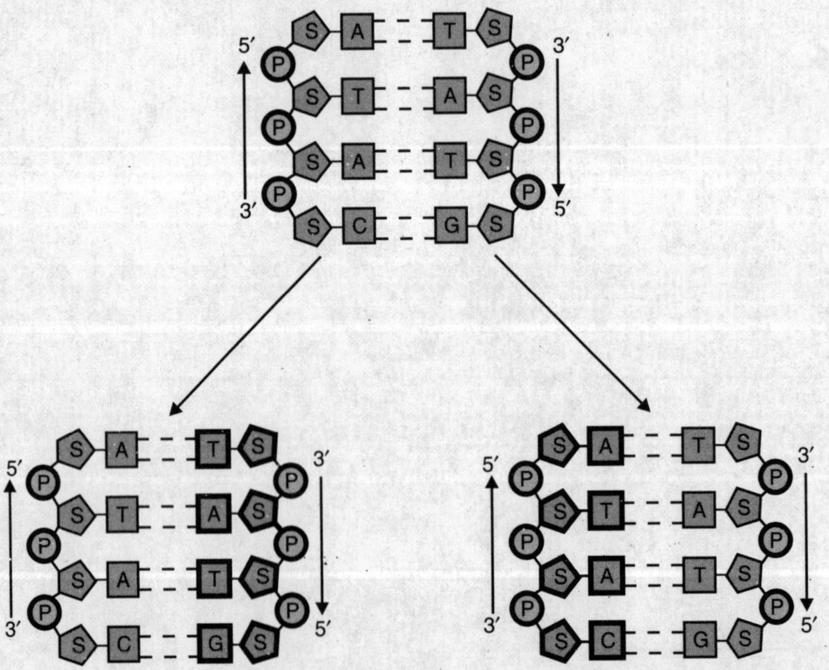

Fig. 7.1. Semi-conservative model of DNA replication.

Once held in place on the parent template chain, the free nucleotides were sewn together by formation of the phosphate diester bonds that linked adjacent deoxyribose residues, forming a new polynucleotide molecule of predetermined base sequence. Thus, two double helical molecules, identical with each other, are formed (Fig. 7.1).

1. Experimental Evidence for Semiconservative DNA Replication in *E.coli*

The Watson–Crick's model of DNA structure and replication suggested that once DNA replication is initiated, the two original polynucleotide strands of the duplex of helix will unwind, at least locally, so that each can serve as a template for a new strand.

An immediate prediction follows from this proposal: both duplexes that result from replication should be **hybrid** in nature, each containing an old strand derived from the original molecule and a new strand which has been formed during the replication process. Since each of the two double helices or duplexes conserves only one of the parent polynucleotide strands, the process is said to be **semiconservative**. This prediction is diagrammed in Fig. 7.2 in which old DNA is shown in black and new DNA in white. Figure 7.2 also outlines what would be predicted if these two hybrid duplexes went on to replicate themselves. Four duplexes would result, two of which would contain a single strand derived from the original chromosome and two of which would contain totally new DNA strands.

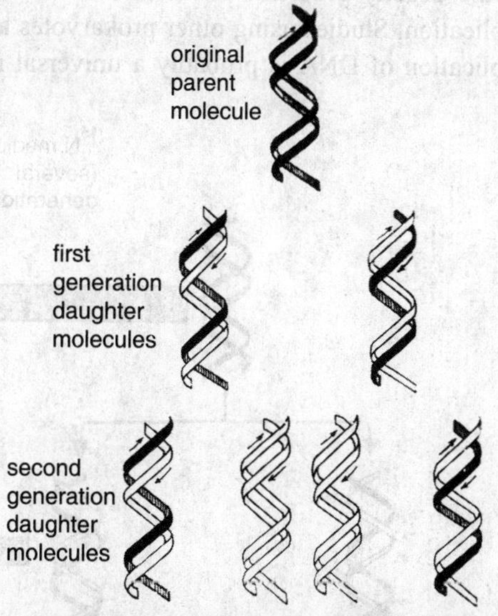

Fig. 7.2. Semiconservative DNA replication.

Besides the semiconservative mode of DNA replication, the following two methods of DNA replication were deemed equally feasible: (*i*) **conservative replication**, in which both strands of parent double helix would be conserved and the new DNA molecule would consist of two newly synthesized strands; and (*ii*) **dispersive replication**, in which replication would involve fragmentaion of the parent double helix and the intermixing of pieces of the parent strands with newly sythesized pieces, thereby forming the two new double helices.

***(i)* Meselson and Stahl's experiment. M. Meselson** and **F.W. Stahl** (1958) verified the semiconservative nature of DNA replication in a series of elegant experiments using isotopically labelled DNA and a form of isopycnic density gradient centrifugation. They cultured *Escherichia coli* cells in a medium in which the nitrogen was ^{15}N (a 'heavy' isotope of nitrogen, but not a radioisotope) instead of commonly occurring and lighter ^{14}N. In time, the purines and pyrimidines of DNA in new cells contained ^{15}N (where ^{14}N normally occurs) and, thus, the DNA molecules were denser. DNA in which the nitrogen atoms are heavy (^{15}N) can be distinguished from DNA containing light nitrogen (^{14}N), because during isopycnic centrifugation, the two different DNAs band at different density positions in the centrifuge tube.

Depending on its content of ^{15}N and ^{14}N, the DNA bands at a specific position in the density gradient. Because the DNA synthesized by *E.coli* cells grown in ^{15}N would be denser than ^{14}N containing DNA, it would band further down the tube.

E.coli cells growth for sometime in the presence of ^{15}N-medium were washed free of the medium and transferred to ^{14}N-containing medium and allowed to continue to grow for specific lengths of time (*i.e.,* for various numbers of generation time). DNA isolated from cells grown for one generation of time in the ^{14}N medium had a density intermediate to that of the DNA from cells grown only in ^{15}N-containing medium (identified as *generation O*; Fig. 7.3) and that of DNA from cells grown only in ^{14}N-containing medium (*the controls*). Such a result immediately ruled out the possibility that DNA replication was conservative, because the conservative replication would have yielded two DNA bands in the density gradient for **generation 1** (*i.e.,* F_1 cells). The single band of intermediate density (identified as "**hybrid DNA**") consisted of DNA molecules in which one strand contained ^{15}N and the other contained ^{14}N. When the incubation in the ^{14}N-medium was carried out for two generations of time (*i.e.,* **generation 2**), two DNA bands were formed – one at the same density position as the DNA from cells grown exclusively in ^{14}N medium (*i.e.,* light controls) and the other of intermediate density. Subsequent generations produced greater numbers of DNA molecules that banded at the "light" (^{14}N-containing DNA) position in the density gradient. These results are consistent only with the model of semiconservative replication. Studies using other prokaryotes as well as eukaryotes indicates that semiconservative replication of DNA is probably a universal mechanism.

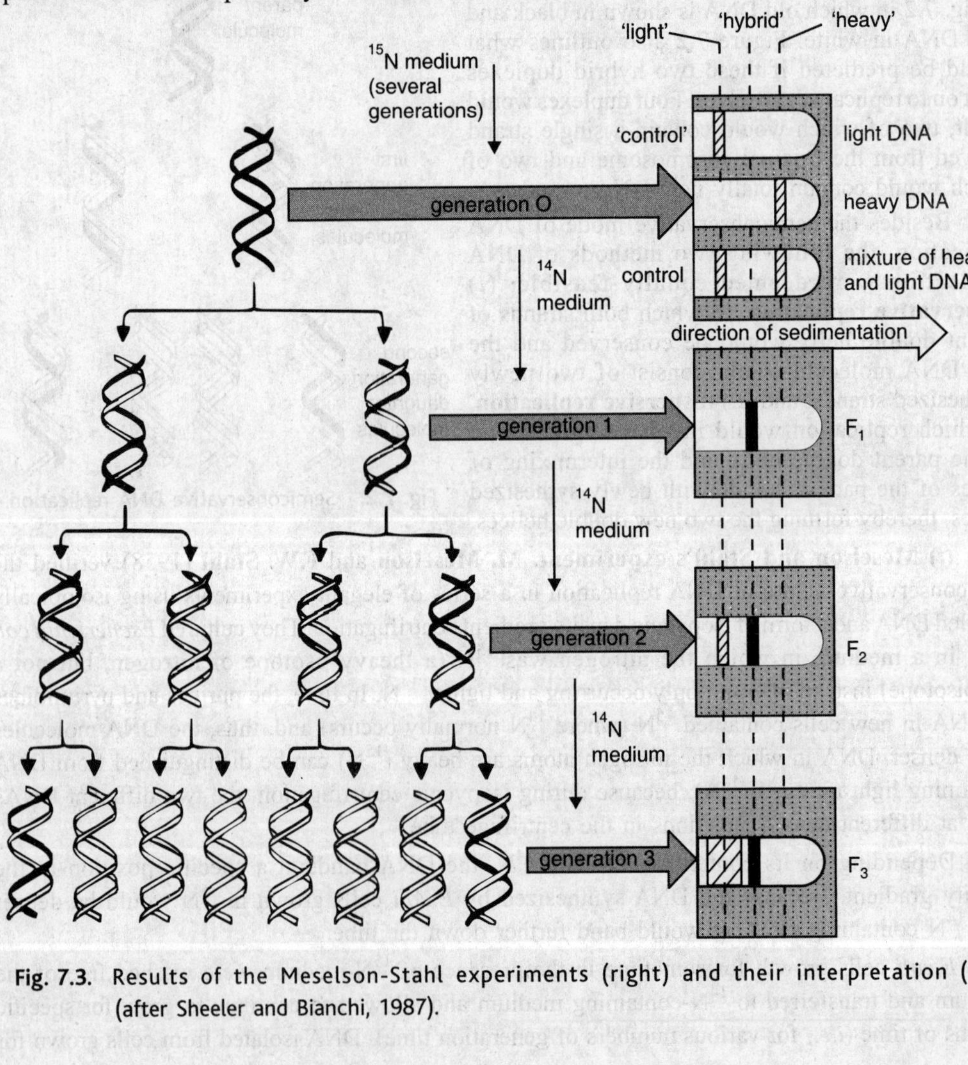

Fig. 7.3. Results of the Meselson-Stahl experiments (right) and their interpretation (left) (after Sheeler and Bianchi, 1987).

(ii)* Visualization of replication in *E.coli. In 1963, **J.Cairns** developed a technique employing a combination of microscopy and autoradiography that made it possible to visualize the replication of the chromosome of *E.coli.* Cairns placed *E.coli* cells in a medium containing ^{3}H-thymidine (tritiated-thymidine) for various periods of timc so that the radioactive thymidine was incorporated into the DNA as the chromosome was replicated in successive generations of cells. *E.coli* cells were removed from the medium after various periods of incubation and gently lysed to release the chromosome from the cell (since the shear forces created by harsh lysis break the chromosome into small pieces). The chromosomes were then transferred to glass slides and coated with a photographic emulsion sensitive to the low energy beta-particles emitted by the ^{3}H-thymidine. After exposing the emulsion to the beta rays, the emulsion was developed and examined by light microscopy. Wherever decay of labelled thymidine had occurred in a chromosome, the emulsion was exposed and created visible grains. A chromosome not engaged in replication appeared as a circular structure formed from a close succession of explosure spots. Chromosomes "caught in the act" of replication gave rise to what are called **theta configurations** because they have the appearance of the Greek letter theta (*i.e.,* θ, Fig. 7.4). The theta structures reveal the positions of the replication forks in the circular chromosomes. Cairns' observations also clearly supported the semiconservative nature of replication.

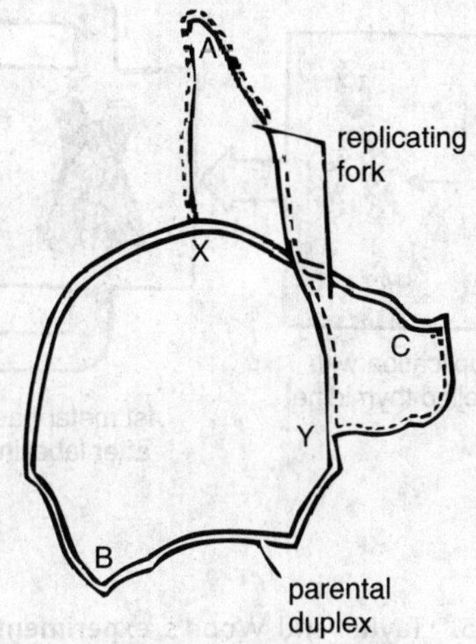

Fig. 7.4. Theta (θ) configuration of replication in *E.coli* chromosome obtained by Cairns. Loops A and B have completed replication, whereas loop C remains to replicated. X and Y are the replication forks.

The rate at which the replication proceeds could also be worked out by measuring the length of DNA undergoing replication in a known interval of time. Cairns worked out generation time of *E.coli* as 30 minutes. The length of the chromosome was worked out to about 1 mm. The rate of replication, thus, would be approximately 30 μm – 40 μm per minute (1mm = 1000 μm).

2. Evidence for Semiconservative Replication of Chromosomes (or DNA) in Eukaryotes

J.H. Taylor and **P. Woods** (1957) provided evidence in support of semiconservative mode of DNA replication in eukaryotes by using the technique of autoradiography and light microscopy in dividing root tip cells of the bean, *Vicia faba.* After incorporation of tritiated thymidine, when root tips were transferred to unlabelled culture medium (and colchicine was added to the medium to prevent anaphase separation of sister chromatids), in the first generation of duplication both chromatids were labelled (this is interpreted as one DNA double helix in each chromatid and only one of the two strands labelled). In the second cycle of duplication (in the unlabelled medium) in each chromosome, one of the two chromatids was found to be labeiled (Fig. 7.5). This was interpreted as showing semiconservative mode duplication.

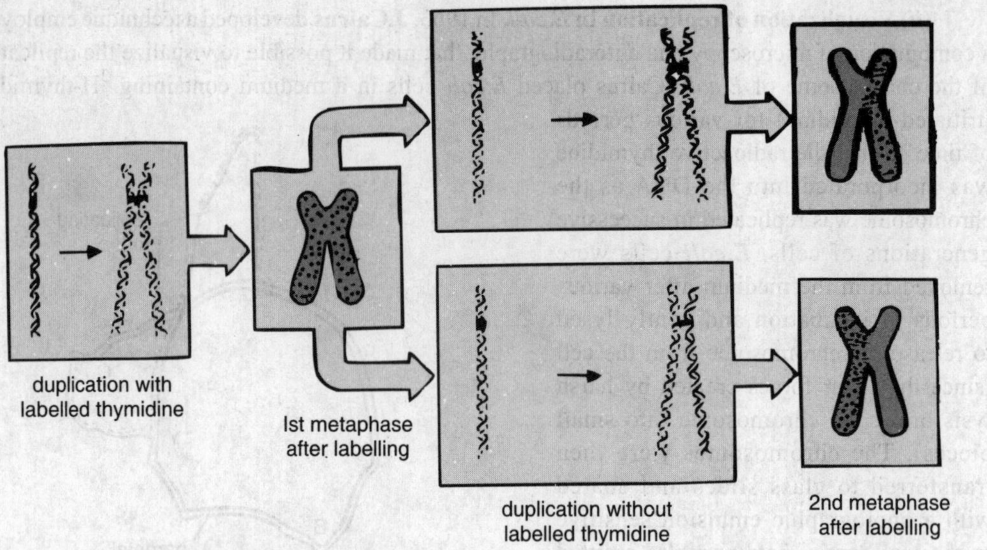

Fig. 7.5. Taylor and Wood's experiment on *Vicia faba* root tips using autoradiography technique (after Sheeler and Bianchi, 1987).

3. Semidiscontinuous DNA Replication

Various experimental evidences have suggested that DNA synthesis is continuous on one strand (3′ to 5′ strand), called **leading strand** and discontinuous on the other strand (5' to 3' strand), called **lagging strand**. Since DNA synthesis always proceeds in 5′ to 3′ direction, so, on the lagging strand, synthesis takes place discontinuously in pieces, called **Okazaki fragments** (after the name of discoverer **R. Okazaki**, 1968). Okazaki fragments average about 1,500 nucleotides in prokaryotes and 150 in eukaryotes (see **Tamarin**, 2002). Later on, these pieces are fused with the help of ligase enzyme to form an intact lagging strand. Such a DNA replication, where the leading strand is synthesized continuously and the lagging strand is synthesized discontinuously, is called **semidiscontinuous replication**. Once initiated, continuous DNA replication can proceed indefinitely. DNA polymerase III on the leading – strand template has what is called **processivity**: once it is attaches, it does not release until the entire strand is replicated. Discontinuous replication, however, requires the repetition of four steps: primer synthesis, elongation, primer removal with gap filling and ligation.

4. Unidirectional and Bidirectional DNA Replication

Regarding the direction of replication, DNA replication may be of the following two types:

(*a*) Bidirectional replication. All known DNA molecules, with only few exceptions, replicate as circles (or bubbles/eyes) and, hence, initiate within the helix. In the electron microscope, eukaryotic chromosomes are found to contain multiple expanding replication eyes, in contrast to a single eye of the prokaryotic DNA. Further DNA synthesis within a given replication unit eye is initiated somewhere at or near the midpoint of the unit at a site termed as the origin (O; Fig. 7.9); prokaryotes contain a solitary origin, while eukaryotes have multiple (up to several thousands) origins for DNA replication. Both ends of the eye are moving and serve as **replication forks** (Fig. 7.6).

Two replicating forks are then believed to travel in opposite directions until they reach either end of that unit, the two end points being called **termini (T)**. A given replication unit may or may not undergo bidirectional synthesis in synchrony with continuous units.

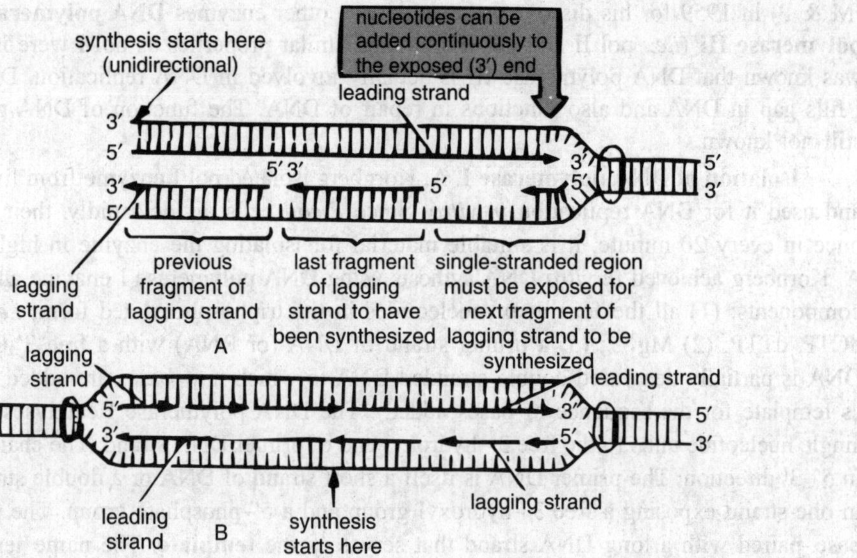

Fig. 7.6. A – Unidirectional DNA replication; B – Bidirectional DNA replication.

In either case the newly replicated strands in adjacent units will eventually meet. These strands are then linked, perhaps, by a DNA ligase to form long, continuous daughter DNA strands (Fig. 7.6).

***(b)* Unidirectional replication.** In case of this type of DNA replication, one of two ends of the replication eye remains stationary and the other end serves as the replication fork and moves with replication (Fig. 7.7). An example of unidirectional replication is the replication of mitochondrial DNA (mt DNA) by D-loop (or displacement loop) in vertebrates.

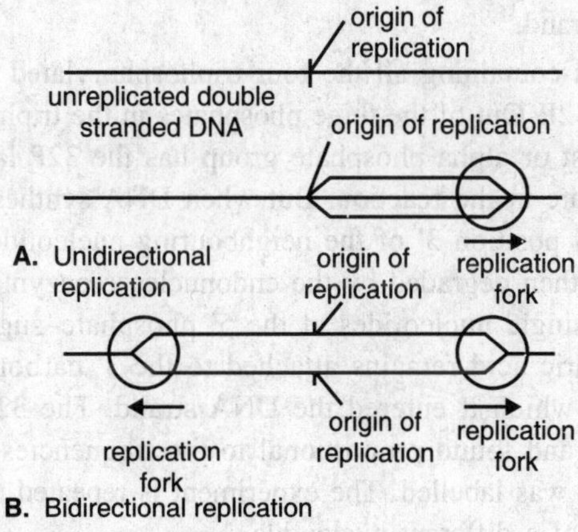

Fig. 7.7. The difference between unidirectional and bidirectional replications in the movement of replication forks.

7.2 DISCOVERY OF DNA POLYMERASES

Studies on molecular mechanisms underlying DNA replication were initiated only after **Arthur Kornberg** in 1957 discovered the enzyme which is involved in DNA replication. This enzyme today is known as **DNA polymerase I** or **pol I**. A. Kornberg was awarded Nobel Prize

(M & P) in 1959 for his discovery. Later on two other enzymes **DNA polymerase II** and **DNA polymerase III** (*i.e.,* pol II and pol III) having similar properties as pol I were isolated. Later it was known that DNA polymerase III is actually involved in DNA replication. DNA polymerase I fills gap in DNA and also functions in repair of DNA. The function of DNA polymerase II is still not known.

Isolation of DNA polymerase I. A. Kornberg isolated pol I enzyme from living *E.coli* cells and used it for DNA replication *in vitro*. Since *E.coli* cells divide rapidly, their DNA duplicate once in every 20 minute, it is suitable material for isolating the enzyme in high concentration. A. Kornberg achieved *in vitro* DNA sythesis using DNA polymerase I enzyme and the following components: (1) all the four deoxynucleotides in the triphosphorylated form, *i.e.,* dATP, dGTP, dCTP, dTTP; (2) Mg^{++} ; (3) a primer strand of DNA (or RNA) with a free 3′ OH end (primer DNA is partially denatured double stranded DNA to which new bases are added and also serves as template for the sequence of bases added). The DNA polymerase I catalyses the addition of single nucleotide units to the free 3′–hydroxyl end of primer DNA strand. The chain is synthesized in 5′–3′ direction. The primer DNA is itself a short strand of DNA or a double strand with a nick in one strand exposing a free 3′–hydroxyl group and a 5′–phosphate group. The primer strand is base–paired with a long DNA strand that serves as the **template**. The name **template–primer** is now given to this structure.

Kornberg's Technique of Nearest Base Frequency Analysis. DNA polymerase I isolated from *E.coli* has a single polypeptide chain of about 4000 amino acids and a molecular weight of 109,000. The enzyme molecule is roughly spherical in shape with a diameter of about 6.5 nm. In the purified state one molecule of the enzyme, at 37°C can add about 1,000 nucleotides per minute to a growing DNA chain. To show that the DNA synthesized *in vitro* by the purified DNA polymerase I has the same base sequence as the DNA used in the primer and template strand, A. Kornberg devised a method known as **nearest–neighbour base frequency analysis**. This method determines the frequency with which any two bases occur as adjacent or nearest neighbours in the DNA strand.

The *in vitro* systems containing all the four triphosphorylated nucleotides has one of the nucleotides labelled with 32P. Out of the three phosphates in the triphosphate linked to 5′ carbon of the sugar, the innermost or alpha phosphate group has the 32P label. This phosphate is not eliminated as pyrophosphate in the reaction. But when DNA synthesis occurs, it forms second linkage with the carbon at position 3′ of the neighbouring nucleotide in the newly synthesized DNA chain. The DNA is then degraded by the endonuclease enzyme spleen phosphodiesterase which breaks DNA into single nucleotides at the 5′ phosphate–sugar linkages only. The 32P containing alpha phosphoric acid remains attached to the 3′ carbon of the nearest neighbour nucleotide to the one on which it entered the DNA strand. The 32P contents of all the four nucleotides are measured and found proportional to the frequencies with which they occurred next to the nucleotide that was labelled. The experiment is repeated three more times each time labelling the triphosphate of a different nucleotide.

Kornberg's experiments established beyond doubt the DNA polymerase I lead to *in vitro* DNA synthesis. However, in 1967 **J. Cairns** provided evidence that this may not always be so. Cairns isolated a mutant strain of *E.coli* known as *pol AI* which was lacking in DNA polymerase, but still was able to grow and replicate its DNA normally. When these cells were exposed to UV and X-radiation, it was found that they could not repair damage to their DNA. From this evidence, it follows that DNA polymerase I is a repair enzyme and may not be responsible for replication.

Later on DNA polymerase I was shown to have exonuclease activity in both 3′ to 5′ and 5′ to 3′ directions (Fig. 7.8A). It performs exonuclease activity mainly for removing a wrongly placed nucleotide. Further experiments proved that DNA polymerase I molecule actually consists of two subunits: a larger polypeptide chain with molecular weight of about 76,000 daltons, a smaller chain with molecular weight of 36,000 daltons. The two subunits dissociate on treatment with proteolytic enzymes. The larger subunit of pol I possess the repair and 3′ to 5′ exonuclease activities. The smaller subunit of pol I has 5′ to 3′ exonuclease activity.

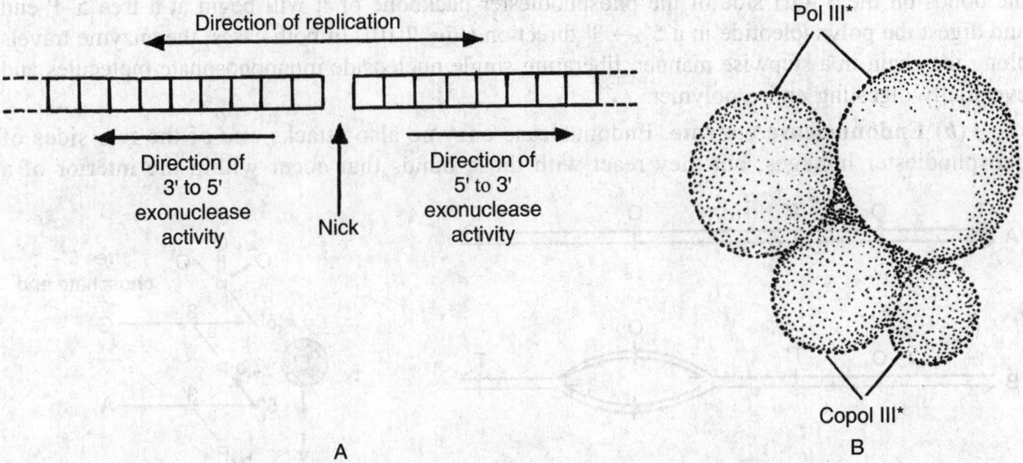

Fig. 7.8. A–The exonuclease activities of DNA polymerase I. B–DNA polymerase II.

DNA Polymerase II and III

From the pol AI mutants of *E.coli* two other enzymes were isolated namely DNA polymerase II and DNA polymerase III (pol II and pol III). The wild type *E.coli* cells contain all the three polymerase enzymes, *i.e.,* pol I, pol II and pol III which are coded by the structural genes pol A, pol B and pol C. The exact function of **DNA polymerase II** in replication is not known with certainty. It has exonuclease activity in 3′ to 5′ direction. Mutants defective in pol II (*pol B*) have been isolated but they are phenotypically indistinguishable from the wild type. It is, therefore, likely that pol II enzyme is not necessary for DNA replication. .

DNA polymerase III was discovered in 1972 by **Roger D. Kornberg** (son of Arthur Kornberg; he was also awarded the 2006 Nobel Prize in Chemistry for this discovery) and **Gefter**. Pol III enzyme is most actively involved in DNA replication. Pol III enzyme also performs exonuclease activity in both 3′ to 5′ and 5′ to 3′ directions. Later on Kornberg showed that pol III enzyme is actually a dimer of pol III and pol III* and is known as DNA **polymerase III*** (pol three star; Fig. 7.8B). It requires a protein, **copolymerase III*** (copol III*) to function. Like pol I, DNA polymerase III* replicates a template-primer strand of DNA. Thus, pol III*–Copol III* complex along with ATP, a DNA template and an RNA primer are necessary to initiate replication.

7.3 ENZYMES AND PROTEINS OF DNA METABOLISM

Different prokaryotic and eukaryotic cells have been found to contain three kinds of nuclear enzymes or enzymatic activities that act on DNA, namely **nucleases, polymerase** and **ligases**.

A. Enzymes Involved in DNA Replication

(I) Nuclease enzymes. The nuclease enzymes act to hydrolyze or break down a polynucleotide chain into its component nucleotides. A polynucleotide is held together by 3′, 5′ phosphodiester bonds and a nuclease enzyme will attack either the 3′ or the 5′ end of this linkage. The nuclease enzymes may be of the following two kinds.

***(a)* Exonuclease enzymes.** A nuclease enzyme which begins its attack from a free end of a polynucleotide is called **exonuclease**. Therefore, depending on the specificity of the enzyme, an exonuclease will either begin at a free 3′-OH end of a polynucleotide and progressively cleave the bonds on the 3′-OH side of the phosphodiester backbone or it will begin at a free 5′-P end and digest the polynucleotide in a 5′ → 3′ direction (Fig. 7.10). In both cases, the enzyme travels along the chain in a stepwise manner, liberating single nucleoside monophosphate molecules and eventually digesting entire polymer.

***(b)* Endonuclease enzyme**. Endonuclease enzyme also attacks one of the two sides of phosphodiester linkages, but they react with those bonds that occur within the interior of a

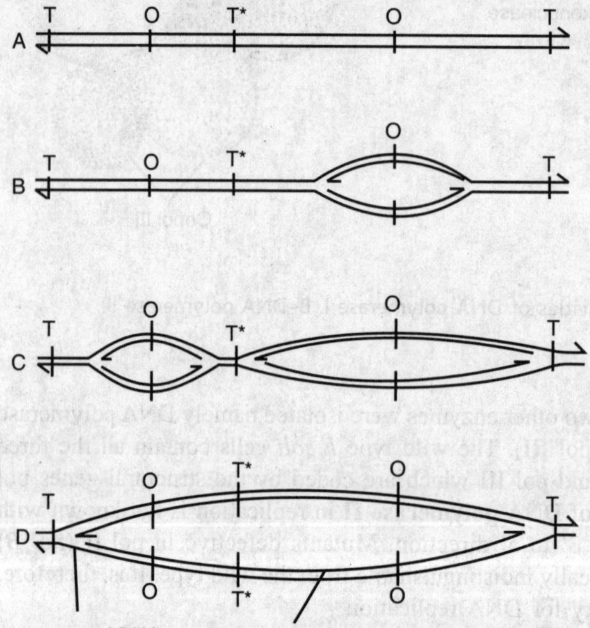

Fig. 7.9. Bidirectional mode for mammalian chromosomal DNA replication. Origins are indicated by O and termini by T. A – DNA prior to replication; B – replication started in right-hand replication unit; C – replication started in left-hand replication unit and completed at termini of right-hand unit; D – replication completed, sister duplexes joined at common terminus T.

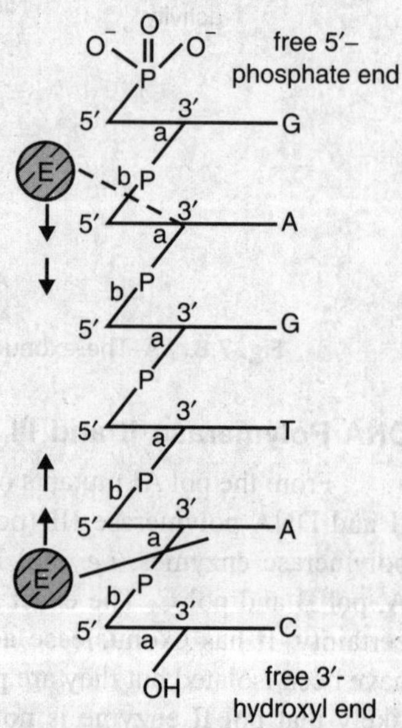

Fig. 7.10. Exonuclease action on a nucleotide chain. The 3′ side of each phosphodiester linkage is labelled a, the 5′ side is labelled b. An exonuclease that cleaves a linkages, starting from a free 3′-OH end (*e.g.*, snake venom phosphodiesterase) is shown in check. An exonuclease that cleaves b linkages, starting from a free 5′-P end (*e.g.*, bovine spleen phosphodiesterase) is shown in line shade (after Lehninger, 1970).

polynucleotide chain. If the polynucleotide chain is single-stranded (*e.g.,* viral DNA), such an attack will obviously cut the chain into two pieces. If, however, the polynucleotide strand is a member of a DNA helix (*e.g.,* prokaryotic and eukaryotic DNA), a single endonucleolytic cut will create a **nick** in the helix; the helix remains in one piece but it now possesses a gap that contains two free ends, which can serve as substrates for exonucleases. A nicked double helix suffers a localized disruption of its secondary structure. The molecules becomes free to bend or rotate around its intact strand and the two broken ends are free to dangle. The increased molecular motion in the region of the nick will very likely interrupt hydrogen bonding between bases in the vicinity of the nick, thus, effecting a limited "untravelling" of the helix.

(II) Polymerase or replicase enzymes. A polymerase enzyme catalyses the formation of a polymer and cellular polymerase enzymes of genetic interest are those that bring about the synthesis of one polynucleotide chain that is a copy of another. A polymerase enzyme is called **replicase** enzyme when the copy of polynucleotide chain so produced is inherited by daughter cells or viruses, that is, when the enzyme brings about chromosome replication (see **Goodenough** and **Levine**, 1974).

***In vitro* DNA polymerization.** To understand the mechanism of DNA-replication inside the living cell (*in vivo*), molecular biologists tried *in vitro* polymerization of DNA and found that in addition to DNA polymerase enzyme, three classes of organic molecules are essential for an *in vitro* reaction. The first are **deoxynucleoside triphosphates**. These are the familiar deoxynucleotide monophosphates (*i.e.,* dAMP, dCMP, dGMP and dTMP) with two additional phosphate groups attached to the initial or α – phosphate group (*i.e.,* dATP, dCTP, dGTP and dTTP). The second class of essential molecules for many DNA polymerase enzymes are polynucleotide chains with free **3′-OH** ends (often called **primer strands**), meaning that they cannot initiate the *de novo* synthesis of a new strand. The third class of essential molecules for DNA polymerase enzymes are **template strands**. All biologically important DNA polymerases possess a critical property: they will add nucleotides to a primer strand only in response to the base sequence found on a second template strand. Just as **Watson** and **Crick** suggested in their original model of DNA replication, the polymerase enzymes observe the rule of complementary base pairing. Thus, if a free 3'-OH group on a primer strand lies opposite a thymine on a template strand (see Fig. 7.11), a polymerase enzyme will add only an adenine group to the primer, even when dCTP, dTTP, dGTP are also present in the reaction mixture.

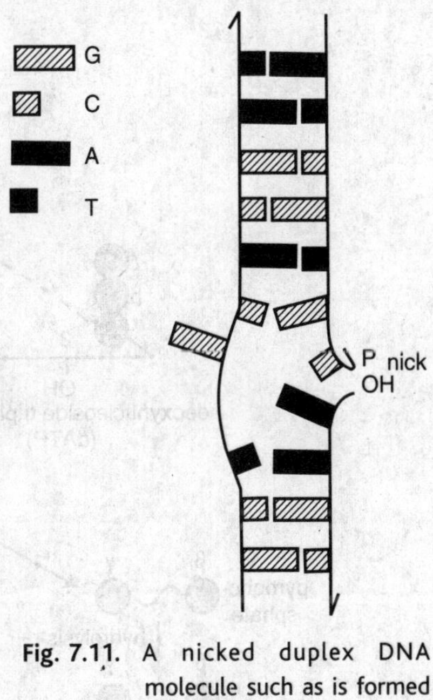

Fig. 7.11. A nicked duplex DNA molecule such as is formed by an endonuclease enzyme.

Once the appropriate molecules are present and the appropriate ionic conditions are maintained, *in vitro* polymerization reaction summarized in Fig. 7.11 will occur. The α-phosphate of a nucleoside triphosphate molecule forms a 3′, 5′ phosphodiester bond with a free 3′–OH in the growing polynucleotide chain, and a molecule of **pyrophosphate** (P ~ P) is simultaneously released. Pyrophosphate contains a "high-energy" or " ~ " bond, meaning that when the released pyrophosphate is hydrolyzed into two phosphate molecules, energy is liberated which drives the polymerization process forward. The resultant polymerization will always proceed in a net 5′ → 3′ direction, meaning that the nucleotide at the 3′ end is always the most recently added to the chain.

(1) Prokaryotic DNA polymerases. Three different DNA polymerases are known in *E.coli* and other prokaryotes (see Table 7.1), of which **DNA polymerase I** and **II** are meant for DNA repair and **DNA polymerase III** is meant for actual DNA replication.

1. DNA polymerase I. This enzyme was isolated around 1960 by **Arthur Kornberg** and was the first enzyme suggested to be involved in DNA replication. It is also called **Kornberg enzyme**. DNA polymerase I enzyme is now considered to be a DNA repair enzyme rather than a replication enzyme. This enzyme is known to have five active sites, namely **template site**, **primer site**, **5′ → 3′ cleavage** or **exonuclease site, nucleoside triphosphate site** and **3’ → 5’ cleavage site** (or **3′ → 5′ exonuclease site**) (Fig. 7.13; Fig. 7.14). DNA polymerase I is mainly involved in removing RNA primers from okazaki or precursor fragments and filling the resultant gaps due to its 5′ → 3′ polymerizing capacity. DNA polymerase I enzyme can also remove thymine dimers produced due to UV – irradiation and fill the gap due to excision. Both polymerization (= chain elongation) and exonuclease activity of DNA polymerase I have been shown in Fig. 7.16 and Fig. 7.17. This is called **proof reading** or **editing function** of this enzyme.

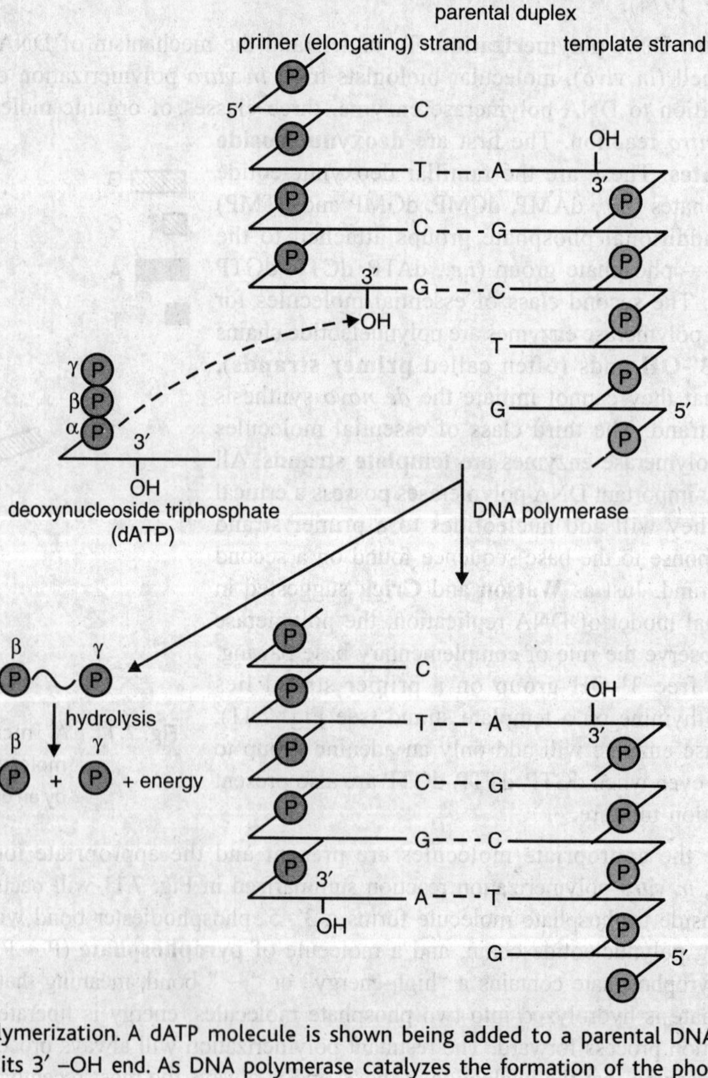

Fig. 7.12. DNA polymerization. A dATP molecule is shown being added to a parental DNA duplex or helix at its 3′ –OH end. As DNA polymerase catalyzes the formation of the phosphodiester bond, pyrophosphate (P ~ P) is hydrolyzed into two phosphate molecules.

Klenow fragment/polymerase. T. Steitz and coworkers have done much X-ray crystallography work that has given us a good look at the structure of a DNA polymerase. In fact, most of their work has been done on a fragment of DNA polymerase I called **Klenow fragment.**

Klenow fragment/polymerase was initially prepared by cutting the natural *E.coli* DNA polymerase I enzyme into two segments with a protease. One of these segments retained the polymerase and 3′ → 5′ exonuclease activity, but lacked the 5′ → 3′ exonuclease. Klenow enzyme is shaped like a cupped right hand with enzymatic activity taking place in two places, separated by a distance of about two or three nucleotides (Fig. 7.12). It is proposed that when the polymerization site senses a mismatch, the DNA is moved so that the 3′ end enters the exonuclease site, where the incorrect nucleotide residue is then cleaved. Polymerization then continue (see **Tamarin,** 2002).

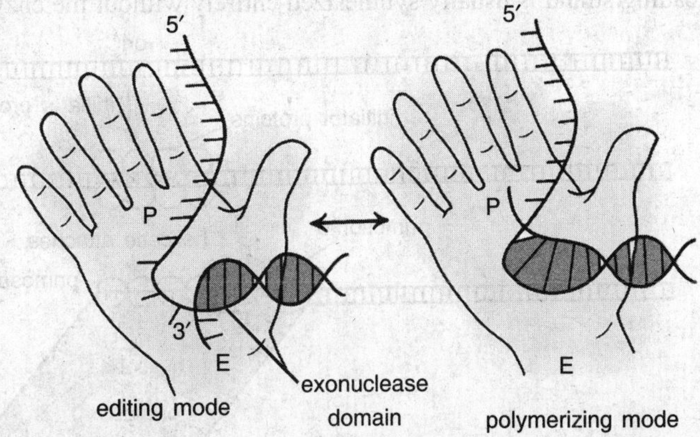

Fig. 7.13. DNA polymerase (P) and exonuclease (E) activities of Klenow fragment of DNA polymerase I in *E.coli*. On the right (A), 5′ → 3′ polymerization occurring. On the left, 3′ end of nascent strand has been backed up into the exonuclease site, presumably when a mismatch was detected (after Tamarin, 2002).

2. DNA polymerase-II. This enzyme resembles DNA polymerase-I in its activity, but is a DNA repair enzyme. It brings about the growth in 5′ → 3′ direction, using free 3′-OH groups.

3. DNA polymerase-III. DNA polymerase-III or Pol III enzyme plays an essential role in DNA replication. It is a multimeric enzyme or holoenzyme having ten subunits such as alpha

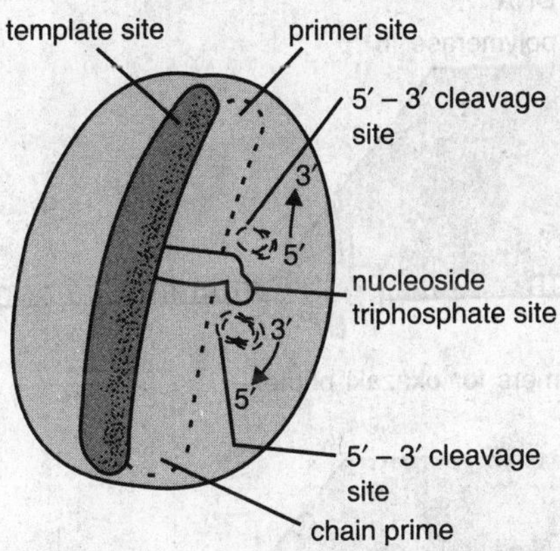

Fig. 7.14. A model of DNA polymerase-I enzyme, showing its five-different active sites

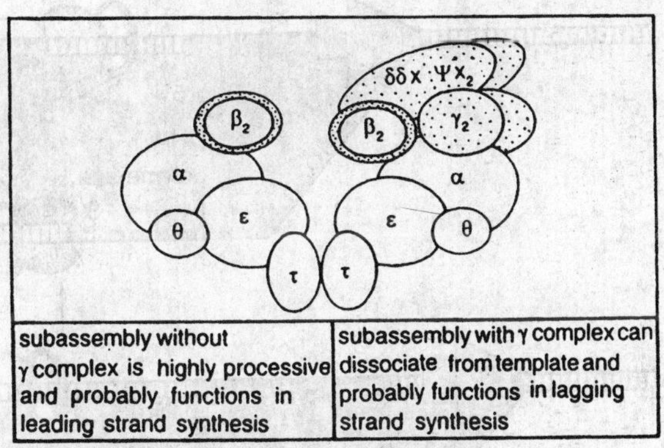

Fig. 7.15. Heterodimeric assembly of *E.coli* polymerase III holoenzyme to facilitate simultaneous leading and lagging strand synthesis.

($\propto$), beta (β), epsilon (ε), theta (θ), tau (τ), gamma (γ), delta (δ), delta dash (δ'), chi (χ) and psi (Ψ) (Table 7.2). All these ten subunits are needed for DNA replication *in vitro;* however, all having different functions. For example, three of the subunits α, ε and θ form the **polymerization core**, with both 5′ → 3′ polymerase activity and 3′ → 5′ exonuclease activity. One subunit, the β

subunit, is a "**processivity clamp**". As a dimer (two identical copies attached head to tail), the protein forms "doughnut" around the DNA so it can move freely on the DNA. When it is attached to the core enzyme, the polymerase is held tightly to DNA and shows high processivity: the leading strand is usually synthesized entirely without the enzyme leaving the template (Fig. 7.15).

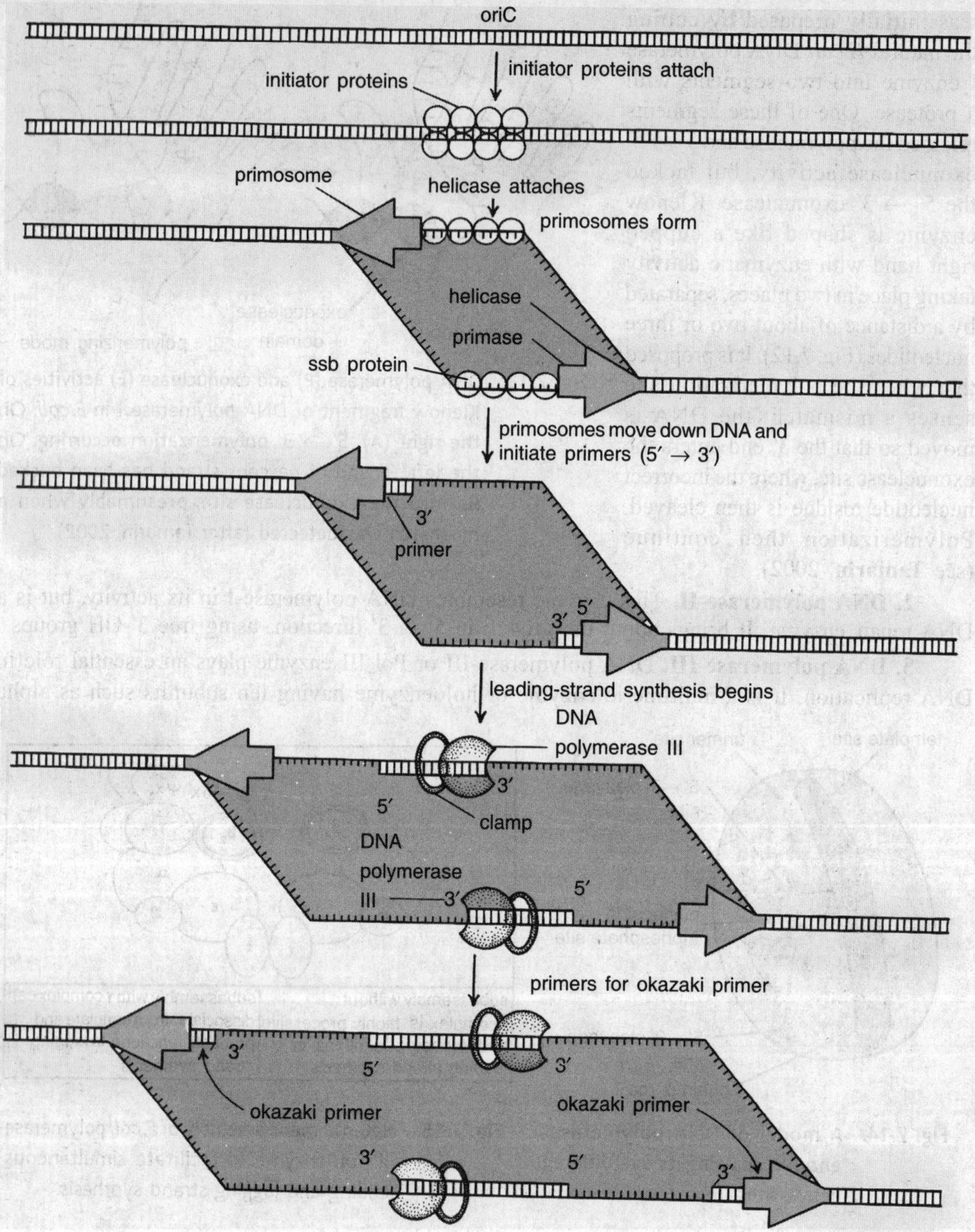

Fig. 7.16. Events at the origin of DNA replication in *E.coli*. The DNA opens up at *oric* to create two moving Y-junctions (replication forks). Initiator proteins attach and then bind helicase. The helicase then binds primase, forming a primosome. After the primer forms and two copies of DNA polymerase III are bound, the polymerization begins (after Tamarin, 2002).

The remaining subunits are involved in processivity control and replisome formation. They allow polymerase to move off and on DNA of lagging-strand template as Okazaki fragments are completed (a process known as **polymerase cycling**).

Table 7.1 Comparison of different characters of three types of DNA polymerases of *E.coli.*

S. No.	*Character*	*DNA polymerase* I	II	III
1.	Polymerization 5′ → 3′	Yes	Yes	Yes
2.	Exonuclease 3′ → 5′	Yes	Yes	Yes
3.	Exonuclease 5′ → 3′	Yes	No	No
4.	Use of primer single strands	Yes	No	No
5.	Use of nicked duplex or helix	Yes	No	No
6.	Molecular weight	109,000 (single chain)	120,000 (single chain)	> 250,000 (hetero-multimeric chain)
7.	Molecules per cell	400	Not known	10–12
8.	Gene	*pol A*	*pol B*	*pol C (dna E), dna N, dna X, dna Q, dna T*
9.	Nucleotides polymerized at 37°C molecules/minute	up to 1,000	up to 50	up to 15,000
10.	Affinity for nucleoside triphosphate (NTPs) precursors	low	low	high

(2) Eukaryotic DNA polymerases. Eukaryotes (*e.g.,* yeast, rat liver, human tumour cells) are found to contain the following five types of DNA polymerases:

(i) **DNA polymerase α (=alpha).** This relatively high molecular weight enzyme is also called **cytoplasmic polymerase** or **large polymerase**. It is found in nucleus and cytoplasm.

(ii) **DNA polymerase β (=beta).** This enzyme is also called **nuclear polymerase** or **small polymerase** and is found only in vertebrates.

(iii) **DNA polymerase γ (=gamma).** This enzyme is called **mitochondrial polymerase** and is encoded in the nucleus.

(iv) **DNA polymerase δ (=delta).** This enzyme is found in mammalian cells and is PCNA dependent for DNA-synthesis processivity (PCNA = proliferating cell nuclear antigen).

(v) **DNA polymerase ε (=epsilon).** It was previously known as **DNA polymerase II**. This enzyme is PCNA independent and occurs in mammalian HeLa cells and budding yeast.

The large DNA polymerase α is the predominant DNA polymerase enzyme in eukaryotic cells and was believed for long time to be only enzyme involved in DNA replication. But now one more polymerase, namely DNA polymerase δ is also found to be involved in eukaryotic DNA replication.

(III) DNA ligases. DNA ligase enzymes are capable of catalyzing phosphodiester bond formation between free 3′–OH and free 5′–P groups of a nick of DNA which is created by endonuclease enzyme, thereby restoring an intact DNA duplex. Many DNA ligases have already

Table 7.2 Summary of the enzymes and proteins involved in DNA replication in *E.coli* (*Source:* **Tamarin,** 2002).

S. No.	*Enzyme or Protein*	*Genetic locus*	*Function*
1.	DNA polymerase I	*polA*	Gap filling and primer removal
2.	DNA polymerase II	*polB*	Replicating damaged templates
3.	DNA polymerase III (Holoenzyme)		
	(i) α subunit	*dnaE*	Polymerization core; 5′ → 3′ polymerase
	(ii) ε subunit	*dnaQ*	Polymerization core; 3′ → 5′ exonuclease
	(iii) θ subunit	*holE*	Polymerization core
	(iv) β subunit	*dnaN*	Processivity clamp (as a dimer)
	(v) τ subunit	*dnaX*	Preinitiation complex; dimerization of core
	(vi) γ subunit	*dnaX*	Preinitiation complex: loads clamp
	(vii) δ subunit	*holA*	Processivity core
	(viii) δ′ subunit	*holB*	Processivity core
	(ix) χ subunit	*holC*	Processivity core
	(x) Ψ subunit	*holD*	Processivity core
4.	Helicase	*dnaB*	Primosome; unwinds DNA
5.	Primase	*dnaG*	Primosome; creates Okazaki fragment primers
6.	Initiator protein	*dnaA*	Binds at origin of replication
7.	DNA ligase	*lig*	Closes Okazaki fragment
8.	Ssb protein	*ssb*	Binds single stranded DNA
9.	DNA topoisomerase I	*topA*	Relaxes supercoiled DNA
10.	DNA topoisomerase type II (DNA gyrase)		
	(i) α subunit	*gyrA*	Relaxes supercoiled DNA; ATPase
	(ii) β subunit	*gyrB*	Relaxes supercoiled DNA
11.	Topoisomerase IV	*parE*	Unconcatenates DNA circles
12.	Termination protein	*tns*	Binds at termination sites

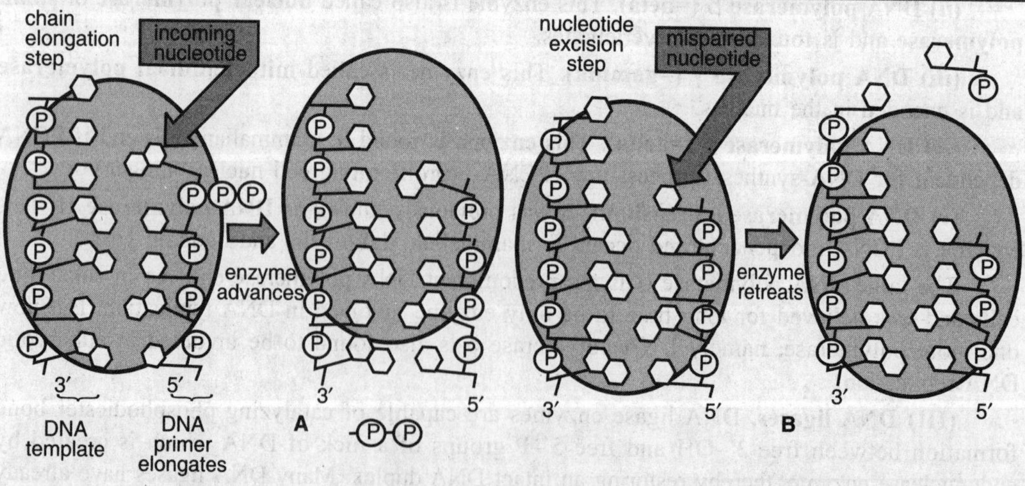

Fig. 7.17. Functions of DNA polymerase-I enzyme. A-Mode of polymerization of nucleotides, on the primer site of the enzyme; B-Removal of mispaired nucleotide activity of the enzyme.

been discovered. The ligase enzyme from *E.coli* requires the presence of oxidized nicotinamide adenine dinucleotide (NAD^+) a cofactor, whereas the ligase enzyme specified by T_4 bacteriophage requires ATP to bring about the joining reaction.

***(iv)* Restriction enzymes.** In the 1970s, a new class of endonucleases has been isolated from microorganisms whose action is limited only to specified nucleotide sequences. They are known as **restriction enzymes** and will produce breaks only within sequences which have two identical bases in adjacent positions such as *CCT*, *TTA*, *GGC* and *AAG*. Since at these positions the complementary strand also has identical bases (*i.e., GGA*, *AAT*, *CCG*, *TTC*), restriction enzymes act on both strands and produce a break. A number of different restriction enzymes have been isolated (such as **Hind II**, **Hind III**, **EcoR1** and others) which specifically break particular nucleotides. They are thus useful for breaking a DNA molecule into specific known types of smaller fragments.

***(v)* Swivelases.** The two helical strands of DNA are wound around each other as if two parallel strands had been coiled around a central axis. This is called **plectonemic coiling** and is revealed by X-ray diffraction data. This winding does not allow rotation or movement of helices about each other. DNA replication requires rotation at specific sites called **swivels** within the helical strands. Swivels allow free rotation of one part of the molecule relative to the adjacent part. In this way, only the DNA between the growth point and the nearest swivel would have to rotate during replication. Enzymes that produce transient swivels are known as **swivelases**. These DNA enzymes have endonuclease action on one of the strands that allowing free rotation within DNA molecule. The site is then immediately sealed leaving an intact strand for replication.

***(vi)* Unwinding enzymes and proteins.** The unwinding of DNA during replication requires an endonuclease enzyme that makes a single break in a strand. This allows DNA to unwind in that region instead of the whole DNA molecule having to twist. The further opening up of local regions is aided by specific **unwinding proteins** which bind preferentially to single-stranded DNA, resulting in separation of strands. The binding of one unwinding protein to a DNA strand promotes the binding of a second, so that the helix continue to unwind. Another class of proteins called **strand separation proteins** (**SSP**) have also been isolated from both prokaryotes and eukaryotes.

Roles of DNA Primers in DNA Replication

No known DNA polymerase can initiate synthesis of DNA without the availability of a **primer RNA strand**. So, before actual DNA replication starts, short RNA oligonucleotide segments, called **RNA primers** or simply the **primers,** have to be synthesized by **DNA primase** enzyme utilizing ribonucleoside triphosphates (Fig. 7.18). This RNA primer is synthesized by copying a particular base sequence from one DNA strand and differs from a typical RNA molecule in that after the synthesis the primer remains hydrogen-bonded to the DNA template (**Freifelder**, 1985). The primers are about 10 nucleotides long in eukaryotes and they are made at intervals on the lagging strand where they are elongated by the DNA polymerase enzyme to begin each Okazaki fragment. These RNA primers are later excised and filled with DNA with the help of DNA repair system in eukaryotes (or DNA polymerase I in *E.coli*).

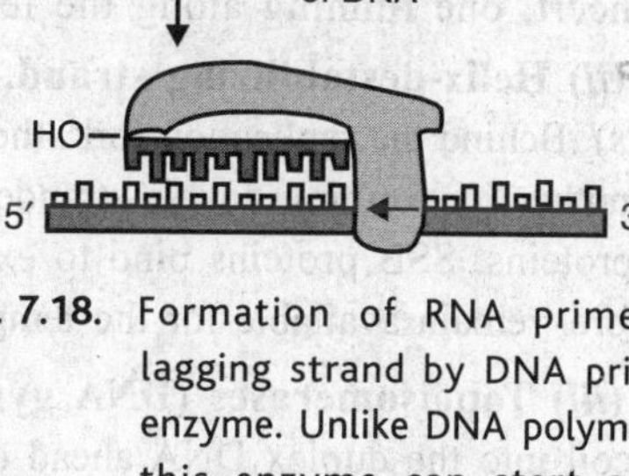

Fig. 7.18. Formation of RNA primer on lagging strand by DNA primase enzyme. Unlike DNA polymerase, this enzyme can start a new polynucleotide chain by joining two nucleotide triphosphates together (after Alberts *et al.*, 1989).

In bacteria, two different enzymes are known to synthesize primer RNA oligonucleotides – **RNA polymerase** (on the leading strand) and **DNA primase** (on the lagging strand).

Replicons

DNA replication in prokaryotes and eukaryotes is attained in discrete units, called **replicons**. The number of replicons may vary in a genome from one in bacteria (*E.coli*) and 500 in yeast to several thousands in plants and animals (Table 7.3). For example, in the *E.coli* there is a single replication with the **origin**, identified as a genetic locus *ori C* (245 bp). The origin is A: T rich, a feature that is related to unwinding of DNA to initiate replication. In *E.coli*, there are also termination sites (*ter A-F*), each consisting of ~ 23 bp. The process of termination of DNA replication requires the product of *tus* gene (**Tus protein** or *TBP*, *i.e.,* ter binding protein) which recognizes *ter* or termination sites.

Table 7.3 Prokaryotic and eukaryotic replicons.

S. No.	Organism	Number of replicons	Average length (kb)	Fork movement (bp/min)
1.	Bacteria (*E.coli*)	1	4200	50,000
2.	Yeast (*S.cervisiae*)	500	40	3,600
3.	Fruit fly (*D.melanogaster*)	3,500	40	2,600
4.	Toad (*Xenopus laevis*)	15,000	200	500
5.	Mouse (*Mus musculus*)	25,000	150	2,200
6.	Bean (*Vicia faba*)	35,000	300	Not known

B. Proteins Involved in Opening of DNA Helix

The following three types of proteins are needed to help the DNA double helix to open and to provide exposed DNA template for the DNA polymerase to copy (Table 7.2):

***(i)* DNA helicases**. DNA helicases are ATP-dependent upwinding enzymes which promote separation of the two parental strands and establish replication forks that will progressively move away from the origin. DNA helicase hydrolyze ATP when they are bound to single strands of DNA. Hydrolysis of ATP can change the shape of a protein molecule in a cyclic manner that allows the protein to perform mechanical work. DNA helicases utilize this principle to move rapidly along a DNA single strand; when they encounter a region of double helix, they continue to move along their strand, thereby unwinding the helix (Fig. 7.19). Unwinding of the template DNA helix at a replication fork could in principle be catalyzed by two DNA helicases, acting in concert, one running along the leading strand and the other along the lagging strand.

***(ii)* Helix-destabilizing strand.** (also called **single strand DNA-binding proteins** or **SSBPs**). Behind the replication fork, the single DNA strands are prevented from rewinding about one another (or forming double-stranded hair-pin loops in each single strands) by the action of SSB proteins. SSB proteins bind to exposed DNA strands without covering the bases, which, therefore, remain available for the templating process.

***(iii)* Topoisomerases (DNA gyrases).** The action of a helicase introduces a positive supercoil into the duplex DNA ahead of the replication fork. Enzymes, called **topoisomerases**, relax the supercoil by attaching to the transiently supercoil duplex, nicking one of the strands and rotating it through the unbroken strand. The nick is then released. Thus, a DNA topoisomerase can be viewed as a "reversible nuclease" that adds itself covalently to a DNA phosphate, thereby breaking a phosphodiester bond in the DNA strand. Because the covalent linkage that joins a topoisomerase to DNA phosphate retains the energy of the broken phosphodiester bond, the breakage reaction is reversible; resealing is rapid and does not require additional energy input. The rejoining mechanism is different from DNA ligase enzyme.

One type of topoisomerase (*i.e.,* **topoisomerase I**) causes a single-strand break or nick which allows the two sections of DNA helix on either side of nick to rotate freely relative to each other, using the phosphodiester bond in the strand opposite the nick as a swivel point. A second type of topoisomerase (*i.e.,* **topoisomerase II**) forms a covalent bond to both strands of DNA helix at the same time, making transient **double-strand break** in the helix.

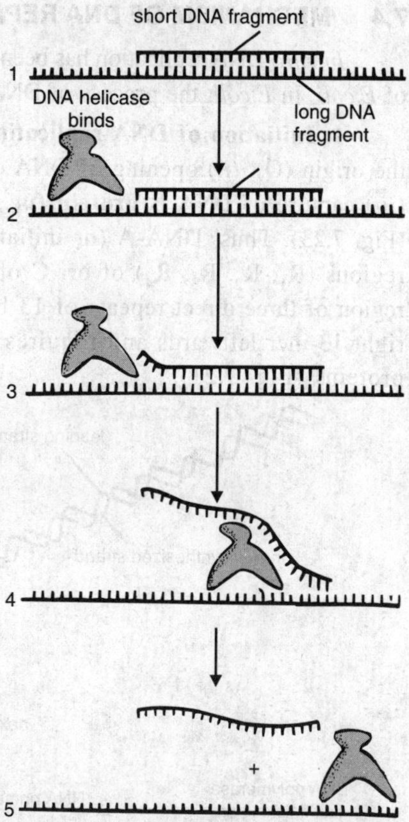

Fig. 7.19. Mode of action of DNA helicase (after Alberts *et al.,* 1989).

Replisome and Primosome

N.K. Sinha and **A. Kornberg** have suggested that the DNA polymerases, RNA primases and helicases may be associated with one another to form a multienzyme complex – the **replisome** that carries out the synthesis of leading and lagging strands in a **coordinated fashion**. Such a complex would be highly processive and assure rapid replication of the DNA (Fig. 7.20).

Likewise, the proteins at a replication fork cooperate to form replication machine, *i.e.,* the primase molecule is linked directly to the helicase to form a unit on the lagging strand called a **primosome** which moves with the fork, synthesizing RNA primers as it moves (Fig. 7.21).

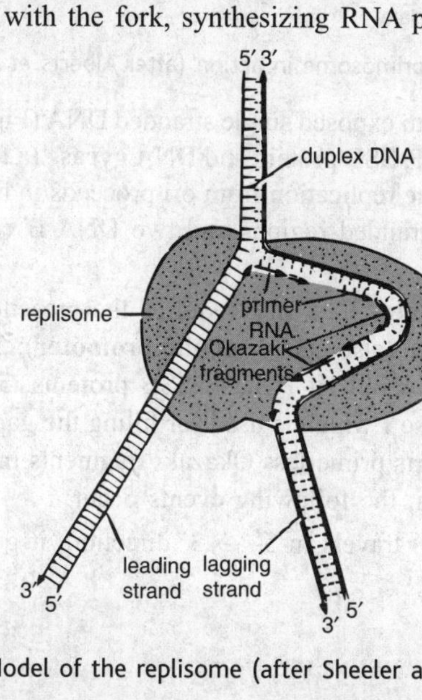

Fig. 7.20. Model of the replisome (after Sheeler and Bianchi, 1987).

7.4 MECHANISM OF DNA REPLICATION IN PROKARYOTES

In vitro DNA replication has been extensively studied in *E.coli* and in the phages and plasmids of *E.coli*. In *E.coli*, the process of DNA replication involves the following three steps (Fig. 7.25).

1. Initiation of DNA replication. This process comprises three steps: (*i*) recognition of the origin (O), (*ii*) opening of DNA duplex to generate a region of single stranded DNA, and (*iii*) capture of **DNA B protein** (*i.e.*, 5′ → 3′ helicase; also acts as the activator of primase (Fig. 7.23). Thus, DNA-A (or initiator protein)-ATP complex binds at 9 bp inverted repeat regions (R_1, R_2, R_3, R_4) of ori C of *E.coli* and promotes opening of the DNA duplex in a region of three direct repeats of 13 bp sequence (called **13-mers**). The opening occurs from right 13-mer leftwards and requires negatively supercoiled DNA and HU or IHF initiator proteins (Fig. 7.21).

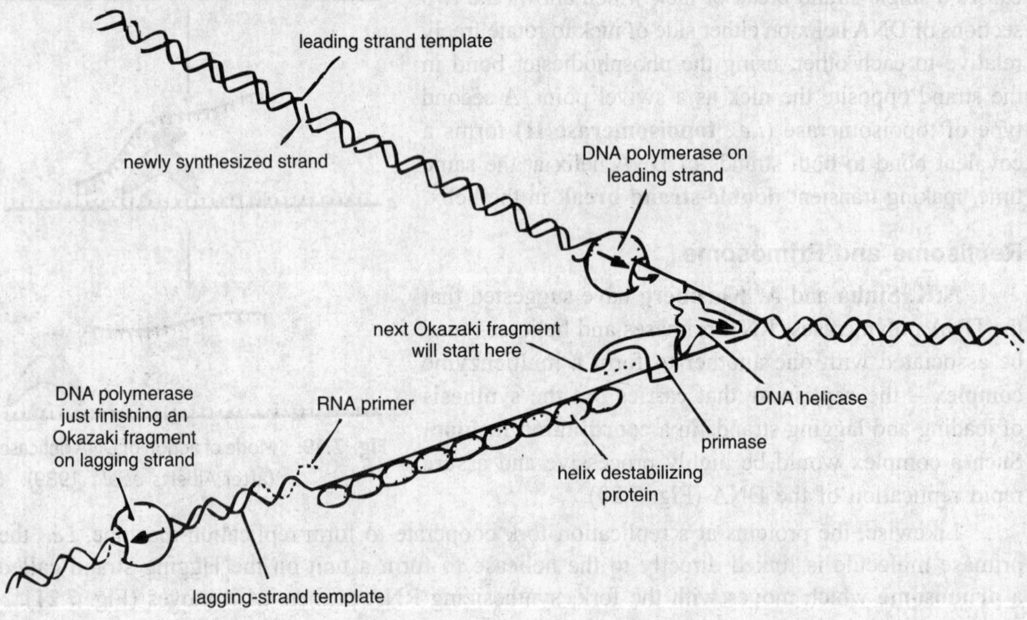

Fig. 7.21. A primosome in action (after Alberts *et al.*, 1989).

DNA B (=helicase) is transferred to exposed single stranded DNA (Fig. 7.22) and causes unwinding of the DNA in the presence of ATP, SSB protein and **DNA gyrase** (a topoisomerase). This results in unwinding of DNA duplex and the replication from ori proceeds in both directions (bidirectional); SSB binding occurs on single stranded regions and two DNA B complexes (=primosomes) are loaded one on each strand.

2. Elongation of DNA chain. This step requires the presence of the following enzymes and factors: 1. DNA B or helicase (also called **mobile promoter**); 2. Primase (DNA G); 3. DNA polymerase holoenzyme (or DNA pol III HE); 4. SSB proteins; 5. RNAase H which removes RNA primers; 6. DNA polymerase I which is used for filling the gap created due to RNA primers and 7. DNA ligase (which converts primerless Okazaki fragments into continuous strand). During initiation to elongation transition, the following events occur:

1. As helicase (or DNA B) travels in 5′ → 3’ direction, it generates a replication fork by opening the DNA duplex.

2. The DNA strand having helicase becomes the lagging strand. DNA primase associates with DNA B helicase, forming the primosome which synthesizes multiple primers for lagging strand and single RNA primer for the leading strand.

3. For the synthesis of lagging strand, the DNA pol III HE has to work on the same strand to which DNA B helicase is bound, but it travels in opposite direction.

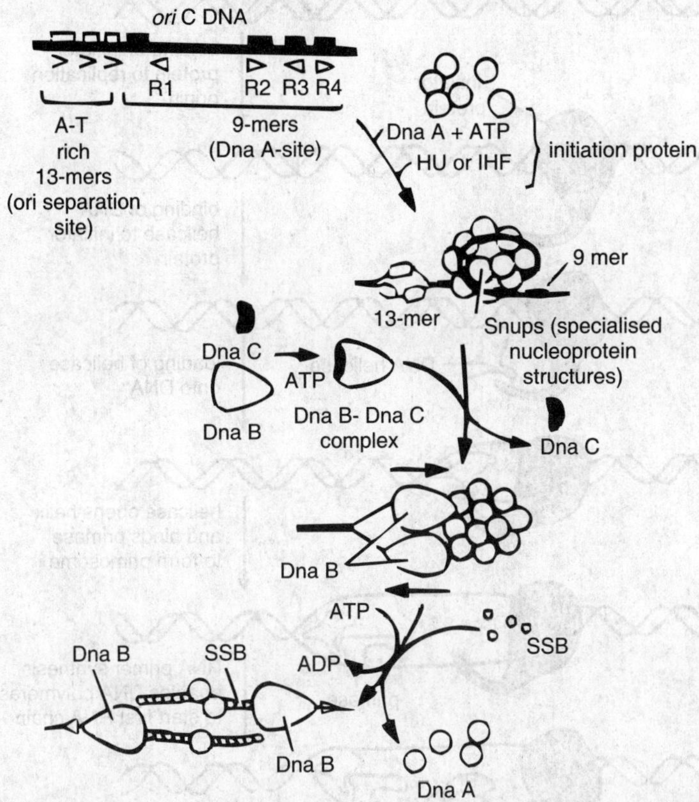

Fig. 7.22. Mode of the initiation of DNA replication at *ori C* of *E.coli.*

4. DNA B helicase, DNA G primase and DNA pol III HE work together in strand elongation. Helicase and DNA polymerase assembly remains *processive, i.e.,* they remain tightly bound to the fork and stay bound through the reaction.

Synthesis (=elongation) of lagging and leading strands takes place by somewhat different methods; it is far more complex for lagging strand than for the leading strand:

A. Discontinuous synthesis on lagging strand. 1. Primase is taken up from solution and is activated by helicase (DNA B) to synthesize a RNA primer (10 to 20nt or nucleotides long) on the lagging strand. 2. The RNA primers are recognized by DNA pol III HE on the lagging strand and are utilized for synthesis of precursor or Okazaki fragments. In fact, each new RNA primer is recognized by the gamma (γ) subunit of DNA pol III HE and loaded with a subunit of the same polymerase. This preloaded β subunit may then capture the *core* of DNA poly III HE when it becomes available after finishing its synthetic job on the preceding Okazaki fragments. 3. On completion of the Okazaki fragments, the RNA primers are excised by DNA polymerase I, which then fills the resulting gaps with DNA. 4. After DNA **polymerase I** adds the final deoxyribonucleotides in the gap left by the excised primer, the enzyme **DNA ligase** forms the phosphodiester bond that links the free 3′ end of the primer replacement to the 5′ end of the Okazaki fragment.

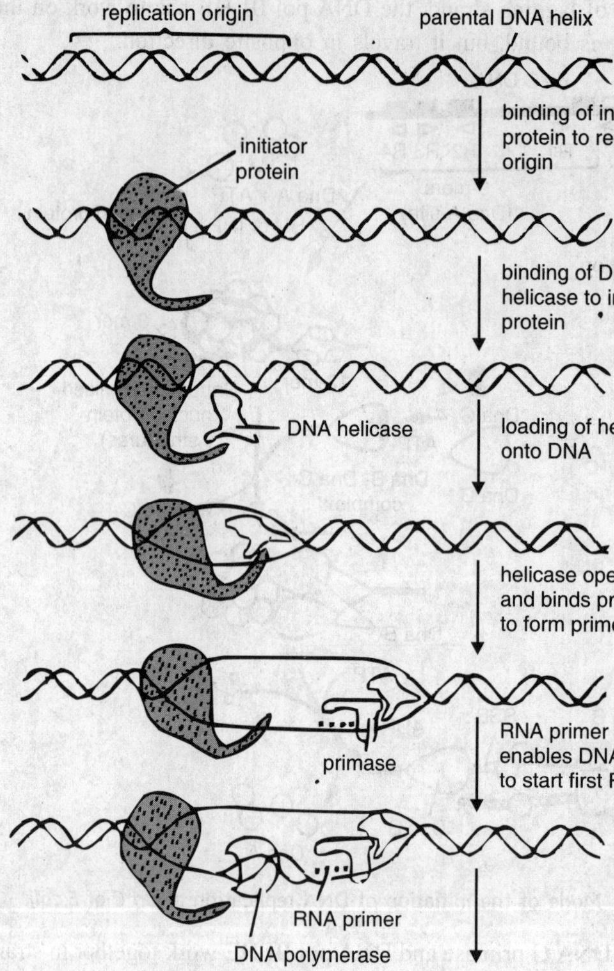

Fig. 7.23. Initial steps leading to formation of replication forks in the *E.coli* (after Alberts *et al.*, 1989).

B. Continuous synthesis on leading strand. (1) In bidirectional DNA replication, the leading strand is primed once on each of the parental strands. (2) The RNA primer of the leading strand is synthesized by RNA polymerase enzyme. (3) DNA pol III HE causes elongation of the leading strand and finally DNA pol I and ligase enzymes give final touch to the leading strand as in case of the lagging strand (Fig. 7.24).

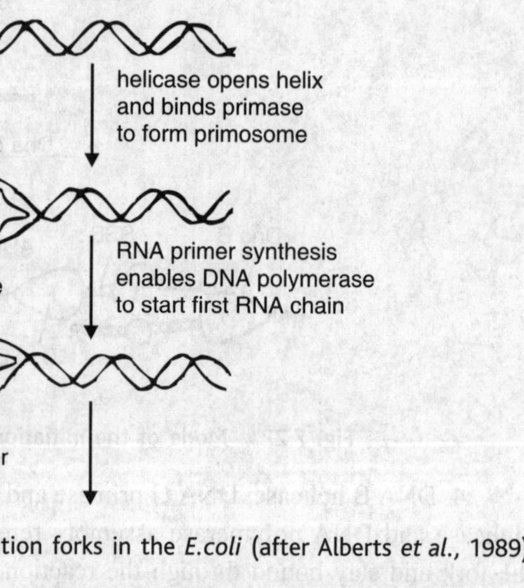

Fig. 7.24. Model of a DNA replication fork in prokaryotes during elongation.

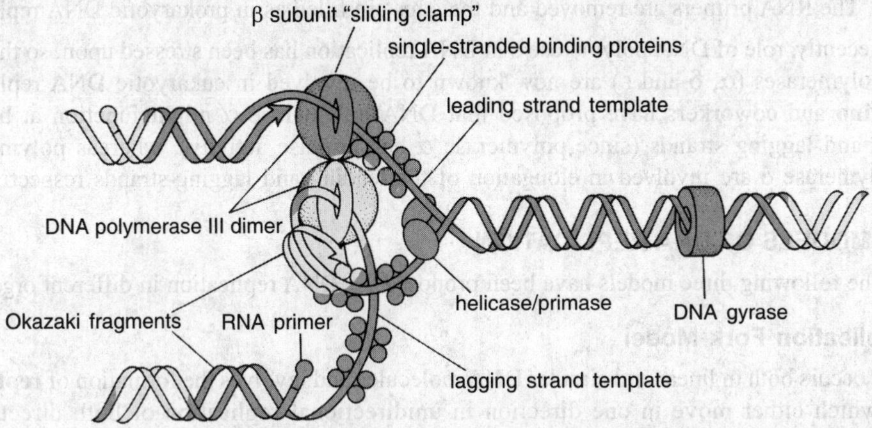

Fig. 7.25. Summary of prokaryotic DNA synthesis at a single replication fork. Various enzymes and proteins essential to the process are shown (after Klug and Cummings, 2003).

7.5 DNA REPLICATION IN EUKARYOTES

Eukaryotic DNA replication requires two different DNA polymerase enzymes, namely **DNA polymerase** α and **DNA polymerase** δ (Fig. 7.26). DNA polymerase δ synthesizes the DNA on the leading strand (continuous synthesis), whereas DNA polymerase α synthesizes the DNA on the lagging strand (discontinuous DNA synthesis). Besides these two enzymes, six more factors are involved in eukaryotic DNA replication: (1) T antigen; (2) replication factor A or RF-A (also called RP-A or eukaryotic SSB); (3) topoisomerase I; (4) topoisomerase II; (5) proliferating – cell nuclear antigeñ (PCNA), also called **cyclin**), and (σ) replication factor C or RF-C.

The process of eukaryotic DNA replication involves the following steps:

1. Before the onset of DNA synthesis, there is a presynthetic stage of 8 – 10 minutes duration for the formation of unwound DNA complex. This step needs only three purified proteins, namely T antigen (T-ag or tumour antigen), RF-A and topoisomerase I and II.

2. The T-antigen, using its DNA-binding domain, forms a multi-subunit complex with site I and site II in the presence of ATP and caused local unwinding.

3. More extensive duplex unwinding occurs due to association of RF-A and a topoisomerase with the help of DNA helicase. Topoisomerases help in unwinding of DNA by altering topology of DNA at the replication fork.

4. RF-A or SSB proteins bind to unwound single stranded DNA.

5. The primer RNA synthesis is performed by primase which is tightly associated with DNA polymerase α.

6. DNA polymerase α helps in synthesis of an Okazaki fragment in 5′ to 3′ direction. In eukaryotes, Okazaki fragments are much smaller – only about 135 bases long, about the size of the DNA on a nucleosome (see **Mathews**, 2000).

7. Replication factor C (or RF-C) and PCNA (cyclin) help in switching of DNA polymerases so that pol α is replaced by pol δ which then continuously synthesized DNA on the leading strand.

8. Another Okazaki fragment is then synthesized from the replication fork on the lagging strand by pol α-primase complex and this step is repeated again and again, till the entire DNA molecule is covered.

9. The RNA primers are removed and the gaps are filled as in prokaryotic DNA replication.

Recently, role of DNA polymerase ε in DNA replication has been stressed upon, so that three DNA polymerases (α, δ and ε) are now known to be involved in eukaryotic DNA replication. **A. Sugino** and coworkers have proposed that DNA polymerase α might function at both the leading and lagging strands (since polymerase α has primase activity), whereas polymerase ε and polymerase δ are involved in elongation of the leading and lagging-strands respectively.

7.6 MODELS OF DNA REPLICATION

The following three models have been proposed for DNA replication in different organisms:

1. Replication Fork Model

It occurs both in linear and circular DNA molecules and involves the formation of **replication forks** which either move in one direction in unidirectional replication or both directions in bidirectional replication.

In *E.coli*, replication of circular DNA is of bidirectional type involving intermediate theta (θ) type structures.

2. Rolling Circle Model (Rolling Circle Replication)

In the rolling circle mode of DNA replication, a **nick** (a break in one of the phosphodiester bonds) is made in one of the strands of the circular DNA and this nick has 3′ – OH and 5′ – P termini. Due to a helicase and *Ssb* protein α, Y–fork is generated. Ultimate result is replication of a circle and a tail (Fig. 7.27). This form of replication occurs in F plasmid or *E.coli* Hfr chromosome during conjugation. The F^+ or Hfr bacterial cell retains the circular daughter while passing the linear tail into the F^- cell, where replication of the tail takes place. Several phages (including l phage) also use this DNA replication method, filling their heads (protein coats) with linear DNA replicated from the circular molecule. Rolling circle model is also called **sigma (s) replication** and it is an efficient mechanism for rapid synthesis of multiple copies of a circular genome (**Novick**, 1998).

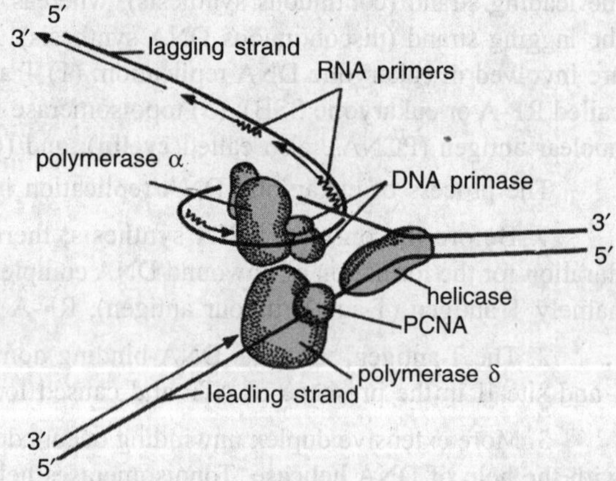

Fig. 7.26. Two different DNA polymerases, synthesizing leading and lagging strands at a eukaryotic DNA replication fork.

3. D-loop Model (Displacement Replication)

Chloroplasts and mitochondria (in eukaryotic cells) have their own circular DNA molecules that replicate by a slightly different mechanism. The **origin** of replication is at a different point on each of the two parental template strand. Replication begins on one strand, displacing the other while forming a **displacement loop** or **D–loop** structure (Fig. 7.28; Fig. 7.29). Replication continues until the process passes the origin of replication on the other strand. Replication then initiates on the second strand, in the opposite direction. Y– junction (or replication fork) replication, also occurs in mitochondrial DNA under some growth condition.

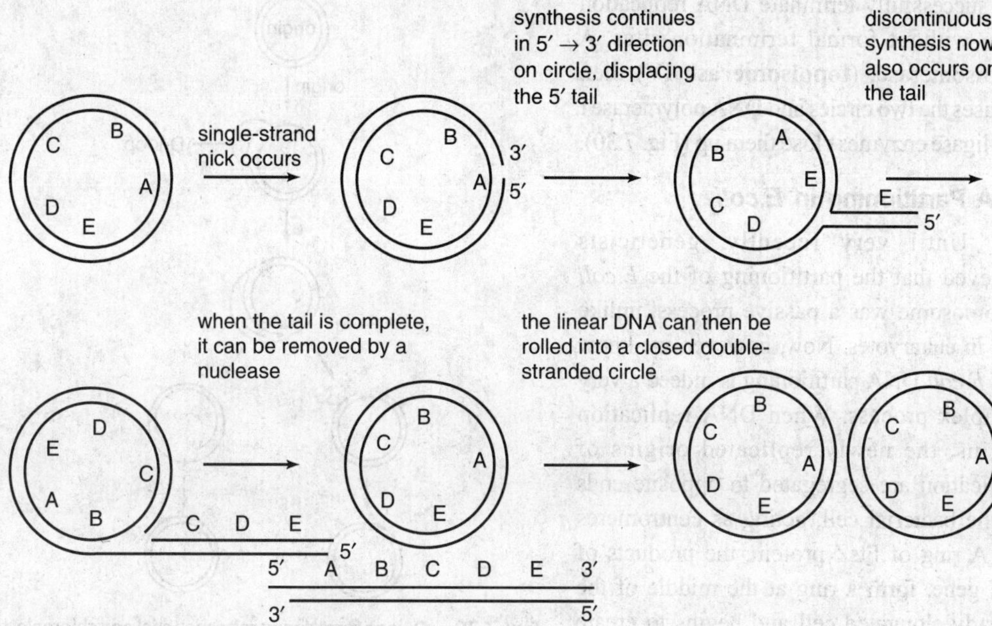

Fig. 7.27. Rolling circle model of DNA replication. The letters A to E provide landmarks on the chromosome (after Tamarin, 2002).

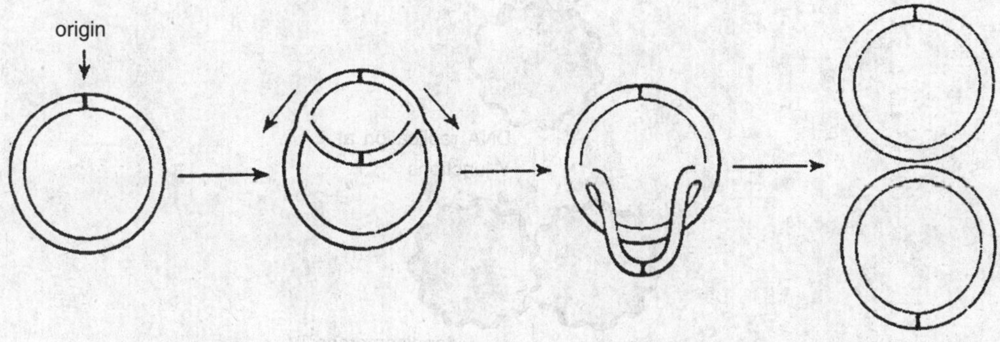

Fig. 7.28. Observable stages in the DNA replication of a circular chromosome, assuming bidirectional DNA synthesis. The intermediate figures are called theta structures (after Tamarin, 2002).

7.7 TERMINATION OF REPLICATION

The termination of the replication of a circular chromosome DNA (of bacteria, mitochondria and chloroplasts) presents no major topological problems. At the end of the theta structure replication, both Y-junctions have proceeded around the molecule. The region of termination on the *E.coli* chromosome, the **terminus** (*Ter*) is 180 degrees from *OriC* on the circular chromosome, between minutes 28 and 36. There are six terminator sites; three arrest the Y-junction from the left, and three arrest the one from the right when bound by a termination protein, the product of the *tus* gene. (*Tus* stands for terminus utilization substance; each *ter* site is about twenty base pairs). One interesting aspect of the termination of *E.coli* DNA replication is that the cells are viable even if the whole terminator region is deleted. There are fewer viable cells and some growth problems, but in general, *E.coli* can successfully terminate DNA replication even without formal termination sites. A topoisomerase, (**topoisomerase IV**) then releases the two circles and DNA polymerase I and ligase enzymes close them up (Fig. 7.30).

DNA Partitioning in *E.coli*

Until very recently, geneticists believed that the partitioning of the *E.coli* chromosome was a passive process, unlike that in eukaryotes. Now, however, we know that *E.coli* DNA partitioning is indeed a very complex process. When DNA replication begins, the newly replicated origins of replication are segregated to opposite ends of the bacterial cell, acting as centromeres do. A ring of FtsZ protein, the products of *ftsZ* gene, form a ring at the middle of the already elongated cell and begins to create septum that will divide the already elongated cell into two daughter cells (Fig. 7.31).

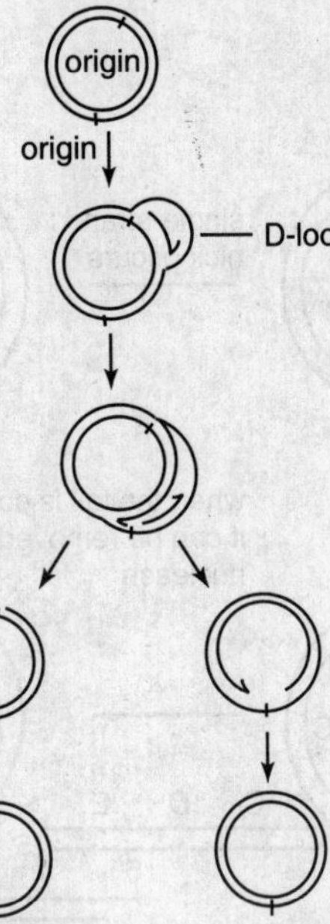

Fig. 7.29. D–loop form during mitochondrial and chloroplast DNA replication because the origins of replication are located at different places on the two strands of the double helix (after Tamarin, 2002).

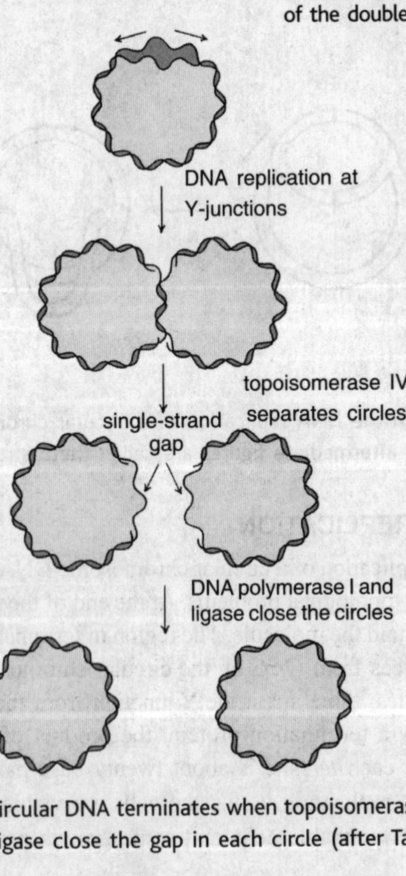

Fig. 7.30. The replication of circular DNA terminates when topoisomerase separates the circles and DNA polymerase I and ligase close the gap in each circle (after Tamarin, 2002).

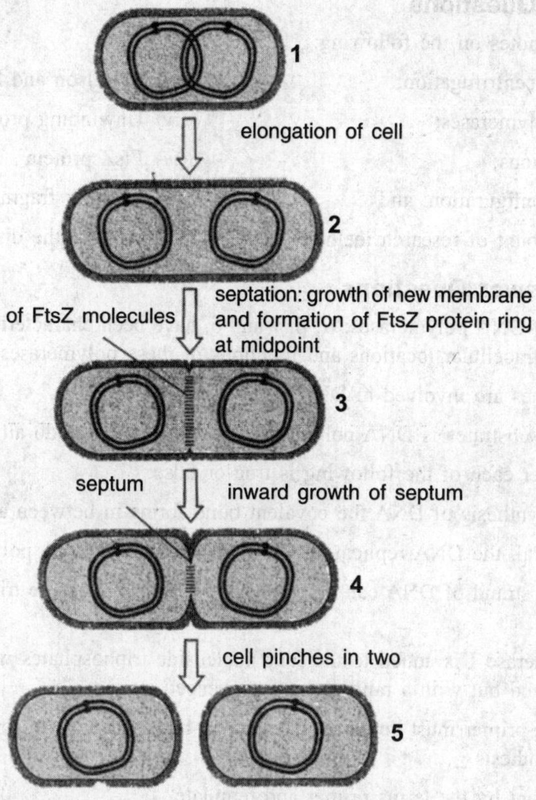

Fig. 7.31. Binary fission in *E.coli.* After the DNA is replicated, the cell elongates and the DNA is partitioned in the cell. Separation then begins, in which new cell membrane material begins to grow and form a septum at approximately the midpoint of the cell. A protein molecule, called FtsZ, (It is a tubulin-like protein) facilitates this processs, *i.e.*, FtzZ forms a ring at approximately the midpoint of the cell. When the septum is complete, the cell pinches in two (after Raven *et al.*, 2005).

QUESTIONS

Long Answer Questions

1. Describe the two classical experiments which demonstrated the semiconservative mode of DNA replication.
2. Describe the different steps involved in the initiation of DNA synthesis in *E.coli.* Discuss the roles of different enzymes or proteins in this process.
3. Describe various steps of DNA replication in eukaryotes. Elaborate the specific role of two different DNA polymerases for the leading and the lagging strands.
4. Describe rolling circles and D-loops for DNA replication.
5. Distinguish between the terms "dispersive", "semiconservative" and "conservative", as applied to DNA replication.
6. Explain why DNA replication must be primed and describe how the priming problem is solved by *Escherichia coli* and by eukaryotes.
7. Give a detailed description of the events occurring at the Y-fork or replication fork in *Escherichia coli.*

Short Answer Questions

1. Write short notes on the following:

   (*i*) Density centrifugation; (*ii*) Meselson and Stahl experiment;

   (*iii*) DNA polymerases; (*iv*) Unwinding proteins;

   (*v*) Replications; (*vi*) FtsZ protein

   (*vii*) Theta configuration; and (*viii*) Okazaki fragments.

2. Give an account of research method of Arthur Kornberg for the discovery of DNA polymerase I.

Very Short Answer Questions

1. Four distinct DNA polymerases: α, β, γ and δ, have been characterized in several higher eukaryotes. What are intracellular locations and functions of these polymerases?
2. What enzymes are involved in DNA replication in *E. coli*?
3. From what substrates is DNA polymerized? What properties do all known polymerase share?
4. State whether each of the following is true or false:

   (*a*) In the synthesis of DNA the covalent bond forms in between a 3' – OH and 5' – P group.

   (*b*) In general, the DNA replicating enzyme in *E. coli* is DNA polymerase I.

   (*c*) A single strand of DNA can be copied if the four nucleoside triphosphates and polymerase I are provided.

   (*d*) If polymerase I is added to the four nucleoside triphosphates without a DNA template, DNA is synthesized but with a random base sequence.

   (*e*) An RNA primer must be complementary in base sequence to some region of the DNA to initiate DNA synthesis.

5. What is meant by the terms primer and template?
6. What is the role of RNA in DNA replication?
7. Distinguish the roles of helicases and SSB proteins in DNA replication?
8. If one strand of DNA is found to have the sequence 5' A A C G T A C T G C 3', what is the sequence of nucleotides on the 3', 5' strand?
9. For a molecule of n-deoxyribonucleotide pairs, give a mathematical expression that can be used to calculate the number of possible sequences of those nucleotide pairs if only the "usual" bases are present.
10. Develop a formula for determining the length in micrometers of a DNA molecule whose number of deoxyribonucleotide pairs is known.
11. Phage T2 DNA is estimated to consist of about 200,000 deoxyribonucleotide pairs. What is the length in micrometers of its DNA complement?
12. Who devised the technique of nearest– neighbour base frequency analysis for pol I enzyme?

Fill in the Blanks

1. M. Meselson and .... in 1958 verified the .... nature of DNA replication.
2. DNA replication in *E. coli* was visualised by .... in 1963 by using microscopy and ....
3. .... and P. Woods in 1957 experimentally supported .... replication of DNA by working on dividing root tip cells of the plant ....
4. DNA polymerase I enzyme was discovered in 1957 by .... from .... *E. coli* cells.
5. Name the discovers of DNA polymerase III enzyme: .... and .... in 1972.
6. DNA polymerase III is a .... and it has .... subunits.

7. Pieces of DNA segments formed on lagging strand of DNA are called ....

8. Okazaki fragments are joined together to produce a continuous DNA strand with the help of .... enzyme.

9. The idea of replisome was suggested by .... and ....

10. Rolling circle DNA replication of *E.coli* is also called .... replication.

Multiple Choice Questions

1. DNA replication occurs in

(*a*) G_1 phase (*b*) G_2 phase

(*c*) S phase (*d*) interphase

2. Duplication of DNA is called

(*a*) transduction (*b*) transcription

(*c*) replication (*d*) translation

3. The process of DNA replication is

(*a*) dispersive (*b*) conservative

(*c*) semi-conservative (*d*) non-conservative

4. Semiconservative DNA replication was first demonstrated by

(*a*) Taylor (*b*) Watson and Crick

(*c*) Meselson and Stahl (*d*) Khurana

5. Taylor demonstrated that DNA replication is semi-conservative in

(*a*) *Vicia faba* (*b*) mouse liver cells

(*c*) peanut (*d*) HeLa cells

6. Isotopes used for proving semiconservative replication are

(*a*) N^{14} and N^{31} (*b*) N^{14} and C^{14}

(*c*) H^3 and N^{15} (*d*) C^{14} and P^{31}

7. The unwinding of DNA duplex is performed by an enzyme called

(*a*) gyrase (*b*) lactase

(*c*) maltase (*d*) ligase

8. The enzyme that reduces the tension during the unwinding of DNA helix in front of the replication form is

(*a*) topoisomerase (*b*) helicase

(*c*) ligase (*d*) polymerase

9. During DNA replication the strands separate by

(*a*) DNA polymerase (*b*) gyrase

(*c*) DNA helicase (*d*) DNA primase

10. During DNA replication, the term leading strand is applied to the one which replicates in

(*a*) 5′ – 3′ direction continuously (*b*) 3′ – 5′ direction continuously

(*c*) 5′ – 3′ direction discontinuously (*d*) 3′ – 5′ direction discontinuously

11. Which of the following enzymes makes short RNA chains using a DNA template?

(*a*) DNA polymerase I (*b*) DNA helicase

(*c*) DNA primase (*d*) DNA gyrase

12. Okazaki fragments are formed in
    (*a*) the splicing of mRNA
    (*b*) the synthesis of the lagging strand of DNA
    (*c*) the neurons of vertebrates
    (*d*) the oocytes of amphibians
13. The enzyme that removes the RNA primer after DNA replication is
    (*a*) primase
    (*b*) DNA polymerase I
    (*c*) DNA polymerase II
    (*d*) ligase
14. The enzyme that joins DNA fragments is
    (*a*) restriction endonuclease (*b*) lipase
    (*c*) ligase (*d*) peroxidase

ANSWERS

Very Short Answer Questions

1. DNA polymerases α, β and γ are located in the nuclei of cells; polymerase γ is located in mitochondria and chloroplasts. Current evidence suggests that polymerases α and δ are both required for the replication of nuclear DNA. Polymerase δ is thought to catalyze the continuous synthesis of the leading strand and polymerase α is believed to catalyze discontinuous synthesis of the lagging strand since it contains primase activity required for the repeated initiation of "Okazaki fragments". Polymerase γ presumbly catalyzes replication of organellar chromosomes. The function of polymearase β remains unknown, perhaps it functions in DNA repair pathways like polymerase I of *E. coli*.
2. A replisome, consisting of a primosome (primase and a helicase) and two polymerase III holoenzymes, forms at a Y-junction (*i.e.,* replication fork) on DNA. One polymerase acts processively, synthesizing the leading strand, while the other forms Okazaki fragments initiated by primers created by the primase. DNA polymerase I completes the Okazaki fragments, eliminating the RNA primer of the previous Okazaki fragment and replacing it with DNA. Finally, DNA ligase connects the fragments. Also involved in the process are single-stranded binding proteins and topoisomerases that relieve the DNA's supercoiling. Initiation involves initiation proteins at *ori C*, and termination requires termination proteins bound to the termination sites and a topoisomerase.
3. Deoxynucleoside triphosphates. Addition to 3′ –OH group; requirement for a template.
4. (*a*) and (*e*) are true.
5. Primer means nucleotide bound to DNA and having a 3′ –OH group; template means a polynucleotide strand whose base sequence can be copied.
6. RNA serves as a primer.
7. A helicase unwinds a helix. The SSB proteins prevent the helix from rewinding and prevent intramolecular base pairing from occurring.
8. 3′ T T G C A T G A C G 5′ because of pairing qualities of deoxyribonucleotides.
9. 4^n.
10. $L\,\mu\,m = 3.4 \times 10^{-4}$ P, where $L\,\mu\,m$ represents the length in micrometers, and P represents the number of nucleotide pairs.
11. 68 μ *m*.
12. Arthur Kornberg.

Fill in the Blanks

1. F.W. Stahl, semiconservative.
2. J. Cairns, autoradiography.
3. J.H. Taylor, semiconservative, *Vicia faba*.
4. Arthur Kornberg, living.
5. Roger G. Kornberg, Gefter.
6. Holoenzyme, ten.
7. Okazaki fragments.
8. DNA ligase.
9. N.K. Sinha, A. Kornberg.
10. Sigma.

Multiple Choice Questions

1. (*c*)	**2.** (*c*)	**3.** (*c*)	**4.** (*c*)	**5.** (*a*)	**6.** (*c*)
7. (*a*)	**8.** (*a*)	**9.** (*c*)	**10.** (*a*)	**11.** (*c*)	**12.** (*b*)
13. (*b*)	**14.** (*c*).				

8

Transcription

In the organisms (*viz.*, prokaryotes and eukaryotes), where coded genetic informations are contained in the DNA molecule, different genetically controlled functions of their cells are performed by a different kind of nucleic acid, called **non-genetic ribonucleic acid (RNA).** In the cells of such organisms, the DNA molecule occurs in the chromosomes of nucleus or nucleoids of eukaryotes or prokaryotes respectively, and the process of protein synthesis occurs in the cytoplasm. It is investigated that a DNA molecule does not leave the nucleus to participate directly in the process of protein synthesis but employs different types of non-genetic ribonucleic acid molecules for carrying its genetic informations from the nucleus to the site of protein synthesis, *i.e.*, ribosomes. Thus, gene expression accomplished by the transfer of genetic information from DNA to RNA molecules and then from RNA to protein molecules. RNA molecules are synthesized by using the base sequence (triplet codons) of one strand of DNA as a template in a polymerization reaction that is catalized by enzymes called **DNA-dependent RNA polymerases** or simply **RNA polymerases.** The process by which RNA molecules are initiated, elongated, and terminated is called **transcription.**

8.1 CHEMICAL COMPOSITION OF NON-GENETIC RIBONUCLEIC ACID (RNA)

Chemically, the non-genetic RNA is closely related with DNA. However, the non-genetic RNA is a single stranded polymer of ribonucleotide units and is composed of phosphoric acid, ribose (pentose) sugar and nitrogen bases, which are purines (adenine and guanine) and pyrimidines (cytosine and uracil). The molecular weights of the various RNA molecules vary from about 25,000 to over one million. Its base composition does not follow the base composition of DNA, although, in general, the proportion of A+C roughly equals G+U, as has been shown in Table 8.1.

Table 8.1 Base ratios of RNA from various sources (as molar percentages) (From, **Hall** *et al.,* 1974).

S. No.	*Source*	*Adenine*	*Guanine*	*Cytosine*	*Uracil*
1.	*Allium cepa* (onion seed)	24.9	29.8	24.7	20.6
2.	*Phaseolus vulgaris* (bean seed)	24.9	31.4	24.1	19.6
3.	*Cucurbita pepo* (pumpkin seed)	25.2	30.6	24.8	19.4
4.	Ox liver	17.1	27.3	33.9	21.7
5.	Yeast	25.4	24.6	22.6	27.4
6.	*E. coli*	25.3	28.8	24.7	21.2

Further, some RNA molecules contain significant proportions of some methylated and an unusual nucleoside known as pseudouridine (ψU), in which the glycosidic bond is associated with position 5 of uracil rather than position 3.

Types of RNA

In a cell, there are three different kinds of RNA which serve in three different roles in the process of protein synthesis. The first type is **messenger RNA (mRNA)** which carries the DNA sequence information to particles in the cytoplasm known as *ribosomes*, where the messenger RNA is translated. The second type is *transfer RNA* (tRNA), which brings the amino acids to the ribosomes, where protein synthesis takes place. The third type of RNA is a structural and functional part of the ribosome called ribosomal RNA (rRNA). The general relationship of the roles of these three types of RNA is diagrammed in Figure 8.1. In addition, small RNAs play other roles in cellular metabolism.

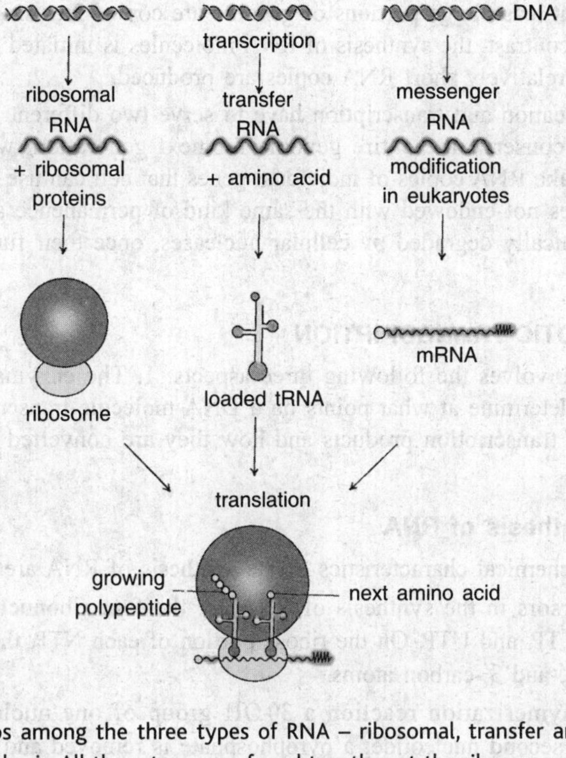

Fig. 8.1. Relationships among the three types of RNA – ribosomal, transfer and messenger – during protein synthesis. All three types are found together at the ribosome during protein synthesis (after Tamarin, 2002).

Further, DNA does not take part directly in protein synthesis because, in eukaryotes, translation occurs in the cytoplasm, whereas DNA remains in the nucleus. Molecular biologists suspected for a long time that genetic intermediate in bacteria and eukaryotes was RNA because the cytoplasmic RNA concentration increases with increasing protein synthesis, and the cytoplasmic RNAs carry nucleotide sequences complementary to the cell's DNA. Proof of an RNA intermediate came when it was shown that mRNA directs protein synthesis.

8.2 COMPARISON BETWEEN DNA REPLICATION AND TRANSCRIPTION

The replication and transcription are two chief activities of DNA and they should be compared at the outset in the following manner:

1. In DNA replication an enzyme (or enzyme complex) matches up complementary deoxyribonucleoside triphosphates with a DNA template according to Watson-Crick pairing rules and the bases are polymerized to form a daughter DNA molecule. In DNA transcription a

different enzyme, called **DNA-depenent RNA polymerase** medicates a similar process, except that complementary ribonucleoside-triphosphates are matched with the DNA template and RNA polymers are formed. The base uracil (U) replaces thymine in RNA but, otherwise, the DNA strand is faithfully copied. The RNA copy has the opposite polarity from the template, so that, for example, the sequence $\xrightarrow[\text{TACAAC}]{}$ in DNA is transcribed as $\xleftarrow[\text{AUGUUG}]{}$ in RNA.

2. The RNA transcripts do not ordinarily remain hydrogen-bonded with their DNA templates as DNA daughter strands do. Instead, they "peel off" the template as they are formed, thus, becoming available to participate in protein synthesis (Fig. 8.1).

3. DNA of prokaryotes (*E.coli*) and eukaryotes has one or few initiation points for DNA replication, so that, at least, large portions of genome are copied into single, enormous daughter DNA molecules. In contrast, the synthesis of RNA molecules is initiated at close intervals along the DNA and, thus, relatively short RNA copies are produced.

4. Lastly, replication and transcription have to serve two different functions. The purpose of replication is to conserve the entire genome for next generation, whereas the purpose of transcription is to make RNA copies of individual genes that cell can use in the metabolism. The copies are themselves not endowed with the same kind of permanence as the genetic material; instead, they are typically degraded by cellular nucleases, once their functional usefulness has been spent.

8.3 PROKARYOTIC TRANSCRIPTION

Transcription involves the following three aspects: 1. The enzymatic synthesis of RNA; 2. The signals that determine at what points on a DNA molecule transcription starts and stops; and 3. The types of transcription products and how they are converted to the RNA molecules needed by the cell.

1. Enzymatic Synthesis of RNA

The essential chemical characteristics of the synthesis of RNA are following:

(1) The precursors in the synthesis of RNA are the four ribonucleotide 5′-triphosphates (rNTP)-ATP, GTP, CTP, and UTP. On the ribose portion of each NTP, there are two OH groups – one each on the 2′-and 3′-carbon atoms.

(2) In the polymerization reaction a 3′-OH group of one nucleotide reacts with the 5′-triphosphate of a second nucleotide; a pyrophosphate is removed and a phosphodiester bond results (Fig. 8.2).

(3) The sequence of bases in RNA molecule is determined by the base sequence of the DNA. Each base, added to the growing end of the RNA chain, is chosen by its ability to base-pair with the DNA strand used as template; thus, the bases C,T,G and A in a DNA strand cause G,A,C, and U respectively, to appear in the newly synthesized RNA molecule.

(4) The DNA molecule being transcribed is double-stranded, yet in many particular regions only one strand serves as a template (Fig. 8.3).

(5) The RNA chain grows in 5′→ 3′ direction; that is, nucleotides are added only to the 3′-OH end of the growing chain. (This is the same as the direction of chain growth in DNA synthesis). The RNA molecule is terminated by a 5′-triphosphate at the non-growing end. The RNA molecule is antiparallel to the DNA strand being copied. Once intiated, RNA chains grow at a rapid rate – 40 to 50 nucleotides per second in *E.coli* at 37°C.

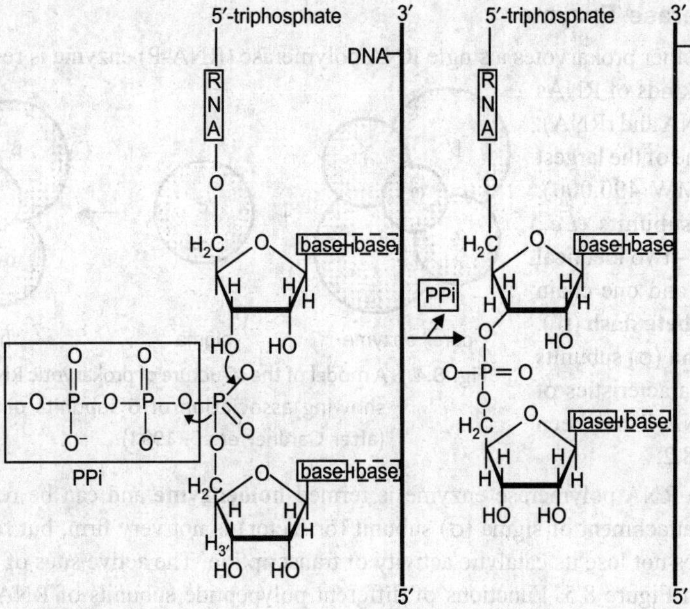

Fig. 8.2. Mechanism of the chain-elongation reaction catalyzed by RNA polymerase. The broken arrows join the reacting groups. The pyrophosphate group (PP) and the bold hydrogen do not appear in the RNA strand. The DNA template and the RNA strands are antiparallel as in double-stranded DNA (after Freifelder, 1985).

(6) RNA polymerases, in contrast with DNA polymerases, are able to initiate chain growth; that is, no primer is needed.

(7) Only ribonucleoside 5′-triphosphates participate in RNA synthesis and the first base to be laid down in the initiation event is a triphosphate. Its 3′-OH group is the point of attachment of the subsequent nucleotide. Thus, the 5′ end of a growing RNA molecule terminates with a triphosphate.

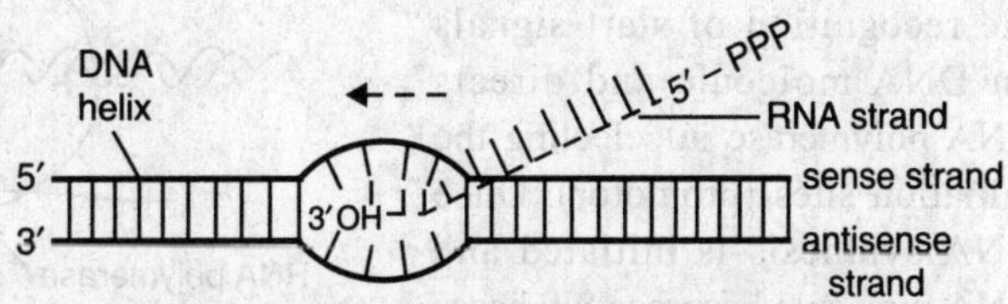

Fig. 8.3. An RNA strand is copied only from strand A (sense strand) of a segment of a DNA molecule. No RNA is copied from strand B (antisense strand) in that region of the DNA molecule. However, elsewhere, for example, in a different gene, strand B might be copied; in that case strand A would not be copied in that region of the DNA. The arrow shows the direction of RNA chain growth (after Freifelder, 1985).

The overall polymerization reaction can be written as follows:

$$_{n}NTP \;+\; XTP \xrightarrow[Mg^{2+}]{DNA,\ RNA\text{–}P} (NMP)_n\ XTP \;+\; {}_{n}PP_1$$

In which XTP represents the first nucleotide at the 5′ terminus of the RNA chain, NMP is a mononucleotide in the RNA polymerase, and PP_1 is the pyrophosphate released each time a nucleotide is added to the growing chain. The Mg^{2+} is required for all nucleic acid polymerization reactions.

The RNA Polymerase Enzyme

In *E.coli* and other prokaryotes a single RNA polymerase (RNA-P) enzyme is responsible for the synthesis of all kinds of RNAs (such as mRNA, tRNA and rRNA). RNA-P is, in fact, one of the largest enzymes known (MW 490,000). It consists of six subunits (*i.e.*, polypeptide chains) – two identical **alpha** (α) subunits and one chain of each of **beta** (β), **beta dash** (β'), **omega** (ω) and **sigma** (σ) subunits (Fig. 8.4). Some characteristics of each subunit of RNA-P has been tabulated in Table 8.2.

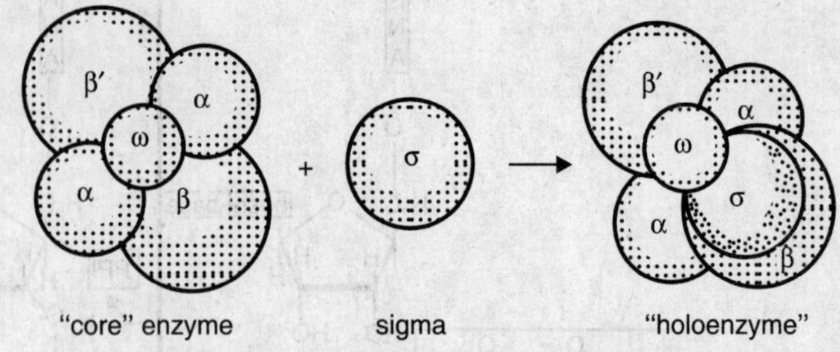

Fig. 8.4. A model of the structure of prokaryotic RNA polymerase showing association of 6 subunits or polypeptides (after Gardner *et al.*, 1991).

The complete RNA polymerase enzyme is termed **holoenzyme** and can be represented as $\alpha_2\beta\beta'\omega\sigma$ in which attachment of sigma (σ) subunit (or factor) is not very firm, but resultant **core enzyme** ($\alpha_2\beta\beta'$) does not lose its catalytic activity of transcription. The active sites of core enzyme have been shown in Figure 8.5. Functions of different polypeptide subunits of RNA polymerase are now known, but not in any detail. Thus, β and β' subunits form the catalytic centre of RNA-P and help RNA polymerase in unwinding of DNA molecule for the purpose of transcription. The sigma (σ) factor helps in the recognition of start signals on DNA molecule and directs RNA polymerase in selecting the initiation sites (promotor). Once RNA synthesis is initiated and RNA molecule becomes 8-9 bases long, the σ factor dissociates from the holoenzyme and then the core enzyme brings about elongation of mRNA (or any other RNA).

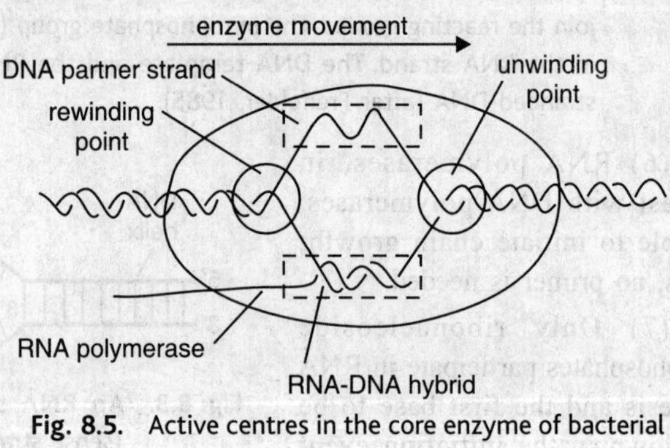

Fig. 8.5. Active centres in the core enzyme of bacterial RNA polymerase enzyme.

2. Binding of RNA Polymerase to Promoter and Initiation

Unlike replication, transcription does not progress along the entire length of a chromosome. Instead, certain parts of the chromosome are transcribed. Only one of the two strands of a DNA duplex is transcribed; this strand is called the **sense strand.** The other strand of DNA which is not transcribed at the moment, is called **anti-sense strand**. Like DNA, RNA is synthesized in the 5→3' direction from the single-stranded region of the DNA template. This localized unwinding moves along the molecule followed by single-recoiling of the helix behind the newly synthesized RNA. The region of sense strand of DNA which is actually transcribed into RNA, is called the **coding region**.

The DNA region that RNA polymerase enzyme associates with immediately before beginning transcription is known as the *promoter.* The promoter is a significant part of gene expression in both prokaryotes and eukaryotes. Promoters contain the information for transcription initiation and are the major sites in which gene expression is regulated.

Table 8.2 Some basic characteristics of different subunits (=polypeptides) of RNA polymerase of *E.coli* (*Source:* **Gardner,** *et al.*, 1991).

S. No.	*Subunit*	*Gene*	*Molecular weight (Daltons)*	*Location*	*Function*
1.	α_2	*rpo A*	41,000 each	Core enzyme	Promoter binding
2.	β	*rpo B*	155,000	Core enzyme	Nucleotide binding
3.	β'	*rpo C*	165,000	Core enzyme	DNA template binding
4.	ω	–	12,000	Core enzyme	–
5.	σ	*rpo D*	95,000	Sigma factor	Initiation

The molecule of enzyme of RNA polymerase covers a region of about sixty base pairs of DNA. This point was determined by causing the polymerase to bind to DNA and then digesting the mixture with nuclease enzymes, in a technique known as *footprinting* (Fig. 8.6). The polymerase "protects" or prevents degradation of the region it covers. The undigested DNA is then isolated and its size determined. Geneticists have gained much new information about the nature of recognition regions within promoters through recombinant DNA technology and nucleotide sequencing techniques. Sequencing of numerous promoters has revealed that they contain common sequences.

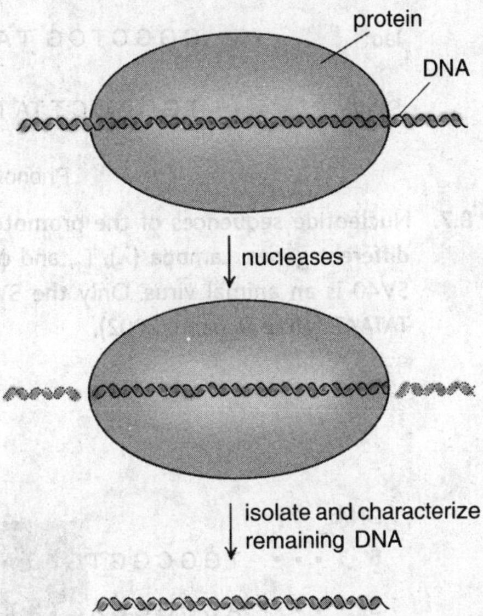

Fig. 8.6. Technique of footprinting. DNA in contact with a protein (e.g., RNA polymerase) is protected from degradation by nuclease enzyme. In next step, the protected DNA is isolated and characterized (after Tamarin, 2002).

If the promoter nucleotide sequences align with each other, and each has exactly the same series of nucleotides in a given segment, then sequence of that segment is regarded as *conserved sequence.* If, however, there is some variation in the sequence, but certain nucleotides occur at a high frequency (significantly greater than by chance), then these nucleotides are said to form a *consensus sequence.* Surrounding a point in prokaryotic promoters about ten nucleotides before the *first transcribed base* is just such a consensus sequence – TATAAT. This sequence is called *Pribnow box* after one of its discoverers (Fig. 8.7).

The nucleotides in the Pribnow box are mostly adenines (A) and thymines (T), so the region is primarily held together by only two hydrogen bonds per base pairs. Since local DNA denaturation occurs during transcription by RNA polymerase (*i.e.*, the DNA is opened to allow transcription), fewer hydrogen bonds make this process easier energetically. When the RNA polymerase is bound at the promoter region (Fig. 8.7), it is in position to begin polymerization six to eight nucleotides down from the Pribnow box.

In Fig. 8.7. have been shown the sequences which represent the *coding strand* of DNA. It is general convention to show the coding strand because both that strand and mRNA have the same sequences, substituting U for T in RNA, they are both complementary to the same *template strand* (which is also known as *anticoding strand* or *noncoding strand*; Fig. 8.8). Another convention is to indicate the first base transcribed by number +1 and to use positive numbers to count farther

	Promoter region	First base transcribed	
λP_R 5′ • • •	TGGCGGTGATAA TGGTTGCATGT		• • • 3′
T7A1 • • •	CC TATAGGA TACTTACAGGCAT		• • •
T7A2 • • •	CATGCAGTA AGAT ACAAATCGCTA		• • •
ϕ×174A • • •	TGTATGTTTTCATGCCTCCAAAT		• • •
lac • • •	CGGCTCG TATGTTGTGTGGAAT		• • •
SV40 • • •	TGCAGCTTATAATGGTTACAAATA		• • •

Pribnow box

Fig. 8.7. Nucleotide sequences of the promoter region and the first base transcribed from several different genes. Lambda (λ), T_7, and ϕ × 174 are bacteriophages. *Lac* is an *E.coli* gene, and SV40 is an animal virus. Only the SV40 promoter has the actual consensus sequence of TATAAT (after Tamarin, 2002).

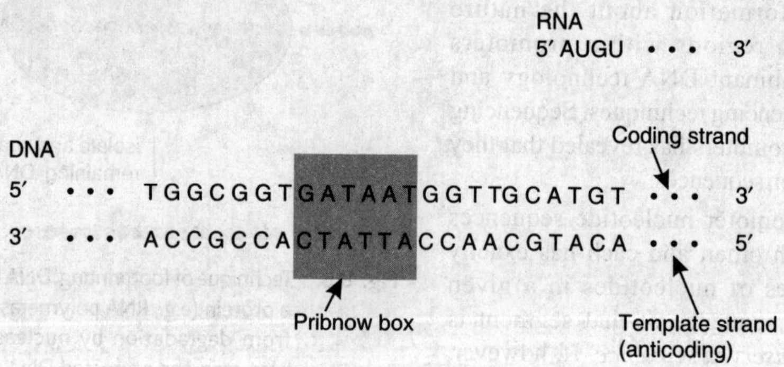

Fig. 8.8. The template (anticoding) strand of DNA is complementary to both the coding strand and the transcribed RNA. The sequences are from the promoter of the bacteriophage λP_R region (after Tamarin, 2002).

down the DNA in the *downstream* direction of transcription. If transcription is proceeding to the right, the direction to the left is called *upstream,* with bases indicated by negative numbers (Fig. 8.9) Under this convention, the Pribnow box is often referred to as the – 10 sequence.

Figure 8.9 also indicates another region with similar sequences among many promoters centred near – 35 and referred to as the – *35 sequence.* The consensus sequence at – 35 is TTGTCA. Mutation studies have tried to determine the relative roles of the – 10 and – 35 sequences in transcription. In other words, mutations of bases in the – 10 and – 35 regions of promoter were examined to determine how they affected transcription initiation. The conclusion from these studies are that both regions contribute to the efficiency of binding of polymerase enzyme to promoter. In other words, the more each sequence differs from the consensus sequence, the less frequently that promoter initiates the transcription. The sigma factor recognizes both the –35 and the –10 sequences. The sigma factor is also sensitive to the spacing between these sequences, preferring (*i.e.*, being most efficient at seventeen base pairs).

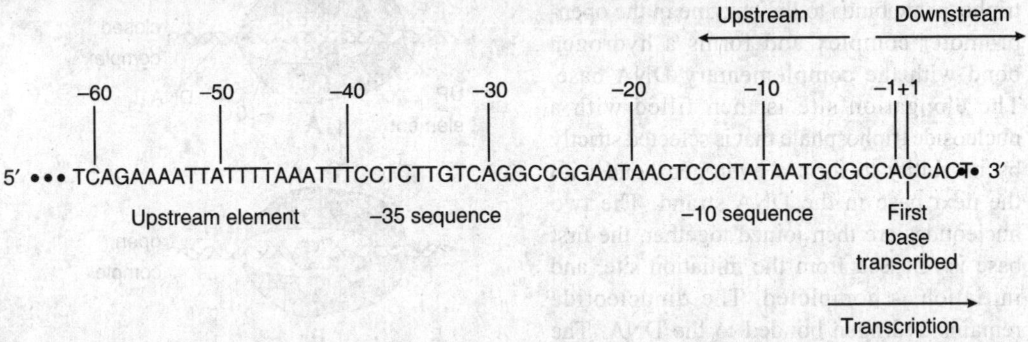

Fig. 8.9. Promoter of the *Escherichia coli* ribosomal RNA gene, rrnB. Note the – 10 and – 35 sequences and the upstream element. The first base transcribed (the transcription start site) is noted (+1), as well as, the upstream, downstream, and transcription direction (after Tamarin, 2002).

Moreover, in bacterial chromosomes, farther upstream from the – 35 sequence is a recognition element that are very strongly expressed, e.g., ribosomal RNA genes (Fig. 8.8). This *upstream element* or *UP element*, is about twenty base pairs long, is centered at – 50 and is rich in A and T. By mutational studies, it has been shown that adding UP element to promoters that do not normally have it greatly increases the rate of transcription. There are other recognition sites in prokaryotes, both upstream and downstream, at which various proteins attach that can enhance or inhibit transcription by direct contact with the polymerase (the α and σ subunits).

Because the holoenzyme (RNA polymerase) recognizes consensus sequences in a promoter, it is not surprising that some promoters are bound more efficiently than others or that different sigma factors exist within a cell. For example, in *E.coli*, the major sigma factor is a protein of 70,000 Daltons, referred to as σ_{70}. There exist about five less common sigma factors, due to which a bacterial cell remains able to transcribe different genes under different circumstances. For instance, in an *E.coli* cell subjected to elevated temperatures, a group of new proteins, called *heat shock proteins*, which protect the cell to some extent against the elevated temperatures. These heat shock proteins all appear at once because they have promoters that a different sigma factor recognizes, one with a molecular weight of 32,000 Daltons (σ_{32}); this new sigma factor is produced by the cell after heat shock.

According to the current picture of mode of action of polymerase holoenzyme, this enzyme sets down on a DNA promoter because the sigma factor recognizes the –10 and – 35 elements, the α proteins recognize the UP element and the α and σ subunits recognise proteins to bound to various other upstream elements, when present (Fig. 8.10A). This initiation complex is initially known as a *closed complex* because the DNA has not melted, which is the next step in transcription initiation (Fig. 8.10B). After the transcription of 5 to 10 bases, the sigma factor is released (Fig. 8.10 C and D).

Once an open promoter complex has formed, RNA polymerase is ready to initiate RNA synthesis. RNA polymerase contains two nucleotides binding sites, called the **initiation site** and the **elongation site**. The initiation site binds only purine triphosphates, namely ATP and GTP, and one of these (usually ATP) is the first nucleotide in the growing RNA chain. Thus, the first DNA base that is transcribed is usually thymine (T). The initiating nucleoside triphosphate binds to the enzyme in the open-promoter complex and forms a hydrogen bond with the complementary DNA base. The elongation site is then filled with a nucleoside triphosphate that is selected strictly by its ability to form a hydrogen bond with the next base in the DNA strand. The two nucleotides are then joined together, the first base is released from the initiation site, and initiation is completed. The dinucleotide remains hydrogen bonded to the DNA. The **elongation phase** begins when the polymerase releases the base and then moves along the DNA chain.

3. Elongation phase. Since there is just one bacterial RNA polymerase, the general mechanism of transcription is the same for all bacterial genes. The following description of elongation and termination, given in the context of mRNA synthesis, therefore apply equally well to the synthesis of non-coding RNA (*i.e.*, rRNA and tRNA).

During transcription, ribonucleotides are added one after another to the growing 3′ end of the RNA transcript, the identity of each nucleotide specified by the base-pairing rules: A base pairs with T or U; G base pairs with C. During each nucleotide addition, the β- and γ- phosphates are removed from the incoming nucleotide (Fig. 8.11) and the hydroxyl group is removed from the 3′ – carbon of the nucleotide present at the end of the chain.

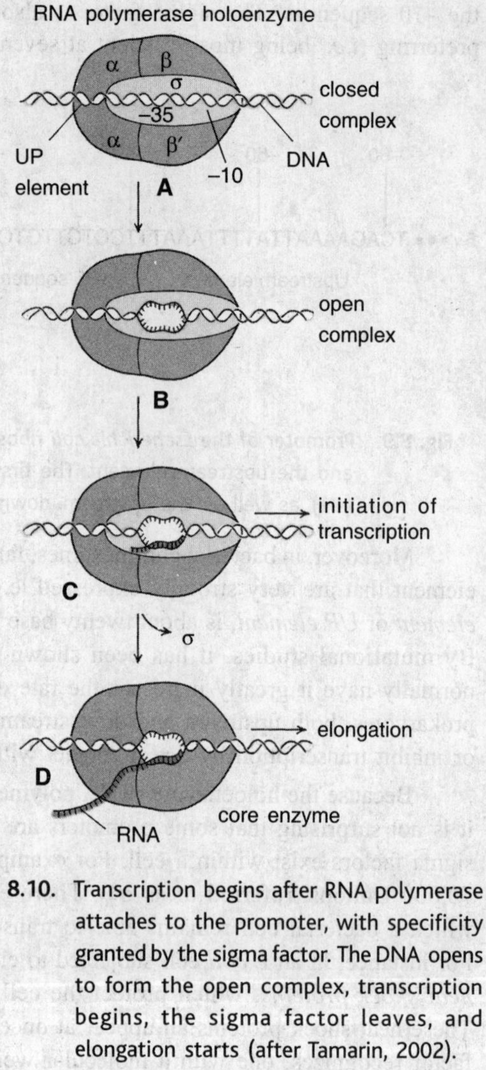

Fig. 8.10. Transcription begins after RNA polymerase attaches to the promoter, with specificity granted by the sigma factor. The DNA opens to form the open complex, transcription begins, the sigma factor leaves, and elongation starts (after Tamarin, 2002).

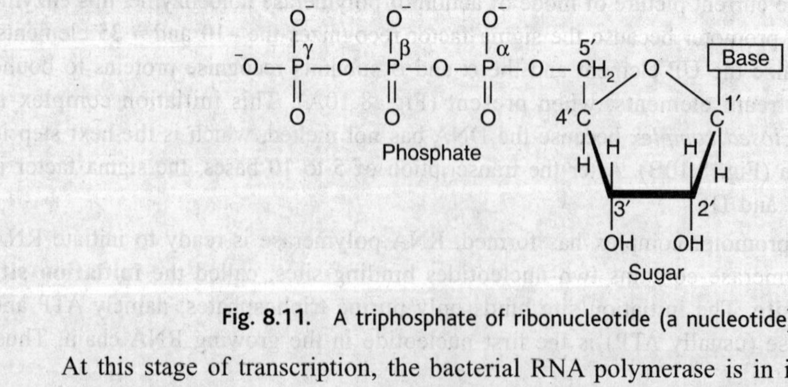

Fig. 8.11. A triphosphate of ribonucleotide (a nucleotide).

At this stage of transcription, the bacterial RNA polymerase is in its core enzyme form, comprising four proteins, two relatively small (approximately 35 k Da) α subunits and one each of the related subunits β and β' (both approximately 150 k Da): the α subunit that has the key role in initiation has now left the complex. The RNA polymerase covers about 30 bp of the template DNA, including the *transcription bubble* of 12 – 14 bp, within which the growing transcript is held to the template strand of the DNA by approximately eight RNA – DNA base pairs (Fig. 8.12). The polymerase has to keep a tight grip on both the DNA template and the RNA that it is making in order to prevent the transcription complex from falling apart before the end of the gene is reached. However, this grip must not be so tight as to prevent the polymerase from moving along the DNA.

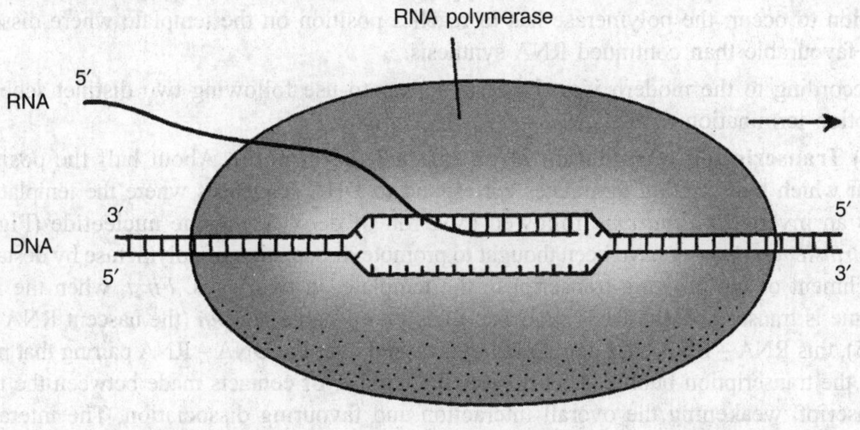

Fig. 8.12. Transcription elongation complex of *E. coli* (a schematic representation). The arrow indicates the direction in which the RNA polymerase moves along the DNA (after Brown, 2002).

In a recent model, which has been produced by X-ray crystallographic studies and crosslinking experiments, the DNA template lies between the β and β′ subunits, within a trough on the enclosed surface of β′. The active site for RNA synthesis also lies between these two subunits, with the non-template strand of DNA held within the β subunit and RNA transcript extruded from the complex via a channel formed partly by the β and partly by the β′ subunit (Fig. 8.13; **Korzheva** *et al.* 2000).

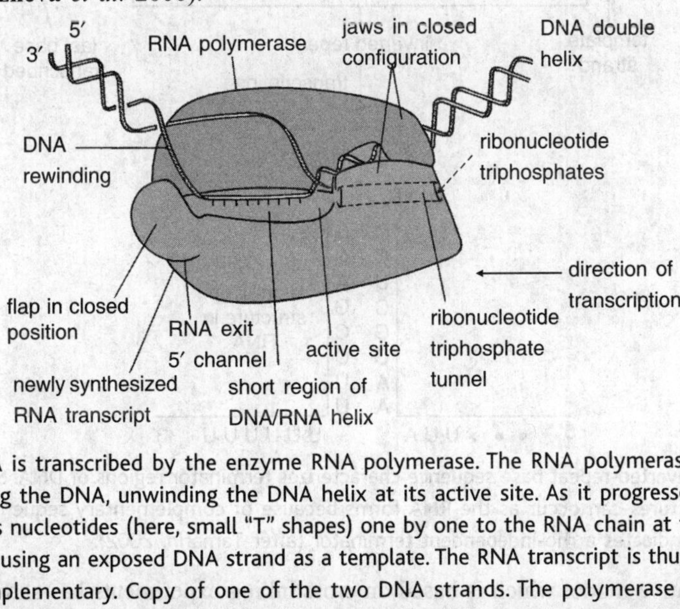

Fig. 8.13. DNA is transcribed by the enzyme RNA polymerase. The RNA polymerase moves stepwise along the DNA, unwinding the DNA helix at its active site. As it progresses, the polymerase adds nucleotides (here, small "T" shapes) one by one to the RNA chain at the polymerization site using an exposed DNA strand as a template. The RNA transcript is thus a single-stranded complementary. Copy of one of the two DNA strands. The polymerase has a rudder that displaces the newly formed RNA, allowing the two strands of DNA behind the polymerase to rewind (after Alberts *et al.*, 2002).

4. Termination phase. Exact mechanism involved in termination of transcription is still not known. However, molecular biologists believed that transcription is a stepwise addition process of nucleotide-by-nucleotide, with the polymerase pausing at each position and making a choice between continuing elongation by adding another ribonucleotide to the transcript, or terminating by dissociating from the template. Which choice is selected depends on which alternative is more favourable in thermodynamic terms (**Von Hippel**, 1998). This model emphasizes that in order for termination to occur, the polymerase has to reach a position on the template where dissociation is more favourable than continued RNA synthesis.

According to the modern view bacteria appear to use following two distinct schemes for transcription termination.

(*a*) Transcription termination at an intrinsic terminator. About half the positions in *E. coli* at which transcription terminates correspond to DNA sequences where the template strand contains an inverted *palindrome* followed by a run of deoxyadenosine nucleotide (Fig. 8.14). These *intrinsic terminators* have been thought to promote dissociation of polymerase by destabilizing the attachment of the growing transcript to the template, in two ways. *First*, when the inverted palindrome is transcribed, the RNA sequence folds into a stable *hairpin* (the nascent RNA hairpin; Fig. 8.15), this RNA – RNA base pairing being favoured over the DNA – RNA pairing that normally occur in the transcription bubble. This reduces the number of contacts made between the template and transcript, weakening the overall interaction and favouring dissociation. The interaction is further weakened when the run of As (= adenine bases) in the template is transcribed, because the resulting A-U base pairs have only two hydrogen bonds each, compared with three for each G – C pair. The net result is that termination is favoured over continued elongation (**Von Hippel**, 1998).

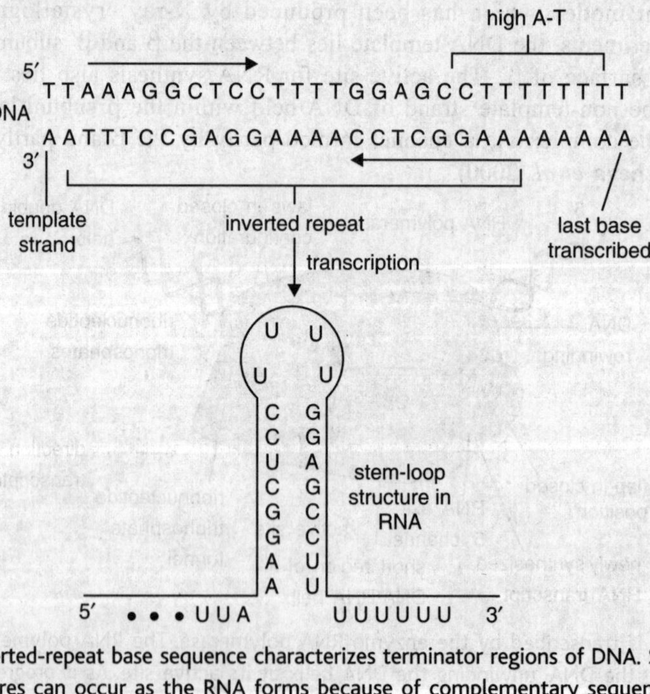

Fig. 8.14. An inverted-repeat base sequence characterizes terminator regions of DNA. Stem-loop (hairpin) structures can occur as the RNA forms because of complementary sequences. The 3′ poly-U tail indicates a rho-independent terminator (after Tamarin, 2002).

An alternative model which is based on crosslinking experiments has shown that the **RNA hairpins** makes contact with flap structure on the outer surface of the RNA polymerase β subunit, adjacent to the exit point of the channel through which the RNA emerges from the complex.

Although the flap structure is quite distant (some 6.4 nm) from the active site of the polymerase, a direct connection is made between the two by a segment of β-sheet within the β-subunit. Movement of the flap could therefore affect the positioning of amino acids within the active site, possibly leading to breakage of the DNA-RNA base pairs and termination of transcription (**Korzheva** *et al.,* 2000; **Toulokhonov** *et al*., 2001). **Toulokhonov** *et al*., (2001) have demonstrated the presence of a protein called **NusA** which enhances termination at intrinsic promoters, interacts with the hairpin loops and flap structure and may stabilize the contact between the two.

***(b)* Rho dependent transcription termination.** The second type of bacterial termination signal is **Rho dependent.** These signals usually retain the hairpin feature of intrinsic terminators, although the hairpin is less stable and there is no run of As in the template. Termination requires the activity of a protein called **Rho,** which attaches to the transcript and moves along the RNA towards the polymerase. If the polymerase continues to synthesize RNA then it keeps ahead of the pursuing Rho, but at the termination signal the polymerase stalls (Fig. 8.16). Rho is a helicase, which means that it actively breaks base pairs, in this case between the template and transcript, resulting in termination of transcription.

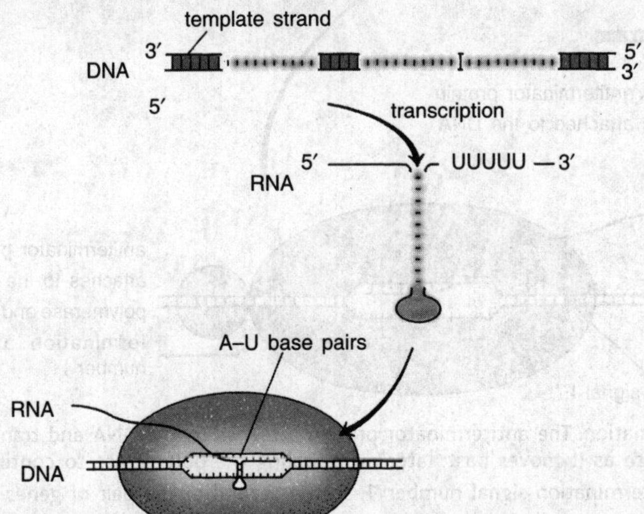

Fig. 8.15. Termination at an intrinsic terminator. The presence of an inverted palindrome in the DNA sequence result in formation of a hairpin loop in the transcript (*e.g.*, mRNA) (after Brown, 2002).

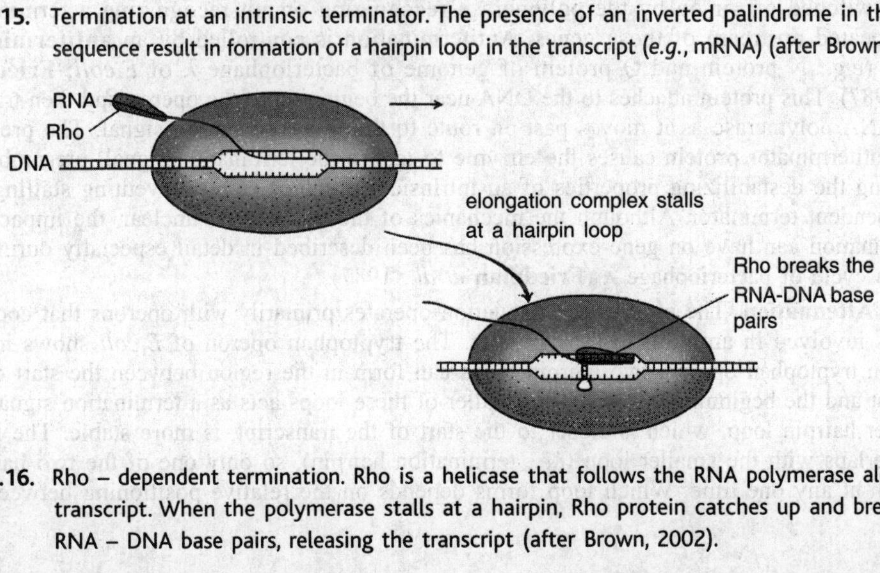

Fig. 8.16. Rho – dependent termination. Rho is a helicase that follows the RNA polymerase along the transcript. When the polymerase stalls at a hairpin, Rho protein catches up and breaks the RNA – DNA base pairs, releasing the transcript (after Brown, 2002).

5. Antitermination and attenuation. In bacteria, the following two mechanisms have evolved for influencing the repeated choice that the polymerase has to make between elongation and termination when copying a template (DNA).

A. Antitermination. This process occurs when the RNA polymerase ignores a termination signal and continues elongating its transcript until a second signal is reached (Fig. 8.17). Antitermination provides a mechanism whereby one or more of the genes at the end of an operon

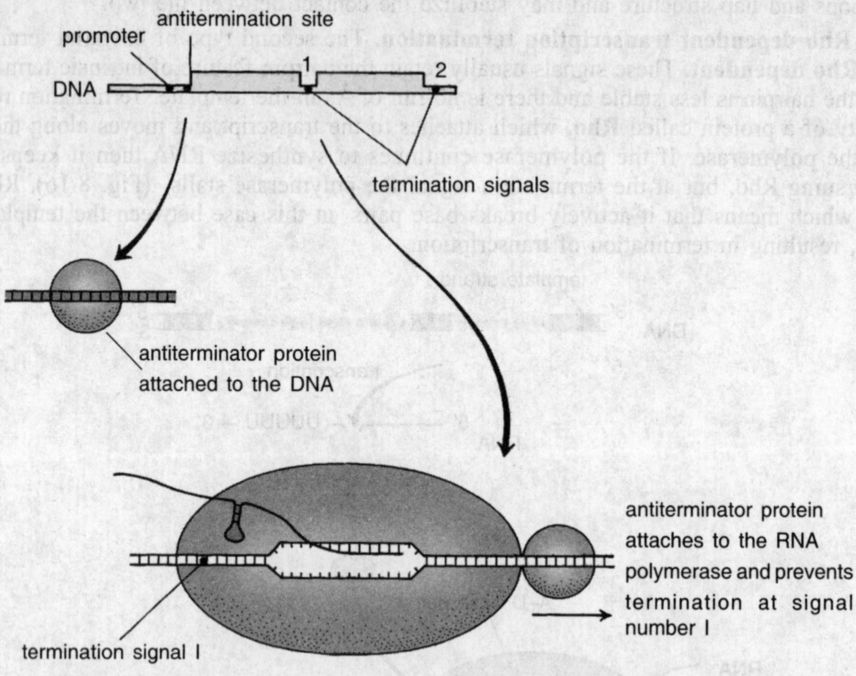

Fig. 8.17. Antitermination. The antiterminator protein attaches to the DNA and transfers to the RNA polymerase as it moves past; later on, enabling the polymerase to continue transcription through termination signal number 1, so the second of the pair of genes in this operon is transcribed (after Brown, 2002).

can be switched off or on by the polymerase recognizing or not recognizing a termination signal located upstream of those genes. Antitermination is controlled by an **antiterminator protein** (*e.g.,* N protein and Q protein of genome of bacteriophage λ of *E.coli*; **Friedman** *et al.,* 1987). This protein attaches to the DNA near the beginning of the operon and then transfer to the RNA polymerase as it moves past en route to the first termination signal. The presence of the antiterminator protein causes the enzyme to ignore the termination signal, presumbly by countering the destabilizing properties of an intrinsic terminator or by preventing stalling at a Rho-dependent terminator. Although the mechanics of the process are unclear, the impact that antitermination can have on gene expression has been described in detail especially during the infection cycle of bacteriophage λ (**Friedman** *et al.,* 1987).

B. Attenuation. The process of attenuation operates primarily with operons that code for enzymes involved in amino acid biosynthesis. The tryptophan operon of *E.coli* shows how it works. In tryptophan operon, two hairpin loops can form in the region between the start of the transcript and the beginning of *trpE.* The smaller of these loops acts as a termination signal, but the larger hairpin loop, which is closer to the start of the transcript, is more stable. The larger loop overlaps with the smaller loop (*i.e.,* termination hairpin), so only one of the two hairpins can form at any one time. Which loop forms depends on the relative positioning between the

RNA polymerase and a ribosome which attaches to the 5′ end of the transcript as soon as it is synthesized in order to translate the genes into protein (Fig. 8.18).

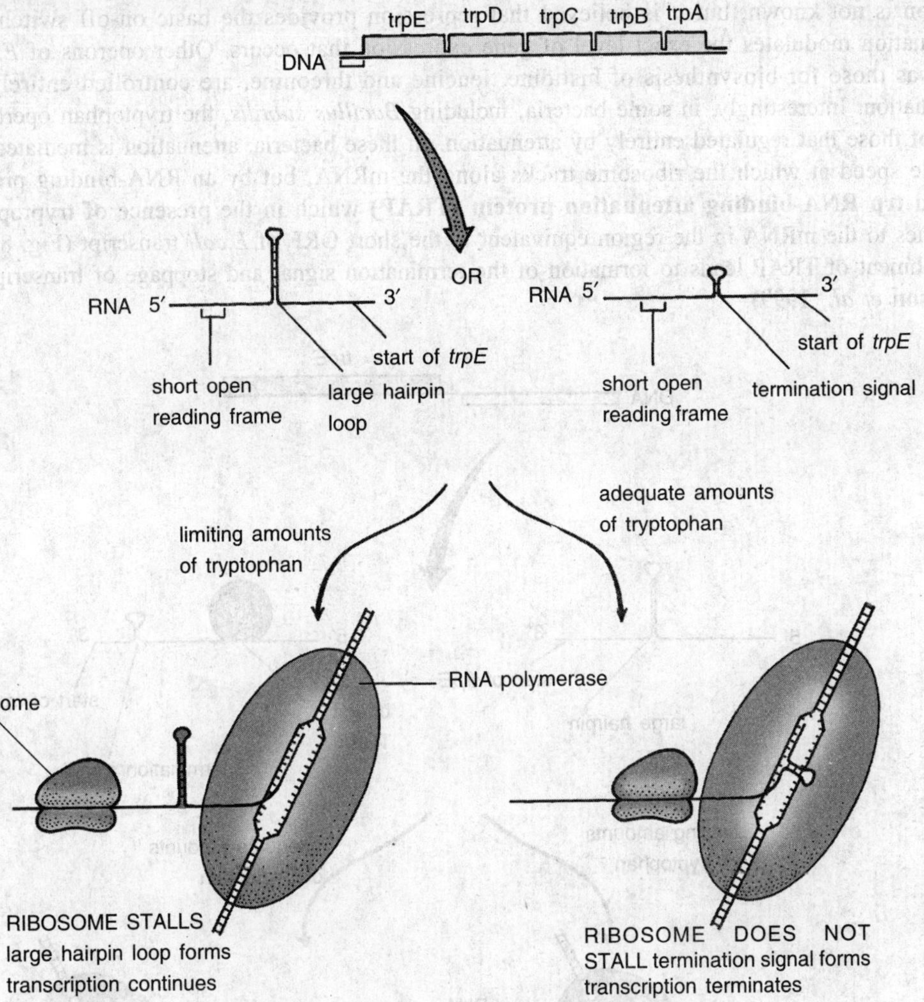

Fig. 8.18. Attenuation at the tryptophan operon of *E.coli.*

Thus, if the ribosome stalls so that it does not keep up with the polymerase, then the larger hairpin forms and transcription continues. However, if the ribosome keeps pace with the RNA polymerase then it dislocates the larger hairpin by attaching to the RNA that forms part of the stem of this hairpin. When this happens the termination hairpin is able to form and transcription stops.

Stalling of the ribosome can occur because upstream of the termination signal is a short open reading frame (ORF) coding for a 14-amino-acid peptide that includes two tryptophans. Again, if the amount of free tryptophan is limiting, then the ribosome stall as it attempts to synthesize this peptide, while the polymerase continues, to make its transcript (mRNA). Because this transcript contains copies of genes coding for the biosynthesis of tryptophan, its continued elongation undertakes the requirement that the cell has for this amino acid. When the amount of tryptophan in the cell reaches a satisfactory level, the attenuation system prevents further transcription of the tryptophan operon, because now the ribosome does not stall while making the short peptide, and instead keeps pace with the polymerase, allowing the termination signal to form.

In fact, tryptophan operon of *E.coli* is controlled not only by attenuation but also by a **repressor**. Exactly how attenuation and repression work together to regulate expression of the operon is not known, but it is believed that repression provides the basic on-off switch and attenuation modulates the exact level of gene expression that occurs. Other operons of *E.coli*, such as those for biosynthesis of histidine, leucine and threonine, are controlled entirely by attenuation. Interestingly, in some bacteria, including *Bacillus subtilis*, the tryptophan operon is one of those that regulated entirely by attenuation. In these bacteria, attenuation is mediated not by the speed at which the ribosome tracks along the mRNA, but by an RNA-binding protein called **trp RNA-binding attenuation protein (TRAP)** which in the presence of tryptophan, attaches to the mRNA in the region equivalent to the short ORF of *E.coli* transcript (Fig. 8.19). Attachment of TRAP leads to formation of the termination signal and stoppage of transcription (**Antson** *et al.*, 1999).

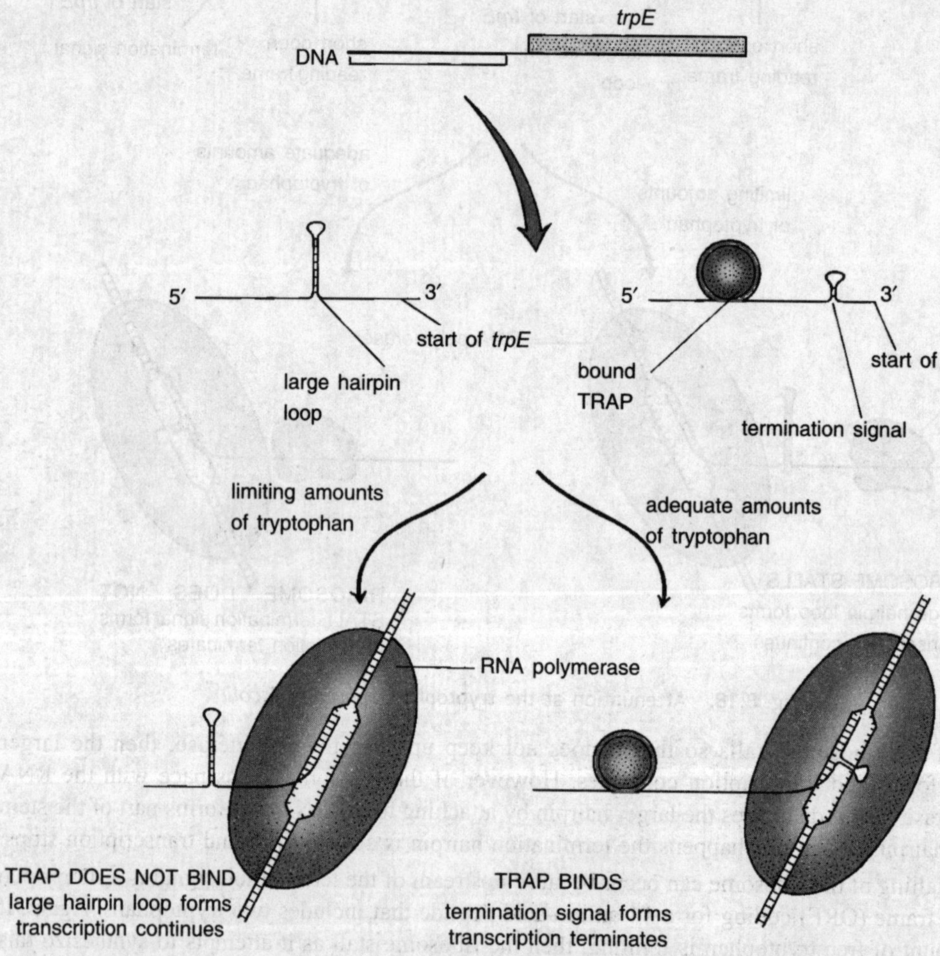

Fig. 8.19. Regulation of the tryptophan operon of *Bacillus subtilis*. Regulation centres on the protein TRAP, which attaches to the leader region of the transcript when tryptophan level are adequate. When attached, TRAP prevents formation of the large hairpin loop and so allows formation of the termination signal (after Brown, 2002).

8.4 INITIATION OF TRANSCRIPTION IN EUKARYOTES

The mRNA molecules make up the **transcriptosome** (*i.e.,* the entire mRNA content of a cell) and specify the protein content of the cell. As the central players in genome expression, mRNAs have received the greatest attention from the molecular biologists. Transcriptional events in prokaryotes (bacteria) are different in many respects from those in eukaryotes. In bacteria, mRNA do not undergo any significant forms of processing: the primary transcript that is synthesized by the RNA polymerase in itself the nature mRNA and its translation usually begins before transcription is complete (Fig. 8.20). The messenger RNA is synthesized in the $5' \rightarrow 3'$ direction, and it is near the $5'$ end of mRNA that translation begins. As soon as the $5'$ end of mRNA is available, a ribosome can attach to the mRNA and move along it in the $5' \rightarrow 3'$ direction lengthening the growing polypeptide as it moves. When the first ribosome moves away from the $5'$ end of the transcript, a second ribosome can attach and begin translation. These processes evidently are repetitive. In eukaryotes, however, messenger RNA is synthesized in the nucleus, but protein synthesis takes place in the cytoplasm. (Such a regional division of labour is not present in *E.coli* because the bacterium has no nucleus). Before a mRNA leaves the nucleus, it is highly modified by processes that generally do not occurs in prokaryotes. Eukaryotic processing of primary transcript includes 1. 5' capping; 2. RNA splicing and 3. 3' polyadenylation (Fig. 8.21). Here now we will describe initiation of transcription in the eukaryotes.

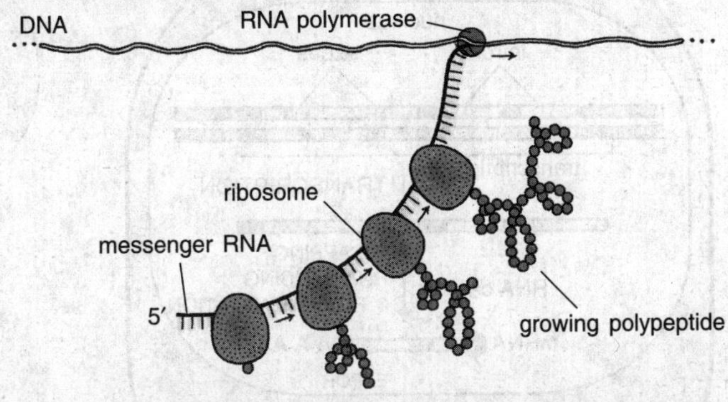

Fig. 8.20. In prokaryotes, translation of messenger RNA (*i.e.,* mRNA) by ribosomes begins before transcription is complete. Ribosomes attach to the growing mRNA strand when the 5′ end becomes accessible. They then move along the RNA as it elongates. When the first ribosome moves from the 5′ end, a second ribosome can attached and so on (after Tamarin, 2002).

1. Promoters

Eukaryotic promoters are somewhat similar to prokaryotic promoters; both are regions of DNA at the beginning of genes with signals that allow RNA polymerase to attach and begin transcription (Fig. 8.26). In eukaryotes, however, more proteins are involved in promoter recognition and more proteins are involved in the control of transcription, many recognizing signals thousands of base pairs away.

In eukaryotes, there are three major classes of RNA polymerases which are designated as I, II and III and are found in the nucleus. The three polymerases have different properties (see Table 8.3) and can be distinguished by the ions required for their activity, the optimal ion strength and their sensitivity to inhibition by various antibiotics (*e.g.,* α-amanitin). Each of these enzymes is a large protein (~500,00 Daltons), with two large and several (8 to 10) smaller subunits.

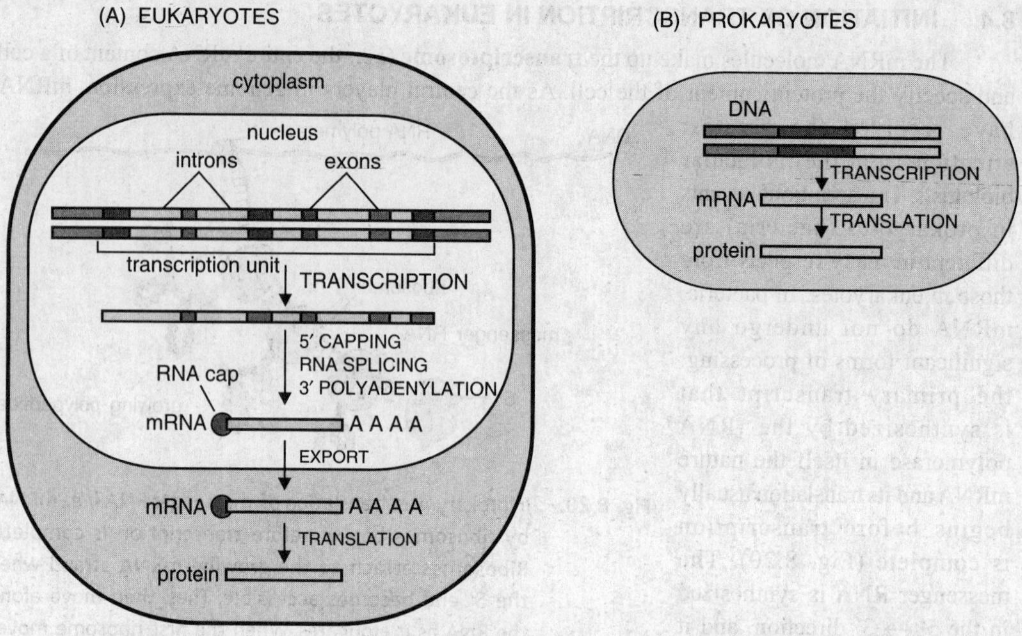

Fig. 8.21. Steps leading from gene to protein in eukaryotes. In eukaryotic cells, the RNA molecule produced by transcription is called primary RNA transcript. It contains both coding (exon) and noncoding (intron) sequences. Before it is translated into protein, the two ends of the RNA are modified, the introns are removed by an enzymatically catalyzed RNA splicing reaction and resulting mRNA is transported from the nucleus to the cytoplasm (after Alberts *et al.*, 2002).

Table 8.3 Properties of three different eukaryotic nuclear RNA polymerases.

S. No.	*Enzymes*	*Location*	*Product and abundance*	*Sensitivity to α-amanitin*
1.	RNA polymerase I	Nucleolus	rRNA (50 – 70%) (except 5S rRNA)	Not sensitive
2.	RNA polymerase II	Nucleoplasm	hnRNA (mRNA) (20 – 40%)	Sensitive
3.	RNA polymerase III	Nucleoplasm	rRNA(~ 10%) (and 5S rRNA)	Inhibited in animals at high levels; not in yeast and insects

All three eukaryotic RNA polymerases (I, II and III) recognize a seven–base sequence, TATAAAA, located at about – 25 on the promoter DNA. It is similar to the – 10 sequence in prokaryotes and is called the **TATA box** (or **Hogness box** after its discover, D. Hogness). Since RNA polymerase II transcribes most of the genes in eukaryotes, so let us describe it in sufficient detail.

Among the large number of promoters that have been sequenced, a few lack the TATA box, yet are still transcribed. Transcription initiation in these promoters appears to be controlled by a CT-rich area, called the **initiator element (INR),** at +1 of the transcript (close to the transcription start site), coupled with a **downstream promoter element (DPE)** at about +28 to the +34 of the transcript. In TATA – less promoters, a protein called **TFIID** requires both these elements to bind. The initiator element has a consensus sequence of TCAGTT or TCATTC, and the downstream promoter element has the consensus sequence of AGACGTG or GGTCGTG (Fig. 8.22).

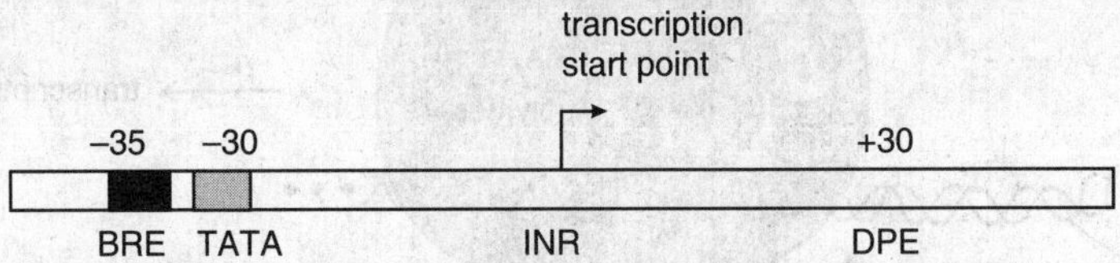

element	consensus sequence	general transcription factor
BRE	G/C G/C G/A C G C C	TFIIB
TATA	T A T A A/T A A/T	TBP
INR	C/T C/T A N T/A C/T C/T	TFIID
DPE	A/G G A/T C G T G	TFIID

Fig. 8.22. Consensus sequences found in the vicinity of eukaryotic RNA polymerase II start points. The name given to each consensus sequence (first column) and the general transcription factor that recognizes it (last column) are indicated. N indicates any nucleotide and two nucleotides separated by a slash indicate an equal probability of either nucleotide at the indicated position. Most polymerase II promoters have a TATA box sequence, and those do not typically have a "strong" INR sequence. Although most of the DNA sequences that influence transcription initiation are located "upstream" of the transcription start point, a few, such as the DPF shown in the figure, are located in the transcribed region (after Alberts *et al.,* 2002).

2. General Transcription Factors

RNA polymerase II of yeast is a protein of twelve subunits. This enzyme cannot locate promoters or attach to DNA in a stable fashion. To attach at the beginnings of genes, RNA polymerase II must interact with several proteins called **general transcription factors** (Table 8.4). The proteins are "general" because they assemble on all promoters used by RNA polymerase II. In eukaryotes, general transcription factors are named after the polymerase they work with. Thus, the transcription factor that recognizes the TATA box for polymerase II genes is called TFIID (D being the fourth letter of the alphabet for the fourth transcription factor so named). TFIID is composed of one subunit that recognizes the TATA sequence, called **TATA binding protein (TBP),** and upto a dozen other proteins called **TBP – associated factors (TAFs)**, which recognize the initiator element, when present and aid in regulating transcription. TFIID is virtually similar to the sigma factors of prokaryotic RNA polymerase. One interesting aspect of the binding of TBP is that it causes a significant bending and opening of the DNA. This bending may be an important signal for other binding proteins (Fig. 8.24).

Box 8.1 Techniques employed in study of initiation of eukaryotic transcription

Much of the information regarding initiation of transcription in eukaryotes has been gathered by foot printing, mutational studies, cloning and isolating the genes and proteins involved, and then reconstituting various purified combinations in test tube. These studies have been combined with kinetic research to determine which arrangements are stable, immunological research to isolate various components with antibodies, and photocrosslinking studies to determine which moieties are in contact with each other.

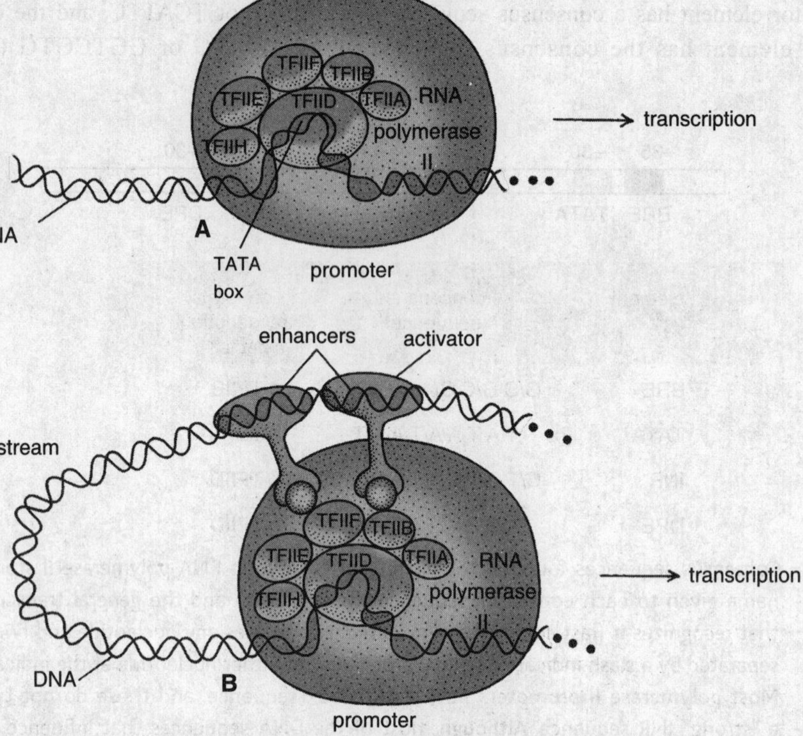

Fig. 8.23. A – Formation of an RNA polymerase II preinitiation complex at a promoter. TFIID binds to the TATA box. The other transcription factors are then recruited with the polymerase. B – Two activators are shown bound at one end (their DNA domains) to enhancers upstream on the DNA. The activators are bound at their other ends (their transcriptional activation domains) to other protein associated with the polymerase machinery. Phosphorylation of the polymerase initiates activated transcription (after Tamarin, 2002).

Table 8.4 Postulated roles of the general transcription factors of RNA polymerase II (*Source:* **Tamarin,** 2002).

S. No.	*General transcription factors*	*Function*
1.	TFIID, TBP	Recognizes TATA box
2.	TFIID, TAFs	Recognizes initiator elements and regulatory proteins
3.	TFIIA	Stabilizes TFIID
4.	TFIIB	Aids in start-site selection by RNA polymerase II
5.	TFIIE	Controls TFIIH functions; enhances promoter melting
6.	TFIIF	Destabilizes nonspecific interactions of RNA polymerase II and DNA
7.	TFIIH	Melts promoter with helicase activity; activates RNA polymerase I with kinase activity

Once TFIID binds to the TATA box (Fig. 8.25), a cascade of recruitment (*i.e.,* binding) of other transcription factors take place. Transcription factors TFIIA, TFIIB and TFIIF bind, as does RNA polymerase II in an unphosphorylated state. In next stage, transcription factors TFIIE and TFIIH bind, forming a **preinitiation complex (PIC),** equivalent to the *E.coli* holoenzyme (Fig. 8.24A).

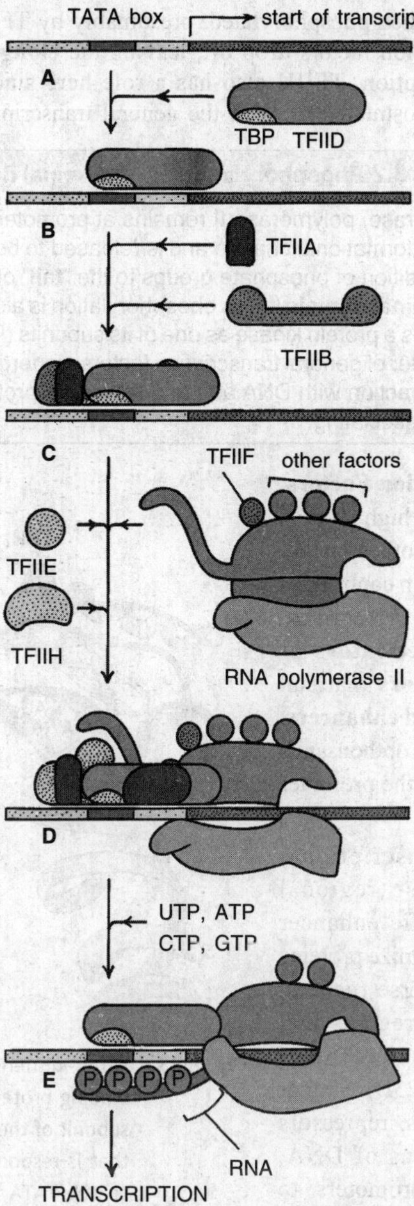

Fig. 8.24. Initiation of transcription of a eukaryotic gene by RNA polymerase II. To begin transcription, RNA polymerase requires a number of general transcription factors (called TFIIA, TFIIB and so on). A – The promoter contains a DNA sequence called the TATA box, which is located 25 nucleotide away from the site at which transcription is initiated. B – The TATA box is recognized and bound by transcription factor TFIID, which then enables the adjacent binding of TFIIB. C – For simplicity the DNA distortion produced by the binding of TFIID is not shown. D – The rest of the general transcription factors, as well as the RNA polymerase itself, assemble at the promoter. E – TFIIH then uses ATP to pry apart the DNA double helix at the transcription start point, allowing transcription to begin. TFIIH also phosphorylate RNA polymerase II, changing its conformation so that polymerase is released from the general factors and can begin the elongation phase of transcription. As shown, the site of phosphorylation is a long C – terminal polypeptide tail that extends from the polymerase molecule (after Alberts *et al.*, 2002).

The RNA polymerase II is then phosphorylated, presumably by TFIIH, which is a kinase; at this point, most of the transcription factors drop off, leaving the **elongation complex,** which carries out a basal rate of transcription. TFIIH also has a role here since it is a helicase (Box 8.2). Table 8.4 summarizes the postulated roles of the general transcription factors.

Box 8.2 Phosphorylation of C–terminal domain

Like the bacterial polymerase, polymerase II remains at promoter synthesizing short lengths of RNA until it undergoes a conformational change and is released to begin transcribing a gene. A key step in this release is the addition of phosphate groups to the "**tail**" of the RNA polymerse (which is known as the **CTD** or **C-terminal domain**). This phosphorylation is also catalysed by TFIIH, which in addition to a helicase contains a protein kinase as one of its subunits (Fig. 8.24). The polymerase can then disengage from the cluster of general transcription factors, undergoing a series of conformational changes that tighten its interaction with DNA and acquiring new proteins that allow it to transcribe for long distances without dissociating.

3. Specific Transcription Factors. For activated transcription, a high level of transcription, to take place, other factors are needed that are involved in controlling which promoters are actively transcribed. These other factors are **activators** or **specific transcription factors** that can bind to DNA sequence, called **enhancers**. Enhancers are often hundreds or thousands of base pairs upstream from the promoter (Fig. 8.23A).

These specific transcriptional activators have domains (regions) that recognize their specific enhancer sequences, regions that recognize proteins associated with the polymerse (general transcription factors) and regions that allow the joint attachment of other transcription factors (Fig. 8. 23B). Similar to activators and enhancers, repressors can bind to silencer regions of DNA, often far upstream of the promoters, to repress transcription. Thus, many genes are associated with numerous and complex arrangements of transcription factors, providing intricate control of transcription.

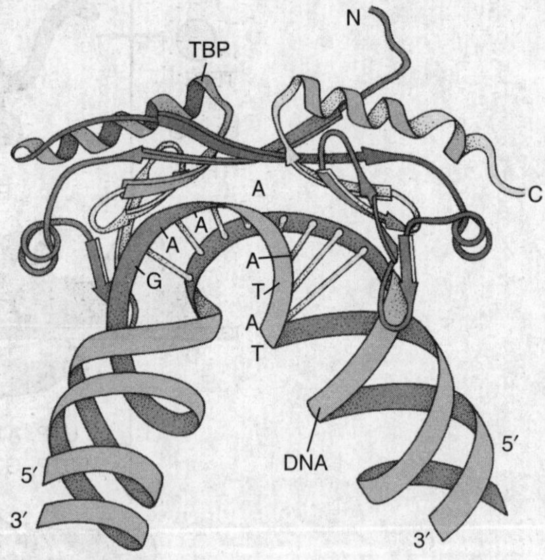

Fig. 8.25. Three–dimensional structure of TBP (TATA – binding protein) bound to DNA. The TBP is the subunit of the general transcription factor TFIID that is responsible for recognizing and binding to the TATA box sequence in the DNA. The unique DNA bending caused by TBP, *i.e.*, two kinks in the double helix separated by partly unwound DNA, may serve as a landmark that helps to attract the other general transcription factors. TBP is a single polypeptide chain that is folded into two very similar domains.

For specific transcription factors to attach to both enhancers and the polymerse machinery, possibly thousands of base pairs apart, the DNA evidently bends to allow them to come into the range of polymerse.

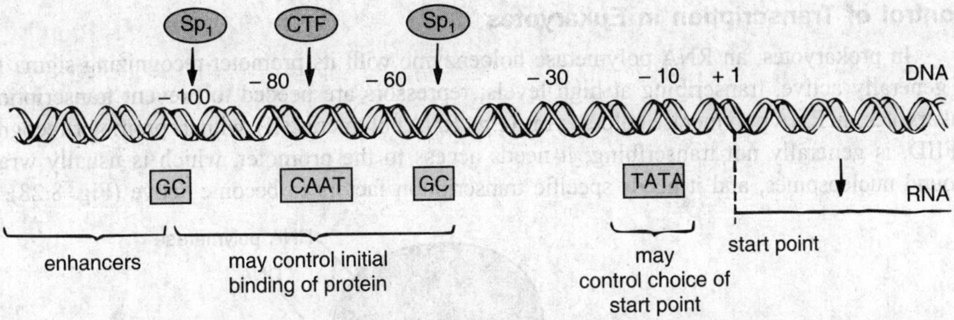

Fig. 8.26. An eukaryotic DNA segment showing promoter sites (TATA, CAT, GC boxes and enhancer sites).

Although RNA polymerase I and II seem to have termination signals similar to rho-independent promoters in prokaryotes, termination of transcription of RNA polymerse II genes is more complex, coupled with further processing of the mRNA.

Further unlike prokaryotic RNA polymerases, eukaryotic RNA polymerases do proof reading (showing 3′ → 5′ exonuclease activity). Moreover, eukaryotic DNA is complexed with histone proteins that can interfere with transcription. In turn, part of the RNA polymerase II complex is made up of proteins that can disrupt the histones bound to the DNA. In addition, the RNA polymerase II complex contains proteins that act as **mediators** between activators and the polymerase holoenzyme (Fig. 8.27). This complex coordination of the initiation of transcription in eukaryotes has been termed **combinatorial control**, the huge initiation complex may contain 100 or more different polypeptides (see **Alberts** *et al.,* 2002). Finally, transcription in archaea (archaebacteria), although under much simpler control than in eukaryotes, resembles transcription in eukaryotes rather prokaryotes. The study of details of the transcription process – its **initiation, control** and **termination** – is one of the most active and exciting areas in modern genetics.

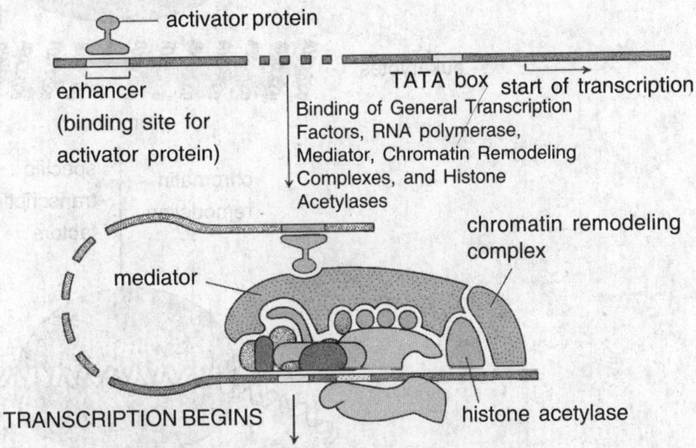

Fig. 8.27. Initiation of transcription by RNA polymerase II in a eukaryotic cell. Transcription initiation *in vivo* requires the presence of transcription activator proteins. These proteins bind activator proteins. These proteins bind to specific short sequences in DNA. A typical eukaryotic gene has many activator proteins (here only one is shown), which together determine its rate and pattern of transcription. Sometimes acting from a distance of several thousand nucleotide pairs (indicated by dashed DNA molecule), these gene regulatory proteins, help RNA polymerase, the general factors, and the mediator all to assemble at the promoter. In addition, activators attract ATP – dependent chromation remodeling complexes and histone acetylases (after Alberts *et al.,* 2002).

Control of Transcription in Eukaryotes

In prokaryotes, an RNA polymerase holoenzyme with its promoter-recognizing sigma factor is generally active, transcribing at high levels; repressors are needed to prevent transcription. In eukaryotes, an RNA polymerase holoenzyme (*e.g.*, RNA polymerase II), with its promoter-recognizing TFIID, is generally not transcribing; it needs access to the promoter, which is usually wrapped around nucleosomes, and it needs specific transcription factors to become active (Fig. 8.28).

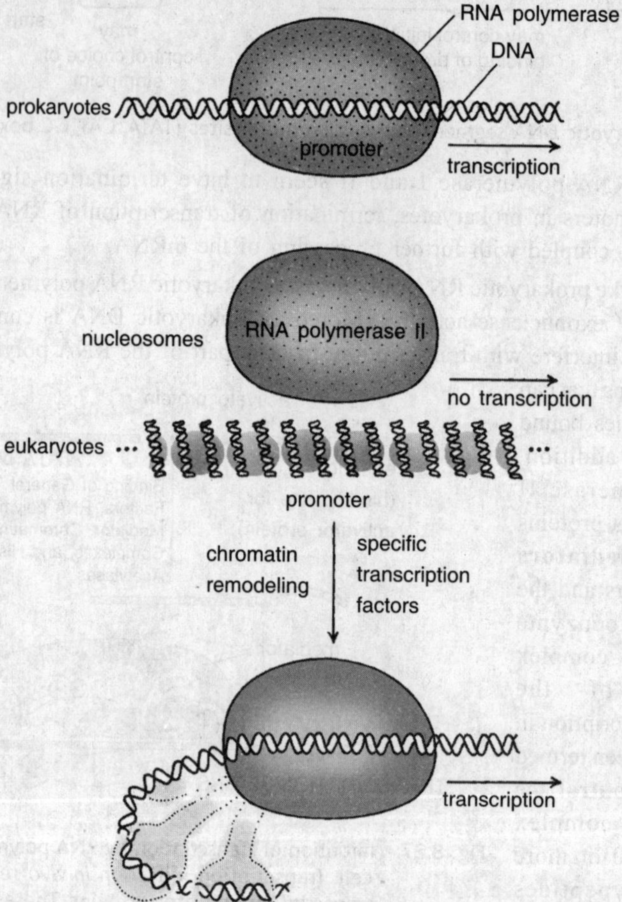

Fig. 8.28. In prokaryotes, the default condition is active transcription. In eukaryotes, the default condition is no transcription since the DNA of promoters is usually wrapped around nucleosomes and specific transcription factors are needed to recruit the polymerase holoenzyme. Transcription in eukaryotes is generally initiated when specific transcription factors bind to enhancer sequences near the promoter, and chromatin is remodeled at the promoter (after Tamarin, 2002).

Thus, although the parts of the transcribing machinery of prokaryotes and eukaryotes are generally similar, the essence of prokaryotic transcription is activity, whereas the essence of eukaryotic transcription is inactivity. In addition, eukaryotes generally do not have operons; however, groups of eukaryotic genes involved in the same pathway or function can be induced simultaneously by having common enhancers that respond to the same specific transcription factors. Such a group of genes is called a **synexpression group.**

1. Chromatin remodeling. For transcription to take place in eukaryotes the DNA must be available for the preinitiation complex to form, with its RNA polymerase and general transcription factors. It appears that DNA wrapped around nucleosomes is often not accessible for the formation of the preinitiation complex, but is available for recognition by transcription-activating proteins, also called *specific transcription factors* (as compared to general transcription factors of the RNA polymerase machinery). One model of initiation of transcription by genes whose promoters are wrapped around nucleosomes is for specific transcription factors recruit chromatin-remodeling proteins. There are two general classes of proteins that remodel nucleosomes: *histone acetyl transferases* and ATP-dependent chromatin remodeling proteins such as the *SWI/SNF complex* in yeast. Thus, the presence of one or more specific transcription factors can begin the process of transcription by recruiting chromatin-remodeling proteins that allow the RNA polymerase access to the promoter.

2. Specific transcription factors. Eukaryotic transcription is started with the formation of a preinitiation complex formed by the amalgamation of a group of general transcription factors (such as TFIID in RNA polymerase II formation). Proteins that exert control over transcription at specific promoters are the specific transcription factors (Fig. 8.23). These proteins generally have two domains: a domain that recognizes specific DNA sequence, and a domain that recognizes another protein such as a protein in the preinitiation complex. Thus, these proteins recognize signals in the vicinity of the promoter of a gene, bind there, and initiate transcription. Currently, we believe that the majority of specific transcription factors act by recruiting the components of the DNA polymerase holoenzyme. Thus, the binding of a specific transcription factor at a promoter is the first step in the formation of a preinitiation complex at the promoter of a gene. Some transcription activating proteins also recruit chromatin-remodeling proteins.

An example of a specific transcription factor is Dorsal protein, the product of the *dorsal* gene in fruit flies, active in development. Dorsal protein controls the transcription of several genes and at several different levels of protein concentration. The ability to have different effects at different concentrations is extremely important, allowing gradients of the same protein to control the expression of different genes. One gene Dorsal protein controls is the *rhomboid*, which has three sites in its promoter that Dorsal protein binds to, initiating transcription. Another gene, *twist*, also has three sites in its promoter that bind Dorsal protein, also initiating transcription. However, the *rhomboid* sites are more efficient in binding Dorsal protein; thus, *rhomboid* is transcribed at lower concentrations of Dorsal protein than *twist* is (Fig. 8.29). One other signal in the control of transcription that is of current interest is methylation.

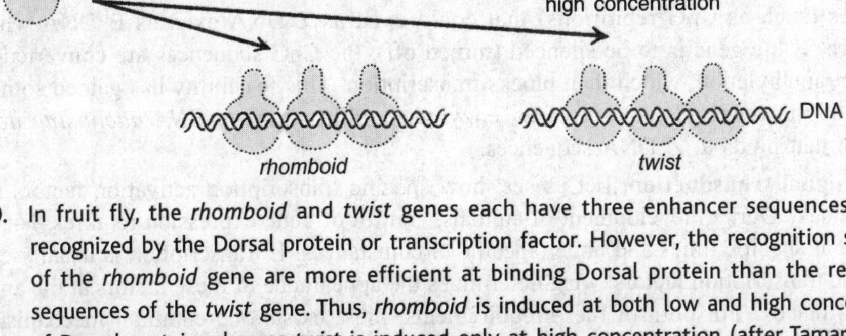

Fig. 8.29. In fruit fly, the *rhomboid* and *twist* genes each have three enhancer sequences that are recognized by the Dorsal protein or transcription factor. However, the recognition sequences of the *rhomboid* gene are more efficient at binding Dorsal protein than the recognition sequences of the *twist* gene. Thus, *rhomboid* is induced at both low and high concentrations of Dorsal protein whereas *twist* is induced only at high concentration (after Tamarin, 2002).

3. Methylation of DNA. The importance of methylation in DNA-protein interactions is well known. Molecular biologists have showed that a particular DNA sequence could be protected from restriction endonucleases if it was methylated. A small percentage of cytosine residues are methylated in many eukaryotic organisms, mainly in CpG sequences; 80% of the cytosines in CpG sequences in human DNA are methylated. (**Note.** Often when we refer to a sequence of two bases on the same strand of DNA, we put a "p" between them –CpG to indicate that they are on the same strand connected by a phosphodiester bond and not on two different strands as a hydrogen-bonded base pair).

The *degree* of methylation of DNA is related to the silencing of gene. Genes that are dormant at one stage of development but active in another, are usually less methylated when active and more fully methylated when inactive. For example, adenovirus, a cancer-causing virus, has been observed in many eukaryotic cell lines. In most lines in which the adenovirus DNA has integrated into the host chromosome, late viral genes are turned off. These genes are highly methylated at their CCGC or GCGC sites.

Fig. 8.30. *Arabidopsis thaliana.*

In addition, chemicals that prevent methylation frequently activate previously dormant genes. For example, a chemical compound, called *5-azacytidine*, inhibits methylation; X chromosomal genes, which are normally deactivated, can be reactivated by treatment with 5-azacytidine. There are numerous other examples of the activation of genes after treatment with this chemical. The activated genes lack methylated cytosines that were previously methylated. Finally, the possibility exists that DNA methylation can affect chromatin structure.

Recent work has also indicated that the methylation itself may not prevent transcription, but rather may be signal for transcriptional inactivity. In the thale cress plant, *Arabidopsis thaliana* a protein named Mom (for *Mo*rpheus molecule) has been discovered that, when mutated, results in genes that have heavy methylation level but are actively transcribed. Thus, the methylation level can be separated from the transcriptional activity of genes, although the two usually occur together. *Arabidopsis* is providing to be a good model in the study of the role of methylation in transcriptional activation because other common model organisms, namely fruit flies, yeast and the nematode, *Caenorhabditis elegans*, do not have methylation of their DNA.

Further interest has been generated in the role of methylation in *regulating* gene expression by the discovery of Z DNA, and the fact that Z DNA can be stabilized by methylation. This observation has led to a model of transcriptional regulation based on alternative DNA structures. Sequences (such as CpG repititions) that could exist as Z DNA exist as B DNA when being transcribed. If the gene is to be silenced (turned off), the CpG sequences are converted to stable Z DNA by methylation, which then blocks transcription. This feasibility has gained some interest because of the recent discovery of an enzyme, double stranded *RNA adenosine deaminase (ADAR1),* that binds to Z DNA sequences.

4. Signal transduction. Let us see, how specific transcription activation factors appear at specific times? During development of animals, control of gene expression requires that genes be expressed at specific times and under specific circumstances. If transcription is usually controlled by specific transcription factors, what determines the appearance of these factors at the appropriate times and places? For both of these requirements, there exists one common mechanism, called

signal transduction pathway, in which signals pass from the external environment through the cytoplasm, into the nucleus.

For example, in a signal transduction pathway involved in development of the fruit fly, the *Toll protein* spans the plasma membrane (Fig. 8.30). It acts as a receptor for the *Spatzle protein,* which when detected, causes a change in the cytoplasmic end of Toll protein, activating it. Activated Toll protein activates *Pelle*, a protein kinase that phosphorylates the *Cactus* protein, causing it to dissociate from Dorsal protein. Once Dorsal protein dissociates from Cactus protein (which acts to repress Dorsal protein), Dorsal protein becomes an active specific transcription factor that can cross the nuclear membrane and activates its target genes (Fig. 8.31). We thus see that Spatzle protein attaching to its receptor protein (*i.e.*, Toll protein) on the cell surface results in activation of the target gene of the Dorsal protein in the nucleus. These pathways can become very complex, with many protein elements. More elements mean more sensitive control of various processes, often requiring that several conditions be met before gene is activated. In addition, these pathways are usually conserved by evolution. A similar pathway, though more complex, occurs in mammals in which the target gene is *interleukin-1,* a protein in the immune system that induces fever. In the mammalian protein, the signal protein is called Toll-like receptor-4 and the specific transcription factor is called NF-kB (see Tamarin, 2002).

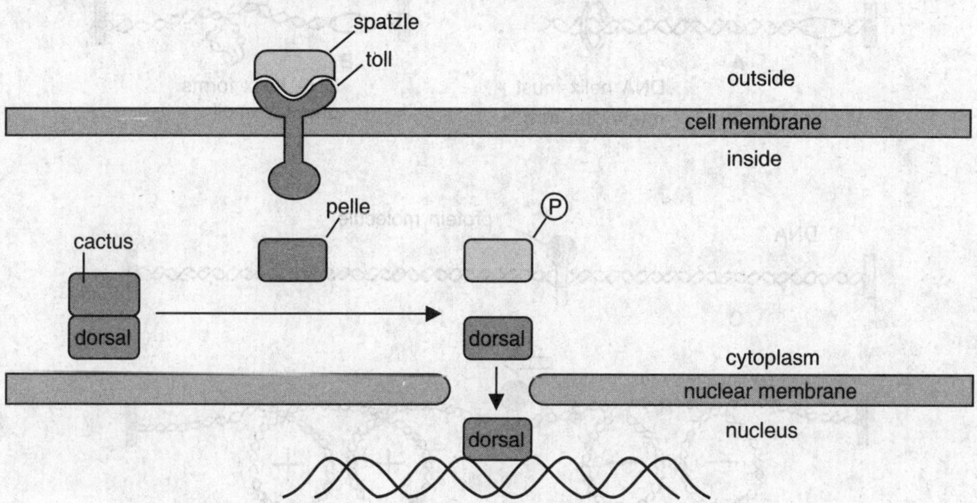

Fig. 8.31. Signal transduction pathway. In this mechanism, the spatzle protein outside of a cell intereacts with the Toll receptor protein, freeing the Dorsal protein to act as a transcription factor in the nucleus. When Spatzle protein, binds to Toll protein which spans the plasma membrane, it changes the configuration of the interior domain of Toll protein, which then interacts with pelle protein, causing it to phosphorylate the Cactus protein. Previously, cactus protein had been bound to Dorsal protein, making Dorsal protein inactive; phosphorylation of Cactus protein releases it from Dorsal protein. Dorsal protein is then free to cross the nuclear membrane and act as a transcription factor (after Tamarin, 2002).

5. Transcription elongation produces super helical tension of DNA. Once it has initiated transcription, RNA polymerase does not proceed smoothly along a DNA molecule, rather it moves jerkily, pausing at some sequences and rapidly transcribing through others. RNA polymerases involved in *RNA-elongation* both bacterial and eukaryotic, are associated with a series of *elongation factors,* proteins that decrease the likelihood that RNA polymerase will dissociate before it reaches the end of a gene. These factors typically associate with RNA polymerase shortly after initiation has occurred and help polymerases to move through the wide variety of different DNA sequences that are found in genes. Eukaryotic RNA polymerases must also cope with chromatin structure as they move along a DNA template. Experiments have shown that bacterial RNA polymerases, which never encounter nucleosomes *in vivo*, can nonetheless transcribe through them *in vitro*, suggesting that a nucleosome in easily traverses. However, eukaryotic polymerases have to move through forms of chromatin that are more compact than a simple nucleosome. It therefore seems likely that they transcribe with the aid of *chromatin remodeling complexes*. These complexes may move with the RNA polymerases or may simply seek out and rescue the occasional stalled RNA polymerase. In addition, some elongation factors associated with eukaryotic RNA polymerase facilitate transcription through nucleosomes without requiring additional energy. It is not yet understood how this is accomplished, but these proteins may help to dislodge parts of the nucleosome core as the RNA polymerase transcribes the DNA of a nucleosome.

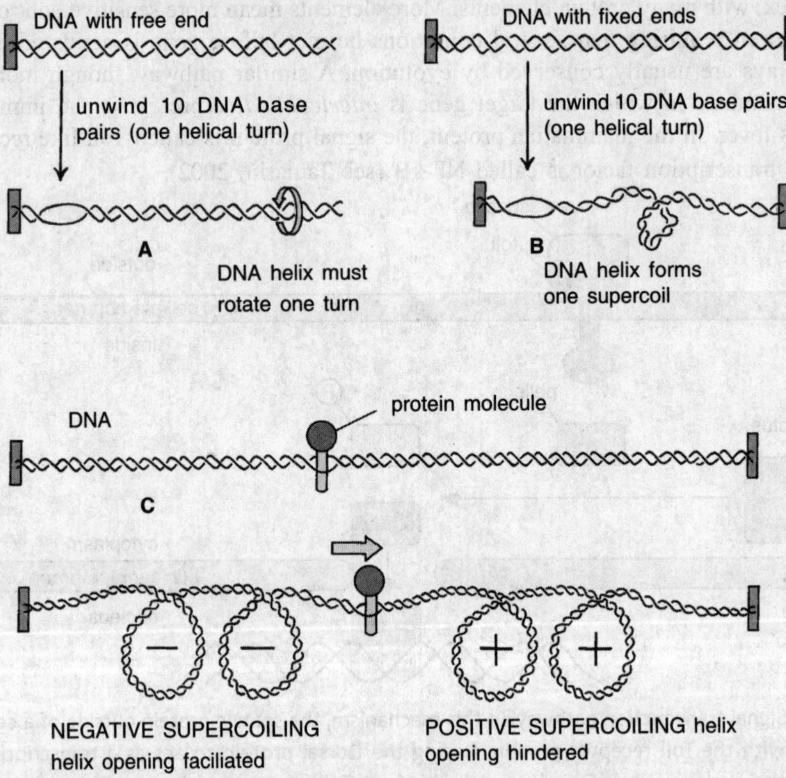

Fig. 8.32. Superhelical tension in DNA causes DNA super coiling A-For a DNA molecule with one free end (or a nick in one strand that serves as a swivel), the double helix rotates by one turn for every 10 nucleotide pairs opened; B-If rotation is prevented, superhelical tension is introduced into the DNA by helix opening. One way of accommodation this tension would be to increase the helical twist from 10 to 11 nucleotide pairs per turn in the double helix that remains in this example; the DNA helix, however, resists such a deformation in a spring-like fashion, preferring to relieve the superhelical tension by bending into super coiled loops. As a result, one DNA supercoil forms in that DNA double helix for every 10 nucleotide pairs opened. The super coil forms in this case is positive super coil. C-Supercoiling of DNA is induced by a protein tracking through the DNA helix. The two ends of the DNA show here are unable to rotate freely relative to each other, and the protein molecule is assumed also to be prevented from rotating freely as it moves. Under these conditions, the movement of the protein causes an excess of helical turns to accumulate in the DNA helix ahead of the protein and a deficit of helical turns to arise in the DNA behind the protein as shown (after Alberts *et al.*, 2002).

Elongating RNA polymerases also have to phase out some another type of barrier, both bacterial and eukaryotic. To discuss this issue, we need first to consider a refined property inherent in the DNA double helix called *DNA supercoiling.* DNA supercoiling represents a conformation that DNA will adopt in response to superhelical tension; conversely, creating various loops or coils in the helix which can create such tension. A simple way of visualizing the topological constraints that cause DNA supercoiling is illustrated in Figure 8.32A. There are approximately 10 nucleotide pairs for every helical turn in a DNA double helix. In a gene, a helix whose two ends are fixed (Fig. 8.32B) with respect to each other (as they are in a DNA circle, such as bacterial chromosome, or in a tightly clamped loop, as thought to exist in eukaryotic chromosomes). In this case, one large DNA supercoil will form to compensate for each 10 nucleotide pairs that are opened (unwound). The formation of this super coil is energetically favourable because it restores a normal helical twist to the base paired regions that remain which woned otherwise need to be overwound because of the fixed ends.

Superhelical tension is also created as RNA polymerase moves along a stretch of DNA that is anchored at its both ends (Fig. 8.32). As long as the polymerase is not free to rotate rapidly (and such rotation is unlikely given the size of DNA polymerases and their attached transcripts), a moving polymerase generates *positive* superhelical tension in the DNA in front of it and *negative* helical tension behind it. For eukaryotes, this situation is thought to provide a bonus: the positive superhelical tension ahead of the polymerase makes the DNA helix more difficult to open but this tension should facilitate the unwrapping of DNA in nucleosomes, as the release of DNA from the histone core helps to relax positive superhelical tension.

Any protein that propels itself along a DNA strand of a double helix tends to generate superhelical tension. In eukaryotes, DNA topoisomerase enzyme rapidly removes this superhelical tension. But, in bacteria, a specialized topoisomerase called *DNA gyrase* uses the energy of ATP hydrolysis to pump super coils continuously into the DNA, thereby maintaining the DNA under constant tension. These are *negative supercoils,* having the opposite handedness from the *positive supercoils* that form when a region of DNA helix opens (see Fig. 8.32C). These negative super coils are removed from bacterial DNA whenever a region of helix opens, reducing the superhelical tension. DNA gyrase therefore makes the opening of the DNA helix energetically favourable compared with helix opening in DNA that is not supercoiled. For this reason, it usually facilitates those genetic processes in bacteria, including the initiation of transcription by bacterial RNA polymerase, that require helix opening.

QUESTIONS

Long Answer Questions

1. Describe the process of initiation, elongation and termination during prokaryotic and eukaryotic transcription.
2. Describe transcription of the information from DNA to mRNA.
3. Describe the structure of *E.coli* RNA polymerase and discuss the roles of different components of their enzyme in RNA synthesis on DNA template.
4. Describe the structure and functions of three RNA polymerases known in eukaryotes.
5. Describe the promoter sites for initiation of transcription in prokaryotes and eukaryotes.
6. What are transcription factors? Describe them for three different RNA polymerases in eukaryotes.
7. What is Pribnow box? Describe its evolutionary and biochemical significance.
8. Give an account of control of transcription in eukaryotes.

Short Answer Questions

1. Given a single strand of DNA ... 3′ TACCGAGTAC 5′...., construct
   (*a*) the complementary DNA chain,
   (*b*) the mRNA chain which would be made from strand.
2. What are the functions of the core enzyme and the holoenzyme *in vivo.?*
3. Describe specific transcription factors.
4. Write short notes on the following:
   (*i*) General transcription factors; (*ii*) Specific transcription factors;
   (*iii*) Methylation of DNA; (*iv*) Signal transduction;
   (*v*) Superhelical tension in DNA.

Very Short Answer Questions

1. What is TATA box?
2. What is consensus sequence?
3. Define Pribnow box?
4. What is preinitiation complex?
5. Define transcription factors.
6. (*a*) From what substrates is RNA made? (*b*) On what template?
   (*c*) By which enzyme? (*d*) Is a primer required?
7. Describe the differences, if any, between the chemical reactions catalyzed by DNA polymerase and RNA polymerase.
8. What is a transcription unit? Is it the same thing as a gene?
9. In what way, relating to polycistronic mRNA, do eukaryotic and prokaryotic protein synthesis differ?
10. What is meant by the term "upstream" and "downstream"?
11. An RNA molecule is isolated having 3′–OH terminus and 5′–P terminus. What information does this fact provide?
12. What is a consensus sequence? What is a conserved sequence?
13. Why do you think that most promoter regions are A-T rich?
14. What is the difference between a σ-70 and a σ-32?
15. What is footprinting? How did it help to define promoter?
16. Can one nucleotide be a conserved sequence?
17. What is a stem-loop or hair-pin structure? An inverted repeat? A tandem repeat? Draw a section of a DNA double helix with an inverted repeat of seven base pairs.

Fill in the Blanks

1. .... is the process by which RNA molecules are initiated, elongated and terminated.
2. In prokaryotes, ... type of RNA polymerase enzyme is responsible for synthesis of all kinds of RNA.
3. Prokaryotic RNA polymerase is a .... and has .... subunits.
4. During transcription, a molecule of RNA polymerase enzyme covers a region of about .... base pairs of DNA.
5. Transcription starts from the .... of sense strand of DNA which is recognized by .... of RNA polymerase enzyme.

6. In transcription, ... bind to nitrogen bases of sense strand of DNA in .... direction.

7. In eukaryotes, these are .... major classes of RNA polymerases.

8. TATA box of eukaryotic promoter is similar to .... in prokaryotes.

9. .... is base sequence in DNA to which RNA polymerase binds when initiating transcription.

Multiple Choice Questions

1. Transcription is the transfer of genetic information from

(*a*) chromosome to cytoplasm (*b*) tRNA to mRNA

(*c*) DNA to mRNA (*d*) mRNA to rRNA

2. Nongenetic RNA is of

(*a*) one type (*b*) two types

(*c*) three types (*d*) non functional type

3. Regarding transcription which one of these is false?

(*a*) uses RNA polymerase

(*b*) one of the mother strands of DNA is the template

(*c*) is bidirectional

(*d*) reverse transcription used RNA dependent DNA polymerase

4. The enzyme required for transcription is

(*a*) DNA polymerase (*b*) RNA polymerase

(*c*) endonuclease (*d*) all the above

5. Pribnow box contribute to

(*a*) protein synthesis (*b*) ATP synthesis

(*c*) RNA synthesis (*d*) none of the above

6. The process involved in the RNA formation on DNA template is

(*a*) translation (*b*) transduction

(*c*) transcription (*d*) transformation

7. Transcription takes place in

(*a*) matrix (*b*) nucleus

(*c*) cytosol (*d*) cytoplasm

8. Essential components of eukaryotic cistron are

(*a*) intron

(*b*) exons

(*c*) operons

(*d*) operator and regulator genes

9. After initiation of transcription with core enzyme RNA polymerase the sigma factor is

(*a*) functionless (*b*) released to take part again

(*c*) used during closing of chain (*d*) retained and it performs special function

10. During transcription, the DNA site at which RNA polymerase binds is called

(*a*) promoter (*b*) regulator

(*c*) receptor (*d*) enhancer

ANSWERS

Very Short Answer Questions

1. TATA box is part of the core promoter for many eukaryotic genes transcribed by RNA polymerase II. It consists of a consensus sequence of TATA followed by two or three more A's.

2. It is a sequence of the common nucleotides found in many different DNA or RNA samples of homologous regions (*e.g.,* promoters).

3. It is consensus sequence of TATA at in prokaryotic promoters centred at position–10 of cistron.

4. Preinitiation complex is the form of the RNA polymerase II enzyme with general transcription factors bound equivalent to the *E. coli* holoenzyme. Phosphorylation of the enzyme then allows transcription to begin.

5. Eukaryotic proteins that aid RNA polymerase enzyme in recognizing promoters.

6. (*a*) Ribonucleoside 5' triphosphates.

(*b*) Double-stranded DNA

(*c*) RNA polymerase.

(*d*) No.

7. The reactions are identical, however, substrates are different, *i.e.*, DNA polymerase joins deoxynucleotides and RNA polymerase joins ribonucleotides.

8. A transcription unit is a section of DNA extending from a promoter to an RNA polymerase termination site. It is usually not a gene, but typically includes many genes.

9. In eukaryotes all mRNA is monocistronic.

10. Upstream and downstream usually refer to regions in the 5′and 3′ directions respectively, from a particular site that is being discussed.

11. Since the triphosphate of the primary transcript is absent, the molecule has been processed.

12. A *consensus sequence* is made up of the nucleotides that appear in a significant proportion of cases when similar sequences are aligned. A *conserved sequence* consists of nucleotides found in all cases in when similar sequences are aligned . For example, the Pribnow box is the consensus sequence TATAAT.

13. A double helix of DNA must unwind in order for transcription to occur. A-T pairs, because they have only two H-bonds, are more easily disrupted than G-C pairs.

14. The superscripts of the sigma factors refers to their molecular weights (e.g., σ-70 is 70,000 Daltons). Different sigma factors usually recognize different prokaryotic promoters.

15. **Footprinting** is a technique in which DNA in contact with a protein is exposed to nucleases; only DNA protected by the protein is undigested. Promoters could be isolated by protection with RNA polymerase in the absence of ribonucleotides-the polymerase will not move and then sequenced.

16. Conserved sequences are invariant sequences of DNA or RNA recognizable to either a protein or a complementary sequence of DNA or RNA. However, in group II introns, an adenine (A) is needed near the 3′ end of the intron for lariat formation. Thus, this single nucleotide, given its relative position in the intron and possible surrounding bases, is a conserved sequence of one.

17. A stem-loop (or hair-pin) structure can form when a single strand of DNA or RNA has a double helical section. An *inverted repeat* is a sequence read outward on both strands of a double helix from a central point. A *tandem repeat* is a segment of nucleic and repeated consecutively; that is, the same sequence repeats in the same directions on the same strand:

5′- TCCGGTCCGGTCCGG-3′

3′- AGGCCAGGCCAGGCC-5′

A DNA sequence with a *seven-base inverted repeat* is

5′- ATTACCGCGGTAAT-3′

3′- TAATGGCGCCATTA-5′

Fill in the Blanks

1. Transcription.

2. Single.

3. Holoenzyme, six.

4. Sixty.

5. Promoter site, sigma (σ) factor.

6. Ribonucleotides, $5' \rightarrow 3'$.

7. Three.

8. –10 sequence.

9. Promoter site.

Multiple Choice Questions

1. (*c*)	**2.** (*c*)	**3.** (*c*)	**4.** (*b*)	**5.** (*c*)	**6.** (*c*)
7. (*b*)	**8.** (*d*)	**9.** (*b*)	**10.** (*a*).		

9

Messenger RNA (Coding RNA) (mRNA: Modification, Processing, Editing, and Splicing)

Messenger RNA (mRNA) or **coding RNA** acts as the intermediate molecule between the gene and the polypeptide translation product. Its existence was postulated by **Crick** and his associates during the 1950s. Among the indirect evidence for mRNA at that time was the knowledge that *in eukaryotes the genes reside on the chromosomes in the nucleus, whereas protein synthesis occurs in ribosomes in the cytoplasm.* The physical separation of genes and ribosomes means that some sort of messenger molecule must carry the biological information from nucleus to cytoplasm. In bacteria, the physical separation of DNA and ribosomes is less distinct but nevertheless a messenger molecule is still necessary.

Evidence that this messenger is in fact RNA came first from **Elliot Volkin** and **Lazarus Astrachan** in 1956, but more convincingly from **Sol Spiegelman** and **Benjamin D.Hall** in 1961. Both groups demonstrated that after infection of a culture of bacteria with a bacteriophage, the new RNA that is synthesized is related in sequence to the phage DNA, suggesting that the phage genes are copied into RNA before synthesis of the phage proteins occurs. Shortly afterward two independent groups-**Sydney Brenner, Francois Jacob** and **Mathew Meselson**, who did the crucial experiment at the California Institute of Technology and **James Watson's** group at Harvard-directly identified mRNA molecules in *E. coli* cells.

9.1 LIFE SPAN OF mRNA

Messenger RNA molecules are not generally long lived in the cell *i.e.*, most of them are unstable. Most bacterial mRNAs have a **half life** of only a few minutes and so are turned over very rapidly. In eukaryotic cells, mRNA molecules have longer half-lives (*e.g.*, 6 hours for many mammalian mRNAs) but are still subject to turnover. Turnover of mRNA is important because it means that the absolute amount of a particular mRNA in the cell can be controlled by adjusting the rate at which the relevant gene is transcribed. If the transcription rate for the gene decreases then the level of the mRNA in the cell also decreases until a new steady state is reached (Fig. 9.1).

In fact, in both bacteria and eukaryotes mRNAs are known that are apparently resistant to cytoplasmic ribonuclease enzyme and survive for long period of time. For example, mRNA with life time of six hours has been detected in the bacterium *Bacillus cerus* at a time when the cells are induced to become long lived spores. Likewise, in differentiating eukaryotic cells mRNAs with a life time of days have been detected. For example, in the immature red blood cells (reticulocytes) of the mammals the mRNA is synthesized originally by the nucleus in early stages and expelled to the cytoplasm. In later stages, the nuclei of maturing reticulocytes degenerate but the mRNA exists up to 2 days for prolong utilization in the synthesis of globin protein of haemoglobin. Further, in extreme cases, such as in the state of dormancy adopted by many animal eggs and plant seeds, mRNA is maintained in a stable form for months or even years.

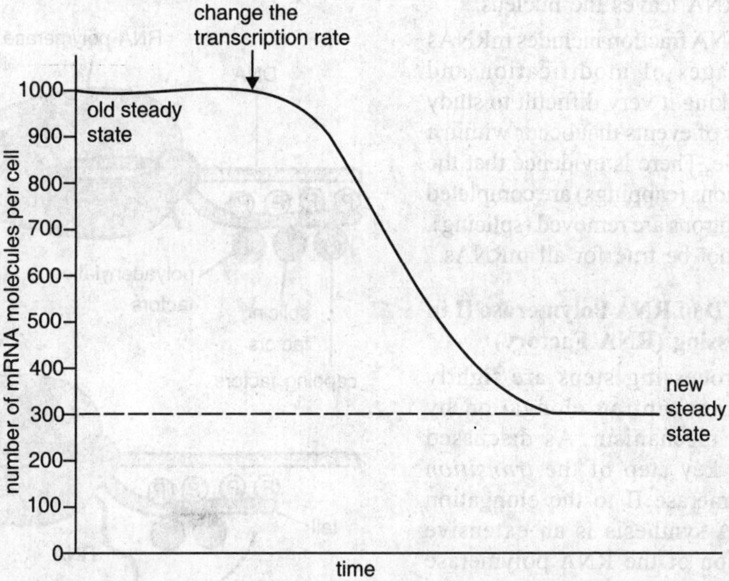

Fig. 9.1. Effect of transcription rate on the mRNA steady state in the cell. In this example, the half-life of the mRNA molecule is 5 minutes. The initial transcription rate is 100 molecules of mRNA synthesized per minute, leading to a steady state of 1000 molecules per cell. The transcription rate then decreases to 30 molecules of mRNA synthesized per minute; the steady state gradually drops until the new value of 300 molecules per cell is reached (after Brown, 1998).

9.2 MODIFICATION AND PROCESSING OF mRNA

Most eukaryotic mRNAs undergo a quite complicated series of modifications and processing events before translation occurs. These include:

1. Chemical modification to the two ends of the mRNA molecule (*Capping;* Fig. 9.2).
2. Polyadenylation
3. Removal of introns (*Splicing*)
4. In a few special cases, alterations to the nucleotide sequence of the mRNA (*RNA editing*).

These events occur only in eukaryotes. The first clue that eukaryotic mRNAs undergo modification and processing came when RNA fractions present in the nucleus and cytoplasm were compared. The nucleus can be divided into two distinct regions:

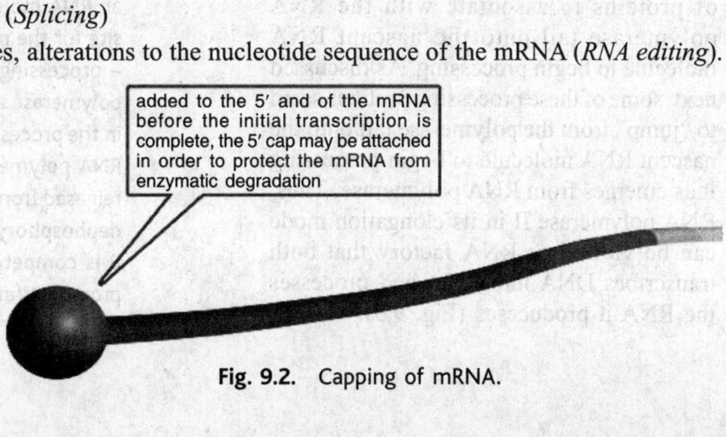

Fig. 9.2. Capping of mRNA.

the *nucleolus,* in which rRNA genes are transcribed, and the *nucleoplasm,* where other genes, including those for mRNA, are transcribed. The nucleoplasmic RNA fraction is called *heterogenous nuclear RNA,* the name indicating that it is made up of a complex mixture of RNA molecules, some over 20 kb in length. The mRNA in the cytoplasm is also heterogenous, but its average length is only 2 kb. If the mRNA, in the cytoplasm is derived from hnRNA, then modification and processing events including a reduction in the length of the primary transcripts, must occur before the mRNA leaves the nucleus.

The hnRNA fraction includes mRNAs at various stages of modification and processing making it very difficult to study the exact order of events that occur within a single molecule. There is evidence that the end-modifications (cappings) are completed before all the introns are removed (splicing), but this may not be true for all mRNAs.

1. Role of CTD of RNA Polymerase II in mRNA Processing (RNA Factory)

RNA processing steps are tightly coupled to transcription elongation by an ingenious mechanism. As discussed previously, a key step of the *transition* of RNA polymerase II to the elongation mode of RNA synthesis is an extensive phosphorylation of the RNA polymerase II tail, called the CTD. This C-terminal domain of the largest subunit consists of a long tandem array of a repeated seven-amino acid sequence, containing two serines per repeat that can be phosphorylated. Because there are 52 repeats in CTD of human RNA polymerase II, its complete phosphorylation would add 104 negatively charged phosphate groups to the polymerase II. This phosphorylation step not only dissociates the RNA polymerase II from other proteins present at the start point of transcription, it also allows a new set of proteins to associate with the RNA polymerase tail onto the nascent RNA molecule to begin processing. As discussed next, some of these processing proteins seem to "jump" from the polymerase tail onto the nascent RNA molecule to begin processing it as emerges from RNA polymerase. Thus, RNA polymerase II in its elongation mode can be viewed as RNA factory that both transcribes DNA into RNA and processes the RNA it produceses (Fig. 9.3).

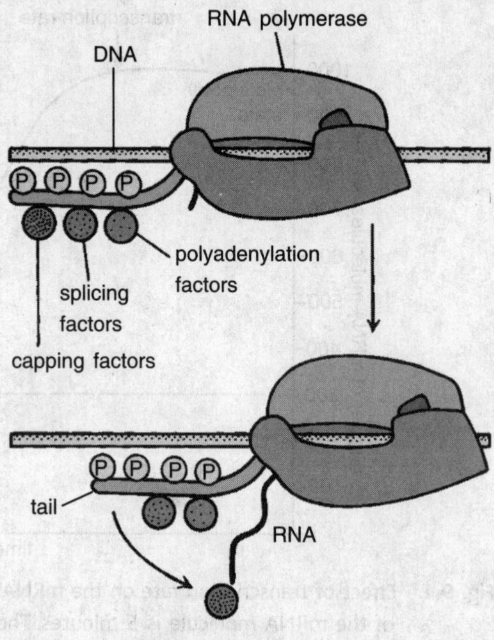

Fig. 9.3. The "RNA factory" concept for eukaryotic RNA polymerase II. Not only does the RNA polymerase II transcribe DNA into RNA, but it also carries pre-mRNA-processing proteins on its tail, which are then transferred to the nascent RNA at the appropriate time. There are many RNA – processing enzymes, and not all travel with the polymerase. For RNA splicing, for example, only a few critical components are carried on the tail, once transferred to an RNA molecule, they serve as a nucleation site for the remaining components. The RNA – processing proteins first bind to the RNA polymerase tail when it is phosphorylated late in the process of transcription initiation. Once RNA polymerase II finishes transcribing, it is released from DNA, the phosphates. Only this dephosphorylated form of RNA polymerase II is competent to start RNA synthesis at a promoter (after Alberts *et al.*, 2002).

2. RNA Capping

As soon as RNA polymerase II has produced about 25 nucleotides of RNA, the 5' end of the new RNA molecule is modified of a "cap" that consists of a modified guanine nucleotide (*i.e.*, 7–methyl G). The capping reaction is performed by *three* enzymes acting in succession: One enzyme called *phosphatase* removes one phosphate from the 5′ end of the nascent RNA, another enzyme (a *guanyl transferase*) adds a GMP in a reverse linkage (5′ to 5′ instead of 5′ to 3′) and a third enzyme (a *methyl transferase*) adds a methyl group to the guanosine (Fig. 9.4). Because all three enzymes bind to the phosphorylated RNA polymerase tail, they are ready to modify the 5′ end of the nascent transcript as soon as it emerges from the polymerase.

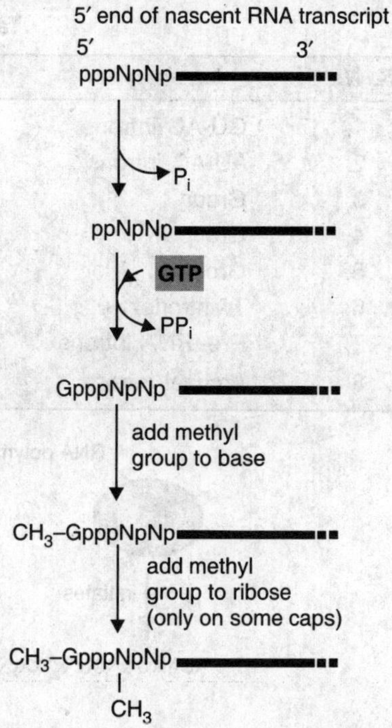

Fig. 9.4. The reactions that cap the 5′ end of each RNA molecule synthesized by RNA polymerase II. The final cap contains a novel 5′- to – 5′ linkage between the positively charged 7–methyl G residue and the 5′ end of the RNA transcript. The latter N represents any one of the four ribonucleotides, although the nucleotide that starts an RNA chain is usually a purine (an A or a G) (after Alberts *et al.*, 2002).

The 5′-methyl cap protects mRNA against the attack of phosphatases and nucleases (**Shatkin**, 1976). Also, the cap signals the 5′ end of eukaryotic mRNA, and this landmark helps the cell to distinguish mRNA, from the other types of RNA molecules present in the cell. For example, RNA polymerase I and III produce uncapped RNAs during transcription, in part because their polymerases lack the tail. In the nucleus, the cap binds a protein complex called *CBC* (*cap binding complex*) which helps the RNA to be properly processed and exported. The 5′ methyl cap also has an important role in translation of mRNA in the cytosol.

3. Polyadenylation

At the other end, the 3′ end of polymerase II transcripts, a sequence of twenty to two hundred adenine containing nucleotides, known as a **poly-A tail,** is added by the enzyme **poly-A polymerase.** This step is called **polyadenylation** which takes place after the 3′ end of the transcript is removed by a nuclease that cuts about twenty nucleotides downstream from the signal 5′–AAUAAA–3′ (Fig. 9.5). The tail adds stability to the molecule and aids in its transportation from the nucleus.

4. Introns and RNA Splicing

The protein coding sequences of eukaryotic genes are typically interrupted by noncoding intervening sequences (introns). Discovered in 1977, this feature of eukaryotic genes came as a surprise to molecular biologists, who had been until that time, familiar only with bacterial genes, which typically consists of a continuous stretch of coding DNA that is directly transcribed into mRNA. In marked contrast, eukaryotic genes were found to be broken up into small pieces of codon sequence (*expressed sequences or exons*) interspersed with much longer *intervening sequences or introns;* thus, the coding portion of a eukaryotic gene is often only a small fraction of the length of the gene (*e.g.*, globin gene, see Fig. 9.6).

Types of introns. Molecular biologists have recognized seven distinct types of introns in eukaryotes, and additional forms in the archaea (Table 9.1). Two of these types, namely GV-AG and AU-AC introns, are found in eukaryotic protein coding genes.

Table 9.1 Types of introns.

S. No.	*Intron type*	*Where found*
1.	GU-AG introns	Eukaryotic nuclear pre-mRNA
2.	AU-AC introns	Eukaryotic nuclear pre-mRNA
3.	Group I	Eukaryotic nuclear pre-rRNA, few bacterial RNAs
4.	Group II	Organelle RNAs, some prokaryotic RNAs
5.	Group III	Organelle RNAs
6.	Twintrons	Organelle RNAs
7.	Pre-tRNA introns	Eukaryotic nuclear pre-tRNA
8.	Archael introns	Various RNAs

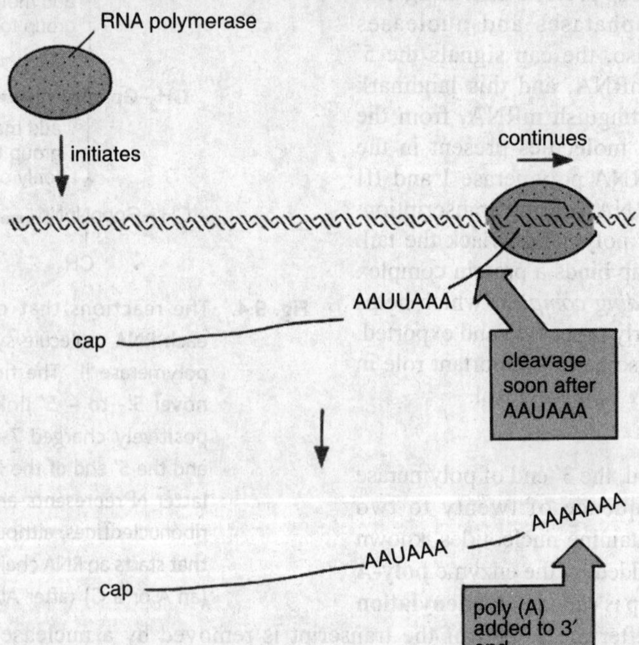

Fig. 9.5. Cleavage of signal AAUAAA sequence and polyadenylation at the cleaved 3′ end.

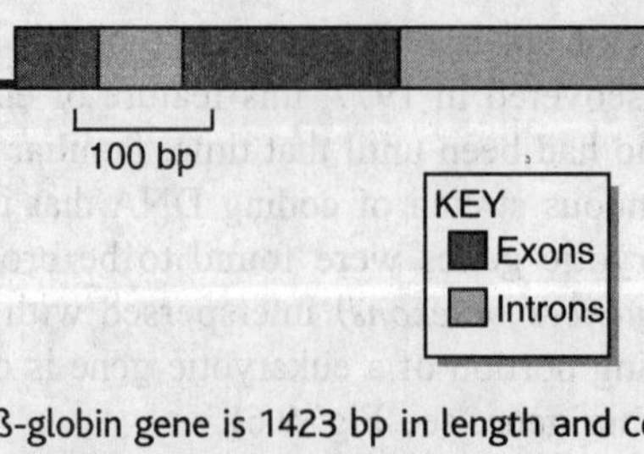

Fig. 9.6. Introns. Human β-globin gene is 1423 bp in length and contains two introns, one of 131 bp and other of 851 bp, which together make up 69% of the length of the gene (after Brown, 2002).

In protein-coding genes, introns are less common in lower eukaryotes: the 6000 genes in the yeast genome contain only 239 introns in total, whereas many individual mammalian genes contain 50 or more introns. A eukaryotic pre-mRNA may contain many introns, perhaps over 100, taking up a considerable length of the transcript (Table 9.2). These introns must be excised and the exons joined together in the correct order before the transcript can function as a mature mRNA (Fig. 9.7).

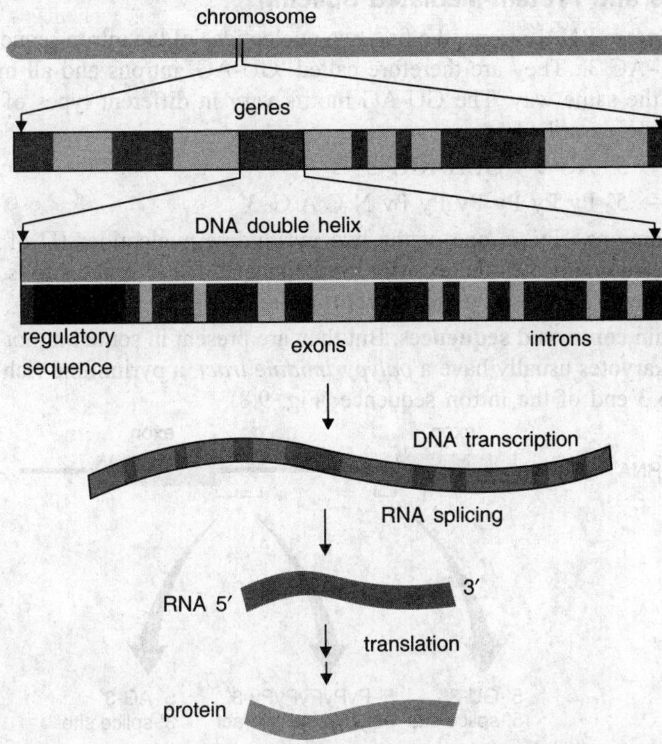

Fig. 9.7. Only exons are included in mRNA which translates into protein.

Table 9.2 Introns in human genes (*Source:* **Brown**, 2002).

S. No.	*Gene*	*Length (kb)*	*Number of introns*	*Amount of the gene taken up by the introns (%)*
1.	Insulin	1.4	2	69
2.	β-globin (of haemoglobin)	1.6	2	61
3.	Serum albumin	18	13	79
4.	Type VII collagen	31	117	72
5.	Factor VIII	186	25	95
6.	Dystrophin*	2400	78	98

*A transmembrane glycoprotein complex of skeletal muscle cells and is absent in Duchenne muscular dystrophy and deficient in Becker muscular dystrophy.

Evolution of introns. When the same gene is compared in related species, molecular biologists find that some of the introns are in identical positions but that each species has one or more unique introns. This implies that some introns remain in place for millions of years. Retaining their positions while species diversify, whereas others appear or disappear during this same period. This leads to two opposing hypothesis for the evolution of introns:

1. 'Intron late' hypothesis. This hypothesis holds that introns evolved relatively recently and are gradually accumulating in eukaryotic genomes.

2. 'Introns early' hypothesis. This hypothesis suggests that introns are very ancient and are gradually being lost from eukaryotic genomes.

A. GU-AG Introns and Protein-Mediated Splicing

In most of the pre-mRNA introns, the first two nucleotides of the intron sequence are 5′-GU-3′and the last two 5′-AG-3′. They are therefore called 'GU-AG' introns and all members of this class are spliced in the same way. The GU-AG motifs vary in different types of eukaryotes; in vertebrates they can be described as:

5′ splice site → 5′–AG ↓ GUAAGU–3′

3′ splice site → 5′–Py Py Py Py Py Py N C A G–3′

In these designations, 'Py' is one of the two pyrimidine nucleotides (U of C); 'N' is any nucleotide, and the arrow indicates the exon-intron boundary. The 5'splice site is also known as the **donor site** and the 3' splice site as the **acceptor site.**

There are certain **conserved sequences**. But they are present in some but not all eukaryotes. Introns in higher eukaryotes usually have a *polypyrimidine tract*, a pyrimidine-rich region located just upstream of the 3'end of the intron sequence (Fig. 9.8).

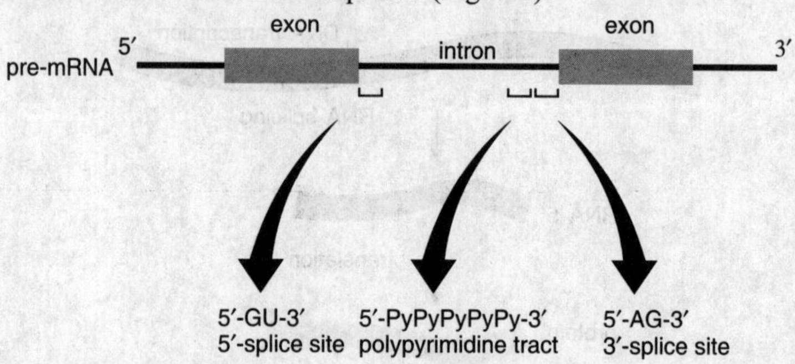

Fig. 9.8. Conserved sequences in vertebrates introns. Py, pyrimidine nucleotide (U or C).

This tract is not common in yeast. However, introns of yeast contain a 5′-UACUAAC-3′ sequence, located between 18 and 140 bp upstream of the 3′splice; these sites are absent in higher eukaryotes. The polypyrimidine tract and the 5′ –UACUAA-3′sequence have different functions.

Two steps of splicing-pathways

1. Cleavage of the 5′ splice site. This step occurs by a **transesterification reaction** promoted by the hydroxyl group attached to the 2′carbon of an adenosine nucleotide located within the intron sequence. In yeast, this adenosine is the last one in the conserved 5′-UACUAA-3′sequence. The result of the hydroxyl attack is cleavage of the phosphodiester bond at the 5′splice site, accompanied by formation of a new 5′-2′ phosphodiester bond linking the first nucleotide of the intron (the G of the 5′-GU-3′ motif) with the internal adenosine. This means that the intron has now been looped back on itself to create a **lariat structure**. Lariat is a rope with running noose, for catching horses and cattle.

2. Cleavage of the 3′ splice site and joining of exons (Fig. 9.9). These two reactions occur due to second transesterification reaction, which is promoted by the 3′-OH group attached to the end of the upstream exon. This hydroxyl group attacks the phosphodiester bond at the 3′ splice site, cleaving and so releasing the intron as the lariat structure. The free lariat structure is converted back to a linear RNA and degraded. At the same time, the 3′ end of the upstream exon joins to the newly formed 5′ end of the downstream exon completing the splicing process.

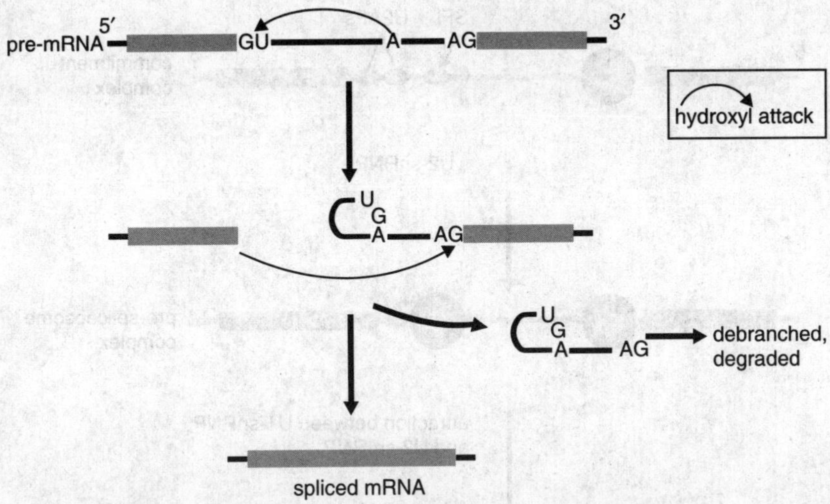

Fig. 9.9. Outline of splicing process. Cleavage of the 5′ splice site is promoted by the hydroxyl (OH) group attached to the 2′-carbon of an adenosine nucleotide within the intron sequence. This results in the lariat structure and is followed by the 3′-OH group of the upstream exon inducing. Cleavage of the 3′ splice site. This enables the two exons to be ligated, with the released intron debranched and degraded.

The splicing apparatus of GU-AG introns. The central components of the splicing apparatus for GU-AG introns are the small nuclear (sn) RNAs called U1, U2, U4, U5 and U6. These are short molecules [between 106 nucleotides (*e.g.*, U6) and 185 nucleotides (*e.g.*, U2) in vertebrates that associate with proteins to form *small nuclear ribonucleoprotein particles* (snRNPs) in the nucleus of eukaryotic cells (Fig. 9.10). The snRNPs, together with accessory proteins, attach to the transcript and form a series of complexes, the last one of which is the **spliceosome**, the structure within which the actual splicing reactions occur (**Smith** and **Valcarcel**, 2000). A completed spiceosome contains 5 RNAs (snRNA molecules such as U1, U2, U4, U5 and U6) and more than 50 proteins (see **Becker** *et al.,* 2006).

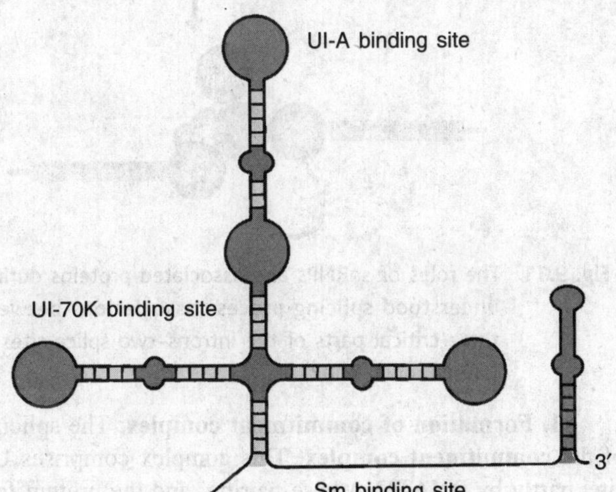

Fig. 9.10. Structure of U1-snRNP. The mammalian U1-snRNP comprises the 165-nucleotide UI-RNA plus ten proteins. Three of these (U1-70K, U1-A and U1-C) are specific to this snRNP, the other seven are Sm proteins that are found in all the snRNPs involved in splicing. U1-RNA forms a base-paired structure as shown. The U1-70K and U1-A proteins attach to two of the major stem-loops of this base-paired structure, the U1-C attaches via a protein-protein interaction. The Sm proteins attach to the Sm site (after Brown, 2002).

Role of Splicing Apparatus in Splicing

Splicing involves the formation of following *three* transition structures (Fig. 9.11).

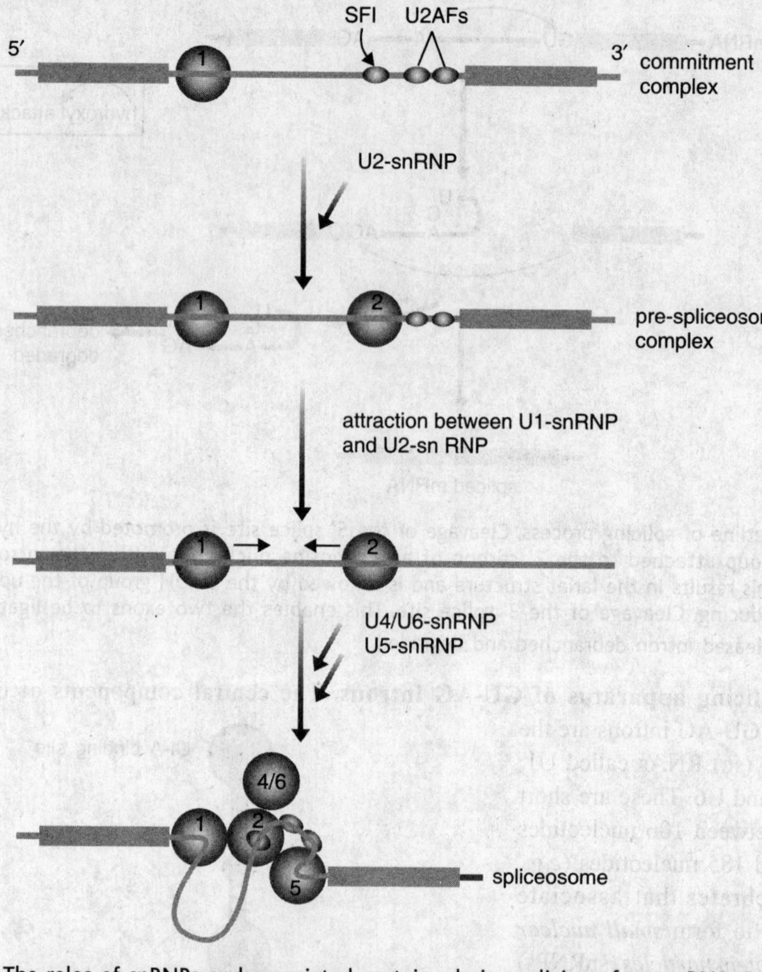

Fig. 9.11. The roles of snRNPs and associated proteins during splicing of pre-mRNA. During this little understood splicing process associations between the snRNPs are thought to bring the three critical parts of the introns–two splice sites and the branch point-into close proximity (after Brown, 2002).

1. Formation of commitment complex. The splicing activity is initiated by the formation of the **commitment complex**. This complex comprises U1-snRNP, which binds to the 5′ splice site, partly by RNA-RNA base pairing, and the protein factors SFI, $U2AF^{35}$ and $U2AF^{65}$, which make protein-RNA contacts with the branch site, the polypyrimidine tract and the 3′ splice site, respectively.

2. Formation of the pre-spliceosome complex. The pre-spliceosome complex comprises the commitment complex plus U2-snRNP, the latter attached to the branch site. At this stage, an association, between U1-snRNP and U2-snRNP brings the 5′ splice site into close proximity with the branch point.

3. Formation of spliceosome. It is formed when U4/U6-snRNP (a single snRNP containing two snRNAs) and U5-snRNP attach to the pre-spliceosome complex. This results in addition interactions that brings the 3′ splice site close to the 5′ splice site and the branch point. All three key positions in the intron are now in proximity and the two transesterifications occur as a linked reaction, possibly catalyzed by U6-snRNP, completing the splicing process (Fig. 9.12).

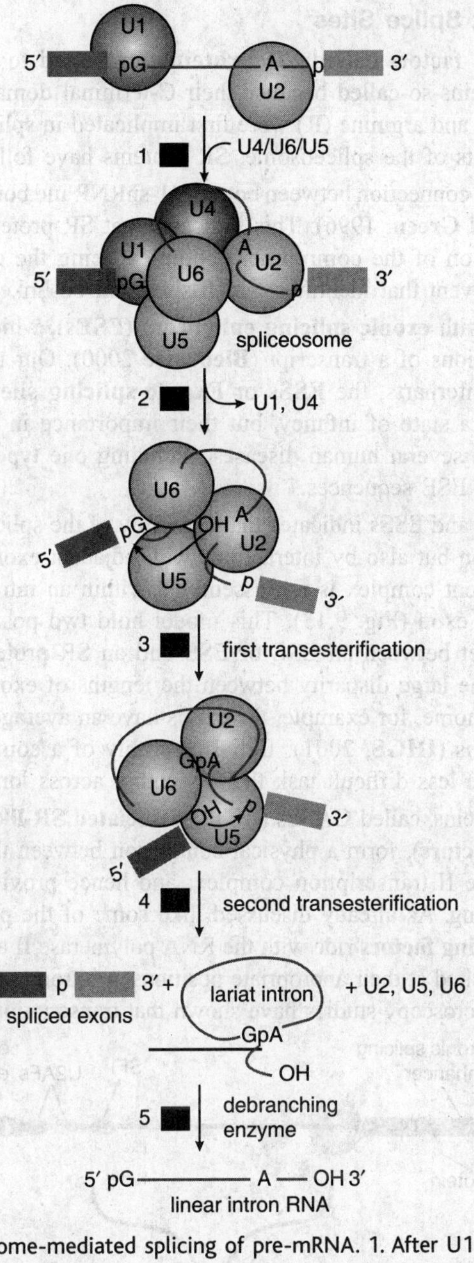

Fig. 9.12. Model of spliceosome-mediated splicing of pre-mRNA. 1. After U1 and U2 snRNPs associate with the pre-mRNA, via base-pairing interactions, a trimeric snRNP complex of U4, U5 and U6 joins the initial complex to form the spliceosome. 2. Rearrangements of base-pairing interactions between snRNAs converts the spliceosome into a catalytically active conformation and destabilizes the U1 and U4 snRNPs, which are released. 3. The catalytic core, thought to be formed by U6 and U2, then catalyzes the first transesterification reaction, forming the intermediate containing a 2′, 5′-phosphodiester bond. 4. Following further rearrangements between the snRNPs, the second transesterification reaction joins the two exons by a standard 3′, 5′-phosphodiester bond and the remaining snRNPs. 5. The excised lariat intron is converted into a linear RNA by a debranching system (after Lodish *et al.*, 2004).

Selection of Correct Splice Sites

A set of splicing factors called **SR Proteins** are found to be important in splice site selection. The SR proteins-so-called because their C-terminal domains contain a region rich in serine (abbreviations S) and arginine (R) were first implicated in splicing when it was discovered that they are components of the spliceosome. SR proteins have following functions:

1. They establish a connection between bound U1-snRNP and bound U2AF in the commitment complex (**Valcarcel** and **Green**, 1996). This indicates that SR proteins have a role in splice-site selection, since formation of the commitment complex being the critical stage of the splicing process, as this is the event that identifies which sites will be linked.

2. They interact with **exonic splicing enhancers (ESEs),** which are purine-rich sequences located in the exon regions of a transcript (**Blencowe** 2000). Our understanding about the role of ESEs and their counterparts, the **ESSs** or **Exonic splicing silencers** (**Del Gatto-Konczak** *et al.,* 1999) still is in a state of infancy, but their importance in controlling splicing is clear from the discovery that several human diseases, including one type of muscular dystrophy, are caused by mutations in ESE sequences.

The sites of ESEs and ESSs indicates that assembly of the spliceosome is driven not simply by contacts within intron but also by interactions with adjacent exons. In fact, it is possible that an individual commitment complex is not assembled within an intron as shown in Figure 9.11, but initially bridges an exon (Fig. 9.13). This model hold two positive points: 1. it provides a means by which contact between an ESE or ESS and an SR protein could influence splicing; 2. it takes account of the large disparity between the lengths of exons and introns in vertebrates genes. In the human genome, for example, the exons have an average length of 145 bp compared with 3365 bp for introns (**IHGS**, 2001). Initial assembly of a commitment complex across an exon might therefore be less difficult task than assembly across longer introns.

3. Certain SR proteins, called **CASPs** (**=CTD-associated SR-like proteins**) or **SCAFs (=SR-like CTD-associated factors)**, form a physical connection between the spliceosome and the CTD of the RNA polymerase II transcription complex, and hence provide a link between transcript elongation and processing. As already discussed, like some of the polyadenylation proteins, it is probable that these **splicing factors** ride with the RNA polymerase II as this enzyme synthesizes the transcript, and are deposited at their appropriate positions at introns splice sites as soon as these are transcribed. Electron microscopy studies have shown that transcription and splicing occur together.

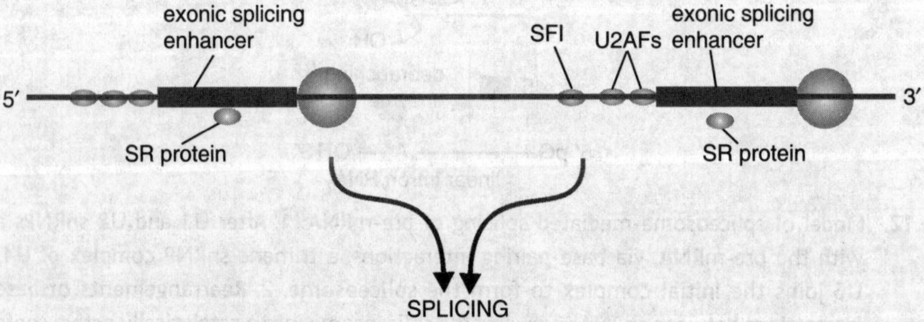

Fig. 9.13. An alternative model for assembly of the commitment complex. In this model, each individual commitment complex is built up across an exon, bringing the complex into close association with an exonic splicing enhancer or silencer and its attached SR proteins. Once complexes are formed over adjacent exons, splicing follows the pathway illustrated in Figure 9.10, the only difference being that the resulting spliceosome is made up of components of adjacent commitment complexes, rather than being derived from the single commitment complex shown in Figure 9.10. (after Brown, 2002).

Alternative Splicing Pathways

With the discovery of introns, it has been the prevailing concept that each gene always give rise to the same mRNA: in other words, there is a single **splicing pathway** for each primary transcripts (Fig. 9.15A). This assumption was found to be incorrect in the 1980s, when it was shown that the primary transcripts of some genes can follow two or more **alternative splicing pathways**, enabling a single transcript to be processed into related but different mRNAs and hence to direct synthesis of a range of proteins (Fig.9.14, 9.15B). The alternative splicing is most common in higher animals. So while there are only three examples of alternative splicing in yeast *Saccharomyces cerevisiae,* there are much more examples in *Drosophila* and humans.

Presence of many alternative splicing pathways were became established when draft *Drosophila* sequence and sequence of microscopic nematode *Caenorhabditis elegans* were compared and found that fruit files have fewer genes than this nematode. Thus, the apparent greater physical complexity of *Drosophila* is not reflected in the far simpler diversity of its genome. The most likely explanation for the lack of parallelism between the number of genes in *Drosophila* genome and the number of proteins in its genome proteome is that a substantial number of the genes give rise to the multiple proteins via alternative splicing (**Note.** Proteome is the collection of functioning proteins synthesized by a living cell). Similarly, when the first human chromosome sequences were obtained and it was recognized that rather than having 80000-100000 genes, as suggested by the size of the human proteome, humans have only 35000 or so genes. It is now believed that at least 35% of the genes in humans genome undergo alternative splicing (**Graveley,** 2001) with this new awakening, the principle 'one gene one protein', serving as biological dogma since the 1940s, has become completely overthrown.

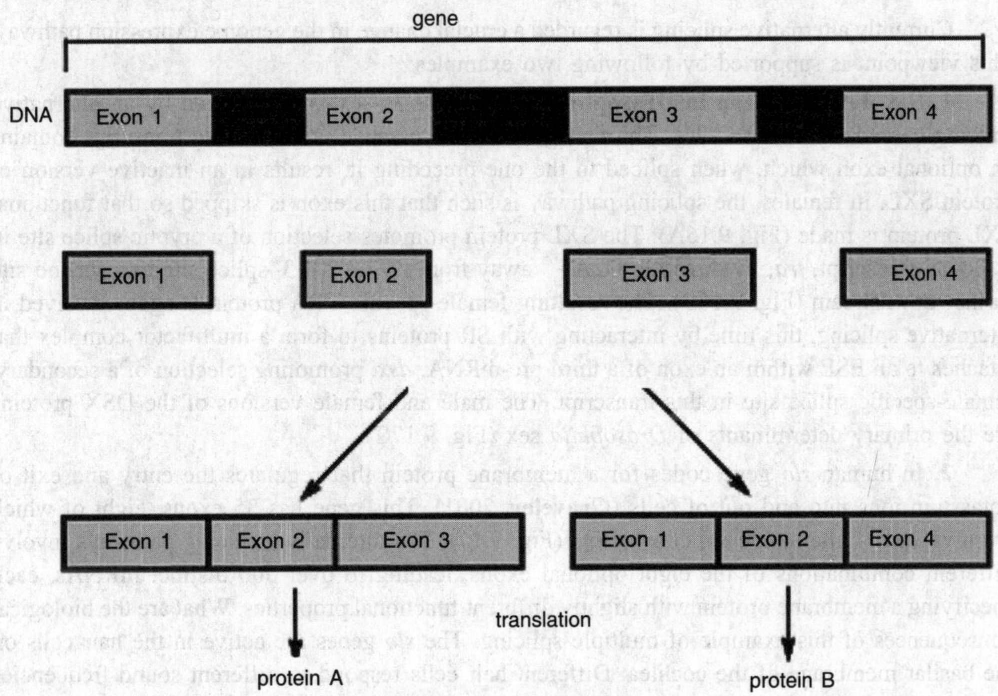

Fig. 9.14. Alternative splicing.

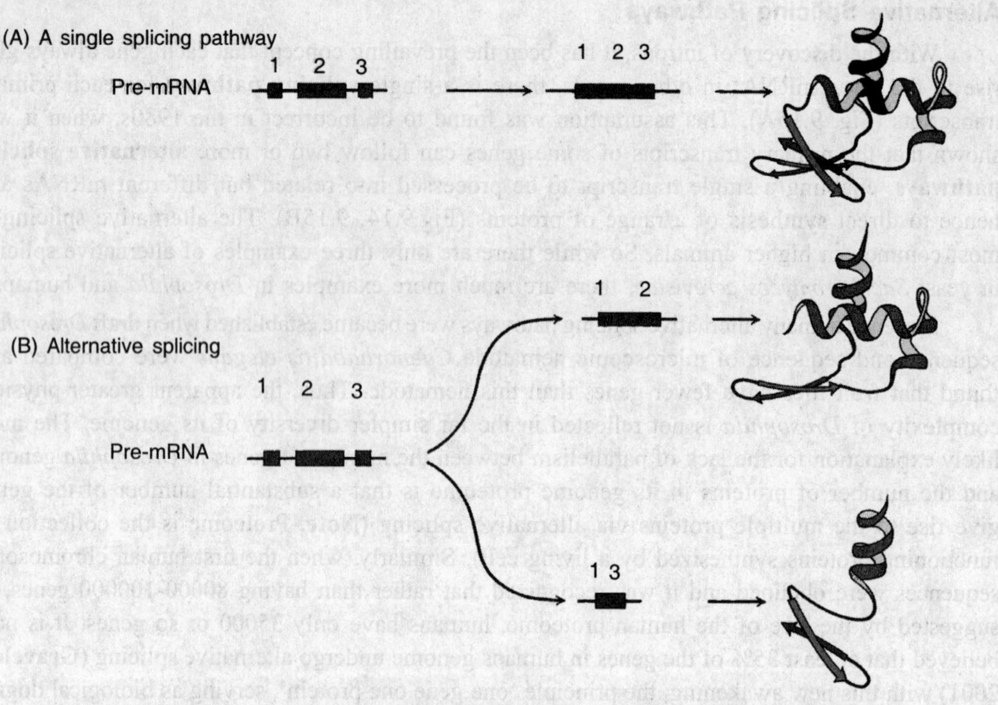

Fig. 9.15. Diagram explaining two modes of alternative splicing (after Brown, 2002).

Currently alternative splicing is regarded a crucial change in the genome expression pathway. This viewpoint is supported by following two examples:

1. Sex determination in Drosophila. In *Drosophila* sex is determined by an alternative splicing cascade (**Chabot** 1996). The first gene in this cascade is *sxl*, whose transcript contains an optional exon which, when spliced to the one preceding it, results in an inactive version of protein SXL. In females, the splicing pathway is such that this exon is skipped so that functional SXL protein is made (Fig. 9.16A). The SXL protein promotes selection of a cryptic splice site in a second transcript, *tra*, by directing $U2AF^{65}$ away from its normal 3' splice site to a second site further downstream (Fig. 9.16B). The resulting female-specific TRA protein is again involved in alternative splicing, this time by interacting with SR proteins to form a multifactor complex that attaches to an ESE within an exon of a third pre-mRNA, *dsx*, promoting selection of a secondary, female-specific splice site in this transcript. The male and female versions of the DSX proteins are the primary determinants of *Drosophila* sex (Fig. 9.17C).

2. In human *slo* gene codes for a membrane protein that regulates the entry and exit of potassium ions into and out of cells (**Graveley,** 2001). This gene has 35 exons, eight of which are involved in alternative splicing events (Fig. 9.17). The alternative splicing pathways involve different combinations of the eight optional exons, leading to over 500 distinct mRNAs, each specifying a membrane protein with slightly different functional properties. What are the biological consequences of this example of multiple splicing. The *slo* genes are active in the hair cells on the basilar membrane of the cochlea. Different hair cells respond to different sound frequencies between 20 and 20000 Hz, their individual capabilities determined in part by the properties of their *slo* proteins. Alternative splicing of *slo* genes in cochlear hair cells therefore determines the auditory range of humans.

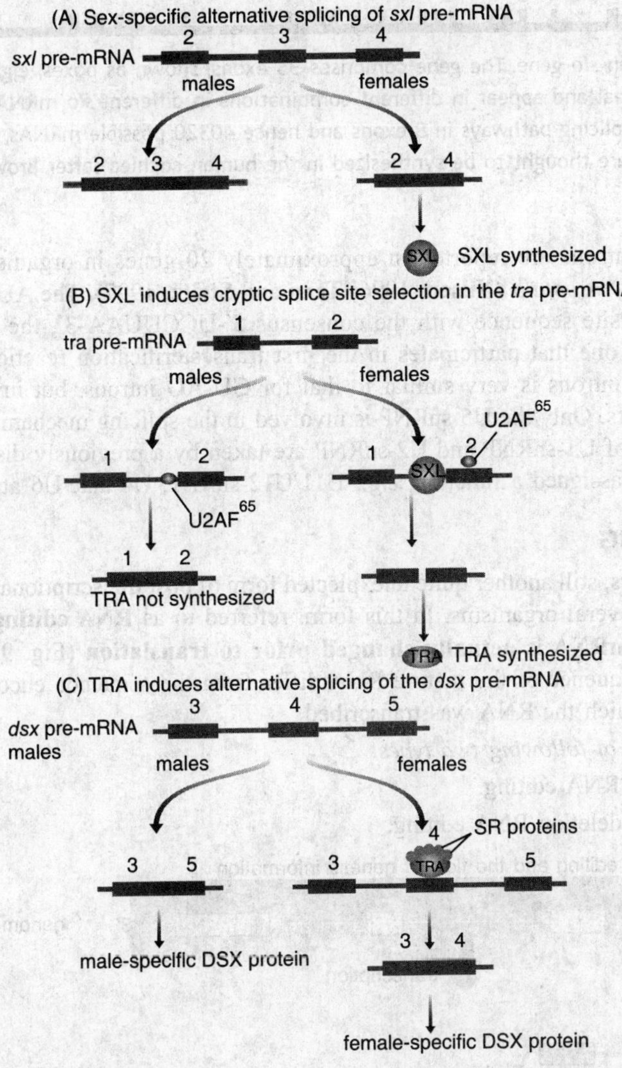

Fig. 9.16. Scheme of **Chabot** (1996) for expression of genes involved in sex determination of *Drosophila*. A–The cascade begins with sex-specific alternative splicing of the *sxl* pre-mRNA. In males all exons are present in the mRNA, but this means that a truncated protein is produced because exon 3 contains a termination codon. In females, exon 3 is skipped, leading to full-length, functional SXL protein. B–In females, SXL blocks the 3′ splice site in the first intron of the *tra* pre-mRNA. $U2AF^{65}$ is unable to locate this site and instead directs splicing to a cryptic site in exon 2. This results in an mRNA that codes for a functional TRA protein. In males, there is no SXL protein so the 3′ splice site is not blocked and a nonfunctional mRNA is produced. C–In males, exon4 of the dsx pre-mRNA is skipped. The resulting mRNA codes for a male-specific DSX protein. In females, TRA protein stabilizes the attachment of SR proteins to an exonic splicing enhancer located within exon 4, so this exon is not skipped, resulting in the mRNA that codes for the female specific DSX protein. The two versions of DSX protein are the primary sex determinants of the male and female physiologies. The female dsx mRNA ends with exon 4 because the intron between exons 4 and 5 has not 5′ splice site, meaning that exons 5 cannot be ligated to the end of exon 4. Instead a polyadenylation site at the end of exon 4 is recognized in females (after Brown, 2002).

Fig. 9.17. The human *slo* gene. The gene comprises 35 exons, shown as boxes, eight of which (i.blue) are optional and appear in different combinations in different *slo* mRNAs. There are 40320 possible splicing pathways in 8 exons and hence 40320 possible mRNAs, but only some 500 of these are thought to be synthesized in the human cochlea (after Brown, 2002).

B. AU-AC Introns

The AU-AC introns are reported in approximately 20 genes in organisms as diverse as humans, plants and *Drosophila* (**Nilsen**, 1996; **Tarn** and **Steitz,** 1997). The AU-AC introns have a conserved branch site sequence with the consensus 5′-UCCUUAA-3′, the last adenosine in this motif being the one that participates in the first transesterification reaction. Thus, splicing pathway of AU-AC introns is very similar to that for GU-AG introns, but involves a different sets of splicing factors. Only the U5-snRNP is involved in the splicing mechanisms of both types of intron. The roles of U1-snRNP and U2-snRNP are taken by a previously discovered complex that had never been assigned a function, *e.g.*, U11/U12-snRNP, U4 atac/U6 atac-snRNP.

9.3 RNA EDITING

In the late 1980s, still another quite unexplected form of post transcriptional RNA processing was discovered in several organisms. In this form, referred to as **RNA editing, the nucleotide sequence of a pre-mRNA is actually changed prior to translation** (Fig. 9.18). As a result, the ribonucleotide sequence of the mature RNA differ from the sequence encoded in the exons of the DNA from which the RNA was transcribed.

RNA editing is of following two types:

1. Substitution RNA editing.
2. Insertion or deletion RNA editing.

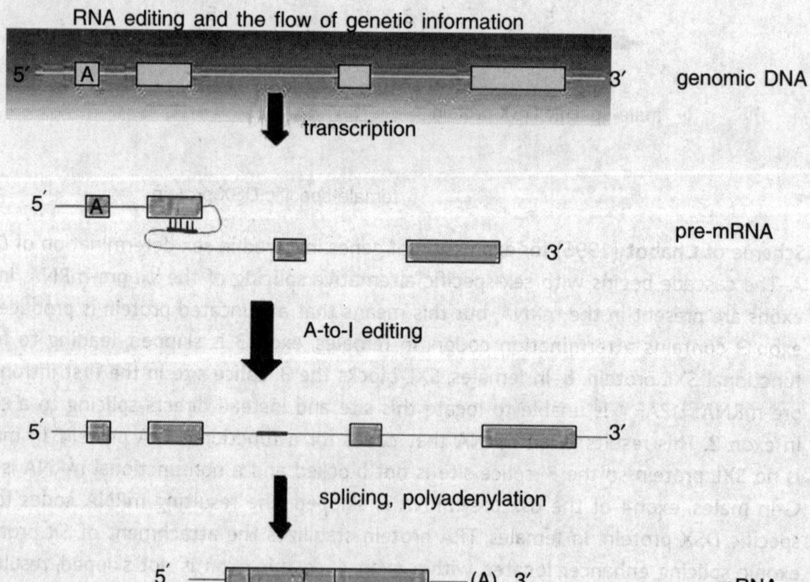

Fig. 9.18. RNA editing.

1. Substitution RNA editing. In this process, the identities of individual nucleotide bases are altered. Substitution editing is used in some nuclear-derived eukaryotic RNAs and is also very prevalent in mitochondrial and chloroplast RNAs transcribed in plants.

2. Insertion or deletion RNA editing. In this process, nucleotides are added to or subtracted from the total number of bases.

Examples of RNA Editing

1. The slime mold, *Physarum polycephalum* uses both substitution and insertion/deletion editing for its mitochondrial mRNAs.

2. RNA editing was particularly evident in mitochondrial proteins of a group of protozoan parasites, the trypanosomes (some of which cause African sleeping sickness); in one case, more than 50% of the nucleotides in the mRNA were added uridines (Us). Uridines were also **deleted** from the original sequence. These parasites had another curious trait-the existence of **minicircles** and **maxicircles** of DNA in specialized mitochondria, called **kinetoplasts**. In the average kinetoplasts, there are about 50 maxicircles and about 5000 minicircles, concentrated like chain links (Fig. 9.19A). The maxicircles contain genes for mitochondrial function. **Larry Simpson** and his colleagues showed in 1990, that both maxicircles and minicircles are templates for **guide RNA (gRNA),** RNA that guides the process of messenger editing.

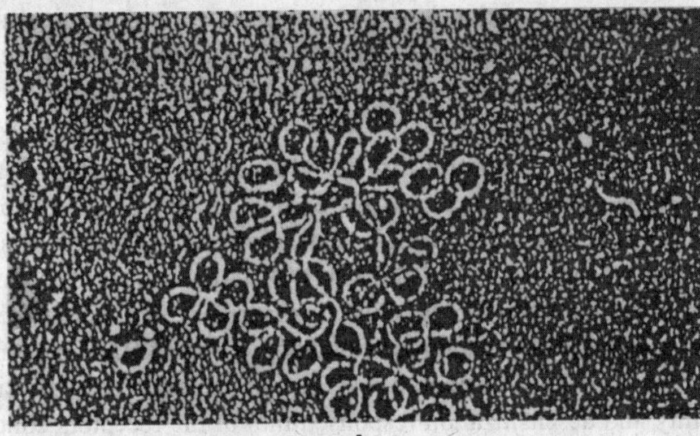

A

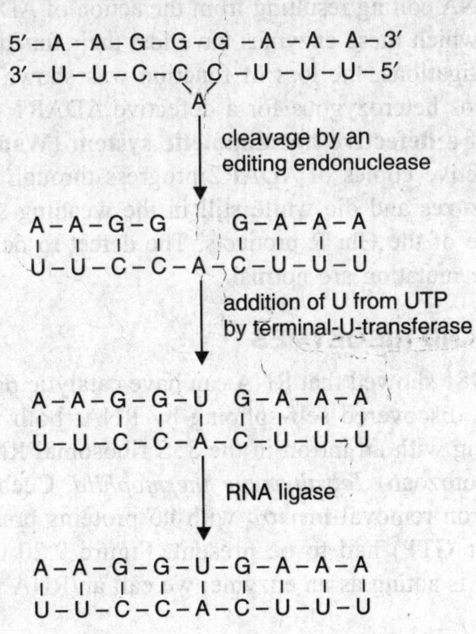

B

Fig. 9.19. RNA editing. A–870 bp minicircles of DNA from *Leishmania tarentolae* (mitochondria). B–Mechanism by which a guide RNA is involved in the editing of a mRNA (after Tamarin, 2002).

The guide RNA forms a complement with the mRNA to be edited; however, the gRNA has the sequence complementary to that of the final mRNA, the one with bases added. Since the bases have not yet been added, a bulge occurs in the gRNA where the complement to be added is (Fig. 9.19). The mRNA is then cleaved opposite the bulge by an enzyme **editing endonuclease**. A uridylate (U) is brought into the mRNA as complement to the adenine (A) with the enzyme **terminal-U-transferase.** An **RNA ligase** then closes the nick in the mRNA, which now has a uridylate added.

An exciting outcome of this research, aside from learning about a novel mechanism of messenger RNA processing, is the possibility of clinical rewards. Anytime there is a specialized pathway in a parasite not found in its host, it is possible to use that pathway to attack the parasite. Thus, this research might lead to new approaches of combating these trypanosome parasites.

3. The most well-studied examples of substitutional editing occur in mammalian nuclear encoded mRNA transcripts. For example, **apolipoprotein-B (apo-B)** exists in both a long and a short form, though a single gene *apo B* encodes both proteins (**Hodges** and **Scott** 1992). In human intestinal cells, apo B mRNA is edited by a single C (called nucleotide 6666) to U change, which converts a CAA glutamine codon into a UA stop codon and terminates the polypeptide at approximately half its genomically encoded length. The editing is performed by a complex of proteins that bind to a "mooring sequence" (**moor** means tie up) on the mRNA transcript just downstream of the editing site.

The synthesis of subunits constituting the **glutamate receiptor channels (Glu R)** in mammalian brain tissue is also affected by RNA editing. In this case, adenosine (A) to inosine (I) editing occurs in pre-mRNAs prior to their translation, during which I is read as guanosine (G). The enzymes produced by a family of three **ADAR** (Adenosine Deaminase Acting on RNA) genes are believed to be responsible for the editing of various sites within the glutamate channel subunits. The double-stranded RNAs required for editing by the **ADAR** enzymes are provided by intron-exon pairing of the GluR mRNA transcripts. The editing changes the physiological parameters (*i.e.*, solute permeability and desensitization response time) of the receptors containing the subunits.

The importance of RNA editing resulting from the action of ADARs is most apparent when one examine situations in which these enzymes have lost their functional capacity as a result of mutation. In several investigations, the loss of function was shown to have a lethal impact in mice. In one study, embryos heterozygous for a defective ADAR1 gene die during embryonic development as a result of a **defective haemopoietic system (Wang** *et al.,* 2000). In another study, mice with two defective copies of ADAR2 progress through development normally, but are prone to **epileptic seizures** and die while still in the weaning stage. Their tissues contain the unedited version of one of the Glu R products. The defect to death is believed to be in the brain. Heterozygote for the mutation are normal.

9.4 SELF-SPLICING AND RIBOZYMES

Sidney Altman in 1981 showed that RNA can have catalytic properties. In 1982, **Thomas Cech** and **Sidney Altman** discovered self-splicing by RNA; both were awarded 1989 Nobel Prize in Chemistry. Working with an intron in the 35S ribosomal RNA precursor in the single-cell eukaryote (ciliated protozoan) *Tetrahymena thermophila*, Cech and his colleagues found that they could induce intron removal *in vitro* with no proteins present. A guanine-containing nucleotide (GMP; GDP or GTP) had to be present. Figure 9.20 diagram depicts how self-splicing occurs. The intron is acting as an enzyme; we call an RNA with catalytic or enzymatic properties a **ribozyme.**

During self-splicing, the U-A bond at the left (5′) side of the intron is transfered to the GTP. The U that is now unbounded displaces the G at the right (3′) side of the intron, reconnecting the

RNA with a U-U connection and releasing the intron (Fig. 9.20). Since all bonds are reversible transfers (transesterifications) rather than new bonds, no external source of energy is required. Self-splicing introns of this type are called **group I introns**.

Although the first enzymatic activity of the ribozyme is its own removal, its secondary structure after removal gives it the ability to further catalyze reactions (Fig. 9.21). The reactions that ribozymes catalyze are transesterifications and the hydrolysis reaction of splitting an RNA molecule into two parts. Ribozymes can also perform other functions, including peptide bond formation. Currently, at least seven different classes of ribozymes are known, based on their enzymatic properties (Table 9.3). A ribozyme that can split other RNAs and that occurs in small plant pathogens (viruses) is called **hammerhead ribozyme** because of its shape.

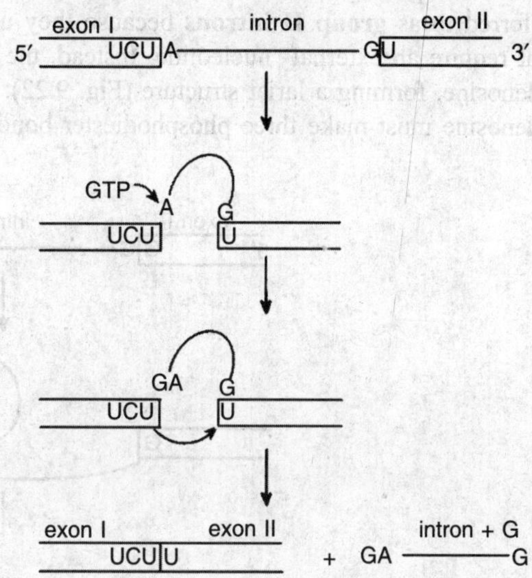

Fig. 9.20. Self-splicing of a ribsosomal RNA precursor in *Tetrahymena*. An external GTP is required. Two bonds transfer produce a shortened RNA and a free intron (after Tamarin, 2002).

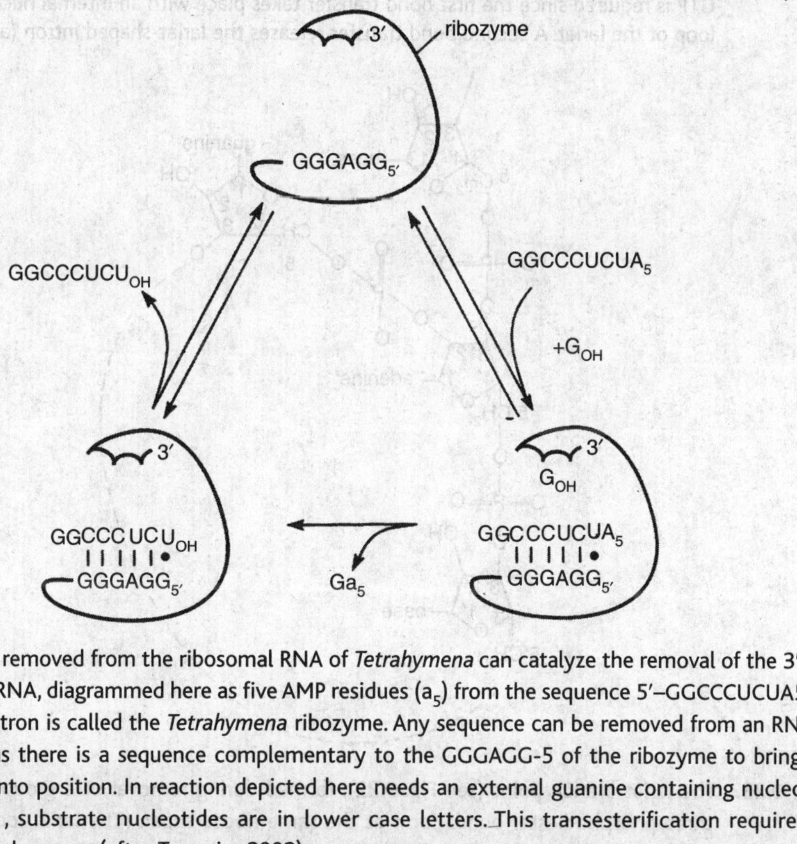

Fig. 9.21. Intron removed from the ribosomal RNA of *Tetrahymena* can catalyze the removal of the 3′ end of an RNA, diagrammed here as five AMP residues (a_5) from the sequence 5′–GGCCCUCUA5′-3′. The intron is called the *Tetrahymena* ribozyme. Any sequence can be removed from an RNA as long as there is a sequence complementary to the GGGAGG-5 of the ribozyme to bring the RNA into position. In reaction depicted here needs an external guanine containing nucleotide (G_{OH}) , substrate nucleotides are in lower case letters. This transesterification requires no external energy (after Tamarin, 2002).

Self-splicing has also been found in genes in the mitochondria of yeast. These introns are referred to as **group II introns** because they use a different mechanism of splicing that does not require an external nucleotide. Instead, the first bond is transferred within the intron to an adenosine, forming a lariat structure (Fig. 9.22). In order for the lariat to form, the ribose of the adenosine must make three phosphodiester bonds (Fig. 9.23).

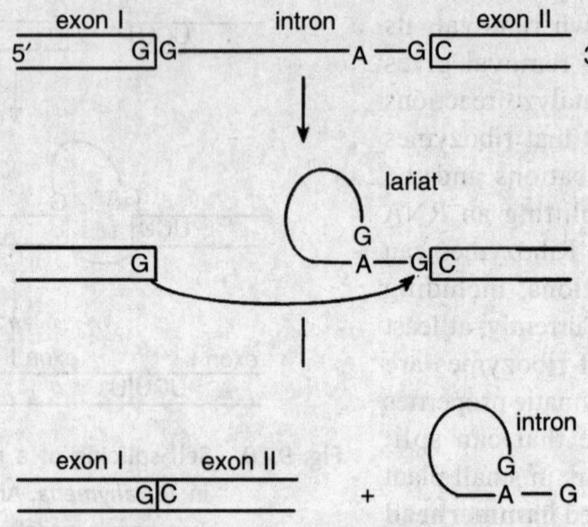

Fig. 9.22. Self-splicing of a group II intron results in a lariat configuration of the released intron. No external GTP is required since the first bond transfer takes place with an internal nucleotide, forming the loop of the lariat. A second bond transfer releases the lariat-shaped intron (after Tamarin, 2002).

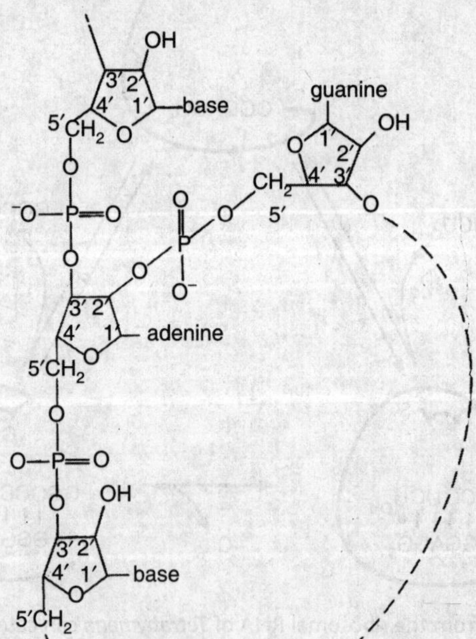

Fig. 9.23. The lariat branch point (see Fig. 9.22), formed during removal of a group II intron, occurs as three phosphodiester bonds form at the same ribose sugar. (The lariat loop is formed by the 2'-phosphodiester bond (after Tamarin, 2002).

Table 9.3 Examples of ribozymes.

S. No.	*Ribozyme*	*Description*
1.	Self-splicing introns	Some introns of Groups I, II and II splice themselves by an autocatalytic process. However, the splicing pathway of GU-AG introns includes at least some steps that are catalyzed by snRNAs (Newman, 2001).
2.	Ribonuclease P	The enzyme that creates the 5′ ends of bacterial tRNAs consists of an RNA subunit and a protein subunit, with the catalytic activity residing in the RNA. (**Note.** The RNA catalyzed reactions are found to follow **Michaelis-Menten Kinetics,** which provides further evidence to the fact RNA component of ribonuclease P is acting like a true enzyme; see Becker *et al.,* 2007).
3.	Ribozomal RNA	The peptidyl transferease activity required for peptide bond formation during protein synthesis is associated with the 23S rRNA of the large subunit of the ribosome.
4.	$tRNA^{Phe}$	Undergoes self-catalyzed cleavage in the presence of divalent lead ions.
5.	Virus genome	Replication of the RNA genomes of some viruses involves self-catlyzed cleavage of chains of newly synthesized genomes linked to head to tail. Examples are the plant viroids and virusoids and the animal hepati is delta virus. These viruses form a diverage group with the self-cleaving activity specified by a variety of different base-paired structure called, *e.g.,* hammerhead.

Structure of Ribozymes

There appear to be at least two kinds of ribozymes. Some ribozymes have folded structures and catalyze reactions on themselves, a process called **intramolecular catalysis**. Other ribozymes act on other molecules, without themselves being changed, a process called **intermolecular catalysis.** The ribozyme consists of a catalytic core made up of two domains, each one comprising two of the base paired regions, with the splice sites brought into proximity by interaction between two other parts of the secondary structure. Although this RNA structure is sufficient for splicing, it is possible that with some introns the stability of the ribozyme is enhanced by non-catalytic protein factors that bind to it. For example, Group I introns in organelle genes, many of these containing an ORF coding for a protien called a **maturase** that appears to play a role in splicing (see **Brown,** 2002).

Functions of Ribozymes

Those ribozymes that are known today carry out three types of biochemical reactions:

1. Self-cleavage, as displayed by the self-splicing group I, II and III introns and by some virus genomes.

2. Cleavage of other RNAs, as carried out by RNase P (discovered by **Altman** *et al.,* 1983)

3. Synthesis of peptide bonds, by the rRNA component of the ribosome (discovered by **Harry Noller** and his colleagues in 1992).

In the test tube, synthetic RNA molecules have been shown to carry out other biologically relevant reactions, such as;

1. Synthesis of ribonucleotides (**Unran** and **Bartel,** 1998).
2. Synthesis and copying of RNA molcculcs (**Exland** and **Bartel,** 1996: **Johnston** *et al.,* 2001).

3. Transfer of an RNA-bound amino acid forming a dipeptide, in a manner analogous to the role of tRNA in protein synthesis (**Lohse** and **Szostak,** 1996).

The discovery of these catalytic properties solved the polynucleotide-polypeptide dilemma by showing that the **first biochemical systems could have been centered entirely** on RNA (**Bartel** and **Unran**, 1990).

Role of Ribozymes in the Origin of Life

Ideas about the RNA world during chemical evolution and origin of life have taken shape in recent years (**Robertson** and **Ellington,** 1998). Biologists now envisage that RNA molecules initially replicated in a slow and haphazard fashion simply by acting as templates for binding of complementary nucleotides which polymerized spontaneously. This process would have been very in accurate so a variety of RNA sequences would have been generated, eventually leading to one or more with incipient ribozyme properties that were able to direct their own, more accurately self-replication. It is possible that a form of natural selection operated so that the most efficient replicating systems began to predominate. A greater accuracy in replication would have enabled RNAs to increase in length without losing their sequence specificity, providing the potential for more sophisticated catalytic properties, possibly culminating in structures as complex as present day Group-I introns (*e.g. Tetrahymena* tRNA intron) and ribosomal RNAs (Fig. 9.3).

9.5 CHANCES OF ERRORS IN GENE SPLICING

There are certain topological problems which might pose chances of errors in gene splicing. Following two types of errors can commonly occur during splicing of a gene:

1. The first of these is the *substantial* distance that might lie between splice sites, possibly a few tens of kb, representing 100 nm or more if the mRNA is in the form of linear chain. A means is therefore needed to bring the splice sites into proximity.

2. The second topological problem concerns **selection** of the correct splice site. All splice sites are similar, so if a pre-mRNA contains two or more introns then there is the possibility that the wrong splice sites could be joined, resulting in **exon skipping,** *i.e.*, the loss of an exon from the mature mRNA (Fig. 9.24A). Equally unfortunate would be selction of a **cryptic splice site,** a site within an intron or exon that has sequence similarity with consensus motifs of real splice sites (Fig. 9.24B). Cryptic sites are present in most pre-mRNAs and must be ignored by splicing apparatus (see **Brown**, 2002).

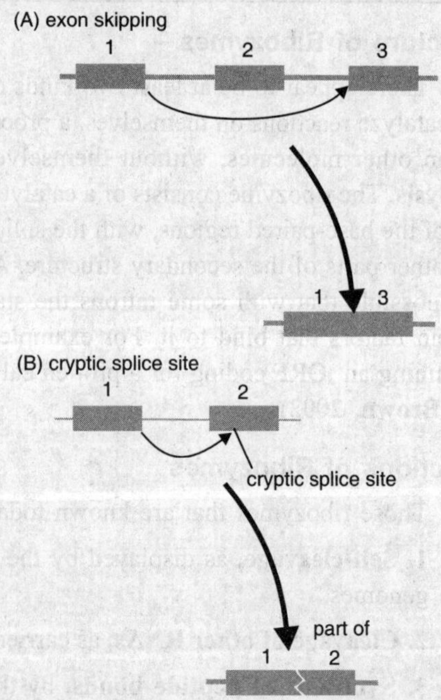

Fig. 9.24. Two aberrant forms of splicing. A-In exon skipping the aberrrant splicing results in an exon being lost from mRNA. B–When a cryptic splice site is selected, part of an exon might be lost from the mRNA (after Brown, 2002).

QUESTIONS

Long Answer Questions

1. Describe the post-transcriptional modification of heterogenous nuclear RNA in eukaryotes.
2. What is heterogenous nuclear messenger RNA (hnRNA)? What are small nuclear ribonucleo-proteins?
3. What chemical groups are present at the origin and terminus of a molecule of mRNA that has just been synthesized?

   (*i*) What is mRNA?

   (*ii*) What is polycistronic mRNA?

Very Short Answer Questions

1. What is pre-mRNA?
2. Enlist three basic functions of known ribozymes.
3. What is RNA editing?
4. Define spliceosome.
5. These questions refer to eukaryotic RNA:

   (i) (*a*) What is a cap?

   (*b*) At which end of mRNA is the poly (A)?

   (*c*) Are there eukaryotic mRNA molecules that do not contain either feature?

   (ii) (*a*) What are intervening sequences or introns?

   (*b*) What is mRNA splicing?
6. Many eukaryotic genes contain noncoding introns that separate the coding sequences or exons of these genes. At what stage, during the expression of these mosaic genes are the noncoding intron sequences removed?
7. In what ways are ribosomes and spliceosomes similar? Different?
8. What would be the effect on the final protein product if an intervening sequence were removed with an extra base? One base? Too few?
9. What product would DNA-RNA hybridization produce in a gene with five introns? No introns? Draw these hybrid molecules.
10. What are differences between group I and group II introns?
11. Enhancers can often exert their effect from a distance; some enhancers are located thousands of bases upstream from the promoter. Propose an explanation to account for this.
12. For several decades, the dogma in biology has been that molecular reactions occurring in living cells are catalyzed by enzymes composed of polypeptides. We now know that the introns of some precursor RNA molecules such as the rRNA precursors in *Tetrahymena* are removed autocatalytically ("self-spliced") with no involvement of any protein. What does the demonstration of autocatalytic splicing indicate about the dogma that biological reactions are always catalyzed by proteinaceous enzymes?
13. What components of the introns of nuclear genes that encode proteins in higher eukaryotes are conserved and required for the correct excision of intron sequences from primary transcripts by spliceosomes?
14. (A) Which of the following nuclear pre-mRNA nucleotide sequences potentially contains and intron?

    (1) 5′–CCAUGGCGCUAACACUGCCAAUUGGCCAAUACUGACCUGAAG-CAUCAGCCAA-3′

    (2) 5′-AGUCUCAUCUGUCCAUUGACUUCGAAACUGAAUCGUAACUCCU-ACGUCUAUGGA-3′

(3) 5′–GCUGUUUGUCAUGACUGACUGGUCACUAUCGUACUAACCUGU-CAUGCAAUGUC-3′

(4) 5′–AGCAGUUCUGUCGCCUCGUGGUGCUGGCUGGCCCUUCGUCG-CUCGGGCUUAGCUA-3′

(5) 5′-UUCGCAUGACGUACUUCUGAGACUACUACUACUAACGCAUCG-AGUCUCAA-3′

(B) One of the five pre-mRNAs shown above is a likely candidate for splicing out an intron sequence. What mRNA nucleotide sequence would one expect to result from this splicing event?

Fill in the Blanks

1. RNA polymerase II enzyme is located in .... and synthesized .... (heterogenous nuclear RNA) which acts precursor or pre-RNA of ....

2. The mRNAs of mammalian reticulocytes exists up to .... for prolong utilization in the synthesis of .... of haemoglobin.

3. Modification and processing of pre-RNA of hnRNA includes four steps, namely .... , ... , ... and RNA editing.

4. RNA polymerase II enzyme both .... DNA into RNA and .... the RNA it produces.

5. Capping of nascent mRNA involves three enzymes, namely, ... , ... and methyl transferase.

6. Polyadenylation of mRNA involves addition of .... at .... of RNA.

7. Size of poly–A tail of mRNA may be .... to .... adenine containing nucleotides.

8. The *slo* gene of humans contains .... exons.

9. RNA editing includes .... and ....

10. Ribozyme is an .... molecule with .... activity.

11. RNA polymerase II synthesizes .... and most of .... (small–nuclear RNAs) which are used in post-transcriptional processing of mRNA.

12. RNA polymerase is sensitive to ....

13. The cap of .... protects mRNAs against the attack of .... and ....

Multiple Choice Questions

1. Nuclear DNA sends information for protein synthesis through
   (*a*) tRNA (*b*) mRNA
   (*c*) rRNA (*d*) All the above

2. The RNA which carries the genetic message from the nucleus to the ribosome is
   (*a*) hn mRNA (*b*) mRNA
   (*c*) tRNA (*d*) rRNA

3. The type of RNA specifically responsible for directing the proper sequence of amino acids in protein synthesis is
   (*a*) chromosomal RNA (*b*) rRNA
   (*c*) tRNA (*d*) mRNA

4. Maximum formation of mRNA occurs in
   (*a*) cytoplasm (*b*) nucleolus
   (*c*) ribosome (*d*) nucleoplasm

5. The ribozyme is a RNA
(*a*) without sugar (*b*) with extra phosphate
(*c*) with enzymatic property (*d*) without phosphate

6. mRNA is complimentary copy of
(*a*) tRNA (*b*) rRNA
(*c*) ribosomal DNA (*d*) a single strand of DNA

7. The cap structure of mRNA is of methyl GTP and tail of
(*a*) methionine (*b*) formylmethionine
(*c*) poly A (*d*) UAG codon

8. Synthesis of mRNA molecule is terminated by ____ factor.
(*a*) alpha (*b*) beta
(*c*) sigma (*d*) rho

9. Which RNA is having least age?
(*a*) mRNA (*b*) tRNA
(*c*) rRNA (*d*) none of these

ANSWERS

Very Short Answer Questions

1. It is primary transcript whose processing yields a mature mRNA.
2. 1. Self-cleavage; 2. cleavage of other RNAs; and 3. synthesis of peptide bonds.
3. RNA editing is the insertion of uridines into messenger RNAs after transcription is completed, it is controlled by guide RNA.
4. Spliceosome is protein-RNA complex that catalyzes the removal of introns from pre-mRNA. It contains 5 RNAs (*e.g.*, U1, U2, U4, U5 and U6) and more than 50 proteins.
5. (*i*) (*a*) A terminal structure in which a methylated guanosine is in 5'-5'-triphosphate linkage at the 5'terminus of mRNA.
   (*b*) The 3' –OH end.
   (*c*) All mRNA molecules (except those of several viruses) are capped. Some mRNA molecules lack the poly (A) tail.
   (*ii*) (*a*) Untranslated sequences that interrupt the coding sequence of a transcript and that are removed before translation begins.
   (*b*) Removal of introns.
6. The entire nucleotide-pair sequences-including the introns-of the genes are transcribed by RNA polymerase to produce primary transcript that still contain the intron sequences. The intron sequences are then "spliced out" off the primary transcripts to produce the mature, functional RNA molecules. In the case of protein encoding nuclear genes of higher eukaryotes, the introns are "spliced out" by complex ribosome-like macro-molecular structures called **spliceosomes.**
7. Ribosomes and spliceosomes both play essential roles in gene expression, and both are complex macromolecular structures composed of RNA and protein molecules. Ribosomes are located in cytoplasm, and spliceosomes in the nucleus. Ribosomes are larger and more complex than spliceosomes.
8. Removing one base too many or too few would result in shift in the reading frame during translation, thus radically altering the protein product.

9. Five introns

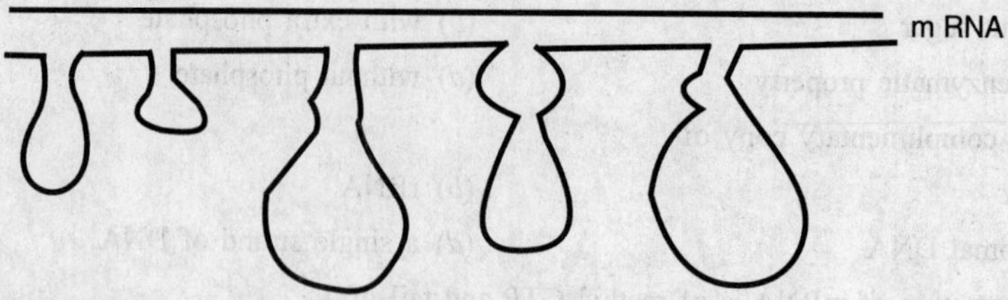

No introns

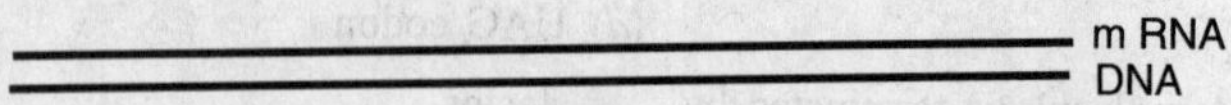

10. Group I introns are self-splicing introns that require a guanine (G)-containing nucleotide for splicing. Group II introns are similar but do not require an external nucleotide for splicing. Group I and II introns are released as linear and lariat-shaped molecules respectively.

11. Enhancers (E) bind activator proteins that also bind proteins of the polymerase at the promoter (P). In some eukaryotic genes, there are many enhancers, allowing numerous levels of control and forcing enhancers further and further upstream. For activators bound to enhancers to bind polymerase proteins, the DNA must lost around.

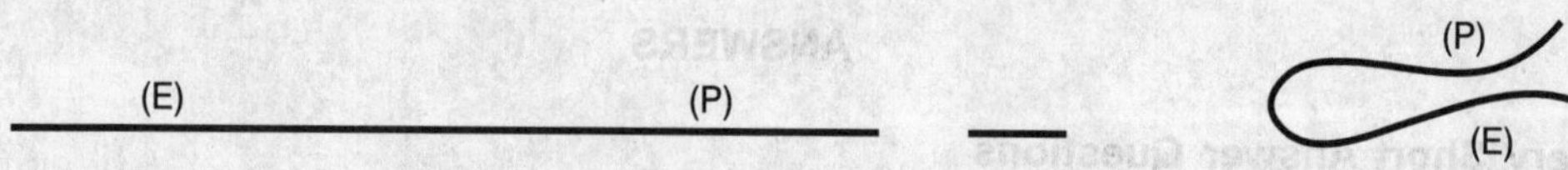

12. "Self splicing" of RNA precursors demonstrates that RNA molecules can also contain catalytic sites; this property is not totally restricted to proteins.

13. The introns of protein encoding nuclear genes of higher eukaryotes almost without exception begin (5′) with GT and end (3′) with AG. In addition, the 3′ subterminal A in the "TACTAAC box" is completely conserved; this A is involved in bond formation during intron excision.

14. (*a*) Sequence 5. It contains the conserved intron sequences: a 5′ GU, a 3′ AG, and UACUAAC internal sequence providing a potential bonding site for intron excision. Sequence 4 has a 5′GU and 3′ AG, but contains no internal A for the bonding site during intron excision.

(*b*) 5′–UAGUCUCAA-3′; the putative intron from the 5′ GU through the 3′AG has been removed.

Fill in the Blanks

1. Nucleoplasm, hnRNA, mRNA.
2. Two days, globin protein.
3. Capping, polyadenylation, splicing.
4. Transcribes, processes.
5. Phosphatase, guanyl transferase and methyl transferase.
6. Poly A-tail, 3′–end.
7. Twenty, two hundred.
8. *slo*, 35.
9. Subsitution, insertion/deletion.
10. RNA, catalytic activity.
11. Pre-mRNA, snRNAs.
12. α-amanitin.
13. 7-methyl guanosine, phosphatases, nucleases.

Multiple Choice Questions

1. (*b*) **2.** (*b*) **3.** (*d*) **4.** (*d*) **5.** (*c*) **6.** (*d*)
7. (*c*) **8.** (*d*) **9.** (*a*).

10

Ribosomal RNA (rRNA) and Transfer RNA (tRNA)

10.1 RIBOSOMAL RNA (rRNA)

Ribosomes are cell organelles, which function as cell's protein factories. They provide for the sites of protein synthesis involving translation of the genetic information encoded in the messenger RNA (mRNA). When not engaged in protein synthesis, the ribosomes in a bacterial cell are always free, but in an eukaryotic cell, these may be found either free in the cytosol or attached to membrane of ER (called rough endoplasmic reticulum).

The initial progress in understanding the detailed structure of the ribosome came not from observing them with the electron microscope but by analyzing their components by **ultracentrifugation**. Ribosomes, as well as other small particles and molecules, are measured in units that describe their rate of sedimentation during density gradient centrifugation in sucrose. This technique gives information on size and shape (due to the speed of sedimentation) while simultaneously isolating the molecules. Isolation by centrifugation in sucrose is a relatively gentle isolation technique; the molecule still retain their biological properties and can be use for further experimentation. In the 1920s, physical chemist **T. Svedberg** developed ultracentrifugation, giving his name to the unit of sedimentation: the **Svedberg unit, S.**

In **sucrose density-gradient centrifugation,** the gradient is formed by layering on decreasingly concentrated sucrose solutions. In a related technique, **cesium chloride density-gradient centrifugation,** the gradient develops during centrifugation. The sucrose centrifugation is stopped after a fixed time, whereas in the cesium chloride technique, the system spins until it reaches equilibrium. The sucrose method tends to be more rapid. Samples can be isolated from a sucrose gradient by punching a hole in the bottom of the tube and collecting the drops in sequentially numbered containers. The first (lowest-numbered) containers will contain the heaviest molecules (with the highest S values).

Components of Ribosome

Intact ribosomes have sedimentation coefficients of 80S for eukaryotes and 70S for bacteria, and each can broken down into smaller components (Fig. 10.1):

1. Each ribosome comprises two subunits of unequal size. In eukaryotes these subunits are 60S and 40S, in bacteria they are 50S and 30S. Note that sedimentation coefficients are not additive because they depend on shape as well as mass; it is perfectly acceptable for the intact ribosome to have a S value less than the sum of its two subunits.

2. The large subunit contains three rRNAs in eukaryotes (the 28S, 5.8S and 5S rRNAs). In bacteria, the equivalent of the eukaryotic 5.8 rRNA is contained within the 23S rRNA.

3. The small subunit contains a single rRNA in both types of organisms: an 18S rRNA in eukaryotes and a 16S rRNA in bacteria.

4. Both subunits contain a variety of **ribosomal proteins,** the numbers have been shown in Figure 10.1. The ribosomal proteins of the small subunit are called S1, S2, S3, etc.; those of the large subunit are called L1, L2, etc. There is just one of each protein per ribosome, except for L7 and L12, which are present as dimers.

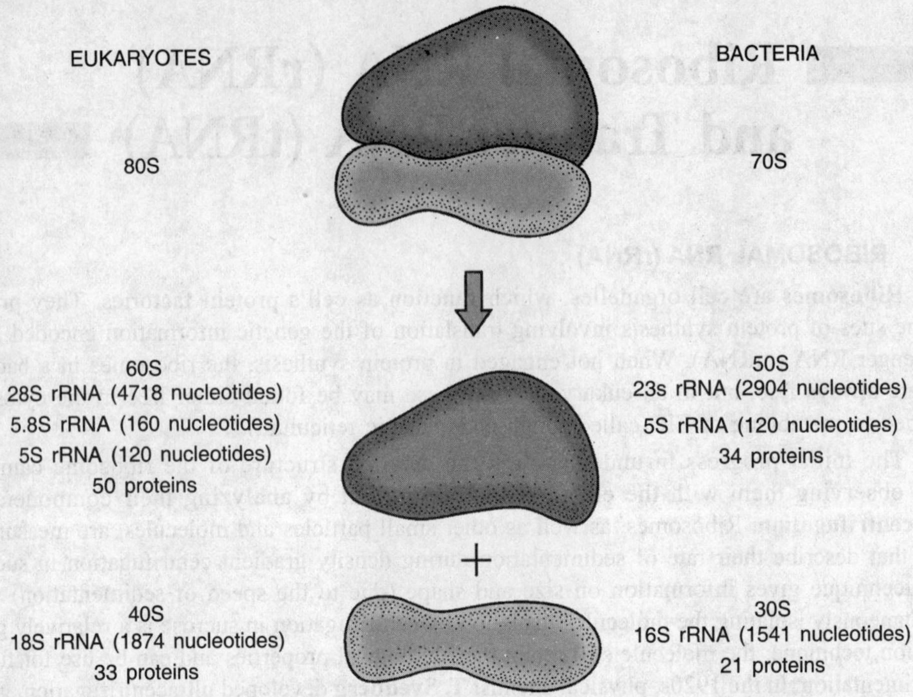

Fig. 10.1. The comparison of eukaryotic and bacterial ribosomes. The details refer to a 'typical' eukaryotic ribosome and the *Escherichia coli* ribosome. Variations between different species mainly concern the numbers of ribosomal proteins (after Brown, 2002).

Investigation of the Fine Structure of Ribosome

Once the basic composition of eukaryotic and bacterial ribosomes had been worked water, attention was focused on the way in which the various molecules of rRNAs and proteins fit together. Important information was provided by the first rRNA sequences, comparisons between these identifying conserved regions that can base-pair to form complex two-dimensional structures (Fig. 10.2; Fig. 10.4). This suggested that the rRNAs provide a scaffolding within the ribosome, to which the proteins are attached, an interpretation that under-emphasizes the active role that rRNAs play in protein synthesis (Fig. 10.5).

Much of the subsequent research has concentrated on the bacterial ribosome, which is smaller than the eukaryotic version and available in large amounts from extracts of cells grown to high density in liquid cultures. Following technical approaches have been used to study the bacterial ribosome.

1. **Nuclease protection studies** enable contacts between rRNAs and proteins to be identified.

2. **Protein-protein crosslinking** identifies pairs or groups of proteins that are located close to one another in the ribosome.

3. **Electron microscopy** has gradually become more sophisticated, enabling the overall structure of the ribosome to be resolved in greater detail. Innovations such as **immunoelectron**

microscopy, in which ribosomes are labeled with antibodies specific for individual ribosomal proteins before examination, have been used to locate the positions of these proteins on the surface of the ribosome.

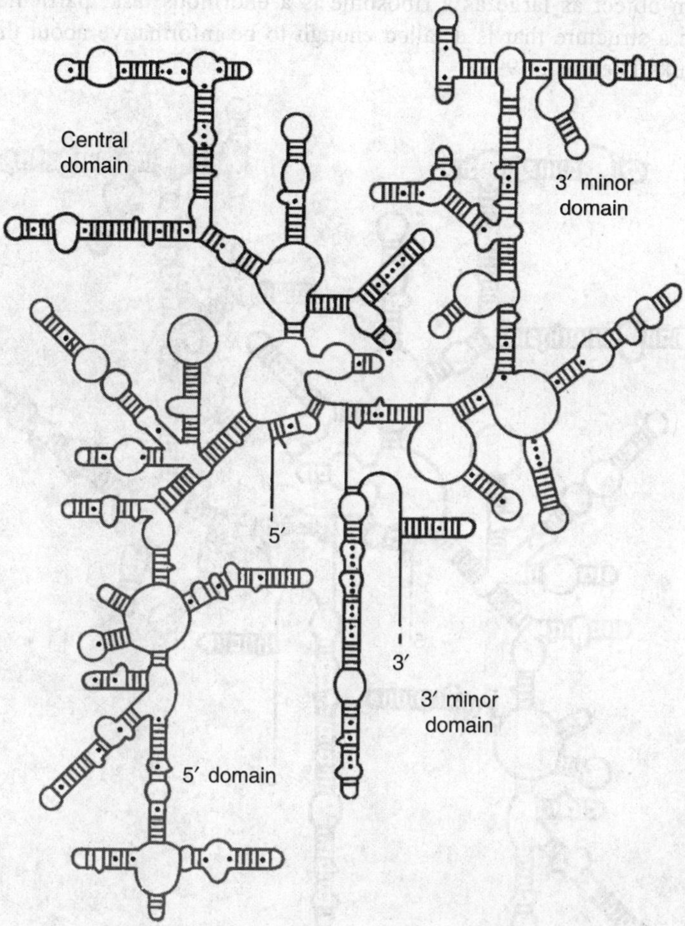

Fig. 10.2. The base-paired structure of 16S rRNA of *E.coli.* In this representation, standard base pairs (G-C, A-U) are shown as bars; non-standard base pairs (*e.g.*, G-U) are shown as dots (after Brown, 2002).

4. Site-directed hydroxyl radical probing makes use of the ability of Fe (II) ions to generate hydroxyl radicals that cleave RNA phosphodiester bonds located within 1 nm of the site of radical production. This technique was used to determine the exact positioning of ribosomal protein S5 in the ribosome of *E.coli.* Different amino acids within S5 were labeled with Fe(II) and hydroxyl radicals induced in the reconstituted ribosomes. The positions at which 16S rRNA was cleaved were then used to infer the topology of the rRNA in the vicinity of S5 protein (Fig. 10.4; **Heilek** and **Noller**, 1996).

Fig. 10.3. X-ray crystallographic apparatus.

5. X-ray crystallography. In recent years, above discussed techniques have been increasingly supplemented by X-ray crystallography which has been responsible for the most exciting insights into ribosome structure. Analyzing the bulk amounts of X-ray diffraction data that are produced by crystals of an object as large as a ribosome is a enormous task, particularly at the level needed to obtain a structure that is detailed enough to be informative about the way in which the ribosome works (**Pennisi**, 1999).

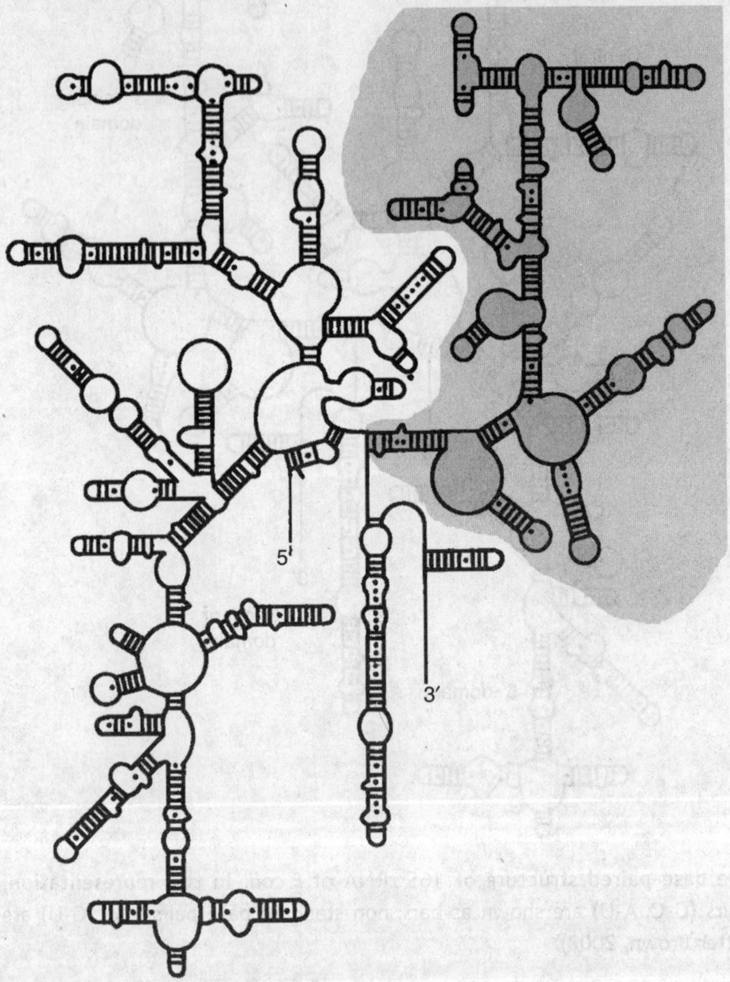

Fig. 10.4. Position within the 16S rRNA of *E.coli* that form contacts with the ribosomal protein S5. The distribution of the contact positions (shaded part) for this single ribosomal protein emphasizes the extent to which the base-paired secondary structure of the rRNA is further folded within the three-dimensional structure of the ribosome (after Brown, 2002).

Box 10.1

By the help of X-ray diffraction, details of following ribosomal features have been revealed by different molecular biologists:

1. Binding of ribosomal proteins to their segments of rRNA: **Conn** *et al.*, 1999; **Agalarov** *et al.*, 2000).
2. Structure of the large and small subunits of ribosome: **Ban** *et al.*, 2000; **Wimberly** *et al.*, 2000.
3. Entire bacterial ribosome attached to mRNA and tRNA: **Yusupov** *et al.*, 2001.

In *E.coli,* all three ribosomal RNA segments are transcribed as a single long piece of RNA that is then cleaved and modified to form the final three pieces of RNA (*i.e.*, 16S, 23S and 5S). The region of DNA that contains the three ribosomal RNA molecules also contains genes for four tRNAs (Fig. 10.6). There occur about ten copies of this region in each chromosome of *E.coli* (see *Albert et al*., 2002). The occurrence of three rRNA segments on the same piece of RNA ensures a final ratio of 1:1:1, the ratio needed for ribosomal construction.

Nucleolus and Transcription in Eukaryotes

In eukaryotes, all rRNAs (*i.e.*, 5.8S, 18S and 28S) but the 5S ribosomal RNA section are transcribed as part of the same piece of RNA. However, eukaryotic cells have many copies of these rRNA genes, depending on the species (Box 10.1). For example, the fruit fly, *Drosophila melanogaster*, has about 130 copies of the DNA region that the larger segments of ribosomal RNA are transcribed from (Box 10.2). These regions occur in tandem on the X and Y sex chromosomes and are known collectively as the *nucleolar organizer*. The smallest ribosomal RNA (*i.e.,* 5S rRNA) is also produced from a duplicated gene, but at a different point in the genome. For example, in *D. melanogaster*, the 5S rRNA is produced on chromosome 2. It is still not known why 5S rRNA is transcribed separately (see **Alberts** *et al*., 2002).

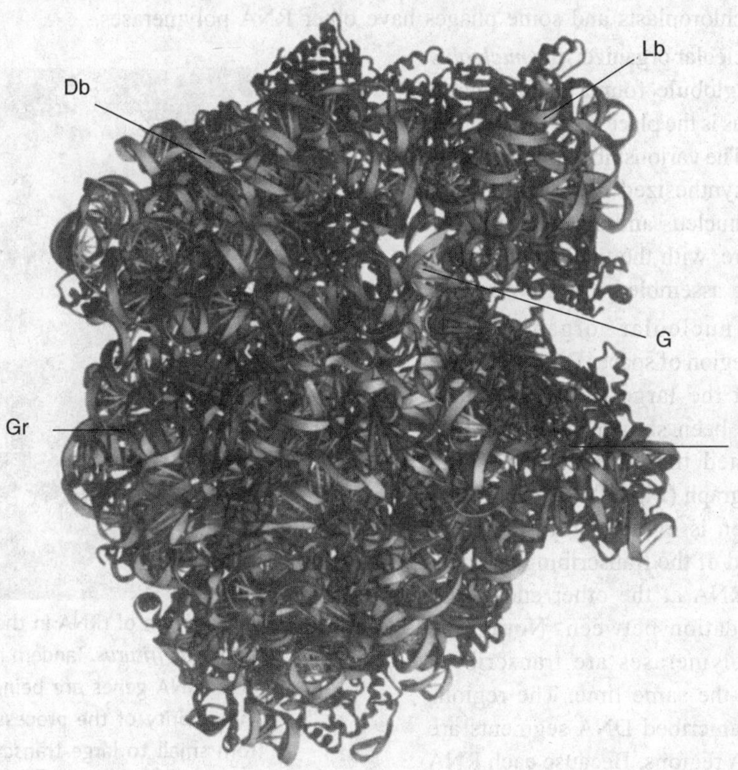

Fig. 10.5. This picture shows the molecular construction of 70S ribosome of the bacterium *Thermus thermophilus*. The small subunit is at the top, with the 16S rRNA in Lb and the small subunit ribosomal proteins in Db. The large subunit rRNAs are in Gr and the proteins in P. The G area is the A site-the point at which aminoacylated tRNAs enter the ribosome during protein synthesis. This site and most of the region within which protein synthesis occurs, is located in the cleft between the two subunits (after Brown, 2002).

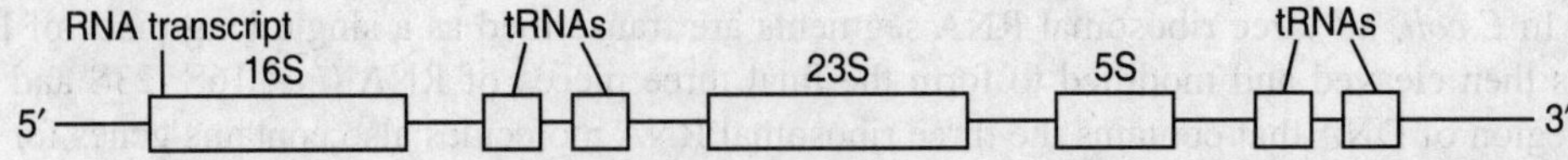

Fig. 10.6. The *E. coli* transcript that contains the three ribosomal RNA segments also contains four tRNAs and some spacer RNA that separates the tRNA are r RNA genes (after Tamarin, 2002).

Box 10.2

Human cells contain about 200 rRNA gene copies per haploid genome. These are spread out in small clusters on five different chromosomes. Cells of the frog *Xenopus* also contain about 600 rRNA gene copies per haploid genome in a single cluster on one chromosome.

As we know that the eukaryotes have three RNA polymerases (enzymes). Eukaryotic RNA polymerase I (or polymerase A) transcribes only the nucleolar organizer DNA (*i.e.*, most of rRNAs except 5S). RNA polymerase II (or polymerase B) transcribes most genes (*i.e.*, genes for mRNAs). RNA polymerase III (or polymerase C) transcribes small genes, primarily the 5S rRNA gene and tRNA genes. Primase do primer synthesis during DNA replication. In addition, mitochondria, chloroplasts and some phages have other RNA polymerases.

At the nucleolar organizer, the *nucleolus* forms as dark globule found in eukaryotic nuclei. Nucleolus is the place where ribosomes are assembled. The various ribosomal proteins that has been synthesized in the cytoplasm migrate to the nucleus and eventually to the nucleolus, where, with the final forms of the rRNA, they are assembled into ribosomes.

In the nucleolar organizer, an untranscribed region of spacer DNA separates each repeat of the large ribosomal DNA gene. This has been shown in Figure 10.7 and diagrammed in Figure 10.8. In the electron micrograph (Fig. 10.7), the polarity of transcription is evident from the short RNA at one end of the transcribing segment and the long RNA at the other ends, with a uniform gradation between. Notice the many RNA polymerases are transcribing each region at the same time. The regions between the tranicribed DNA segments are the spacer DNA regions. Because each RNA gene has a fixed initiation site (promoter) and a fixed termination site, the transcripts adopt the characteristic Christmas tree' configuration (Fig. 10.7).

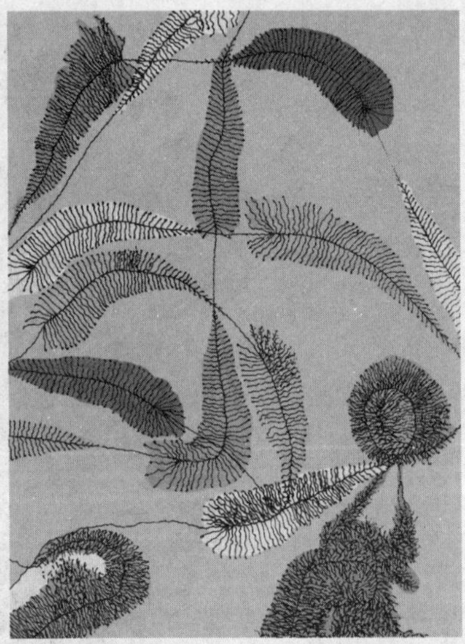

Fig. 10.7. Transcription of rRNA in the nucleolus of the newt, *Triturus*. Tandem repeats of the large rRNA genes are being transcribed. The polarity of the process (progressing from small to large transcripts), as well as the spacer DNA (thin lines between transcribing areas), is clearly visible.

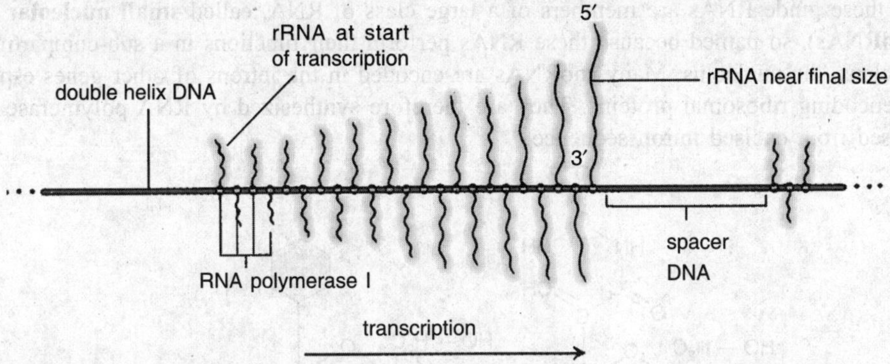

Fig. 10.8. Details of the transcription of large ribosomal RNA gene. Note the polarity of the process and the spacer DNA (after Tamarin, 2002).

Processing of Ribosomal RNAs

Three of four rRNAs (185, 5.8 and 28S) are made by chemically modifying and cleaving a single large (45S) precursor rRNA (Fig. 10.9). Extensive chemical modifications occur in the 13,00 – nucleotide – long precursor rRNA before the rRNAs are cleaved out of it and assembled into ribosomes. (Note. The 5SrRNA does not require chemical modification). Such modifications include about 10 methylations of the 2'- OH positions on nucleotide sugars and 100 isomerizations of uridine nucleotides to pseudouridine (Fig. 10.10A). The functions of these modifications are not

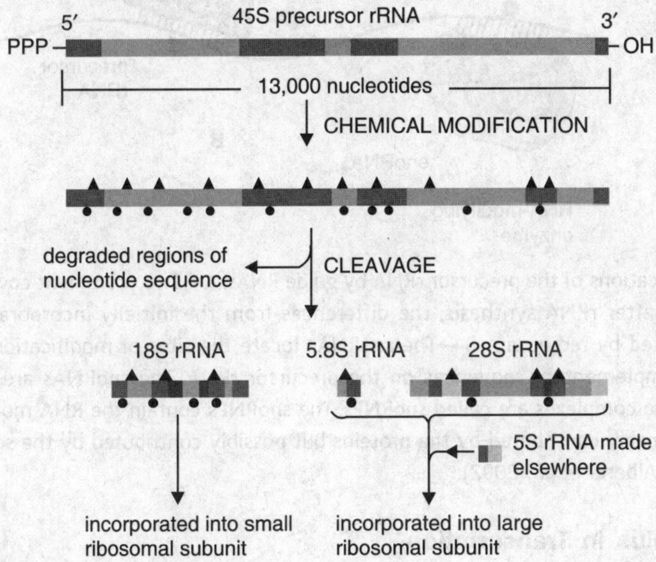

Fig. 10.9. The chemical modification and nucleolytic processing of a eukaryotic 45S precursor rRNA molecule into three separate ribosomal RNAs (*i.e.*, 18S, 5.8S and 28SrRNAs) (Alberts *et al.*, 2002).

understood in detail, but they probably aid in the folding and assembly of the final rRNAs and may also delicately alter the function of ribosomes. Each modification is made at a specific position in the precursor RNA. These positions are specified by several hundred **"guide RNAs"**, which locate themselves through base pairing to the precursor rRNA and thereby brings on RNA modifying enzyme to the appropriate position (Fig. 10.10B). Other guide RNAs promote, probably by causing conformational changes in the precursor RNA, the cleavage of the precursor rRNA into the rRNAs.

All of these guide RNAs are members of a large class of RNA, called **small nucleolar RNAs** (or **snoRNAs**), so named because these RNAs perform their functions in a sub-compartment of the nucleus, the nucleolus. Many snoRNAs are encoded in the introns of other genes especially those encoding ribosomal proteins. They are therefore synthesized by RNA polymerase II and processed from excised intron sequence.

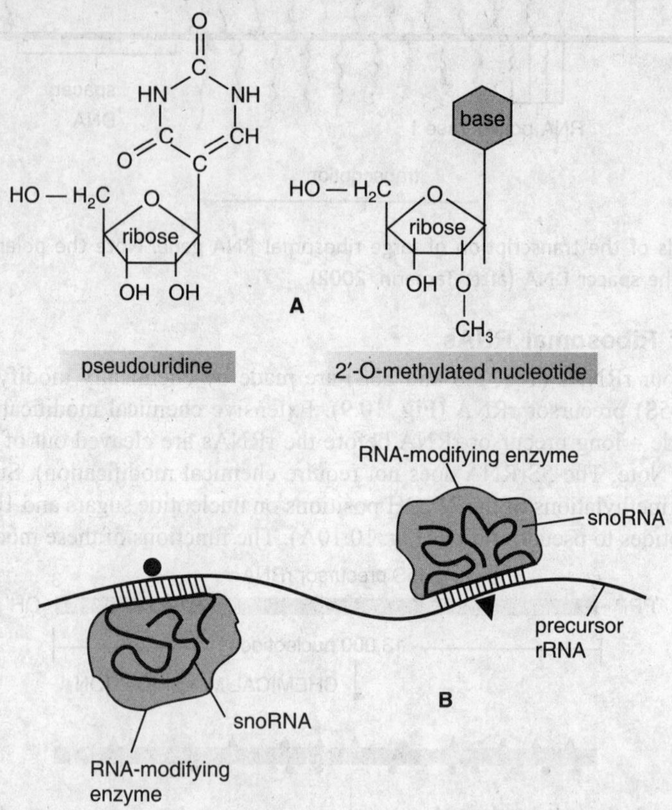

Fig. 10.10. Modifications of the precursor rRNA by guide RNAs. A – Two prominent covalent modifications occur after rRNA synthesis; the differences from the initially incorporated nucleotide are indicated by red atoms. B – The snoRNAs locate the sites of modification by base – pairing to complementary sequences on the precursor rRNA. The snoRNAs are bound to proteins and the complexes are called snoRNPs. The snoRNPs contain the RNA modification activities, presumably contributed by the proteins but possibly contributed by the snoRNAs themselves (after Alberts *et al.*, 2002).

Role of Nucleolus in Transcription

In addition to the important role in ribosome biogenesis (Fig. 10.11), the nucleolus is also the site where other RNAs are produced and other RNA protein complexes are assembled. For example, the U6 snRNP which functions in pre-mRNA splicing, is composed of one RNA molecule and at least seven proteins. The U6 snRNA is chemically modified by snoRNAs in the nucleolus before its final assembly there into the U6 snRNP. Other important RNA protein complexes including **telomerase** and the **signal recognition particle,** are also believed to be assembled at the nucleolus. Finally, the tRNAs are processed as well in the nucleolus. Thus, the nucleolus can be thought of as a large factory at which many different noncoding RNAs are processed and assembled with proteins to form a large variety of ribonucleoprotein complexes.

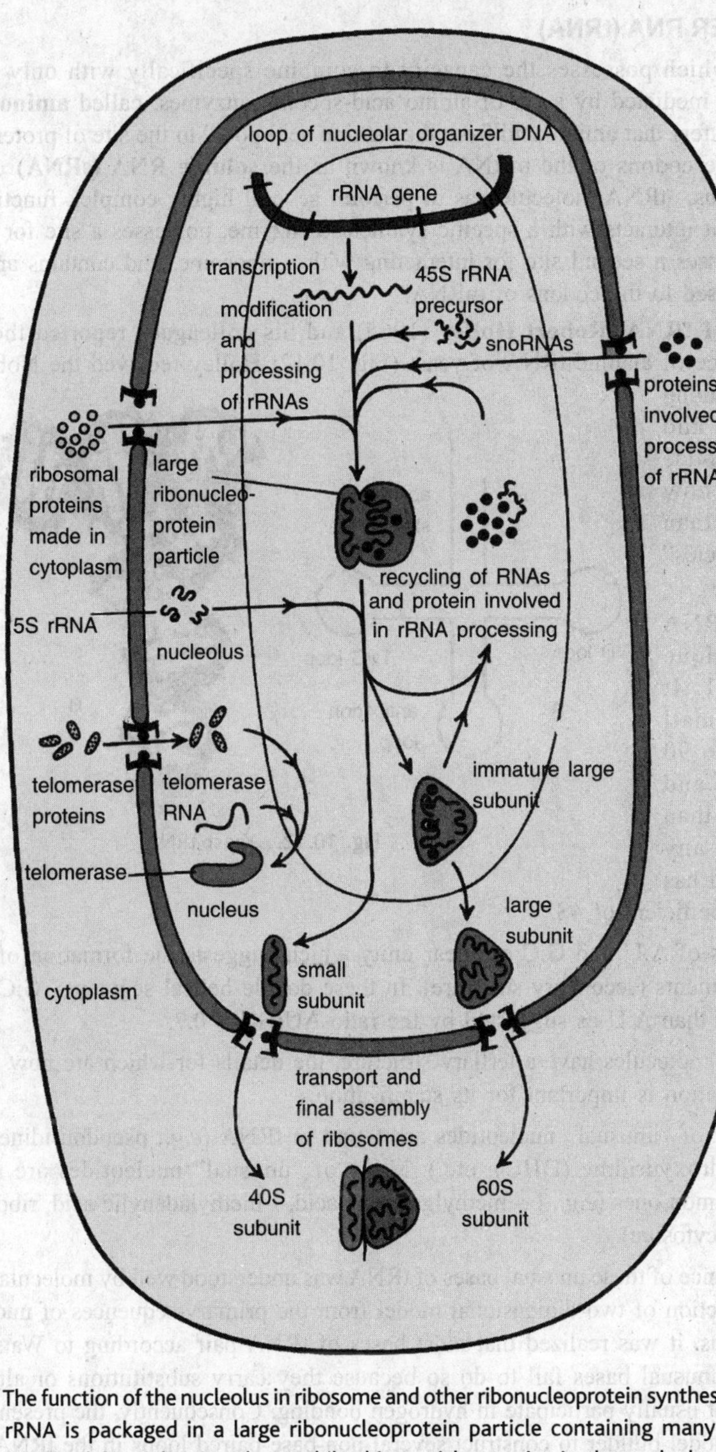

Fig. 10.11. The function of the nucleolus in ribosome and other ribonucleoprotein synthesis. The 45S precursor rRNA is packaged in a large ribonucleoprotein particle containing many ribosomal proteins imported from the cytoplasm while this particle remains in the nucleolus, selected pieces are added and others discarded as it is processed into immature large and small ribosomal subunits. The two ribosomal subunits are thought to attain their final functional form only as each is individually transported through the nuclear pores into the cytoplasm. Other ribonucleoprotein complexes (*e.g.*, telomerase) are also assembled in the nucleolus (after Alberts *et al.*, 2002).

10.2 TRANSFER RNA (tRNA)

The RNA which possesses the capacity to combine specifically with only one amino acid in a reaction mediated by a set of amino acid-specific enzymes, called **aminoacyl-tRNA synthetases;** transfers that amino acid from the "amino acid pool" to the site of protein synthesis and recognizes the codons of the mRNA is known as the **soluble RNA (sRNA)** or **transfer RNA (tRNA).** Thus, tRNA molecule has to perform several highly complex functions during protein synthesis–it interacts with a specific synthetase enzyme, possesses a site for binding an amino acid, possesses a second site for interacting with a ribosome, and contains an anticodon that must be exposed to the codons of mRNA.

Structure of tRNA. Robert Holley (1965) and his colleagues reported the complete nucleotide sequence of alanine tRNA of yeast (Fig. 10.12; Holley received the Nobel Prize in 1968 for his work along with **Khurana** and **Nirenberg**). Nucleotide sequences are now known for more than 100 different "species" of tRNA.

Transfer RNA has several unique characteristics: 1. It is a relatively small molecule of 75 to 90 ribonucleotides and is, thus, smaller than either mRNA or any of the rRNAs, and has a sedimentation coefficient of 4S.

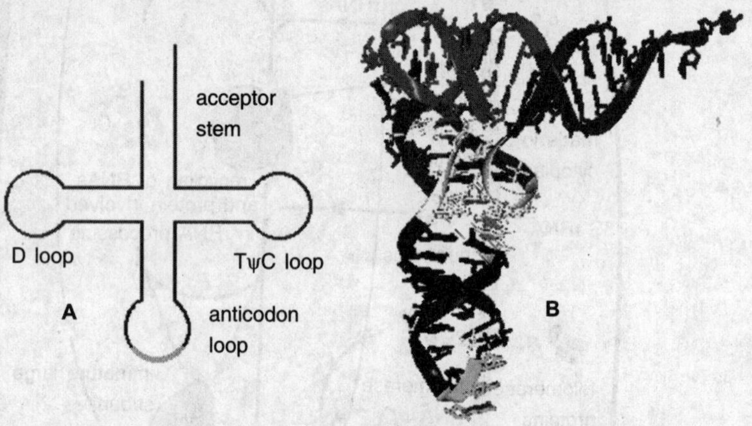

Fig. 10.12. Yeast tRNA.

2. The ratios of A:U and G:C are near unity which suggests the formation of DNA-like double helical segments (secondary structure). In these double helical segments, G:C base pairs are more common than A:U as suggested by the ratio AU:GC = 0.7.

3. All tRNA molecules have a tertiary structure, the details for which are now known and Mg^{2+} ion concentration is important for its stabilization.

4. A number of "unusual" nucleotides are found in tRNA (*e.g.*, pseudouridine ψ or psi), inosine (1), dihydroxyuridine (DHU), etc.). Many of "unusual" nucleotides are methylated derivatives of common ones (*e.g.*, I – methylguanylic acid, I-methyladenylic acid, ribothymidylic acid and 5-methylcytosine).

The significance of these unusual bases of tRNA was understood well by molecular biologists during the construction of two-dimensional model from the primary-sequences of nucleotides of known tRNA. Thus, it was realized that most bases of tRNA pair according to Watson-Crick's pairing rule, but unusual bases fail to do so because they carry substitutions or alterations in those positions that usually participate in hydrogen bonding. Consequently, the presence of these bases forces the model builder to construct several non-base-paired loops in the tRNA molecule. By working on these lines, **R.Holley** (1965) first of all proposed a clover leaf model for yeast $tRNA^{ala}$. The clover leaf model of tRNA because accommodated several of the known functions of tRNA, so, it gained general acceptance. A typical clover-leaf model tRNA (Fig. 10.13) depicts the following structural peculiarities:

(*i*) All tRNA molecules have guanine residue G at the 5′ terminal end and unpaired (single stranded) C-C-A sequence at the 3′ end. This is called **amino acid attachment** site or **carrier end**, because the amino acid becomes covalently attached to adenylic acid or A of CCA sequence during polypeptide synthesis. If a tRNA lacks this sequence at its 3′ end then tRNA is inactive.

(*ii*) The **amino acid stem** or **helix** consists of seven paired bases.

(*iii*) The **T-stem** is composed of five paired bases-the last (*i.e.,* nearest the **T-loop** or TψC loop is C-G. T-loop contains seven unpaired bases and is involved in the binding of rRNA molecules to the ribosomes.

(*iv*) The **anticodon stem** includes five paired bases. The **anticodon loop** consists of seven unpaired bases, the third, the fourth and the fifth of which (from the 3') end of the molecule) constitute the **anticodon**. The anticodon permits temporary complementary pairing with three bases (triplet codon) on mRNA.

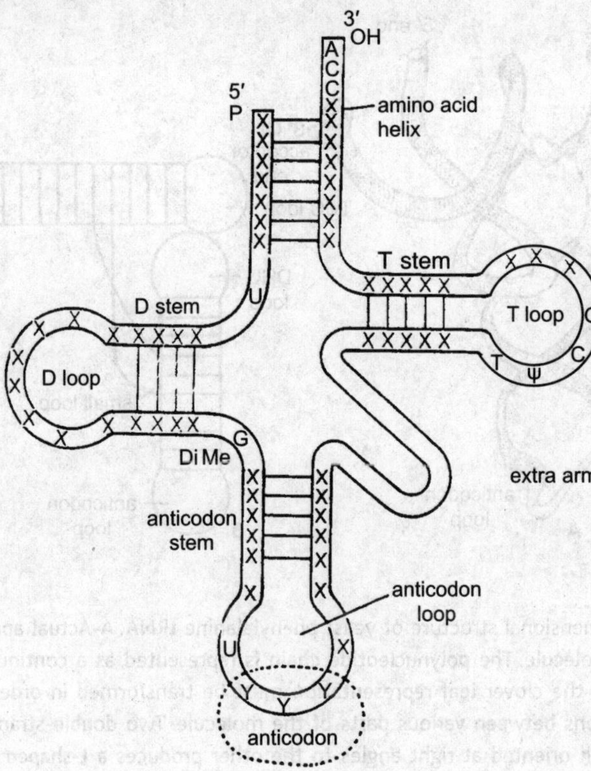

Fig. 10.13. Generalized two-dimensional clover-leaf model of tRNA, based on analyses of several yeast tRNA molecules by various investigators. Note the common 3′ terminal–CCA, the TCG in the T-loop, the anticodon-U (here YYYU) in the anticodon loop, the DiMeG between the anticodon loop and D loop and the U as the first unpaired based on the 5′ strand. As diagrammed here, the anticodon is read from right to left 3′ → 5′ (A=adenosine, C=cytidine, G=guanosine, T=ribothymidine, U=uridine, ψ = pseudouridine, DiMeG=dimethylguanosine, Y=any base of the anticodon (after Burns and Bottino, 1989).

(*v*) The base on the 3′ side of anticodon is a purine.

(*vi*) Immediately adjacent to the 5′ side of the anticodon, uracil and another pyrimidine occurs.

(*vii*) A purine, often dimethylguanylic acid, is located in the "corner" between the anticodon stem and the D stem.

(*viii*) The **D-stem** is composed of three or four base pairs (depending on the "species" of tRNA). The **DHU-loop** or **D-loop** (DHU=dihydrouracil) is also variable in size containing 8 to 12 unpaired bases. The D-loop helps in binding of amino-acyl synthetase.

(*ix*) The extra arm is variable in nucleotide composition and is lacking entirely in some tRNA.

Three-dimensional structure of tRNA. In order to understand the structure-function relationship of tRNA, its three dimensional structure (TDS) was worked out by the help of X-ray crystallography study. **A. Klug,** the Nobel laureate of 1982, has contributed much to the TDS of tRNAs. **S.H.Kim** (1973) proposed a most acceptable TDS model of tRNA of yeast cells (Fig. 10.16). According to **Kim**, TDS of tRNA takes the shape of letter L with a thickness of 10Å. Each arm of the L doubled over by bonds holding complementary base together. Such an L-shape can also easily derived from two-dimensional clover-leaf model (Fig. 10.14).

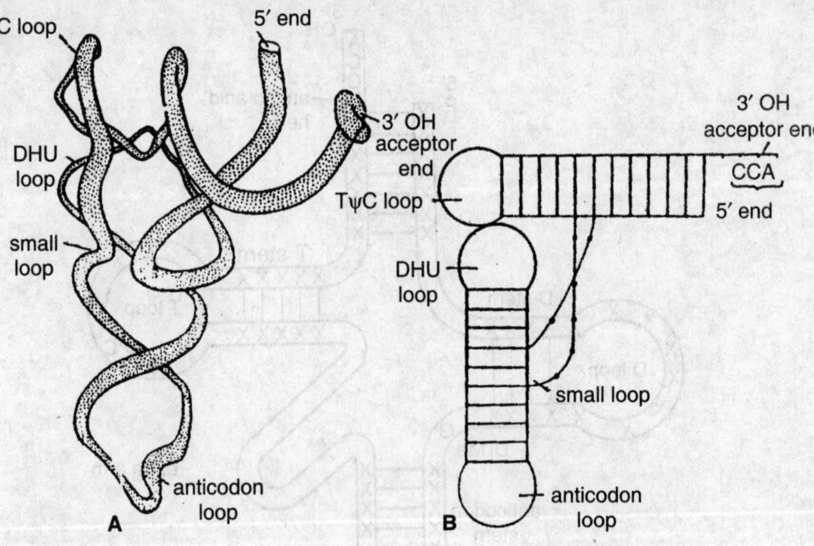

Fig. 10.14. Three-dimensional structure of yeast phenylalanine tRNA. A-Actual appearance of the folding of the molecule. The polynucleotide chain is represented as a continuous coiled tube. B-The way that the clover leaf representation must be transformed in order to show the physical connections between various parts of the molecule. Two double-stranded helical regions are seen, each oriented at right angles to the other produces a L-shaped structure.

Extended anticodon hypothesis. Recently it is reported that the performance of anticodon, when isolated from tRNA, is weak and inaccurate. However, the performance of this anticodon triplet is enhanced if a matching sequence is present on the anticodon loop and on the stem on either side of anticodon triplet. As a result, an **extended anticodon hypothesis** has been proposed which suggests that the structures of anticodon loop and that of the proximal anticodon stem are related to the sequence of anticodon (**Michael Yarus,** 1982). Thus, the anticodon is extended into the nearby sequence and consists of in all 12 nucleotides arranged in the following way: (*i*) two nucleotides (*i.e.*, CU, ψU or UU) at the 5′ side of anticodon loop; (*ii*) three nucleotides of anticodon (3′ nucleotide of anticodon being very important and termed **cardinal nucleotide);** (*iii*) two nucleotides at the 3′ side of anticodon loop; and (*iv*) five pairs of nucleotides in the anticodon stem which can be conveniently written by giving only the bases on 3′ side of the stem (Fig. 10.15).

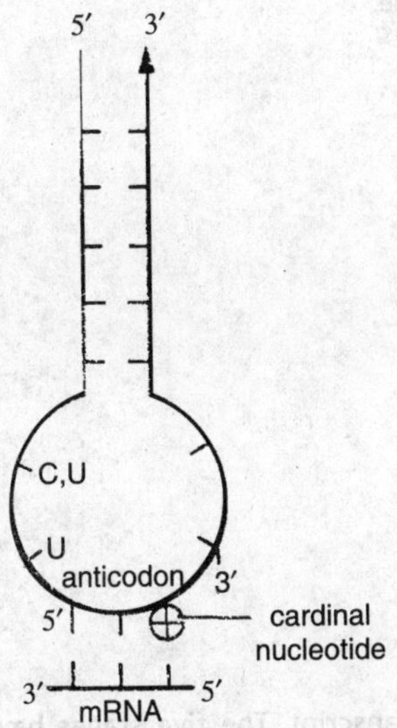

Fig. 10.15. General pattern of organization of nucleotide sequence around the anticodon, *i.e.*, in anticodon and loop and stem of tRNA.

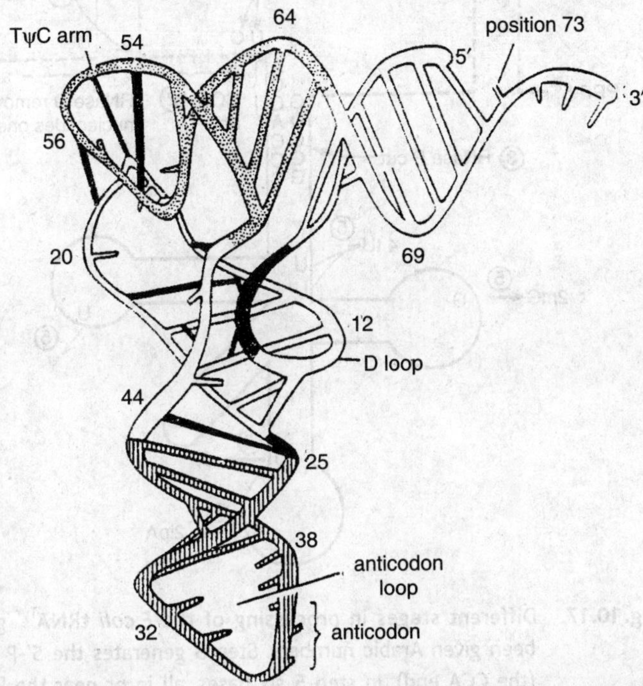

Fig. 10.16. Three-dimensional model proposed by Kim *et al.*, (1974) for yeast phenylalanine tRNA molecule (after Burns and Bottino, 1989).

Processing of tRNA. Transfer RNA is transcribed from several particular sites on template DNA and comprised of about 15 percent of the RNA present at any one time in an *E.coli* cell. Like the synthesis of rRNA and mRNA, synthesis of tRNA molecules is initiated at a promoter site and completed at a terminator sequence. Transfer RNAs are also processed from larger precursors. In prokaryotes, tRNA processing includes removal of nucleotides from the precursor or primary transcript and modification of some internal nucleotides. For example, in *E.coli,* the precursor for the amino acid tyrosine (pre $tRNA^{tyr}$) consists of 128 bases; in processing to $tRNA^{tyr}$ 41 nucleotides are removed from the 5′ end and two are removed from the 3′ end (Fig. 10.17).

In addition to processing, methylation and inclusion of other "unusual" intercalary bases takes place after transcription, as does the addition of the 3′ terminal–C-C-A. The final functional tRNA molecule has 85 bases. Three-dimensional shape is achieved through hydrogen bonding. The processing of eukaryotic tRNAs resembles with that of prokaryotes. However, in them it is far more complex.

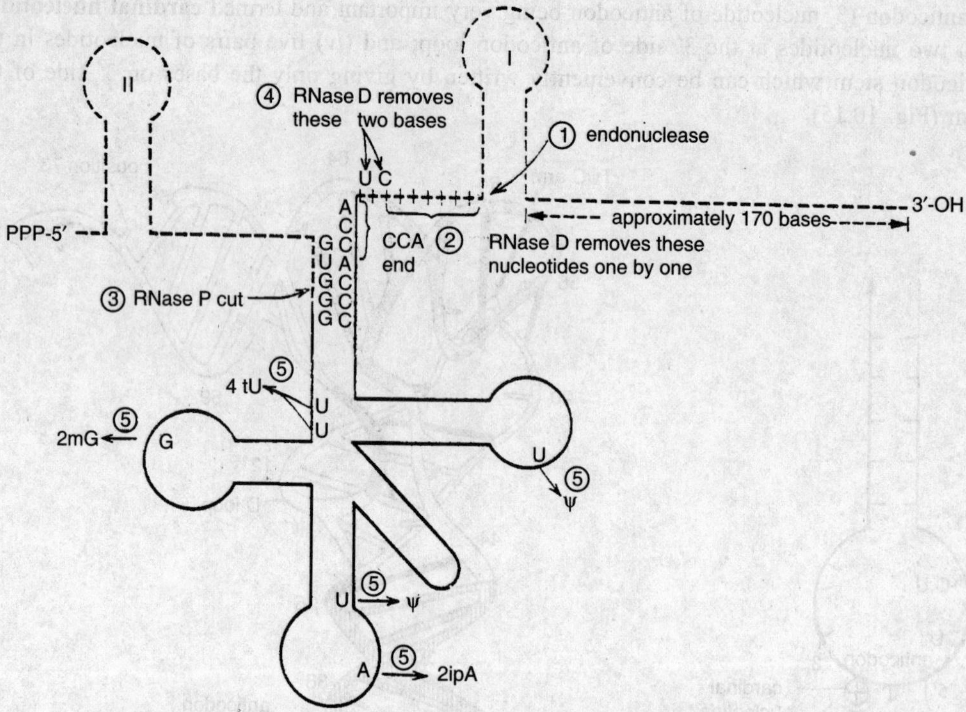

Fig. 10.17. Different stages in processing of the *E.coli* tRNA^{Tyr} gene transcript. The five stages have been given Arabic numbers. Step 3 generates the 5′-P end. Step 4 generates the 3′-OH end (the CCA end). In step 5 six bases, all in or near the loops of tRNA molecule, are modified to form pseudouridine (ψ, psi), 2-isopentenyladenosine (2 ip A), 2-O-methylguanosine (2 mg), and 4-thiouridine (4tU). The continuous sequence that forms the final tRNA molecule is given in solid line (after Freifelder, 1985).

Splicing of tRNA Precursors

The tRNA precursor splicing reaction has been worked out in detail in the yeast *Saccharomyces cerevisiae.* Both *in vitro* splicing systems and temperature-sensitive splicing mutants have been used in dissecting the tRNA splicing mechanism in *S.cerevisiae.* The excision of introns from yeast tRNA precursors occurs in two stages (Fig. 10.18).

1. In the first step, a nuclear membrane-bound splicing **endonuclease** enzyme makes two cuts precisely at the ends of the intron.

2. Then, in a fairly complex set of reactions, a splicing **ligase** enzyme joins the two halves of the tRNA to produce the mature form of the tRNA molecule.

The specificity of these reactions exists in the conserved three dimensional structural features of the tRNA precursors, not in the nucleotide sequences per se (*i.e.*, by itself). Cleavage of the tRNA precurson yields 5′–OH termini and 2′–3′ cyclic phosphate groups at the 3′ termini. The stage II ligation process actually involves four separate reactions:

1. The first reaction is the **addition** of a phosphate group to the 5′-OH terminus; this reaction requires **kinase** (enzyme) and a **phosphate donor** (ATP).

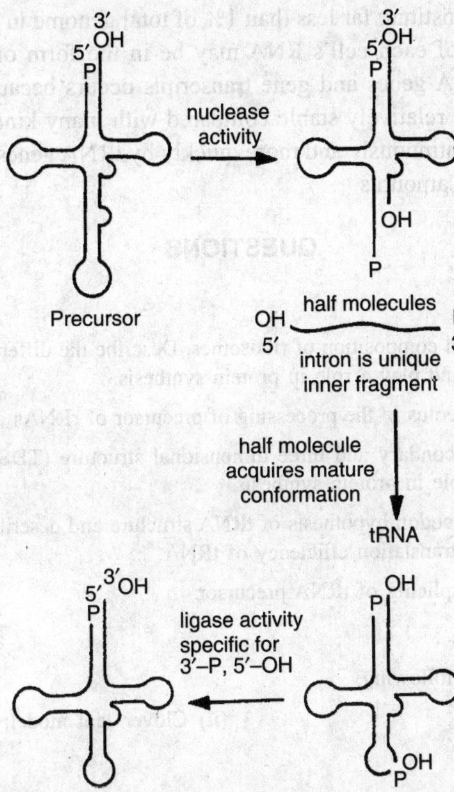

Fig. 10.18. Splicing mechanism of yeast tRNA, due to nuclease and ligase activities.

2. The second reaction, the 5′ phosphate group is activated by the transfer of an AMP group to the terminus from an AMP–ligase intermediate (the AMP originally having been derived from ATP also).

3. The 2′-3′cyclic phosphate is opened by a cyclic phosphodiesterase activity that produces a 2'-phosphate and a free 3′ hydroxyl.

4. The final ligation reaction occurs via a nucleophilic attack of the free 3′-OH on the interior 5′ phosphate with the release of AMP.

All four of these reactions are catalyzed by the splicing ligase. Finally, the 2′-phosphate group (remaining from the 2′-3′cyclic phosphate produced by the original cleavage reaction) is removed by a phosphatase activity to yield the mature tRNA molecule. The overall two stage mode of tRNA intron excision appears to occur in other organisms as well, *e.g.*, plants, mammals, etc., (**Gardner** *et al.*, 2002).

Earlier we considered a simple definition of a gene as a length of DNA that codes for one protein. But we have just encountered a variance: genes code code for both tRNAs and rRNAs, yet neither is eventually translated into a protein. Their transcription function as final products without ever being translated. Thus, tRNA and rRNA are the major exceptions to the general rule that a gene codes for a protein.

Genes for tRNA. There are, probably, at least 30 to 40 different tRNA genes and tRNA molecules in *E.coli.* Higher organisms are found to contain 60 tRNA molecules and 60 tRNA genes. Since the cell uses only 20 amino acids in protein synthesis (and probably ony 20 synthetase enzymes), it follows that several tRNA will often have an affinity for the same amino acid. For example, *E.coli* cells contain five species of tRNA for leucine amino acid.

All the tRNA genes constitute far less than 1% of total genome in both *E.coli* and eukaryotic cells, yet some 10 to 15% of each cell's RNA may be in the form of tRNA. This discrepancy between the number of tRNA genes and gene transcripts occurs because of the following facts: (1) The tRNA molecules are relatively stable compared with many kinds of RNA. (2) The tRNA molecules are transcribed continuously and more quickly by tRNA genes than other RNAs because they are needed in plentiful amounts.

QUESTIONS

Long Answer Questions

1. Describe the structure and composition of ribosomes. Describe the different sites and enzymes which are found in ribosomes and play a role in protein synthesis.
2. Describe the role of nucleolus in the processing of precursor of rRNAs and biogenesis of ribosomes.
3. Describe the primary, secondary and three dimensional structure (TDS) of transfer RNA structure and relate this with its role in protein synthesis.
4. Discuss the extended anticodon hypothesis of tRNA structure and describe how sequences other than anticodon determine the translation efficiency of tRNA.
5. Describe the process of splicing of tRNA precursor.

Short Answer Questions

1. Write short notes on the following:
   (*i*) Protein modification; (*ii*) Clover–leaf model;
   (*iii*) Cardinal nucleotides.

Very Short Answer Questions

1. What are ribosomal RNAs (rRNAs)?
2. What is function of RNA polymerase I?
3. What is the role of RNA polymerase III?
4. Define transfer RNAs (tRNAs).
5. Explain the role of carrier end of tRNA.
6. Who is discoverer of tRNA?

Fill in the Blanks

1. In eukaryotic cells, all ribosomal RNAs (*i.e.,* 5.85, 185 and 285) but .... rRNA are transcribed as part of the same piece of RNA.
2. Because each RNA gene has a fixed initiation site (promoter) and a fixed termination site, the transcripts adopt the characteristic .... configuration.
3. Processing of rRNA is done by .... RNA.
4. The snoRNA stands for .... which function inside the ....
5. Since transfer RNAs are soluble in IM-NaCl so are known as ....
6. Clover leaf model of transfer RNA of yeast was proposed by .... in 1965.
7. .... of tRNA permits temporary complementary pairing with three bases (triplet ....) on ....
8. The tRNA molecules are smallest sized RNA molecules and are formed of .... to .... nucleotides.
9. .... of tRNA is amino acid recognition site.

10. In *Drosophila* about .... cistrons are involved in the biogenesis of tRNA.

11. Ribosomes are measured in units that describe their rate of sedimentation during .... centrifugation in sucrose.

Multiple Choice Questions

1. The function of nucleolus is the synthesis of

(*a*) DNA (*b*) mRNA
(*c*) rRNA (*d*) tRNA

2. Functional unit in synthesis of protein is

(*a*) peroxisome (*b*) dictyosome
(*c*) lysosome (*d*) polysome

3. An anticodon of tRNA represents

(*a*) Wobble hypothesis (*b*) template hypothesis
(*c*) gene flow hypothesis (*d*) Richmond and Long effect

4. Which of the following RNAs picks up specific amino acid from amino acid pool in the cytoplasm to ribosome during protein synthesis?

(*a*) tRNA (*b*) mRNA
(*c*) rRNA (*d*) all of them

5. Which site of the tRNA molecule hydrogen bonds to a mRNA molecule?

(*a*) codon (*b*) anticodon
(*c*) 5′ end of tRNA (*d*) 3′ end of tRNA

6. Clover leaf model belongs to

(*a*) tRNA (*b*) DNA
(*c*) centriole (*d*) flagella

7. An anticodon is complementary to the nucleotide triplet in

(*a*) tRNA (*b*) rRNA
(*c*) mRNA (*d*) cDNA

8. Which of the character is not applicable to tRNA?

(*a*) it is the smallest of the RNAs
(*b*) it acts as an adapter for amino acids
(*c*) it has clover leaf structure
(*d*) it is the largest of the RNAs
(*e*) It bears anticodon

9. According to Wobble hypothesis

(*a*) first base is unstable
(*b*) second base is unstable
(*c*) third base is unstable
(*d*) the process of polypeptide chain elongation has been established

10. The codon for anticodon 3′UUA 5′ is

(*a*) 5′AAU 3′ (*b*) 3′ AAU 5′
(*c*) 5′AAT 3′ (*d*) 3′ AAC 5′

11. tRNA recognizes aminoacyl synthetase enzyme by

(*a*) anticodon (*b*) DHU loop

(*c*) T Ψ C loop (*d*) AA-site

12. Shape of 3–D view of tRNA is

(*a*) Z-shaped (*b*) X-shaped

(*c*) Y-shaped (*d*) L-shaped

13. Amino acid binding site in tRNA is

(*a*) 5′ end (*b*) anticodon loop

(*c*) CCA 3′ end (*d*) DHU loop

14. Genetic information for the synthesis of ribosomal RNA is coded in

(*a*) DNA present in nucleus

(*b*) nucleolar associated chromatin

(*c*) granular zone of nucleolus

(*d*) amorphous zone of nucleolus

ANSWERS

Very Short Answer Questions

1. The RNAs of a ribosome are called ribosomal RNAs (rRNAs). rRNAs recognize and bind other molecules, provide structural support and catalyze the chemical reaction in which amino acids are covalently linked to one another.

2. RNA polymerase I is a transcribing enzyme found in eukaryotic cells that synthesizes the large (285, 185 and 5.8S) ribosomal RNAs.

3. RNA polymerase III is a transcribing enzyme found in eukaryotic cells that synthesizes the various transfer RNAs and 5S ribosomal RNA.

4. Transfer RNAs are a family of small RNAs that translate the information encoded in the nucleotide "alphabet" of an mRNA into the amino acid "alphabet" of a polypeptide.

5. Each tRNA, at its 3′ end, has—CCA base sequence and represents the site for binding of specific activated amino acid as carboxyl group of the acid is attached to 3′ OH of adenosine (A) moeity of tRNA by an ester bond. It is nonspecific end as it is similar in all tRNAs. It is called *carrier end* of tRNA. If a tRNA lacks this sequence at its 3′ end then tRNA is inactive.

6. Mahlon Hoagland in 1958.

Fill in the Blanks

1. 5S. **2.** Christmas tree.

3. Guide. **4.** Small nucleolar RNA, nucleolus.

5. Soluble RNAs (sRNAs). **6.** Robert Holley.

7. Anticodon. **8.** 73, 93.

9. DHU loop. **10.** 55.

11. Density gradient.

Multiple Choice Questions

1. (*c*) **2.** (*d*) **3.** (*a*) **4.** (*a*) **5.** (*b*) **6.** (*a*)

7. (*c*) **8.** (*d*) **9.** (*c*) **10.** (*a*) **11.** (*b*) **12.** (*d*)

13. (*c*) **14.** (*b*).

11

Genetic Code

As DNA is a genetic material, it carries genetic informations from cell to cell and from generation to generation. Inside each cell this genetic information flows and ultimately form the proteins (the bases of the phenotype of the organisms (Fig. 11.1) At this stage, an attempt will be made to determine that in what manner the genetic informations are existed in DNA molecule? Are they written in articulated or coded language on DNA molecule? If in the language of codes what is the nature of genetic code?

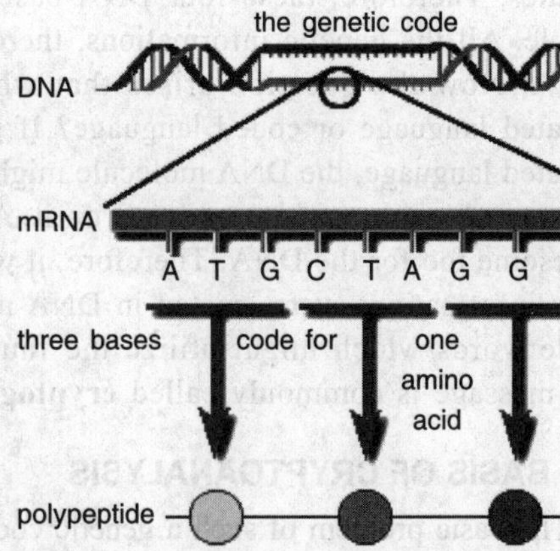

Fig. 11.1. Flow of genetic information.

A DNA molecule is composed of three kinds of moieties: (*i*) phosphoric acid, (*ii*) deoxyribose sugar, and (*iii*) nitrogen bases. The genetic informations may be written in any one of the three moieties of DNA. But the poly-sugar-phosphate backbone is always the same, and it is, therefore, unlikely that these moieties

	U	C	A	G
U	UUU, UUC — Phe UUA, UUG — Lou	UCU, UCC, UCA, UCG — Ser	UAU, UAC — Tyr UAA, UAG — Stop	UGU, UGC — Cys UGA — Stop UGG — Trp
C	CUU, CUC, CUA, CUG — Leu	CCU, CCC, CCA, CCG — Pro	CAU, CAC — His CAA, CAG — Gin	CGU, CGC, CGA, CGG — Arg
A	AUU, AUC, AUA — Ile AUG — Met	ACU, ACC, ACA, ACG — Thr	AAU, AAC — Asn AAA, AAG — Lys	AGU, AGC — Ser AGA, AGG — Arg
G	GUU, GUC, GUA, GUG — Val	GCU, GCC, GCA, GCG — Ala	GAU, GAC — Asp GAA, GAG — Glu	GGU, GGC, GGA, GGG — Gly

Fig. 11.2. The standard genetic code.

of DNA molecule carry the genetic informations. The nitrogen bases, however, vary from one segment of DNA to another, so the informations might well depend on their sequences. The sequences of nitrogen bases of a given segment of DNA molecule, actually, has been found to be identical to linear sequence of amino acids in a protein molecule. The proof of such a **colinearity** between DNA nitrogen base sequence and amino acid sequence in protein molecules has first obtained from an analysis of mutants of head protein of bacteriophage T_4 (**Sarabhai** *et al.,* 1964) and the A protein of tryptophan synthetase of *Escherichia coli* (**Yanofski** *et al*., 1964). The colinearity of protein molecules and DNA polynucleotides has given the clue that the specific arrangement of four nitrogen bases (*e.g.*, A, T, C and G) in DNA polynucleotides chains, somehow, determines the sequence of amino acids in protein molecules. Therefore, these four DNA bases can be considered as four alphabets of DNA molecule. All the genetic informations, therefore, should be written by these four alphabets of DNA. Now the question arises that whether the genetic informations are written in articulated language or coded language? If genetic informations might have occurred in an articulated language, the DNA molecule might require various alphabets, a complex system of grammar and ample amount of space on it. All of which might be practically impossible and troublesome too for the DNA. Therefore, it was safe to conclude for molecular biologists that genetic informations were existed in DNA molecule in the form of certain special language of code words which might utilize the four nitrogen bases of DNA for its symbols. Any coded message is commonly called **cryptogram.**

11.1 BASIS OF CRYPTOANALYSIS

The basic problem of such a genetic code is to indicate how information written in a four letter language (four nucleotides or nitrogen bases of DNA) can be translated into a twenty letter language (twenty amino acids of proteins). The group of nucleotides that specifies one amino acid is a **code word** or **codon**. By the **genetic code**, one means the collection of base sequences (codons) that correspond to each amino acid and to translation signals. We can consider here the classical but logical reasoning done by **George Gamov** (1954) about the possible size of a codon. The simplest possible code is a **singlet** code (a code of single letter) in which one nucleotide amino acid could be specified. A **doublet code** (a code of two letters) is also inadequate, because it could specify only sixteen (4 × 4) amino acids, whereas a **triplet code** (a code of three letters) could specify sixty four (4 × 4 × 4) amino acids (Fig. 11.2). Therefore, it is likely that there may be 64 triplet codes for 20 amino acids. The possible singlet, doublet and triplet codes, which are customarily represented in terms of "**mRNA language**" [mRNA is a complementary molecule which copies the genetic informations (cryptogram of DNA) during its transcription], have been illustrated in Table 11.1.

The first experimental evidence in support to the concept of triplet code is provided by **Crick** and coworkers in 1961. During their experiment, when they added or deleted single or double base pairs in a particular region of DNA of T_4 bacteriophages of *E. coli*, they found that such bacteriophages ceased to perform their normal functions. However, bacteriophages with addition or deletion of three base pairs in DNA molecule, had performed normal functions. From this experiment, they concluded that a genetic code is in triplet form, because the addition of one or two nucleotides has put the reading of the code out of order, while the addition of third nucleotide resulted in a return to the proper reading of the message.

Table 11.1 Possible singlet, doublet and triplet codes of mRNA.

Singlet code (4 words)	*Doublet code (16 words)*	*Triplet code (64 words)*
		AAA AAG AAC AAU
		AGA AGG AGC AGU
		ACA ACG ACC ACU
		AUA AUG AUC AUU
		GAA GAG GAC GAU
		GGA GGG GGC GGU
A	AA AG AC AU	GCA GCG GCC GCU
G	GA GG GC GU	GUA GUG GUC GUU
C	CA CG CC CU	CAA CAG CAC CAU
U	UA UG UC UU	CGA CGG CGC CGU
		CCA CCG CCC CCU
		CUA CUG CUC CUU
		UAA UAG UAC UAU
		UGA UGG UGC UGU
		UCA UCG UCC UCU
		UUA UUG UUC UUU

11.2 CODON ASSIGNMENT (CRACKING THE CODE OR DECIPHERING THE CODE)

The genetic code has been cracked or deciphered by the following kinds of approaches:

A. Theoretical Approach

The physicist **George Gamow** proposed the **diamond code** (1954) and the **triangle code** (1955) and suggested an exhaustive theoretical framework to the different aspect of the genetic code. **Gamow** suggested the following properties of the genetic code:

(*i*) **A triplet codon** corresponding to one amino acid of the polypeptide chain.

(*ii*) **Direct template translation** by codon-amino acid pairing.

(*iii*) Translation of the code in an **overlapping** manner.

(*iv*) **Degeneracy** of the code, *i.e.*, an amino acid being coded by more than one codon.

(*v*) **Colinearity** of nucleic acid and the primary protein synthesized.

(*vi*) **Universality** of the code, *i.e.*, the code being essentially the same for different organisms.

Some of these Gamow's proposals have been contradicted by the molecular biologists. For example, **Brenner** (1957) showed that the overlapping triplet code is an impossibility, and subsequent work has shown that the code is a **non-overlapping** one. Similarly, Gamow's idea of direct template relationship between nucleic acid and polypeptide chain was challenged when **Crick** proposed his **adopter hypothesis**. According to this hypothesis, **adaptor molecules** intervene between nucleic acid and amino acids during translation. In fact, it is now known that tRNA molecules act as adaptors between codons of mRNA and amino acids of the resulting polypeptide chain.

	T	C	A	G
T	TTT Phe (F) TTC" TTA Leu (L) TTG"	TCT Ser (S) TCC" TCA" TCG"	TAT Tyr (Y) TAC TAA Ter TAG Ter	TGT Cys (C) TGC TGA Ter TGG Trp (W)
C	CTT Leu (L) CTC" CTA" CTG"	CCT Pro (P) CCC" CCA" CCG"	CAT His (H) CAC" CAA Gln (Q) CAG"	CGT Arg (R) CGC" CGA" CGG"
A	ATT Ile (I) ATC" ATA" ATG Met (M)	ACT Thr (T) ACC" ACA" ACG"	AAT Asn (N) AAC" AAA Lys (K) AAG"	AGT Ser (S) AGC" AGA Arg (R) AGG"
G	GTT Val (V) GTC" GTA" GTG"	GCT Ala (A) GCC" GCA" GCG"	GAT Asp (D) GAC" GAA Glu (E) GAT "	GGT Gly (G) GGC" GGA " GGG"

Fig. 11.3. Degeneracy of genetic code.

B. The *in vitro* Codon Assignment

1. Discovery and use of polynucleotide phosphorylase enzyme. Marianne Grunberg-Manago and **Severo Ochoa** isolated an enzyme from the bacteria (*e.g., Azobacter vinelandii* or *Micrococcus lysodeikticus)* that catalyzes the breakdown of RNA in bacterial cells. This enzyme is called **polynucleotide phosphorylase. Manago** and **Ochoa** found that outside of the cell *(in vitro),* with high concentrations of ribonucleotides, the reaction could be driven in reverse and an **RNA** molecule could be made (see **Burns** and **Bottino**, 1989). Incorporation of bases into the molecule is random and does not require a DNA template. Thus, in 1955 **Manago** and **Ochoa** made possible the artificial synthesis of polynucleotides (=mRNA) containing only a single type of nucleotides (U, A, C, or G respectively) repeated many times.

Polynucleotide	Configuration
1. Polyuridylic acid or poly (U)	UUUUUU
2. Polyadenylic acid or poly (A)	AAAAAA
3. Polycytilic acid or poly (C)	CCCCCC
4. Polyguanidylic acid or poly (G)	GGGGGG

Thus, the action of polynucleotide phosphorylase can be represented in the following way:

$$(RNA)_n + \text{Ribonucleoside diphosphate} \xrightleftharpoons[]{\text{polynucleotide phosphorylase}} (RNA)_{n+1} + Pi$$

The polynucleotide phosphorylase enzyme differs from RNA polymerase used to transcribe mRNA from DNA in that: (*i*) it does not require a template or primer; (*ii*) the activated substrates are ribonucleoside diphosphates (*e.g.*, UDP, ADP, CDP and GDP) and not triphosphates; and (*iii*) orthophosphate (*Pi*) is produced instead of pyrophosphates (*PPi*).

The deciphering of the genetic code was made possible by the use of synthetic (or artificial) polynucleotides and trinucleotides. The different types of techniques used include the use of polymers containing a single type of nucleotide (called **homopolymers**), the use of mixed polymers (**copolymers**) containing more than one type of nucleotides (**heteropolymers**) in random or defined sequences and the use of trinucleotides (or "**minimessengers**") in ribosome-binding or filter-binding.

2. Codon assignment with unknown sequence. **(*i*) Codon assignment by homopolymer.** The first clue to codon assignment was provided by **Marshall Nirenberg** and **Heinrich Matthaei** (1961) when they used *in vitro* system for the synthesis of a polypeptide using an artificially synthesized mRNA molecule containing only one type of nucleotide *(i.e.*, homopolymer). Prior to performing the actual experiments, they tested the ability of a cell-free protein synthesizing system to incorporate radioactive amino acids into newly synthesized proteins. Their cell-free extracts of *E.coli* contained ribosomes, tRNAs, aminoacyl-tRNA synthetase enzymes, DNA and mRNA. The DNA of this extract was eradicated by the help of **deoxyribonuclease** enzyme, thus, the template which might synthesize new mRNA was destroyed. When twenty amino acids were added to this mixture along with ATP, GTP, K^+ and MG^{2+}, they were incorporated into proteins. This incorporation continued so long as mRNA was present in such a cell-free suspension. It also continued in the presence of synthetic polynucleotides (mRNAs) which could be made with the help of polynucleotide phosphorylase enzyme.

The first successful use of this technique was made by **Nirenberg** and **Matthaei** who synthesized a chain of uracil molecules [poly (U)] as their synthetic mRNA (homopolymer). Poly (U) seemed a good choice, because there could be no ambiguity in a message consisting of only one base. Poly (U) was a good choice for other reasons: it binds well to ribosomes and, as it turned out, the product protein was insoluble and easy to isolate. When poly (U) was presented as the message to the cell free system containing all the amino acids, one amino acid was exclusively selected from the mixture for incorporation into the polypeptide, called **polyphenylalanine**. This amino acid was phenylalanine and it could be concluded that some sequence of UUU coded for phenylalanine. Other homogeneous chains of nucleotides (Poly A, Poly C and Poly G) were inactive for phenylalanine incorporation. The mRNA code word for phenylalanine was, therefore, shown to be UUU. Thus, the first code word to be deciphered was UUU.

This discovery was extended in the laboratories of **Nirenberg** and **Ochoa**. The experiment was repeated using synthetic **poly (A)** and **poly (C)** chains, which gave **polylysine** and **polyproline** respectively. Thus, AAA was identified as the code for lysine and CCC as the code for proline. A poly (G) message was found non-functional *in vitro*, since it attains secondary structure and, thus, could not attach the ribosomes. In this way, three of 64 codons were easily accounted for.

(*ii*) Codon assignment by heteropolymers (Copolymers with random sequences). Further exposition of the genetic code took place by using synthetic messenger RNAs containing two kinds of bases. This technique was used in the laboratories of **Ochoa** and **Nirenberg** and led the deduction of the composition of codons for the 20 amino acids. The synthetic messengers contained the bases at random (called **random copolymers**). For example, in a random copolymer using U and A nucleotides eight triplets are possible, such as UUU, UUA, UAA, UAU, AAA, AAU, AUU and AUA. Theoretically, eight amino acids could be coded by these eight codons. Actual experiments, however, yielded only six phenylalanine, leucine, tyrosine, lysine, asparagine and isoleucine. By varying the relative compositions of U and A in the random copolymer and determining the percentage of the different amino acids in the proteins formed, it was possible to deduce the composition of the code for different amino acids.

3. Assignment of codons with known sequences. (*i*) Use of trinucleotides or minimessengers in filter binding (Ribosome-binding technique). Ribosome technique of **Nirenberg** and **Leder** (1964) made use of the finding that aminoacyl-tRNA molecules specificially bind to ribosome-mRNA complex. The binding does not require the presence of a long mRNA molecule; in fact, the association of a **trinucleotide** or **minimessenger** with the ribosome is sufficient to cause aminoacyl-tRNA binding. When a mixture of such small mRNA molecules-ribosomes and amino acid-tRNA complexes are incubated for a short time and then filtered through a **nitrocellulose membrane**, then the mRNA-ribosome-tRNA-amino acid complex is retained back and rest of the mixture passes through the filter. By using a series of 20 different amino acid mixtures, each containing one radioactive amino acid at a time, it is possible to find out the amino acid corresponding to each triplet by analysing the radioactivity absorbed by the membrane, *e.g.*, the triplet GCC and GUU retain only alanyl-tRNA and valyl-tRNA respectively. All 64 possible triplets have been synthesized and tested in this way. Forty five of them have given clear-cut results. Later on, with the help of longer synthetic messages it has been possible to decipher 61 out of the possible 64 codons.

C. The *in vivo* Codon Assignment

The cell free protein synthetic systems, though have proved of great significance in decipherment of the genetic code, but they could not tell us whether the genetic code so deciphered is used in the living systems of all organisms also. Three kinds of techniques are used by different molecular biologists to determine whether the same code is also used *in vivo* (*a*) amino acid replacement studies (*e.g.*, tryptophan synthetase synthesis in *E.coli* (**Yanofsky** *et al.* 1963) and haemoglobin synthesis in humans), (*b*) frameshift mutations (*e.g.*, investigations of **Terzaghi** *et al.* 1966, on lysozyme enzyme of T_4 bacteriophages, and (*c*) comparison of a DNA or mRNA polynucleotide cryptogram with its corresponding polypeptide clear text (*e.g., comparison* of amino acid sequence of the R_{17} bacteriophage coat protein with the nucleotide sequence of the R_{17} mRNA in the region of the molecule that dictates coat-protein synthesis by **S. Cory** *et al.*, 1970).

Thus, *in vitro* and *in vivo* studies, so far described, gave the way to formulate a code table for twenty amino acids (see Table 11.2 and Table 11.3).

Table 11.2 The genetic dictionary. The trinucleotide codons are written in the 5′→3′ direction.

First base	*Second base* U		C		A		G		*Third base*
U	UUU	Phe	UCU	Ser	UAU	Tyr	UGU	Cys	U
	UUC		UCC		UAC		UGC		C
	UUA	Leu	UCA		UAA	Non-† sense codon	UGA	Non-sense † codon	A
	UUG		UCG		UAG		UGG	Trp	G
C	CUU	Leu	CCU	Pro	CAU	His	CGU	Arg	U
	CUC		CCC		CAC		CGC		C
	CUA		CCA		CAA	Gln	CGA		A
	CUG		CCG		CAG		CGG		G
A	AUU	Ileu	ACU	Thr	AAU	Asn	AGU	Ser	U
	AUC		ACC		AAC		AGC		C
	AUA		ACA		AAA	Lys	AGA	Arg	A
	AUG*	Met	ACG		AAG		AGG		G
G	GUU	Val	GCU	Ala	GAU	Asp	GGU	Gly	U
	GUC		GCC		GAC		GGC		C
	GUA		GCA		GAA	Glu	GGA		A
	GUG		GCG		GAG		GGG		G

*AUG—Met or chain initiation codon; †UAA, UAG, UGA—Stop codon.

Table 11.3 Amino acids and their messenger RNA codons.

Alanine	–	GCA	Glycine	–	GGA	Leucine	–	CUA	Threonine	–	ACA
(Ala)		GCG	(Gly)		GGG	(Lcu)		CUG	(Thr)		ACG
		GCC			GGC			CUC			ACC
		GCU			GGU			CUU			ACU
Arginine	–	AGA	Glutamine	–	CAA			UUA			
(Arg)		AGG	(Gln; Glun)		CAG			UUG	Tryptophan	–	UGG
		CGA	Glutamic			Methionine	–	AUG	(Trp)		
		CGG	acid	–	GAA	(Met)			Tyrosine	–	UAC
		CGC	(Glu)		GAG	(Starting			(Tyr)	–	UAU
		CGU	Histidine	–	CAC	codon)			Valine	–	GUA
Aspartic			(His)		CAU	Phenylalanine	–	UUC	(Val)		GUG*
acid	–	GAC	Isoleucine	–	AUC	(Phe)		UUU			GUC
(Asp)		GAU	(Ile)		AUU						GUU
Asparagine	–	AAC			AUA	Proline	–	CCA			
(Asn; Aspn)		AAU				(Pro)		CCG	Terminator	–	UAA
Cysteine	–	UGC	Lysine	–	AAA			CCC	(Nonsense		(ocher)
(Cys)		UGU	(Lys)		AAG			CCU	codons)		UAG
						Serine	–	AGC			(amber)
						(Ser)		AGU			UGA
								UCA			(opal)
								UCG			
								UCC			
								UCU			

* GUG is also used as a start codon for some proteins.

11.3 CHARACTERISTICS OF GENETIC CODE

The genetic code has the following general properties:

1. The code is a triplet codon. The nucleotides of mRNA are arranged as a linear sequence of codons, each codon consisting of three successive nitrogenous bases, *i.e.*, the code is a triplet codon. The concept of triplet codon has been supported by two types of point mutations: frameshift mutations and base substitution.

***(i)* Frameshift mutations.** Evidently, the genetic message once initiated at a fixed point is read in a definite frame in a series of three letter words. The framework would be disturbed as soon as there is a deletion or addition of one or more bases. When such frameshift mutations were intercrossed, then in certain combinations they produce wild type normal gene. It was concluded that one of them was deletion and the other an addition, so that the disturbed order of the frame due to mutation will be restored by the other (Fig. 11.4).

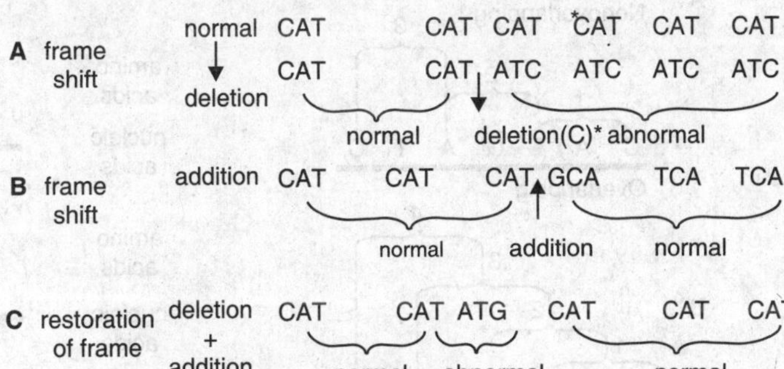

Fig. 11.4. Frame-shift mutations: A—Deletion; B—Addition; C—Restoration of frame.

***(ii)* Base substitution.** If in a mRNA molecule at a particular point, one base pair is replaced by another without any deletion or addition, the meaning of one codon containing such an altered base, will be changed. In consequence, in place of a particular amino acid at a particular position in a polypeptide, another amino acid will be incorporated. For example, due to substitution mutation, in the gene for tryptophan synthetase enzyme in *E.coli*, the GGA codon for glycine becomes a missence codon AGA which codes for arginine. **Missence codon** is a codon which undergoes an alteration to specify another amino acid.

A more direct evidence for a triplet code came from the finding that a piece of mRNA containing 90 nucleotides, corresponded to a polypeptide chain of 30 amino acids of a growing haemoglobin molecule. Similarly, 1200 nucleotides of "satellite" tobacco necrosis virus direct the synthesis of coat protein molecules which have 372 amino acids.

2. The code is non-overlapping. In translating mRNA molecules the codons do not *overlap* but are "read' sequentially (Fig. 11.5). Thus, **a non-overlapping code** means that a base in a mRNA is not used for different codons. In Figure 11.6., it has been shown that an overlapping code can mean coding for four amino acids from six bases. However, in actual practice six bases code for not more than two amino acids. For example, in case of an overlapping code, a single change (of substitution type) in the base sequence will be reflected in substitutions of more than one amino acid in corresponding protein. Many examples have accumulated since 1956 in which a single base substitution results into a single amino acid change in insulin, tryptophan synthetase, TMV coat protein, alkaline phosphatase, haemoglobin, etc.

Fig. 11.5. Bases in mRNA are read sequentially in the 5' to 3' direction, in group of three bases.

Recently, however, it has been shown that in the bacteriophage $\phi \times 174$ there is a possibility of overlapping of genes and codons (**Barrel** and coworkers, 1976; **Sanger**, *et al.*, 1977).

3. The code is commaless. The genetic code is commaless, which means that no codon is reserved punctuations. It means that after one amino acid is coded, the second amino acid will be automatically coded by the next three letters and that no letters are wasted as the punctuation marks (Fig. 11.7).

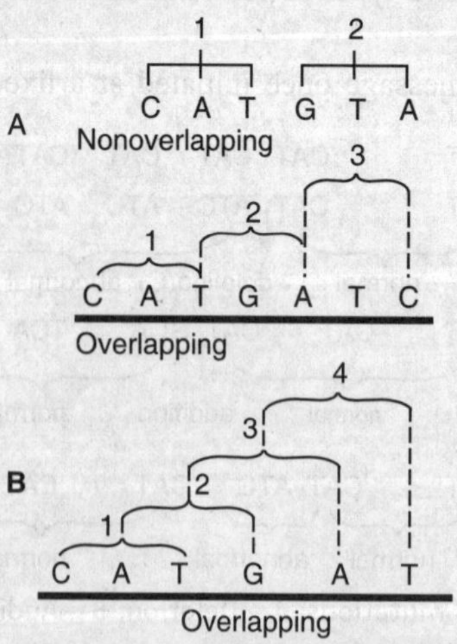

Fig.11.6. A—Nonoverlapping codons; B—Overlapping of codon due to one base; C—Overlapping of codon due to two bases.

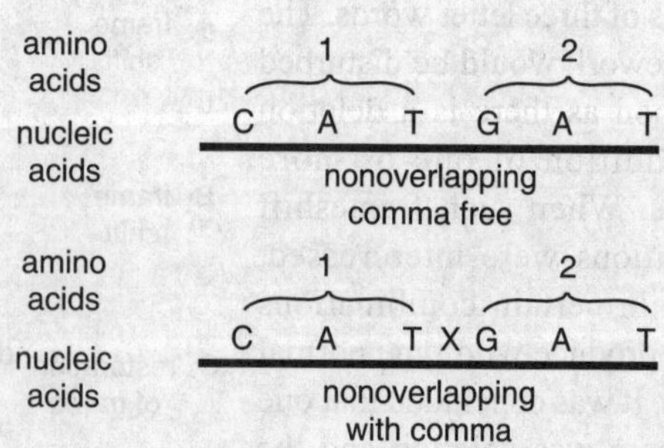

Fig. 11.7. Two forms of genetic code. A—Genetic code without comma; B—Genetic code with comma.

4. The code is non-ambiguous. Non-ambiguous code means that a particular codon will always code for the same amino acid. In case of ambiguous code, the same codon could have different meanings or in other words, the same codon could code two or more than two amino acids. Generally, as a rule, the same codon shall never code for two different amino acids. However, there are some reported exceptions to this rule: the codons AUG and GUG both may code for *methionine* as initiating or starting codon, although GUG is meant for *valine*. Likewise, GGA codon codes for two amino acids *glycine* and *glutamic acid* (Box 11.1).

5. The code has polarity. The code is always read in a fixed direction, *i.e.*, in the 5′ → 3′ direction. In other words, the codon has a **polarity**. It is apparent that if the code is read in opposite directions, it would specify two different proteins, since the codon would have reversed base sequence:

Codon:	UUG	AUC	GUC	UCG	CCA	ACA	AGG
Polypeptide: →	Leu	Ile	Val	Ser	Pro	Thr	Arg
	Val	Leu	Leu	Ala	Thr	Thr	Gly ←

6. The code is degenerate. More than one codon may specify the same amino acid; this is called **degeneracy** of the code (Fig. 11.3). For example, except for *tryptophan* and *methionine*, which have a single codon each, all other 18 amino acids have more than one codon. Thus, nine amino acids, namely *phenylalanine*, *tyrosine*, *histidine*, *glutamine*, *asparagine*, *lysine*, *aspartic acid*, *glutamic acid* and *cysteine*, have two codons each. Isoleucine has three codons. Five amino acids, namely *valine*, *proline*, *threonine*, *alanine* and *glycine*, have four codons each. Three amino acids, namely *leucine*, *arginine* and *serine*, have six codons each (see Table 11.3).

The code degeneracy is basically of two types: partial and complete. **Partial degeneracy** occurs when first two nucleotides are identical but the third (*i.e.*, 3′ base) nucleotide of the degenerate codons differs, *e.g.*, CUU and CUC code for leucine. **Complete degeneracy** occurs when any of the four bases can take third position and still code for the same amino acid (*e.g.*, UCU, UCC, UCA and UCG code for serine).

Degeneracy of genetic code has certain biological advantages. For example, it permits essentially the same complement of enzymes and other proteins to be specified by microorganisms varying widely in their DNA base composition. Degeneracy also provides a mechanism of minimizing mutational lethality.

7. Some codes act as start codons. In most organisms, AUG codon is the **start** or **initiation** codon, *i.e.*, the polypeptide chain starts either with **methionine** (eukaryotes) or **N-formylmethionine**

Box 11.1 Genetic code is degenerate not ambiguous

After the start and stop codons, the remaining 60 codons are far more than enough to code for the other 19 amino acids and indeed there are repeats. Thus we say that the genetic code is **degenerate,** that is, an amino acid may be represented by more than one codon. The degeneracy is not evenly divided among the amino acids. For example, methionine and tryptophan are represented by only one codon each, whereas leucine is represented by six different codons.

The term "degeneracy" should not be confused with "ambiguity". To say that the code was ambiguous would mean that a *single codon could specify either of two (or more) different amino acids;* there would be doubt whether to put in, say leucine or something else. The genetic code is not ambiguous. Degeneracy in the code means that *there is more than one clear*

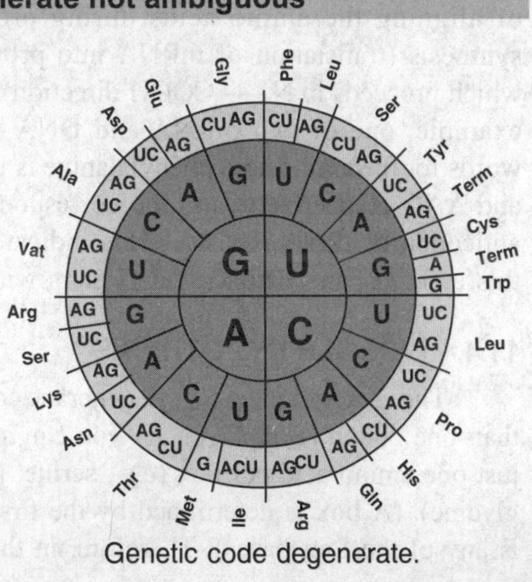

Genetic code degenerate.

way to say "put leucine here". In other words, a given amino acid may be encoded by more than one codon, but a codon can code for only one amino acid. But just as people in different places prefer different ways of saying the same thing —"Good-bye" ! "See you" | "Ciao"! and "So long"! have the same meaning — different organisms prefer one or others of the degenerate codons (Ciao means informal hello or good bye). These preferences are important in genetic engineering.

(prokaryotes). methionyl or N-formylmethionyl-tRNA specifically binds to the **initiation site** of mRNA containing the AUG initiation codon. In rare cases, GUG also serves as the initiation codon, *e.g.*, bacterial protein synthesis. Normally GUG codes for valine, but when normal AUG codon is lost by deletion, only then GUG is used as initiation codon.

8. Some codes act as stop codons. Three codons UAG, UAA, UGA are the chain **stop** or **termination** codons. They do not code for any amino acids. These codon are not read by any tRNA molecules (via their anticodons), but are read by specific proteins, called **release factors** (*e.g.*, RF-1, RF-2, RF-3 in prokaryotes and RF in eukaryotes). These codons are also called as **nonsense codons**, since they do not specify any amino acid.

The UAG was the first termination codon to be discovered by **Sidney Brenner** (1965). It was named **amber** after a graduate student named **Bernstein** (= the German word for 'amber' and amber means brownish yellow) who helped in the discovery of a class of mutations. Apparently, to give uniformity the other two termination codons were also named after colours such as **ochre** for UAA and **opal** or **umber** for UGA. (*Ochre* means yellow red or pale yellow; *opal* means milky white and *umber* means brown). The existence of more than one stop codon might be a safety measure, in case the first codon fails to function.

9. The code is universal. Same genetic code is found valid for all organisms ranging from bacteria to man. Such universality of the code was demonstrated by **Marshall, Caskey** and **Nirenberg** (1967) who found that *E.coli* (bacterium), *Xenopus laevis* (amphibian) and guinea pig (mammal) amino acyl-tRNA use almost the same code. **Nirenberg** has also stated that the genetic code may have developed 3 billion years ago with the first bacteria, and it has changed very little throughout the evolution of living organisms.

Codon and Anticodon

The codon words of DNA would be complementary to the mRNA code words (*i.e.*, DNA codes run in 3′→ 5′ direction and mRNA code words run in 5′ → 3′ direction) (Fig. 11.8) and so thereby the three bases forming the **anticodon** of tRNA (*i.e.*, bases of anticodons run in 3′ → 5′ direction (Fig. 11.10)). Three bases of anticodon pair with the mRNA on the ribosomes at time of aligning the amino acids during protein synthesis (translation of mRNA into proteins which proceeds in N_2 — COOH direction). For example, one of two mRNA and DNA code words for the amino acid phenylalanine is UUC and AAG respectively, and the corresponding anticodon of tRNA is CAA. This indicates that codon and anticodon pairing is antiparallel. In this case, C pairs with G and U pairs with A.

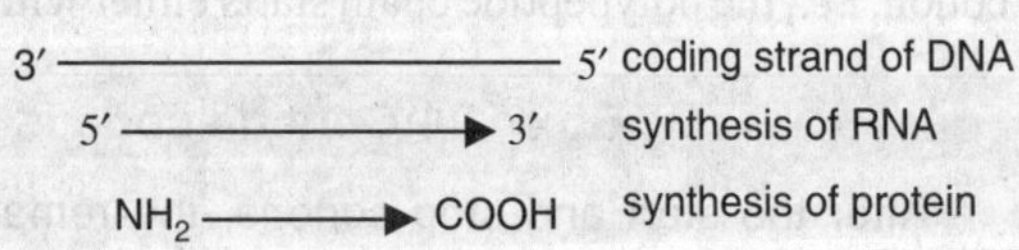

Fig. 11.8. Directions of synthesis of RNA and protein with respect to the coding strand of DNA (after Freifelder, 1985).

11.4 WOBBLE HYPOTHESIS

The genetic code is a degenerate code, meaning that a given amino acid may have more than one codon (Box 11.1). As you can see from Table 11.2, eight of the sixteen boxes contain just one amino acid per box (*e.g.,* serine, leucine, proline, arginine, threonine, valine, alanine and glycine). (A **box** is determined by the first and second positions, *e.g.,* the UUX box in which X is any of the four bases). Therefore, in these eight amino acids, the codon need only be read in the first two positions because the same amino acid will be represented regardless of the third base of the codon. These eight groups of codons are termed **unmixed families** of codons. An unmixed family is the four codons beginning with the same two bases that specify a single amino acid. For example, the codon family GUX codes for valine.

Mixed families of codons code for two amino acids or for stop signals and one or two amino acids. Six of the mixed-family boxes are split in half so that the codons are differentiated by the presence of a purine or a pyrimidine in the third base. For example, CAU and CAC both code for histidine: in both, the third base, U (uracil) or C (cytosine) is a pyrimidine. Only two of the families of codons are split differently.

The lesser importance of the third position is the genetic code is linked with two facts about transfer RNAs. First, although there would seem to be a need for sixty-two transfer RNAs—since there are sixty-one (*i.e.,* one codons specifying amino acids and an additional codon for initiation) there are actually only about fifty different transfer RNAs in an *E.coli* cell. Second, a rare base such as inosine (I) can appear in the anticodon, usually in the position that is complementary to the third position of the codon. These two facts researchers believe that some kind of conservation of transfer RNAs is occurring and that rare base may be involved.

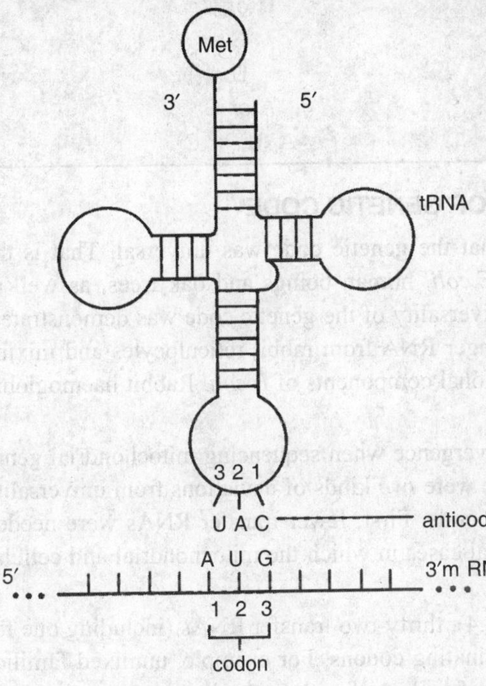

Fig. 11.9. Codon and anticodon base positions are numbered from the 5′ end. The 3′ position in the codon (5′ in the anticodon) is the wobble base (after Tamarin, 2002).

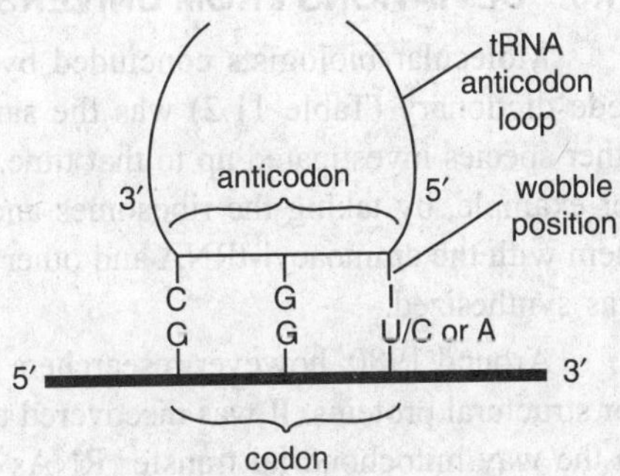

Fig. 11.10. Wobble hypothesis. In the third site (5′ end) of the anticodon I can take either of three wobble positions, thus, being able to pair with U, C or A. This means that a single tRNA species carrying an amino acid can recognize three codons in mRNA: GCU, GCC and GCA (After Suzuki *et al.*, 1986).

At this stage, one point should be very clear that both messenger RNA and transfer RNA bases are usually numbered from the 5′ side. Thus, the number one base of the codon is complementary to the number-three base of the anticodon (Fig. 11.9). Thus, the codon base of lesser importance is the number-three base, whereas its complement in the anticodon is the number-one base.

Since the first position of the anticodon (5′) is not as constrained as the other two positions, a given base at that position may be able to pair with any of several bases in the third position of the codon. **Crick** (1966) characterized this ability as **wobble** (wobble means to sway or move unsteadily) (Fig. 11.9). Table 11.4 shows the possible pairings that would produce a transfer RNA system compatible with the known code. For example, if an isoleucine transfer RNA has the anticodons 3′–UAI–5′, it is compatible with the three codons for that amino acid (see Table 11.4 and Table 11.5): 5′–AUU–3′, 5′–AUC–3′ and 5′–AUA–3′. That is inosine is the first (5′) position of the anticodon can recognise U, C, or A in the third (3′) position of the codon, and thus one transfer RNA complements with all three codon for isoleucine.

Table 11.4 Pairing combinations at the third codon position (*Source:* **Tamarin**, 2002).

Number-one base in tRNA (5′ end)	*Number-three base in mRNA (3′ end)*
G	U *or* C
C	G
A	U
U	A *or* G
I	A, U, *or* C

11.5 DEVIATIONS FROM UNIVERSALITY OF GENETIC CODE

Molecular biologists concluded by 1979 that the genetic code was universal. That is the code dictionary (Table 11.2) was the same for *E.coli,* human beings and oak trees, as well as other species investigated up to that time. The universality of the genetic code was demonstrated, for example, by taking the ribosomes and messenger RNA from rabbit reticulocytes and mixing them with the aminoacyl-tRNA and other translational components of *E.coli.* Rabbit haemoglobin was synthesized.

Around 1980, however researchers noted divergence when sequencing mitochondrial genes for structural proteins. It was discovered that there were *two* kinds of deviations from universality in the way mitochondrial transfer RNAs read the code. First, fewer transfer RNAs were needed to read the genetic code, second, there were several cases in which the mitochondrial and cellular systems interpreted a codon differently.

According to Crick's wobble rules (Table 11.4), thirty-two transfer RNAs (including one for initiation) can complement all sixty-one non-terminating codons. For example, unmixed families of codons require two transfer RNAs, and mixed families of codons require one, two or three transfer RNAs, depending on the family. The yeast mitochondrial coding system apparently needs only twenty-four transfer RNAs. The reduction in numbers is attained primarily by having only one transfer RNA recognize each unmixed family (Table 11.5). Because mitochondrial transfer RNAs for unmixed families of codons have a U in the first (*i.e.,* wobble) position of the anticodon, apparently, given the structure of the mitochondrial transfer RNAs, the U can pair with U, C, A or G. Most probably, evolutionary pressure has minimized the number of transfer RNA genes in the DNA of the mitochondrion, in keeping with its small size. Reduction from thirty-two to twenty-four is a 25% savings. Recent evidence suggests that mammalian mitochondria may need only twenty-two transfer RNAs.

Table 11.5 The genetic code dictionary of yeast mitochondria; anticodons (3′→5′) are given within boxes (*Source:* **Tamarin,** 2002).

First Position (5′ End)	**Second Position** U	C	A	G	**Third Position (3′ End)**
U	Phe AAG	Ser AGU	Try AUG	Cys AGG	U C
	Leu AAU		*stop*	Trp ACU	A G
C	Thr GAU	Pro GGU	His GUG	Arg GCA	U C
			Glon GUU		A G
A	Ile UAG	Thr UGU	Asn UUG	Ser UCG	U C
	Met UAC		Lys UUU	Arg UCU	A G
G	Val CAU	Ala CGU	Asp CUG	Gly CCU	U C
			Glu CUU		A G

It has also been found that yeast mitochondria read the CUX family (*i.e.,* CUU, CUC, CUA, CUG) as threonine rather than as leucine (Table 11.2 and Table 11.5) and the terminator UCA (opal) as tryptophan rather than as termination. However, there have been found differences among mitochondria of different groups of organisms reading the CUX family. Human and *Neurospora* mitochondria appear to read the CUX codons as leucine just as cellular systems do. Of the groups so far analyzed, only yeast reads the CUX family as threonine. Similarly, human and *Drosophila* mitochondria read AGA and AGG as stop signals rather than as arginine (Table 11.6).

Table 11.6 Common and alternative meanings of codons (*Source:* **Tamarin**, 2002).

S. No.	*Codon*	*General meaning*	*Alternative Meaning*
1.	CUX	Leu	Thr in yeast mitochondria
2.	AUA	Ile	Met in mitochondria of yeast, *Drosophila* and vertebrates
3.	UGA	Stop	Trp in mycoplasmas and mitochondria other than higher plants
4.	AGA/AGG	Arg	*Stop* in mitochondria of yeast and vertebrates in mitochondria of *Drosophila*
5.	CGG	Arg	Trp in mitochondria of higher plants
6.	UAA/UAG	*Stop*	Gln in ciliated protozoa
7.	UAG	*Stop*	Ala or Leu in mitochondria of some higher plants

The ciliated protozoan *Paramecium* species read the UAA and UAG stop codons as glutamine within the cell. In addition, a prokaryote (*Mycoplasma capricolum*) reads UGA codon as tryptophan. We do not yet know how general this finding is: scientists have analysed the genetic code of very few species. We can thus conclude that *the genetic code seems to have universal tendencies among prokaryotes, eukaryotes and viruses.* Mitochondria however, read the code slightly differently: different, wobble rules apply, and mitochondria and cells read at least one terminator and one unmixed family of codons differently. Also the mitochondrial deviations are not universal among all types of mitochondria. Further work, involving the sequencing of more mitochondrial DNAs, should explain the pattern of deviations among the mitochondria of diverse species. Further, it has also become clear that not every organism reads all codons in the same way. For example, ciliated protozoa and a mycoplasma read some stop signals as coding for amino acids. Likewise, nuclear variants are known in the following codons: CUG, AUA, UAA, UAG, UGA, CGG and AGA. Mitochondrial variants are known in CUX, AUA, UAA, UAG, AAA, UGA, CGX, AGA and AGG.

Site specific variation. In this type of variation in codon reading, the interpretation of a codon depends on its specific location. For example, codons GUG and, rarely, UUG can serve as prokaryotic initiation codons. This means that they are recognized by $tRNA_f^{Met}$. However, they are not recognised by $tRNA_m^{Met}$ (*i.e.,* GUG and UUG are not misread internally in messenger RNAs). In some cases, two of the termination codons (UGA and UAG, but not UAA) are misinterpreted as codons for amino acids. That is, termination will not occur at the normal place, resulting in a longer-than usual protein. In some cases, these "read-through" proteins are vital — the organism depends on their existence. For example, in the phage θ β, the coat-protein gene is read through about 2% of the time. Without this small number of read through proteins, the phage coat (capsid) cannot be constructed properly.

Selenocysteine codon. Site-specific variation also involves the amino acid selenocysteine (*i.e.,* cysteine with a selenium atom replacing the sulphur). Although many proteins have unusual amino acids, almost all are due to post-translational modifications of normal amino acids. However, the amino acid selenocysteine is inserted directly into some proteins, such as formate dehydrogenase in *E.coli*, which has selenium in its active site. Selenocysteine is inserted into the protein by a new transfer RNA that recognises the termination codon, UGA, if that codon involved in a particular stem-loop secondary structure in the messenger RNA. The selenocysteine transfer RNA is originally charged with a serine that is then modified to a selenocysteine. In addition to the stem-loop structure 3′ (downstream) from the amber codon (UAG), a selenocysteine elongation factor (SELB) is also needed at the ribosome. This same mechanism may occur in eukaryotes, but not all of the components have yet been recognized.

11.6 EVOLUTION OF GENETIC CODE

Certain modern theories have suggested that the genetic code has the wobble in it because it originally arose from a code in which only the first two bases were needed for the small number of amino acids in use several billion years ago. As new amino acids with useful properties became available, they were incorporated into proteins by a code modified by the third base, albeit with less specificity. This view has been supported by the fact that codons starting with the same nucleotide come from the same biosynthetic pathway. This indicates that in early evolution, as biosynthetic pathways were extended to new amino acids, the newcomers were incorporated by use of the second and the third bases of the code.

Let us see whether there is a relationship between the codons and the amino acids they code for, or is the code just one of many random/chance possibilities. That is, whether the genetic code is highly evolved or just a "frozen accident". Recent computer simulations of random codes indicate that the current genetic code is far outside the range of random in its ability to protect the organism from mutation. This suggests that the genetic code is not a frozen accident, but rather is highly evolved. Numerous examples in the current code support this view.

For example, in the unmixed family 5′ – CUX – 3′, any mutation in the third position produces another codon for the same amino acid. Wobble in the third position and codon arrangement ensures that less than half of the mutations in the third codon position result in the specification of a different amino acid.

Some patterns also have been observed in the genetic code. For example, the mutation of one codon to another results in an amino acid of similar properties. A high probability exists that such a mutation will produce a functional protein. All the codons with U as the middle base, for example, are for amino acids that are hydrophobic (*e.g.*, phenylalanine, leucine, isoleucine, methionine and valine). Mutation in the first or third positions for any of these codons still codes a hydrophobic amino acids. Both of the two negatively charged amino acids, aspartic acid and glutamic acid, have codons that start with GA. All of the aromatic amino acids—phenylalanine, tyrosine and tryptophan–have codons that begin with uracil. Such patterns minimize the negative effects of mutation (see **Tamarin,** 2002).

QUESTIONS

Long Answer Questions

1. What do you understand by 'genetic code'? How will you show that the minimum size of a code word should be triplet?

2. A basic concept of molecular biology is the collinearity of gene, DNA, RNA and protein. Discuss the various lines of evidence supporting or providing this concept.

3. Gamow (1954) pointed out that since the genetic language contains only four letters, A, U = (T), C and G, if all code words are of the same size, codons must be at least three bases long.

(*a*) Show why codons cannot consist of one or of two bases.

(*b*) Discuss two lines of research that indicate that codons are three base long.

(*c*) Since codons are three base long, 64 different triplets can exist. Illustrate these using the branching method.

(*d*) How many of the 64 triplets will contain (1) no adenine, (2) at least one adenine?

4. (*a*) Explain what is meant by a degenerate code and illustrate your answer to show degeneracy in translation.

(*b*) Which are the nonsense triplets and why are they so termed?

(*c*) Do codons have the same meaning *in vitro as in vivo*?

5. Describe the general properties of the genetic code.

6. Give a brief account of the experiments which helped in deciphering the genetic code.

7. Discuss the experiments conducted by H.G. Khurana and his coworkers for cracking the genetic code.

Short Answer Questions

Write short notes on the following:

(1) Homopolymers; (2) Copolymers of repetitive sequences; (3) Poly U; (4) Frame shift mutation; (5) Overlapping genes; (6) Wobble hypothesis; (7) Split genes.

Very Short Answer Questions

1. Define triplet code.

2. What is degenerate code?

3. What is advantage of degeneracy of genetic code?

4. Explain wobble hypothesis.

5. The size of haemoglobin gene in humans is estimated to consist of approximately 450 nucleotide pairs. The protein product of the gene is estimated to consist of about 150 amino acid residues. Calculate how many nucleotides are present in a genetic codon for each amino acid.
6. How many triplet codons can be made from the four ribonucleotides A, U, G and C containing (*a*) no uracils; (*b*) one or more uracils?
7. Assume a series of different one-base changes in the codon GGA, which produces following several new codons: (*a*) UGA; (*b*) GAA; (*c*) GGC, (*d*) CGA. Which of these represent (*s*) degeneracy, which missense, and which nonsense?
8. If single-base changes occur in DNA (and, therefore, in mRNA), which amino acid, tryptophan or arginine, is most likely to be replaced by another in protein synthesis?
9. How was the genetic code first decoded? What refinements have since been incorporated in the technique.
10. In what sense and to what extent is the genetic code (*a*) degenerate and (*b*) universal?
11. What evidence supports the hypothesis of colinearity between the nucleotide sequence in a gene and the amino acid sequence in a polypeptide?
12. Why is colinearity between codons and polypeptides significant?
13. **A. Garen** (1968) extensively studied a particular nonsense (Chain-termination) mutation in the alkaline phosphatase gene of *E. coli.* This mutation resulted in the termination of the alkaline phosphatase polypeptide chain at a position where the amino acid tryptophan occurred in the wild-type polypeptide. Garen induced revertants (in this case, mutations altering the same codon) of this mutant with chemical mutagens that induced single base-pair substitutions and sequenced the mutant polypeptides. Seven different types of revertants were found, each with a different amino acid at the tryptophan position of the wild-type polypeptide (termination position of the mutant polypeptide fragment). The amino acids present at this position in the various revertants included tryptophan, serine, tyrosine, leucine, glutamic acid, glutamine and lysine. Was the nonsense mutation studied by Garen an *amber* (UAG), an *ochre* (UAA), or an *opal* (UGA) nonsense mutation? Explain the basis of your deduction.

Fill in the Blanks

1. The sequences of nitrogen bases of a given segment of DNA molecule, actually, has been found to be identical to linear sequences of .... in a protein molecule. This is called ....
2. The first experimental evidence in support of the concept of triplet code is provided by .... and coworkers in 1961.
3. The deciphering of the genetic code was made possible by the use of .... polynucleotides and ....
4. .... and Matthaei synthesized a chain of .... molecules as their synthetic homopolymers.
5. Three codons .... , ... and .... are the chain stop or .... codons.
6. Amber codon stands for .... triplet.
7. .... codon stands for UAA triplet.
8. The bases of anticodon run in .... direction.

Multiple Choice Questions

1. Genetic code is sequence of nitrogenous bases on

(*a*) mRNA (*b*) tRNA

(*c*) rRNA (*d*) DNA

2. The terminator codons are

(*a*) UAA, UAG, UGA (*b*) AUG, UAG, UGA

(*c*) UAC, AUG, UAG (*d*) AUG, ACG, GAG

3. Who discovered the entire genetic code?

(*a*) Nirenberg (*b*) Ochoa

(*c*) Khurana (*d*) Crick

4. A codon consists of how many nitrogen bases in it

(*a*) 1 (*b*) 2

(*c*) 3 (*d*) 4

5. The arrangement of three bases in DNA represents

(*a*) protein (*b*) amino acid

(*c*) plasmid (*d*) nucleic acid

6. Degeneration of a genetic code is attributed to the

(*a*) first member of codon (*b*) second member of codon

(*c*) entire codon (*d*) third member of codon

7. In the genetic code dictionary, how many codons are used to code for all the 20 essential amino acids?

(*a*) 20 (*b*) 64

(*c*) 61 (*d*) 60

8. What would happen if in a gene encoding a polypeptide of 50 amino acids, 25th codon (UAU) is mutated to UAA?

(*a*) a polypeptide of 24 amino acids will be formed

(*b*) two polypeptide of 24 and 25 amino acids will be formed

(*c*) a polypeptide of 49 amino acids will be formed

(*d*) a polypeptide of 25 amino acids will be formed

9. Which of the chemical characteristics is not common to all living beings?

(*a*) ribosomes are sites of protein synthesis (*b*) types of protein present in the body

(*c*) energy is stored by high phosphate bond (*d*) similar triplet codes for amino acid

ANSWERS

Very Short Answer Questions

1. A coding system in which three units of information are read as a unit; a reference to the genetic code which is read from mRNA in units of three bases called codons.

2. Degenerate code is ability of the genetic code to use more than one triplet codes to specify the same amino acid. For each of Arginine, Serine and Leucine (amino acids) is coded by six codons.

3. Most amino acids show multiple coding. This degeneracy of genetic code provides a protection to organisms against harmful mutations, stabilizes the phenotype and decreases chances in base pairing errors.

4. **Crick** (1968) proposed the wobble hypothesis to explain the degeneracy of codons. It states that during codon–anticodon pairing between the codon of mRNA and anticodon of tRNA, complimentary pairing is required for only towo nucleotides while third nucleotide at 3′-end of codons act as a wobbly base and may pair with more than one nucleotide of anticodon. This shows that only first two nucleotides of a codon specify an amino acid while third nucleotide of codon lacks the specificity. This third position is called **wobble position** due to which more than one charged tRNA can leave their amino acid for the same codon during the process of translation. For example, valine is coded by 4 codons, *e.g.*, GUU, GUC, GUA and GUG. In all four codons, first two nucleotides are GU which specify for valine. Third nucleotide may be and of A, U, C or G but does not change the coded amino acid.

5. Approximately 3 nucleotides code for each amino acid.

6. (*a*) Since uracil represents 1 among 4 nucleotides; the probability that uracil will be the first letter of the codon is 1/4 and the probability that U will not be the first letter is 3/4. The same reasoning holds true for the second and third letters of the codon. The probability that none of the three letters of the codon are uracils is $(3/4)^3 = 27/64$.

(*b*) The number of codons containing at least one uracil is $1 - 27/64 = 37/64$.

7. (*a*) Nonsense; UGA is a stop signal and does not translate.

(*b*) Missense; GGA codes for glycine, GAA for glutamic acid.

(*c*) Degeneracy; both GGA and GGC code for glycine.

(*d*) Missense; GGA codes for glycine, CGA for arginine.

8. Tryptophan, which has only one codon. Arginine has six codons.

9. (*a*) Synthetic RNA molecules (polyuridylic acid molecules) containing only the base uracil were prepared when these synthetic molecules were used to activate *in vitro* protein synthesis systems, small polypeptides containing only the amino acid phenylalanine (polyphenyl alanine molecules) were synthesized. Codons composed only of uracil were therefore shown to specify phenylalanine. Similar experiments were carried out using synthetic RNA molecules with different base compositions.

(*b*) Better *in vitro* systems activated with synthetic RNA molecules with known repeated base sequences were developed ultimately, *in vitro* systems in which specific aminoacyl-tRNAs were shown to bind to ribosomes activated with specific mini-mRNAs, which were trinucleotides of known base sequence, were developed and used in codon identification.

10. (*a*) The genetic code is degenerated in that all but 2 of the 20 amino acids are specified by two or more codons. Some amino acids are specified by six different codons. The degeneracy occurs largely at the third or 3′ base of the codons. "*Partial degeneracy*" occurs where the third base of the codon may be either of the two purines or either of the two pyrimidines and the codon still specifies the same amino acid. "*Complete degeneracy*" occurs where the third base of the codon may be any one of the four bases and the codon still specifies the same amino acid.

(*b*) The code appears to be almost completely universal. Known exceptions to universality include strains carrying suppressor mutations that alter the reading of certain codons (with low efficiencies in most cases) and the use of UGA as a tryptophan codon in yeast and human mitochondria.

11. Colinearity received strong support from studies that showed a direct correlation between the linear sequence of mutational sites in a gene (established by genetic mapping experiments) and the linear sequence of mutational defects (amino acid substitutions or chain terminations) in the polypeptide gene-product (established by purification and amino acid sequencing, etc., of mutant polypeptides).

12. Colinearity is an important prediction of our present concepts of noncoding sequences ("introns" or "intervening sequences") within eukaryotic genes does not violate the concept of colinearity; their presence simply means that the colinear structures frequently contain interruptions.

13. *Amber* (UAG). This is the only nonsense codon that is related to tryptophan, serine, tyrosine, leucine glutamic acid, glutamine and lysine codons by a single base pair substitution in each case.

Fill in the Blanks

1. Amino acids, colinearity.
2. Crick.
3. Artificial or synthetic, trinucleotides.
4. Nirenberg, uracil.
5. UAG, UAA, UGA, termination.
6. UAG.
7. Ochre.
8. 5′ → 3′.

Multiple Choice Questions

1. (*a*) 2. (*a*) 3. (*a*) 4. (*c*) 5. (*b*) 6. (*d*)

(*c*) 8. (*c*) 9. (*d*).

12

Translation: Protein Synthesis

DNA, with its correct mechanism of replication, serves to carry genetic information from cell to cell and from generation to generation. This information is translated into proteins that determine the phenotype (Fig. 12.1). Virtually, all the phenotypes examined so far are the result of biochemical reactions that occur in the cell. All of these reactions require enzymes and enzymes are proteins. In fact, more than 2000 types of enzymes have been identified in the living organisms. Each enzyme is a unique molecule catalyzing a specific chemical reaction. Other phenotypes are due primarily to the kinds and amounts of non-enzymatic proteins (including the structural proteins) present, for example, haemoglobin, myoglobin, gamma-globulin (*e.g.*, immunoglobulins or Igs), insulin, cytochrome C, fibroin (silk protein), or collagen. Proteins are composed of one or more, long linear polymers of amino acid residues (polypeptide chains) that are synthesized almost exclusively in the cytoplasm. The topic of this chapter is how the information present in the sequence of bases (=triplet codons) of the mRNA is translated into a sequence of amino acids in proteins. However, at the outset, one point should be clear that the genetic information of mRNA is undirectionally read beginning at the 5′–hydroxyl end by one or more ribosomes (polysomes) and 3′–end signify the C-terminal amino acid. The 5′–end of the mRNA correspond to the N-terminal amino acid and ending at 3′-end in the completed protein.

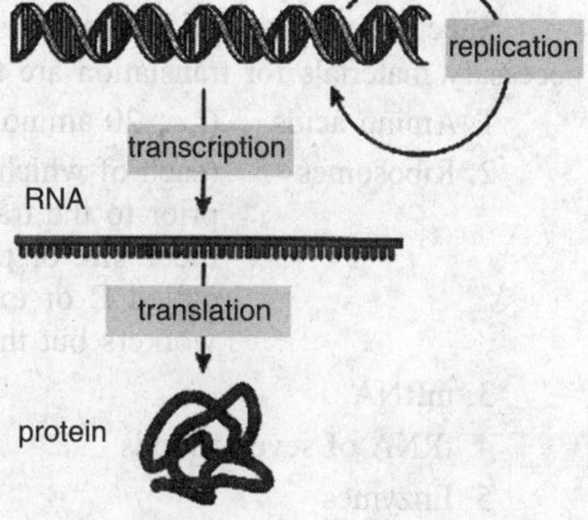

Fig. 12.1. Central Dogma.

12.1 CENTRAL DOGMA AND CENTRAL DOGMA REVERSE

The process of synthesis of protein involves one of the central dogma of molecular biology, which postulates that genetic information flows from nucleic acids to protein (Fig. 12.1). (It was first forwarded by **Crick** in 1958). The first step of this central dogma is known as transcription and does not involve a change of code since DNA and mRNA are complementary. The second step involves a change of code from nucleotide sequences to amino acid sequences and is called translation. It can be illustrated as follows:

$$\text{DNA} \xrightarrow{\text{replication}} \text{DNA} \xrightarrow{\text{transcription}} \text{RNA} \xrightarrow{\text{translation}} \text{Proteins}$$

Thus, according to this central dogma, the flow of information is one way, *i.e.*, from DNA the information is transferred to RNA (mRNA) and from RNA to proteins. In 1968, **Barry Commoner** suggested a circular flow of information, *i.e.*, DNA transcribes RNA, RNA translates into proteins, proteins synthesize RNA and RNA synthesizes DNA, as has been illustrated in following figure:

$$\begin{array}{ccc} \text{DNA} & \longrightarrow & \text{RNA} \\ \uparrow & & \downarrow \\ \text{RNA} & \longleftarrow & \text{Proteins} \end{array}$$

Later on, **Temin** (1970) reported the existence of an enzyme "**RNA-dependent DNA polymerase" (reverse transcriptase)** which could synthesize DNA from a single-stranded RNA template. **D. Baltimore** (1970) also reported the activity of this enzyme in certain tumour viruses. This exciting finding in molecular biology gave rise to the concept of "**central dogma reverse**" or **teminism**, suggesting that the sequence of information flow is not necessarily from DNA to RNA to protein but can also take place from RNA to DNA. The central dogma reverse can be illustrated as follows:

$$\text{DNA} \underset{\text{inverse transcription}}{\overset{\text{transcription}}{\rightleftarrows}} \text{RNA} \xrightarrow{\text{translation}} \text{Proteins}$$

In fact, the step of translation of this central dogma is the time in the flow of information between genes and proteins to change the language being used. In going from DNA to RNA, the language (nucleotide sequences) remained the same. However, in going from RNA to protein the language is changed from a nucleotide sequence to amino acid sequence. Just as in the process of translation from one language to another, this process of using information in RNA to make protein is called **translation**.

12.2 MINIMUM NECESSARY MATERIALS FOR TRANSLATION

Success in polypeptide synthesis in *in vitro* cell free systems shows that the minimum necessary materials for translation are the following:

1. Amino acids (*i.e.*, 20 amino acids forming the pool of amino acids in the cytoplasm)
2. Ribosomes (each of which comprises two subunits which exist as separate subunits prior to the translation of mRNA and contain two tRNA binding sites: the **P site** or **peptidyl site** and an **A** or **aminoacyl site**. One more site called **E** or **exit site** has been recognized in the ribosomes by some workers but this three site model (*i.e.,* A, P and E) still is not popular.
3. mRNA
4. tRNA of several kinds
5. Enzymes
   (*a*) Amino-acid activating system (*e.g.*, aminoacyl-tRNA-synthetase).
   (*b*) Peptide polymerase system
6. Adenosine triphosphate (ATP) as an energy source
7. Guanosine triphosphate (GTP) for synthesis of peptide bonds
8. Soluble protein for initiation and transfer factors
9. Various inorganic cations (*e.g.*, K^+, NH^+ $NH_4{}^+$, Mg^{2+}).

Because, the proteins are the polymers of amino acids therefore, protein synthetic process requires the amino acids as the raw material. All the naturally occurring proteins of living organisms fundamentally are the polymers of about 20 amino acids. These amino acids occur in the matrix forming an "**amino acid pool"** and are readily available for the process of protein synthesis.

All amino acids present in amino acid pool are called α-*amino acids* because one carbon, the α-carbon has four specific groups attached to it: an amino group, a carboxyl (acidic) group, a hydrogen and one of the twenty different R groups (side chains), imparting the specific properties of that amino acid. (Technically, proline is called an **amino acid** because of its structure). Having these four groups attached to C imparts a property known as **chirality** on the amino acid: such as left and right-handed gloves, the mirror images cannot be superimposed. (*Chirality* means handedness or the property by which a form is distinguishable from its mirror image). Because of optical properties, the two forms of each amino acids are referred to as D and L, in which

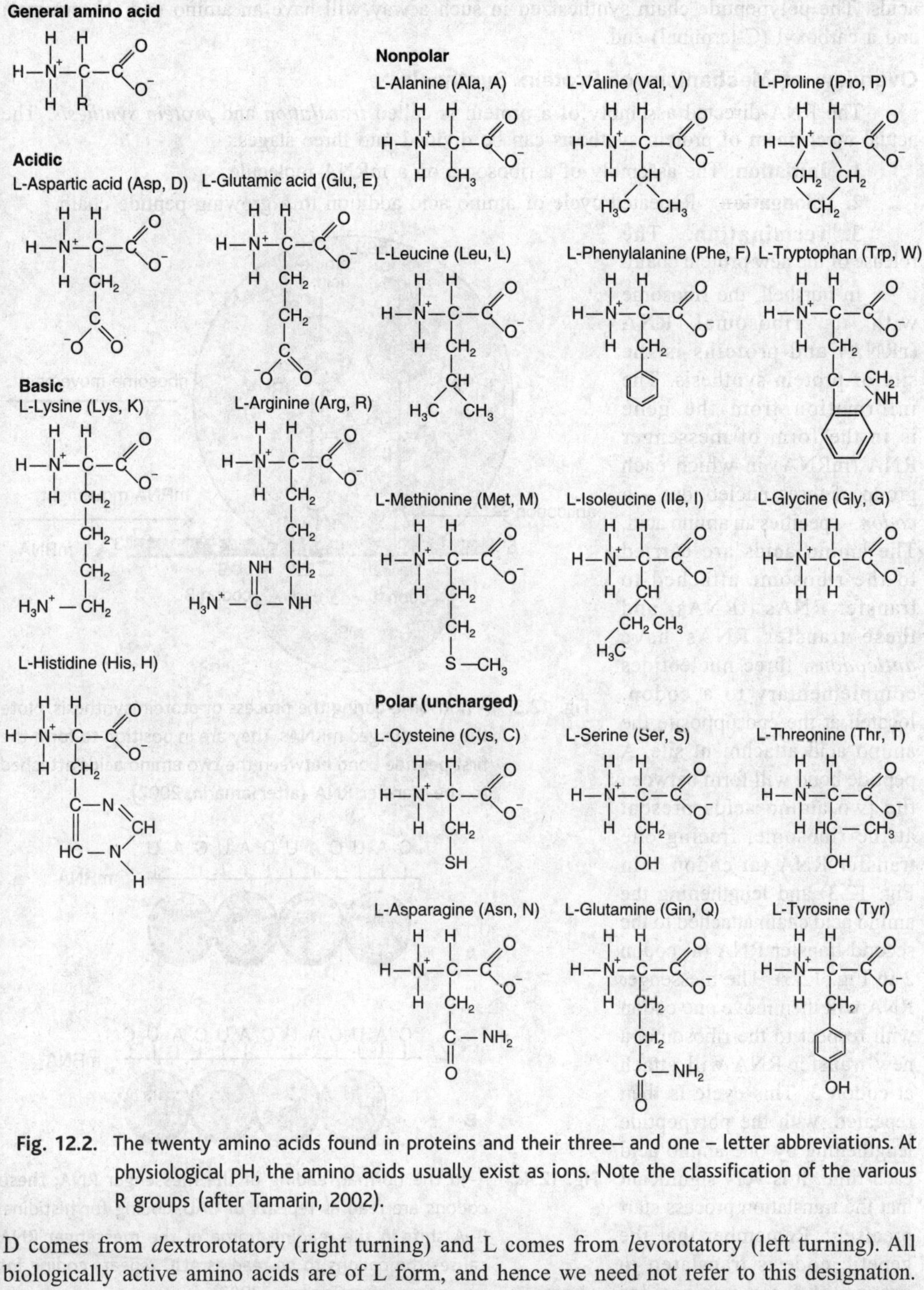

Fig. 12.2. The twenty amino acids found in proteins and their three– and one – letter abbreviations. At physiological pH, the amino acids usually exist as ions. Note the classification of the various R groups (after Tamarin, 2002).

D comes from *d*extrorotatory (right turning) and L comes from *l*evorotatory (left turning). All biologically active amino acids are of L form, and hence we need not refer to this designation. Proteins (polypeptides) are synthesized when peptide bonds form between any two amino acids-called **residues** when incorporated into a protein – can join and all chains will have an amino (N – terminal) end and a carboxyl (C–terminal) end.

Proteins (polypeptides) are synthesized by the formation of peptide bonds between two amino acids. The polypeptide chain synthesized in such a way will have an amino end (N-terminal) and a carboxyl (C-terminal) end.

Overview of Mechanism of Protein Synthesis

The RNA-directed assembly of a protein is called *translation* and *protein synthesis.* The actual mechanism of protein synthesis can be divided into three stages:

1. **Initiation.** The assembly of a ribosome on a mRNA molecule.
2. **Elongation.** Repeated cycle of amino acid addition to a growing peptide chain.
3. **Termination.** The release of the new protein chain.

In nutshell, the ribosome with its ribosomal RNA (rRNA) and proteins is the site of protein synthesis. The information from the gene is in the form of messenger RNA (mRNA) in which each group of three nucleotides – a *codon* – specifies an amino acid. The amino acids are carried to the ribosome attached to transfer RNAs (tRNAs) and these transfer RNAs have *anticodons,* three nucleotides complementary to a codon, located at the end opposite the amino acid attachment site. A peptide bond will form between the two amino acids present at the ribosome, freeing one transfer RNA (at codon 1 in Fig. 12.3) and lengthening the amino acid chain attached to the second transfer RNA (at codon 2 in Fig. 12.3). The messenger RNA will then move one codon with respect to the ribosome, a new transfer RNA will attach at codon 3. This cycle is then repeated, with the polypeptide lengthening by one amino acid each time. It is very significant that the translation process start precisely. Remember that the genetic code is translated in groups of three nucleotides

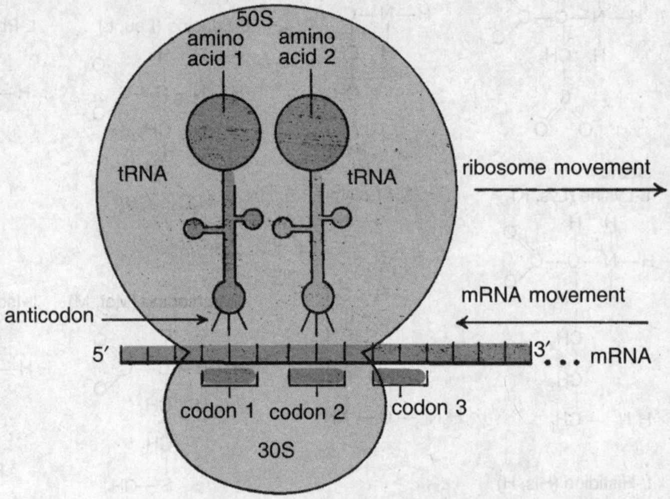

Fig. 12.3. A ribosome during the process of protein synthesis. Note the two charged mRNAs. They are in position to form the first peptide bond between the two amino acids attached to the transfer RNA (afterTamarin, 2002).

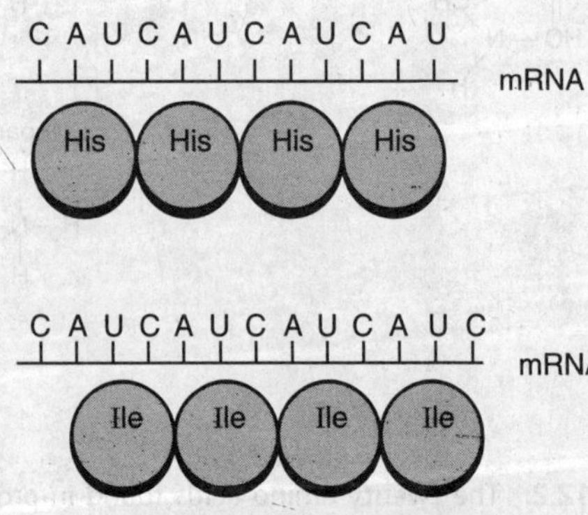

Fig. 12.4. A–In the normal reading of the messenger RNA, these codons are read as repeats of CAU, coding for histidine. B–A shift in the reading frame of the messenger RNA causes the codons to be read as AUC repeats coding for isoleucine (after Tamarin, 2002).

(codons). If the reading of the messenger RNA begins one base too early or too late, the reading frame is shifted so that an entirely different set of codons is read (Fig. 12.4). The protein produced, if any, will probably bear no structural or functional similarity to the protein the gene is coded for.

12.3 MECHANISM OF TRANSLATION OR PROTEIN SYNTHESIS

I. Aminoacylation of tRNA (Attachment of Amino Acid to Transfer RNA)

The function of tRNA (transfer RNA) is to ensure that each amino acid incorporated into a protein corresponds to a particular codon in the messenger RNA. The transfer RNA serves this function through its peculiar structure: it has an anticodon at one end and an amino acid attachment site at the other end. The "correct" amino acid, the amino acid corresponding to the anticodon, is attached to the transfer RNA by enzymes known as **aminoacyl tRNA synthetases** (e.g., arginyl-tRNA synthetase, leucyl-tRNA synthetase). A transfer RNA with an amino acid attached is said to be "charged". The transfer RNAs for each amino acid are designated by the convention $tRNA^{Leu}$ (for leucine), $tRNA^{His}$ (for histidine) and so on.

An aminoacyl – tRNA synthetase joins a specific amino acid to its transfer RNA in a two stage reaction that takes place on the surface of the enzyme (Box 12.1). In the first stage, the amino acid is activated with ATP. In the second stage of the reaction, the amino acid is attached with a

Box 12.1

Each activating enzyme (*i.e.*, aminoacyl – tRNA synthetase) is specific for one amino acid and for a tRNA. The enzyme has a three part active site that recognizes three smaller molecules:

1. a specific amino acid (AA);
2. ATP; and
3. a specific transfer RNA.

high energy bond to the 2′ or 3′ carbon of the ribose sugar at the 3′ end of the transfer RNA: The first step in aminoacyl – tRNA synthesis involves the ***activation*** of the amino acid using energy from ATP:

$$\text{Amino acid + ATP} \xrightarrow{\text{Aminoacyl synthetase}} \text{amino acid} \sim \text{AMP + P + P}$$

The aminoacyl ~ AMP intermediate is not normally released from the enzyme before undergoing the second step in aminoacyl tRNA synthesis, namely the reaction with appropriate tRNA:

$$\text{Amino acid} \sim \text{AMP + tRNA} \xrightarrow{\text{Aminoacyl -tRNA synthetase}} \text{amino acid} \sim \text{tRNA + AMP}$$

The resultant aminoacyl ~ tRNAs (amino acid ~ tRNAs) are immediate precursors of protein (=polypeptide) synthesis.

In the Figure 12.5, we denote high energy bonds, bonds that liberate a lot of free energy when hydrolysed, as "~". Thus, during the process of protein synthesis, the energy for the formation of the peptide bond will be present when it is needed, at the point of peptide bond formation.

Few Characteristics of Aminoacyl-tRNA Synthetases

In bacteria, there are twenty aminoacyl-tRNA synthetases, one for each amino acid. A particular enzyme recognizes a particular amino acid, as well as, all the transfer RNAs that code for that amino acid. In eukaryotes, there are separate sets of twenty cytoplasmic and twenty mitochondrial aminoacyl-tRNA synthetases, all coded in the aminoacyl-tRNA synthetases, all coded in the nucleus.

Aminoacyl-tRNA synthetases are a heterogeneous group of enzymes. In *E.coli*, these enzymes vary from monomeric proteins (one subunit) to tetrameric proteins, made up of two copies each of two subunits. The enzymes fall into two categories based on sequence similarity, structural features, and whether the amino acid is attached at the 2′-OH (in class I enzymes) or 3′-OH (in class II enzymes) of the 3′-terminal adenosine of the transfer RNA (Table 12.1; **Ibba** *et al.,* 2000).

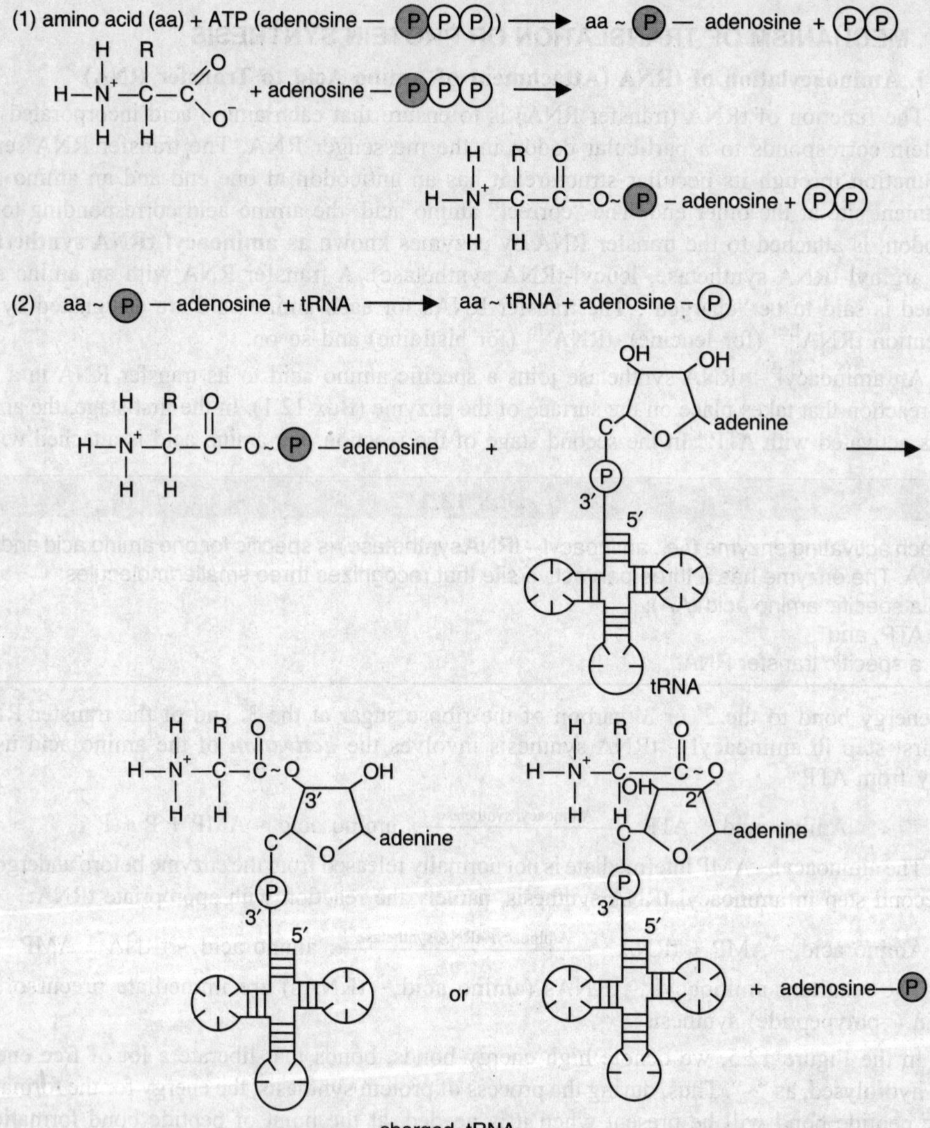

Fig. 12.5. Two steps of aminoacylation that is attachment of a specific amino acid to its transfer RNA by an aminoacyl synthetase enzyme. High–energy bonds are indicated by ~. In the first step, an amino acid is attached to AMP with a high–energy bond. In the second step, the high-energy bond is transferred to the tRNA, which is then referred to "charged". Depending on which class of aminoacyl-tRNA synthetase is involved, the amino acid will be attached to either the 2′ or 3′ carbon of the sugar of the 3′ terminal adenosine. The link is formed between the —COOH group of the amino acid and the —OH group attached to carbon on the sugar of the last nucleotide, which is always an A(= adenine).

To add its appropriate amino acid to the appropriate transfer RNA, a synthetase recognizes many parts of transfer RNA. This can be demonstrated by experiments that introduces specific changes in transfer RNAs by site-directed mutagenesis. In seventeen of the twenty synthetase enzymes of *E.coli*, recognition involves part of the anticodon itself.

Table 12.1 Features of aminoacyl-tRNA synthetases (*Source:* **Brown**, 2002).

S. No.	*Feature*	*Class I enzymes*	*Class II enzymes*
1.	Structure of the enzyme's active sites	Parallel β-sheet	Antiparallel β-sheet
2.	Interaction with tRNA	Minor groove of the acceptor stem	Major groove of the acceptor stem
3.	Orientation of the bound tRNA	V loop faces away from the enzyme	V loop faces the enzyme
4.	Amino acid attachment	To the 2′–OH of the terminal nucleotide of the tRNA	To the 3′–OH of the terminal nucleotide of the tRNA
5.	Enzymes for *	Arg, Cys, Glu, Gln, Ile, Leu, Lysl, Met, Trp, Tyr, Val	Ala, Asn, Asp, Gly, His, Lysl, Phe, Pro, Thr, Ser

* The aminoacyl-tRNA synthetase for lysine in a class I enzyme in some archaea and bacteria and a class II enzyme in all other organisms.

Further, an aminoacyl-tRNA synthetase enzyme can initially make errors and attach the "wrong" amino acid to a tRNA. For example, isoleucine (Ile) synthetase will attach valine about one in 225 times. This type of error occurs because a similar, but smaller, amino acid can sometimes occupy the active site of the enzyme (compare chemical formulae of isoleucine and valine in Fig. 12.2). However, because of a proof-reading step, only 1 in 270 to 1 in 800 of the errors are released intact from the enzyme. The amino acids on the rest of the incorrectly charged transfer RNAs are hydrolyzed before the transfer RNAs are released. Thus, only about one incorrectly charged transfer RNA occurs per 60,000 to 80,000 formed.

In various bacteria, the number of amino acyl-tRNA synthetases in a particular organism is below twenty. For example, in some archaea, there is no cysteinyl-tRNA synthetase (enzyme). However, the prolyl-tRNA synthetase activates the tRNAs for both cysteine and proline with their appropriate amino acids. Similarly, in some eubacteria, there is no glutaminyl-tRNA synthetase; the glutaminyl-tRNA is charged with glutamic acid, rather than glutamine. An amido transferase (enzyme) then converts the glutamic acid to glutamine (Fig.12.2). This process is called **transamidation**. Some archaea also use process of transamidation to synthesise asparagine-$tRNA^{Asn}$ from aspartic acid-$tRNA^{Asn}$ (**Ibba** *et al.*, 2000).

Lastly, three of sixty-four codons are used to terminate translation. Thus, sixty-one transfer RNAs are needed because there are sixty-one different *nonterminator codons.* About fifty transfer RNAs are known in *E.coli*. The number fifty can be explained by **Wobble phenomenon** (see chapter 10, Genetic Code; Box 12.2), which occurs in the third position of the codon.

Box 12.2 Wobble helps in the reduction of number of tRNAs

Aminoacylation represents the first level of specificity demonstrated by a tRNA. The second level is the specificity of the interaction between the anticodon of the tRNA and codon of the mRNA being translated. This specificity ensures that protein synthesis follows the rules of the genetic code.

In principle, codon-anticodon recognition is an easy process involving base-pairing between the anticodon of the tRNA and a codon in the mRNA (Figure 12.6). The specificity of aminoacylation ensures that the tRNA carries the amino acid denoted by the codon that it pairs with, and the ribosome controls the topology of the interaction in such a way that only a single triplet of nucleotides is available for pairing. Because base-paired polynucleotides are always antiparallel and because the mRNA is read in the 5′→3′ direction, the first nucleotide of the codon pairs with nucleotide 36 of the tRNA, the second with nucleotide 35 and the third with nucleotide 34.

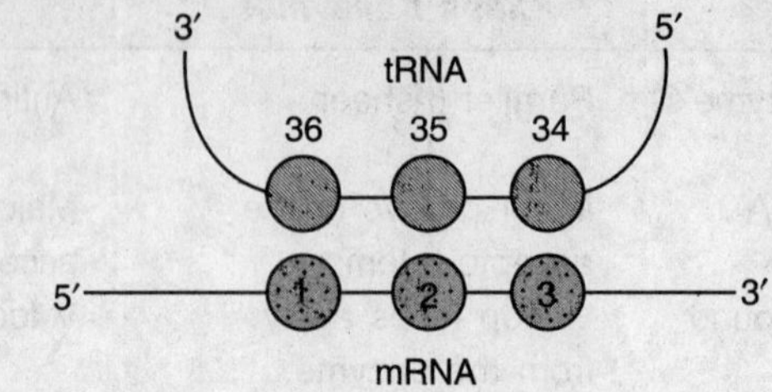

Fig. 12.6. Mode of interaction between a codon of mRNA and an anticodon of tRNA. The numbers indicate the nucleotide positions in the tRNA.

All three nucleotides (*i.e.*, 36, 35 and 34) form an anticodon.

In practice, codon recognition is complicated by the possibility of **Wobble.** Because the anticodon is in a loop of tRNA, the triplet of nucleotides is slightly curved and so cannot make an entirely uniform alignment with the codon. As a result, a non-standard base pair can form between the third nucleotide of the codon and the first nucleotide (number 34) of the anticodon. This is called "wobble". A variety of pairings is possible, especially if the nucleotide at position 34 is modified. In bacteria, the two main features of wobble are (**Ikemura,** 1981).

1. G-U base pairs are permitted. This means that an anticodon with the sequence 3′ – XX G – 5′ can base-pair with both 5′ – XX C – 3′ and 5′ – XX U – 3′. Similarly, the anticodon 3′ – XX U – 5′ can base-pair with both 5′ – XX A – 3′ and 5′ – XXG – 3′. The consequence is that, rather than needing a different tRNA for each codon, the four members of codon family [*e.g.*, 5′ – GCN – 3′ (*i.e.*, GCC, GCA, GCU and GCG) all coding for alanine] can be decoded by just two tRNA.

2. Inosine, abbreviated to I, is a modified purine that can base pair with A, C and U. Inosine can occur only in the tRNA because mRNA is not modified in this way. The triplet 3′ – UAI – 5′ is sometimes used as the anticodon in a $tRNA^{Ile}$ because it pairs with 5′ – AUA – 3′, 5′ – AUC – 3′ and 5′ – AUU – 3′, which form the three codon family for this amino acid (*i.e.* Isoleucine) in the standard genetic code.

Thus, *wobble tends to reduce the number of tRNAs in a cell by enabling one tRNA to read two or possibly three codons.* Hence, bacteria can decode their mRNAs with as few as 30 tRNAs. Eukaryotes also make use of wobble but in a restricted way. For example, human genome has 48 tRNAs. Of these, 16 tRNAs are predicted to use wobble to decode two codons each, with the remaining 32 tRNAs being specific for just a single triplet (see **Brown,** 2002).

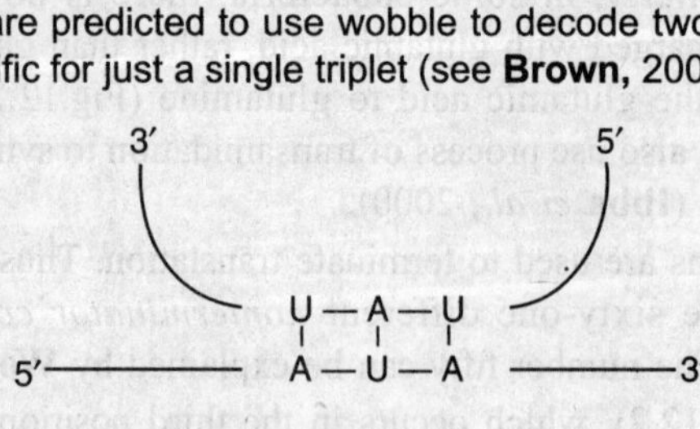

Isoleucine codon

Methionine codon

Fig. 12.7. A tRNA with the anticodon 3′ – UAU – 5′ could read the isoleucine codon 5′ – AUA – 3′ as well as the methionine codon.

In eukaryotes, the bacterial G – U wobble is used with eight tRNAs but in every case the wobble involving an anticodon with the sequence 3′ – XXG – 5′. The alternative version of G – U wobble, where the anticodon sequence is 3′ – XXU – 5′, appears not be used in eukaryotes, possibly because this could result in a $tRNA^{Ile}$ with the anticodon 3′ – UAU – 5′ reading the methionine codon 5′ –AUG – 3′ (Fig. 12.7). Eukaryotes may therefore have a means of preventing this type of wobble from occurring (**Percudani**, 2001).

Eight other human tRNAs have anticodons containing inosine (3′ – XXI – 5′) but, these decode only 5′ – XXC- 3′ and 5′ – XXU – 3′ codons. The base pairing between I and A is weak, which means that 5′ – XXA – 3′ codons are only inefficiently recognized by an 3′ – XXI – 5′ anticodon. To avoid this inefficiency, in every example of wobble involving inosine in the human tRNA set, the 5′ – XXA – 3′ codon is recognized by a separate tRNA. Note, however that recognition by a separate tRNA does not preclude the 5′ – XXA – 3′ codon from also being decoded by the tRNA 3′ – XXI – 5′, albeit inefficiently. This does not compromise the specificity of the genetic code, because wobble involving inosine is limited to those codon families in which all four triplets specify the same amino acid.

Superwobble. Human mitochondria use only 22 tRNAs, with some of these tRNAs the nucleotide in the wobble position of the anticodon is virtually redundant because it can base-pair with any nucleotide, enabling all four codons of a family to be recognized by the same tRNA. This phenomenon has been called **superwobble.**

Experiment to show that the transfer RNA, not the amino acid, is recognized during protein synthesis.

F. Chapeville and colleagues, in 1962, isolated transfer RNA with cysteine attached. They chemically converted the cysteine to alanine by using Raney nickle, a catalytic form of nickle that

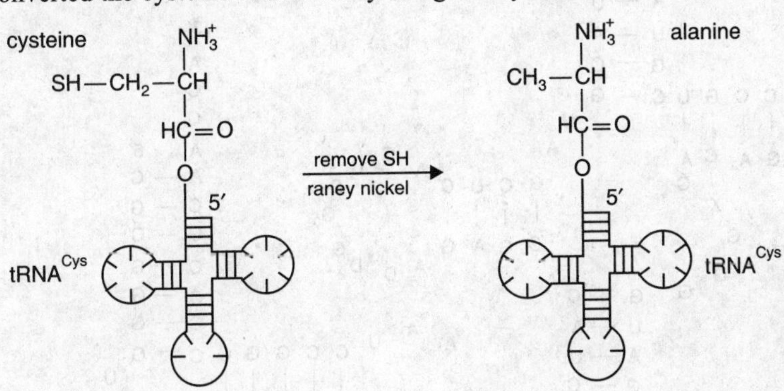

Fig. 12.8. Cysteine – $tRNA^{cys}$ treated with Raney nickel becomes alanine – tRNAcys by the removal of the SH group of cysteine.

removes the SH group of cysteine (Fig. 12.8). When such modified tRNAs were used in protein synthesis, alanine was incorporated where cysteine should have been. This result demonstrated that the transfer RNA, not the amino acid, was recognized during protein synthesis. Thus, the synthetase puts a specific amino acid on a specific tRNA; then, during protein synthesis, the anticodon on the tRNA – not the amino acid itself – determines which amino acid is incorporated.

A. Initiation of Translation

I. Initiation in prokaryotes

1. Interaction between initiation codon and formylmethionine – tRNA. An important feature of initiation of polypeptide (=protein) synthesis is the use of specific initiating tRNA molecule. In *Escherichia coli* and other prokaryotes, the synthesis of every protein begins with the modified amino acid N–formyl methionine (Fig. 12.9).

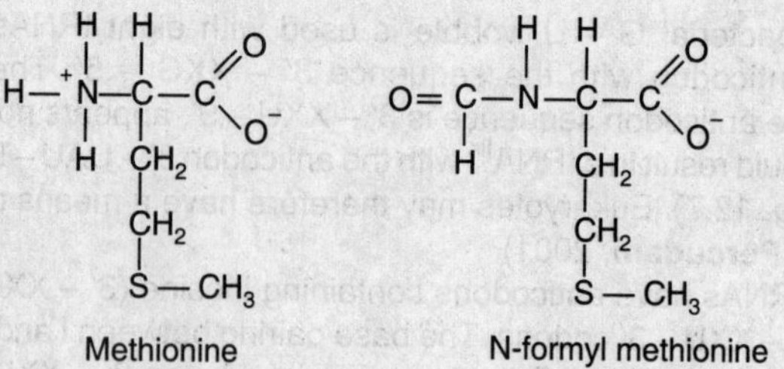

Fig. 12.9. The structure of the amino acid methionine and N-formyl methionine.

However, none of the completed protein in *E.coli* contains N-formyl methionine. Many of these proteins do not even have methionine as their first amino acid. Evidently, before a protein becomes functional, the initial amino acid is modified or removed (called **processing**). In eukaryotes, the initial amino acid is also methionine but it does not have a N-formyl group.

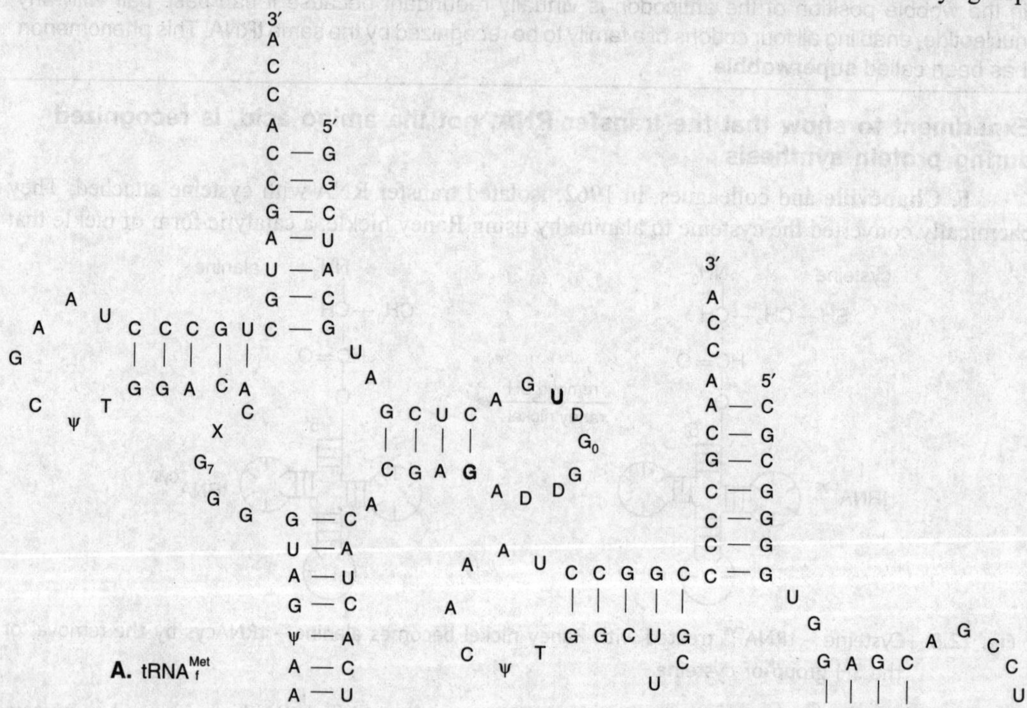

Fig. 12.10. Molecular structure of two tRNAs for methionine in *E.coli*. A – The initiator tRNA; B – The interior or internal tRNA (after Tamarin, 2002).

The codon for methionine is 5′ – AUG – 3′ and is called **initiation codon**. Methionine has two tRNAs with the same anticodon (3′ – UAC – 5′) but different structures (Fig. 12.10). One of these tRNAs ($tRNA_f^{met}$) serves as a part of the initiation complex. Before the initiation of translation, this tRNA will have its methionine chemically modified to N-formyl methionine (fMet). The other transfer RNA will not have its methionine modified ($tRNA_m^{met}$). The translation process will use it to insert methionine into proteins, where it is required, in all but the first position. The cell thus has a device to make use of methionine in the normal way as well as to use a modified form of it (*i.e.,* formyl methionine) to initiate protein synthesis. Because of the structure of the prokaryotic initiation transfer RNA, it can recognize AUG, GUG and rarely UUG as initiation codons. In eukaryotes, CUG as well as AUG can serve as an initiation codon. Since the initiation methionine is not formylated in eukaryotes, the eukaryotic transfer RNA is designated $tRNA_i^{met}$; there is a separate internal methionine transfer RNA, termed $tRNA_m^{met}$, in eukaryotes as well as prokaryotes (Fig. 12.10).

Since in protein synthesis the peptide chain always grows in a sequence from the free terminal amino ($-NH_2$) group towards the carboxyl (—COOH) end, so the function of formylmethionine – tRNA is to ensure that protein are synthesized in that direction. In the formylmethionine – tRNA, the amino ($-NH_2$) group is blocked by the formyl group leaving only the –COOH group available to react with the $-NH_2$ group of the second amino acid (AA_2). In this way, the synthesis of protein chain follows in the correct sequence.

2. Formation of initiation complex

The 30S and 50S ribosomal subunits of bacteria usually dissociate from each other when not involved in translation. To begin translation, an **initiation complex** forms (Fig. 12.11), consisting of following components in prokaryotes (*i.e., E.coli*): the 30S subunit of the ribosome, a messenger RNA, the charged N-formyl methionine tRNA (fmet – $tRNA_f^{Met}$), and three **initiation factors (IF1, IF2, IF3)**. Initiation factors (as well as elongation and termination factors) are proteins loosely associated with the ribosome (Table 12.2). They were discovered when ribosomes were isolated and then washed, losing the ability to perform translation (*i.e.,* polypeptide synthesis).

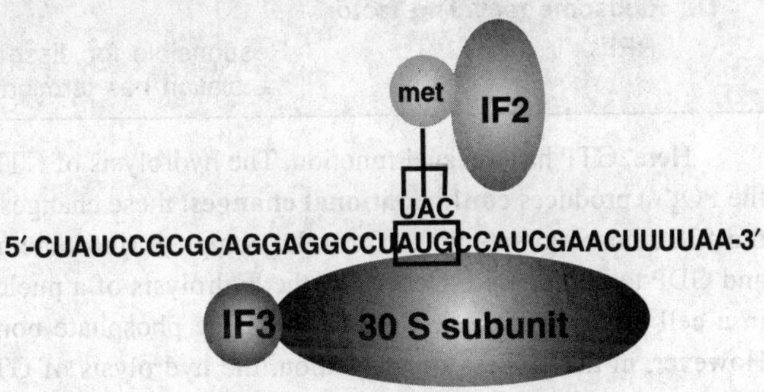

Fig. 12.11. Initiation complex.

Various components of the initiation complex interact in a series of steps. It is known that IF3 binds to the 30S ribosomal subunit, allowing the 30S subunit to bind to messenger RNA (Fig. 12.12, step 1). Meanwhile, a complex forms with IF2, the charged N-formylmethionine tRNA (fmet – $tRNA_f^{Met}$) and GTP (guanosine triphosphate; Fig. 12.12, step 2).

It is IF_2 that brings the initiator transfer RNA to the ribosome. IF2 binds only to the charged initiator transfer RNA, and without IF2, the initiator tRNA cannot bind to the ribosome.

The final step in initiation-complex formation is bringing together the first two components (Fig. 12.12; step 3).

Table 12.2 Functions of the bacterial (=prokaryotic) translation factors (*Source:* **Brown,** 2002).

S. No.	*Factor*	*Function*
A.	**Initiation factors**	
	IF–1	It may prevent premature entry of tRNA into A site and also may cause conformational changes that prepare the small subunit for attachment to the large subunit.
	IF–2	Directs the initiator $tRNA_f^{Met}$ to its correct position in the initiation complex.
	IF–3	Prevents premature reassociation of the large and small subunits of the ribosome.
B.	**Elongation factors**	
	EF–Tu	Directs the next tRNA to its correct position in the ribosome
	EF–Ts	Regenerates EF-Tu after the latter yielded the energy contained in its attached GTP molecule
	EF–G	Mediates translocation
C.	**Release factors**	
	RF–1	Recognizes the termination codons 5′–UAA–3′ and 5′–UAG–3′.
	RF–2	Recognises 5′–UAA–3′ and 5′–UGA–3′
	RF–3	Stimulates dissociation of RF1 and RF2 from the ribosome after termination
D.	**Ribosome recycling factor**	
	RRF	Responsible for dissociation of ribosomal subunits after translation has terminated

Here, GTP has unusual function. The hydrolysis of GTP to GDP + Pi (inorganic phosphate, the PO_4^{-3}) produces **conformational changes**; these changes allow the initiation complex to join the 50S ribosomal subunit to form the complete ribosome and then allow the initiation factors and GDP to be released. Generally, the hydrolysis of a nucleoside triphosphate (*i.e.,* ATP, GTP) in a cell occurs to release the energy in the phosphate bonds for use in a metabolic process. However, in the process of translation, the hydrolysis of GTP apparently changes the shape so that it and the initiation factors can be released from the ribosome after 70S particle has been formed. Thus, *hydrolysis of GTP in translation is for conformational change rather than covalent bond formation.* IF1 helps the other two initiation factors bind to the 30S initiation complex.

How does 70S ribosome recognizes mRNA? The ribosome recognizes the prokaryotic mRNA through complementarity of a region at the 3′ end of the 16S ribosomal RNA and a region slightly upstream from the initiation sequence (AUG) on the mRNA (Fig. 12.15A, Fig. 12.13). This consensus sequence is 5′ – AGGAGGU – 3′ and is called **Shine-Dalgarno sequence** (Fig.12.13). It is located between 3 to 10 nucleotides upstream of the initiation codon.

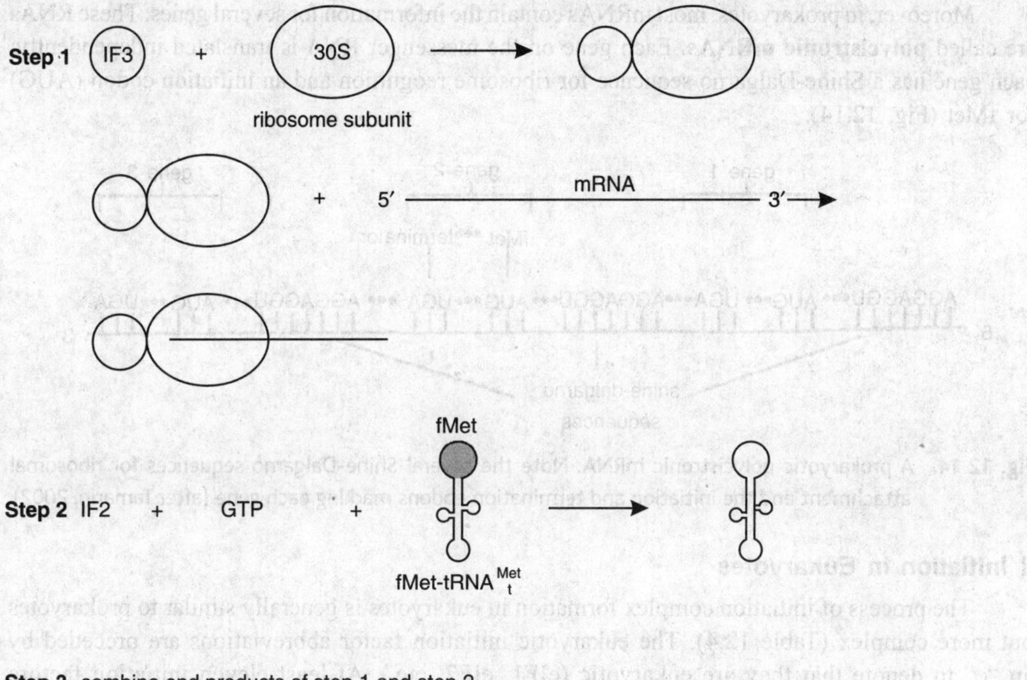

Fig. 12.12. The prokaryotic 70S ribosome forms in a three step process. In the first step, the 30S ribosomal subunit and the mRNA combine. In the second step, the initiator tRNA combines with IF2. In the final step, the components from steps 1 and step 2 combine to form the initiation complex, followed by the formation of the 70S ribosome (after Tamarin, 2002).

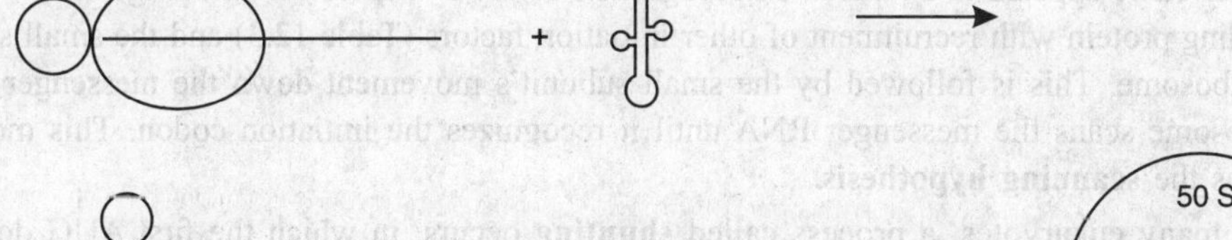

Fig. 12.13. The Shine-Dalgarno sequence or ribosome binding site for bacterial (*E.coli*) translation.

Moreover, in prokaryotes, most mRNAs contain the information for several genes. These RNAs are called **polycistronic mRNAs**. Each gene on the messenger RNA is translated independently: each gene has a Shine-Dalgarno sequence for ribosome reognition and an initiation codon (AUG) for fMet (Fig. 12.14).

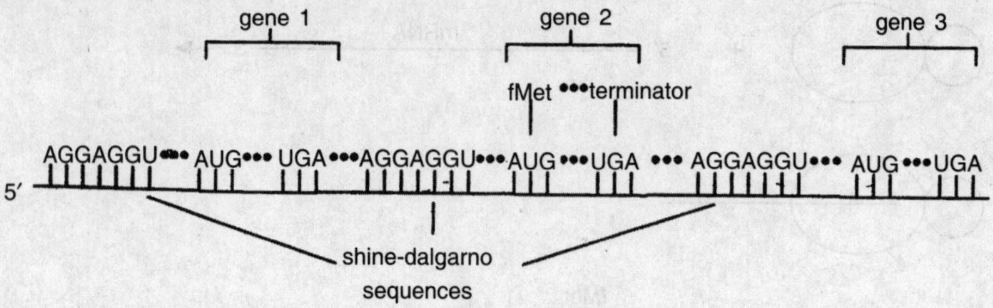

Fig. 12.14. A prokaryotic polycistronic mRNA. Note the several Shine-Dalgarno sequences for ribosomal attachment and the initiation and termination codons marking each gene (after Tamarin, 2002).

II Initiation in Eukaryotes

The process of initiation complex formation in eukaryotes is generally similar to prokaryotes but more complex (Table 12.4). The eukaryotic initiation factor abbreviations are preceded by an "e" to denote that they are eukaryotic (eIF1, eIF2, etc.). At least eleven initiation factors are involved, including a specific cap-binding protein, eIF4E. The Shine-Dalgarno sequence is altogether absent in eukaryotes. The actual mechanism for recognizing the 5′ end of eukaryotic messenger RNA appears to be based on recognition of the **5′ cap** of the messenger RNA by the cap-binding protein with recruitment of other initiation factors (Table 12.3) and the small subunit of the ribosome. This is followed by the small subunit's movement down the messenger RNA. The ribosome scans the messenger RNA until it recognizes the initiation codon. This model is known as the **scanning hypothesis.**

In many eukaryotes, a process called **shunting** occurs, in which the first AUG does not serve as the initiation codon; rather, scanning begins, but it bypasses a region of the messenger RNA upstream of the initiation codon, called the **leader** or **5′ untranslated region** (5′ UTR), in favour of an AUG further down the messenger RNA. The cause of shunting seems to be secondary structure in the messenger RNA, upstream from the AUG codon that actually serves as the initiation codon.

In some eukaryotes, very small genes called **open reading frames (ORFs)** are present in this region of the messenger RNA and play some role in shunting. In the genes of some plant and animal viruses, ORFs are translated, and then the main gene is translated by the same ribosome in a process called **reinitiation** (Fig. 12.15B).

In some eukaryotes, ribosomes can initiate protein synthesis within the mRNA if that mRNA contains a sequence called an **IRES** (*i.e.*, internal ribosome entry site). These sequences were discovered in RNA genome of the human poliovirus and rhinovirus (for cold) and in several cellular messenger RNAs. They are at least four hundred nucleotides long.

***(i)* Initiation of eukaryotic translation with scanning.** The first step involves assembly of the *pre-initiation complex.* This structure comprises of the 40S ribosomal subunit, a 'ternary complex' made up of the initiation factor eIF-2 bound to the initiator tRNA ^{Met} and a molecule of GTP, and three additional initiation factors, eIF-1, eIF-1A and eIF-3. As in bacteria, the initiator

Table 12.3 Eukaryotic translation factors (*Source:* **Brown**, 2002).

A. Initiation factors	
eIF-1	Component of pre-initiation complex
eIF-1A	Component of pre-initiation complex
eIF-2	Binds to initiator tRNAMet within the ternary complex component of the pre-initiation complex; phosphorylation of eIF-2 results in a global repression of translation
eIF-3	Component of the pre-initiation complex; makes direct contact with eIF-4G and forms the link with the cap binding complex
eIF-4A	Component of the cap binding complex, a helicase that aids scanning by breaking intramolecular base pairs in the mRNA
eIF-4B	Aids scanning possible by acting as a helicase that breaks intramolecular base pairs in the mRNA
eIF-4E	Component of the cap-binding complex, possibly the component that makes direct contact with the cap structure at the 5′ end of mRNA
eIF-4F	The cap-binding complex, comprising eIF-4A, eIF-4E and eIF-4G, which makes the primary contact with the cap structure at the 5′ end of the mRNA
eIF-4G	Component of the cap-binding complex; forms a bridge between the cap binding complex and eIF-3 in the preinitiation complex; in at least some organisms, eIF-4G also forms an association with the poly (A) tail, via the polyadenylate-binding protein.
eIF-5	Aids release of the other initiation factors at the completion of initiation
eIF-6	Associated with the large subunit of the ribosome, prevents large subunits from attaching to small subunits in the cytoplasm
B. Elongation factors	
eEF-1	Complex of four subunits (eEF-1a, eEF-1b, eEF-1d and eEF-1g); directs the next tRNA to its correct position in the ribosome
eEF-2	Mediates translocation
C. Release factors	
eRF-1	Recognizes the termination codon
eRF-2	Possibly stimulates dissociation of eRF-1 from the ribosome after termination; possibly causes the ribosome subunits to dissociate after termination of translation

tRNA is distinct from the normal $tRNA^{Met}$ that recognizes internal 5′–AUG-3′ codons but, unlike bacteria, it is aminoacylated with normal methionine, not the formylated version.

After assembly, the pre-initiation complex associates with the 5′ end of the mRNA. This step requires the **cap binding complex** (sometimes called eIF-4F) which comprises the initiation factors eIF-4A, eIF-4E and eIF-4G. The contact with the cap might be made by eIF-4E alone (Fig. 12.16A) or involve a more general interaction with the cap binding complex (**Pestova** and **Hellen**, 1999). The factor eIF-4G serves as a bridge between eIF-4E, bound to the cap, and eIF-3, attached to pre-initiation complex gets attached to the 5′ region of the mRNA. Attachment of the pre-initiation complex to the mRNA is also influenced by the poly (A) tail, at the distant 3′ end of the mRNA. This interaction is thought to be mediated by the polyadenylate-binding protein (PADP) which is attached to the poly (A) tail. In yeast and plants, PADP is found to form an association with eIF-4G and for such an association mRNA has to bend back on itself.

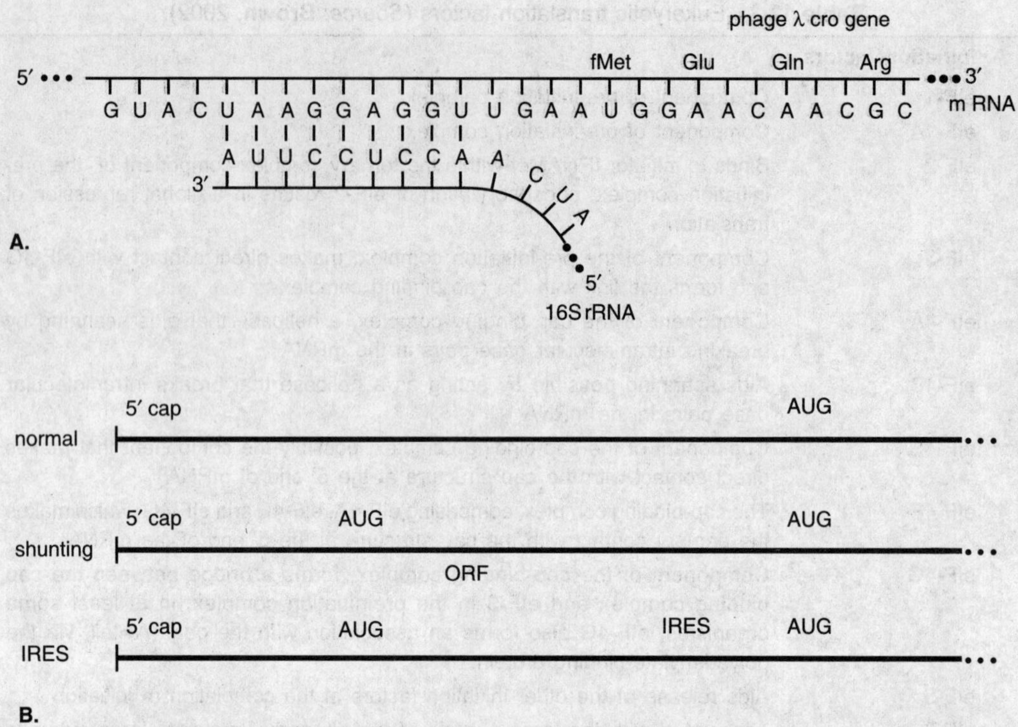

Fig. 12.15. Initiation signals for translation. A-The Shine-Dalgarno hypothesis for prokaryotic translation. The Shine-Dalgarno sequence (AGGAGGU) is on the prokaryotic messenger RNA just upstream from the initiation codon AUG. Complementarity exists between this sequence and a complementary sequence (UCCUCCA) on the 3′ end of the 16S ribosomal RNA. B-Scanning, shunting and internal ribosome entry in eukaryotic messenger RNA. In the scanning model, the initiation codon of the gene is the first AUG encountered. In shunting, an open reading frame (ORF) may or may not be present to provide secondary structure in the messenger RNA to shunt scanning to the main gene. If the open reading frame (ORF) is translated, reinitiation of translation at the same ribosome may occur at the main gene. Finally, an internal ribosome entry site (IRES) allows translation to begin with the messenger RNA without scanning (after Tamarin, 2002).

With its attachment to the 5′ end of the mRNA, the **initiation complex** has to scan along the mRNA molecule and find the initiation codon. The leader regions of eukaryotic mRNAs can be several tens or even hundreds of nucleotides in length and often contain regions that form hairpins and other base paired structures. These are probably removed by a combination of eIF-4A and eIF-4B. The eIF-4A, and possibly also eIF-4B, has a helicase activity and is able to break intra-molecular base pairs in the mRNA, freeing the passage for the initiation complex (Fig. 12.16B). The initiation codon (AUG) is recognizable because it is contained in a short consensus sequence: 5′-ACCAUGG-3′, which is called **Kozak consensus** (see **Brown**, 2002).

As soon as, the initiation complex is positioned over the initiation codon, the large subunit (60S) of the ribosome attaches. As in bacteria, this requires hydrolysis of GTP and leads to release of the initiation factors. Two final initiation factors are involved at this stage: eIF-5 which aids release of the other factors and eIF-6, which is associated with the unbound large subunit and prevents it from attaching to a small subunit in the cytoplasm.

(*ii*) Initiation of eukaryotic translation without scanning. Transcripts (mRNA) of picornaviruses (*i.e.*, RNA viruses such as poliovirus and rhinovirus) are not capped but instead have an internal ribosome entry site (IRES). The presence of IRES on their transcripts means that picorna viruses can block protein synthesis in the host cell by inactivating the cap binding complex, without affecting translation of their own transcripts.

Quite interestingly, no virus proteins are required for the recognition of an IRES by a host ribosome. In other words, the normal eukaryotic cell possesses proteins and/or other factors that enable it to initiate translation by the IRES method (**Holick** *et al.,* 2000). Some nuclear genes are reported to have IRES. For example, IRES are present on mRNAs for the mammalian immunoglobulin heavy chain binding protein and the *Drosophila* antennapedia protein. IRES are also found on several mRNAs whose protein products are translated when the cell is put under stress, for example by exposure to heat, irradiation, on low oxygen conditions. Under these circumstances, cap dependent translation is globally suppressed. The presence of IRESs on the 'survival' mRNAs therefore enables these to undergo preferential translation at the time when their products are needed (see **Brown,** 2002).

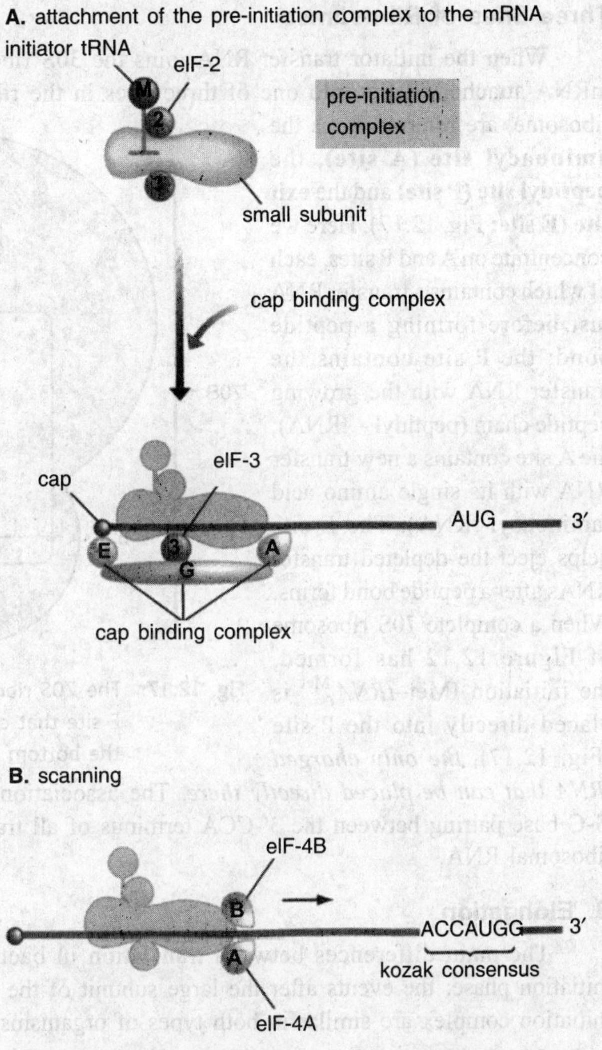

Fig. 12.16. Initiation of eukaryotic translation. A-Assembly of the pre-initiation complex and its attachment to the mRNAs (according to Hentze 1997). B-The pre-initiation complex scans along the mRNA until it reaches a initiation codon, which is recognisable because it is located within Kozak consensus sequence. Scanning is aided by eIF-4A, eIF-4B, which are thought to have helicase activity. Most probably, eIF-3 remains attached to the pre-initiation complex during scanning, as shown here. Scanning is an energy-dependent process that requires hydrolysis of ATP. M=methionine (after Brown, 2002).

Three Sites of Ribosomes

When the initiator transfer RNA joins the 30S ribosomal subunit of prokaryotes with its mRNA attached, it fits into one of three sites in the ribosome. These sites or cavities in the ribosome, are referred to as the **aminoacyl site (A site),** the **peptidyl site (P site)** and the **exit site (E site;** Fig. 12.17). Here we concentrate on A and P sites, each of which contains a transfer RNA just before forming a peptide bond: the P site contains the transfer RNA with the growing peptide chain (peptidyl – tRNA); the A site contains a new transfer RNA with its single amino acid (aminoacyl-tRNA). The E site helps eject the depleted transfer RNAs after a peptide bond forms. When a complete 70S ribosome of Figure 12.12 has formed, the initiation fMet–$tRNA_f^{Met}$ is placed directly into the P site (Fig. 12.17), *the only charged tRNA that can be placed directly there.* The association of tRNA and ribosome is aided by a G-C base pairing between the 3′-CCA terminus of all transfer RNAs and a guanine in the 23S ribosomal RNA.

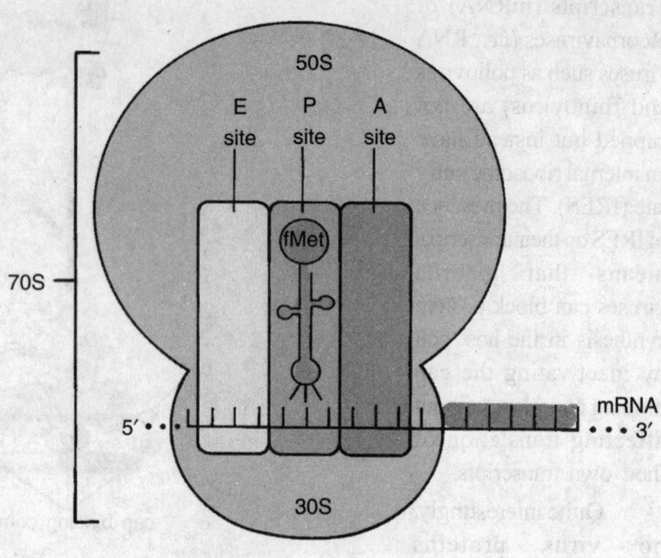

Fig. 12.17. The 70S ribosome contains an A site, a P site and an E site that can receive tRNAs. The mRNA runs through the bottom of the sites (after Tamarin, 2002).

B. Elongation

The main differences between translation in bacteria and eukaryotes occur during the initiation phase; the events after the large subunit of the ribosome becomes associated with the initiation complex are similar in both types of organisms.

1. Positioning a second transfer RNA

The second charged tRNA binds (due to a codon-directed binding) to the first ribosome at the latter's 'A' site with the help of the proteins, called elongation factors (e.g. EF-Tu). EF-Tu carries a molecule of GTP. Correct hydrogen bonding with the mRNA template dictates the selection of a new tRNA, and the activity of the EF-Tu ensures the proper positions of the tRNA in the A site. Such a placement activity needs energy that is provided by the hydrolysis of the molecule of GTP to GDP and phosphate. After performing its function, the **EF-Tu** protein dissociates from the ribosome, and in the cytoplasm is subsequently regenerated to its active form by another elongation factor, the **EF-Ts**. At this point, both sites of the ribosome are occupied by tRNA's each of which carries an amino acid, and each of which is hydrogen bonded to the template mRNA (Fig. 12.18).

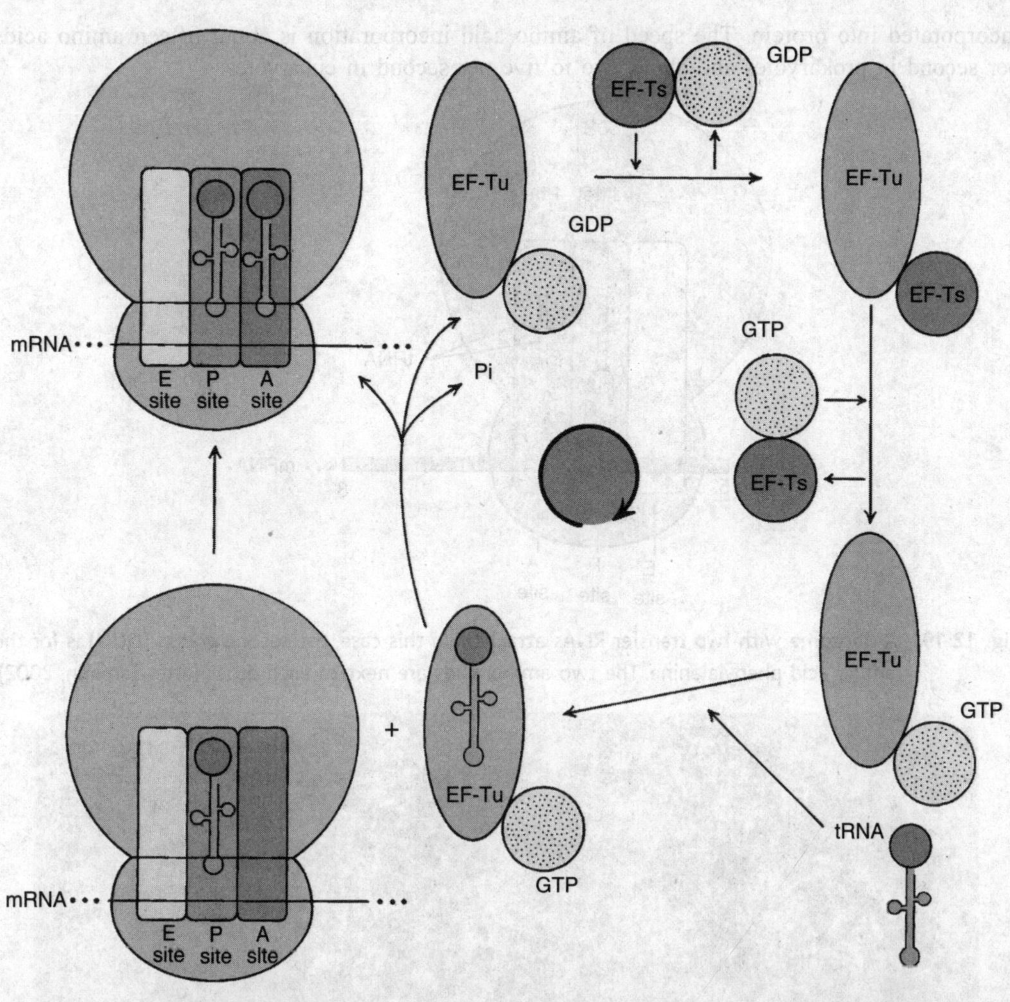

Fig. 12.18. The EF–Ts/EF–Tu cycle. EF–Ts and EF–Tu are required for a transfer RNA to attach to the A site of the ribosome. At *top centre,* there is EF–Tu attached to a GDP. The GDP is then displaced by EF–Ts, which in turn is displaced by GTP. A transfer RNA attaches and is brought to the ribosome. If the codon-anticodon fit is correct, the transfer RNA attaches at the A site with the help of the hydrolysis of GTP to GDP + Pi, allowing EF–Tu to release. The EF–Tu is now back where we started. Since EF–Tu has a strong affinity for GDP, the role of EF–Ts is to displace the GDP, and later to be replaced by GTP (after Tamarin, 2002).

The structures revealed by X-ray crystallography (Fig. 12.20) show that A, P and E sites are located in the cavity between the large and small subunits of the ribosome, the codon-anticodon interaction being associated with the small subunit and the aminoacyl end of the mRNA with the large subunit (**Yusupov** *et al.* 2001; Fig. 12.18).

Further, it takes several milliseconds for the GTP to be hydrolysed, and another few milliseconds for the EF-Tu/GDP to actually leave the ribosome. During those two intervals of time, the codon-anticodon fit of the transfer RNA is scrutinized. If the correct tRNA is in place, a peptide bond forms. If not, the charged tRNA is released and a new cycle of EF-Tu/GTP-mediated testing of tRNAs begins. The error rate is only about one mistake in ten thousand amino acids incorporated into protein. The speed of amino acid incorporation is about fifteen amino acids per second in prokaryotes and about two to five per second in eukaryotes.

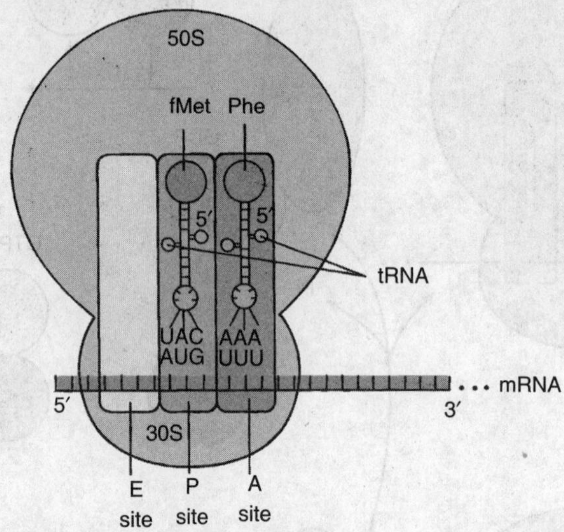

Fig. 12.19. A ribosome with two transfer RNAs attached. In this case, the second codon (UUU) is for the amino acid phenylalanine. The two amino acids are next to each other (after Tamarin, 2002).

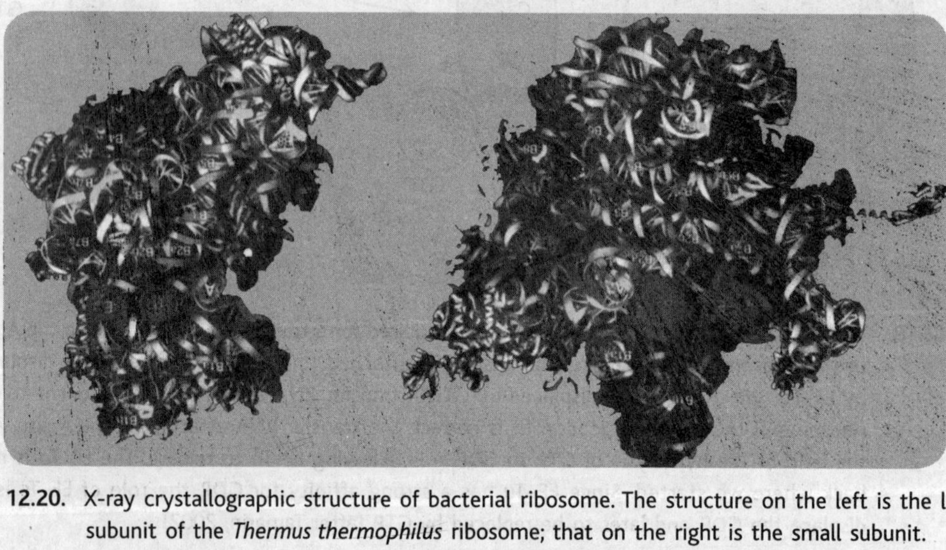

Fig. 12.20. X-ray crystallographic structure of bacterial ribosome. The structure on the left is the large subunit of the *Thermus thermophilus* ribosome; that on the right is the small subunit.

***(ii)* Peptide bond formation.** The next step is the formation of a peptide bond between the two amino acids. To accomplish this job, the first amino acid (N-formylmethionine) is removed from its attachment to its tRNA and transferred to the free-NH_2 terminus of the second amino acid. The first amino acid is, thus, placed "on top of" the second. The ensuing peptide bond, thus, joins the carboxyl group of the first amino acid with the amino group of the second amino acid (Fig. 12.21). The resulting compound is a dipeptide whose carboxyl end is still bonded to the second tRNA, but whose amino end is free. The reaction is catalyzed by an enzyme associated with 50S subunit and called **peptidyl transferase.** This enzymatic centre, an integral part of the

50S subunit, was originally believed to be composed of parts of several of the 50S proteins. Now, however, it is believed to have ribozymic activity (enzymatic activity) of 23S rRNA of 50S ribosomal subunit.

The enzymatic activity involve a bond transfer from the carboxyl end of N-formyl methionine to the amino end of the second amino acid (phenylalanine in Fig. 12.19). Every subsequent peptide bond is identical, regardless of the amino acids involved. The energy used is contained in the high-energy ester bond between the transfer RNA in the P site and its amino acid (Fig. 12.21). Immediately after the formation of the peptide bond, the transfer RNA with the dipeptide is in the A site, and a depleted tRNA is in the P site.

***(iii)* Translocation**

The next stage in elongation is *translocation* of the ribosome in relation to the tRNA and the mRNA. Elongation factor **EF-G**, earlier called ***translocase***, catalyses the translocation process. The ribosome must be converted from *pretranslocational state* to the *post-translocational* state by the action of EF-G, which physically moves the mRNA and its associated transfer RNA (Fig. 12.22).

Thus, during translocation following three things happen at once:

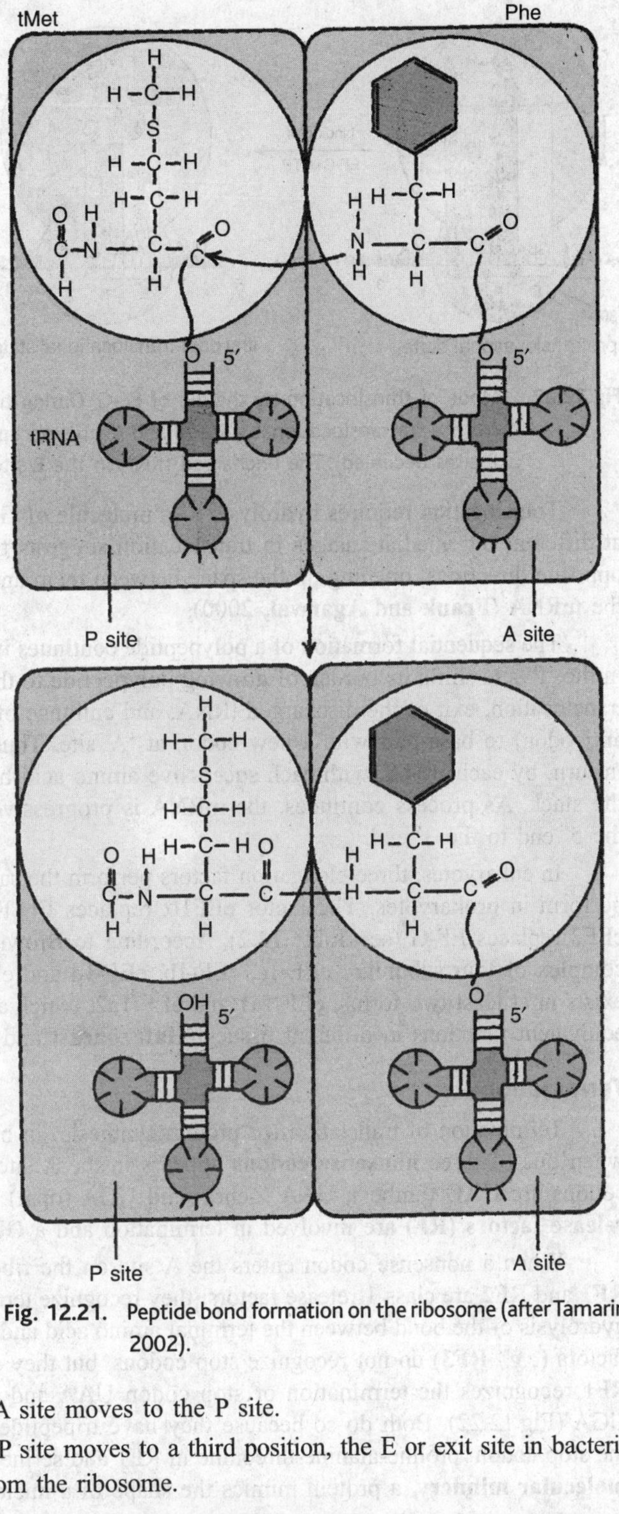

Fig. 12.21. Peptide bond formation on the ribosome (after Tamarin, 2002).

1. The ribosome moves along the three nucleotides, so that the next codon enters the A site.
2. The dipeptide-tRNA in the A site moves to the P site.
3. The deacylated tRNA in the P site moves to a third position, the E or exit site in bacteria or in eukaryotes, is simply ejected from the ribosome.

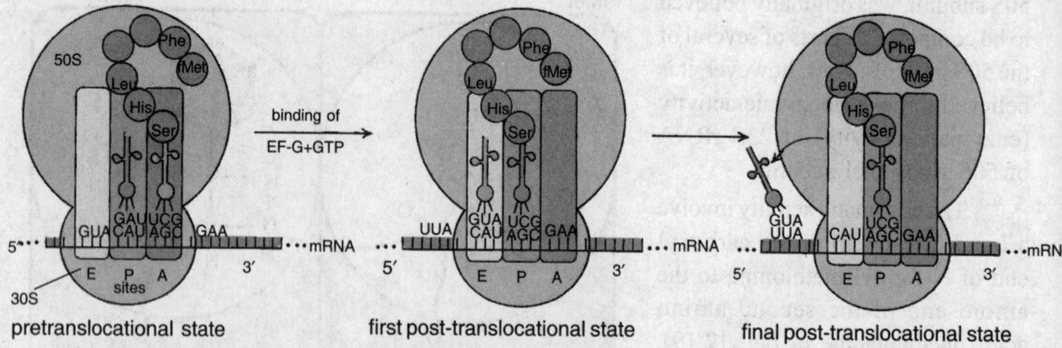

Fig.12.22. Mode of translocation by the aid of EF-G. During translocation, the ribosome converts istself from a pretranslocational state (P and A sites occupied) to a post-translocational state (E and P sites occupied). The uncharged tRNA in the E site is then ejected (after Tamarin, 2002).

Translocation requires hydrolysis of a molecule of GTP. Electron microscopy of ribosomes at different intermediate stages in translocation suggests that the two subunits rotate slightly in opposite directions, opening up the space between them and enabling the ribosome to slide along the mRNA (**Frank** and **Agarwal,** 2000).

The sequential formation of a polypeptide continues in the manner described above. A tRNA in the 'P' site shifts its burden of growing polypeptide to the next succeeding tRNA, followed by translocation, exit of the discharged tRNA, and entrance of a new charged tRNA (having correct anticodon) to base pair with a new codon at 'A' site. Thus, the growing polypeptide is adopted in turn, by each tRNA, with each successive amino acid being added in effect, to the bottom of the stack. As process continues, the mRNA is progressively translated, codon by codon, from the 5′ end to the 3′ end.

In eukaryotes, three elongation factors perform the same tasks that EF-Tu, EF-Ts and EF-G perform in prokaryotes. The factor eEF1α replaces EF-Tu, factor eEF1ßγ replaces EF-Ts and eEF2 replaces EF-G (see Table 12.2). According to Brown (2002), elongation factor eEF-1 is a complex of four subunits : eEF-1a, eEF-1b, eEF-1d and eEF-1g (Table 12.3). The first of these exists in at least two forms, eEF-1a1 and eEF-1a2, which are similar proteins that probably have equivalent functions in different tissues (**Hafezparast** and **Fisher,** 1998).

Termination

Termination of translation (or protein synthesis) in both prokaryotes and eukaryotes occur when one of three **nonsense codons** appears in the A site of the ribosome (Fig. 12.23). These codons are UAG (amber), UAA (ochre) and UGA (opal). In prokaryotes, three proteins called **release factors (RF)** are involved in termination and a GTP is hydrolyzed) to GDP + Pi.

When a nonsense codon enters the A site on the ribosome, a release factor recognizes it. **RF1** and **RF2** are class 1 release factors: they recognize termination/stop codon and then promote hydrolysis of the bond between the terminal amino acid and its tRNA in the P site. Class 2 release factors (*e.g.,* **RF3**) do not recognize stop codons, but they stimulate class1 release factors to act. RF1 recognizes the termination or stop codon UAA and UAG and RF2 recognizes UAA and UGA (Fig.12.22). Both do so because they have tripeptides that mimic anticodons to recognize the stop codon: proline-alanine-threonine in RF1 and serine-proline-phenylalanine in RF2. In this **molecular mimicry,** a protein mimics the shape of a nucleic acid in order to function properly.

The next base in mRNA past the stop codon is usually an adenine, required for efficient termination. After the release factors act, with the hydrolysis of a GTP, the ribosome has completed its task of translating mRNA into a polypeptide. Final release of all factors and dissociation of the two subunits of the ribosome take place with the help of IF3, which rebinds to the 30S subunit, and a **ribosome recycling factor (RRF).** RRF has a tRNA-like structure like the eRF-1 (**Selmer** *et al.,* 1999). RRF probably enters the P or A site and "unlocks" the ribosome. Dissociation requires energy, which is released from GTP by EF-G and also requires the initiation factor IF-3 to prevent the subunits from attaching together again. The disassociated ribosome subunits enter the cytoplasmic pool, where they remain until used again in another round of translation.

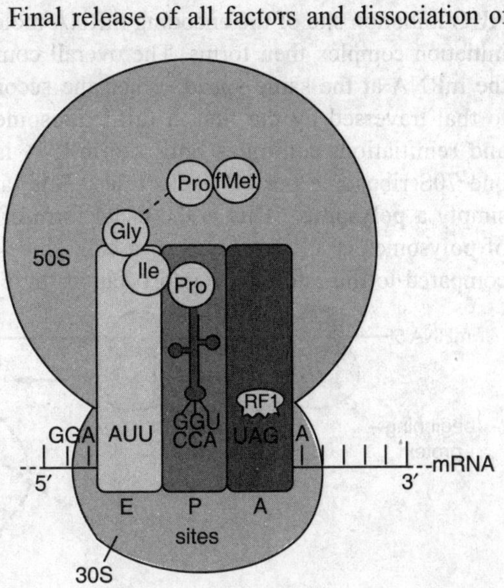

Fig. 12.23. Chain termination at the ribosome. One of two release factors recognizes a termination or stop codon in the A site in ribosome. In this case, RF1 recognizes a UAG. The complex then falls apart releasing the peptide (after Tamarin, 2002).

Eukaryotes have just two release factors: eRF-1, which recognizes the termination codon, and eRF-3, which might play the same role as RF-3 although this has not proven (see **Brown,** 2002). The structure of eRF-1 has been solved by X-ray crystallography, showing that the shape of this protein is very similar to that of a tRNA (**Kisselev** and **Buckingham,** 2000).

Further, in contrast to the often polycistronic mRNA of bacteria, mRNA of eukaryotes are monocistronic, containing the coding sequence only for one polypeptide. The initiation codon for this sequence is located near the 5′ end of the message. Thus, the association of mRNA takes place on 5′ end and not at the initiation codon AUG as in prokaryotes.

Comparison of Prokaryotic and Eukaryotic Translation (Table 12.4)

Table 12.4 Comparison between prokaryotic and eukaryotic translation (*Source:* **Tamarin,** 2000).

S. No.		*Prokaryotes*	*Eukaryotes*
1.	Initiation codon	AUG, occasionally GUG, UUG	AUG, occasionally GUG, CUG
2.	Initiation amino acid	N-formyl methionine	Methionine
3.	Initiation tRNA	$tRNA_f^{met}$	$tRNA_i^{Met}$
4.	Internal methionine tRNA	$tRNA_{in}^{met}$	$tRNA_m^{Met}$
5.	Initiation factors	IF1, IF2, IF3	eIF factors
6.	Elongation factor	EF-Tu	eIF1α
7.	Elongation factor	F-Ts	eIF 1 βγ
8.	Translocation factor	EF-G	eIF 2
9.	Release factors	RF1, RF2, RF3, RRF	eRF1, eRF3

Polysomes and Coupled Transcription – Translation

The unit of translation is never simply a ribosome traversing a mRNA molecule, but is a more complex structure. After about 25 amino acids have been joined in a polypeptide chain, the

AUG initiation site of the encoding mRNA molecule is completely free of the ribosome. A second initiation complex then forms. The overall configuration is of two 70S ribosomes moving along the mRNA at the same speed. When the second ribosome has moved along a distance similar to that traversed by the first, a third ribosome is able to attach. This process, *i.e.,* movement and reinitiation, continues until the mRNA is covered with ribosomes at a density of about one 70S ribosome per 80 nucleotides. This large translation unit is called a **polyribosome** or simply a **polysome.** This is the usual form of translation unit in all cells (Fig. 12.24). The use of polysomes is advantageous to a cell, since the overall rate of protein synthesis is increased compared to the rate that would occur if there were no polysomes.

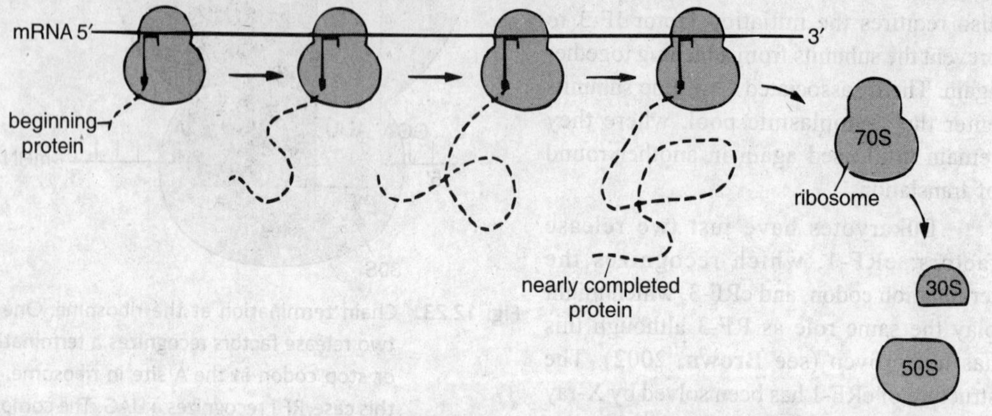

Fig. 12.24. Polysome formation: diagram shows the relative movement of the 70S ribosome and the mRNA, and growth of the protein chain (after Freifelder, 1985).

Further, a mRNA molecule being synthesized has a free 5′ terminus and translation also occurs in the 5′ → 3′ direction, so each cistron contained in the mRNA immediately starts its translation. As a result, the ribosome binding site is transcribed first, followed in order by the AUG codon, the region encoding the amino acid sequence, and finally the stop codon. Thus, in bacteria in which no nuclear membrane separates the DNA and the ribosomes, there is no

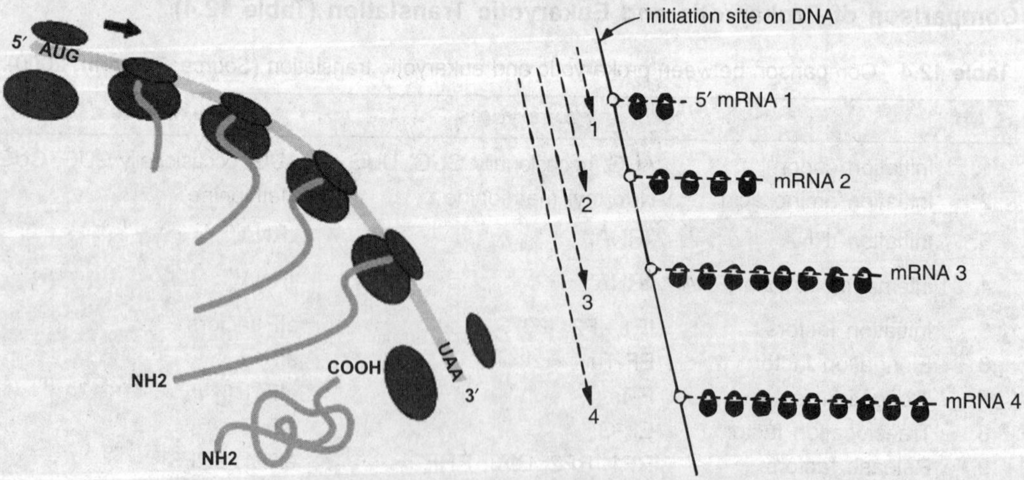

Fig. 12.25. Coupled transcription-translation. Transcription of a section of the DNA of *E.coli* and translation of the nascent mRNA. The dashed arrows show the distances of each RNA polymerase from the transcription initiation site, mRNA 4 probably becomes shorter due to partial digestion off its 5′ end by a RNase (after Freifelder, 1985).

obvious reason why the 70S initiation complex should not form before the mRNA is released from the DNA. With prokaryotes (*E.coli)* this does indeed occur; this process is called **coupled transcription-translation** (Fig. 12.25). This coupled activity does not occur in eukaryotes, because the mRNA is synthesized and processed in the nucleus and later on transported through the nuclear membrane to the cytoplasm where the ribosomes are located. Coupled transcription-translation, too, speeds up protein synthesis in the sense that translation does not have to await release of mRNA from the DNA.

12.4 UNUSUAL EVENTS DURING TRANSLATION ELONGATION

The simple codon-by-codon translation of an mRNA is viewed as the standard way in which proteins are synthesized. But following three unusual translation elongation events have been reported:

1. Frameshifting;
2. Translation slippage; and
3. Translation bypassing

1. Frameshifting

This event occurs when a ribosome pauses in the middle of an mRNA, moves back one nucleotide or, less frequently, forward one nucleotide, then continues translation (**Farabaugh,** 1996). The result is that the codons that are read after the pause are not contiguous with the preceding set of codons, *i.e.*, they lie in a different reading frame (Fig. 12.26A). Such frame shifts may be spontaneous or programmed.

***(i)* Spontaneous frameshifts.** They occur randomly and are deleterious because the polypeptide synthesized after the frameshift has the incorrect amino acid sequence.

***(ii)* Programmed frameshifts.** They occur only in some mRNAs. A few mRNAs utilize programmed frameshifting to induce the ribosome to change frame at a specific point within transcript. Programmed frameshifting occurs in all types of organisms, from bacteria through to humans, as well as during expression of a number of viral genomes. An example occurs during synthesis of **DNA polymerasc III** (enzyme) in *E.coli*, the main enzyme involved in replication of DNA. Two of the subunits of DNA polymerase III, γ (gamma) τ and (tau), are coded by a single gene, *dnaX*. Subunit τ is the full-length translation product of the *dnaX* mRNA, and subunit γ is a shortened version. Synthesis of γ subunit invoves a frameshift in the middle of *dnaX* mRNA, the ribosome encountering a termination codon immediately after frameshift and so producing the truncated γ version of the translation product. It is thought that the frameshift is induced by *three* features of the *dnaX* mRNA:

1. A hairpin loop, located immediately after the frameshift position, which stalls the ribosome.

2. A sequence similar to a ribosome binding site immediately upstream of the frameshift position, which is thought to base-pair with the 16S rRNA, again causing the ribosome to stall.

3. The codon 5′–AAG–3′ at the frameshift position. The presence of a modified nucleotide at the wobble position of the $tRNA^{lys}$ that decodes 5′–AAG–3′ means that the codon-anticodon interactions is relatively weak at this position, enabling the frameshift to occur.

2. Translation slippage

The phenomenon of translation slippage enables a single ribosome to translate an mRNA that contains copies of two or more genes (Fig.12.26B). Due to this phenomenon, a single ribosome can synthesize each of the five proteins coded by the mRNA transcribed from the tryptophan operon of *E.coli* (Fig. 12.27). When the ribosome reaches end of one series of codons (*e.g.*, UAA), it releases protein it has just made, slips to the next initiation codon (*e.g.*, AUG), and begins synthesizing the next protein (Fig.12.26B).

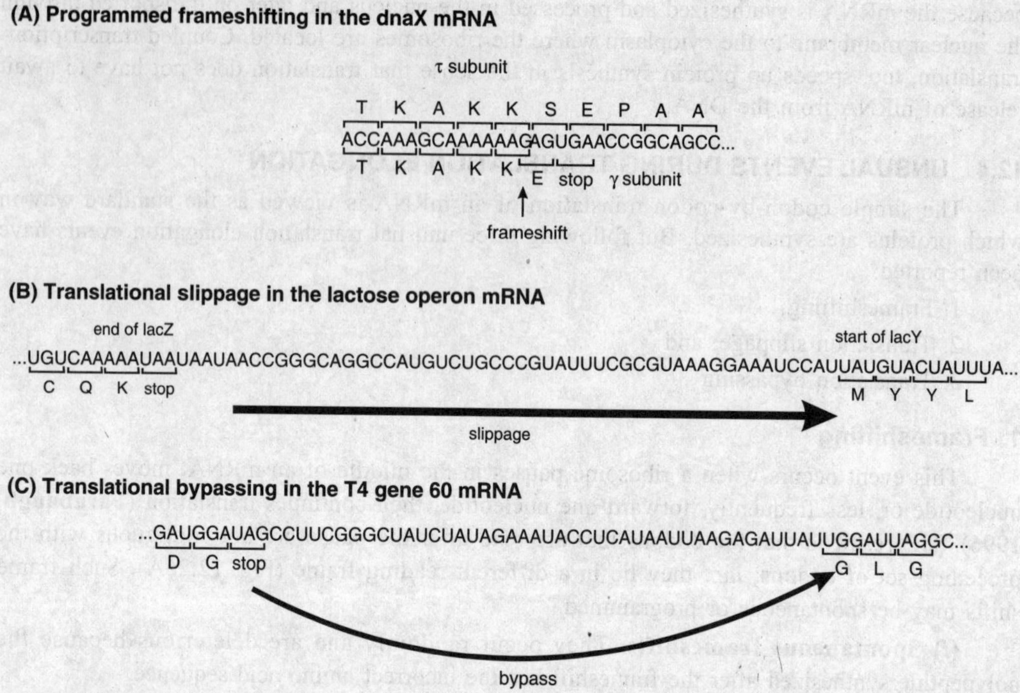

Fig. 12.26. Three unusual translation elongation events occurring in *Escherichia coli*. A-Programmed frameshifting during the translation of the *dnaX* mRNA. During the synthesis of γ subunit the ribosome shifts back one nucleotide, immediately after a series of As coding for two lysine amino acids (K, K). The ribosome inserts a glutamic acid (E) into the polypeptide and then encounters a termination codon. B-Slippage between the *lac Z* and *lac Y* genes of the lactose operon mRNA. C-Bypassing during translation of the T4 gene 60 mRNA involves a jump between two glycine (G) codons (**Note.** T, K, A, etc., stand for respective amino acids; Refer Fig. 11.3) (after Brown, 2002).

3. Translation bypassing

It is an extreme form of translation slippage (**Herr** *et al.*, 2000). In translation bypassing, a larger part of the transcript, possibly a few tens of base pairs, is skipped and elongation of the original protein continues after the bypassing event (Fig. 12.26C). The bypass starts and ends either at two identical codons or at two codons that can be translated by the same tRNA by wobble. This suggests that the jump is controlled by the tRNA attached to the growing polypeptide, which scans the mRNA as the ribosome tracks along and halts the bypass when a new codon to which it can base-pair is reached. Translation bypassing of 44 nucleotides occurs in *E.coli* during translation of the mRNA for *gene 60* of T4 bacteriophage, which codes for the DNA topoisomerase subunit. Similar events have also been identified in a variety of other bacteria. Bypassing could result in two different proteins being synthesized from one mRNA-one protein from normal translation and one from bypassing, but whether this is its general function is not yet known (see **Brown**, 2002).

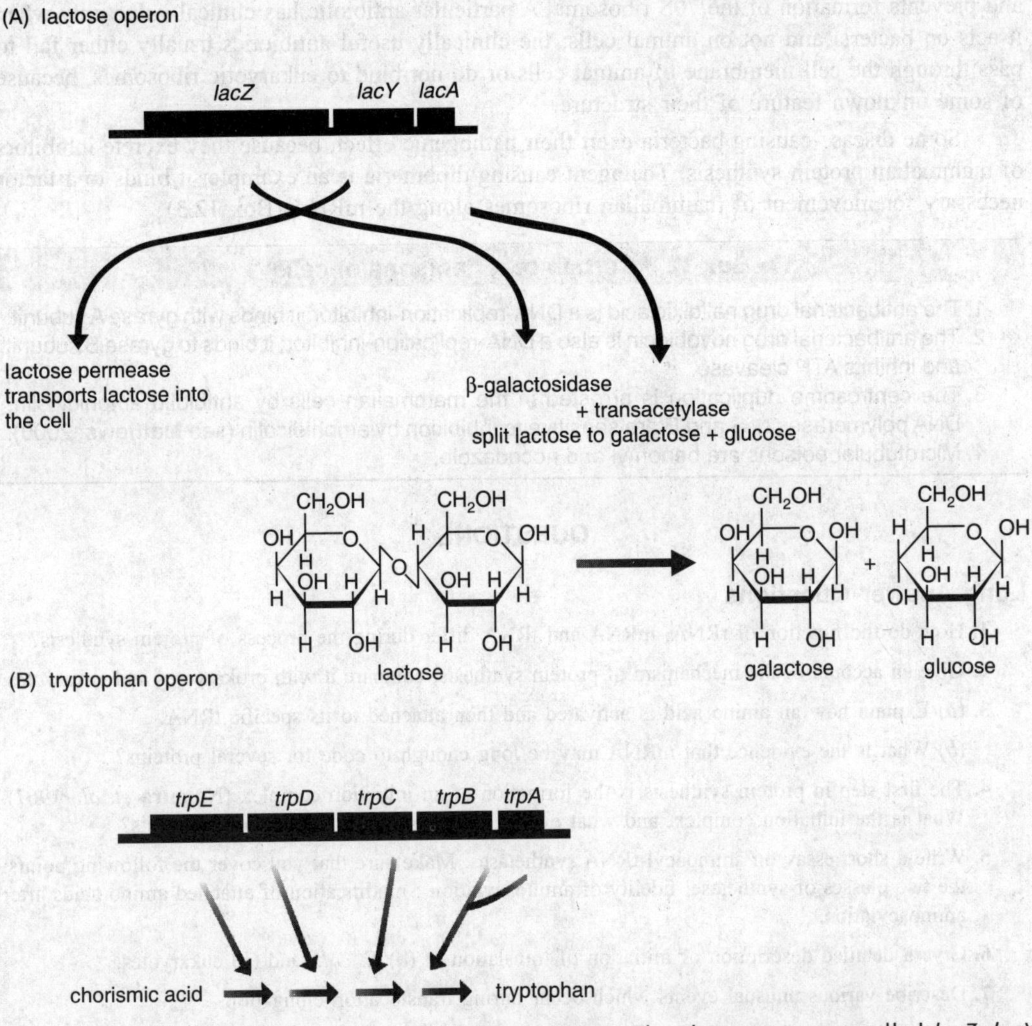

Fig. 12.27. Two operons of *E.coli*. *A*-The lactose operon. The three genes are called *lacZ, lacY* and *lacA*, the first two genes are separated by 52 bp and the second two genes by 64 bp. All three genes are expressed together, *lacY* coding for the lactose permease which transports lactose into the cell, and *lacZ* and lacA coding for enzymes that split lactose into its component sugars—galactose and glucose. *B*-The tryptophan operon, which contains five genes coding for enzymes involved in the multistep biochemical pathway that converts chorismic acid into the amino acid tryptophan. The genes in the tryptophan operon are closer together than those in the lactose operon: *trpE* and *trpD* overlap by 1bp, as do *trpB* and *trpA; trpD* and *trpC* are separated by 4bp and *trpC* and *trpB* by 12bp (after Brown, 2002).

12.5 ANTIBIOTICS AND PROTEIN SYNTHESIS

Many antibacterial agents (called antibiotics) have been isolated from fungi. Most of these are inhibitors of protein synthesis. For example, streptomycin and neomycin bind to a particular protein in the 30S particle and thereby prevent binding of $tRNA^{fmet}$ to the 'P' site; the tetracyclines inhibit binding of charged tRNA; lincomycin and chloramphenicol inhibit the peptidyl transferase: and puromycin causes premature chain termination; erythromycin binds to a free 50S particle and prevents formation of the 70S ribosome. A particular antibiotic has clinical value only when it acts on bacteria and not on animal cells; the clinically useful antibiotics usually either fail to pass through the cell membrane of animal cells or do not bind to eukaryotic ribosomes, because of some unknown feature of their structure.

Some disease-causing bacteria exert their pathogenic effect, because they excrete inhibitors of mammalian protein synthesis. The agent causing diphtheria is an example; it binds to a factor necessary for movement of mammalian ribosomes along the mRNA (Box 12.3)

Box 12.3 Certain other poisons of cells

1. The antibacterial drug nalidixic acid is a DNA-replication-inhibitor; it binds with gyrase A subunit.
2. The antibacterial drug novobiocin is also a DNA-replication-inhibitor; it binds to gyrase B subunit and inhibits ATP cleavase.
3. The centrosome duplication is arrested in the mammalian cells by antibiotic amphidicolin. DNA polymerases L, S and E are sensitive to inhibition by amphidicolin (see **Mathews,** 2000).
4. Microtubular poisons are benomyl and nocodazole.

QUESTIONS

Long Answer Questions

1. How do the function of rRNA, mRNA and tRNA differ during the process of protein synthesis?
2. Give an account of the mechanism of protein synthesis; compare it with prokaryotes and eukaryotes.
3. (*a*) Explain how an amino acid is activated and then attached to its specific tRNA.

   (*b*) What is the evidence that mRNA may be long enough to code for several proteins?
4. The first step in protein synthesis is the formation of an initiation complex (**Nomura** *et. al.,* 1967). What is the initiation complex, and what events follow to begin polypeptide synthesis?
5. Write a short essay on aminoacyl-tRNA synthetases. Make sure that you cover the following points: the two classes of synthetase; fidelity of aminoacylation ; modification of attached amino acids after aminoacylation.
6. Give a detailed description of initiation of translation in (*a*) *E. coli,* and (*b*) eukaryotes.
7. Describe various unusual events which occur during translocation elongation.

Short Answer Questions

1. Write short notes on the following:

   (*i*) Central dogma; (*ii*) Teminism; (*iii*) Reverse transcription; (*iv*) Kozak's scanning hypothesis; (*v*) Shine-Dalgarno sequence; (*vi*) rho factor; (*vii*) Aminoacyl RNA synthetase; (*viii*) Elongation factors; (*ix*) Chaperones; (*x*) Peptidyl transferase.
2. Differentiate between transcription and translation.

Very Short Answer Questions

1. How many different mRNAs could specify the amino acid sequence met-phe-ser-pro ?
2. Using the information in Table 10.2 convert the following mRNA segments into their polypeptide equivalents:

   (*a*) .... 5′GAAAUGGCAGUUUAC3′....;

   (*b*) .....3ÚUUCGAGAUGUCAA 5′......; and

   (*c*) ..... 5′AAA ACC UAG AAC CCA3′.....

3. A certain codon is determined to be AUG.
   (*a*) Of what nucleic acid molecule is this codon a part?
   (*b*) What is the corresponding anticodon?
   (*c*) Of what nucleic acid molecule is this anticodon a part?
   (*d*) What is the deoxyribonucleotide responsible for this codon?
4. Alpha (α) polypeptide chain of human haemoglobin consists of 141 amino acid residues. Would you expect the DNA segment responsible for the ultimate synthesis of this chain to be shorter than, longer than, or about the same length as the functional mRNA molecule for this chain?
5. An antibiotic is a microbial product of low molecular weight that specifically interferes with the growth of microorganisms when it is present in exceedingly small amounts. Specify some of the physiological activities that might be interrupted by an appropriate antibiotic and the reason why human cells are not harmed.
6. Which processes in protein synthesis require hydrolysis of GTP?

Fill in the Blanks

1. During the process of translation, the peptide formation at P site of 50S ribosomal subunit is catalyzed by the ....
2. The initiating formyl methionine tRNA can bind only with .... of 50S ribosomal subunit.
3. The translocation step of translation require _______ which is also called ....
4. The three chain termination codons are UAG (.................), UAA (.................) and UGA (opal).
5. In protein synthesis of prokaryotes, the three releasing factors are ....,................. and ....
6. The formyl group of methionine is removed by the enzyme ....
7. .... such as streptomycin prevents binding of .... to P site of ribosome.
8. The .... (*i.e.*, AUG) in eukaryotes is recognizable because it is contained in a short consensus sequence: 5'-ACCAUGG-3' which is called ....
9. Ribosomes contain three sites: aminoacyl site (A site), ... site (P site) and .... (.................).
10. Ribosomal translocation in relation to tRNA and mRNA requires hydrolysis of a molecule of ....

Multiple Choice Questions

1. The 'central dogma' states that biological information flaws in which of these patterns?
   (*a*) DNA → Protein → RNA (*b*) RNA → DNA → Protein
   (*c*) Protein → RNA → DNA (*d*) DNA → RNA → Protein
2. One of these cannot be prepared directly from DNA
   (*a*) another DNA (*b*) mRNA
   (*c*) Protein (*d*) tRNA
3. Amino acid sequence, in protein synthesis is decided by the sequence of
   (*a*) *r*RNA (*b*) *t*RNA
   (*c*) mRNA (*d*) *c*RNA
4. One of the following is not involved in translation
   (*a*) mRNA (*b*) tRNA
   (*c*) DNA (*d*) rRNA

5. Codes of mRNA and amino acid residues of proteins are

(*a*) colinear (*b*) coplanar

(*c*) zig-zag (*d*) none of these

6. The name of Temin and Baltimore is associated with

(*a*) photorespiration (*b*) RNA synthesis

(*c*) reverse transcription (*d*) all of these

7. Enzyme involved in reverse transcription

(*a*) primase (*b*) reverse transcriptase

(*c*) DNA polymerase (*d*) terminase

8. RNA retrovirus has a special enzyme that

(*a*) polymerizes host DNA (*b*) disintegrates host DNA

(*c*) translates host DNA (*d*) transcribes viral RNA to DNA

9. The starting tRNA of prokaryotes is loaded with

(*a*) valine (*b*) methionine

(*c*) formylated methionine (*d*) tryptophan

10. Initiation of polypeptide chain formation is always brought about at the site of codon coding for an amino acid called

(*a*) isoleucine (*b*) cysteine

(*c*) phenylalanine (*d*) methionine

11. Which of the following is associated with protein synthesis?

(*a*) aminoacyl tRNA (*b*) initiation factors IF3, IF2, IF1

(*c*) peptidyl transferase (*d*) transposons

12. The site of protein synthesis in a cell is

(*a*) mitochondria (*b*) chloroplasts

(*c*) ribosomes (*d*) plasma membrane

13. During translation, initiation in prokaryotes, a GTP molecule is needed in

(*a*) formation of formyl-met-tRNA

(*b*) binding of 30S subunit of ribosome with mRNA

(*c*) association of 50S subunit of ribosome with initiation complex

(*d*) association of 30S-mRNA with formyl-met-tRNA

14. Which step of translation does not consume high energy phosphate bond?

(*a*) translocation

(*b*) amino acid activation

(*c*) peptidyl transferase reaction

(*d*) aminoacyl tRNA binding to A site

15. Translocase is an enzyme required for

(*a*) DNA replication

(*b*) RNA synthesis

(*c*) initiation of protein synthesis

(*d*) chain elongation in protein synthesis

ANSWERS

Very Short Answer Questions

1. $1 \times 2 \times 6 \times 2 = 48$

2. (*a*) - glu-met-ala-val-tyr-.

(*b*) - phe-ala-arg-cys-asn-(since genetic code is always read in $5' \rightarrow 3'$ direction).

(*c*) - lys-thr- (nonsense), chain terminates prematurely.

3. (*a*) By definition, codon occurs only in mRNA

(*b*) UAC

(c) By definition, anticodons occur only in tRNA

(*d*) TAC

4. Longer.

5. Cell wall formation is interfered with by penicillins and cephalosporins. DNA replication is prevented by the bleomycins and anthracyclines. Rifamycins interrupt the transcription of DNA into RNA. Translation is disrupted by erythromycin, tetracyclines, chloramphenicol, neomycin, lincomycin, puromycin, and streptomycin. Antibiotics are toxic for microorganisms but safe for humans because all of these metabolic processes are subtly different in bacteria and humans.

6. Formation of the 70S initiation complex and translation.

Fill in the Blanks

1. Peptidyl transferase.

2. P site

3. Elongation factor G(EF-G), translocase.

4. Amber, Ochre.

5. RF1, RF2 and RF3(S).

6. Deformlyase.

7. Antibiotics, $tRNA^{fmet}$.

8. Initiation codon, Kozak consensus.

9. Peptidyl, exit site, E site.

10. GTP.

Multiple Choice Questions

1. (*d*)	**2.** (*c*)	**3.** (*c*)	**4.** (*c*)	**5.** (*a*)	**6.** (*c*)
7. (*b*)	**8.** (*d*)	**9.** (*c*)	**10.** (*d*)	**11.** (*a*)	**12.** (*c*)
13. (*d*)	**14.** (*d*)	**15.** (*d*).			

13

Post-Translational Processing of Proteins

The process of gene expression is not over when the genetic code has been used to create the sequence of amino acids that constitutes a protein. To be useful to the cell, this new polypeptide chain must fold up into its unique three-dimensional conformation, bind any small molecule cofactors required for its activity, be appropriately modified by protein kinases or other protein-modifying enzymes, and assemble correctly with the other protein subunits with which it functions. Thus, the polypeptide that emerges from the ribosome is inactive, and before taking on its functional role in the cell, it must undergo at least the first of following four types of post-translational processing.

1. Protein folding
2. Proteolytic cleavage
3. Chemical modification
4. Inteine splicing

13.1 PROTEIN FOLDING

The proteome is the final product of genome expression and comprises all the proteins present in a cell at a particular time. A 'typical' mammalian cell, for example a hepatocyte, is thought to contain 10,000 – 20,000 different proteins, about 8×10^9 individual molecules in all, representing approximately 0.5ng of protein or 18 – 20% of the total cell weight (**Brown,** 2002). A protein, like a DNA molecule is a linear unbranched polymer. The monomeric subunits of a protein are called **amino acids** and the resulting polymers, or polypeptides, are rarely more than 2000 units in length.

Four levels of Protein Structure

Proteins are found to have four distinct levels of structure. These levels are hierarichal, the protein being built up stage by stage, with each level of structure depending on the one below it.

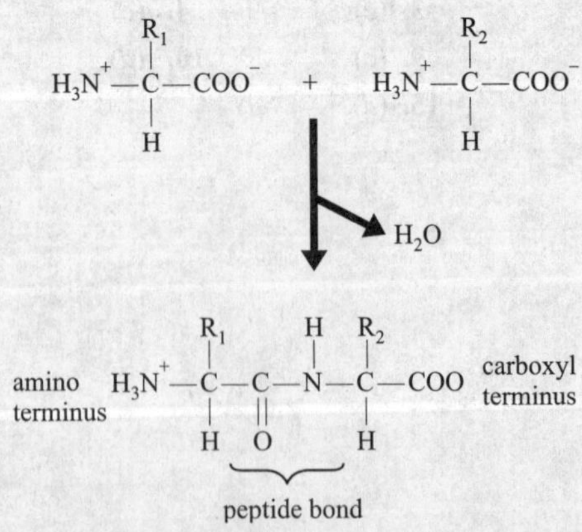

Fig. 13.1. In polypeptides, amino acids are linked by peptide bonds.

1. Primary structure: The primary structure of protein is formed by joining amino acids into a polypeptide. The amino acids are linked by

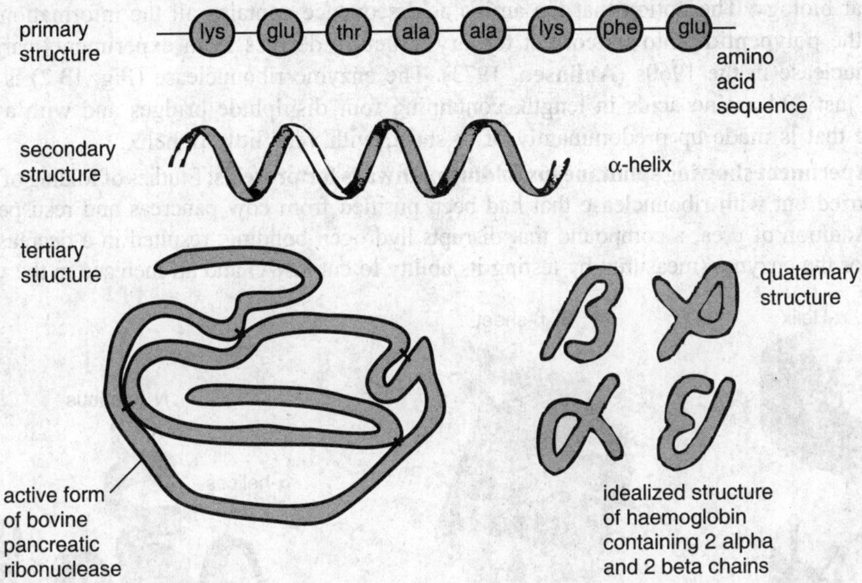

Fig. 13.2. Stages in the development of a functional protein (after Stansfield, 1969).

peptide bonds which are formed by a condensation reaction between the carboxyl group of one amino acid and the amino group of a second amino acid (Fig. 13.1). The two ends of the polypeptide are chemically distinct: one has a free amino group and is called the **amino**, **NH_2, - or N – terminus**; the other has a free carboxyl group and is called the **carboxyl, COOH** – or **C – terminus**. The direction of the polypeptide can therefore be expressed as either N → C (left to right) or C → N (right to left).

2. Secondary structure: The secondary structure of protein refers to the different conformations that can be taken up by the polypeptide. The two types of secondary structure are the **β–helix** and **α–sheet** (Fig. 13.3), both of which are stabilized by hydrogen bonds that form between different amino acids in the polypeptide. Most polypeptides are long enough to be folded into a series of secondary structures one after another along the molecule.

3. Tertiary structure: The tertiary structure of protein results from folding the secondary structural components of the polypeptide into a three-dimensional configuration (Fig. 13.4). The tertiary structure is stabilized by various chemical forces, notably hydrogen bonding between individual amino acids, and hydrophobic forces, which dictate that amino acids with non-polar (*i.e.*, water hating or hydrophobic) side groups must be shielded from water by embedding within the internal regions of the protein. There may also be covalent linkages called **disulphide bridges** between cysteine amino acids at various places in the polypeptide.

4. Quaternary structure: The quarternary structure of proteins involves the association of two or more polypeptides, each folded into its tertiary structure, into a multisubunit protein. Not all proteins form quaternary structures, but it is a feature of many proteins with complex functions, including several involved in genome expression. Some quaternary structures held together by disulphide bridge between different polypeptides, but many proteins comprise looser associations of subunits stabilized by hydrogen bonding and hydrophobic effects. However, according to the functional requirements, quaternary proteins can revert to their component polypeptides, or change their subunit composition.

All the information that a polypeptide needs in order to adopt its correct three-dimensional structure is contained within its amino acid sequence. This is one of the central principles of molecular biology. The notion that the amino acid sequence contains all the information needed to fold the polypeptide into its correct tertiary structure derives from experiments carried out on ribonuclease in the 1960s (**Anfinsen**, 1973). The enzyme ribonuclease (Fig. 13.2) is a small protein, just 124 amino acids in length, containing four disulphide bridges and with a tertiary structure that is made up predominantly of α–sheet, with very little β–helix.

Experiment showing spontaneous folding pathways for proteins: Studies of folding of proteins were carried out with ribonuclease that had been purified from cow pancreas and resuspended in buffer. Addition of urea, a compound that disrupts hydrogen bonding, resulted in a decrease in the activity of the enzyme (measured by testing its ability to cut RNA) and an increase in the viscosity

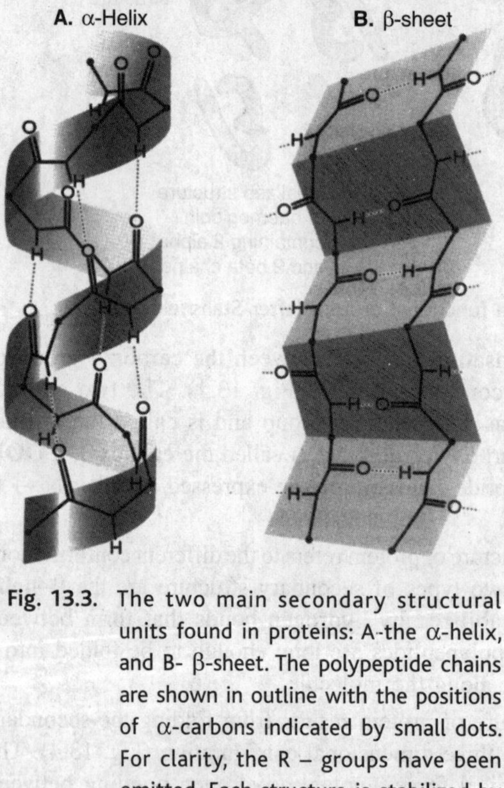

Fig. 13.3. The two main secondary structural units found in proteins: A-the α-helix, and B- β-sheet. The polypeptide chains are shown in outline with the positions of α-carbons indicated by small dots. For clarity, the R – groups have been omitted. Each structure is stabilized by hydrogen bonds between the C = O and N– H groups of different peptide bonds. The β-sheet conformation that is shown is anti- parallel, the two chains running in opposite directions. Parallel β-sheets also occur (after Brown, 2002).

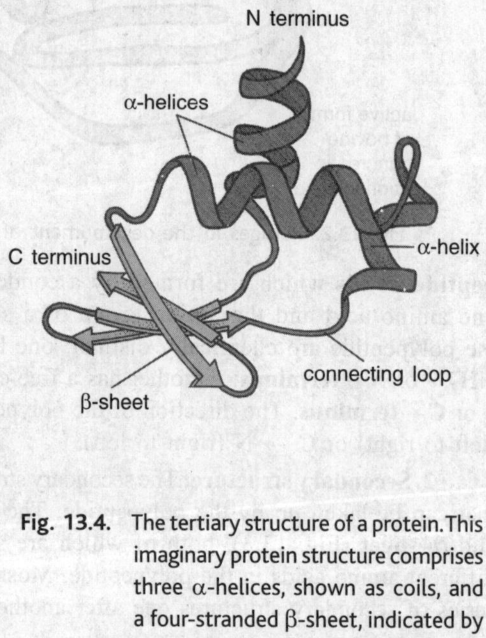

Fig. 13.4. The tertiary structure of a protein. This imaginary protein structure comprises three α-helices, shown as coils, and a four-stranded β-sheet, indicated by the arrows (after Brown, 2002).

of the solution (Fig. 13.5), indicating that the protein was being **denatured** by unfolding into an unstructured polypeptide chain. Critical observation was that when the urea was removed from the experimental solution, the viscosity decreased and the enzyme activity reappeared. The conclusion is that the protein refolds spontaneously when the denaturant (*i.e.,* urea) is removed. In these initial experiments the four disulphide bonds remained intact because they were not disrupted by urea but same result occurred when the urea treatment was combined with addition of a reducing agent to break the disulphide bonds: the activity was still regained on renaturation. This shows that the disulphide bonds are not critical to the protein's ability to refold, they merely stabilise the tertiary structure once it has been adopted.

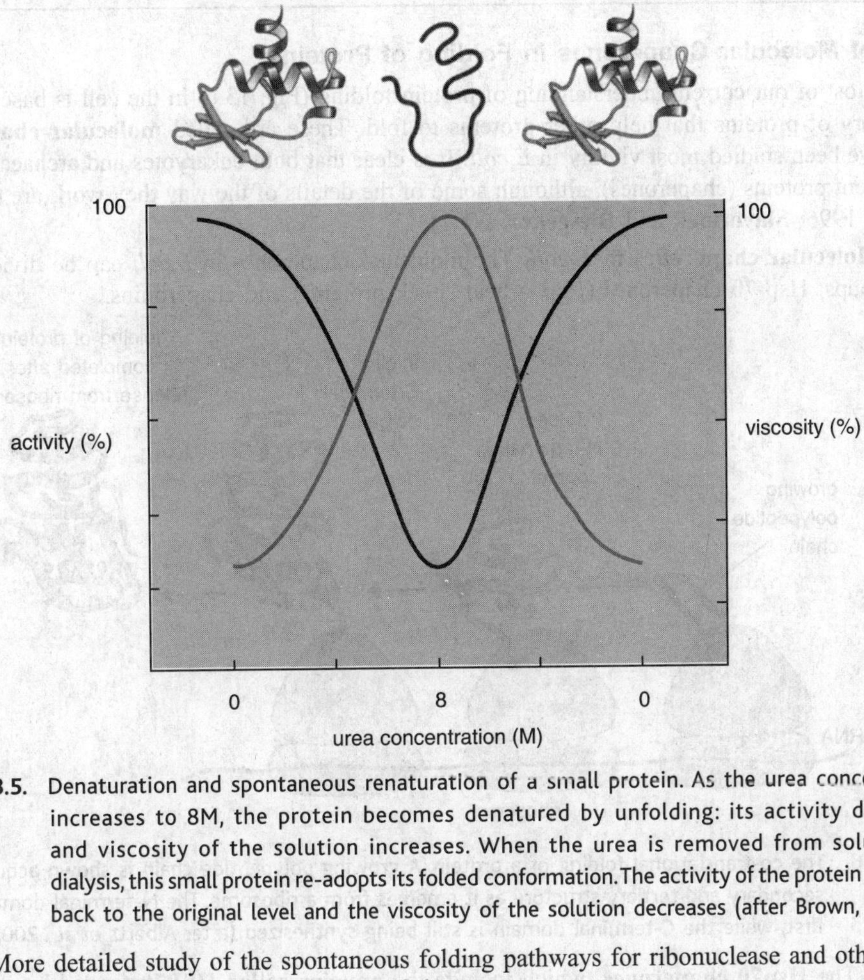

Fig. 13.5. Denaturation and spontaneous renaturation of a small protein. As the urea concentration increases to 8M, the protein becomes denatured by unfolding: its activity decreases and viscosity of the solution increases. When the urea is removed from solution by dialysis, this small protein re-adopts its folded conformation. The activity of the protein increases back to the original level and the viscosity of the solution decreases (after Brown, 2002).

More detailed study of the spontaneous folding pathways for ribonuclease and other small proteins has led to the following two sets of the process:

1. The secondary structural motifs along the polypeptide chain form within a few milliseconds of the denaturant being removed.

2. During the next few seconds or minutes, the secondary structural motifs interact with one another and the tertiary structure gradually takes shape, often *via* a series of intermediate conformations (Box 13.1). In other words, the protein follows a **folding pathway.** However, there may be more than one possible pathway that a protein can follow to reach its correctly folded structure (**Radford**, 2000). Such *in vitro* folding pathways were found true for only the small proteins; large proteins fail to refold.

Box 13.1

Experiments have demonstrated that once a protein domain in a multidomain protein emerges from the ribosome, it forms a compact structure within a few seconds that contains most of the final secondary structure (β-helices and α-sheets) aligned in roughly the right way. For many protein domains, this unusually open and flexible structure, which is called a **molten globule**, is the starting point for a relatively slow process in which many side chain adjustments occur that eventually form the correct tertiary structure. Nevertheless, because it takes several minutes to synthesize a protein of average size, a great deal of folding process is completed by the time the ribosome releases the C – terminal end of a protein (**Federov** and **Baldwin,** 1997).

Role of Molecular Chaperones in Folding of Proteins

Most of our current understanding of protein folding (Fig. 13.6) in the cell is based on the discovery of proteins that help other proteins to fold. These are called **molecular chaperones** and have been studied most vividly in *E.coli*. It is clear that both eukaryotes and archaea possess equivalent proteins (chaperones), although some of the details of the way they work are different (**Hartl**, 1996; **Slavotinek** and **Biesecker**, 2001).

Molecular chaperones in *E.coli*. The molecular chaperones in *E.coli* can be divided into two groups: Hsp 70 chaperones (Hsp = heat shock proteins) and chaperonins.

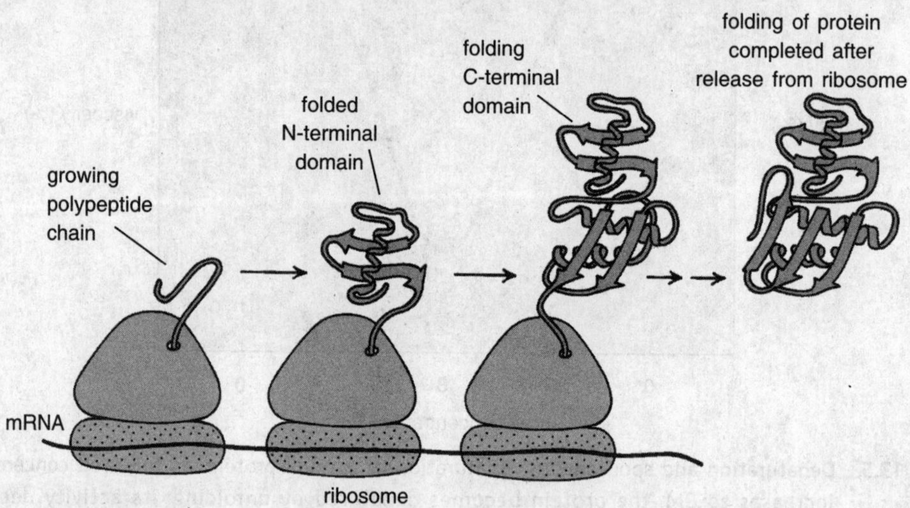

Fig. 13.6. The co-translational folding of a protein. A growing polypeptide chain is shown acquiring its secondary and tertiary structure as it emerges from a ribosome. The N-terminal domain folds first, while the C-terminal domain is still being synthesized (after Alberts *et al.*, 2002).

1. The **Hsp70 chaperones**, which include the proteins called **HSP70** (coded by the *dnak* gene and sometimes called **Dnak protein**), **Hsp40** (coded by *dnaJ*) and Grp E.
2. The **chaperonins**, the main version of which in *E.coli* is the **GroEL – GroES** complex.

Molecular chaperones do not specify the tertiary structure of a protein, they merely help the protein find that correct structure. The two types of chaperones do this in different ways. For example, the Hsp70 family (of chaperones) bind to hydrophobic regions of proteins, including proteins that are still being translated (Fig. 13.7A). They prevent protein aggregation by holding the protein in an open conformation until it is completely synthesized and ready to fold. The Hsp70 chaperones are also involved in other processes that require shielding of hydrophobic regions in proteins, such as transport through membranes and disaggregation of proteins that have been damaged by heat stress.

The chaperonins work in quite a different way. GroEL and GroES form a multi-subunit structure that looks like a hollowed out bullet with a central cavity (Fig. 13.7B; **Xu** *et al.,* 1997).

A single unfolded protein enters the cavity and emerges folded. The mechanism for this is not known but it is postulated that GroEL/GroES acts as a cage that prevents the unfolded protein from aggregating with other proteins, and that the inside surface of the cavity changes from hydrophobic to hydrophilic in such a way as to promote the burial of hydrophobic amino acids within the protein. It is later found that the cavity unfolds proteins that have folded incorrectly, passing these unfolded proteins back to the cytoplasm so that they can have a second attempt at adopting their correct tertiary structure (**Shtilerman** *et al.*, 1999).

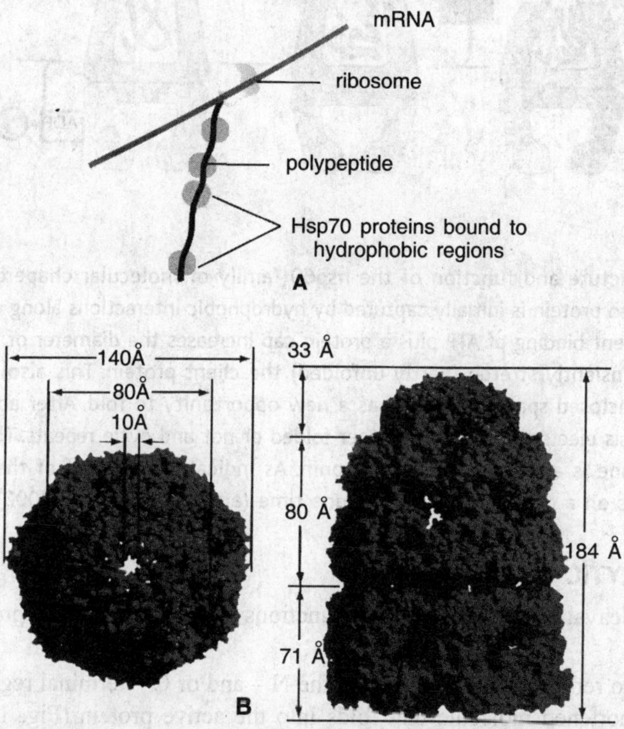

Fig. 13.7. Molecular chaperones of *E.coli*. A – Hsp70 chaperones bind to hydrophobic regions in unfolded polypeptides, including those that are still being translated and hold the protein in an open conformation until it is ready to be folded. B – The structure of the GroEL/GroES chaperonin (*i.e.*, GroEL bound to its GroES cap). On the left is a view from the side. 1Å is equal to 0.1nm. The GroES part of the structure is made up of seven identical protein subunits and is shown in gold. The GroEL components consist of 14 identical proteins arranged into two rings (shown in red and green), each containing seven subunits. The main entrance into the central cavity is thought at the bottom of the structure shown on the right. The images were creations of **Dr. Zhaohni Xu** and coworkers in 1977 (after Brown, 2002).

Molecular chaperones and chaperonins in eukaryotes. Eukaryotic cells have at least two families of molecular chaperones, known as the **hsp60** and **hsp70** proteins. Different family members function in different organelles. For example, mitochondria contain their own hsp60 and hsp70 molecules that are distinct from those that function in the cytosol (where hsp60 is called **TCP – 1** in vertebrate cells), and a special hsp70 (called **BIP**) helps to fold proteins in the endoplasmic reticulum. The hsp70 machinery act early in the life of many proteins, binding to a string of about seven hydrophobic amino acids before protein leaves the ribosome.

The hsp60 – like proteins form a large barrel-shaped structure that acts later in proteins life, after it has been fully synthesized. This type of chaperone forms an "*isolation chamber*" into which misfolded proteins are fed, preventing their aggregation and providing them with a favourable environment in which they attempt to refold (Fig. 13.8).

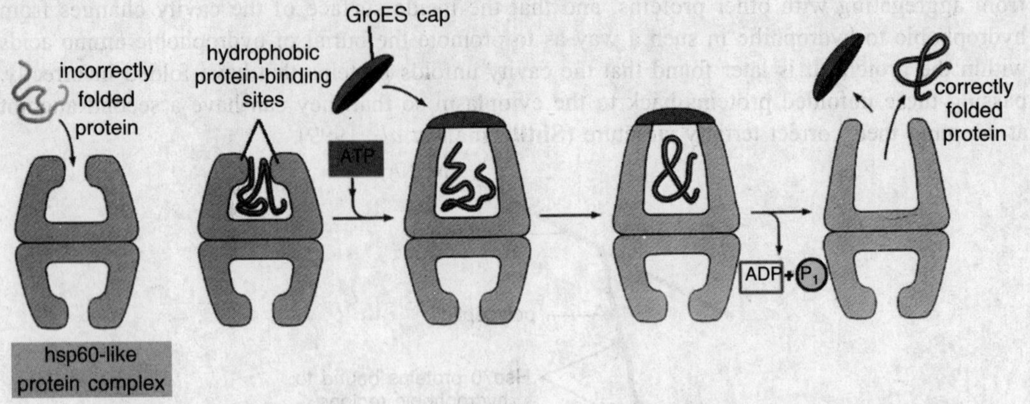

Fig. 13.8. The structure and function of the hsp60 family of molecular chaperones. As indicated, a misfolded protein is initially captured by hydrophobic interactions along one rim of barrel. The subsequent binding of ATP plus a protein cap increases the diameter of the barrel rim, which may transiently stretch (partly unfolded) the client protein. This also confines the protein in an enclosed space, where it has a new opportunity to fold. After about 15 seconds, ATP hydrolysis ejects the protein whether folded or not and cycle repeats. This type of molecular chaperone is also known as chaperonin. As indicated only half of the symmetrical barrel operates on a client protein at any one time (after Alberts *et al.*, 2002).

13.2 PROTEOLYTIC CLEAVAGE

Proteolytic cleavage has following *two* functions in post-translational processing of proteins (Fig. 13.9).

1. It is used to remove short pieces from the N – and/or C – terminal regions of polypetides, leaving a single shortened molecule that folds into the active protein (Fig. 13.9).

2. It is used to cut **polyproteins** into segments, all or some of which are active proteins.

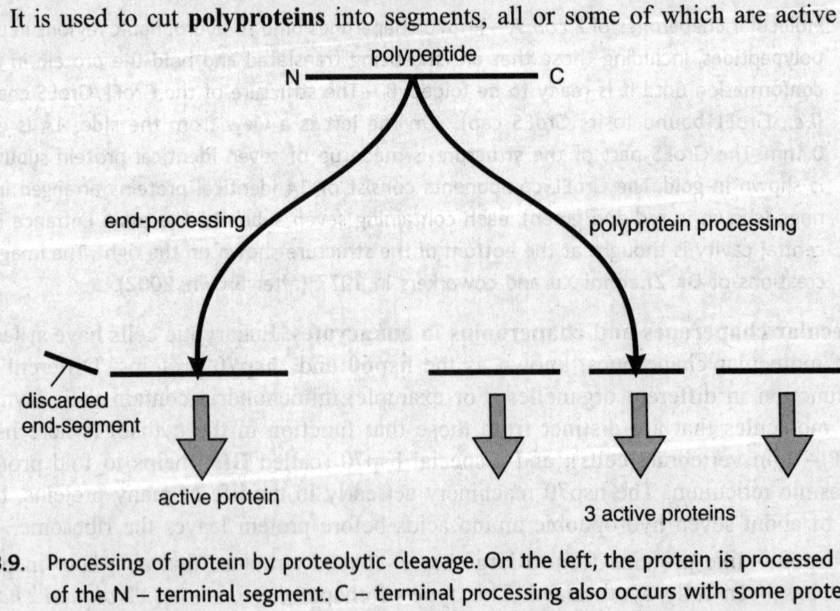

Fig. 13.9. Processing of protein by proteolytic cleavage. On the left, the protein is processed by removal of the N – terminal segment. C – terminal processing also occurs with some proteins. On the right, a polyprotein is processed to give three different proteins (after Brown, 2002).

These events are relatively common in eukaryotes but less frequent in bacteria. Not all proteins undergo proteolytic cleavage.

1. Cleavage of the Ends of Polypeptides

Processing by proteolytic cleavage is common with secreted polypetides whose biochemical activities might be injurious to the cell producing the protein. An example is provided by **melittin**, the most abundant protein in bee venom and the one responsible for causing cell lysis after injection of the bee sting into the person or animal being stung. Melttin (protein) lyses cells in bees as well as animals so have to be initially synthesized as an inactive precursor. This precursor, called **promelittin**, has 22 amino acids at its N – terminus. The pre-sequence is removed by an extracellular protease that cuts it at all eleven positions (Fig. 13.10A), releasing the active venom protein. The protease does not cleave within the active sequence because its mode of action is to release dipeptides with the sequence X – Y, where X is alanine, aspartic acid or glutamic acid, and Y is alanine or proline, these motifs are not found in the active sequence (Fig. 13.10A).

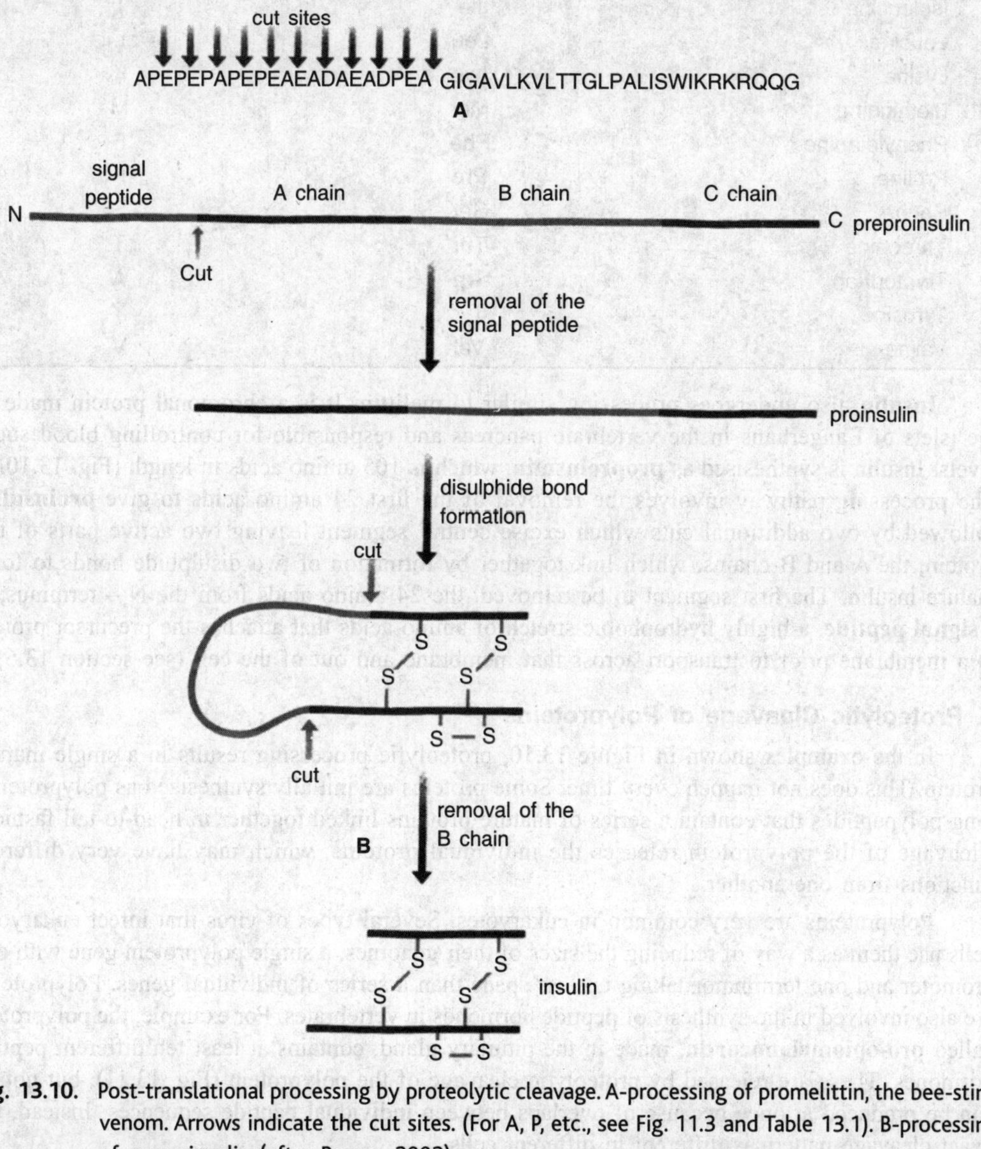

Fig. 13.10. Post-translational processing by proteolytic cleavage. A-processing of promelittin, the bee-sting venom. Arrows indicate the cut sites. (For A, P, etc., see Fig. 11.3 and Table 13.1). B-processing of preproinsulin (after Brown, 2002).

Table 13.1 Amino acid abbreviations.

Amino acid	*Abbreviation* *Three-letter*	*One-letter*
Alanine	Ala	A
Arginine	Arg	R
Asparagine	Asn	N
Aspartic acid	Asp	D
Cysteine	Cys	C
Glutamic acid	Glu	E
Glutamine	Gln	Q
Glycine	Gly	G
Histidine	His	H
Isoleucine	Ile	I
Leucine	Leu	L
Lysine	Lys	K
Methionine	Met	M
Phenylalanine	Phe	F
Proline	Pro	P
Serine	Ser	S
Threonine	Thr	T
Tryptophan	Trp	W
Tyrosine	Tyr	Y
Valine	Val	V

Insulin also undergoes processing similar to melittin. It is a chromonal protein made in the islets of Langerhans in the vertebrate pancreas and responsible for controlling blood sugar levels. Insulin is synthesised as **proproinsulin**, which is 105 amino acids in length (Fig. 13.10B). The processing pathway involves the removal of the first 24 amino acids to give **proinsulin,** followed by two additional cuts which excise central segment leaving two active parts of the protein, the A and B chains, which link together by formation of two disulphide bonds to form mature insulin. The first segment to be removed, the 24 amino acids from the N – terminus, is a **signal peptide**, a highly hydrophobic stretch of amino acids that attaches the precursor protein to a membrane prior to transport across that membrane and out of the cell (see section 13.6).

2. Proteolytic Cleavage of Polyproteins

In the examples shown in Figure 13.10, proteolytic processing results in a single mature protein. This does not happen every time. Some proteins are initially synthesized as polyproteins, long polypeptides that contain a series of mature proteins linked together in head-to-tail fashion. Cleavage of the polyprotein releases the individual proteins, which may have very different functions from one another.

Polyproteins are very common in eukaryotes. Several types of virus that infect eukaryotic cells use them as a way of reducing the sizes of their genomes, a single polyprotein gene with one promoter and one terminator taking up less space than a series of individual genes. Polyproteins are also involved in the synthesis of peptide hormones in vertebrates. For example, the polyprotein called **pro-opiomelanocortin**, made in the pituitary gland, contains at least ten different peptide hormones. These are released by proteolytic cleavage of the polyprotein (Fig. 13.11), but not all can be produced at once because of overlaps between individual peptide sequences. Instead, the exact cleavage pattern is different in different cells.

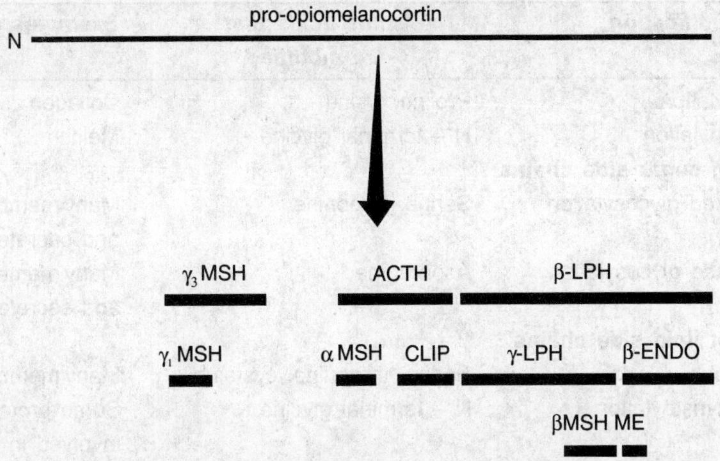

Fig. 13.11. Processing of pro-opiomelanocortin polyprotein. Abbreviations: ACTH, adrenocorticotropic hormones; CLIP, corticotropin-like intermediate lobe protein; ENDO, endorphin; LPH, lipoprotein; ME, met-encephalin; MSH, melanotropin (after Brown, 2002).

13.3 CHEMICAL MODIFICATION

The genome has the capacity to code for 21 different amino acids: the 20 specified by the standard genetic code, and selenocysteine, which is inserted into polypeptides by the context-dependent reading of a 5′ – UGA – 3′ codon. This range is increased dramatically by post-translational chemical modification of proteins, which results in an immense supply of different types of amino acids.

The most simple sort of chemical modifications involve addition of a small chemical group (*e.g.*, an acetyl, methyl or phosphate group; Table 13.2) to an amino acid side chain, or to the amino or carboxyl groups of the terminal amino acids in a polypeptide (**Bradshaw** *et al.*, 1998). More than 150 different modified amino acids, have been reported in different proteins; each modification is found to be carried out in a highly specific manner, *i.e.*, each amino acid being modified in a highly specific manner and the same amino acids being modified in same way in every copy of the proteins, *e.g.*, histone H3 (Fig. 13.12).

```
  Me   Me   Ac       Ac   Me
  ↓    ↓    ↓        ↓    ↓
ARTKQTARKSTGGKAPRKQLATKAARKSAP
```

Fig. 13.12. Post-translational modification of calf histone H3. The first 30 amino acids of this 135 – amino acid protein are listed using the one – letter abbreviations (see Table 13.1). Five modifications occur-three methylations and two acetylations.

Table 13.2 Examples of post-translational chemical modifications (Source: **Brown,** 2002).

Modification	*Amino acids that are modified*	*Examples of proteins*
I. Addition of small groups		
1. Acetylation	Lysine	Histones
2. Methylation	Lysine	Histones
3. Phosphorylation	Serine, threonine, tyrosine	Some proteins involved in signal transduction

Modification	Amino acids that are modified	Examples of proteins
4. Hydroxylation	Proline, lysine	Collagen
5. N - formylation	N – terminal glycine	Melittin
II. Addition of sugar side chains		
6. O – linked glycosylation	Serine, threonine	Many membrane proteins and secreted proteins
7. N – linked glycosylation	Asparagine	Many membrane proteins and secreted proteins
III. Addition of lipid side chains		
8. Acylation	Serine, threonine, cysteine	Many membrane proteins
9. N - myristoylation	N – terminal glycine	Some protein kinases involved in signal transduction
IV. Addition of biotin		
10. Biotinylation	Lysine	Various carboxylase enzymes

We have observed how acetylation and methylation of histone H3 and other histones have an important influence on chromatin structure and hence on genome expression. Other types of chemical modification have important regulatory roles, *e.g.*, phosphorylation is used in transduction.

Glycosylation is a complex type of modification of amino acids. It involves the attachment of large carbohydrate side chains to polypeptides (**Drickamer** and **Taylor,** 1998). There are following two general types of glycosylation (Fig. 13.13).

1. O – linked glycosylation. It is the attachment of a sugar side chain *via* the hydroxyl group of a serine or threonine amino acid.

2. N – linked glycosylation. It involves attachment through the amino group on the side chain of asparagine.

Glycosylation results in attachment to the proteins of grand structures comprising branched networks of 10 to 20 sugar units of various types. These side chains help to target proteins to particular sites in the cells and determine the stability of proteins circulating in the bloodstream.

Another category of large scale modification is **acylation** which involves attachment of long chain lipids, often to serine or cysteine amino acids. Acylation occurs with many proteins that become associated with membranes. A less common modification is **biotinylation,** in which a molecule of biotin is attached to a small number of enzymes that catalyze the carboxylation of organic acids such as acetate and propionate (**Chapman**, **Smith** and **Cronan,** 1999).

13.4 INTEINE SPLICING

The final type of post-translational processing is **inteine splicing**. It is a protein version of the more extensive intron splicing that occurs in context to pre-RNAs. Inteins are internal segments of proteins that are removed soon after translation, the two external segments or **exteins** becoming linked together (Fig. 13.14). The first intein was discovered in 1990 in *S. cerevisiae* and more than 100 inteins have been reported till today. Most inteins are known in bacteria, archaea and lower eukaryotes. In a few cases, there is more than one intein in a single protein.

Most inteins are approximately 150 amino acids in length and like pre-mRNA introns, the sequences at the splice junctions of inteins have some similarity in most of the known examples. The first amino acid of the downstream extein is cysteine, serine or threonine. A few other amino acids within the intein sequence are also conserved. These conserved amino acids are involved in the splicing process, which is catalyzed by the intein itself (**Paulus,** 2000).

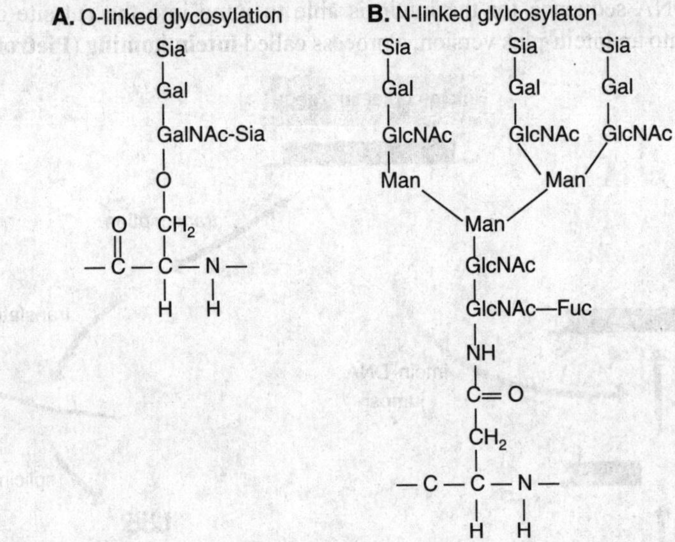

Fig. 13.13. Glycosylation of proteins. A – O – linked glycosylation. The structure shown is found in a number of glycoproteins. It is drawn here attached to a serine amino acid but it can also be linked to a threonine. B – N – linked glycosylation usually results in larger sugar structures than are seen with O – linked glycosylation. The figure shows a typical example of a complex glycan attached to an asparagine amino acid. Abbreviations: Fuc, fucose; Gal, galactose; GalNAc, N – acetylgalactosamine; GlnNAc, N – acetyl glucosamine; Man, mannose; Sia, sialic acid (after Brown, 2002).

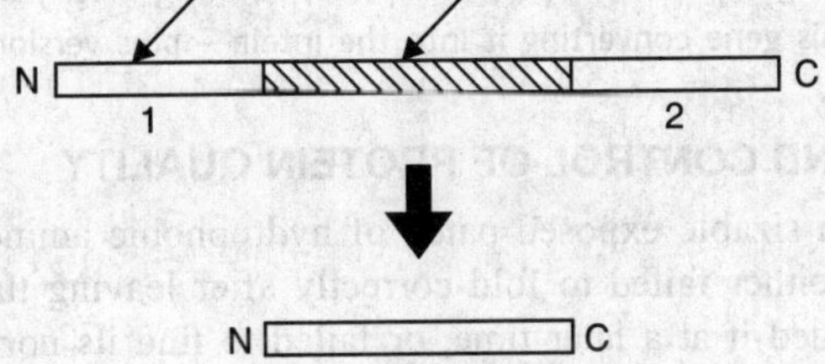

Fig. 13.14. Intein splicing.

Characteristics of Inteins

1. Self – catalysis. The first interesting feature of inteins was discovered when the structure of two inteins were determined by X-ray crystallography (**Duan** *et al.*, 1997; **Klabunde** *et al.*, 1998). These structures are similar in some respects to that of a *Drosophila* protein called **Hedgehog**, which is involved in the development of the segmentation pattern of the embryo of fruit fly. Hedgehog is an autoprocessing protein that cuts itself into two. The structural similarity with inteins lies in the part of the Hedgehog protein that catalyzes its self-cleavage. Molecular biologists believe that possibly the same protein structure has evolved twice, or possibly inteins and Hedgehog protein shared a common link at some stage in the evolutionary past.

2. Intein homing. The second interesting feature of intein is that with some inteins the excised segment is a sequence-specific endonuclease. The intein cuts DNA at the sequence corresponding to its insertion site in a gene coding for intein-free version of the protein from which it is derived (Fig. 13.15). If the cell also contains a gene coding for the intein containing protein, then the DNA sequence for the intein is able to jump into the cut site converting the intein-minus gene into an intein-plus version, a process called **intein homing** (**Pietrokovski,** 2001).

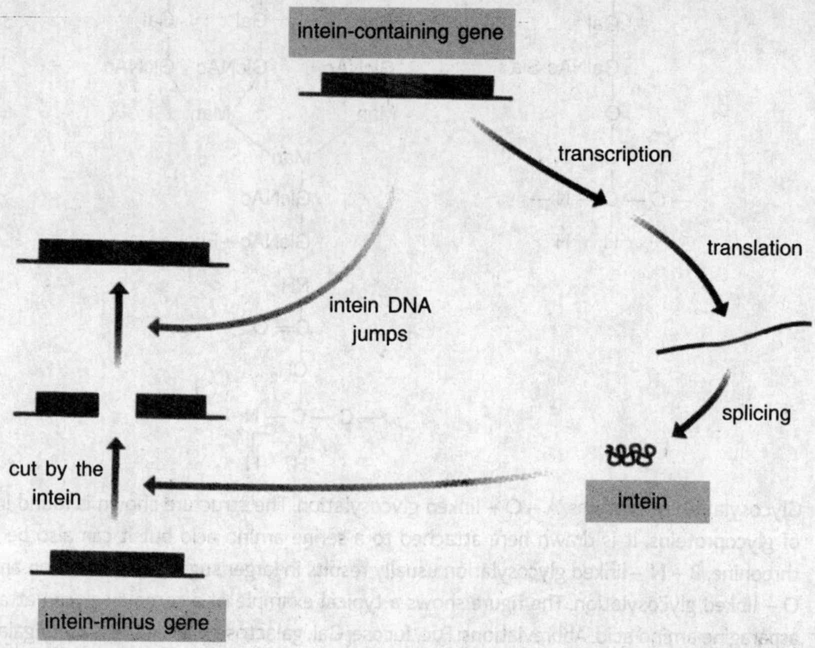

Fig. 13.15. Process of intein homing. The cell is heterozygous for the intein-containing gene, possessing one gene with the intein and one allele without the intein. After protein splicing, the intein cuts the intein minus gene at the appropriate place, allowing a copy of the intein DNA sequence to jump into this gene converting it into the intein – plus version (after Brown, 2002).

13.5 MONITORING AND CONTROL OF PROTEIN QUALITY

A protein that has a sizable exposed patch of hydrophobic amino acids on its surface is usually abnormal: it has either failed to fold correctly after leaving the ribosome, suffered an accident that partly unfolded it at a later time, or failed to find its normal partner subunit in a larger protein complex. Such a protein is useless to the cell. Further, such protein complex can be dangerous. Many proteins with an abnormally exposed hydrophobic region can form large aggregates, precipitating out of solution. We shall see that in rare cases, such aggregates to form and cause severe human diseases. But in the vast majority of cells, powerful protein quality control mechanisms prevent such disasters.

Cells have evolved elaborate mechanisms that recognize and remove the hydrophobic patches on proteins. Two of mechanisms depend on the molecular chaperones just discussed, which bind to the patch and attempt to repair the defective protein by giving it another chance to fold. At the same time, by covering the hydrophobic patches, these chaperones transiently prevent protein aggregation. Proteins that very rapidly fold correctly on their own do not display such patches and are therefore bypassed by chaperones.

All of the options of quality control that a cell makes for a difficult-to-fold newly synthesized protein have been outlined in Figure 13.16. As indicated, when attempts to refold a protein fail, a third option is called into play that completely destroys the protein by proteolysis.

The proteolytic pathway begins with the recognition of an abnormal hydrophobic patch on a proteins surface and it ends with the delivery of the entire protein to a protein destruction machine, a complex protease called **proteosome**. As described next, this process depends on an elaborate protein marking system that also carries out other central functions in the cell by destroying selected normal proteins.

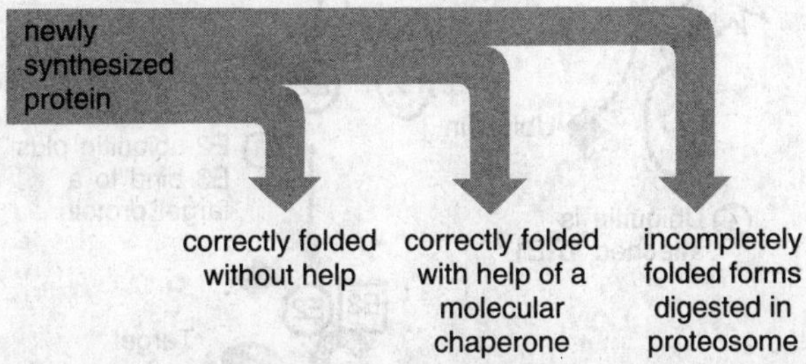

Fig. 13.16. The cellular mechanisms that monitor protein quality after protein synthesis. A newly synthesized protein sometimes folds correctly and assembles with its partners on its own, in which case it is left alone. Incompletely folded proteins are helped to refold by molecular chaperons: first by a family of hsp70 proteins, and if this fails, then by hsp70- like proteins. In both cases the client proteins are recognized by an abnormally exposed patch of hydrophobic amino acids on their surface. These processes complete with a different system that recognizes an abnormally exposed patch and transfers proteosome for complete destruction. The combination of all of these processes is needed to prevent massive protein aggregation in a cell, which can occur when hydrophobic regions on protein clump together and precipitate the entire mass out of solution (after Alberts *et al.*, 2002).

1. Protein Degradation

The most common method for targeting proteins for destruction is to link them to **ubiquitin**, a small protein chain containing 76 amino acids. Ubiquitin is joined to target proteins by a process that involves three components: 1. a *ubiquitin-activating enzyme (E1)*, 2. a *ubiquitin-conjugating enzyme (E2)*, and 3. a *substrate recognition protein (E3)*. As shown in Fig. 13.17, ubiquitin is first activated by attaching it to E1 in an ATP-dependent reaction. The activated ubiquitin is then transferred to E2 and subsequently linked, in a reaction facilitated by E3, to a **lysine** residue in a target protein. Additional molecules of ubiquitin are then added in sequence, forming short chains.

These ubiqutin chains serve as targeting signals that are recognized by large protein-degrading structures called **proteasomes** (Fig. 13.18; Box 13.2). Proteasomes are the predominant proteases (protein-degrading enzymes) of the cytosol and are often present in high concentration, accounting for up to 1% of all cellular proteins. Each proteasome has a mass of roughly two million daltons and consists of half a dozen proteases associated with several ATPases and a binding site for the ubiquitin chains. Shaped like a short cylinder, the proteasome binds to ubiquitylated proteins and removes their ubiquitin chains. The proteins are then fed into a central channel of the proteasome and their peptide bonds are hydrolyzed in ATP-dependent process, generating small peptide fragments that are released from the other end of the cylinder.

The key to regulating protein degradation lies in the ability to select particular proteins for ubiquitylation. This selectivity stems in part from the existence of multiple forms of E3, each directing the attachement of ubiquitin to different target proteins. One feature recognized by the various forms of E3 is the amino acid present at the N-terminus of a potential target proteins. Some N-terminal amino acids cause proteins to be rapidly ubiquitylated and degraded; others make proteins less susceptible.

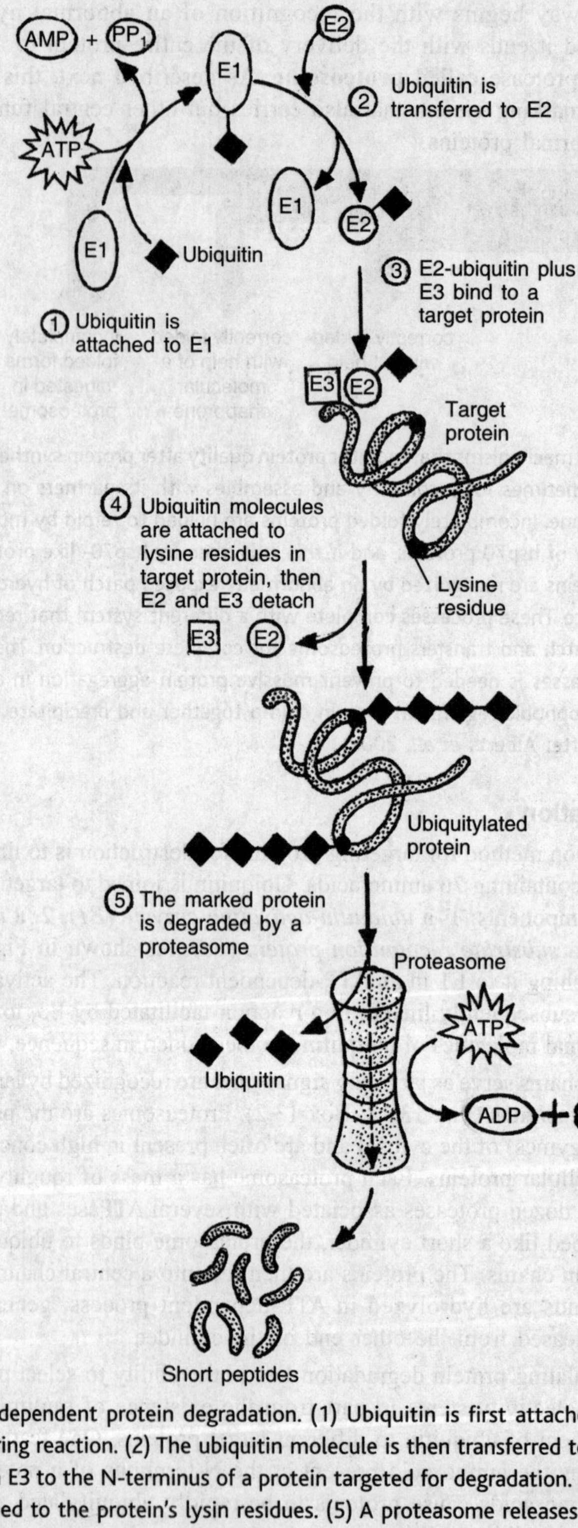

Fig. 13.17. Ubiquitin-dependent protein degradation. (1) Ubiquitin is first attached to enzyme E1 in an ATP-requiring reaction. (2) The ubiquitin molecule is then transferred to enzyme E2, which (3) binds with E3 to the N-terminus of a protein targeted for degradation. (4) Ubiquitin molecules are attached to the protein's lysin residues. (5) A proteasome releases the ubiquitin for reuse and degrades the targeted protein into short peptides in another ATP-requiring reaction (after Becker *et al.*, 2007).

Internal amino acid sequences called **degrons** also allow particular proteins to be selected for destruction. For example, progression through the final stages of mitosis is controlled by the anaphase-promoting complex, which targets selected proteins, such as mitotic cyclic, for degradation. The anaphase-promoting complex accomplishes this task by functioning as a substrate recognition protein (E3) that binds to target proteins containing a particular type of degron, promoting ubiquitylation of these target proteins by E2 and their subsequent degradation by proteasomes. In addition to such **selective mechanisms** for degrading specific proteins at appropriate times, proteasomes also play a role in general, ongoing mechanisms for eliminating defective proteins from cells. Recent observation suggest that upto 30% of newly synthesized proteins are defective in some way and are immediately tagged with ubiquitin, triggering their destruction by proteasomes.

Microautophagy. Lysosomes can take up and degrade cytosolic proteins by an infolding of the lysosomal membrane, creating small vesicles that are internalized within the lysosome and broken down by the organelle's hydrolytic enzymes. This process of **microautophagy** tends to be rather **nonselective** in the proteins it degrades. The result is the slow, continual recycling of the amino acid building blocks found in most of a cells protein molecules (**Becker** *et al.*, 2007).

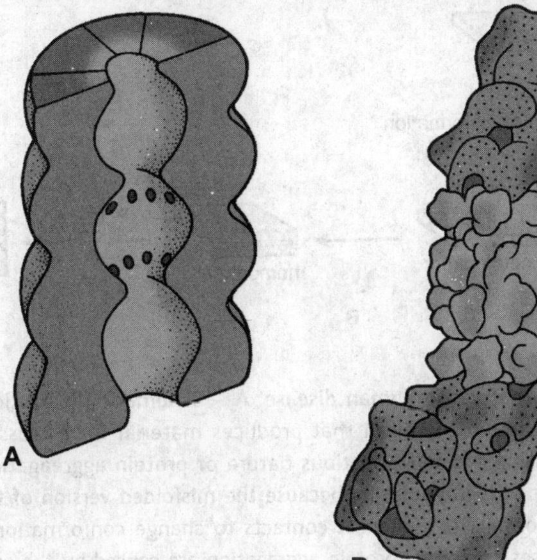

Fig. 13.18. The proteasome. A – A cut-away view of the structure of the central 20S cylinder, as determined by X-ray crystallography, with the active sites of the proteases indicated by *red dots*. B – The structure of the entire proteasome, in which the central cylinder (*yellow*) is supplemented by a 19S cap (*blue*) at each end. The complex cap structure selectively binds these proteins that have been marked for destruction (after Alberts *et al.*, 2002).

Box 13.2

In eukaryotes, the proteosome is a large, multi-subunit structure with a sedimentation coefficient of 26S, comprising a hollow cylinder of 20S and two "caps" of 19S (**Groll** *et al.*, 1997; **Ferrell** *et al.*, 2000). Archaea also have proteasomes of about the same size but these are less complex, being composed of multiple copies of just two proteins; eukaryotic proteasomes contain 14 different types of protein subunits. The entrance into the cavity within the proteasome is narrow, and a protein must be unfolded before it can enter. This unfolding probably occurs through an energy dependent process and may involve structure similar to chaperonins (see **Brown,** 2002). After unfolding, the protein can enter the proteosome within which it is cleaved into short peptides 4 – 10 amino acids in length. These are released back into the cytoplasm where they are broken down into individual amino acids which can be reutilized in protein synthesis.

2. Abnormal Protein Folding Causes Destructive Human Diseases

When all of a cell's protein quality controls fail, large protein aggregates tend to accumulate in the affected cell. Some of these aggregates, by adsorbing critical macromolecules to them, can severly damage cell and even cause cell death. The protein aggregates released from dead cells tend to accumulate in the extracellular matrix that surrounds the cells in a tissue, and in extreme cases they can also damage tissues. Since the brain is composed of a highly organised collection of nerve cells, it is especially vulnerable. These protein aggregates primarily tend to causes diseases of neurodegeneration. Prominent among these are **Huntington's disease** and **Alzheimer's disease** – the latter causing age-related dementia in more than 20 million people in today's world (**Alberts** *et al.*, 2002).

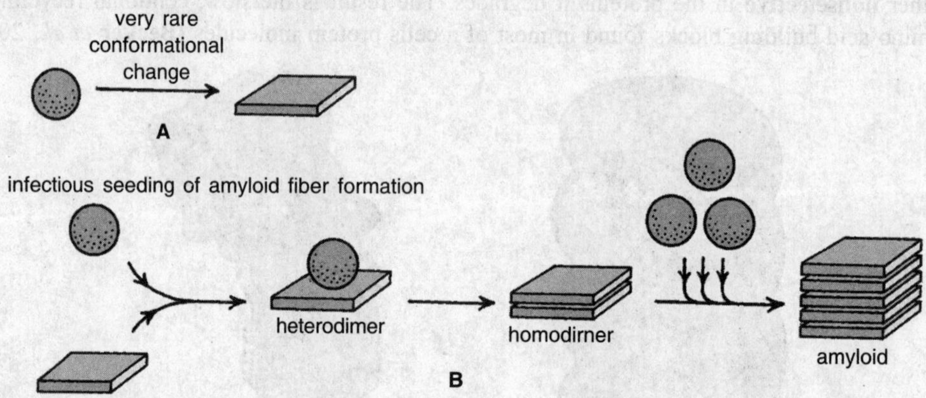

Fig. 13.19. Protein aggregates that cause human disease. A – Schematic illustration of the type of conformational change in a protein that produces material for a cross – beta filament. B– Diagram illustrating the self infectious nature of protein aggreagation that is central to prion disease. Prp is highly unusual because the misfolded version of the protein, called Prp*, induces the normal Prp protein it contacts to change conformation, as shown. Most of the human diseases caused by protein aggregation are caused by the over- production of a variant protein that is especially prone to aggregation, but because this structure is not infectious in this way, it cannot spread from one animal to another. (Drawing of a cross beta filament, a common type of protease-resistant protein aggregate found in a variety of human neurological diseases. Because the hydrogen-bond interactions in a β-sheet form between polypeptide backbone atoms, a number of different abnormally folded proteins can produce this structure (after Alberts *et al.*, 2002).

A protein aggregate, which have to survive, grow and damage an organism, must be highly resistant to proteolysis both inside and outside the cell. Many of the protein aggregates that cause problems from fibrils built from a series of polypeptide chains that are layered one over the other as a continuous stack of β-sheets. This is so-called **cross beta filaments** (Fig. 13.19), tends to be highly resistant to proteolysis. This resistance presumably explains why this structure is observed in so many of the neurological disorders caused by protein staining deposits, called **amyloid**.

One particular variety of these neurological diseases is quite notorious. These are **prion diseases**. Unlike Huntington's or Alzheimer's disease, a prion disease can spread from one organism to another, providing that the second organism eats a tissue containing the protein aggregate. A set of diseases – called **scrapie** in sheep, **Creutzfeldt – Jacob disease (CJD)** in humans, and **bovine spongiform encephalopathy (BSE)** in cattle – are caused by a misfolded, aggregated form of protein called **PrP** (for prion protein). The PrP is normally located on the outer surface of the plasma membrane, most prominently in neurones. Its normal function is not known. However, PrP has the unfortunate property of being convertible to a very special abnormal conformation (Fig. 13.19A). This conformation not only forms protease-resistant, cross-beta filaments; it also is "infectious" because it converts normally folded molecules of PrP to the same form. This property creates a positive feedback loop that produces the abnormal form of prp, called **PrP*** (Fig. 13.19B) and thereby allows PrP to spread rapidly from cell to cell in the brain, causing death of both animals and humans. It is dangerous to eat the tissues of animals that contain PrP*, as witnessed most recently by spread of BSE (commonly referred to as the "mad cow disease") from cattle to humans in great Britain.

13.6 SIGNAL HYPOTHESIS (PROTEIN TARGETING)

Ribosomes are either free in the cytoplasm or associated with membranes, depending on the type of protein being synthesised. Membrane-bound ribosomes, which are indistinguishable from free ribosomes, synthesise proteins that enter membranes. These proteins either become a part of the membrane or in eukaryotes, either pass into membrane bound organelles (*e.g.*, the Golgi apparatus, mitochondria, chloroplasts, vacuoles) or transported outside the plasma membrane. The **signal hypothesis** of **Gunter Blobel**, a 1999 Nobel laureate, and his colleagues, explains the mechanism for membrane attachment. The mechanism applies to both prokaryotes and eukaryotes. Here, signal hypothesis has been described for mammals.

The signal for membrane insertion is in the form of polypeptide. It is coded into the first one to three dozen amino acids of membrane-bound proteins. This **signal peptide** takes part in a chain of events that leads the ribosome to attach to the membrane and to insertion of the protein. The *first step* occurs when the signal peptide becomes exposed on outside of the ribosome. A ribonucleoprotein particle, called **signal recognition particle (SRP)**, which consists of six different proteins and a 7S RNA (about 300 nucleotide long), recognizes the signal peptide. In *second step*, the complex of signal recognition particle, ribosome, and signal peptide then passes to a membrane, where the SRP binds to a receptor called a **docking protein (DP) or signal recognition particle receptor** (Fig. 13.20; Fig. 13.21). During this time, protein synthesis or translation halts. The ribosome is brought into direct contact with the membrane, and other proteins which help anchor the ribosome. In *third step*, the protein synthesis resumes, with the nascent protein usually

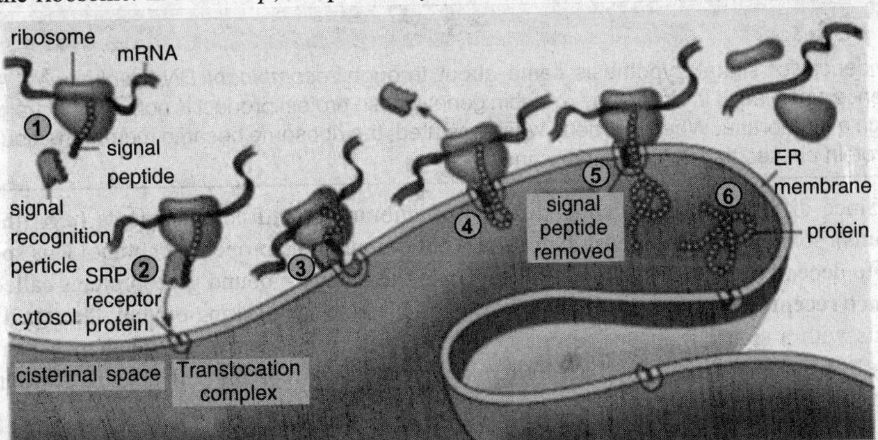

Fig. 13.20. Diagram showing the mode of action of signal peptide (protein) during protein synthesis.

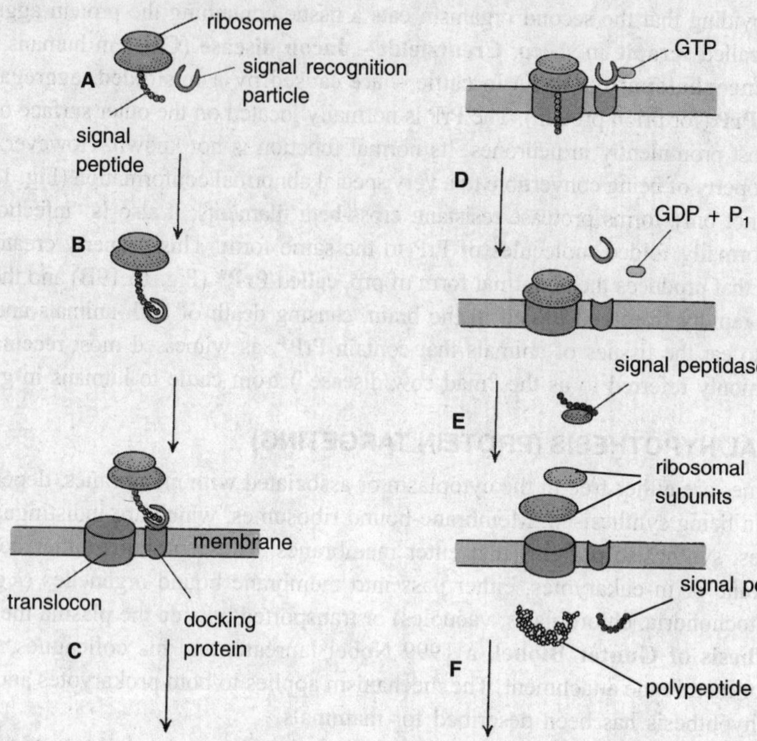

Fig. 13.21. Bobel's signal hypothesis for protein transportation. A signal recognition particle recognizes a ribosome with a signal peptide (A), then draws the ribosome to docking protein located near a translocon in the membrane. With the addition of GTP, the signal recognition particle releases the signal peptide; the hydrolysis of GTP to GDP + Pi causes the signal recognition particle to leave the docking protein (B). Peptide synthesis then resumes, with the newly synthesized peptide passing through the translocon in the membrane. A signal peptide on the other side of the membrane removes the signal peptidase (enzyme). When translation is completed, the ribosome dissociate and drops free of the translocon (after Tamarin, 2002).

passing directly into a **translocation channel** (called **translocon**). Once through the membrane, the signal peptide is cleaved from the protein by an enzyme called **signal peptidase** (Box 13.3).

Box 13.3

Evidence for signal hypothesis came about through *recombinant DNA techniques*. A signal sequence was placed in front of the α-globin gene, whose protein product is normally not transported through a membrane. When this gene was translated, the ribosome became membrane bound, and the protein passed through the membrane.

Since different proteins enter different membrane-bound compartments (*e.g.,* the Golgi apparatus), some mechanism must direct a nascent protein to its proper membrane. This specificity seems to depend on the exact signal sequence and membrane – bound glycoproteins called **signal sequence receptor**. Apparently, after the ribosome binds to the docking protein, the signal peptide interacts with a signal sequence receptor, which presumably determines whether that protein is specific for that membrane. If it is, the remaining processes continue. If not, the ribosome may be released from the membrane.

The signal peptide does not seem to have a consensus sequence similar to transcription or translation recognition boxes. Rather similarities (at least for the endoplasmic reticulum and bacterial membrane bound proteins) include a positively charged (basic) amino acid (commonly lysine or arginine) near the beginning (N – terminal end), followed by about a dozen hydrophobic (nonpolar) amino acids, commonly alanine, isoleucine, leucine, phenylalanine and valine (**Sasavage** *et al.*, 1982; Table 13.3).

Table 13.3 The signal peptide of the bovine prolactin protein* (*Source:* **Tamarin**, 2002).

NH_2 – Met Asp Ser Lys Gly Ser Ser Gln Lys Gly Ser Arg Leu Leu Leu Leu Leu Val Ser Asn Leu Leu Leu Cys Gln Gly Val Val Ser/Thr Pro Val ....Asn Asn Cys - COOH

*The slash line separates the signal peptide from the rest of the protein which consists of 199 residues.

QUESTIONS

Long Answer Questions

1. Describe four levels of protein structure.
2. Describe the experiments which showed that a small protein such as ribonuclease can fold spontaneously *in vitro*. Why are larger proteins unable to fold spontaneously?
3. Distinguish between the activities of Hsp70 chaperones and chaperonins in protein folding.
4. Give examples of proteins that are processed by proteolytic cleavage.
5. What is an intein? How are inteins spliced?
6. Describe the process thought to be responsible for protein degradation in eukaryotes.
7. Are the similarities between inteins and introns purely coincidental?

Very Short Answer Questions

1. Define molecular chaperones.
2. What is ubiquitin?
3. What is protesome?
4. What is signal peptide? What role does it play in eukaryotes? What is its fate?

Fill in the Blanks

1. .... is the final product of genome expression and comprises all the .... present in the cell at a particular time.
2. β-helix is .... structure of a protein.
3. The protein folding is assisted by enzymes, called ....
4. Molecular chaperones in *E.coli* includes .... and ....
5. Eukaryotic cells have mainly two families of molecular chaperones, known as .... and ....
6. Glycosylation involves attachement of large .... side chains to polypeptides.
7. .... are internal segments of proteins that are removed soon after translation, the two external segments or .... becoming linked together.
8. .... is the amino acid sequence within a protein that is used in targeting the ______ for destruction.

Multiple Choice Questions

1. Peptide linkage is

(*a*) —CO—NH— (*b*) —CO—NH_2

(*c*) —COOH—NH_2 (*d*) —CH—N

2. Tertiary structure of proteins having amino acids cystein is achieved through

(*a*) ionic bonds (*b*) covalent bonds

(*c*) disulphide bonds (*d*) hydrogen bonds

ANSWERS

Very Short Answer Questions

1. Molecular chaperone is a protein that facilitates the folding of other proteins but is not a component of the final folded structure.

2. Ubiquitin is a small protein (76 amino acid) that is linked to other proteins as a way of marking the targeted proteins for degradation by proteasomes.

3. Proteosome is a strucutre within which ubiquitinated proteins are broken down. It contains several protein-degrading enzymes.

4. A signal peptide is a sequence of amino acids at the amino terminal end of a protein that signals that the protein should enter a membrane. Although the concept is the same, the situation in eukaryotes is somewhat more complex because there are many different membrane- bound organelles, each having their own membrane specific requirements. Signal peptides are usually cleaved off the protein after the protein enters or passes through the membrane.

Fill in the Blanks

1. Proteome, proteins.

2. Secondary.

3. Chaperons.

4. Hsp70, chaperonins.

5. hsp60 proteins, hsp70 proteins.

6. Carbohydrate.

7. Inteins, exteins.

8. Degron, protein.

Multiple Choice Questions

1. (*a*) **2.** (*c*).

14

Multigene Families (Globin genes, Actin genes, Histone gene clusters and tRNA genes)

The majority of genes are spaced out more or less randomly along the length of DNA molecule. In some cases, however, they are grouped into distinct clusters. Sometimes the individual genes in a cluster are unrelated and there is no apparent reason or advantage to having them organized in this way. More commonly, clusters are made of genes that contain related units of biological information, *e.g.*, operons and multigene family.

14.1 OPERONS

Operons are quite common aspects of the organization of genes in bacteria. A operon is a cluster of genes coding for a series of enzymes that work in concert in an integrated biochemical pathway (Fig. 12.27; Fig. 14.1). The first operon to be discovered was the lactose or lac operon of *E.coli*

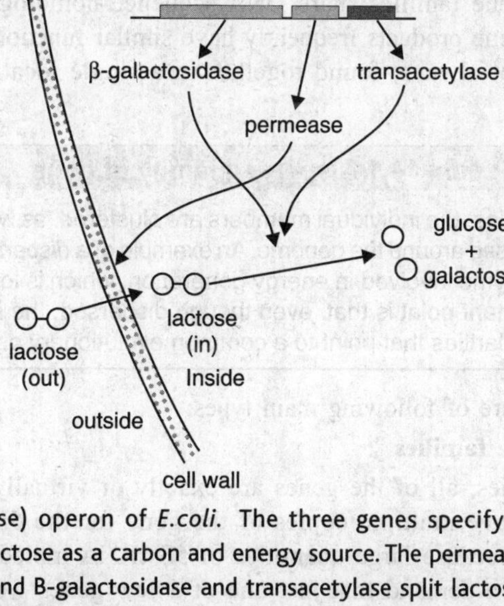

Fig. 14.1. The lac (lactose) operon of *E.coli*. The three genes specify three enzymes involved in utilization of lactose as a carbon and energy source. The permease transports lactose into the bacterial cell, and B-galactosidase and transacetylase split lactose into glucose and galactose (after Brown, 1998).

which is a cluster of three genes each coding for one of the three enzymes involved in conversion (catabolism) of the disaccharide lactose into its monosaccharide units glucose and galactose. These enzymes are not required all the time, only when lactose is available. But when needed they are required together, *i.e.*, the three units of biological information carried by *lac* must be read at the same time. A sophisticated system enabling the *lac* genes to be expressed together and only when needed has evolved round the fact that the genes are clustered.

14.2 MULTIGENE FAMILIES

Operons have no direct counterparts in organisms other than bacteria. In contrast, multigene families are found in many organisms, mainly eukaryotic organisms. A multigene family or gene family is also a cluster of related genes, but in multigene families the individual genes have identical or similar nucleotide sequences, and therefore contain identical or similar pieces of biological information.

Why do Multigene Families are Required?

Some proteins in the cell are required in small quantities, but other proteins may be required in large quantities, so that the demands can not be easily met. These demands can be met in either of the following two ways: (1) In some cases, a solitary gene is present in the cell but is transcribed repeatedly to produce large number of messenger RNA molecules and these mRNA molecules act as templates for many rounds of protein synthesis. For example, although the gene coding for silk worm protein **fibroin** is present in only one copy per haploid complement, but it keeps on making silk actively. Similarly, in a cell (*e.g.*, erythroblasts) engaged in exclusive synthesis of haemoglobin, a solitary copy of the gene produce large quantities of α and β polypeptide chains of haemoglobin. However, although a whole family of related α globin genes are found in each cell, different but solitary α globin genes function at different stages of development. In other cases, multiple copies of a gene are found in the same cell, all taking part in transcription within a cell, giving rise to large quantities of the gene product as in case of ribosomal RNA (rRNA) genes, histone genes, tRNA genes, heat shock protein genes, etc.

Types of Multigenes

Members of multigene families share DNA sequence homology, descend from a single ancestral gene and their gene products frequently have similar functions. Members of multigene families are often, but not always, found together in a single location along a chromosome (Box 14.1).

Box 14.1 Dispersed family of gene

In some multigene families, the individual members are clustered, as with the globin genes, but in others the genes are dispersed around the genome. An example of a *dispersed family* is the five human genes for *aldolase*, an enzyme involved in energy generation, which is located on chromosomes 3, 9,10,16 and 17. The important point is that, even though dispersed, the members of the multigene family have sequence similarities that point to a common evolutionary origin.

Multigene families are of following main types:

1. Simple multigene families

In these gene families, all of the genes are exactly or virtually the same. For example, most higher organisms have multiple copies of the gene for the 5S rRNA (a component of the ribosome), probably because large quantities of 5S rRNA molecules must be synthesized at certain times, creating a demand that just one of a few genes would be unable to satisfy.

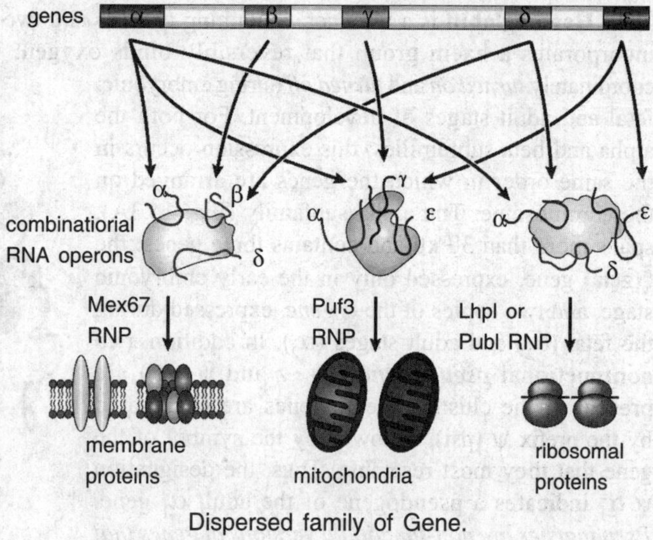

Rather than being spread randomly throughout the chromosomes, the 5S rRNA genes are clustered into a multigene family. In humans, for example, there is a family of about two thousand 5S rRNA genes in chromosome 1.

The 5S rRNA genes are present independently of the rDNA which is localized at the NOR in eukaryotes. However, in prokaryotes and yeast, 5S rRNA genes are present in close vicinity of rDNA (Fig. 14.2). The 5S rRNA genes are also organized in tandem repeats, each repeat consisting of a gene 120 bp long and a spacer region. The length of complete repeat is 375 bp in *Drosophila*. In wheat, there are two loci for 5S rRNA having repeats of different lengths, 480 bp and 500 bp. These repeat units may sometimes carry pseudogenes (instead of functional genes), which are not used for synthesis of any RNA.

2. Complex Multigene Families

A multigene family is a group of similar, but not necessarily identical sequences, each sequence representing a gene, so that the gene is present in multiple copies. Complex multigene families are of following types:

A. Multigene families with divergent members

(i) Globin gene family in human genome

The human α-globin and β-globin genes are subfamilies of the globin gene *superfamily* and are two of the most intensively studied regions of human genome. There is a cluster of three α - globin genes on the short arm of chromosome 16 and another with five β - globin genes on the short arm of chromosome 11. Members of both subfamilies share nucleotide sequence similarity, but members of the same subfamily have the greatest amount of sequence similarity.

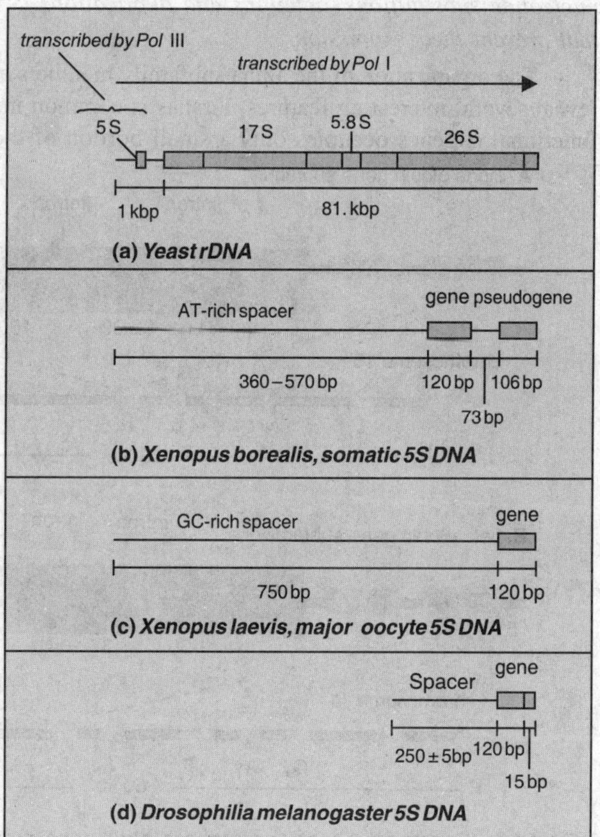

Fig. 14.2. Organization of 5S DNA (produces 5S r RNA for ribosomes) in several organisms; only in yeast it is linked with repeat units of r DNA; in other cases it is independent on a separate chromosome.

Haemoglobin is a tetramer, containing two α - and two β - polypeptides. Each polypeptide incorporates a haem group that reversibly binds oxygen. Within each subfamily, genes are coordinately *turned on* and *turned off* during embryonic, fetal and adult stages of development. For both the alpha and beta subfamilies, this expression occurs in the same order in which the genes are arranged on the chromosome. The alpha subfamily (Fig. 14.3A), spans more than 30 kb and contains three genes: the ζ(zeta) gene, expressed only in the early embryonic stage, and two copies of the α *gene,* expressed during the fetal (α_1) and adult stages (α_2). In addition, two nonfunctional *pseudogenes* (ψ z and ψ α_1) are present in the cluster. Pseudogenes are designated by the prefix ψ (psi), followed by the symbol of the gene that they most resemble. Thus, the designation ψ α_1 indicates a pseudogene of the adult α_1 gene. *Pseudogenes are non-functional versions of genes that resemble other gene sequences, but contain significant nucleotide substitutions, deletions and duplications that prevent their expression.*

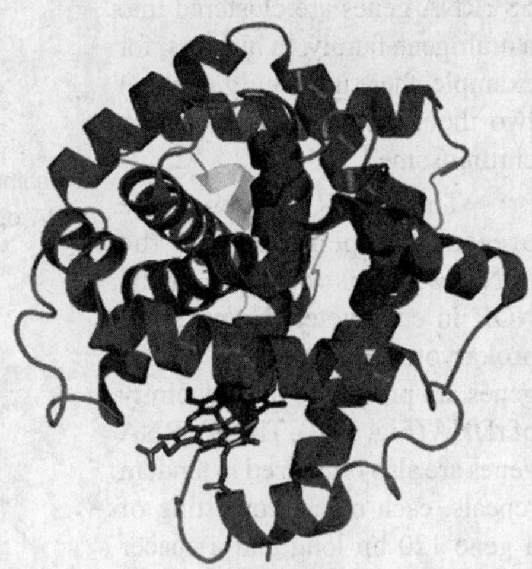

Haemoglobin.

The organization of the alpha subfamily members and the location of their introns and exons reveal several interesting features. First,as is common in eukaryotes, the DNA encoding the three functional α genes occupies only a small portion of the region containing the subfamily. Most

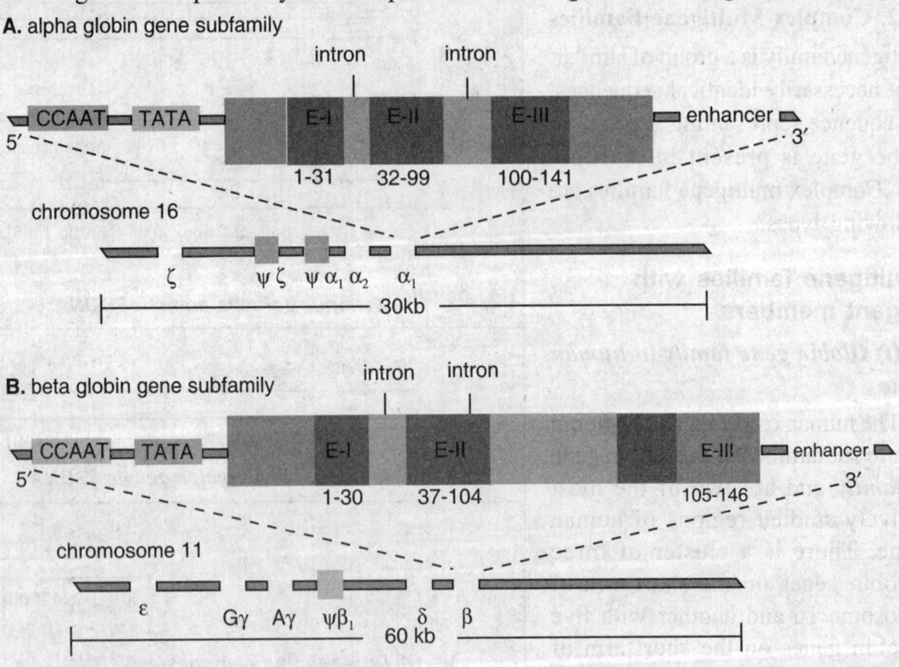

Fig. 14.3. Organization of the alpha-globin gene subfamily (A) on chromosome 16 and the β - globin gene subfamily on chromosome 11 (B). Also shown the internal organization of the α_1 gene and the β gene. Each gene contains three exons (E-I, E-II, E-III) and two introns. The numbers below the exons in dictate the amino acids in the gene product encoded by each exon (after Klug and Cummings, 2003).

of the DNA in this region is intergenic spacer. Second, each functional gene in this subfamily contains two *introns* at precisely the same positions. Third, the nucleotide sequences within corresponding *exons* are nearly identical in the z and α genes. Both genes encode polypeptide chains of 141 amino acids. However, their intron sequences are highly divergent, even though they are about the same size. Significantly, much of the nucleotide sequence of each gene is contained in these noncoding introns.

The human β - globin gene cluster is longer than the α - globin cluster and contains five genes spaced over 60 kb of DNA (Fig. 14.3B). As with the alpha subfamily, the order of genes on the chromosome parallels their order of expression during development. Of the five genes, three are expressed prior to birth. The ε (epsilon) gene is expressed only during embryogenesis, while the two nearly identical γ (gamma) gene (Gγ and Aγ) are expressed only during fetal development. The polypeptide products of the two γ-genes differ only by a single amino acid. The two remaining genes, δ (delta) and β (beta), are expressed following birth. Finally a single pseudogene $\psi\beta_1$ is present within the subfamily. All five functional genes encode proteins with 146 amino acids and have two similarly sized introns at exactly the same positions. The second intron in the β gene is significantly larger than its counterpart in the functional α genes. These similarities reflect the *evolutionary history* of each subfamily and the events such as gene duplication, nucleotide substitution and chromosome translocations that produced the present day globin superfamily.

It has been shown that in these sequences (α cluster genes and β cluster genes), the coding regions show little divergene, but spacers in between the genes show considerable diversity as in the case in many other gene families. The only explanation for this is that the protein function puts a constraint on the evolution of coding sequence, while no such constraint is exercised in spacer region which is fast evolving. Based on the degree of similarity in amino acid sequences of globin of different mammals, evolutionary trees have been prepared in the past (Fig. 14.4 and Fig. 14.5).

(ii) Actin gene family in eukaryotes

Actin is a cytoskeleton protein. Actin genes represent a multigene family, where the members are homologous, but non-identical, giving rise to slightly different variants so that different members function either at different times in the same tissue or in different tissues at the same time. For example, a minimum of six closely related polypeptides are synthesized in different cells at different times in mammals; the number may be larger in some flowering plants. The actins may be broadly classified as α,β and γ chains. Of these α actins make contractile apparatus of the muscle, while β and γ actins are found together in almost all nonmuscle cells. There may be variants in each of these three classes, making upto 20-30 actin related sequences in human genome; humans have six genes for actin (see **Alberts** *et al.*, 2002). In a number of plant systems (*e.g.*, tobacco, potato, etc.) also, more than 20 related but variant actin sequences have been reported.

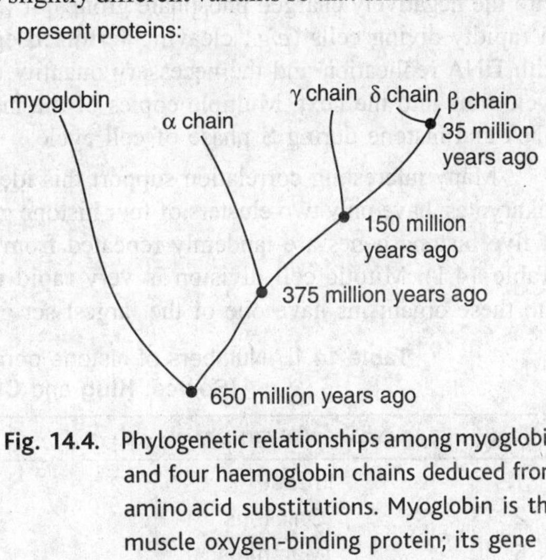

Fig. 14.4. Phylogenetic relationships among myoglobin and four haemoglobin chains deduced from amino acid substitutions. Myoglobin is the muscle oxygen-binding protein; its gene is present on chromosome 22.

Actin is extraordinarily well conserved among eukaryotes. The amino acid sequence of actins from different species are usually about 90% identical. Yeast actin and *Drosophila* muscle actin are 89% identical.

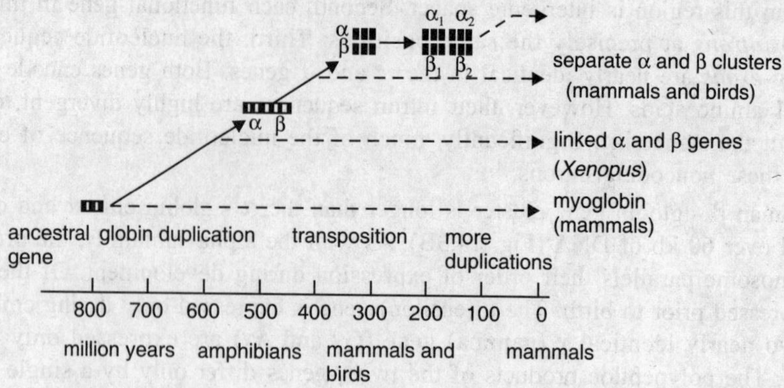

Fig. 14.5. Myoglobin differentiation into α and β chains from earliest period in the history.

(iii) Other multigene families with divergent members. Other multigene families have divergent members; *e.g.,* (*i*) albumin, ; α feroprotein (major components of blood plasma), serine proteases, the interferons (for defence against viral infection) and the immunoglobulins; (*ii*) chorion proteins making the major components of egg shells in insects.

B. Multigene families with identical genes.

In some cases, for producing large quantities of a gene product, many copies of identical genes are required. These genes may occur as repeat units, each of which comprises (*i*) a *coding region,* which is highly conserved to produce exactly the same gene product and (*ii*) a *spacer region,* which may show divergence. Examples of these multigene families are the following:

(*i*) Multiple copies of histone genes. Organization of the histone genes is a variation on the theme established in the globin gene family. Here a cluster of five related, but nonidentical genes, separated from each other by highly divergent intergenic spacer regions, is tandemly repeated many times.

As we already know that histones are positively charged (basic) proteins which interacts with the negatively charged phosphate groups of DNA to form nucleosomes, the chromatin units. In rapidly diving cells (*e.g.*, cleaving fertilized egg of frog), histone synthesis must keep pace with DNA replication and the necessary quantity of histones must appear each time, as one cell cycle runs into the next. Multiple copies of the histone cluster allow the coordinate synthesis of DNA and histone during S phase of cell cycle.

Many interesting correlation support this idea. Yeast cells, with much less DNA than other eukaryotes, have only two clusters of four histone genes(they lack histone H1). In contrast, clusters of five histone genes are tandemly repeated from 10 to 800 times in many complex organisms (Table 14.1). Mitotic cell division is very rapid during development in sea urchins and newts and these organisms have one of the largest set of histone gene clusters.

Table 14.1 Numbers of histone gene clusters in selected eukaryotes (Source: **Klug** and **Cummings**, 2003).

S. No.	*Organism*	*Repeats*
1.	Yeast	2
2.	Chickens	10
3.	Mammals	20
4.	*Xenopus*	40
5.	*Drosophila*	100
6.	Sea urchin	300-600
7.	Newt (amphibian)	600-800

Polarity differences of transcription. Beyond its tandem arrangement, the histone gene family shows several other differences from other families.

First, almost all histone genes lack introns. Second, individual genes within clusters of a given species often orient in opposite directions with respect to transcription. As shown in Figure 14.6,

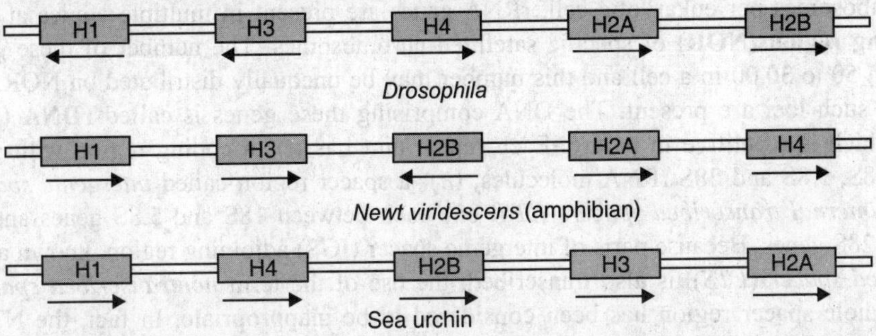

Fig. 14.6. The histone gene cluster in *Drosophila*, *Newt* and sea urchin. Arrows indicate the direction of transcription.

the arrangement of genes and their polarity of transcription, shown by the direction of arrow, vary in the sea urchin and *Drosophila*, two organisms where this cluster has been extensively studied. Studies also show polarity differences in other organisms.

Despite polarity differences the amino acid sequences of the various histones (reflecting the DNA encoding each one) are remarkably similar in highly divergent organisms. The homology of histone genes is one of the best examples of sequence conservation through evolution.

Another interesting feature of the multigene family of histone genes observed in *Drosophila* is the presence of an '*attachment site*' per repeat unit (Fig. 14.7); 100 such repeat units being present in each haploid genome. With the help of attachment site, the genes (DNA) remain attached to *nuclear scaffold*. These attachment sites help in transcription.

In mice and humans, members of the histone gene family are clustered, but do not form tandem arrays. A region at

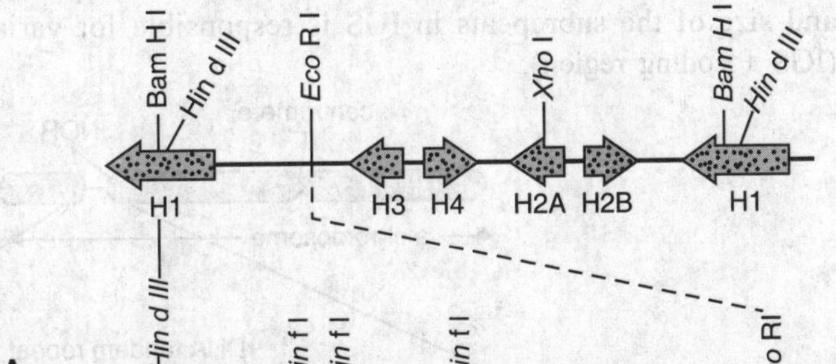

Fig. 14.7. A – A restriction map of major histone gene cluster in *Drosophila*, each repeat unit has one specific attachment site to the nuclear scaffold, this attachment site is Hin1/ ECO RI DNA segment, 657 bp long; B – Schematic representation showing how *Drosophila* histone repeats might be attached to the nuclear scaffold.

the distal tip of the long arm of human chromosome 7 contains all copies of the histone gene family, interspersed with other nonhistone genes. Therefore, no single pattern of histone gene organization applies to all organism (see **Klug** and **Cummings,** 2003).

***(ii)* Ribosomal RNA (rRNA) genes in tandem arrays.** For the production of about 10 million ribosomes per eukaryotic cell, rRNA genes are present in multiple copies at **nucleolar organizing regions(NOR)** of specific satellited chromosomes. The number of these genes may vary from 50 to 30,00 in a cell and this number may be unequally distributed on NOR's, if more than one such loci are present. The DNA comprising these genes is called **rDNA** (*ribosomal DNA*), which is repetitive in nature. Each repeat unit has (*i*) a coding region with genes that specify 18S, 5.8S and 28S rRNA molecules; (*ii*) a spacer region called *intergenic spacer* (*IGS*) and (*iii*) *internal transcribed spacers (ITS)* one each between 18S and 5.8S genes and between 5.8S and 28S genes. Because parts of intergenic spacer (IGS) adjoining region, known as *external transcribed spacer* (*ETS*), is also transcribed, the use of the term *nontranscribed spacer* (*NTS*) for the whole spacer region has been considered to be inappropriate. In fact, the NTS makes only a part of IGS (Fig. 14.8), the remaining part of IGS being ETS.

The IGS region in its turn has a region consisting of a tandem array of variable number of *subrepeats* ranging from 100 – 300 bp (bp = base pairs) in length. The variation in number and size of the subrepeats in IGS is responsible for variation in length of the repeat units (IGS + coding region).

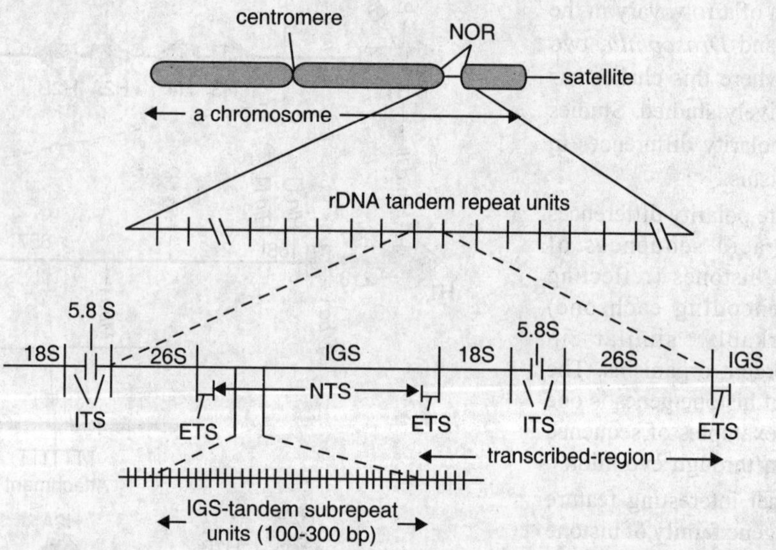

Fig. 14.8. Organization of ribosomal DNA (rDNA) in nucleolar organizing region (NOR) of a eukaryotic chromosome. Abbreviations: IGS = intergenic spacer; NTS = non-trascribed spacer; ETS = external transcribed spacer; ITS = internal transcribed spacer; 18S, 5.8S, 26S = different genes responsible for noncoding RNAs such as 18S rRNA, 5.8S rRNA and 26S rRNA.

The rRNA repeat units are usually looped off from the main chromosome fibres, in the form of extended threads at nucleolar organizing region. These loops, in association with specific proteins form the nucleoli, where rRNA synthesis and processing actually takes place. The number of nucleoli and the corresponding number of nucleolar organizing chromosomes in an organism may vary from one to several. At each locus (NOR), rDNA repeat units may evolve independently both in length and also in the nucleotide sequence of the spacer region. As a result of this, the length of the repeat unit including coding and spacer regions (nontranscribed spacer or NTS; external transcribed spacer or ETS and internal transcribed spacers or ITS) varies from about 7 kb (kilobase) pairs to 14 kb. In wheat and barley, it is usually 9 kb to 10 kb (Fig. 14.9).

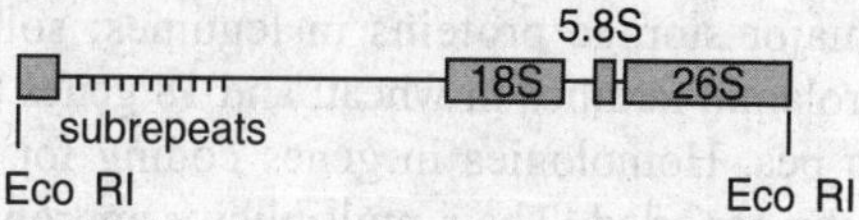

Fig. 14.9. An outline of the structure of rRNA in wheat and barley.

***(iii)* Transfer RNA (tRNA) multigene families.** The genes for each of the different tRNAs are also found in multiple copies to meet the heavy demands of the cell for the production of tRNA molecules. Ten to several hundred genes for each tRNA are present in each haploid genome. While in some cases with fewer copies of a gene, these copies are dispersed, in other cases such as in *Xenopus* (a toad, an amphibian), tandem repeats of long sequences (each repeat having genes for several different tRNAs) are found. In *Xenopus,* for example, a 3.2 kb repeat length has 8 genes, 2 for tRNA$_f^{met}$ and six other, for six different tRNA molecules (Fig. 14.10). In still other cases (*e.g., Drosophila)* genes for completely different tRNA species are clustered over a length of several thousand base pairs. Sometimes, tRNA pseudogenes are also found.

***(iv)* Small nuclear RNA (Sn RNA) genes.** Small nuclear RNAs (SnRNAs) are found in abundance in the nuclei of all eukaryotes and represent neither tRNA, nor rRNA. Six SnRNAs, that are usually found, range from 100 (U6) to 215 (U3) nucleotides in length and are involved in RNA processing through formation of SnRNP (small nuclear ribonucleoproteins). SnRNAs are encoded by multiple copies of identical genes organized in tandem arrays of repeat units, each SnRNA gene being flanked on either side of by spacer DNA ranging from 800 bp to 45,000 bp in length in different organisms.

DNA of genes of Sn RNAs is reported to have high ratio(10:1) of pseudogenes than functional genes. In humans, for example, there are 30 genes, for Ul (an SnRNA) on chromosome 1 and 500 to 1000 pseudogenes distributing throughout the genome.

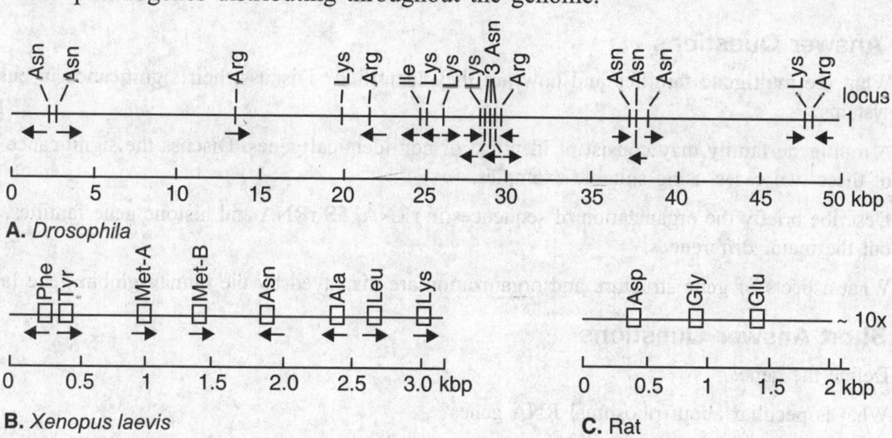

Fig. 14.10. Arrangement of tRNA genes in higher eukaryotes. A – One of the 50 – 60 sites on *Drosophila* chromosomes where tRNA genes are clustered. B – A 3.18 kilobase – pairs unit from *Xenopus* DNA, that is repeated about 150 times; C – Rat DNA has about 10 repeat units per haploid genome, here only one of them is shown.

C. Multigene families for storage proteins in crop plants. Storage proteins of several crop plants have been reported to be encoded by multigene families and are represented mainly as *prolamins* (major storage proteins in cereals; soluble in aqueous alcohol) or *globulins* (major storage proteins in legumes; soluble in salt). Three to ten genes for each of the ten prolamin families in wheat; and 18 genes belonging to three families of globulin genes are known in pea. Homologies in genes coding for prolamins and globulins in different crop plants have been reported. Their multiplicity appropriately meets the demand for rapid synthesis of these proteins in the developing seeds.

14.3 CONCERTED EVOLUTION OF MULTIGENE FAMILIES

Multigene families are found to have mechanisms that prevents the individual copies from accumulating mutations and hence diverging away from the functional sequence. This is called *concerted evolution.* Thus, if one copy of the family aquires an advantageous mutation then it is possible for that mutation to spread throughout the family until all members possess it. Following genetic mechanisms may be involved in such homogenization of concerted evolution:

1. Gene conversion. Gene conversion can result in the sequence of one copy of a gene being replaced with all or part of the sequences of a second copy. Multiple gene conversion events could therefore maintain identity among the sequences of the individual members of a multigene family (see **Brown,** 2002).

2. Unequal crossing over. Unequal crossing over may lead to *duplications or deletions* leading to homogenization, *i.e.*, either mutant member will be deleted or multiply and spread throughout the gene family. In yeast, it has been reported that the rates of gene conversion and unequal crossing over are higher than the rate of mutations so that homogeneity can be maintained. Further, certain specific sequences are also located in the multiple gene families which stimulate higher rates of recombination in a multigene family. In rDNA, NTS (nontranscribed spacer) region has been shown to be hot spot of recombination thus facilitating homogenization.

QUESTIONS

Long Answer Questions

1. What are multigene families and how are they identified? Discuss their significance in eukaryotic systems.
2. A multigene family may consist of identical or non-identical genes. Discuss the significance of each of these two cases using suitable examples.
3. Describe briefly the organization of sequences in rRNA, 5S rRNA and histone gene families, pointed out the major differences.
4. What aspects of gene structure and organization are displayed by the human globin gene families?

Very Short Answer Questions

1. Define the gene.
2. What is peculiar about ribosomal RNA genes?
3. Write about histone genes.
4. What is unique in 5S rRNA genes?
5. The β-globin gene family consists of 60 kb of DNA, yet only 5 percent of the DNA encodes β-globin gene products. Account for as much of the remaining 95 per cent of DNA as you can.

Fill in the Blanks

1. Clusters are made of genes that contain related units of biological information, *e.g.*, ... and .... family.
2. The *lac* operon of *E. coli* is a cluster of .... genes.

3. Members of multigene families share .... homology, descend from a single ancestral gene and their gene products have ....
4. Human beings contain a family of about .... 5SrRNA genes in ....
5. There is a cluster of .... α-globin genes on the short arm of .... of human being.
6. Pseudogenes contain significant nucleotide .... , ... and duplications that prevent their expression.
7. Yeast actin and *Drosophila* muscle actin are .... identical.
8. The homology of histone genes in divergent organisms is one of the best examples of .... through evolution.
9. Spacer regions of rRNA genes include NTS (..................), ... (external transcribed spacer) and .... (internal transcribed spacer).

ANSWERS

Very Short Answer Questions

1. Gene is hereditary factor that specifies an inherited trait, it consists of a DNA sequence that codes for the amino acid sequence of one or more polypeptide chains, or alternatively, for one of several types of RNA that perform functions other than coding for polypeptide chains (*e.g.*, rRNA, tRNA, snRNA, or microRNA).
2. Ribosomal RNA (*i.e.*, rRNA) genes are a cluster of genes associated with the nucleolus. They have been studied most in the load *Xenopus laevis*. The cluster consists of 450 tandemly repeated (arranged one behind the other) units. Each unit is made up of two rigions, one which is transcribed to produce rRNA molecule, and a spacer rigion which is not transcribed.
3. The histone genes are repeated 300-1000 times in *Xenopus*, but only 10–20 times in human beings. By use of restriction endonucleases, the genes for the five types of histones have been determined. Histone genes are clustered together and the repeat units is about 6 to 7 kilobase. Out of the 6–7 kilobase repeat units only 2kb are required to code for the five histones. It was on the basis of this finding that spacers were discovered in histone genes. The transcribed region is rich in G + C nucleotides and the spacer is A + T rich.
4. The genes for 5S rRNA are sepeated 10,000 to 25,000 times in *Xenopus laevis* and are located in the telomere regions of 15 out of 18 chromosomes.
5. While the β-gene family is a relatively large (60 kb) sequence and restriction analyses show that it is composed of six genes, one is a pseudogene and therefore does not produce a product. The five functional genes each contain, two similarly sized intron, which when included with non-coding blanking region (5' and 3'), and spacer DNA between gene, accounts for the 95% mentioned in the question.

Fill in the Blanks

1. Operons, multigene.
2. Three.
3. DNA sequence.
4. 2000, chromosome 1.
5. Three, chromosome 16.
6. Substitutions, deletions.
7. 89%
8. Sequence conservation.
9. Nontranscribed spacer, ETS, ITS.

15

Regulation of Gene Action in Bacteria and Viruses

Bacteria are exposed to a wide range of environmental conditions. For example, *E.coli* may encounter rapidly changing growth conditions as they pass from mammalian intestinal tract to sewer systems to polluted rivers, lakes, ponds and so on. Each of these ecological niches will provide different organic molecules for use as energy sources. From this fact, it can be concluded that natural selection will have preserved those organisms that have evolved ways of adapting to the wide range of environmental conditions encountered during their evolution. Indeed, the available information indicates that most prokaryotes such as *E.coli* exhibit remarkable capacities to adapt to diverse environmental conditions.

To a great extent, the adaptability of bacteria and other prokaryotes depends on their ability to "turn on" and "turn off" the expression of specific sets of genes in response to the specific demands of the environmental milieu (*i.e.,* surroundings). In other words, prokaryotes exhibit a outstanding ability to **regulate the expression of specific genes in response to environmental signals** (Box 15.1). The expression of particular genes is "turned on" when the products of these genes are needed for growth in a given environment. Their expression is "turned off" when their products are no longer needed for growth in the existing environmental surroundings. Clearly, the ability of an organism to regulate gene expression in this way will increase its overall "fitness" (*i.e.,* its ability to grow and leave progeny under a variety of environmental conditions). The synthesis of gene transcripts (*i.e.* various types of RNA molecules) and translation products (*i.e.* proteins) require the expenditure of great amount of energy. By "turning off" the expression

Box 15.1

The chromosome of the bacterium *Escherischia coli*, a single-celled organism, consists of a single circular DNA molecule of about 4.6 × 10^6 nucleotide pairs. This DNA encodes approximately 4300 proteins, although only a fraction of these are made at any one time. The expression of many of them is regulated according to the available food in the environment (see **Alberts** *et al.*, 2002).

of genes when their products are not needed, an organism can avoid wasting energy to synthesis products that maximize the growth rate in the existing environmental surroundings.

Here, one can ask three sort of questions:

1. What are the mechanisms by which these organisms regulate gene expression in respons to changes in the environment?

2. Is there a single mechanism by which the expression of different genes or sets of gene are regulated?

3. Or are different genes controlled by different mechanisms?

Some genes, for example, the genes specifying ribosomal RNAs, ribosomal proteins and transfer RNAs, are certainly expressed at some time in virtually all cells regardless of the environment conditions. The products of these genes are required for growth of all cells in all environments. However, the products of many other genes are required for growth only in certain environments, and *the expression of these genes is regulated such that the products are synthesized only when they are needed*. As a result, the expression of these genes is continually being "turned on" and "turned off" in response to changes in the environment. From this discussion, one can conclude that gene expression can be (and is) regulated at several different levels: for example, **transcription, mRNA processing, mRNA turnover, translation** and **enzyme function**. However, extensive data indicate that the *regulation of transcription is the most important mode of control of gene expression, at least in prokaryotes*.

The regulatory mechanisms of transcription in both prokaryotes and eukaryotes are of two basic types. The *first*, and best understood, category includes mechanisms involved in rapid turn-on and turn-off of gene expression in response to environmental changes. Such sort of mechanisms are very significant in microorganisms (bacteria) because these organisms are exposed to sudden changes in environment. They provide microorganisms with a great deal of "plasticity", an ability to rapidly adjust their metabolic processes in order to achieve maximal growth and reproduction under highly variable environmental conditions. These quick responding on–off switches seem to be less important in higher eukaryotes. This might be expected since the circulatory systems of higher eukaryotes buffer their cells against many sudden environmental changes.

The second major category of regulatory mechanisms includes the so called **preprogrammed circuits of gene expression**. In these cases, some event (*e.g.,* infection by a virus) triggers the expression of one set of genes. The product (or products) of one (or more) of these genes function by turning off the transcription of the first set of genes and/or turning on the transcription of the second set of genes. In turn, one or more of the products of the second set acts by turning on a third set, and so on. In these cases, *the sequential expression of genes is genetically preprogrammed,* and the genes usually cannot be turned on out of sequence. Such preprogrammed sequences of gene expression in viral infections are well worked out. In most of these preprogrammed sequences, it seems the circuit is cyclical. For example, in viral infections some event associated with the packaging of the viral DNA or RNA inside the protein coat somehow seems to reset the program so that the first set of genes will again be expressed when a progeny virus subsequently infects another host cell.

Discovery of Prokaryotic Gene Regulatory Proteins

Genetic analyses in bacteria carried out in the 1950s provided the first evidence for the existence of **gene regulatory proteins** that turn specific sets of genes on or off. One of these regulators, the **lambda repressor**, is encoded by a bacterial virus, the bacteriophage lambda. The repressor shuts off the viral genes that code for the protein components of new virus particles and thereby enables the viral genome to remain a silent passenger in the bacterial chromosome, multiplying with the bacterium when conditions are favourable for bacterial growth. The lambda repressor was among the first gene regulatory proteins to be characterised. Other bacterial regulators respond to nutritional conditions by shutting off genes encoding specific sets of metabolic enzymes when they are not needed. For example, **lac repressor**, the first of these bacterial proteins to be recognised, turns off the production of the proteins responsible for lactose metabolism when this sugar is absent from the medium.

The first step toward understanding gene regulation was the isolation of mutant strains of bacteria and bacteriophage lambda that were unable to shut off specific sets of genes. It was proposed at the time, and later proven, that most of these mutants were deficient in proteins acting as specific repressors for these sets of genes. Because, these proteins, like most gene regulator proteins, are present in small quantities, it was difficult and time consuming to isolate them. They were eventually purified by fractionating cell extracts. Once isolated, the proteins were shown to bind to specific DNA sequences close to the genes that they regulate. The precise DNA sequences that they recognized were then determined by a combination of classical genetics, DNA sequencing and DNA footprinting experiments.

15.1 CONSTITUTIVE GENES AND INDUCIBLE GENES

Certain gene products, such as tRNA molecules, rRNA molecules, ribosomal proteins, RNA polymerase components (polypeptides), and other enzymes catalyzing metabolic processes that are commonly known for cellular "house keeping" functions, are essential components of almost all living cells. Genes that specify products of this type are continually being expressed in most cells. Such genes are said to be expressed **constitutively** and are called **constitutive genes**.

Rest of the gene products are needed for cell growth only under certain environmental conditions. Constitutive synthesis of such gene products would clearly be wasteful, using energy that could otherwise be utilized for more rapid growth and reproduction under the existing environmental conditions. The *evolution* of regulatory mechanisms that would provide for the synthesis of such gene products only when and where they were needed would clearly provide organisms with a selective advantage over organisms lacking these mechanisms. This undoubtedly explains why presently existing organisms, including the "primitive" bacteria and viruses, exhibit highly developed and very efficient mechanisms for the control of gene expression.

E.coli and most other bacteria are capable of growth using any one of several carbohydrates (*e.g.,* glucose, sucrose, galactose, arabinose, lactose) as an energy source. If glucose is present in the environment, it will be preferentially metabolished by *E.coli* cells. In the absence of glucose, however, *E.coli* cells can grow very well on other carbohydrates. Cells of *E.coli* growing in medium containing the sugar lactose, for example, as the sole carbon source synthesise two enzymes, **β–galactosidase** and **β–galactoside permease**, that they uniquely required for the catabolism of lactose (Note: A third enzyme, α-**galactoside transacetylase** is also synthesized). The enzyme β-galactosidase cleaves lactose into glucose and galactose, and enzyme β-galactoside permease pumps α-galactosides into the cell. Neither of these enzymes is of any use to *E.coli* cells when present in an environment not containing lactose. The synthesis of these two enzymes, of course, requires the utilization of great amount of energy (in the form of ATP and GTP). Thus, *E.coli* cells have evolved a regulatory mechanism by which the synthesis of these lactose catabolic enzymes is turned on in the presence of lactose and turned off in its absence.

In natural environments (such as intestinal tracts and sewers), *E.coli* cells probably encounter an absence of glucose and the presence of lactose relatively rarely. Most of the time, therefore, *E.coli* cells growing on a carbohydrate other than lactose are transferred to medium containing lactose as the only carbon source, they rapidly begin sythesising the enzymes required for lactose utilization. This process, by which the expression of genes is turned on in response to a substance in the environment, is called **induction**. Genes whose expression are so regulated are called **inducible genes**. The substances or molecules responsible for induction are known as **inducers** (*e.g.,* in this case lactose is inducer).

Enzymes that are involved in **catabolic** (*i.e.,* degradative) **pathway**, such as in lactose, galactose or arabinose utilization, are characteristically inducible. Researches made it evident that *induction occurs at the level of transcription*. Induction alters the rate of synthesis of enzymes, not the activity of existing enzyme molecules. Induction should not be confused with **enzyme activation**, in which the binding of a small molecule to an enzyme increases the **activity** of the enzyme (but does not affect its rate of synthesis).

Further, *E.coli* and other bacteria possess the metabolic capacity to synthesise most of the organic molecules (such as amino acids, purines and vitamins) required for their growth. For example, *E.coli* has five genes coding for enzymes that are required in the synthesis of tryptophan (an amino acid). These five genes must be expressed in *E.coli* cells growing in an environment devoid of tryptophan in order to provide adequate amounts of this amino acid for ongoing protein synthesis. When *E.coli* cells are present in an environment containing concentrations of tryptophan sufficient to support optimal growth, the continued synthesis of the tryptophan biosynthetic enzymes would be a waste of energy, because these bacteria have the capacity to take in external tryptophan. Thus, a regulatory mechanism has evolved in *E.coli* by which the synthesis of the tryptophan is present in the external milieu (surrounding environment). This process of "turning off" the expression of sets of genes is called **repression**. A gene whose expression has been turned off in this way is said to be repressed; when its expression is turned on, a gene of this type is said to be **depressed**.

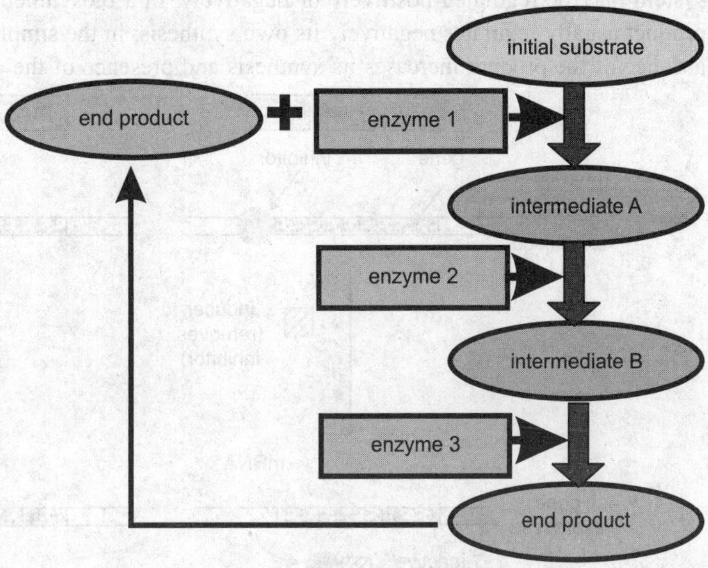

Fig. 15.1. Feedback inhibition.

Enzymes that are components of **anabolic** (biosynthetic) **pathways** are frequently subject to **repression** (are **repressible**). *Repression, like induction occurs at the level of transcription.* However, repression should not be confused with **feedback inhibition** (Fig. 15.1), in which the binding of an end product to the first enzyme in a biosynthetic pathway **inhibits the activity** of the enzyme (but does not affect its synthesis).

15.2 TRANSCRIPTIONAL CONTROL MECHANISMS

In bacteria, there occur several mechanisms of gene regulation at the level of transcription. A notable method depends on whether the enzymes being regulated act in catabolic (degradative) or anabolic (synthetic) metabolic pathways. For example, in a multistep catabolic system, the availability of the molecule to be degraded commonly determines whether the enzymes in the pathway will be synthesised. In contrast, in a biosynthetic pathway the final product is often the regulatory molecule. Even when a single protein molecule is translated from a monocistronic mRNA molecule, the protein may be **autoregulated**, *i.e.,* the protein itself may inhibit initiation of transcription and high concentrations of the protein may cause less transcription of the mRNA that encodes the protein. The molecular mechanisms for each of the regulatory patterns differ greatly and are of the following two types–negative regulation and positive regulation (Fig. 15.2). In a **negative regulated system**, an inhibitor is present in the cell and prevents transcription. An antagonist of the inhibitor, called an **inducer**, is needed to allow initiation of transcription. In a **positively regulated system**, an **effector** molecule (which may be protein, a small molecule, or a molecular complex) activates a promoter; no inhibitor must be abolished. Negative and positive regulation are not mutually exclusive, and some systems are both positively and negatively regulated, utilizing two regulators to respond to different conditions in the cell. Thus, a catabolic system may be regulated positively or negatively. In a biosynthetic (anabolic) pathway, the final product usually regulates negatively its own synthesis; in the simplest type of negative regulation, absence of the product increases its synthesis and presence of the product decreases its synthesis.

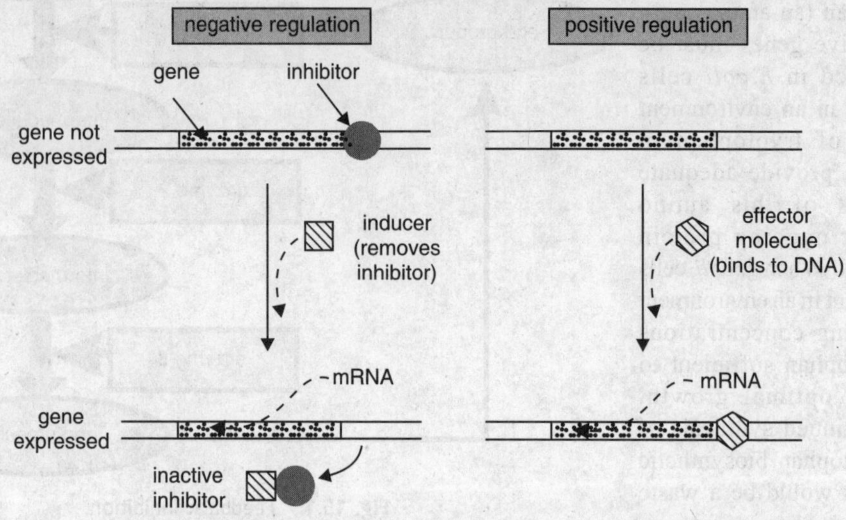

Fig. 15.2. The distinction between negative and positive regulation. In negative regulation an inhibitor, bound to the DNA, must be removed before transcription can occur. In positive regulation, an effector molecule must bind to the DNA (after Freifelder, 1985).

The Operon Model

Induction and repression of gene expression can be accomplished by essentially the same mechanism. This mechanism was first accurately described in 1961 when **Francois Jacob** and **Jacques Monod**, both 1965 Nobel Prize recipients, proposed the **operon model** to explain the regulation of genes encoding the enzymes required for lactose utilization in *E.coli*. **Jacob**

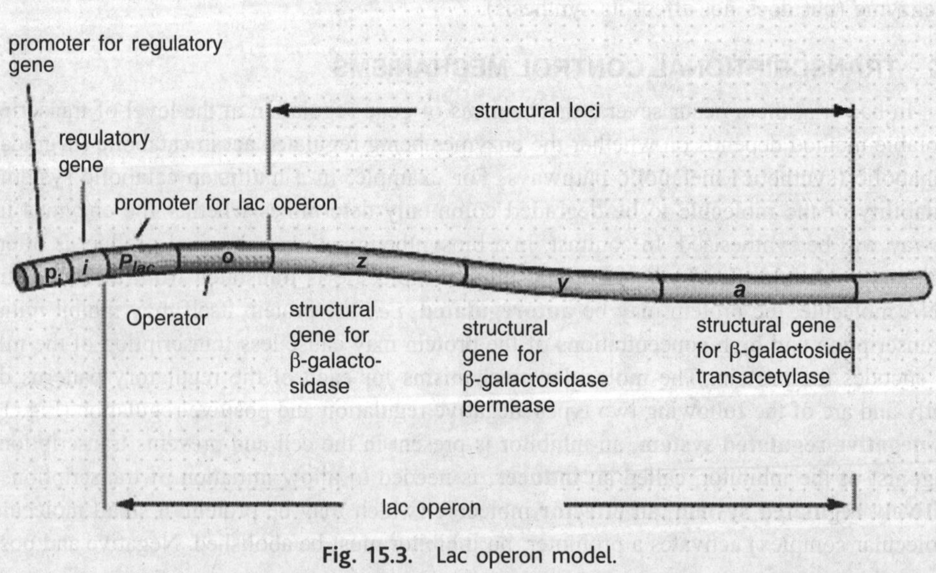

Fig. 15.3. Lac operon model.

and **Monod** proposed that the transcription of one or a set of contiguous structural genes (*i.e.,* genes coding for polypeptides) is regulated by two controlling elements (Fig. 15.3; Fig. 15.4). One of these controlling elements, called the **regulator gene** (or **repressor gene**), codes for a protein called the **repressor**; under the appropriate conditions, the repressor binds to the second element, the **operator**. The operator is always located contiguous to the structural gene or genes whose expression it regulates. *When the repressor is bound to the operator, transcription of the structural genes cannot occur.* We now know that this results because the binding of the repressor to the operator stearically prevents RNA polymerase (enzyme) from binding at the **promoter site** (the RNA polymerase binding site), which is always located contiguous with the operator sequence. The **operator** is usually located between the promoter and structural genes. (Note: The promoter was not recognized at the time of Jacob and Monod's proposal, but has since been shown to be as essential component of an operon). Thus, the complete contiguous unit, including the structural genes, the operator and the promoter is called the operon (**Gardner** *et al.*, 2002). In other words, an operon is a unit of prokaryotic gene expression which includes co-ordinately regulated (structural) genes and control elements (promoter operator) which are recognized by regulatory gene products (**Turner** *et al.,* 2000).

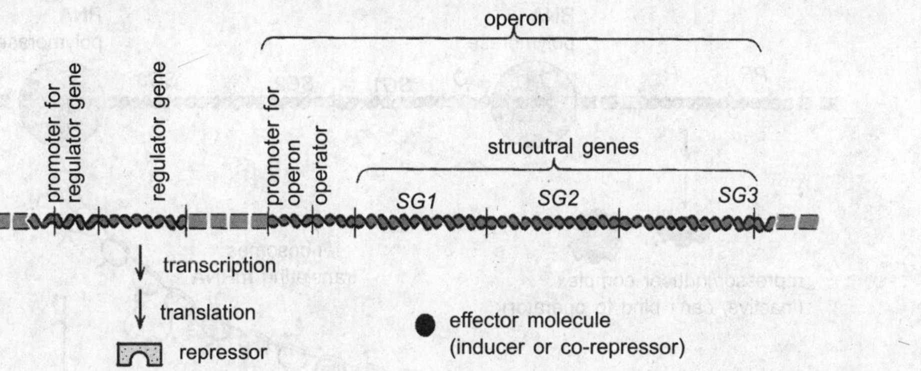

Fig. 15.4. The operon model for regulation of gene expression. The operon consists of one or more structural genes (three–*SG1*, *SG2* and *SG3* are arbitrarily shown) and the adjoining operator and promoter sequences. The promoter for the operon (*PO*) is the site at which RNA polymerase enzyme must bind to initiate transcription of structural genes. The operator (*O*) is the site at which protein repressor (the products of the regulator gene or repressor gene) binds. The regulator gene need not be closely linked to the operon; in fact, it can be located at any position in the genome. The transcription of the regulator gene is initiated by RNA polymerase, which binds to its promoter (labeled *PR*, for promoter for regulator gene). When the repressor is bound to the operator, it stearically prevents RNA polymerase from binding to adjoining promoter (*PO*) and from initiating transcription of the structural genes. Whether the repressor binds to the operator or not, depends on the presence or absence of a metabolite called an effector molecule (after Gardner *et al.*, 2002).

Whether the repressor will bind to the operator and turn off the transcription of the structural genes in an operon is determined by the presence or absence of **effector molecules** (small molecules such as amino acids and sugars) in the environment. In the case of **inducible operons** (*e.g.,* lac operon), these effector molecules are called **inducers**. Those active on repressible operons (*e.g.,* *trp* operon) are called **co–repressors** act by binding to (or forming a complex with) the repressors.

The only essential difference between inducible operons and repressible operons is whether the naked repressor or the repressor effector molecule complex is active in binding to the operator.

1. In case of an inducible operon, the free repressor binds to the operator, turning off (Fig. 15.5). When the effector molecule or the inducer (*e.g.,* allolactose in *lac* operon) is present, it binds to the repressor, that is, the **repressor-inducer complex cannot bind to the operator**. Thus, the addition of inducer turns on (or induces) the transcription of the structural genes in the operon (Fig. 15.6).
2. In case of a repressible operon (*e.g., trp* operon), the situation is just reversed. The free repressor-cannot bind to the operator. Only the repressor-effector molecule (co-repressor) complex is active

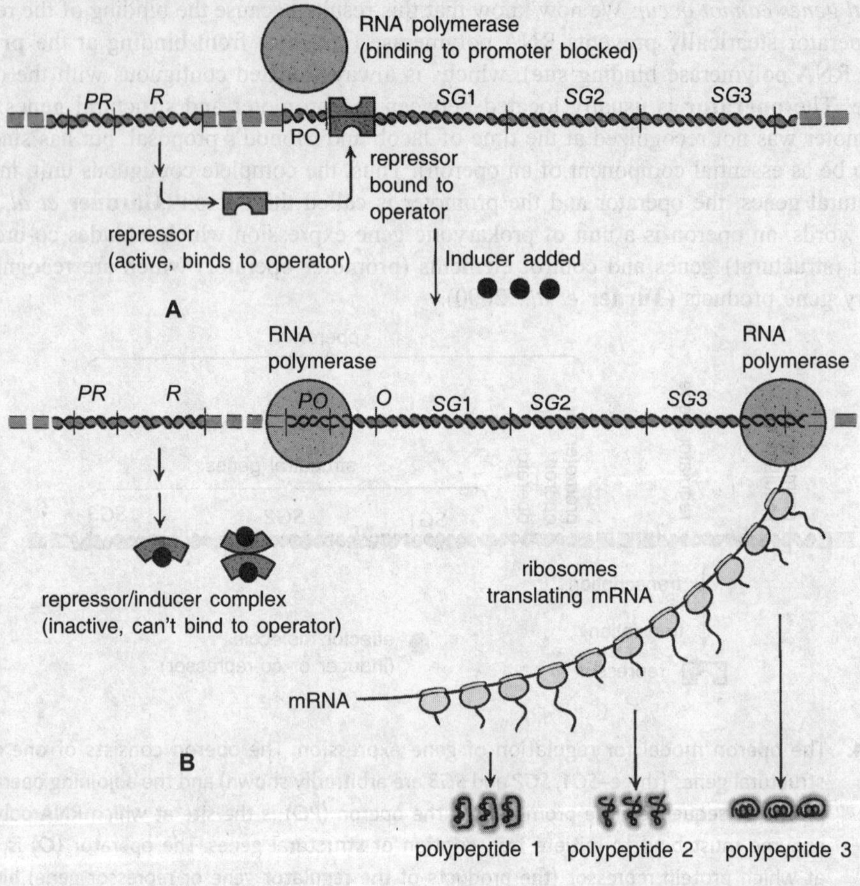

Fig. 15.5. Mode of regulation of gene expression for an inducible operon (*e.g., lac* operon). A–The product of the regulator gene (*R*), the repressor, in the absence of the effector molecule (called an inducer, for inducible operon, *e.g.,* allolactose in *lac* operon), binds to the operator, preventing RNA polymerase from binding to the promoter for the operon (*PO*). Thus, transcription of the structural genes cannot occur. B–When inducer is added, it binds to the repressor, causing it to be released from the operator (*O*). This, in turn, allows RNA polymerase to bind to the promoter (*PO*) and initiate transcription of the structural genes. The resulting multigenic (=polycistronic) mRNA is rapidly translated by ribosomes, producing the three polypeptide products of the structural genes (after Gardner *et al.,* 2002).

in binding to the operator (Fig. 15.6). Thus, transcription of the structural genes in repressible operon is turned on in the absence of and turned off in the presence of the effector molecule (co-repressor). Except for this difference in the operator repressible operons are comparable.

A single mRNA transcript carries the coding information of an entire operon. Thus, the mRNAs of operons consisting of more than one structural gene are polygenic or polycistronic. For example, the tryptophan operon mRNA of *E.coli* is a huge macromolecule carrying the coding sequences that specify five different polypeptides because of their cotranscription, all the structural genes in an operon are coordinately expressed.

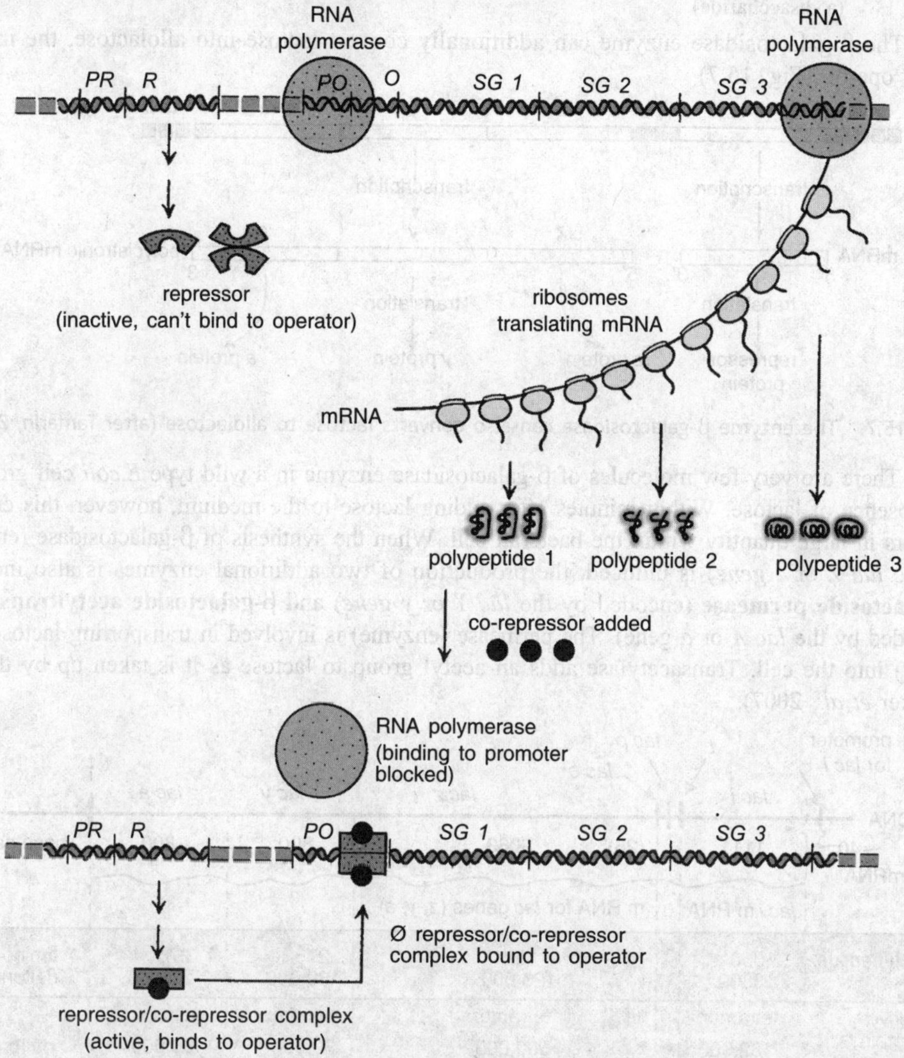

Fig. 15.6. Mode of regulation of gene expression for a repressible operon (*e.g., trp* operon). In this case, the repressor can only bind to the operator in the presence of effector molecule (called a co-repressor, for repressible operons; *e.g.,* tryptophan amino acid in *trp* operon). In its absence, the operator is free, permitting RNA polymerase to bind at the adjoining promoter (*PO*) and to initiate transcription of the structural genes. When co-repressor is added, it forms a complex with the repressor. This repressor – co-repressor complex then binds to the operator (*O*). This, in turn, prevents RNA polymerase from binding at *PO* and transcribing the three structural genes (after Gardner *et al.,* 2002).

Examples of Operons

1. *Lac* Operons (Inducible System)

Lactose (milk sugar, a disaccharide) is a β-galactoside that *E.coli* can use for energy and as a carbon source after it is broken down into glucose and galactose. The enzyme that performs the breakdown is **β–galactosidase:**

$$\text{Lactose} \xrightarrow{\beta\text{-galactosidase} + H_2O} \text{Galactose + Glucose}$$
(a disaccharide)

The β-galactosidase enzyme can additionally convert lactose into allolactose, the inducer of *lac* operon (Fig. 15.7)

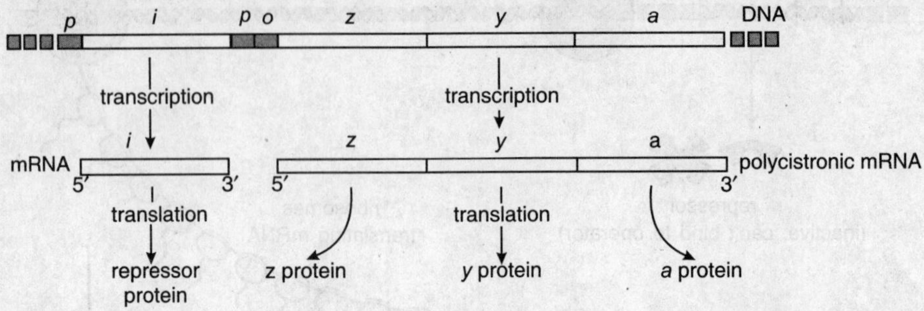

Fig. 15.7. The enzyme β-galactosidase can also converts lactose to allolactose (after Tamarin, 2002).

There are very few molecules of β-galactosidase enzyme in a wild type *E.coli* cell grown in the absence of lactose. Within minutes after adding lactose to the medium, however, this enzyme appears in large quantity within the bacterial cell. When the synthesis of β-galactosidase (encoded by the *lac Z* or *z gene*) is induced, the production of two additional enzymes is also induced, **β-galactoside permease** (encoded by the *lac Y* or *y gene*) and **β-galactoside acetyltransferase** (encoded by the *lac A* or *a* gene). The permease (enzyme) is involved in transporting lactose (Fig. 12.27) into the cell. Transacetylase adds an acetyl group to lactose as it is taken up by the cell (Becker *et al.*, 2007).

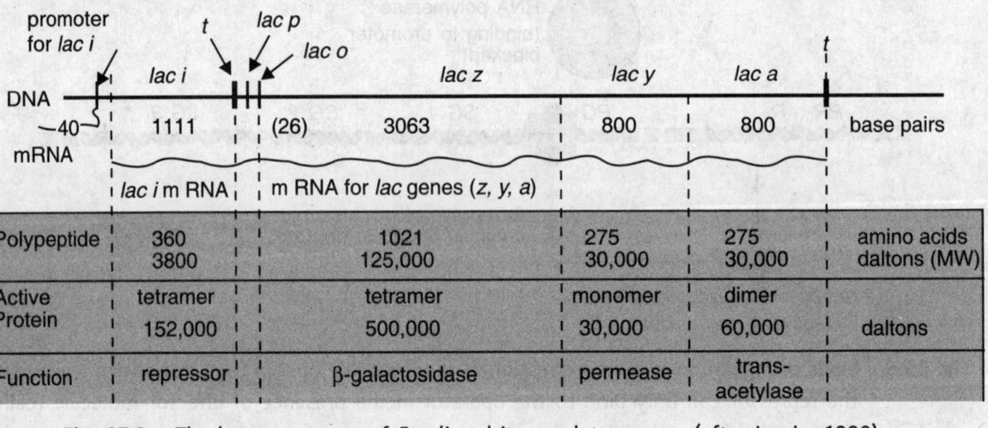

	lac i		lac z	lac y	lac a	
Polypeptide	360 3800		1021 125,000	275 30,000	275 30,000	amino acids daltons (MW)
Active Protein	tetramer 152,000		tetramer 500,000	monomer 30,000	dimer 60,000	daltons
Function	repressor		β-galactosidase	permease	trans-acetylase	

Fig. 15.8. The lactose operon of *E.coli* and its regulatory gene (after Lewin, 1990).

The regulator gene. Not only are three *lac* genes (*z*, *y* and *a*) induced together, but they are adjacent to one another in the *E.coli* chromosome; they are, in fact, transcribed on a single polycistronic messenger RNA (Fig. 15.8 and Fig. 15.9). Induction involves the protein product of another gene, the *regulator* gene, or *i* gene (*lac I*). Although the regulator gene is located adjacent to the three other *lac* genes (*i.e.,* three structural genes), it is totally independent transcriptional entity. The regulatory gene specifies a protein, the **repressor,** that interferes with the transcription of the genes involved in lactose metabolism.

The operator. For the repressor protein to exert its influence over transcription, there must be a control element (receptor site) located near the beginning of the β-galactosidase (*lac Z*) gene.

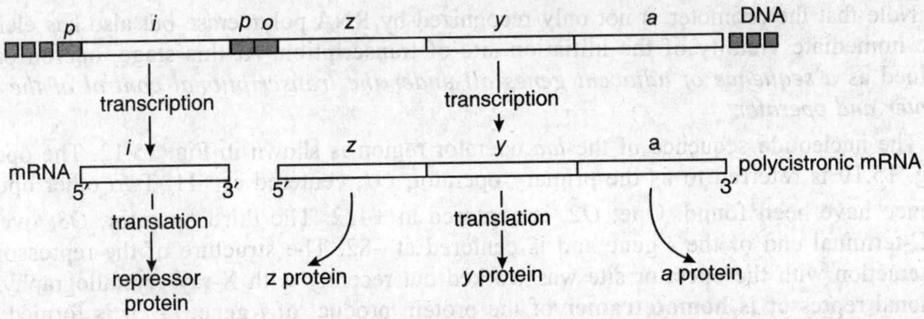

Fig. 15.9. The *lac* operon is transcribed as a multigenic (polycistronic) mRNA. The *z*, *y* and *a* indicate the *lac Z*, lac *Y* and *lac A* loci. The mRNA transcript is then translated as individual proteins. The regulator gene of *lac* operon is denoted as *i*; the *o* stands for operator and *p* for promoter. Both the operon and the regulator gene have their own promoters (*p*) (after Tamarin, 2002).

This control element is a region referred to as the **operator,** or operator site (Fig. 15.9). The operator site is a sequence of DNA that the product of the regulator gene, the repressor, recognizes. When the repressor is bound to the operator, it interferes with RNA polymerase binding or prevents

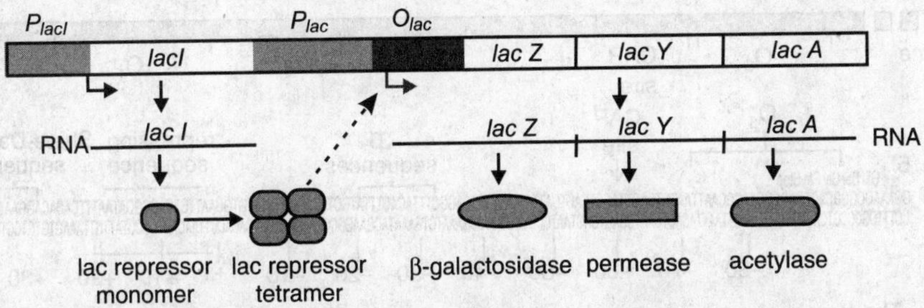

Fig. 15.10. Structure of the lactose operon.

the RNA polymerase from achieving the open complex. In either case, transcription of the operon is prevented (Fig. 15.11). The repressor is released when it combines with an **inducer,** a derivative of lactose called **allolactose** (Fig. 15.10).

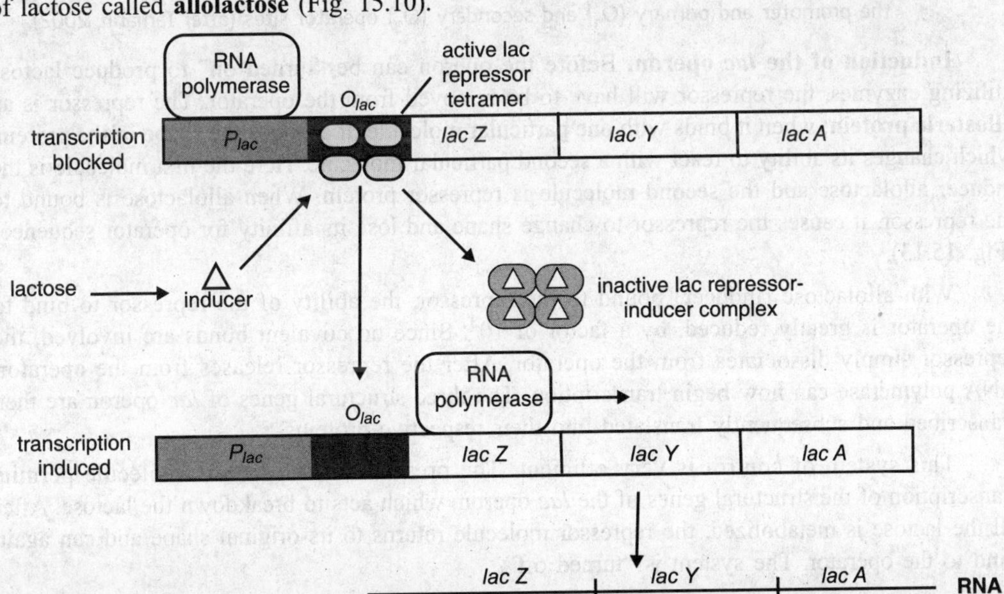

Fig. 15.11. Binding of inducer (allolactose) inactivates the *lac* repressor.

Note that the promoter is not only recognized by RNA polymerase but also has elements in the immediate vicinity of the initiation site of transcription. At this stage, operon can be redefined as *a sequence of adjacent genes all under the transcriptional control of the same promoter and operator.*

The nucleotide sequence of the *lac* operator region is shown in Fig. 15.12. The operator in Fig. 15.10 is referred to as the primary operator, *O1*, centered at +11. Two other operator sequence have been found. One, *O2*, is centered at +412. The third operator, *O3*, overlaps the C-terminal end of the *i* gene and is centered at –82. The structure of the repressor and its interaction with the operator site was worked out recently with X-ray crystallography. The functional repressor is **homotetramer** of the protein product of *i* gene; *i.e,* it is formed from four identical copies of the repressor protein. Since each operator site has two-fold symmetry, two repressor monomer proteins bind to each operator site. The monomer is shaped so that it fits into the major groove of the DNA to locate the exact base sequence of the operator; it then binds at the point through electrostatic forces. A tetramer can bind to two of the operator sites at the same time, presumably *O1 and O3* or *O1* and *O2*. In the process, the DNA is formed into a loop (Fig. 15.13).

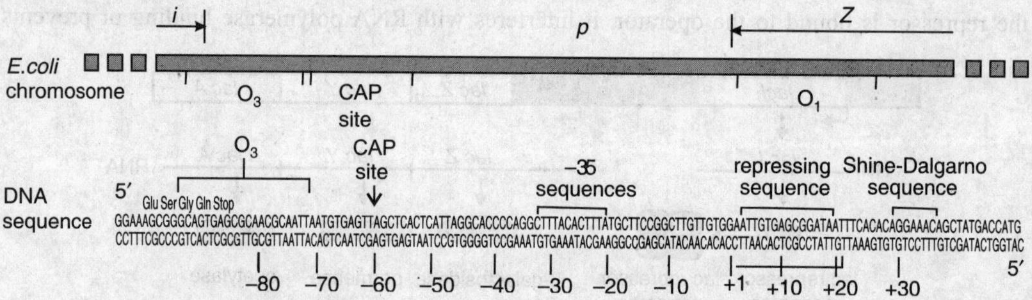

Fig. 15.12. The *lac* operon promoter and operator regions. The base sequence corresponds to the diagram above it. The terminal amino acids of the *i* gene are shown, as well as the initial amino acids of *lac Z* gene. In addition, have been depicted the Shine-Dalgarno sequence of the DNA, the repressor binding region (centered around +10 of the gene), the –10 and –35 sequences of the promoter and primary (O_1) and secondary (O_3) operator sites (after Tamarin, 2002).

Induction of the *lac* operon. Before the operon can be "turned on" to produce lactose utilizing enzymes, the repressor will have to be removed from the operator. The repressor is an **allosteric protein**; when it binds with one particular molecule, it changes the shape of the protein, which changes its ability to react with a second particular molecule. Here the first molecule is the inducer allolactose and the second molecule is repressor protein. When allolactose is bound to the repressor, it causes the repressor to change shape and lose its affinity for operator sequences (Fig. 15.13).

With allolactose (inducer) bound to the repressor, the ability of the repressor to bind to the operator is greatly reduced, by a factor of 10^3. Since no covalent bonds are involved, the repressor simply dissociates from the operator. After the repressor releases from the operator, RNA polymerase can now begin transcription. The three structural genes of *lac* operon are then transcribed and subsequently translated into their respective proteins.

This system of control is very efficient. The presence of the lactose molecule permits transcription of the structural genes of the *lac* operon, which acts to breakdown the lactose. After all the lactose is metabolized, the repressor molecule returns to its original shape and can again bind to the operator. The system is "turned off".

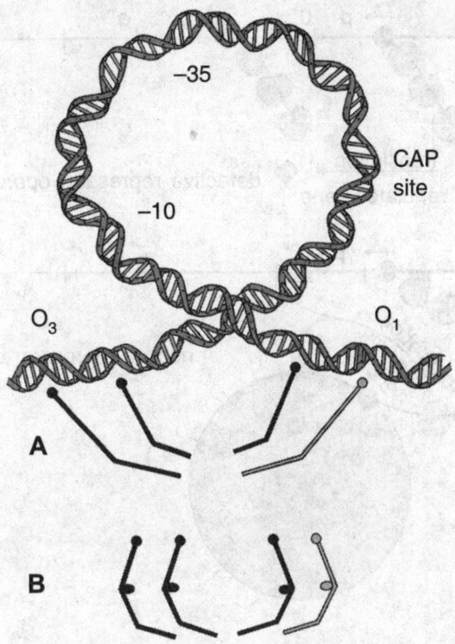

Fig. 15.13. Because the *lac* operator DNA sequences are palindromes (A *palindrome* has the same DNA sequence when one strand is read left to right in a 5′ to 3′ direction and the complementary strand is read right to left in a 5′ to 3′ direction), each half can bind one repressor subunit. A–The tetrameric repressor binds to O_1 and O_3, causing the DNA in between to form a loop. Each of the subunits has been shown in a different shade. The round portion of the subunit in touch with the DNA is the N-terminal end of the repressor subunit; the C–terminal ends form tails that bind the subunits together. Also indicated are the CAP site and the –10 and –35 sequences. B–When each of the subunits binds an allolactose molecule (black circles), the shape of the middle portion of the subunit changes, causing the subunit to fall free of the operators (after Tamarin, 2002).

Lac Operon Mutants (Genetic Manipulations of *Lac* Operon)

Verification of *lac* operon system came about through the use of mutants and partial diploids (merozygotes) of the *lac* operon well before DNA sequencing techniques had been developed. The structural (*i.e.,* enzyme specifying) genes of the *lac* operon, *z, y* and *a*, all have known mutant forms in which the particular enzyme does not perform its function. These mutant forms of the enzymes are z^+, y^+ and a^+.

Merozygotes or partial diploids in *E.coli* can be created through **sexduction** because some strains of *E.coli* have the *lac* operon incorporated into an F′ factor. Since F^+ strain can pass the F′ particle into F^- strain, *lac* operon diploids can be formed. By careful manipulation, various combinations of mutations can be looked at in diploid state.

Constitutive mutants are mutants in which the three *lac* operon structural genes are transcribed at all times, *i.e.*, they are not turned off even in the absence of lactose. Constitutive production of enzymes can come about in several ways. A defective repressor, produced by a mutant regulator gene, will not turn the system off, nor will a mutant operator that will no longer bind the normal repressor. The regulator constitutive mutants are designated i^-; the operator constitutive mutants are designated O^c. Both types of mutants produce the same phenotype: constitutive expression of the three *lac* operon genes.

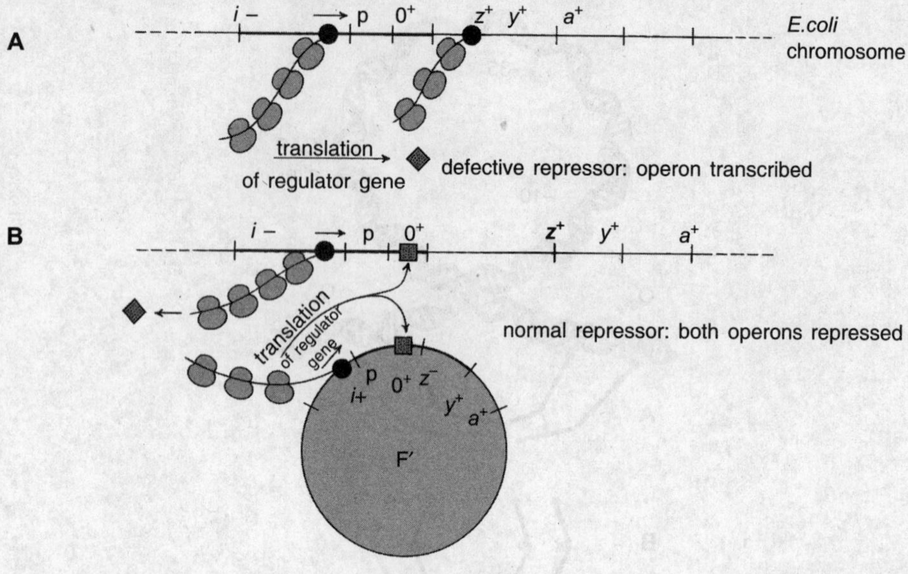

Fig. 15.14. A–A lac operon in *E.coli* with a mutation in the regulator gene (i^-). Transcription and translation of this gene yield a defective repressor; the cell, thus, has a constitutive production of the lac operon. B—The wild-type regulator gene (i^+) is introduced in an F′ factor; there is both a bacterial chromosome and an F′ factor each containing a regulator gene. (The F′ operon carries a mutant z^- allele, allowing geneticists to keep track of the transcriptional control of the chromosomal operon only). In this case, the phenotype is normal (inducible) because enough repressor is produced by the F′ allele (i^+), by transcription and translation, to bind to both operators. RNA polymerase enzyme is shown as solid spheres on the DNA; the wild-type repressor is shown as a square; the mutant repressor, which cannot bind to the operator, is shown as a diamond (after Tamarin, 2002).

When a new mutant is isolated, it is possible to determine whether it is caused by a regulator or operator mutation. For example, we can determine the exact location of a mutation on the bacterial chromosome by standard mapping techniques or, more recently, by DNA sequencing. Alternatively, the **Jacob** and **Monod** model predicts different modes of actions for the two types of mutations. In merozygotes, a constitutive operator mutation affects only the operon it is physically a part of. Operator mutations are therefore called **cis-dominant**. However, a constitutive *i*–gene mutation, since it works through an altered protein, is recessive to a wild–type regulator gene in the same cell, regardless of which operon (chromosomal or F′ factor) the mutation is on. Constitutive regulator mutations are, therefore, **transacting**. (*Note:* If two mutations are on the same piece of DNA, they are in the *cis* configuration. If they are on different pieces of DNA, they are in the *trans* configuration). *Trans*-acting mutations usually work through a protein product that diffuses through the cytoplasm. *Cis*-acting mutants are changes in recognition sequences on the DNA.

Thus, when the bacterium has a regulator constitutive mutation (i^-); the cell has constitutive production of the operon (Fig. 15.14A). If the wild-type regulator is introduced in an F′ plasmid (Fig. 15.14B), the normal (*i.e.,*inducible) phenotype is restored because the F′ i^+ allele is dominant to the chromosomal mutation – the i^+ gene regulates both the chromosomal and F′ operons. Hence, both operons are inducible. We do not need to be bothered about the other components of the F′ plasmid, because it carries a z^- allele; only the activity of the chromosomal operon will be observed. When, the chromosomal operon carries an operator constitutive mutation; the cell also

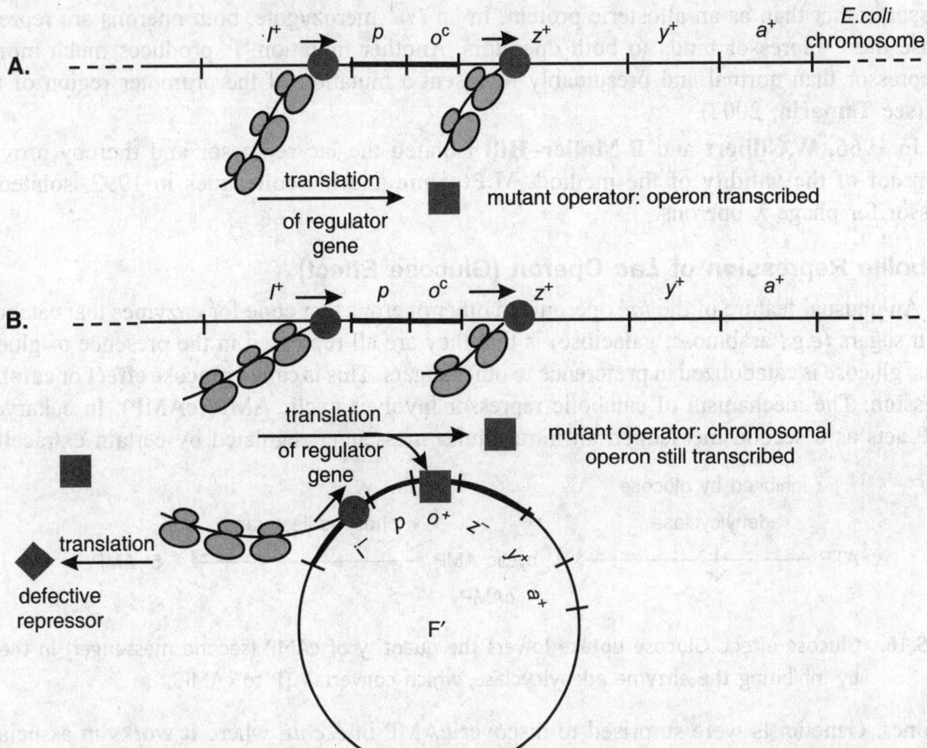

Fig. 15.15. A–A *lac* operon in *E.coli* with a mutation of the operator (O^c). The cell has a constitutive phenotype; the operator cannot bind the wild type repressor protein, and thus, transcription is continuous, even in the absence of lactose. B–The phenotype is unchanged even when a wild-type operator is introduced into the cell in a F′ factor; there is both a bacterial chromosome and a F′ factor; there is both a bacterial chromosome and a F′ factor, each containing an operator. (The F′ operon carries mutant regulator (i^-) and z^-alleles, allowing a geneticist to keep track of the chromosomal operon only). The F′ operator does not change the phenotype of the cell because the wild-type operator exerts no control over the chromosomal operator, which exerts *cis–dominant effect*, another operator on another operon has no effect. RNA polymerase enzyme is shown as solid spheres on the DNA; the wild-type repressor is shown as a square; the mutant repressor, which cannot bind to the operator is shown as diamond (after Tamarin, 2002).

has constitutive production of the operon (Fig. 15.15A). When a wild-type operator is introduced into the cell in an F′ plasmid (Fig. 15.15B), the cell still has the constitutive phenotype because the operator allele on the F′ plasmid does not control the bacterial operon; the *lac* operon on the bacterial chromosome will be continually transcribed. The chromosomal operon has a **cis-dominant** operator mutation that has a constitutive phenotype. One should also note that only the bacterial chromosome determines the phenotype because the introduced F′ plasmid has a z^- allele.

Other Types of *Lac* Operon Control Mutations

Jacob and Monod's operon model has been supported by some other types of mutations. For example, a super-repressed mutation, i^s, was discovered. This mutation represses the operon even in the presence of large quantities of the inducer. Thus, the repressor seems to have lost the ability to recognize the inducer. Basically, the *i* gene product is acting as a constant repressor, rather than as an allosteric protein. In an i^s/i^+ merozygote, both operons are repressed because the i^s repressor binds to both operators. Another mutation, i^Q,produces much more of the repressor than normal and presumably represent a mutation of the promoter region of the *i* gene (see **Tamarin**, 2002).

In 1966, **W.Gilbert** and **B.Muller–Hill** isolated the lac repressor and thereby provided final proof of the validity of the method. **M.Ptashne** and his colleagues in 1992 isolated the repressor for phage λ operons.

Catabolite Repression of *Lac* Operon (Glucose Effect)

An unusual feature of the *lac* operon and other operons that code for enzymes that catabolize certain sugars (*e.g.,* arabinose, galactose) is that they are all repressed in the presence of glucose. That is, glucose is catabolized in preference to other sugars. This is called **glucose effect** or **catabolic repression**. The mechanism of catabolic repressor involves cyclic AMP (cAMP). In eukaryotes, cAMP acts as a second messenger, an intracellular messenger regulated by certain extracellular

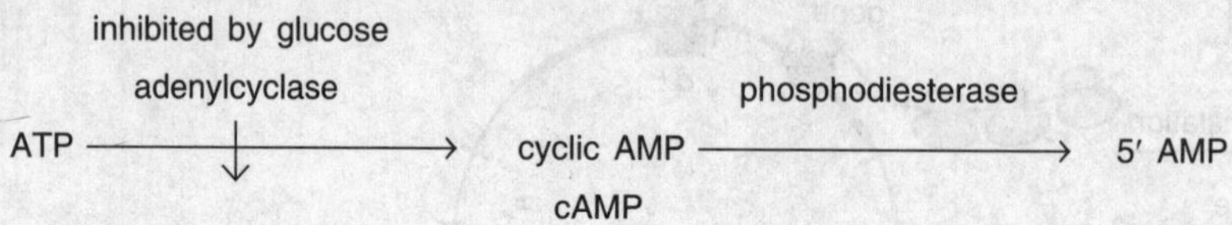

Fig. 15.16. Glucose effect. Glucose uptake lowers the quantity of cAMP (second messenger) in the cell by inhibiting the enzyme adenylcyclase, which converts ATP to cAMP.

hormones. Geneticists were surprised to discover cAMP in *E.coli*, where it works in association with another regulatory protein, the **catabolite activator protein** (**CAP**), to control the transcription of certain operons.

When glucose is not present, cAMP combines with CAP, and the CAP-cAMP complex binds to a distal part of the promoter of operons with CAP sites (*e.g.,* the *lac* operon; see Fig. 15.17A). This binding clearly increases the affinity of RNA polymerase for the promoter, because without the binding of the CAP-cAMP complex to the promoter, the transcription rate is very low. The uptake of glucose by *E.coli* cells causes the loss of cAMP from the cell, probably by inhibiting adenylcyclase (Fig. 15.16, and thus lowers the CAP-cAMP level. The transcription rate of operons with CAP sites will be reduced (Fig. 15.17B). The same reduction of transcription rates has been observed in mutant strains of *E.coli* when this part of the distal end of the promoter is deleted. The binding of CAP-cAMP to the CAP site causes the DNA to bend more than 90 degrees. This bending by itself, may enhance transcription, making the DNA more available to RNA polymerase enzyme.

During process of initiation of transcription, the CAP is in direct contact with RNA polymerase. This was shown by photocross-linking studies in which CAP was treated with a cross-linking agent that bound the subunit of RNA polymerase when irradiated with UV light. (**Photocrosslinking** is a technique used to determine which moieties (proteins, DNA) are in close proximity during a particular process). For the two proteins to cross-link, they must be in direct contact during the initiation of transcription.

Catabolite repression is an example of positive regulation: Binding of the CAP-cAMP complex at the CAP site enhances the transcription rate of the transcriptional unit. Thus, the *lac* operon is both positively and negatively regulated; the repressor exerts negative control, and the CAP-cAMP complex exerts positive control of transcription.

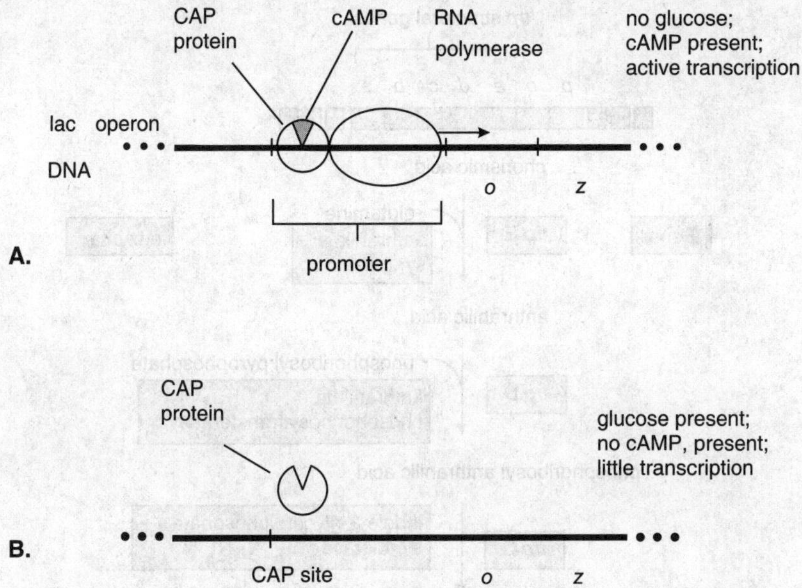

Fig. 15.17. Catabolite repression. When cAMP is present in the cell (no glucose is present), it binds with CAP protein and together they bind to the CAP site in various sugar metabolizing operons, such a *lac* operon shown here. The CAP-cAMP complex increases the transcription of the operon. When glucose is present, it inhibits the formation of cAMP. Thus no CAP-cAMP complex forms, and transcription of the same operon is reduced (after Tamarin, 2002).

2. *TRP* Operon

The **catabolic operons** or inducible operons are activated when the substrate that is to be catabolized enter the cell. **Anabolic operons** function in the reverse manner. They are turned off (repressed) when their end product accumulates beyond the needs of the cell. Two entirely different mechanisms seem to control the transcription of repressible operons. The *first* mechanism follows the basic scheme of inducible operons and involves the end product of the pathway. The *second* mechanism involves secondary structure in messenger RNA transcribed from an attenuator region of the operon.

(1) Tryptophan synthesis. Tryptophan operon is an example of a repressible system. It contains the five genes that code for the synthesis of the enzymes that build tryptophan amino acid, starting from **chorismic acid** (Fig. 15.18). It has a promoter operator sequence (*p, o*) as well as its own regulator gene (*trpR*).

(2) Operator control. In *trp* operon, the product of *trpR* gene is called **repressor**. It is inactive by itself and it does not recognize the operator sequence of *trp* operon. The repressor only become active when it combines with tryptophan. Thus, when tryptophan builds up, enough is available to bind with and activate the repressor. Tryptophan is, thus, called **corepressor**. The corepressor-repressor complex then recognizes the operator, binds to it, and prevents transcription by RNA polymerase.

After the available tryptophan in the cell is used up, the diffusion process causes tryptophan to leave the repressor, which then detaches from the *trp* operator. Since, the transcription process no longer is blocked, so proceeds normally (*i.e.,* the operon is now derepressed). Transcription continues until enough of the various enzymes are synthesized to again produce an excess of tryptophan. Some tryptophan molecules become available to bind to the repressor and make a

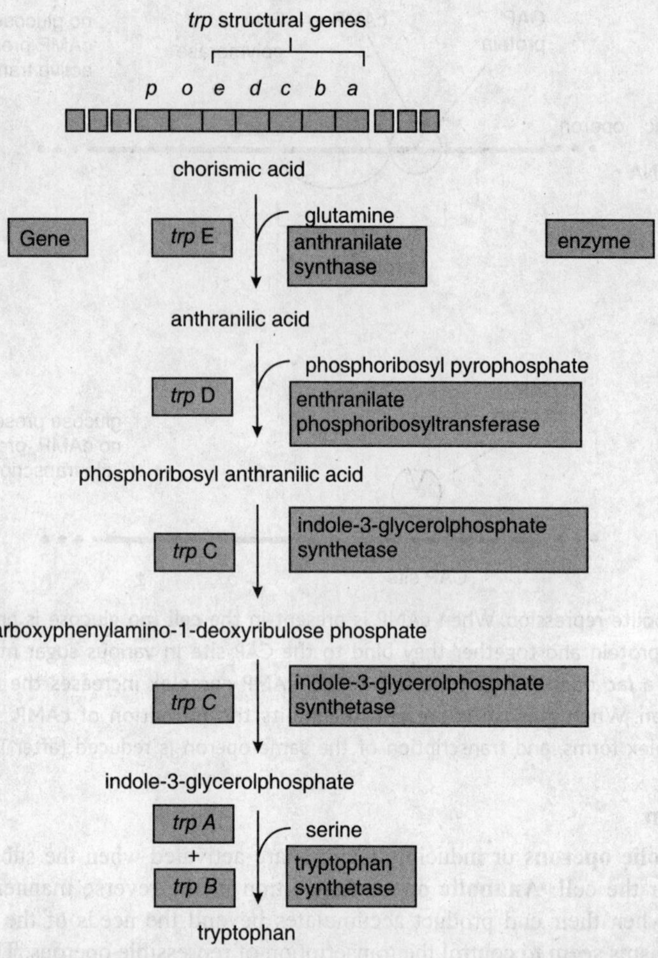

Fig. 15.18. Genes of tryptophan operon (*trp* operon) in *E.coli*. The enzymes they produce control the conversion of chorismic acid to tryptophan. The symbol *o* on the chromosome refers to the *trp* operator, which has its own repressor, the product of the *trpR* gene (after Tamarin, 2002).

functional complex, and the operon is again shut off. This process is repeated to ensure that tryptophan is being synthesized as needed (Fig. 15.19). This regulation is modified, however, by the existence of the second mechanism for regulating repressible operons–the attenuation.

(3) Attenuator–controlled trp operon: Details of the second control mechanism of repressible operons have been explained primarily by **C.Yanofsky** and his colleagues (1984), who worked with the tryptophan operon in *E.coli*. This type of operon is controlled by its own **attenuator region** and similar system is also found in operons involved in synthesis of other amino acids such as leucine and histidine. (**Attenuation** means "premature termination").

***(i)* Leader transcript**: In the *trp* operon, an attenuator region lies between the operator and the first structural gene (Fig. 15.20). The messenger RNA (mRNA) transcribed from the attenuator region, termed the **leader transcript,** has been sequenced, revealing two surprising and interesting facts.

1. Four subregions of the mRNA have base sequences that are complementary to each other so that three different stem-loop structures can form in the mRNA (Fig. 15.21). Depending on conditions, regions 1-2 and 3-4 can form two stem loop structures, or region 2-3 can form a single stem-loop. When one stem-loop structure is formed, the others are acquired. As we will observe, the particular combination of stem-loop structures determines whether transcription continues.

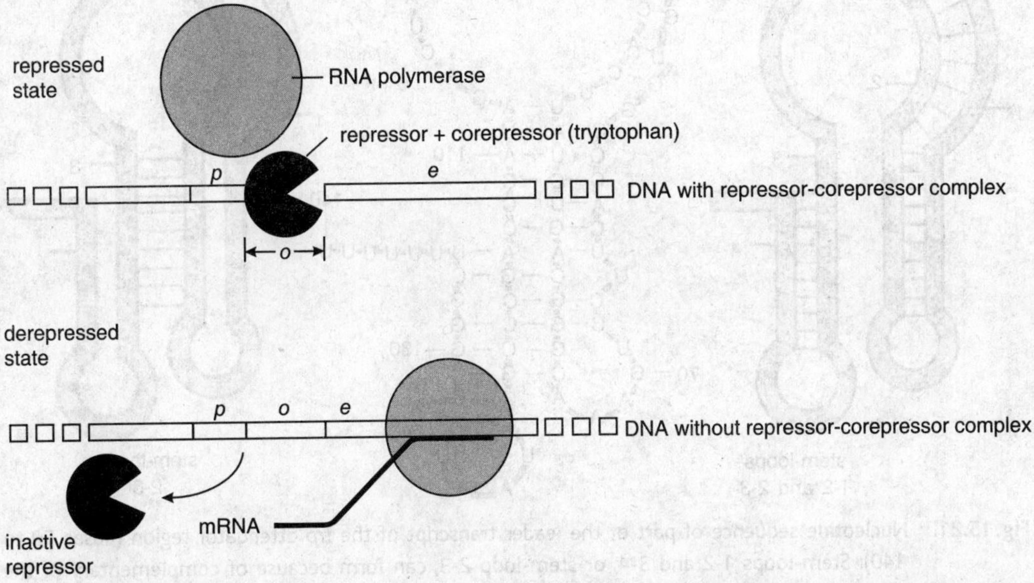

Fig. 15.19. The repressor-corepressor complex binds at the operator and prevents the transcription of the *trp* operon in *E.coli.* Without the co-repressor, the repressor cannot bind, and therefore transcription is not prevented (after Tamarin, 2002).

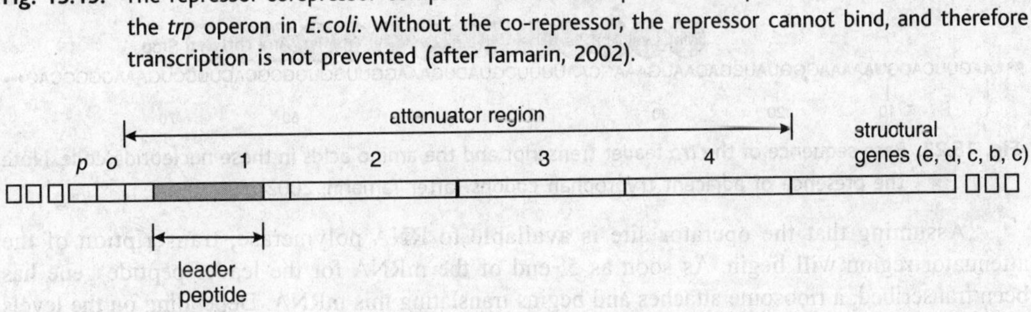

Fig. 15.20. Attenuator region of the *trp* operon, which contains the leader peptide gene. This region is transcribed into the leader transcript (after Tamarin, 2002).

***(ii)* Leader peptide gene**: Another fact has been obtained by sequencing the leader transcript and it has been found that there is a small gene coding infomation for a peptide from base 27 to 68 (Fig. 15.22). The gene for this peptide is called **leader peptide gene**. It codes for fourteen amino acids including two adjacent tryptophans. These adjacent tryptophan codon are critically important in attenuator regulation. The proposed mechanism for this regulation follows.

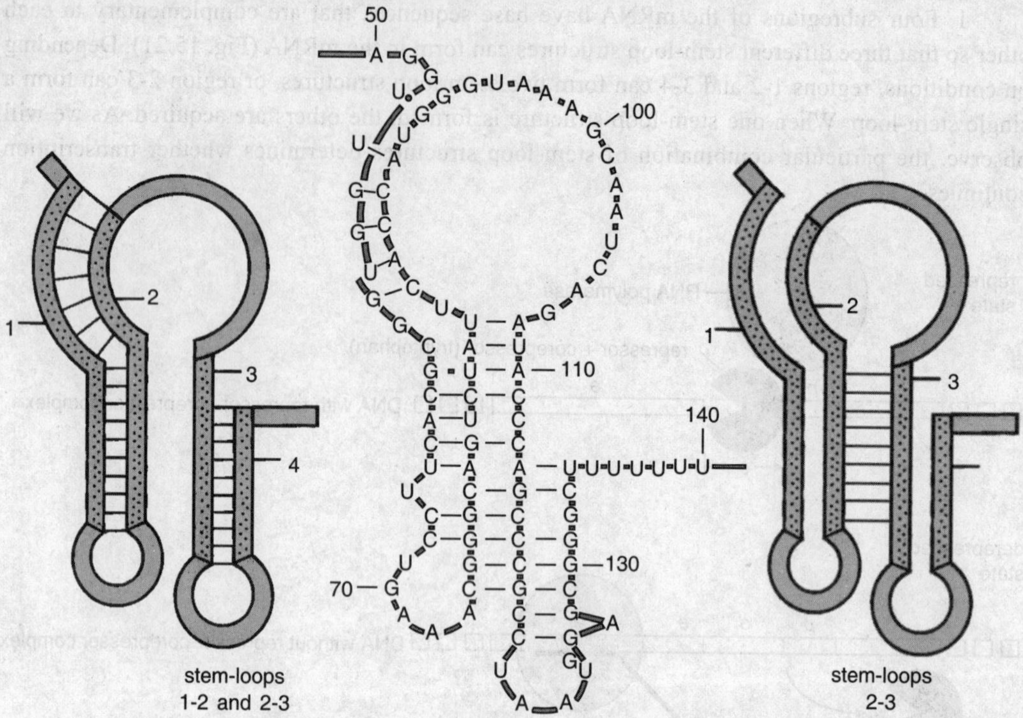

Fig. 15.21. Nucleotide sequence of part of the leader transcript of the *trp* attenuator region (bases 50 to 140). Stem-loops 1-2 and 3-4, or stem-loop 2-3, can form because of complementary of the nucleotides. All possible base pairings are shown in the middle of the figure (after Tamarin, 2002).

Fig. 15.22. Base sequence of the *trp* leader transcript and the amino acids in these nucleotide code. Note the presence of adjacent tryptophan codons (after Tamarin, 2002).

Assuming that the operator site is available to RNA polymerase, transcription of the attenuator region will begin. As soon as 5′ end of the mRNA for the leader peptide gene has been transcribed, a ribosome attaches and begins translating this mRNA. Depending on the levels of amino acids in the cell, *three* different outcomes can take place.

(*a*) Excess tryptophan: If the concentration of tryptophan in the cell is such that abundant tryptophenyl-tRNAs exist, translation proceeds down that leader peptide gene. The moving ribosome overlaps region 1 and 2 of the transcript and allows stem-loop 3-4 to form (Fig. 15.23A). This stem-loop structure is called **terminator** or **attenuator stem,** causes transcription to be terminated. Note that stem-loop 3-4, the terminator stem, followed by a series of uracil (U)–containing bases, is a *rho* independent transcription terminator. Hence, when existing quantities of tryptophan, in the form of tryptophenyl-tRNA, are adequate for translation of the leader peptide 'gene', the transcription is terminated.

(*b*) Tryptophan starvation: If the quantity of tryptophenyl tRNA is lowered, the ribosome must wait at the first tryptophan codon until it acquires a Trp-tRNA^{Trp} (Fig. 15.23B). The stalled ribosome will permit stem-loop 2-3 to form, which precludes the formation of the terminator stem-loop (3-4). In this configuration, transcription is not terminated, so that eventually, the whole

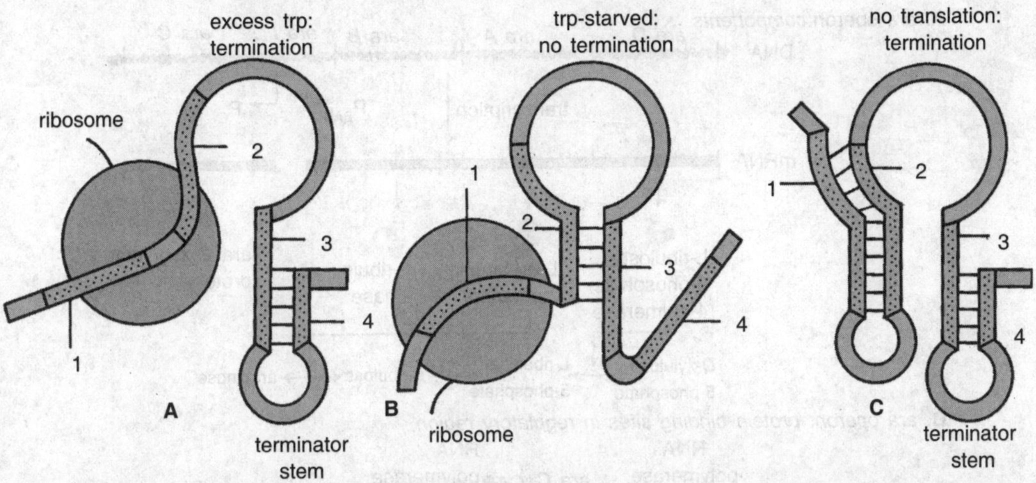

Fig. 15.23. Model for attenuation in the *E.coli trp* operon (**Oxender** *et al.*, 1979). The circle represents the ribosome attempting to translate the leader transcript of Figure 15.21. Under conditions of *excess tryptophan*, the 3-4 stem-loop forms (the terminator stem), terminating transcription. Under conditions of *tryptophan starvation*, the ribosome is stalled, and stem-loop 2-3 forms, allowing continued transcription. Under conditions of *general starvation*, there is no translation, resulting in the formation of stem-loops 1-2 and 3-4, which again results in the termination of transcription (after Tamarin, 2002).

operon is transcribed and translated raising the level of tryptophan in the cell. The stem-loop 2-3 structure is referred to as the **pre-emptor stem**. This is important that the pre-emptor stem is not a rho-independent transcription terminator and thus, without the rho protein present, will not terminate transcription.

***(c)* General starvation**: A final configuration (Fig. 15.23C) is possible. Here, no ribosome interferes with stem formation and presumable stem-loops 1-2 and 3-4 (terminator) form. This configuration also terminates transcription because of the terminator stem. It is believed that this configuration occurs if the ribosome is stalled on the 5′ side of the *trp* codons, which happens when the cell is starved for other amino acids. Presumably, it makes no sense to manufacture tryptophan when other amino acids are in short supply. Hence, the cell can carefully bring up the levels of the various amino acids in the most efficient manner.

The tryptophan operon in bacilli such as *Bacillus subtilis*, is also controlled by attenuation, but secondary structure in the mRNA transport is induced by binding not the ribosome, but a **trp RNA-binding attenuation protein (TRAP) (Antson** *et al.*, 1999).

3. The *ara* Operon

Almost complete mechanism by which the *lac* and *trp* operon of *E.coli* are regulated, are known and supported by an extensive body of experimental data. However, *arabinose (ara)* operon of *E.coli* exhibits much more complex patterns of regulation that are still not completely understood. In the *lac* and *trp* operons, the product of the regulator gene, the repressor, functions in a *negative* manner, turning off transcription of the operon. On the other hand, the catabolic activator protein (CAP) exerts a *positive* control over the *lac* operon by stimulating transcription of the operon. The major regulatory protein of the *ara* operon exhibits both negative and positive regulatory effects on the transcription of the structural genes of the operon, depending on the environmental conditions. Moreover, the regulatory components that control transcription of the *ara* operon include one element that acts from a distance of over 200 nucleotide pairs away from the promoter that it helps to control.

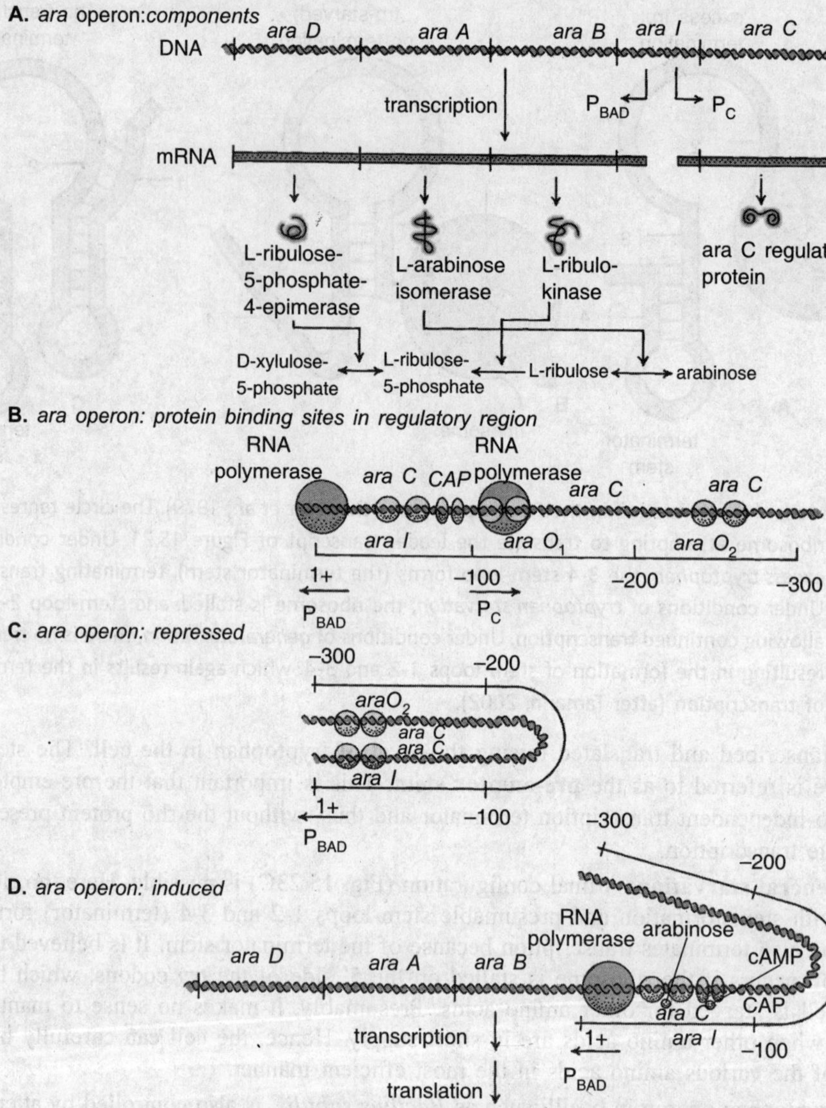

Fig. 15.24. Structure and complex regulation of the arabinose (*ara*) operon of *E.coli*. A–The *ara B*, *ara A* and *ara D* structural genes encode the three enzymes that catalyze the catabolism of arabinose to D-xylulose-5-phosphate. Cotranscription of these three structural genes is initiated at a promoter called P_{BAD}. The *ara C* gene encodes a protein that regulates the transcription of the *ara BAD* operon. The *ara C* protein acts as a repressor of the operon in the absence of arabinose and cAMP; however, it functions as an activator of transcription in the presence of arabinose and cAMP. Transcription of the *ara C* gene is initiated at a promoter called Pc. Transcript extension occurs in opposite directions from the P_{BAD} and Pc promoters. The *ara I* region contains the binding sites for the *ara C* regulator protein and the cAMP–CAP complex. B–Enlargement of the *ara I* region showing the positions of the binding sites for *ara C* protein, cAMP-CAP and RNA polymerase, and the location of ara *C* protein binding site at ara O_2. C–Model for the repression of the *ara* operon. *Ara C* protein dimers bind at the *ara I* and *ara* D_2 sites, and then bind to each other to form a DNA loop. RNA polymerase cannot bind at P_{BAD} to initiate transcription when the DNA is in the looped conformation. D–Induction of the *ara* operon takes place when the arabinose–*ara C* protein and cAMP–CAP complexes both bind at their *ara I* sites. When these complexes are bound at their *ara I* sites, the DNA loop structure cannot form, and RNA polymerase is able to bind at P_{BAD} and initiate transcription of the *ara B*, *ara A* and *ara D* structural genes (after Gardner *et al.*, 2002).

The **arabinose (*ara*) operon** of *E.coli* contains three structural genes (*ara B, ara A* and *ara D*) that encode the three enzymes involved in the catabolism of arabinose (Fig. 15.24A). These three genes are cotranscribed on a single mRNA that is initiated at a promoter called $\mathbf{P_{Bad}}$. Active transport of arabinose into cells is carried out by the products of genes *ara E, ara F,* and *ara G*. These genes are located at sites quite distant from the *ara BAD* operon of interest here and will not be further discussed. The major regulatory protein of the *ara* operon (*i.e.*, the *ara C protein*) is produced from a transcript that is initiated at a promoter called **Pc**. The Pc promoter is only slightly over 100 nucleotide pairs away from P_{BAD}, but the two promoters initiate transcription in opposite directions (Fig. 15.24 A and B). The *ara C* protein acts as a *negative regulator* (a repressor) of transcription of the *ara B, ara A* and *ara D* structural genes from the P_{BAD} promoter in the absence of arabinose and cAMP. It acts as a positive regulator (an activator) of transcription of these genes from the P_{BAD} promoter when arabinose and cAMP are present. Thus, depending on the presence or absence of the effector molecules arabinose and cAMP, the *ara C* regulatory gene products may exert either a positive or a negative effect on transcription of the *ara B, ara A* and *ara D* structural genes. Since the *ara* operon is subject to catabolite repression like the *lac* operon and thus to positive control by CAP and cAMP, induction of the *ara* operon depends on the positive regulatory effects of two proteins, the *ara C* protein and CAP. The binding sites for these two proteins and for RNA polymerase all appear to be located in a region of the *ara* operon historically called *ara I* (*I* for induction), located between the three structural genes of the operon and the regulatory gene (*ara C*) (Fig. 15.24A).

Once geneticists working on the problem of regulation of the *ara* operon thought that all the binding sites for the *ara C* regulatory protein and cAMP-CAP complex were in the *ara I* region. The surprising discovery was that repression of the *ara* operon depended on the binding of *ara C* protein at a site called *ara* O_2 (*O* for operator, *2* because it was the *second ara operator* identified) located 211 nucleotide-pairs upstream (relative to the direction of the transcription from P_{BAD}) from the *ara C* protein binding site in *ara I* (Fig. 15.24). (*Note:* operator *ara* O_1–the first *ara* operator to be identified–controls the transcription of the *ara C* regulator gene initiated at Pc). The currently accepted model for the repression of the *ara* operon is that the *ara C* protein must bind (as a dimer) at both the *ara I* site and the *ara* O_2 site, and that these proteins then bind to each other to form a DNA loop (Fig. 15.24C).

When the loop structure is formed, it must prevent or interfere with the binding of RNA polymerase at the adjacent promoter (P_{BAD}) of the operon (Fig. 15.24C). In the presence of arabinose and cAMP, the *ara* operon is induced. Moreover, under these conditions, the *ara C* protein has been shown to become an activator of transcription of the operon. The arabinose–*ara C* protein complex and the cAMP-CAP complex must open the loop by binding at their *ara I* sites. This, in turn must permit RNA polymerase to bind at the P_{BAD} site and initiate transcription of the *ara* structural genes (Fig. 15.24D).

Mode of Working of Gene Regulatory Proteins

Gene regulatory proteins must recognize specific nucleotide sequences embedded within its structure. It was originally thought that these proteins might require direct access to the hydrogen bonds between base pairs in the interior of the double helix to distinguish between one DNA sequence and another. It is now clear, however, that the outside of the double helix is studded with DNA sequence information that gene regulatory proteins can recognize without having to open the double helix.

Molecular Recognition

Gene regulatory proteins are found to contain structural motifs that read DNA sequences. Molecular recognition in biology generally relies on an exact fit between the surfaces of two molecules, and the study of gene regulatory proteins has provided some of the clearest example of this principle (see **Alberts** *et al.,* 2002). A gene regulatory protein recognizes a specific DNA sequence because the surface of the protein is extensively complementary to the specific surface features of the double helix in that region. In most cases, the protein makes a large number of contacts with the DNA, involving hydrogen bonds, ionic bonds and hydrophobic interactions. Although individual contact is weak, the 20 or so contacts that are typically formed at the protein – helix interface add together to ensure that the interaction is both highly specific and very strong (Fig. 15.25). In fact, DNA-protein interactions include some of the highest and most specific molecular interactions known in biology. DNA-protein interactions are of the following two types.

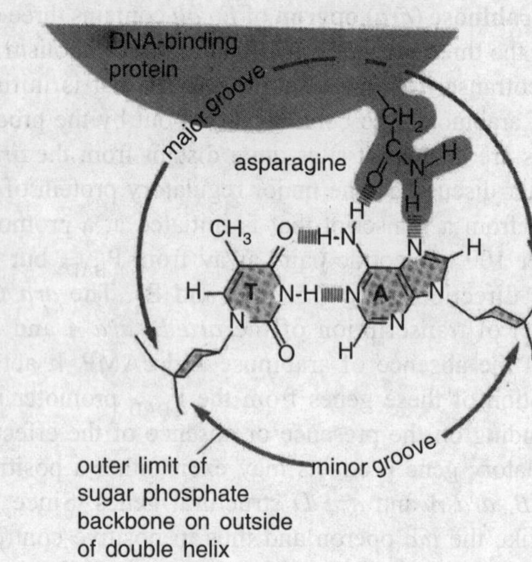

Fig. 15.25. The binding of a gene regulatory protein to the major groove of DNA. Only a single contact is shown. Typically, the protein DNA interface would consist of 10 to 20 such contacts, involving different amino acids, each contributing to the strength of the protein DNA interaction (after Alberts *et al.,* 2002).

Helix-turn-Helix Motif

The first DNA-binding protein motif to be recognized was the helix-turn-helix. This was originally identified in bacterial proteins; but later on, this motif has been reported in hundreds of DNA-binding proteins from both eukaryotes and prokaryotes. This protein is constructed from two α-helices connected by a short extended chain of amino acids, which constitutes the "turn" (Fig. 15.26). The two helices are held at a fixed angle, primarily through interactions between the two helices. The C-terminal helix is called the recognition helix because it fits into the major grooves of DNA; its amino acid side chains, which differ from protein to protein, play an important role in recognizing the specific DNA sequence to which the protein binds. Examples of helix-turn-helix proteins include tryptophan repressor, lambda cro, lambda repressor fragment, CAP fragment, etc.

Homeodomain Protein

Soon after the discovery of gene regulatory proteins in bacteria, genetic analyses in the fruitfly *Drosophila* led to the characterisation of an important class of genes, the **homeotic selector genes**, that play a critical part in coordinating fly development. Mutations in these genes cause one body part in the fly to be converted into another, showing that the proteins they encode control critical developmental decisions.

When the nucleotide sequence of several homeotic selector genes of *Drosophila* and higher animals was determined in the early 1980s, each proved to contain an almost identical stretch of 60 amino acids that defines this class of proteins and is termed as the **homeodomain.** When the three–dimensional structure of homeodomain was determined, it was shown to contain a helix-turn-helix motif related to that of the bacterial gene regulatory proteins, providing one of the first indications that the principle of gene regulation established in bacteria are relevant to higher organisms as well. More than 60 homeodomain proteins now have been discovered in *Drosophila* alone, and homeodomain proteins have been identified in virtually all eukaryotic organisms that have been studied, from yeasts to plants to humans.

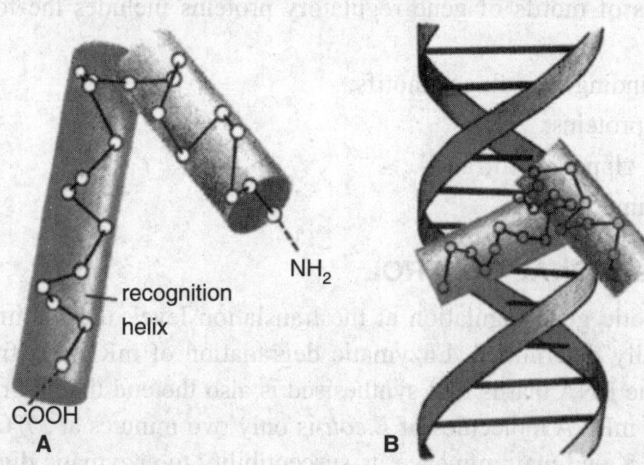

Fig. 15.26. The DNA-binding helix-turn-helix motif. A–Motif with circle which represents the central carbon of an amino acid. The C-terminal α-helix is called the recognition helix because it participates in sequence-specific recognition of DNA. B–The helix of motif fits into the major groove of DNA, where it contacts the edges of the base pairs. The N–terminal α-helix functions primarily as a structural component that helps to position the recognition helix (after Alberts *et al.*, 2002).

The structure of a homeodomain that is bound to its specific DNA sequence, has been shown in Figure 15.27. Where as the helix-turn-helix motif of bacterial gene regulatory proteins is often

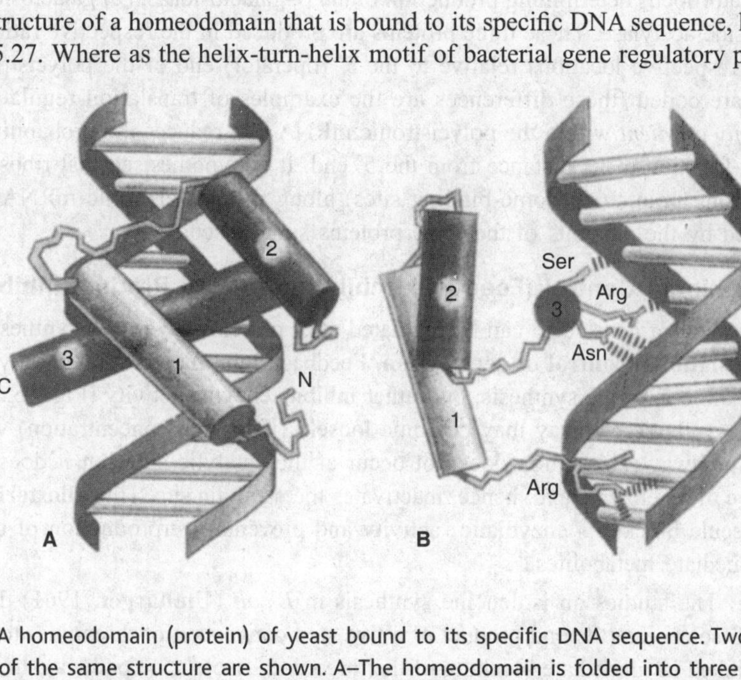

Fig. 15.27. A homeodomain (protein) of yeast bound to its specific DNA sequence. Two different views of the same structure are shown. A–The homeodomain is folded into three α-helices which are packed tightly together by hydrophobic interactions. The part containing helix 2 and 3 closely resembles the helix-turn-helix motif. B–The recognition helix (helix 3) makes important contacts with the major groove of DNA. The asparagine (Asn) of helix 3, for example, contacts an adenine. Nucleotide pairs are also contacted in the minor groove by a flexible arm attached to helix I (after Alberts *et al.*, 2002).

embedded in different structural contexts, the helix-turn-helix motif of homeodomains is always surrounded by the same structure, suggesting that the motif is always presented in the same way.

Other types of motifs of gene regulatory proteins includes the following (see **Alberts** *et al.,* 2002).

1. DNA–binding zinc finger motifs;
2. β-sheet proteins;
3. Leucine zipper motif; and
4. Heterodimerization.

15.3 TRANSLATIONAL CONTROL

In prokaryotic gene regulation at the translation level, the lifetime of a mRNA molecule may be genetically determined. Enzymatic degradation of mRNA is from the 5′ to the 3′ end, *i.e.,* the end of the RNA that is first synthesised is also the end that is first degraded. The average lifetime of many mRNA molecules of *E.coli* is only two minutes at 37°C. The specific nucleotide sequences at the 5′ end may influence its susceptibility to enzymatic digestion. Further, catabolic enzymes are denied access to the mRNA when the ribosome are coated at their 5′ ends (*i.e.,* in case of polyribosomes). Hence, the lifetime of mRNAs may also be correlated with the number of free ribosomes available at any given moment to translate mRNA molecules. Bacteria vary their rates of proteins synthesis by varying their ribosomal content rather than by varying the translational rate.

Example: In the lactose system of *E.coli*, there are three structural genes under control of a common operator locus determining production of (1) β–galactosidase, (2) galactoside permease and (3) galactoside acetylase. These three proteins are produced in the respective ratios 1:1/2:1/5, reflecting their respective locations relative to the 5′ (operator) end of the polycistronic mRNA in which they are coded (these differences are the examples of translation regulations). Thus, there is a *polarity gradient* within the polycistronic mRNA that reduces the probability of cistron translation as a function of its distance from the 5′ end. It is hypothesised that ribosomes attach to different starting points (ribosome-binding sites) along the polycistronic mRNA at different rates as reflected by the amounts of the three proteins synthesised.

1. Post-translational Control (Feedback Inhibition or End Product Inhibition)

The expression of genes also can be regulated after proteins have been synthesised. This is called **post-translational control of gene action**. Feedback inhibition is a regulatory mechanism which does not affect enzyme synthesis, but rather inhibits enzyme activity (Fig. 15.28). The end product of a biosynthetic pathway may combine loosely (if in high concentration) with the first enzyme in the pathway. This union does not occur at the catalytic site, but it does modify the tertiary structure of an enzyme and, hence, inactivates the catalytic site. This **allosteric transition** of protein molecule blocks its enzymatic activity and prevents overproduction of end products and their intermediate metabolites.

Example: The studies on isoleucine synthesis in *E.coli* (**Umbarger,** 1961) demonstrated that addition of isoleucine (the end product of a five step conversion of threonine) to a culture of the bacteria resulted in immediate blocking of the threonine→isoleucine pathway. In the presence of added isoleucine, the cells preferentially use this exogenous end product (i.e., isoleucine) and their own isoleucine synthesis becomes ceased. Moreover, the production of each of the five enzymes is not interfered with, but action of an enzyme responsible for deamination of threonine to α-ketobutyrate is inhibited by the end product, isoleucine.

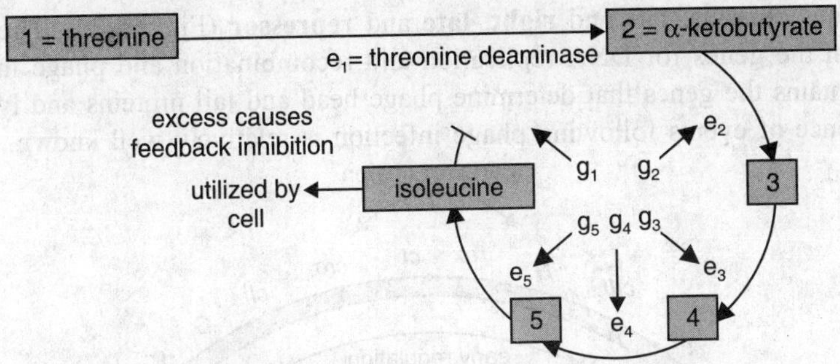

Fig. 15.28. Feedback inhibition in *E.coli.*

15.4 OPERONS OF PHAGE LAMBDA (REGULATION OF PHAGE λ LIFE CYCLES)

When a bacteriophage infects a bacterial cell, it has to express its genes in an orderly fashion; some gene products are needed early in infection, and other products are required during later part of infection. **Early genes** usually control phage DNA replication; **late genes** usually determine phage coat proteins and the lysis of bacterial cell. A bacteriophage is considered most efficient if it expresses early genes first and the late genes last in the infection process. Also, temperate phages have the option of entering into lysogeny with the cell (see Box 15.2), here, also, control processes determine which path is taken. One generalisation has been made regarding the phages that their genes are clustered into early and late operons, with separate transcriptional control mechanisms for each.

Box 15.2

Lysis: Bursting of a cell by the destruction of plasma membrane following an infection by a virus.
Lysogenic: The state of a bacterial cell that has an integrated phage (prophage) in its chromosome.
Lysogenic bacteria: Bacteria harbouring temperate bacteriophages.
Temperate phage: A phage (virus) that invades but may not destroy (lyse) the host (bacterial cell); however, it may continue into the lytic cycle.
Virulent phage: A phage (virus) that destroys the host (bacterial) cell.

Phage λ (=lambda) is one of the best studied bacteriophage. It has a double stranded DNA (chromosome) of about 48,500 base pairs. Since it is a temperate phage, it can exist either vegetatively or as a prophage, integrated into the host chromosome. Life cycle of this phage is suitable for our analysis since its choice of pattern of life cycle is determined by an interesting and complex way. It provides a model system of operon controls. Here, complexity arises from having two conflicting life-cycle choices: one choice favours lytic cycle and other choice prefers nonlytic (*i.e.,* lysogenic) phage of life cycle.

The expression of one of the two life-cycle alternatives, lysogenic or lytic cycles, depends on whether two repressors, *CI* and *cro*, have access to operator site. The **CI repressor** acts to favour lysogeny: it expresses lytic cycle and represses lysogeny. The operator sites, when bound by either *CI* or *cro*, can either enhance or repress transcription. There are certain other control mechanisms which are involved in determining aspects of λ life cycle such as **antitermination** and **multiple promoters** for the same genes (see **Ptashne** and coworkers 1982; **Ptashne,** 1987, 1989).

Structure of Phage λ Operons

Phage lambda (λ) shows a complex system of controls of both early and late operons, as well as controls for the decision of lytic infection versus lysogenic integration. The λ genes are grouped into four operons: **left** and **right**, **late** and **repressor** (Fig. 15.29). The *left* and *right operons* contain the genes for DNA replication and recombination and phage integration. The *late operon* contains the genes that determine phage head and tail proteins and lysis of the host cell. The sequence of events following phage infection is relatively well known.

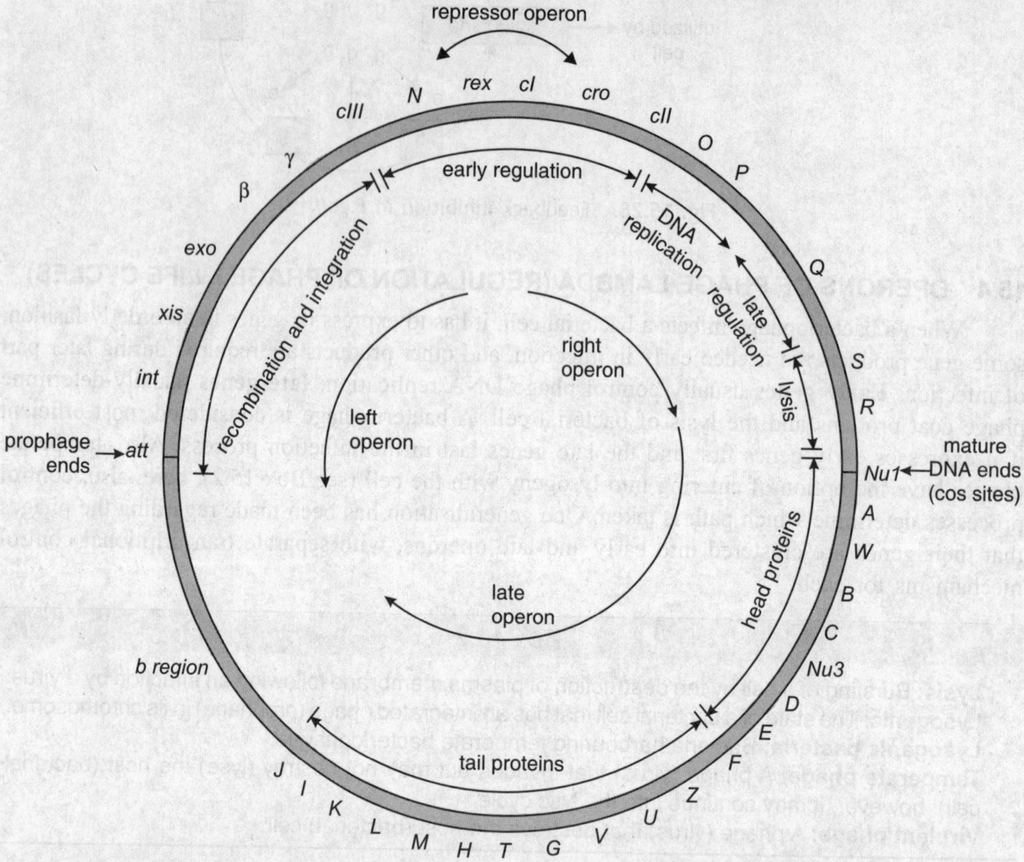

Fig. 15.29. Genetic map of phage λ. There are four operons present: the repressor, left, right and late. The prophage, a linear form integrated in the bacterial chromosome, begins and ends at *att* locus. In mature phage, another linear form of DNA found packed into the phage heads, begins and ends at *Nu1* locus (cos sites) (after Tamarin, 2002).

The genetic map of phage λ (Fig. 15.29) is a circle, but the λ chromosome has two linear stages in its life cycle (Fig. 15.30). (Note: circular form of DNA is infective cellular form). In one linear form, its DNA is packed within the phage head and in another linear form, DNA of the phage becomes integrated into host chromosome to form a **prophage**. These two linear forms of DNA do not have the same ends (Fig. 15.29 and 15.30). The mature DNA which is packed within the phage heads before lysis of the cell, is flanked by *cos* sites. It results from a break in the circular map between the *A* and *R* loci. The prophage is integrated at the *att* site and the circular map is thus broken there at integration.

The homologus integration sites on both λ and the *E.coli* chromosome consists of a 15bp core sequence (called "O" in both), flanked by different sequences on both sides in both the bacterium and the phage (Fig. 15.31). In the phage, the region is referred to as POP′ where P and P′ (P for phage) are two different regions flanking the O core on the phage DNA.

In the bacterium, the region is called BOB′ where B and B′ (B for bacterium) are two different regions flanking the O core on the *E.coli* chromosome. Integration, which is a part of the lysogenic life cycle, requires the product of the λ *int* gene, a protein known as *integrase*, and is referred to as **site-specific recombination.** Later excision of the prophage, during introduction, when the phage leaves the host chromosome to enter the lytic cycle, requires both the integrase and the protein product of the neighboring *xis* gene, the **excisionase** (enzyme).

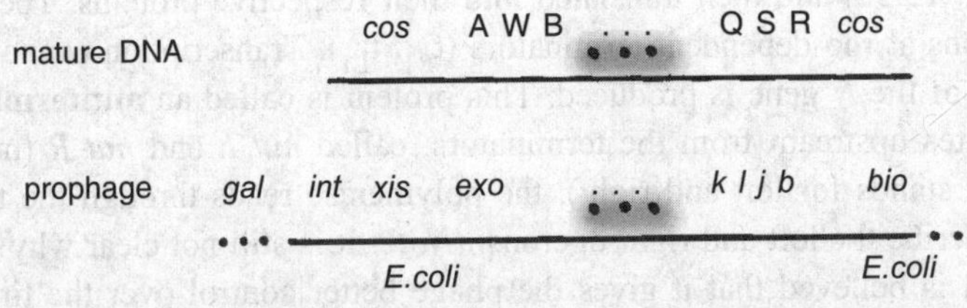

Fig. 15.30. The mature linear DNA (found within phage protein coat) is flanked by *cos* sites. The prophage is flanked by *E.coli* DNA *bio* and *gal* loci (after Tamarin, 2002)

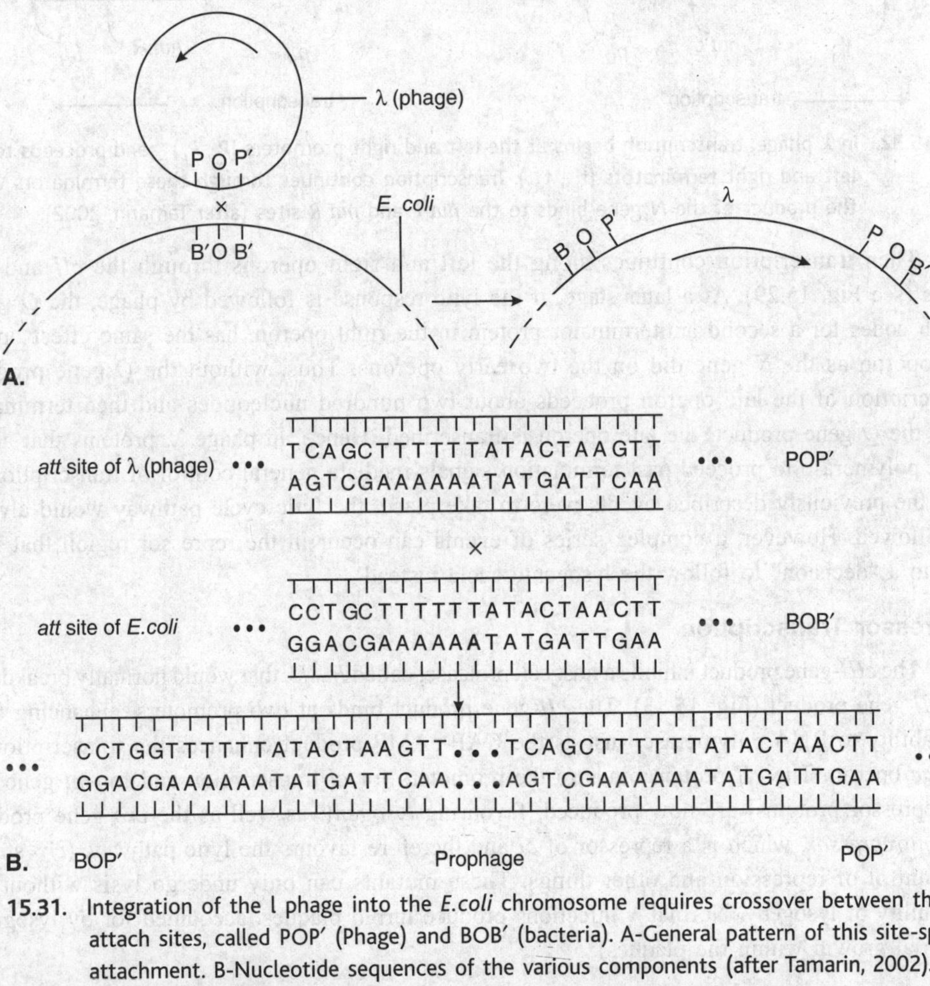

Fig. 15.31. Integration of the l phage into the *E.coli* chromosome requires crossover between the two attach sites, called POP′ (Phage) and BOB′ (bacteria). A-General pattern of this site-specific attachment. B-Nucleotide sequences of the various components (after Tamarin, 2002).

Following infection of the *E.coli* cell by a λ phage DNA circularizes, using the complementarity of the *cos* sites. Transcription begins, and within a very short time the phage is producing virus progeny or entering the lysogenic cycle and integrating into the host chromosome.

Early and Late Transcription

When the phage first infects an *E.coli* cell, transcription of the left and right side operons is started at the left (p_L) and right (p_R) promoters, respectively: The *N* (left) and *cro* (right) genes are transcribed (Fig. 15.32) and then translated into their respective proteins. Then, transcription stops on both operons at rho-dependent terminators (t_{R1}, t_{L1}). Transcription cannot continue until the protein product of the *N* gene is produced. This protein is called an **antiterminator protein**. When it binds at sites upstream from the terminators, called *nut L* and *nut R* (nut stands for *N* utilization; *L* and *R* stands for left and right), the polymerase reads through the terminators and continue on to transcribe the left and right operons. (*Note*: It is still not clear why antitermination has evolved here, it is believed that it gives the phage better control over the timing of events; see **Tamarin,** 2002).

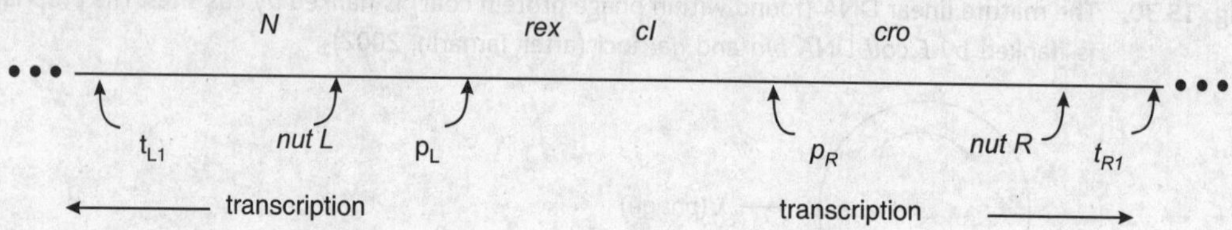

Fig. 15.32. In λ phage, transcription begins at the left and right promoters (P_L, P_R) and proceeds to the left and right terminators (t_{L1}, t_{R1}). Transcription continues through these terminators when the product of the *N* gene binds to the *nut L* and *nut R* sites (after Tamarin, 2002).

Then transcription continues along the left and right operons through the *cII* and *cIII* genes (see Fig. 15.29). At a later stage, if the lytic response is followed by phage, the *Q* gene, which codes for a second antiterminator protein in the right operon, has the same effect on the late operon as the *N* gene did on the two early operons. Thus, without the *Q* gene product, transcription of the late operon proceeds about two hundred nucleotides and then terminates. With the *Q* gene product, the late operon is transcribed. Hence, in phage λ, proteins that allow RNA polymerase to proceed past termination signals mediate general control of transcription. If only the previously described events were to take place, the lytic cycle pathway would always be followed. However, a complex series of events can occur in the repressor region that may lead to a "decision" to follow the lysogenic cycle instead.

Repressor Transcription

The *cIII*-gene product inhibits a host cell protease, called *Fts H*, that would normally breakdown the *cII*-gene product (Fig. 15.33). The *cII*-gene product binds at two promoters, enhancing their availability to RNA polymerase, just as the CAP-cAMP product enhances the transcription of the *lac* operon. The *cII* protein binds to the promoters for *cI* transcription and for *int* gene and *cI* (repressor proteins) are now produced, favouring *lysogeny,* as well as the *cro* gene product, the *antirepressor*, which is a repressor of *cI* and therefore favours the lytic pathway, (*cro* stands for control of repression and other things. These mutants can only undergo lysis without the possibility of lysogeny. Normal λ infections produce turbid plaques, accounted for by lysogenic bacterial growth within the plaques).

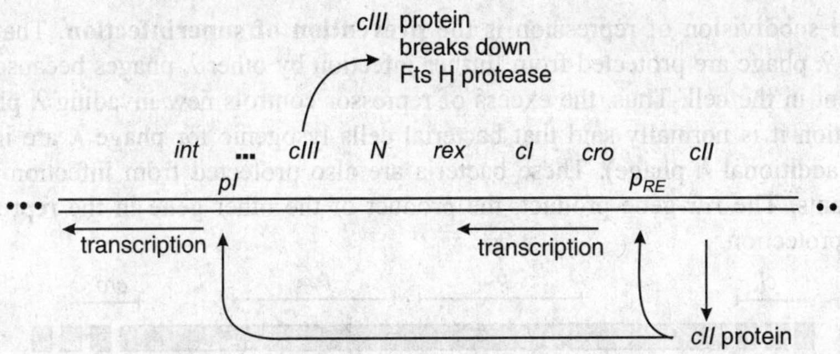

Fig. 15.33. The *cII*-gene product of phage λ binds to the *cI* promoter (p_{RE}) and the *int* promoter (pI), enhancing transcription of those genes. The *cIII* protein breaksdown the Fts H protease (enzyme) that would normally breakdown the *cII* protein (after Tamarin, 2002).

Maintenance of Repression

The *cI* gene, with the aid of the *cII*-gene product, is transcribed from a promoter known as p_{RE} (the *RE* stand for repression establishment; Fig. 15.34). Once *cI* is transcribed, it is translated into a protein called the λ *repressor,* which interacts at the left and right operators, O_L and O_R of the left and right operons. When these operators are bound by *cI* protein, transcription of the left and right operons (and therefore also the late operons) ceases. There are several subdivisions of the repression. **First**, lysogeny can be initiated because the *int gene* has been transcribed at the early stage of infection. **Second,** since cII and cIII are no longer synthesised, *cI* transcription from the

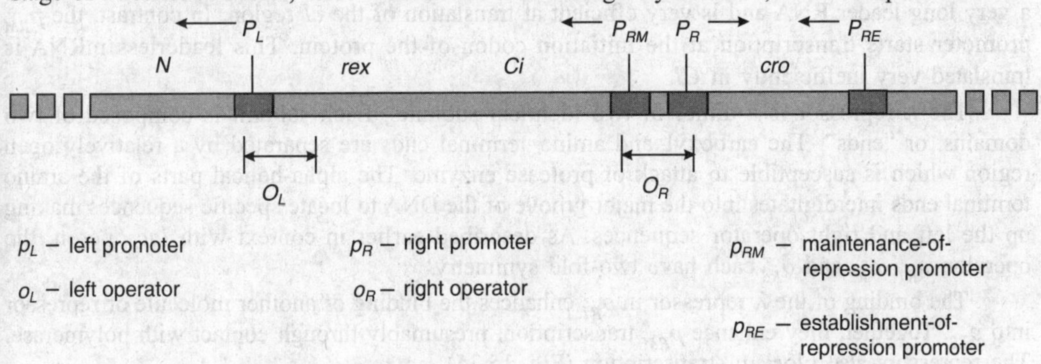

Fig. 15.34. Early regulation region of phage λ. Two promoters, p_{RE} and p_{RM}, transcribe the *cI* and rex genes. The left operator overlaps the left promoter, and the right operator overlaps both the right promoter and the maintenance of repression promoter (after Tamarin, 2002).

p_{RE} promoter stops. However, p_{RM} (RM stands for repression maintenance), that allows low levels of transcription of the *cI* gene.

In fact, the *cI* gene is found to control its own concentration in the cell. When the right and left operators were sequenced, each was discovered to have three sites of repressor recognition (Fig. 15.35). On the right operator, for example, the right most site (Og1) was found to be most efficient at binding repressor. When repressor was bound at this site, the right operon was repressed, and transcription of *cI* was enhanced (it is similar to enhancement of transcription by the binding of CAP-cAMP at the CAP site in *lac* operon). Excess repressor, when present, however, was also bound by the other two sites within *OR*. The above process results in the repression of the *cI* gene itself. Hence, maintenance levels of cI can be kept within very narrow limit.

A **third** subdivision of repression is the **prevention of superinfection**. That is, bacteria lysogenic for λ phage are protected from further infection by other λ phages because repressor is already present in the cell. Thus, the excess of repressor controls new invading λ phages. (*Note*: for this situation it is normally said that bacterial cells lysogenic for phage λ are immune from infection by additional λ phage). These bacteria are also protected from infection by T_4 phage with *rII* mutants. The *rex*-gene product, the product of the other gene in the repressor operon, controls the protection.

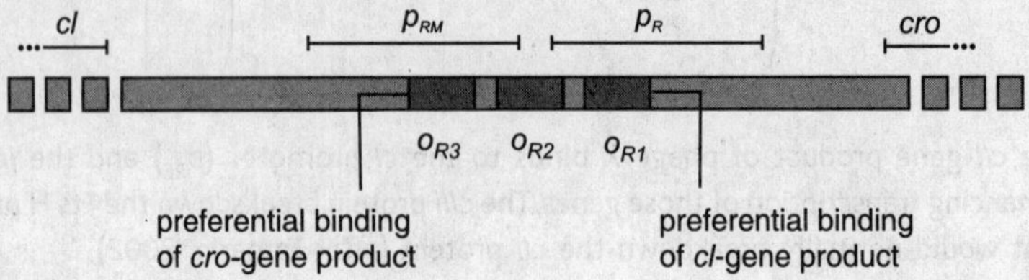

Fig. 15.35. The right operator on the phage λ chromosome overlaps the P_{RM} and P_R promoters. There are three repressor recognition sites within the operator: o_{R1}, o_{R2}, and o_{R3}. Preferential binding by the Cro repressor to o_{R3} and the CI repressor to o_{R1} determines whether transcription occurs to the left or the right (after Tamarin, 2002).

Further the promoters for maintenance and establishment of repression differ markedly in their control of repressor gene expression. When p_{RE} is active, a very high level of repressor is present whereas p_{RM} produces only a low level of repressor. The level of repressor is due to the length of the leader RNA transcribed on the 5′ side of the *cI* gene. The p_{RE} promoter transcribes a very long leader RNA and is very efficient at translation of the *cI* region. In contrast, the p_{RM} promoter starts transcription at the initiation codon of the protein. This leaderless mRNA is translated very inefficiently in *CI*.

The λ repressor is a dimer of two identical subunits. Each subunit is composed of two domains, or "ends". The carboxyl–and amino–terminal ends are separated by a relatively open region which is susceptible to attack of protease enzyme. The alpha-helical parts of the amino terminal ends interdigitates into the major groove of the DNA to locate specific sequences making up the left and right operator sequences. As described earlier in context with *lac* operon, the operator, o_{R1}, o_{R2} and o_{R3} each have two-fold symmetry.

The binding of the λ repressor in o_{R1} enhances the binding of another molecule or repressor into o_{R2}. Together, they enhance p_{RM} transcription, presumably through contact with polymerase. The repressors also block p_R transcription (Fig. 15.35).

Lysogenic versus Lytic Response

Now let us see how does λ phage turn toward the lytic cycle? In this context, the control is exerted by the *cro*-gene product, another repressor molecule that works at the left and right operators in a manner antagonistic to the way the *CI* repressor works. In other words, using the right operator as an example, *cro*-gene product binds preferentially to the left most of the three sites within o_R and represses *cI* but enhances the transcription of *cro* (Fig. 15.35).

The *cro*-gene product can direct the cell toward a lytic response if it occupies the o_R and o_L sites before the λ repressor, or if the **λ repressor** is removed. From the point of view of phage λ, when would be good time for the *CI* repressor to be removed? From evolutionary point of view, it is expected that a prophage might be at an advantage if it left a host's chromosome and began the lytic cycle when it "sensed" damage to the host. In fact, one of the best ways to induce a prophage to enter the lytic cycle is to direct ultraviolet (UV) light at the host bacterium. UV light causes damage to DNA and induces several repair systems. One, called **SOS repair** makes use of the protein products of the *rec A* gene. Among the activities of this enzyme is to cleave the λ repressor in the susceptible region between domains. The cleaved repressor falls free of the DNA, making the operator sites available for the *cro*-gene product. The lytic cycle then follows. Various elements involved in phage λ infection have been tabulated in Table 15.1.

Table 15.1 Elements in phage λ infection (*Source:* **Tamarin**, 2002).

A. Gene Products	
cI	Repressor protein whose function favours lysogeny
cII	Enhances transcription at the p_I and p_{RE} promoters
cIII	Inhibits the *Fts H* protease
cro	Antirepressor protein that favours lytic cycle
N	Antiterminator acting at *nut R* and *nut L*
rex	Protects bacterium from infection by T4 *rII* mutants
int	Integrase for prophage integration
Q	Antiterminator of late operon
Fts H	Bacterial protease that degrades *cII* protein
B. Promoters of	
p_R	Right operon
p_L	Left operon
p_{RE}	Establishment of repression at repressor region
p_{RM}	Maintenance of repression at repressor region
p_I	*int* gene
C. Terminators	
t_{RI}	Terminates after *cro* gene
t_{LI}	Terminates after *N* gene
D. Antiterminators	
nut R	In *cro* gene
nut L	In *N* gene

Under initial infection, the choice between lysogeny and lytic cycle depends primarily on the *cII* protein, which *gauges* the health activity of the host. After lysogeny is established, it can be reversed by the processes that inactivate the *cI* protein, indicating genetic damage to the bacterium (the SOS response), or an abundance of other hosts in the environment (zygotic induction). In zygotyic induction, the lytic cycle is induced during conjugation, presumably when an Hfr cell sends a copy of the F^- prophage into a F^--cell. At that point, without repressor present, the prophage can reassess whether to continue lysogeny or enter the lytic cycle.

QUESTIONS

Long Answer Questions

1. What is the fate of a λ phage entering an *E.coli* cell that contains quantities of λ repressor?
2. What is the fate of a λ phage entering an *E.coli* cell that contains of the *cro*-gene product?
3. Describe the fate of λ phages during the infection process with mutants in the following genes: *cI, cII, cIII, N, cro, att, Q*.
4. What is the fate of λ phages during the infection process with mutants in the following areas: o_{R1}, o_{R3}, p_L, p_{RE}, p_{RM}, p_R, $t_{L1}t_{R1}$, *nut L, nut R*?
5. How does ultraviolet light (UV) damage induce the phage λ chromosome?

Very Short Answer Questions

1. What is inducer?
2. Define inducible enzyme.
3. What is operon?
4. What is inducible operon?
5. Define operator gene.
6. What is repressor protein?
7. Define repressible operon.
8. What is corepressor?
9. Enlist operons of *E.coli*.
10. Define attenuation.
11. What is catabolic repression.
12. How repressor of *lac* operon is a allosteric protein?
13. How can inducible and repressible enzymes of microorganisms be distinguished?
14. Distinguish between (*a*) repression and (*b*) feedback inhibition caused by the end product of biosynthesis pathway. How do these two regulatory mechanisms complement each other to provide for the efficient regulation of metabolism?
15. In the *lac* operon of *E.coli*, what is the function of each of the following genes or sites: (*a*) regulator, (*b*) operator, (c) promoter, (*d*) structural gene *z*, and (*e*) structural gene *y*?
16. What would be the result of inactivation by mutation of the following genes or sites in the *E.coli lac* operon: (*a*) regulator; (*b*) operator; (*c*) promoter; (*d*) structural gene *z* and (*e*) structural gene *y*?
17. Of what biological significance is the phenomenon of catabolite repression?
18. Describe the role of cyclic AMP in transcriptional control in *E.coli.*
19. **J. Beckwith** isolated point mutations that were simulataneously uninducible for the *lac, ara, mal* and *gal* operons, even in the absence of glucose. Provide two different functions that could be missing in these mutations.
20. Describe the interaction of the attenuator and the operator control mechanisms in the *trp* operon of *E.coli* under varying concentrations of tryptophan in the cell. How does attenuator control reacts to shortage of other amino acids?
21. What are the three different physical forms that the phage λ chromosome can take?
22. From an evolutionary point of view, why do you think *Escherichia coli* (bacterium) evolved a CAP system of positive control of gene expression? Why not just metabolize any and all sugars in the environment as they appear.
23. Why would the Rec A protein of *E.coli* cleave the λ repressor?
24. What are the differences in action of the λ promotes p_{RE} and p_{RM}?

Fill in the Blanks

1. Jacob and .... investigated regulation of the activities of genes which control .... of lactose through synthesis of the enzyme .... in *E.coli.*
2. The *lac* enzymes of *E.coli* are β-galactoside permease, ... and ....
3. The tryptophan operons mRNA in *E.coli* is .... carrying coding sequences of .... different polypeptides.
4. The three strucutral genes of *lac* operon of *E.coli* include .... , ... and ....
5. Operon is a sequence of adjacent structural genes all under the transcriptional control of the same .... and ....
6. The repressor of *lac* operon is an .... protein.

Multiple Choice Questions

1. In operon concept, regulator gene functions as

(*a*) inhibitor (*b*) regulator

(*c*) repressor (*d*) all of these

2. The operator gene of lac operon is "turned on" when lactose molecule binds to

(*a*) operator gene (*b*) repressor gene

(*c*) promotor site (*d*) mRNA

3. The example of inducible operon is

(*a*) lac-operon (*b*) try-operon

(*c*) arginine operon (*d*) both (*a*) and (*b*)

4. How many promoters control the transcription of *E.coli* lac operon?

(*a*) one (*b*) two

(*c*) three (*d*) four

5. Which one of the following is not a component of lac operon model?

(*a*) promoter (*b*) structural gene

(*c*) primer gene (*d*) regulator gene

6. In *E.coli*, presence of tryptophan amino acid causes

(*a*) activation of operon (*b*) closure of operon

(*c*) both of above (*d*) none

ANSWERS

Very Short Answer Questions

1. Inducer is any effector molecule that activates the transcription of an inducible operon (*e.g.*, allolactose in case of *lac* operon of *E.coli*).

2. Inducible enzyme is an enzyme whose synthesis is regulated by the presence or absence of its substrate.

3. Operon is a cluster of genes with related functions that is under the control of a single operator and promoter, thereby allowing transcription of these genes to be turned on and off together.

4. Inducible operon is a group of adjoining genes whose transcription is activated in presence of an inducer.

5. Operator (gene) is a base sequence in an operon to which a repressor protein can bind.

6. In prokaryotes, the repressor protein is a protein that binds to the operator site of an operon and prevent transcription of adjacent structural genes (SG).

7. Repressible operon is a group of adjoining genes that are normally transcribed but whose transcription is inhibited in the presence of a corepressor [*e.g.*, tryptophan (*trp*) operon of *E.coli*].

8. It is a effector molecule required along with the repressor to prevent transcription of a bacterial operon (*e.g.*, tryptophan in trp operon of *E.coli*).

9. (*i*) *lac* operon for lactose; (*ii*) *trp* operon for tryptophan and (*iii*) *ara* operon for arabinose.

10. Attenuation is a mechanism for regulating bacterial gene expression based on the premature termination of transcription.

11. Catabolic repression is ability of glucose to inhibit the synthesis of catabolic enzymes produced by inducible bacterial operons (*e.g.*, *lac* operon).

12. When repressor binds with one particular molecule, it changes its shape which changes its ability to react with second particular molecule, it changes its shape which changes its ability to react with second particular molecule. Here the first molecule is the inducer **allolactose** and the second molecule is repressor protein. When allolactose is bound to the repressor, it causes the repressor to change its shape and lose its affinity for operator sequences *lac* operon of *E.coli*.

13. Both enzyme types are distinguished by studying the synthesis or lack of synthesis of the enzyme in cells grown on chemically defined media. If the enzyme is synthesized only in the presence of a certain metabolite or a particular set of metabolites, it is probably inducible. If it is synthesized in the absence but not in the presence of a particular metabolite or group of metabolites, it is probably repressible.

14. Repression occurs at the level of transcription during enzyme synthesis. The end product, or a derivative of the end product, of a repressible system acts as an effector molecule that usually, if not always, combines with the product of one or more regulator genes to turn off the synthesis of the enzymes in the biosynthesis.

15.

Gene or Regulatory element	Function
(*i*) Regulator gene	Codes for repressor
(*ii*) Operator	Binding site repressor
(*iii*) Promoter	Binding site of RNA polymerase and CAP–cAMP complex
(*iv*) Structural gene *z*	Encodes β-galactosidase
(*v*) Structural gene *y*	Encodes β-galactoside permease

16. (*a*) Constitutive synthesis of the *lac* enzymes;
(*b*) Constitutive synthesis of the *lac* enzymes;
(*c*) Uninducibility of the *lac* enzymes;
(*d*) No β-galactosidase activity;
(*e*) No β-galactoside permease activity.

17. Catabolite repression has apparently evolved to assure the use of glucose as a carbon source when this carbohydrate is available, rather than less efficient energy sources.

18. Cyclic AMP, combined with CAP protein, attaches to CAP sites enhancing transcription of nonglucose, sugar metabolizing operon in *E.coli*. Glucose inhibits its formation by inhibiting adenylcyclase.

19. One should have an idea about how these operons are controlled. Each operon requires inducer and also the catabolic repression activation system. These mutants could be unable to make cAMP because the adenylcyclase gene is defective. Alternatively, they could be making a defective catabolite activating protein (CAP).

20. The *E.coli trp* operon functions as a normal repressible operon. In addition, attenuation, based on secondary structure and stalling of the ribosome on the leader transcript, can further prevent transcription. Attenuator control can be exerted based on other amino acids if their codons appear in the leader transcript, causing ribosome stalling.

21. The λ chromosome has one circular and two linear forms. The circular form is the infective cellular form. A break at one point (*cos* site) takes place during packaging into the phage head, and a break at another point forms the linear integrative prophage.

22. One might think that a bacterium should metabolize any sugar in its environment. However, glucose is the most efficient sugar to metabolize and if it is present, it should be metabolized first. Even under maximal growth, it takes an *E.coli* cell about 20 minutes to grow and divide. If the cells, were to begin to take up and metabolize other sugars, there would be a cost for biosynthesis of new enzymes and the inefficiency of initial steps of the sugar metabolism. In fact, the growth of the cell would slow down under these circumstances and be at an evolutionary disadvantage.

23. Since the cleaving of the λ repressor is a signal to begin the lytic phase of the life cycle of the phage, it seems odd that the lysogenized bacterial cell would be an ally to its own destruction. However, the phenomenon makes much more sense if we realize that the Rec A protein has several other functions critically important to the bacterial cell. The λ phage has evolved the ability to take advantage of the existence of the Rec A protein by evolving a repressor sensitive to it. Evolutionary biologists view this as "coevolution", two interacting organisms evolving to take advantage of or minimize properties of the other. The bacterium, however, might be at disadvantage. Since Rec A has many functions involving interactions with other proteins, it may be highly limited in how it can change.

24. The promoters p_{RE} and p_{RM} are both promoters of the repressor operon of phage λ. Transcription from these promoters allows production of the *cI* repressor protein, the repressor that favours lysogeny. Initially, the promoter p_{RE} is activated. For it to be a transcription site, it must be activated by the product of the ell gene, which lies in the right operon. This promoter, p_{RE}, produces a mRNA with a long leader sequence that is translated very efficiently. Once the repressor binds at the operators of the left and right operons, the *cII* gene is no longer transcribed, and therefore p_{RE} is no longer a site for transcription. However, the repressor gene can still be transcribed from the p_{RM} promoter, which does not need the product of the *cII* gene. This promoter produces a transcript with no leader and thus is translated very inefficiently. At that point, however, only a very small quantity of repressor is needed to maintain lysogeny. Thus, the two promoters are the sites for the initiation of the repressor operons under different circumstances: one early in the infection stage and one after lysogeny is under way.

Fill in the Blanks

1. Monod, fermentation, β-galactosidase.

2. β-galactosidase, β-galactoside transacetylase.

3. Polycistronic, five.

4. *lac z, lac y, lac a*.

5. Promoter, operator.

6. Allosteric.

Multiple Choice Questions

1. (*c*) **2.** (*b*) **3.** (*a*) **4.** (*b*) **5.** (*c*) **6.** (*b*).

16

Regulation of Gene Action in Eukaryotes

16.1 REGUALTION OF TRANSCRIPTION IN EUKARYOTES

Initially, researchers assumed that transcriptional regulation would be similar in eukaryotes. However, extensive work over the last 32 years has revealed a surprisingly complex situation, as we shall see in this chapter. The components and steps involved in eukaryotic transcription can be summarized as follows:

1. There are three different gene-class-specific RNA polymerases, each of which consists of at least 10 subunits.

2. A number of additional proteins the so-called **general** or **basal**, **transcription factors** (GTFs), together with the RNA polymerase, form a minimal complex that recognizes a promoter and can initiate transcription *in vitro*.

3. The relative level of transcriptional initiation is regulated by various **specific transcription** factors. These factors bind upstream regions of promoters, or to other regulatory sequences called **enhancers** that are often located very far away from the coding regions of a given gene. Furthermore, additional factors, called **coactivators**, which themselves do not bind DNA, may interact with DNA-bound transcription factors.

4. The specific transcription factors bound to distal promoter or enhancer sequences most likely stimulate transcription by directly interacting with the basal transcription apparatus while looping out the intervening DNA. In additon, at least some of these factors seem to also help reorganize chromatin into more accessible, "loose" form that is experimentally characterized by a higher sensitivity toward DNase I. Conversely, **repressing factors** binding so-called silencer DNA sequences may instead disturb interactions within the transcription apparatus and/or recognize the chromatin into a tight, nonaccessible form (*i.e.*, heterochromatin).

Apparently, this increased degree of complexity in eukaryotes is necessary for the fine-tuned transcriptional regulatory circuits that exist in multicellular organisms. During the differentiation of a precursor cell into a specific cell type, for example, specific sets of genes are turned on, while others are turned off (Fig. 16.1). Such genes include regulatory genes that control the establishment and maintenance of a cell type, as well as genes that encode cell type specific proteins, which may be needed in massive amounts. Examples of the latter type of proteins are the **haemoglobins** of erythroid cells, **insulin** of pancreatic beta cells, **immunoglobulins** of B lymphocytes, and **actin** and **myosin** of muscle cells (**Malacisky 2005**).

Britten-Davidson model or gene-battery model of transcription regulation. This model of eukaryotic gene regulation at the level of transcription was proposed by **Britten** and **Davidson**

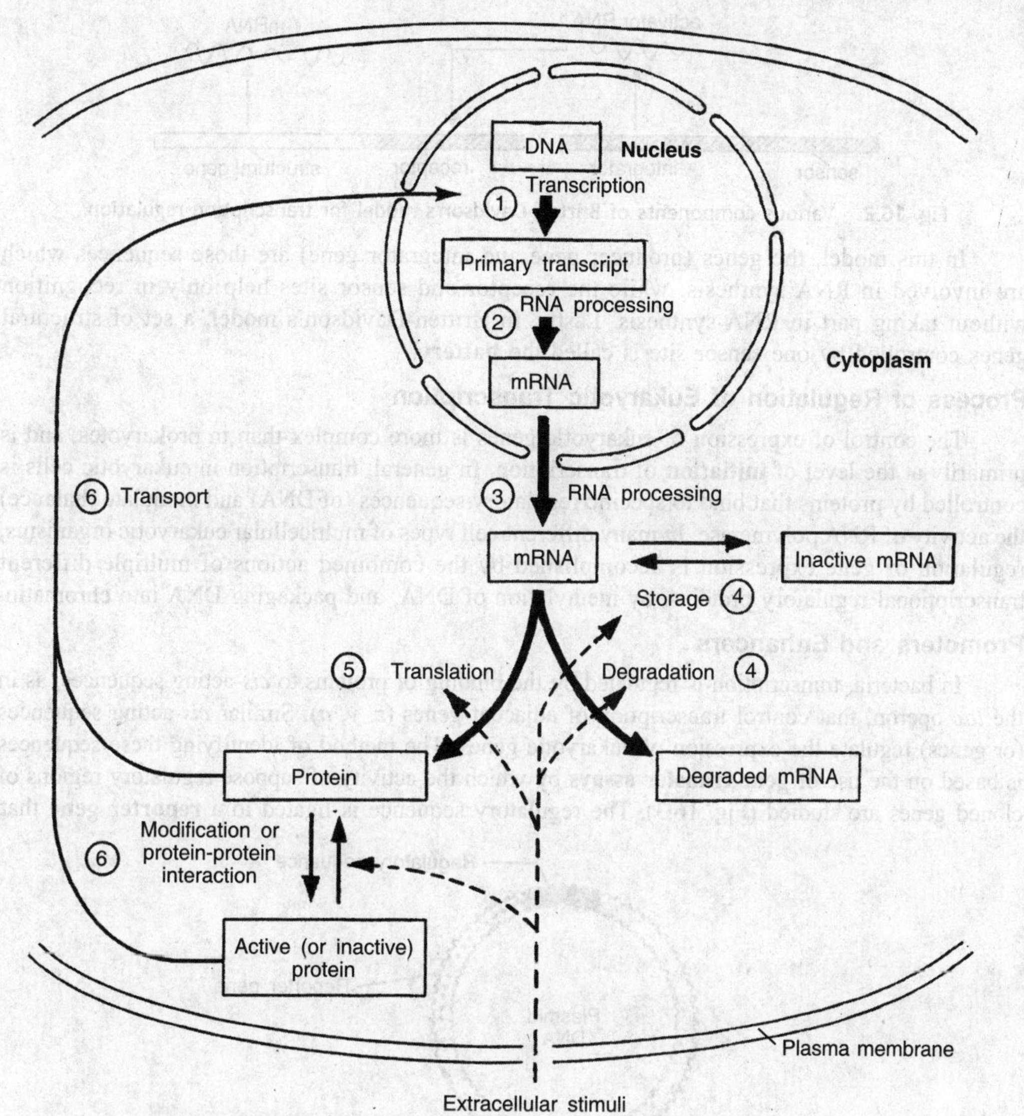

Fig. 16.1. Potential control levels for the regulation of gene expression in eukaryotes. Levels 1 and 2 are localized in the nucleus and 4 and 5 in the cytoplasm; 3 and 6 indicate translocation betweeen those two compartments. Note that with 6 only one possible route of protein translocation (from cytoplasm to nucleus) is indicated (after Malacinsky 2005).

in 1969 and later on elaborated by them in 1973. The gene-battery model (Fig. 16.2) assumes the presence of four classes of sequences: 1. **producer gene** which is comparable to a structural gene of prokaryotic operon; 2. **receptor site** is located adjacent to each producer gene and is comparable to operator gene of prokaryotic operon; 3. **integrator gene** which is comparable to regulator gene and is responsible for synthesis of an **activator RNA** that may or may not give rise to proteins before it activates the receptor site; 4. **sensor site** regulates the activity of integrator gene, which can be transcribed only when the sensor site is activated. The sensor sites are recognised by agents which, like hormones and proteins, change the pattern of gene expression. For instance, hormone-protein complex or a transcription factor may bind to a sensor site and cause the transcription of integrator.

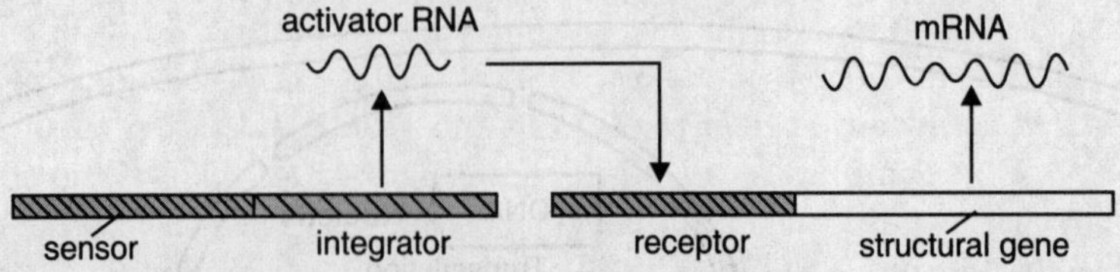

Fig. 16.2. Various components of Britten-Davidson's model for transcription regulation.

In this model, the genes (producer gene and integrator gene) are those sequences which are involved in RNA synthesis, while the receptor and sensor sites help only in recognition without taking part in RNA synthesis. Lastly, in Britten-Davidson's model, a set of structural genes controlled by one sensor site is called the **battery**.

Process of Regulation of Eukaryotic Transcription

The control of expression of eukaryotic genes is more complex than in prokaryotes, and is primarily at the level of **initiation** of transcription. In general, transcription in eukaryotic cells is controlled by proteins that bind to specific regulatory sequences (of DNA) and modulate (balance) the activity of RNA polymerase. In many different cell types of multicellular eukaryotic organisms, regulation of gene expression is accomplished by the combined actions of multiple different transcriptional regulatory proteins, by methylation of DNA, and packaging DNA into chromatin.

Promoters and Enhancers

In bacteria, transcription is regulated by the binding of proteins to *cis*-acting sequences, as in the *lac* operon, that control transcription of adjacent genes (*z*, *y*, *a*). Similar *cis*-acting sequences (or genes) regulate the expression of eukaryotic genes. The method of identifying these sequences is based on the use of **gene transfer assays** by which the activity of suppose regulatory regions of cloned genes are studied (Fig. 16.3). The regulatory sequence is ligated to a **reporter gene** that

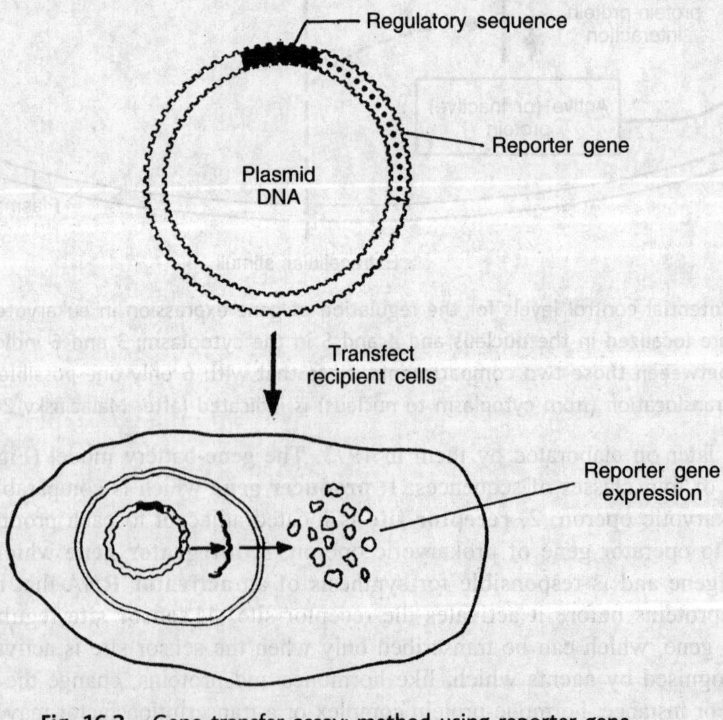

Fig. 16.3. Gene transfer assay; method using reporter gene.

encodes an easily detectable enzyme. The reporter gene is transferred into cultured cells (**transfection**). The expression of the reporter gene indicate biological activity of the regulatory sequence and provides a sensitive assay for the ability of the cloned regulatory sequences to direct transcription.

The two core promoter elements, the **TATA box** and the **InR sequence** in genes transcribed by RNA polymerase II enzyme serve as specific binding sites for transcription factors (Fig. 16.4). Inr is the initiatior sequence spanning the transcription start site in promoters of many genes transcribed by RNA polymerase II. Other *cis*-acting sequences function as binding sites for a variety of regulatory factors that control expression of individual genes. These *cis*-acting regulatory sequences are usually located upstream of TATA box. Interestingly, two regulatory sequences commonly found in eukaryotic gene were found to be present in the promoter of the herpes simplex virus gene that encodes **thymidine kinase**. These two sequences are located about 100 base pairs upstream of the TATA box and their consensus sequences are CCAAT and GGGCGG (called **GC box**). The binding of specific proteins to these sequences has been shown to initiate transcription.

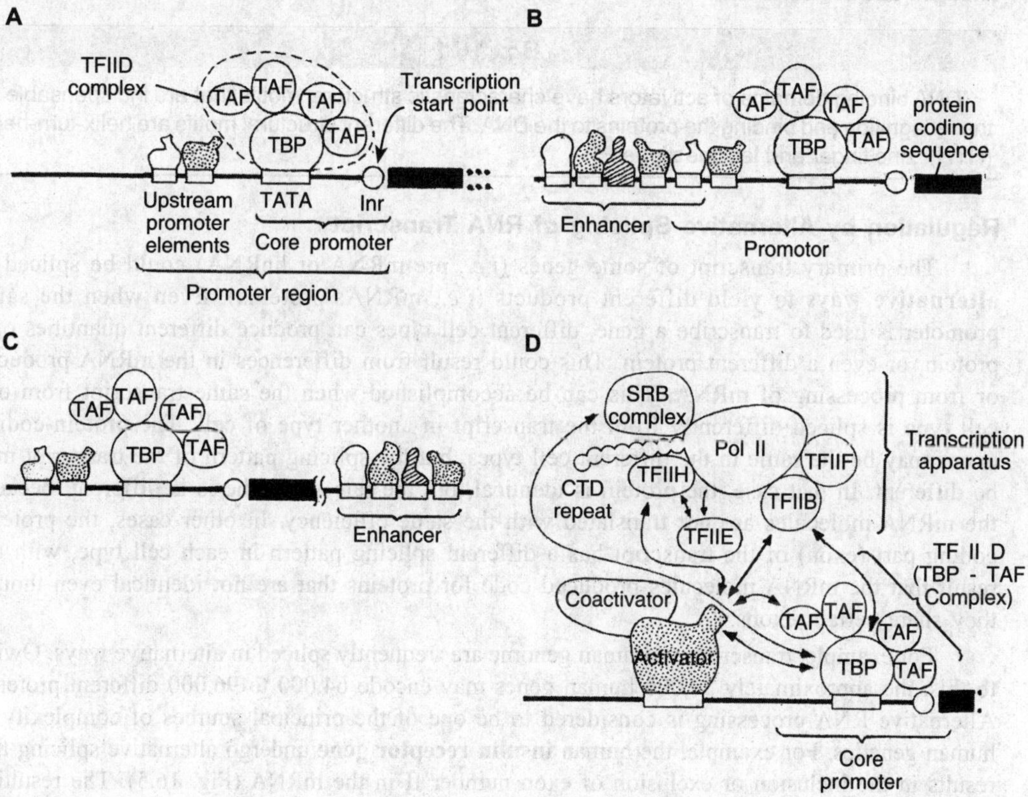

Fig. 16.4. Models for regulation of RNA synthesis from various genes that are transcribed by RNA polymerase II enzyme. Regulatory sequences (enhancers and promoters), as well as the multitude of proteins comprising a transcription complex are illustrated (after Malacinski 2005).

In contrast to CCAAT and GC boxes in the promoter of thymidine kinase of herpes simplex virus, the regulatory sequenecs of several mammalian genes are located further away, upto 10 kilobases, from transcription start point. These sequences are called **enhancers** and were first described in the virus SV4O. Like promoters, enhancers function by binding transcripton factors that act by regulating RNA polymerase. Transcription factors bound to distant enhancers function by the same mechanisms as those bound adjacent to promoters, that is, the *cis*-acting regulatory sequences. The binding of specific transcriptional regulatory proteins to enhancers is responsible for the control of gene expression during development, differentiation and in response of cells to hormones and growth factors.

Transcriptional Activators

Transcriptional activators are well-s..died transcription factors. They bind to regulatory DNA sequences are stimulate transcription. These activators consist of two domains, one region binding to DNA to anchor the factor to the proper site on DNA; the other activates transcription by interacting with other components of the transcriptional system. DNA-binding domains of many of these proteins are found to be related to one-another. The zinc-finger domains contain repeat of cysteine and histidine residues that bind zinc ions and fold into finger-like loops that bind DNA (Box 16.1). Transcription factors of the **steroid hormone** receptors contain zinc-finger domains. The steroid hormone receptors regulate gene transcription in response to hormones **estrogen** and **testosterone**.

Box 16.1

DNA binding domains of activators have characteristic structural motifs that are indispensable in the recognition and binding the proteins to the DNA. The different structural motifs are helix-turn-helix (HTH), zinc finger and leucine zipper.

Regulation by Alternative Splicing of RNA Transcript

The primary transcript of some genes (*i.e.*, pre-mRNA or hnRNA) could be spliced in **alternative ways** to yield different products (*i.e.*, mRNAs/proteins). Even when the same promoter is used to transcribe a gene, different cell types can produce different quantities of a protein, or even a different protein. This could result from differences in the mRNA produced or from processing of mRNA. This can be accomplished when the same transcript from one cell type is spliced differently from the transcript in another type of cell. The protein-coding exons may be the same in the different cell types, but the splicing pattern of the transcript may be different. In that case, the protein is identical, but the rate of synthesis is different, because the mRNA molecules are not translated with the same efficiency. In other cases, the protein-coding part (exon) of the transcript has a different splicing pattern in each cell type, with the result that the mRNA molecules produced code for proteins that are not identical even though they share certain exons.

For example, transcripts of human genome are frequently spliced in alternative ways. Owing to this, the approximately 30,000 human genes may encode 64,000 to 96,000 different proteins. Alternative RNA processing is considered to be one of the principal sources of complexity of human genetics. For example, the human **insulin receptor gene** undergo alternative splicing that results in the inclusion or exclusion of exon number II in the mRNA (Fig. 16.5). The resulting forms of the polypeptide chain differ in length by 12 amino acids. In liver cells, all the 20 exons are present in the mRNA for the long form of the receptor protein (Fig. 16.5A), whereas in skeletal muscle exon II is eliminated along with the flanking introns and excluded from the mRNA for the short form (Fig. 16.5B). The long form of the receptor show low affinity for insulin and is expressed in tissues such as the liver that are exposed to relatively higher concentrations of insulin. The short form of the protein has a high affinity for insulin and is expressed preferentially in tissues such as the skeletal muscle that are normally exposed to lower level of insulin.

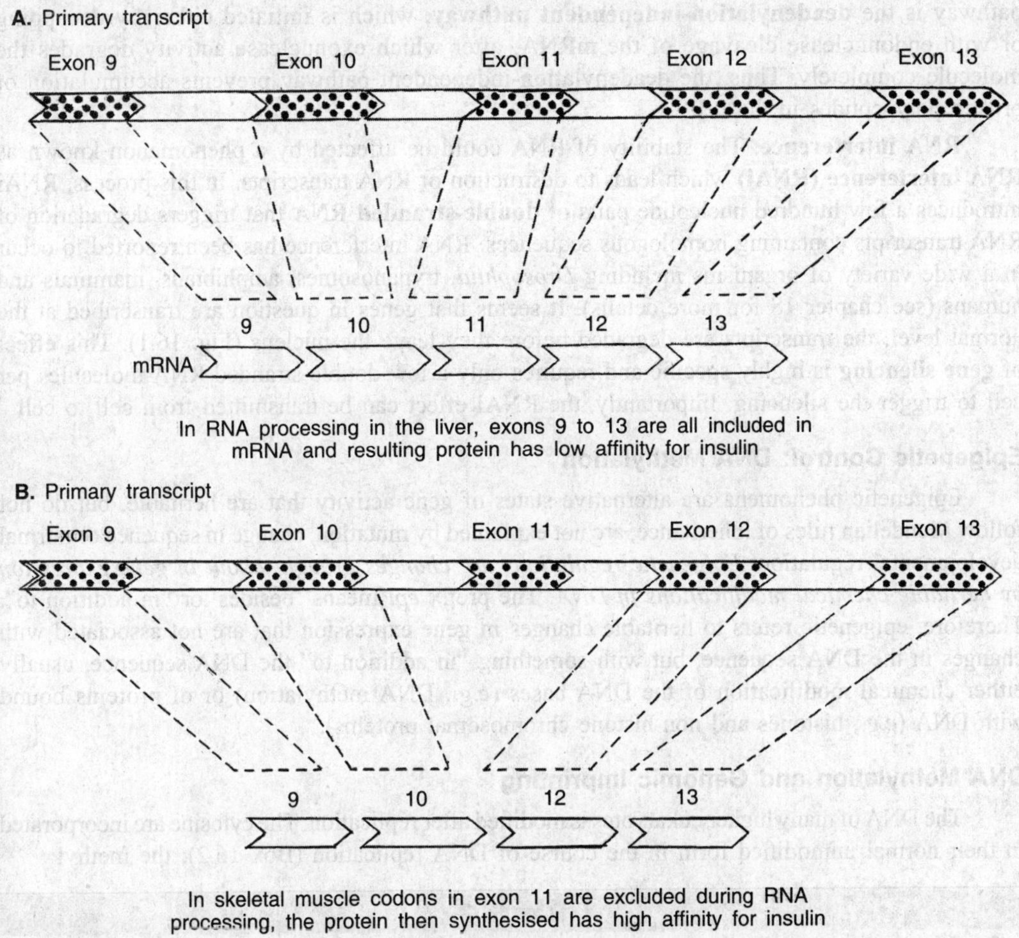

Fig. 16.5. Alternative splicing of human insulin receptor gene mRNA.

Thus, alternative splicing provides a mechanism for generating proteins with different properties from the same gene. The **Dscam** gene of *Drosophila* could give rise to approximately 38016 different proteins by alternative splicing.

The human genes are distributed over a large region of the genome, and the primary RNA transcripts are very long. Alternative splicing of most human genes leads to multiple protein products. About one third of *all* human genes are believed to undergo alternative splicing. Among genes that are alternatively spliced, the average number of distinct mRNAs produced from the primary transcript ranges between 2 and 7.

Stability of Messenger RNA

Different mRNAs persist for different time periods in a cell. Since short-lived mRNAs would produce fewer protein molecules as compared with long-lived mRNAs, factors affecting the **stability of mRNA** determine the level of gene expression. One route of degradation is the **deadenylation-dependent pathway** which begins with enzymatic cutting down of the length of the poly-A tail. When the poly-A tail is trimmed to a length of 25 to 60 nucleotides, the mRNA becomes prone to attack of a decapping enzyme. This enzyme removes the 5′ cap so that the mRNA in this state is rapidly degraded by exonuclease (enzyme). The other degradative pathway is the **deadenylation-independent pathway**, which is initiated either by decapping or with endonuclease cleavage of the mRNA, after which exonuclease activity degrades the molecule completely. Thus, the deadenylation-independent pathway prevents accumulation of broken polypeptides in the cell.

RNA interference. The stability of RNA could be affected by a phenomenon known as **RNA interference (RNAi)** which leads to destruction of RNA transcripts. In this process, RNAi introduces a few hundred nucleotide pairs of **double-stranded RNA** that triggers degradation of RNA transcripts containing homologous sequences. RNA interference has been reported to occur in a wide variety of organisms including *Drosophila*, trypanosomes, amphibians, mammals and humans (see chapter 18 for more details). It seems that genes in question are transcribed at the normal level, the transcripts are degraded before they leave the nucleus (Fig. 16.1). This effect of **gene silencing** is highly specific and requires only a few double-stranded RNA molecules per cell to trigger the silencing. Importantly, the RNAi effect can be transmitted from cell to cell.

Epigenetic Control: DNA Methylation

Epigenetic phenomena are alternative states of gene activity that are heritable, but do not follow Mendelian rules of inheritance, are not explained by mutation, change in sequence or normal developmental regulation. *Epigenetic regulations are changes brought about in gene expression by heritable chemical modifications in DNA*. The prefix *epi* means 'besides' or "in addition to". Therefore, epigenetic refers to heritable changes in gene expression that are not associated with changes in the DNA sequence, but with something "in addition to" the DNA sequence, usually either chemical modification of the DNA bases (*e.g.*, DNA methylation) or of proteins bound with DNA (*e.g.*, histones and non-histone chromosomal proteins).

DNA Methylation and Genomic Imprinting

The DNA of many higher eukaryotes is modified after replication. The cytosine are incorporated in their normal unmodified form in the course of DNA replication (Box 16.2); the methyl

Box 16.2

Methylation in eukaryotes is of two kinds: *de novo* and maintenance methylation. The former is methylation in newly synthesized DNA strand to retain the methylation inheritance pattern. The maintenance methylation is an addition of methyl groups at noval positions.

group (CH_3) is then added by an enzyme **DNA methylase** (Fig. 16.6). In mammals DNA is methylated specifically at the cytosines that precede guanosine residues, that is CpG dinucleotides (5′-CG-3′). Many mammalian genes have CG-rich regions upstream of the coding region that provide multiple sites for methylation. These are called **CpG islands** where "p" represents the phosphate group in the polynucleotide backbone. In higher plants 3–7% of the bases contain methylated cytosine, while in animal DNA the proportion is less. In some protozoa, the adenine groups are methylated, while in the green alga *Chlamydomonas* both adenine and cytosine are methylated. There is a higher concentration of methylated cytosines in highly repetitious eukaryotic DNA. *Methylation is found to correlate with reduced transcriptional activity of genes*. Genes that contain high frequency of CpG dinucleotides in the vicinity of their promoters show a slow rate of transcriptional activity. Methylation inhibits transcription of these genes through the action of a protein, called **MeCP2**, that specifically binds to methylated DNA and brings about repression of transcription.

The inactive X chromosome in mammalian cells has reduced transcription since it is extensively methylated. In fact, in adult mammals, the majority of CpG dinucleotides in all chromosomes are methylated in adult cells. The unmethylated CpGs are usually associated with the promoters of housekeeping genes that are transcriptionally active (**Wolf** *et al.*, 1984).

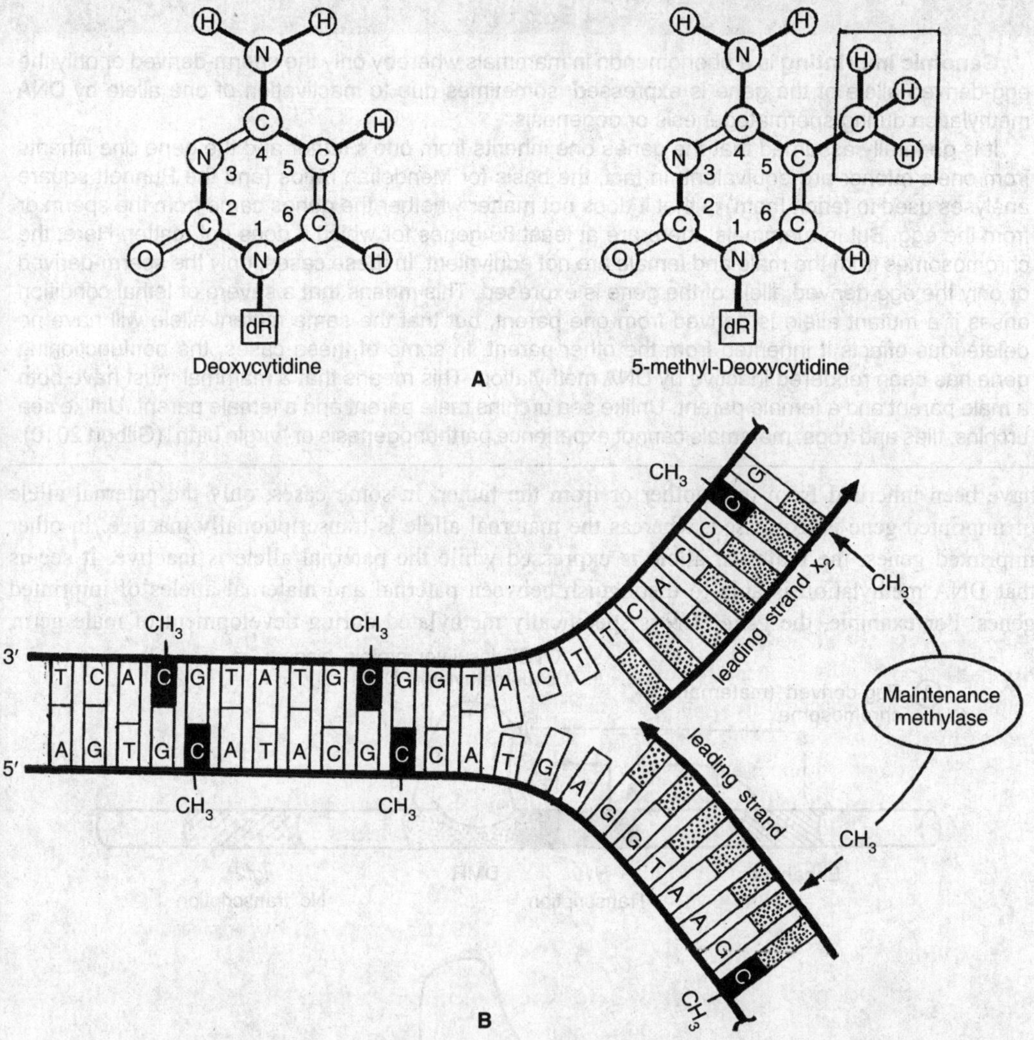

Fig. 16.6. DNA modification by methylation. A-Deoxycytidine and 5-methyl-deoxycytidine ($5^{m}C$) are indicated (dR=deoxyribose). B-The methylation, in which the hydrogen atom at carbon number 5 of deoxycytidine is replaced by a maintenance methylase enzyme that recognized hemimethylated CpG dinuleotides and fully methylates them shortly after replication (after Malacinski, 2005).

When the cells are treated with the analog of cytosine, the **azacytidine**, the latter reverses methylation and the transcriptional activity of such cells is restored. For example, in cell culture, some lineages of rat pituitary tumour cells are able to express the gene for **prolactin**, whereas other related lineages cannot do so. The reason being that the gene is methylated in the non-producing cells but is not methylated in cells that produce prolactin. Treatment of non-producing cells with azacytidine reverses methylation, resulting in prolactin expression in these cells.

Differential methylation and its epigenetic effects. The regulatory role of DNA methylation has been determined through an unusual type of epigenetic silencing known as **genomic imprinting** (Box 16.3), which controls expression of some genes involved in development of embryos in mammals. Usually, both paternal and maternal alleles of a gene are expressed in diploid cells. However, a few imprinted genes have been found in mouse and humans with expression depends upon whether they

Box 16.3

Genomic imprinting is a phenomenon in mammals whereby only the sperm-derived or only the egg-derived allele of the gene is expressed, sometimes due to inactivation of one allele by DNA methylation during spermatogenesis or oogenesis.

It is generally assumed that the genes one inherits from one's father and the gene one inherits from one's mtoher are equivalent. In fact, the basis for Mendelian ratios (and the Punnett square analyses used to teach them) is that it does not matter whether the genes came from the sperm or from the egg. But in mammals, there are at least 80 genes for which it does not matter. Here, the chromosomes from the male and female are not equivalent. In these cases, only the sperm-derived or only the egg-derived allele of the gene is expresed. This means that a severe or lethal condition arises if a mutant allele is derived from one parent, but that the same mutant allele will have no deleterious effects if inherited from the other parent. In some of these cases, the nonfunctioning gene has been rendered inactive by DNA methylation. This means that a mammal must have both a male parent and a female parent. Unlike sea urchins male parent and a female parent. Unlike sea urchins, flies and frogs, mammals cannot experience parthenogenesis or "virgin birth" (Gilbert 2010).

have been inherited from the mother or from the father. In some cases, only the paternal allele of imprinted gene is expressed, whereas the maternal allele is transcriptionally inactive. In other imprinted genes, the maternal allele is expressed while the paternal allele is inactive. It seems that DNA methylation is able to distinguish between paternal and maternal alleles of imprinted genes. For example, the *gene H19* is specifically methylated during development of male germ

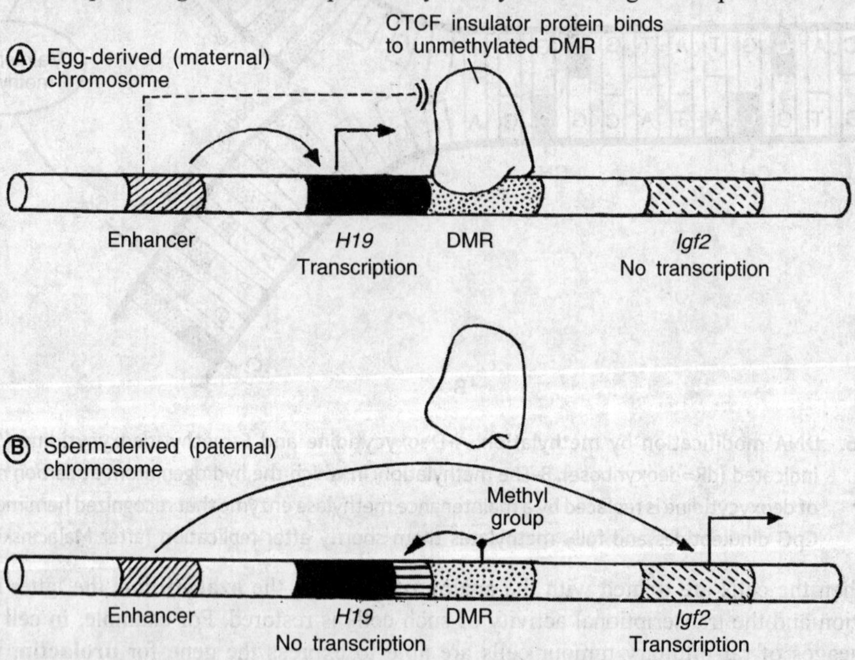

Fig. 16.7. Regulation of the imprinted *Igf2* gene in the mouse. A-This gene is activated by an enhancer element it shares with the H19 gene. The differentially methylated region (DMR) is a sequence located between the enhancer and the *Igf2* gene and is found on both sperm- and egg-derived chromosome, the DMR is unmethylated. The CTCF insulator protein binds to the DMR and blocks the enhancer sequence. B-In the sperm-derived chromosome, the DMR is methylated. The CTCF insulator protein cannot bind to the methylated sequence, and the signal from the enhancer is able to activate *IgF2* transcription (after Gilbert 2010).

cells, but not female germ cells (Fig. 16.7). Thus, *H19* is transcribed only from the maternal copy. After fertilization, the zygote develops into an embryo containing a methylated paternal allele and an unmthylated maternal allele of this gene. The paternal H19 allele therefore, remains methylated and transcriptionally inactive in embryonic cells and somatic tissues.

The process of genomic imprinting displays following characteristics: 1. Imprinting occurs in the germ line. 2. It affects at the most a few hundred genes, many of them located in clusters. 3. It is accompanied by heavy methylation. 4. Imprinted genes are differentially methylated in the female and male germ lines. 5. Once imprinted and methylated, a **silenced gene** remains transcriptionally inactive during embryogenesis.

Although mammalian gametes are extensively methylated, most of the DNA is demethylated in preimplantation development, except for imprinted genes that have a sex-specific **gene-silencing pattern**. After implantation, embryonic DNA is remethylated to the heavy methylation levels that are present in adult somatic cells. In the germ line of both males and females, all remethylated genes acquire methylation to an identical level. Except for a few genes that have sex-specific patterns of imprinting and different levels of methylation in female and male. In females, imprinted genes undergo methylation during growth of oocytes prior to ovulation. Because methylation associated with imprinting is retained throughout embryonic development, any gene that is imprinted in either the female or male germ line has, therefore, only one active copy in the embryo.

Epigenetic sex-specific gene silencing is observed in a pair of human syndromes called **Prader-Willi syndrome** (Box 16.4) and **Angelman syndrome** that display neuromuscular defects, mental retardation and some other abnormalities. Both conditions result from rare spontaneous deletions involving the chromosomal region *15q11*. If the deletion takes place in the father, it results in

Box 16.4 Prader-Willi Syndrome

It is a genetic disorder characterized by short stature, mental retardation, hypotonia (the state of having hypotonic muscle tone), abnormally small hands, and feet, hypogonadism (small gonads), and uncontrolled appetite leading to extreme obesity. It is caused due to deletion in chromosome 15 of sperm.

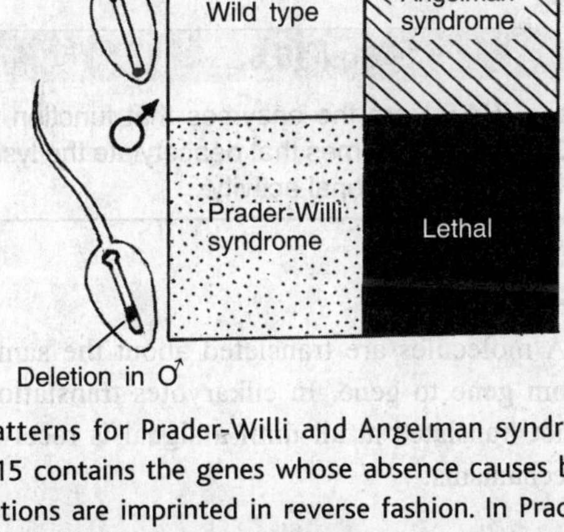

Fig. 16.8. Inheritance patterns for Prader-Willi and Angelman syndromes. A region in long arm of chromosome 15 contains the genes whose absence causes both these syndromes. However, the two conditions are imprinted in reverse fashion. In Prader-Willi syndrome, the paternal genes are active, in Angelman syndrome, the maternal genes are active (after Gilbert 2010).

Prader-Willi syndrome, whereas if the deletion occurs in the mother, it produces Angelman syndorme (Fig. 16.8). The reason is that the DNA in the deleted region *15q11* includes at least three genes that are imprinted and differentially methylated in the male and female gametes.

Nucleosome Remodelling

It is a dynamic process of changing the configuration of the condensed chromatin into one that is accessible to the transcription machinery of the cell. Different proteins act as repressors of the histones and thus activators of transcription.

Box 16.5

Methylation of histone acts as a signal for activation as well as repression of the transcription. Methylation of histone H3 at lysine 9 (H3K9) and at lysine 27 (H3K27) is a signal for gene repression whereas methylation H3K4, H3K36 and H3K79 is a signal for gene activation.

Modifications of histones. Histone component of chromatin is subject to three different post-synthetic modifications which have either direct or indirect effect on eukaryotic gene regulation: 1. **Histone methylation** affects only histones H3 and H4 and involves the irreversible methylation of a few **lysine** residues which alters the hydrophobic nature of the side chain of these histones (Box 16.5). 2. **Histone phosphorylation** involves histone H1, phosphorylation affects **serines** and **threonines**, changing them from a state of neutral charge to one of negative charge and is a reversible reaction. The state of phosphorylation of H1 protein varies through the eukaryotic cell cycle, and after H1 phosphorylation, chromatin becomes much more strongly condensed, as it does in mitotic chromosomes. Evidently, activation of the **histone kinase** enzyme that is responsible for H1 phosphorylation may be the first step in the chain of events that leads to eventual chromatin condensation prior to mitotic cell division. 3. **Histone acetylation** (Box 16.6) is of two types. The first is the irreversible acetylation of the amino terminal **serines** of histones H1, H2A and H4. These modifications seem to be associated with histone synthesis. The second is reversible acetylation of **lysine** residues in the amino-terminal regions of histones H2A, H2B, H3 and H4. Acetylation converts the normally basic lysine side chain to a neutral acetyl lysine, and, thus, reduces the net basic charge of the amino-terminal ends of the affected histones. Both H3 and H4 can have up to four lysines in the acetyl form, and there is a strong correlation between this type of histone acetylation, especially **tetracetylation** of all available lysines, and transcriptionally active chromatin. Evidently, the acetylation of the core histone lysines would tend to loosen the nucleosomal structure which brings about the transition from a silent condensed gene to a transcriptionally active and extended one.

Box 16.6

Histone acetyl transferases (HATs) are the enzymes that function to acetylate the histone. **Histone deacetylases (HDACs)** are the enzymes that deacetylate the lysine residue in the histone proteins and thus repress the gene transcriptional activity.

16.2 TRANSLATIONAL CONTROL

In bacteria, most mRNA molecules are translated about the same number of times with only fairly small variation from gene to gene. In eukaryotes translational regulation occurs in which a mRNA molecule is not translated at all until a signal is received. Translational control may involve the following mechanisms:

1. Extension of lifetime of the mRNA. An important example of translational regulation is that of **informosomes** or **masked mRNA**. Unfertilized eggs are biologically static, but shortly after fertilization many new proteins must be synthesized, for example, the proteins of the mitotic apparatus, the cell membranes, histones for nucleosome formation as well as others. Unfertilized sea urchin eggs store large quantities of mRNA for many months in the form of mRNA-protein particles (=masked mRNA) made during formation of the egg. This mRNA is translationally inactive, but within minutes after fertilization, translation of these molecules begins. Here, the timing of translation is regulated; the mechanism for stabilizing the mRNA, for protecting it against RNases, and for activation are still unknown.

2. Regulation of rate of protein synthesis. This type of regulation also occurs in mature unfertilized eggs. These cells need to maintain themselves but do not have to grow or undergo a change of state. Thus, the rate of protein synthesis in eggs is generally low. This is not due to inadequate supply of mRNA but of a limitation of an as-yet-undefined element, called the **recruitment factor** which apparently interferes with formation of the ribosome-mRNA complex. A good example of translational control is the extension of the lifetime of silk fibroin mRNA in the silkworm *Bombyx mori*. During cocoon formation on the silk gland of the silk worm predominantly synthesises a single type of protein, **silk fibroin**. Since the worm takes several days to construct its cocoon, it is the total amount and not the rate of fibroin synthesis that must be great; the silk worm achieves this by synthesising a fibroin mRNA molecule that is very long lived.

Transcription of the fibroin gene is initiated at a strong promoter by an unknown signal and about 10^4 fibroin mRNA molecules are made in a period of several days (such a synthesis forms an example of transcriptional regulation). A typical eukaryotic mRNA molecule has a lifetime of about three hours before it is degraded. However, the fibroin mRNA survives for several days during which each mRNA is translated repeatedly to yield 10^5 fibroin molecules. Thus, each gene is responsible for the synthesis of 10^9 protein molecules in four days. Altogether the silk gland makes 300 μg or 10^{15} molecules of fibroin during this period. If the lifetime of mRNA were extended, cither 25 times as many genes would be needed or synthesis of the required fibroin would take about 100 days.

Further, free iron ions (Fe^{2+}) within a mammalian cell are bounded by a storage protein, **ferritin**. When iron is in excess, ferritin synthesis rises dramatically. Yet the amount of ferritin mRNA remains constant. This increase in ferritin synthesis is due to an increased rate of mRNA translation. *When the iron level in the cell is low,* a **translational repressor** protein binds to ferritin mRNA and prevents its translation by blocking its attachment to the ribosome. *When iron level rise,* the excess iron binds to the repressor and alters the three-dimensional structure of the translational repressor, causing it to detach from the mRNA and translation proceeds (**Purves** *et al.,* 1998).

16.3 POST-TRANSLATIONAL MODIFICATION OF PROTEINS TO MAKE THEM ACTIVE ONES

Some proteins are altered after synthesis, usually by partial degradation or trimming, as for example, by the enzymatic removal of the central section of the **proinsulin** molecule to yield the active protein, **insulin**. For their activity, many proteins also depend on being complexed into compound proteins together with other subunits, either the same or different in nature. Such post-translational control mechanisms do play a significant role in determining the activities of differential cells. For example, **haemoglobin** production is highly dependent on the availability of haem to complex with globin protein subunits which may be deficient in cases of iron-dependent anaemia.

16.4 HORMONAL CONTROL OF GENE EXPRESSION

In higher plants and animals, intercellular communication is a very important phenomenon. Signals originating in various glands and/or secretory cells somehow stimulate **target tissue** or **target cells** to undergo dramatic changes in their metabolic patterns. These changes frequently include altered pattern of differentiation that are generally dependent on altered patterns of gene expression. **Peptide hormones** such as insulin, epinephrine, etc., and **steroid hormones** such as estrogen, progesterone, testosterone (in higher animals, *e.g,* mammals) and ecdysone (Fig. 16.10) (in insects)

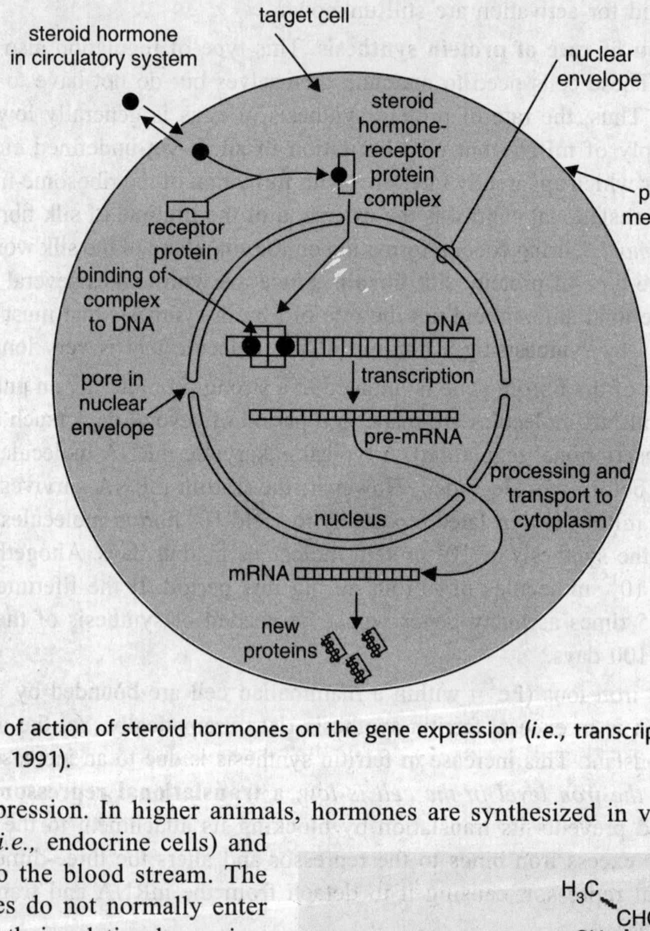

Fig. 16.9. Mode of action of steroid hormones on the gene expression (*i.e.*, transcription) (after Gardner, *et al.*, 1991).

control gene expression. In higher animals, hormones are synthesized in various specialized secretory cells (*i.e.,* endocrine cells) and are released into the blood stream. The peptide hormones do not normally enter cells because of their relative large size. Their effects appear to be mediated by receptor proteins located in target-cell membranes and by the intracellular levels of **cyclic AMP (cAMP)** (called **secondary messenger**). The cAMP activates a protein kinase (*e.g.,* A-kinase) which phosphorylates (activates) many specific enzymes. The steroid hormones, on the other hand, are small molecules that readily enter cells through the plasma membrane (Fig. 16.9). Once inside the appropriate target cells, the steroid hormones become tightly bound to **specific receptor proteins** which are present only in the

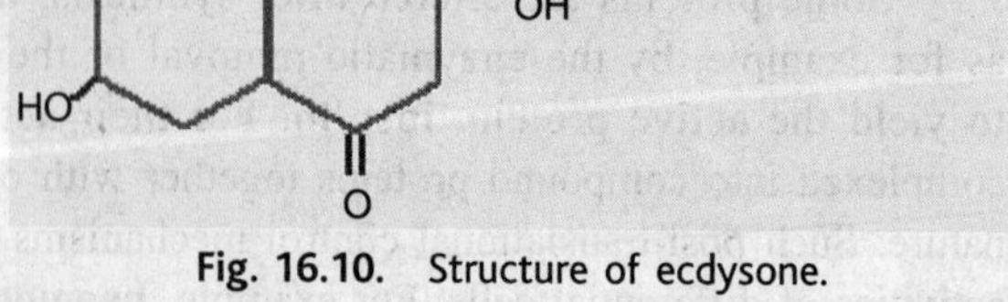

Fig. 16.10. Structure of ecdysone.

cytoplasm of target cells. The hormone-receptor protein complexes activate the transcription of specific genes or sets of genes according to following two methods (=hypotheses): 1. The hormone receptor protein complexes interact with specific non-histone chromosomal protein and this interaction stimulates the transcription of the correct genes (**J. Stein, G. Stein** and **L.Kleinsmith**, 1975). 2. The hormone receptor protein complexes activate transcription of target genes by binding to specific DNA sequences present in the *cis*-acting regulatory regions (the enhancers and promoter regions) of the genes (**R.M. Evans**, 1988). In both of these cases, these hormone-receptor protein complexes would function as positive regulators (or "activators") of transcription much like the CAP-cAMP complexes in prokaryotes.

During development of dipteran flies (*Drosophila melanogaster* and *Chironomus tentans*) the steroid hormone **ecdysone** is released and triggers moulting. If larvae of these insects are treated with ecdysone at stages of development prior to or between moultings, patterns of chromosome puffing occur that are identical to those occurring during natural moultings. Ecdysone tends to affect the gene expression at the level of transcription.

QUESTIONS

Long Answer Questions

1. Describe the Britten-Davidson's model of regulation of gene activity in eukaryotes.

2. Describe regulation of gene action in eukaryotes at the level of DNA or genetic code.

3. Discuss the relative roles of histones and non-histone proteins in regulation of gene activity in the eukaryotes.

4. In an eukaryotic system, what barriers would an extracellular repressor have to pass through before ultimately binding to DNA?

5. In what way is the regulation of a gene that is active in a differentiating cell inherently different from regulation of a bacterial gene?

Short Answer Questions

1. Write short notes on the following:

(*i*) Alternative splicing;

(*ii*) Repression;

(*iii*) Steroid hormones and gene expression;

(*iv*) Genomic imprinting;

(*v*) Heterochromatin.

Very Short Answer Questions

1. What kinds of mutations would completely eliminate translation of the entire sequence encoding a polypeptide?

2. What is enhancer?

3. Define imprinting.

4. What are acetyltransferases (HATs)?

5. Define the histone deacetylases (HDACs).

6. What is alternative splicing?

7. Define activator.

8. What are CpG islands?

9. Define the epigenetic effect.

Fill in the Blanks

1. Eukaryotic transcriptional factors are of two types: .... and ....
2. .... are located faraway from the operator, either upstream or ....
3. In .... a single gene can encode two or more related proteins.
4. Gene-battery model of transcription regulation in eukaryotes was proposed by .... and .... in 1969.
5. In mammals, DNA is methylated specifically at the .... that preced guanosine residues.
6. Methylation of DNA is correlated with .... transcriptional activity of genes.
7. In mammals, genomic imprinting is found to occur in about .... genes.
8. Prader-Willi __________ in humans is caused due to deletion in chromosome 15 of ....

Multiple Choice Questions

1. Methylation of DNA most commonly occurs in the sequence of
   (*a*) Cp G (*b*) Cp A
   (*c*) Cp T (*d*) Cp C
2. "Gene-battery model" of gene regulation in eukaryotes is proposed by
   (*a*) Jacob and Monod
   (*b*) Britten and Davidson
   (*c*) Beadle and Tatum
   (*d*) Kornberg and Ochoa
3. Fisher and Krebs got Nobel Prize in 1992 for their discovery of
   (*a*) organ and cell transplantation
   (*b*) reversible protein phosphorylation as biological regulatory mechanism
   (*c*) single ion channels
   (*d*) all of the above

ANSWERS

Very Short Answer Questions

1. A promoter mutation, which would prevent transcription; or mutation in the start codon, which would prevent initiation.
2. Enhancer is a regulatory site in the DNA that may be located at considerable distance either upstream or downstream from the promoter that it regulates. Binding of one or more transcriptional factors to the enhancer can dramatically increase the rate of transcription.
3. Imprinting is differential expression of genes depending solely on whether they were contributed to the zygote by the sperm or the egg.
4. Enzymes that transfer acetyl groups to lysine and arginine residues of core histones. Histone acetylation is associated with activation of transcription.
5. Enzymes that catalyze the removal of acetyl groups from core histones. Histone deacetylation is associated with the repression of transcription.
6. Alternative splicing is a widespred mechanism by which a single gene can encode two or more related proteins.
7. Activator is eukaryotic specific transcription factor that binds to an enhancer, often far upstream of a promother.

8. CpG islands are stretches of CG repeats (in which CpG indicates sequential bases on the same strand of DNA, rather than a C-G base pair). These repeats, found in imprinted centers, are important in regulation of gene actions.

9. Epigenetic effect is an environmentally induced change in the genetic material that does not cause a change in base pairs of alleles of a locus depending on the parent of origin.

Fill in the Blanks

1. General transcription factors, specific transcription factors.

2. Enhancers, downstream.

3. Alternative splicing.

4. Britten, Davidson.

5. Cytosines.

6. Reduced.

7. 80.

8. Syndrome, sperm.

Multiple Choice Questions

1. (*a*) **2.** (*b*) **3.** (*b*).

17

DNA Repair Mechanisms

Radiation, chemical mutagens, heat, enzymatic errors, and spontaneous decay constantly damage DNA. For example, it is estimated that several thousand DNA bases are lost each day in every mammalian cell due to spontaneous decay (**Tamarin**, 2002). Some types of damage interfere with DNA replication and transcription. In the long evolutionary challenge to minimize mutation, cell have evolved numerous mechanisms to repair incorrectly replicated DNA. Many enzymes, acting alone or in concert with other enzymes repair DNA. DNA repair systems are generally placed in four broad categories: 1. *damage reversal*, 2. *excise repair*, 3. *double strand breakage repair* and 4. *post replicative repair*. Enzymes that process DNA repair steps have been conserved during evolution. That is, enzymes found in *E.coli* have homologues in yeast, fruit flies, and human beings.

17.1 DAMAGE REVERSAL

Ultraviolet (UV) rays cause linkage or **dimerization** of adjacent pyrimidines in DNA (Fig. 17.1). Although cytosine-cytosine and cytosine-thymine dimers are occasionally produced, the

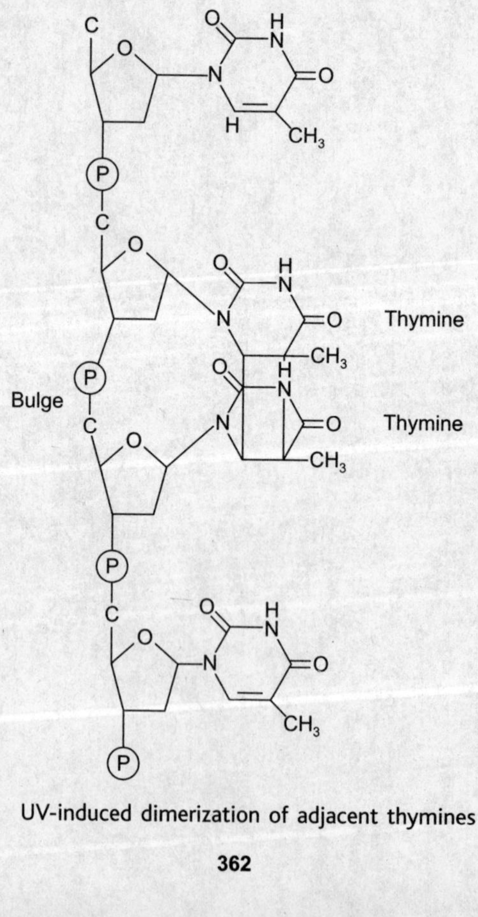

Fig. 17.1. UV-induced dimerization of adjacent thymines in DNA.

principal products of UV irradiation are thymine-thymine **dimers**. These can be repaired in several different ways. The simplest is to reverse the dimerization process and restore the original unlinked thymines.

In *E.coli*, an enzyme called **DNA photolyase**, the product of the *phr* gene (for photoreactivation), binds to dimerized thymines. When light shines on the cell, the enzyme breaks the dimer bonds with light energy. The enzyme then falls free of the DNA. Another example of an enzyme that performs direct DNA repair is **O^6-mGua DNA methyltransferase**, which removes the methyl groups from O^6-methylguanine, the major product of DNA-methylating agents. Photoreactivation does not occur in human beings.

17.2 EXCISION REPAIR

Excision repair refers to the general mechanism of DNA repair that works by removing the damaged portion of a DNA molecule. Various enzymes can detect damage or distortion in the DNA double helix. During excision repair bases and nucleotides are removed from the damaged strand. The gap is then patched using complementarity with the remaining strand. These systems are of three types: base excision repair, nucleotide excision repair and mismatch repair.

***(i)* Base excision repair.** A base can be removed from a nucleotide within DNA in several ways: by direct action of an agent such as radiation, by spontaneous hydrolysis, by an attack of oxygen free radicals, or by **DNA glycosylases** which are enzymes that detect damaged bases and removes them. According to **Tamarin** (2002), there are five glycoylases (enzymes) for excision repairs. For example, **uracil-DNA glycosylase**, the product of the *ung* gene in *E.coli*, recognizes uracil within DNA and cleaves it out at the base-sugar (glycosidic) bond. The resulting site is called an **AP** (=apurinic-apyrimidenic) **site**, because of the lack of a purine or pyrimidine at the site. An **AP endonuclease** then senses the minor distortion of the DNA double helix and initiates excision of the single AP nucleotide in a process known as **base excision repair**. The AP endonuclease nicks the DNA at the 5′ side of the base-free AP site. A **DNA polymerase** then inserts a nucleotide at the AP site; an **exonuclease, lyase** or **phosphodiesterate** enzyme then removes the base-free nucleotide. (Lyases are enzymes that can break C-C, C-O and C-N bonds.) **DNA ligase** then closes the nick (Fig. 17.2). The replacement of just one base occurs 80–90% of the time. In the remaining 10–20% of cases, several nucleotides may be removed, depending probably on which DNA polymerase (I or III) first repairs the site (**Lindahl** 1996).

In mammals, **DNA polymerase** β performs two roles in base excision repair: it both inserts a new base where the AP site was and also eliminates the AP nucleotide residue by exonuclease activity.

***(ii)* Nucleotide excision repair (NER).** Nucleotide excision repair is initiated by enzymes that sense distortions in the DNA backbone and replace a short stretch of nucleotides. For example, six enzymes in *E.coli* excise a short stretch of DNA containing thymine dimers if the dimerization is not reversed by photoreactivation. Two copies of the protein product of the *Uvr A* gene (for ultraviolet light UV-repair) combine with one copy of the product of the *Uvr B* gene to form a **Uvr A_2 Uvr B complex** that moves along the DNA, looking for damage (Fig. 17.3). (The complex has 5′ helicase activity). When the complex finds damage such as a thymine dimer, with moderate to large distortion of the DNA double helix, the Uvr A_2 dimer dissociates, leaving the Uvr B subunit alone. This causes the DNA to bend and attracts the protein product of the *uvr C* gene, *Uvr C*. The Uvr B subunit first nick (hydrolyzes), the DNA four to five nucleotides on the 3′ side of the lesion; next, the Uvr C subunit nicks the DNA eight nucleotides on the 5′ side

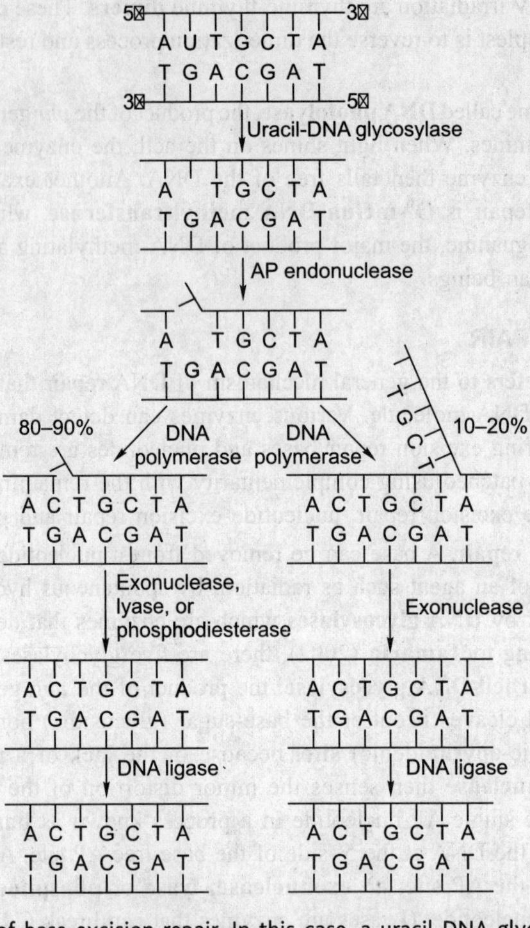

Fig. 17.2. Mechanism of base excision repair. In this case, a uracil-DNA glycosylase enzyme removes a uracil from DNA. An AP endonuclease nicks the DNA on the 5′ side of the base-free site. Between 80% and 90% of the time, a DNA polymerase will replace the single nucleotide, an exonuclease, lyase or phosphodiesterase will remove the base-free nucleotide. The final nick is sealed with DNA ligase. Between 10% and 20% of the time, the DNA polymerase will extend polymerization beyond the single nucleotide. In these cases, an exonuclease and DNA ligase finish the repair (after Tamarin 2002).

of the lesion. (The three components, Uvr A, Uvr B, and Uvr C, are together called the **ABC exonuclease**, for excision endonuclease). The enzyme helicase II, the product of the *Uvr D* gene, then removes the twelve-to-thirteen base oligonucleotide as well as Uvr C. DNA polymerase I fills in the gap and in the remaining nick (Fig. 17.3).

Like the excision repair, nucleotide excision repair is present in all organisms. In yeast, approximately twelve genes are involved, what is called the **RAD 3 group**. In human beings, twenty-five proteins are involved; they remove twenty-seven to twenty-nine nucleotides, as compared to twelve-to-thirteen in *E.coli*.

Transcription and nucleotide excision repair are linked in eukaryotes. Transcription factor TFIIH is involved in repair of UV damage. In human beings, the autosomal recessive trait **xeroderma pigmentosum** is caused by an inability to repair thymine dimerization induced by UV light. Persons with this trait freckle heavily when exposed to UV rays of the sun, and they have a high incidence of skin cancer. Defect in XPD component of TFIIH is main cause of this ailment.

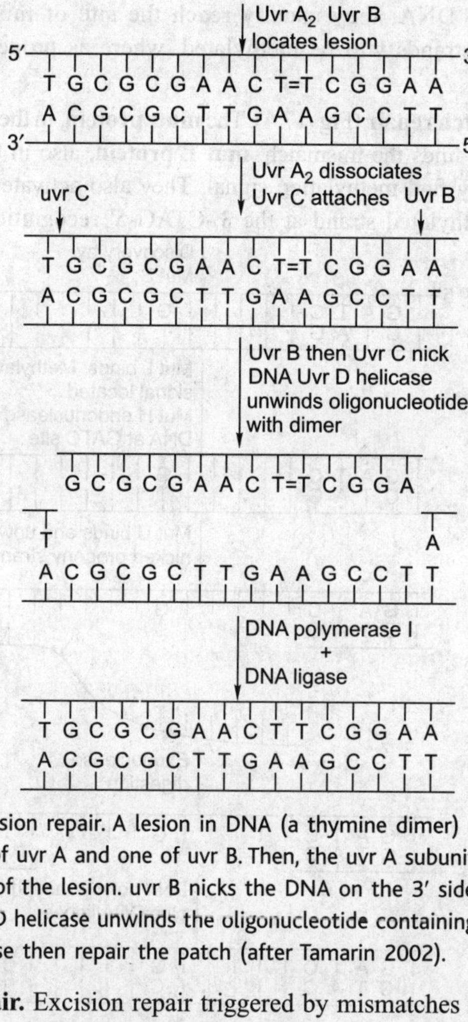

Fig. 17.3. Nucleotide excision repair. A lesion in DNA (a thymine dimer) is located by a protein made of two copies of uvr A and one of uvr B. Then, the uvr A subunits detach, and uvr C attaches on the 5′ side of the lesion. uvr B nicks the DNA on the 3′ side and uvr C on the 5′ side of the lesion; uvr D helicase unwinds the oligonucleotide containing the lesion. DNA polymerase I and DNA ligase then repair the patch (after Tamarin 2002).

***(iii)* Mismatch repair.** Excision repair triggered by mismatches is referred to as **mismatch repair**, which encompasses about 99% of all DNA repairs. As DNA polymerase replicates DNA, some errors are made that the proofreading polymerase does not correct. For example, a template G can be paired with a T rather than a C in the progeny strand. The GT base pair does not fit correctly in the DNA duplex. The mismatch repair system, which follows behind the replicating fork, recognizes this problem. This system, whose members in *E.coli* are specified by the *mut H*, *mut L*, *mut S* and *mut U* genes, is responsible for the removal of the incorrect base by an excision repair process. (The genes are called *mut* for mutator because mutation of these genes cause high levels of spontaneous mutation in the cell. The *mut U* gene is also known as *uvr D*). The mismatch repair enzymes initiate the removal of the incorrect base by nicking the DNA strand on one side of the mismatch.

Let's know how the mismatch repair system recognizes the progeny rather than the template, base as the wrong one. After all in a mismatch, there are defective bases—theoretically, either partner could be the "wrong" base. In *E.coli*, the answer lies in the methylation state of the DNA. **DNA methylase**, then product of the *dam* gene, methylates 5′-GATC-3′ sequences, which are relatively common in the DNA of *E.coli*, at the adenine residue. Since the mismatch repair enzymes follow the replication fork of the DNA, they usually reach the site of mismatch before the DNA methylase does. Template strands will be methylated, where as progeny strands, being newly synthesized, will not be.

Mechanism of mismatch repair (Fig. 17.4). The **muts protein**, in the form of homodimer—two copies of the same protein—finds the mismatch. **mut L protein**, also in the form of a homodimer, then binds, and together they find methylation signal. They also activate the endonuclease mut H, which then nicks the unmethylated strand at the 3′-CTAG-5′ recognition site which can be one

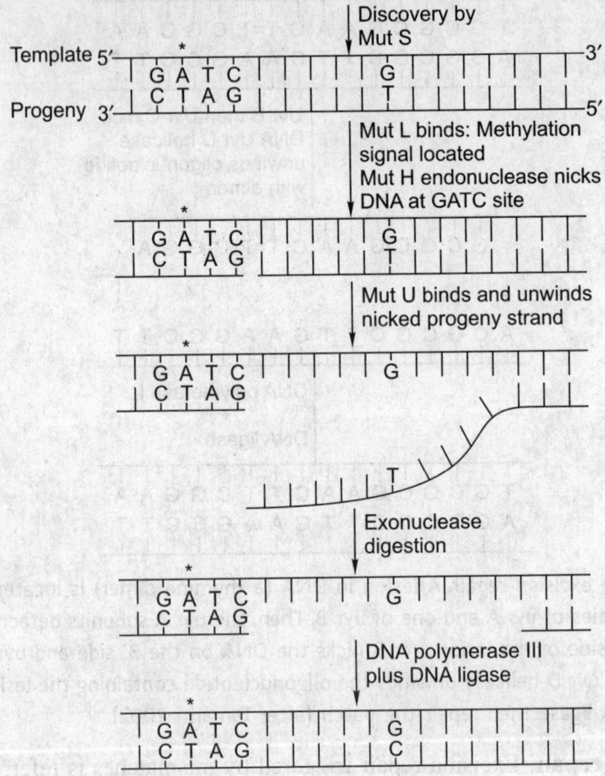

Fig. 17.4. Mismatch repair. The Mut S protein discovers mismatches; Mut L binds and Mut H endonuclease nicks the progeny strand at the 3′-CTAG-5′ sequence. Mut U helicase unwinds the nicked oligonucleotide with this mismatch. Exonuclease digestion, followed by DNA polymerase III and DNA ligase repair, completes the operation (after Tamarin 2002).

thousand to two thousand bases away from the mismatch. At the recognition site, the mut S–mut L tetramer loads the helicase mut U (uvr D), which then unwinds the nicked strand. Any one of at least four different exonucleases then attacks the unwound oligonucleotide. DNA polymerase III then repairs the gap, and DNA ligase seals it. This sequence of events highlights a common theme in DNA repair: once a lesion is found, the damaged DNA has some protein bound to it until the repair is finished.

At this juncture, we can answer an evolutionary question—*why does DNA have thymine while RNA has uracil*? If we live in an RNA world, in which RNA evolved first, why don't DNA and RNA both contain uracil? One answer is that a common damage to cytosine, spontaneous deamination, result in uracil. If uracil were a normal base in DNA, the conversion of cytosine to uracil by deamination would not leave any clue to a mismatch repair system that a mutation had occurred. Thus, thymine replaces uracil in DNA, since thymine is not confused with any other normal base in DNA by common spontaneous changes. Infact cytosine, guanine, adenine and thymine are not converted simply to any other of the bases in DNA. Hence, changes of these bases leave clues for the repair systems.

***(iv)* Double-strand break repair.** Some damage to DNA, such as that caused by ionizing radiation, is capable of breaking both strands of the double helix. When that happens, the cell uses following two mechanisms to repair the broken ends: (1) It can simply bring the ends back together. This is called **nonhomologous end joining**. (2) It can use a mechanism that relies on the nucleotide sequences of a homologous piece of DNA, such as a sister chromatid or a homologous chromosome. This method is called **homology-directed recombination**.

***(v)* Post-replication repair.** When DNA polymerase III encounters certain damage in *E.coli*, such as thymine dimers, it cannot proceed. Instead, the polymerase stops DNA synthesis and leaving a gap, skips down the DNA to restart replication as far as eight hundred or more bases away. If allowed to remain, this gap will result in deficient and broken DNA. Since part of one strand is absent and the other has damage, there appears to be no viable template for replicating new DNA. However, the cell has two mechanisms to repair the gap: one uses polymerases that can replicate these lesions and the other is a repair process that uses homologous DNA.

Initially, several proteins were known to facilitate the replication of DNA with lesions; they were believed to interact with the polymerase to make it capable of using damaged DNA as a template, we now know that these proteins are, in fact, polymerases that have the ability to replicate damage DNA. In *E.coli*, polymerase V can copy damaged DNA. In yeast, polymerases η (eta) and ζ (zeta), also called REV 3/7 and RAD 30 polymerase, respectively, can also copy damaged DNA. Some of these polymerases are relatively error free: polymerase V and polymerase η put adenine-containing nucleotides opposite dimerized thymine. However, polymerase ζ and the *E.coli* polymerase IV, which also appears at the times of damage, are error prone in their replicative roles (Box 17.1).

Box 17.1

One possible reason for error-prone polymerase is that they create mutation at a time when the cell might need variability. That is, DNA damage can occur when the environment is stressful for the cell; variability might help the cell survive (**Tamarin** 2002).

In addition, by using repair polymerases the cell can use a second repair mechanism to replicate damaged DNA when the polymerase leaves a gap. A replication fork creates two DNA duplexes. Thus, an undamaged copy of the region with the lesion exists on the other daughter duplex. A group of enzymes, with one specified by the *rec A* locus having central importance, repairs the gap. Since the repair takes place at a gap created by the failure of DNA replication, the process is called **post-replicative repair**.

The Rec A protein. The Rec A protein has two major properties: 1. It coats single-stranded DNA and causes that coated, single-stranded DNA to invade double-stranded DNA. By invasion, we mean that the single-stranded DNA attempts to form complementary base pairs with the antiparallel strand of the double stranded DNA while displacing the other strand of that double helix. Rec A protein continues to move the single-stranded DNA along the double stranded DNA until a region of homology is found. 2. When stimulated by the presence of single-stranded DNA, Rec A protein causes **autocatalysis** of another repressor called **lex A** and thus initiates several sequences of reactions.

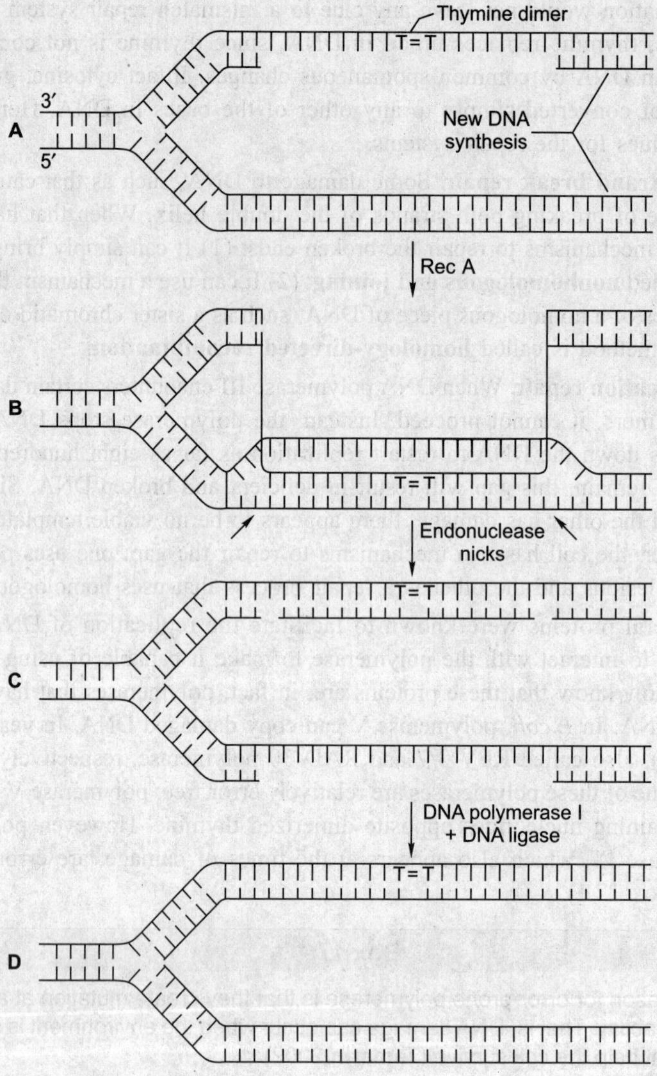

Fig. 17.5. Rec A protein dependent postreplicative DNA repair. A—DNA polymerase III skips past a thymine dimer during DNA replication; it produces a normal duplex and abnormal duplex (with gap against the thymine-dimer); B—With the help of Rec A protein, the single strand with the thymine dimer invades the normal sister duplex; C—An endonuclease nicks the new duplex at either side of the thymine dimer site, freeing the new duplex, with the thymine dimer and leaving the sister duplex single-stranded; D—Repair enzymes (DNA polymerase I and DNA ligase) then create two intact daughter duplexes (after Tamarin 2002).

The Rec A protein is important for filling a postreplicative gap in newly replicated DNA with a strand from the undamaged sister duplex. Gap-filling process then complete both strands. In Fig. 17.5A have been shown a replication fork with a gap in the progeny strand in the region of a thymine dimer. The Rec A protein is responsible for the damaged single strand invading the sister duplex (Fig. 17.5B). Endonuclease activity then frees the double helix containing the thymine dimer (Fig. 17.5C). DNA polymerase I and DNA ligase return both daughter helix to the intact state (Fig. 17.5D). The thymine dimer still exists, but now its duplex is intact, and another cell cycle is available for photoreactivation or excision repair to remove the dimer.

The SOS Response

The postreplicative repair is part of a cell reaction called the **SOS response**. When an *E.coli* cell is exposed to excessive quantities of UV light, other mutagens, or agents that damage DNA (such as alkylating or cross-linking agents), or when DNA replication is inhibited, gaps are created in the

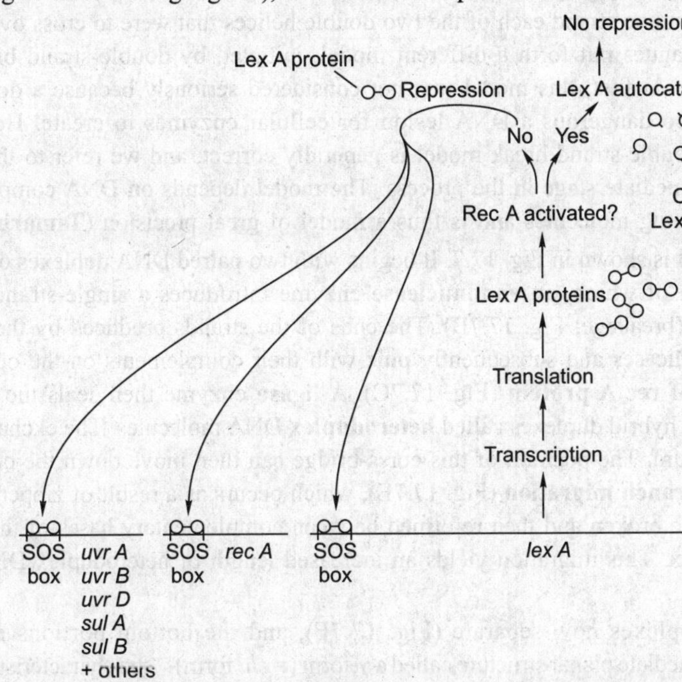

Fig. 17.6. The Lex A protein represses its own gene: *rec A* and several other loci (*uvr A*, *uvr B*, *sul A* and *sul B*) by binding at the SOS box in each of the loci. Activated Rec A protein causes autocatalysis of Lex A, eliminating the repression of all loci, which are then transcribed and translated (after Tamarin 2002).

DNA. In the presence of this single-stranded DNA, the Rec A protein interacts with the **Lex A protein** normally represses about eighteen genes include *rec A*, *uvr A*, *uvr B*, and *uvr D*; two genes that inhibit cell division, *sul A* and *sul B*; and several others. Each of these genes has a consensus sequence in its promoter called the **SOS box**; 5′-CTG X_{10} CAg (where X_{10} refers to any ten bases). The Lex A protein normally binds at the SOS box, limiting the transcription of these genes. When single stranded DNA activates Rec A; Rec A interacts with the Lex A protein to trigger the autocatalytic properties of Lex A (Fig. 17.6). Transcription then follows from all the genes, presumably increase the amount of time the cell has to repair the damage before the next round of DNA replication.

17.3 RECOMBINATION

Although recombination, the nonparental arrangement of alleles in progeny, can come about both by independent assortment and crossing over, we are concerned here with recombination due to crossing over between homologous pieces of DNA, called **homologous recombination**.

Recombination is a **breakage-and-reunion** process. Homologous parts of chromosomes come into apposition and then reconnected in a crosswise fashion. This general model fits what we know about the concordance of recombination and repair: both involve breakage of the DNA and a small amount of repair synthesis and both involve some of the same enzyme. (Concordance means the amount of phenotypic similarity between individuals).

Double Strand Break Model of Recombination

In 1964, **R. Holliday** suggested a model of homologous recombination that involved simultaneous breaks in one strand each of the two double helices that were to cross over. In 1983, **J. Szostak** and colleagues put forth a different model, initiated by double-strand break in one of the double helices. At first, this model was not considered seriously because a double-strand break was thought too dangerous a DNA lesion for cellular enzymes to create. However, we now know that the double-strand break model is generally correct, and we refer to the **Holliday junction** for an intermediate stage in the process. The model depends on DNA complementarity between the recombining molecules and is thus a model of great precision (**Tamarin** 2002).

One such model is shown in Fig. 17.7. It begins with two paired DNA duplexes or homologs (Fig. 17.7A), in each of which, an **endonuclease** enzyme introduces a single-stranded **nick** at an identical position (breakage; Fig. 17.7B). The ends of the strands produced by these cuts are then **displaced** by helicases and subsequently **pair** with their complements on the oppsite DNA duplex by the help of **rec A protein** (Fig. 17.7C). A **ligase** enzyme then seals the loose ends (Fig. 17.7D), creating hybrid duplexes called **heteroduplex** DNA molecules. The exchange creates a **cross-bridge** structure. The position of this corss-bridge can then move down the chromosome by a process called **branch migration** (Fig. 17.7E), which occurs as a result of zipper-like action as hydrogen bonds are broken and then reformed between complementary bases of the displaced strands of each duplex. This migration yields an increased length of heteroduplex DNA on both homologs.

If the DNA duplexes now separate (Fig. 17.7F), and the bottom portions rotate 180° (Fig. 17.7G), an intermediate planar structure called a χ form (= *chi* form)—the characteristic **Holliday structure** or **Holliday junction**—is created. Separation, or **resolution**, of the Holliday structure back into individual double stranded molecules occurs by cleavage across the branch point. *This is key to the entire process* because the cut can be made in either of two orientations, as becomes apparent when the three-dimensional configuration or chi-form of the Holliday structure is examined. These two cuts produce very different results. If the cut is made left-right across the chi-form (*i.e.*, **horizontal resolution**), then all that happens is that a short segment of polynucleotides corresponding to the distance migrated by the branch of the Holliday structure, is transferred between the two molecules. On the other hand, an up-down (*i.e.*, **vertical resolution**) results in **reciprocal strand exchange**, double-stranded DNA being transferred between the two molecules so that the end of one molecule is exchanged for the end of the other molecule. This type of DNA transfer is seen as crossing over.

Evidences supporting this model includes the electron microscopic visualisation of χ-form (chi-form) planar molecules from bacteria in which four duplex arms are joined at a single point of exchange. Further important evidence comes from the discovery of the **rec A protein** in *E.coli*. The molecule promotes the exchange of reciprocal single-stranded DNA molecule as occurs in step C of Fig. 17.7. Rec A protein also enhances the hydrogen-bond formation during strand displacement, thus, initiating heteroduplex formation. Finally, many other enzymes essential to the nicking and ligation process have also been discovered and investigated. The products of the *rec B*, *rec C* and *rec D* genes are thought to be involved in the nicking and unwinding of DNA. Numerous mutations that prevent genetic recombination have been found in viruses and bacteria. These mutations represent genes whose products play an essential role in the process.

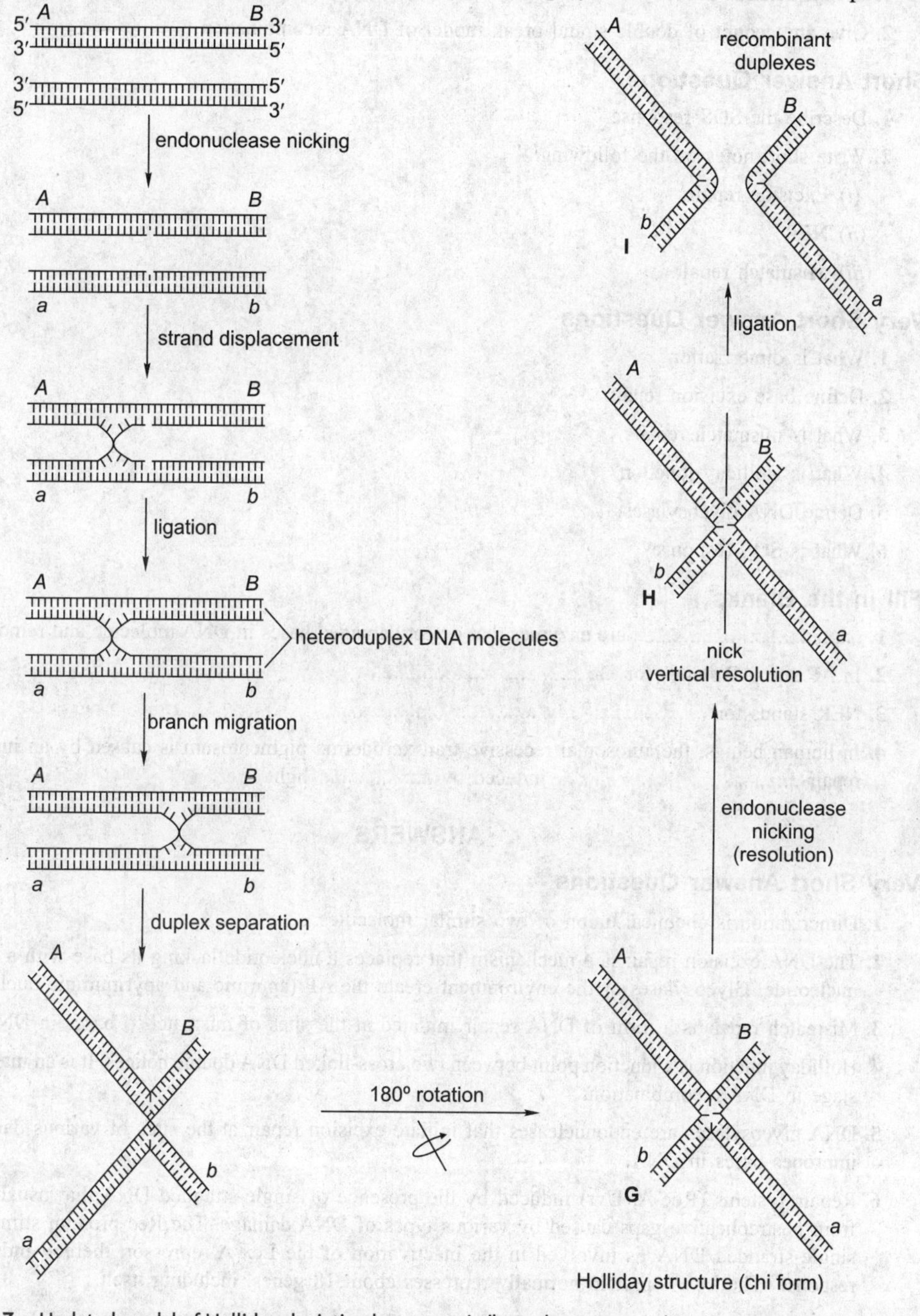

Fig. 17.7. Updated model of Holliday depicting how genetic/homologous recombination can occur as a result of the breakage and rejoining of heterologous DNA strands (after Klug and Cummings, 2003).

QUESTIONS

Long Answer Questions

1. Give a detailed account of various ways of DNA repair.
2. Give an account of double strand break model of DNA recombination.

Short Answer Questions

1. Describe the SOS response.
2. Write short notes on the following:
   (*i*) Excision repair;
   (*ii*) NER;
   (*iii*) Mismatch repair.

Very Short Answer Questions

1. What is dimerization?
2. Define base excision repair.
3. What is mismatch repair?
4. What is Holliday junction?
5. Define DNA glycosylases.
6. What is SOS response?

Fill in the Blanks

1. .... are enzymes that detect damaged bases in DNA molecule and remove them.
2. In AP site, AP stands for ....
3. NER stands for ....
4. In human beings, the autosomal recessive trait xeroderma pigmentosum is caused by an inability to repair .... induced by .... light.

ANSWERS

Very Short Answer Questions

1. Dimerization is chemical union of two similar molecules.
2. The DNA excision repair is a mechanism that replaces a nucleotide lacking its base with a complete nucleotide. Glycosylases or the environment create the AP (apurinic and apyrimidinic nucleotides).
3. Mismatch repair is a form of DNA repair initiated at the sites of mismatched bases in DNA.
4. Holliday junction is a junction point between two cross-linked DNA double helices. It is an intermediate stage in DNA recombination.
5. DNA glycosylases are endonucleases that initiate excision repair at the sites of various damaged or impropes bases in DNA.
6. Repair systems (Rec A, Uvr) induced by the presence of single-stranded DNA that usually occurs from postreplicative gaps caused by various types of DNA damage. The Rec protein, stimulated by single-stranded DNA, is involved in the inactivation of the Lex A repressor, thereby inducing the response. The Lex A protein normally represses about 18 genes, including itselt.

Fill in the Blanks

1. DNA glycosylases.
2. Apurinic-apyrimidenic.
3. Nucleotide excision repair.
4. Thymine dimerization, UV.

18

Regulatory RNAs: Riboswitches, RNA Interference (RNAi), siRNAs and miRNAs

18.1 RIBOSWITCHES

Riboswitch is a site in mRNA to which a small metabolite molecule can bind, triggering changes in mRNA conformation that impact either transcription or translation (**Becker** *et al.*, 2006). Over the past few years a different type of mechanism has captured the interest of researchers studying **bacterial gene regulation**. It has been found that proteins, such as the *lac* and *trp* repressors, are not the only gene regulatory molecules that are influenced by interaction with small metabolites. A number of messenger RNA (mRNAs) have been identified that can bind a **small metabolite** such as **glucosamine** or **adenine** with remarkable specificity. The metabolite binds to a highly strucutred 5′ noncoding region of the mRNA. Once bound to the metabolite, these mRNAs are called **riboswitches**. They undergo a change in their folded conformation that allows them to alter the expression of a gene involved in production of that metabolite. Most riboswitches suppress gene expression by blocking either termination of transcription or initiation of translation. Like the repressors that function in conjuction with operons, riboswitches allow cells to adjust their level of gene expression in response to changes in the available level of certain metabolites. Given that they act without the participation of protein cofactors, riboswitches are likely another legacy from ancestral RNA world. One type of riboswitch has been discovered in plants and fungi and a search for this new type of gene regulation in animal cells is underway (**Karp** 2010).

One of the first riboswitches to be discovered is in the mRNA transcribed from the **riboflavin (rib) operon** of *Bacillus subtilis*. The *rib* operon consists of five genes coding for enzymes that are involved in the pathway for synthesizing the essential coenzymes, FMN and FAD. RNA transcribed from the *rib* operon possesses a leader sequence that, like the *trp* leader discussed earlier (Fig. Genetics 15.19), contains a region that can fold into a **hairpin loop** that will terminate transcription. As shown in Figure 18.1, binding of FMN to the leader sequence promotes the formation of this hairpin loop. Transcription of the *rib* operon is therefore terminated when FMN is present, halting the production of enzymes that are not needed because they are involved in the synthesis of FMN itself.

Besides regulating transcription, the binding of small molecules to riboswitches can also control mRNA translation. One example occurs in mRNAs coding for enzymes involved in the pathways for synthesizing FMN and FAD in *E. coli*. Unlike the situation in *B. subtilis*, the *E. coli* genes coding for these enzymes are not **clustered** in a single operon, but some of them are still controlled by riboswitches. In this case, binding of FMN to its mRNA riboswitch promotes the formation of a hairpin loop that includes the sequences required for binding of mRNA to ribosomes. By sequestering away these sequences, the bound FMN prevents the mRNA from interacting with ribosomes and initiating translation (Fig. 18.1B).

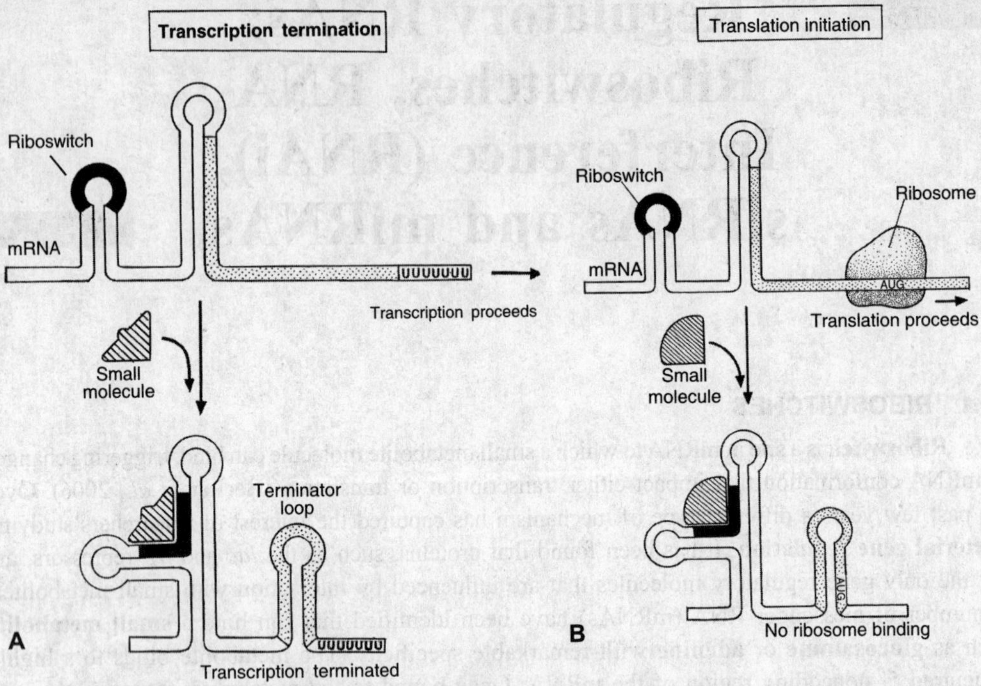

Fig. 18.1. Riboswitch-mediated control of gene expression. This model shows two examples of how the binding of small molecules to riboswitches of mRNA can exert control over transcription and translation. A-Binding of a small molecule to a riboswitch in the leader sequence of some mRNAs triggers the formation of a hairpin loop that terminates transcription. FMN binding to the leader sequence of the mRNA transcribed from the *rib* operon of *B. subtilis* works in this way. B-In other mRNAs, small molecule binding to a riboswitch triggers formation of a hairpin loop containing the site where ribosomes would normally bind, thereby interfering with the initiation of translation. In *E.coli* this type of control is used by FMN to inhibit translation of mRNA, that code for enzymes involved in FMN synthesis (after Becker *et al.*, 2006).

18.2 RNA INTERFERENCE

The idea that RNA molecules are directly involved in the regulation of gene expression began with a puzzling observation. The petals of a petunia plant are normally light purple. In 1990, two groups of investigators reported on an attempt to deepen the colour of the flowers by introducing extra copies of a gene encoding a pigment-producing enzyme. To the surprise of the researchers the presence of the extra genes caused the petals to lose their pigmentation rather than become more darkly pigmented as expected. Subsequent studies indicated that under these experimental conditions, both the added genes and their normal counterparts within the genome were being transcribed, but the resulting mRNAs were somehow degraded. This phenomenon became known as **post-transcriptional gene silencing (PTGS)**.

It was not unitl 1998 that the molecular basis for this form of gene silencing became understood. In that year, **Andrew Fire** and **Craig Mello** and their colleagues conducted an experiment on the nematode *Caenorhabditis elegans*. They injected these worms with several different preparations of RNA hoping to stop production of a particular muscle protein. One of the preparations contained "sense" RNA, that is, an RNA having the sequence of the mRNA that encoded the protein being targeted; another preparation contained "anti-sense" RNA, that is, an RNA having the complementary sequence of a double-stranded RNA containing both the sense and antisense sequences bound to one another. Neither of the single-stranded RNAs had much of an effect, but the double-stranded RNA was very effective in stopping production of encoded protein. Fire and Mello described the phenomenon as **RNA interference (RNAi)**. They demonstrated that double-stranded RNAs (dsRNAs) were taken up by cells where they induced a response leading to the selective destruction of mRNAs having the same sequence as the added dsRNA. In other words, we can say that if one sought to stop the production of the enzyme **glycogen phosphorylase** within the cells of a nematode worm so that the effect of this enzyme deficiency on the phenotype of the worm could be determined. Remarkably, this result can be attained by simply placing the worm in a solution of dsRNA that shares the same sequence as the target mRNA.

The phenomenon of dsRNA-mediated RNA interference is an example of the broader phenomenon of **RNA silencing**, in which small RNAs, typically working in conjuction with protein machinery, act to inhibit gene expression in various ways. RNAi is thought to have evolved as a type of "**genetic immune system**" to protect organisms from the presence of foreign or unwanted genetic material. To be more specific, RNAi porbably evolved as a mechanism to block the replication of viruses and/or to suppress the movements of transposons within the genome because both of these potentially dangerous processes can involve the formation of double-stranded RNAs. Cells can recognize dsRNAs as "undesirable" because such molecules are not produced by the cell's normal genetic activities.

For the process of RNA silencing or RNA interference, the following four main types of RNAs are produced in the eukaryotic cells: 1. si RNAs; 2. miRNAs; 3. piRNAs and 4. nc RNAs.

18.3 SMALL INTERFERING RNAs (siRNAs)

The steps involved in the RNAi pathway are shown in Figure 18.2. The double-stranded RNA that initiates the response is first cleaved into small (21-23 nucleotide), double-stranded fragments, called **small interfering RNAs (siRNAs)**, by a particular type of ribonuclease, called **Dicer** (Fig. 18.3). Enzymes of the class which Dicer belongs act specifically on dsRNA substrates and generate small dsRNAs that have 3′ overhangs (Fig. 18.2). These small dsRNAs are then loaded into a complex (identified as pre-RISC in Fig. 18.2) that contain a member of Argonaute family of proteins. Argonaute proteins play a key role in all of the known RNA silencing pathways. In the RNAi pathway, one of the strands of the RNA duplex (called the **passenger strand**), is cleaved in two and then dissociate from the pre-RISC. The other strand of the RNA duplex is called the **guide strand** and is incorporated together with its Argonaute partner, into a related protein complex named RISC. Several different varieties of RISCs have been identified, distinguished by the particular Argonaute protein they contain. In animal cells, RISCs that contain an siRNA typically include the

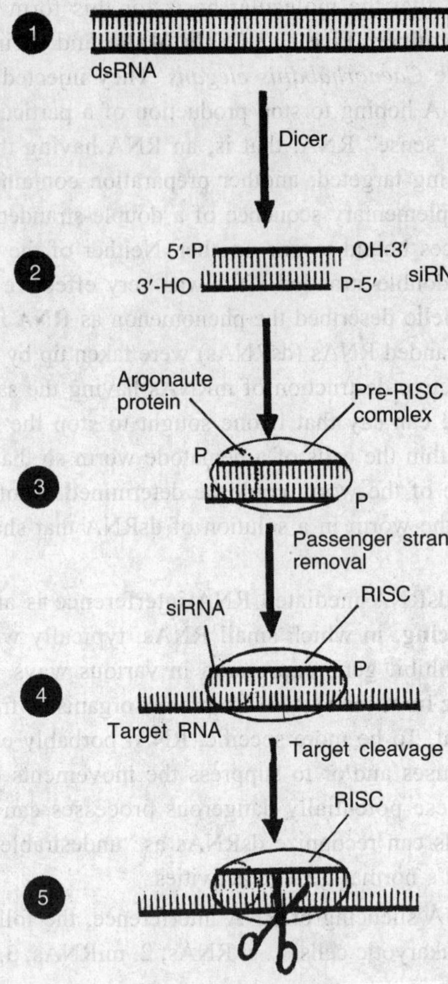

Fig. 18.2. The formation and mechanism of action of siRNAs. In *step 1*, both strands of a double-stranded RNA are cleaved by the endonuclease Dicer to form a small (21-23 nucleotide) siRNA, which has overhanging ends (*step 2*). In *step 3*, the siRNA becomes associated with a protein complex, a pre-RISC, that contains an Argonaute protein (typically Ago2) capable of cleaving and removing the passenger strand of siRNA duplex. In *step 4*, the single-stranded guide RNA, in association with proteins of the RISC complex binds to a target RNA that has a complementary sequence. The target RNA might be a viral RNA, a transcript from a transposon, or an mRNA, depending on circumstances. In *step 5*, the target RNA is cleaved at a specific site by the Argonaute protein and subsequently degraded (after Karp 2010).

Argonaute protein Ago2. The RISC provides the machinery for the tiny single stranded siRNA to bind to an RNA having a complementary sequence. Once bound, the target RNA is cleaved at a specific site by the ribonuclease activity of the Ago protein. Thus, the siRNA acts like a guide to direct the complex toward a complementary target RNA, which is then destroyed

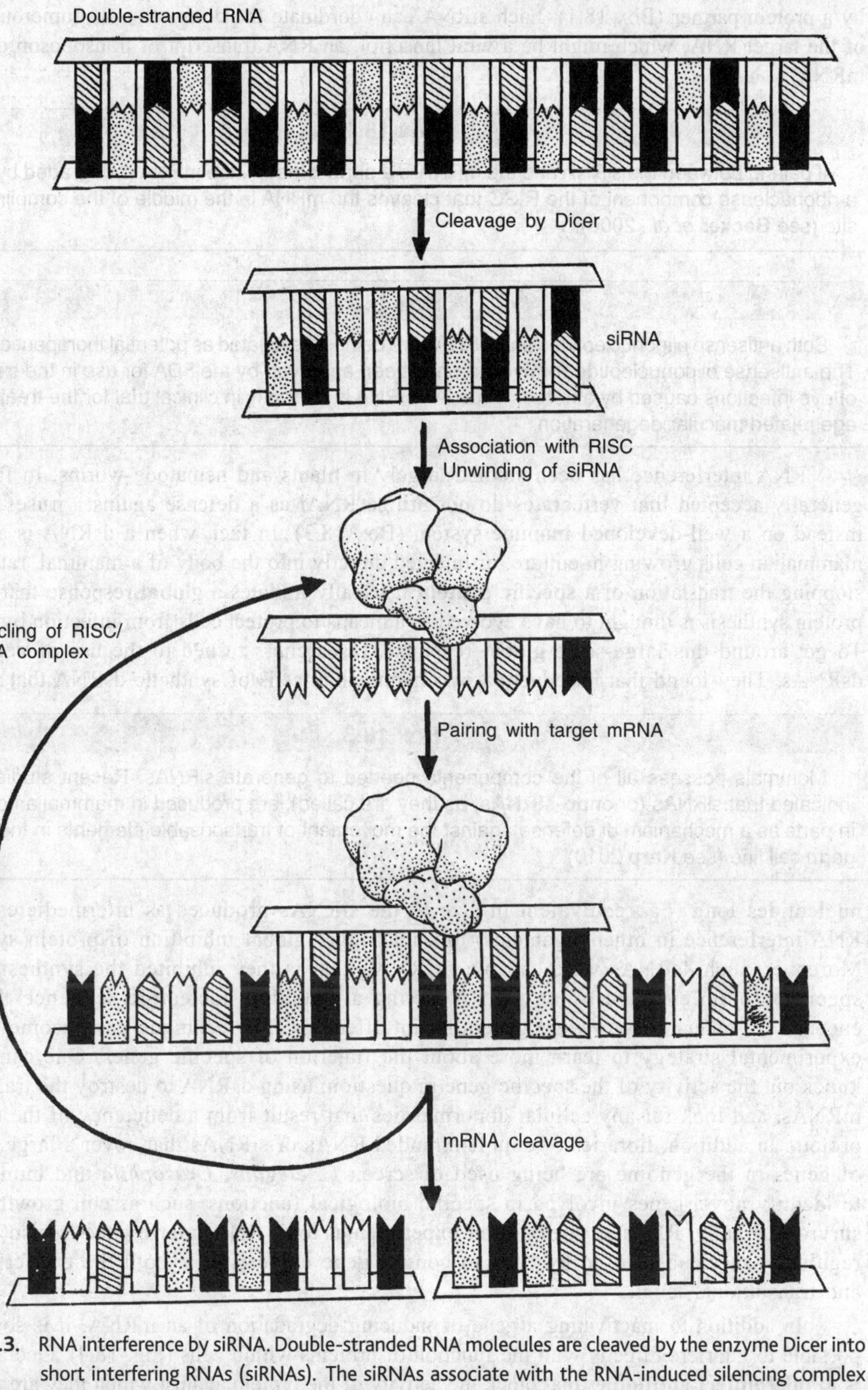

Fig. 18.3. RNA interference by siRNA. Double-stranded RNA molecules are cleaved by the enzyme Dicer into short interfering RNAs (siRNAs). The siRNAs associate with the RNA-induced silencing complex (RISC) and are unwound. The antisense strand of siRNA then targets RISC to a homologous mRNA, which is cleaved by one of the RISC proteins. The RISC-siRNA complex is released and can participate in further cycles of mRNA degradation (after Cooper and Hausman 2007).

by a protein partner (Box 18.1). Each siRNA can coordinate the destruction of numerous copies of the target RNA, which might be a viral tanscript, an RNA transcript of transoposon or a host mRNA.

Box 18.1

If pairing between the siRNA and the mRNA is a perfect match, the mRNA is degraded by **slicer**, a ribonuclease component of the RISC that cleaves the mRNA in the middle of the complimentary site (see **Becker** *et al.*, 2006).

Box 18.2

Both antisense oligonucleotides and siRNAs are being investigated as potential therapeutic agents. The antisense oligonucleotide **fomiversen** has been approved by the FDA for use in the treatment of eye infections caused by a herpes-virus. An siRNA is currently in clinical trial for the treatment of age related macular degeneration.

RNA interference has been studied largely in plants and nematode worms. In fact, it is generally accepted that vertebrates do not utilize RNAi as a defense against viruses, relying instead on a well-developed immune system (Box 18.3). In fact, when a dsRNA is added to mammalian cells growing in culture, or injected directly into the body of a mammal, rather than stopping the translation of a **specific protein** it usually initiates a **global** response that inhibits protein synthesis is thought to have evolved as a means to protect cells from infection by viruses. To get around this large-scale global response, researchers turned to the use of very small dsRNAs. They found that introduction into mammalian cells of synthetic dsRNA that were 21

Box 18.3

Mammals possess all of the components needed to generate siRNAs. Recent studies have indicated that siRNAs (or **endo-siRNAs**, as they are called), are produced in mammalian oocytes, in parts as a mechanism of defense against the movement of transposable elements in the female germ cell line (see **Karp** 2010).

nucleotides long (*i.e.*, equivalent in size to the siRNAs produced as intermediates during RNA interference in other organisms) did not trigger global inhibition of protein synthesis. Moreover, such dsRNAs were capable of RNAi, that is they inhibited the synthesis of the specific protein encoded by an mRNA having a matching nucleotide sequence. Proteins encoded by other mRNAs were generally not affected. This technique has become a major experimental strategy to learn more about the function of specific genes. One can simply knock out the activity of the specific gene in question, using dsRNA to destroy the transcribed mRNAs, and look for any cellular abnormalities that result from a deficiency of the encoded protein. In addition, libraries of double-stranded RNAs or siRNAs that cover a large fraction of genes in the genome are being used to screen *C. elegans*, *Drosophila* and human cells to identify novel genes involved in specific biological functions, such as cell growth or cell survival. Finally, RNAi is not only an experimental tool, RNA interference is also a major regulatory mechanism used by cells to control gene expression at both the transcripitional and translational level.

In addition to inactivating a gene or inducing degradation of an mRNA, it is sometimes possible to interfere directly with the function of proteins within cells (Fig. 18.4). One approach is to **microinject antibodies** that block the activity of the protein against which they are directed.

Alternatively, some **mutant proteins interfere** with the function of their normal counterparts when they are expressed within the same cell-for example, by competing with the normal protein for binding to its target molecule. Cloned DNA encoding such mutant proteins (called **dominant inhibitory mutants**) can be introduced into cells by **gene transfer** and used to study the effects of blocking normal gene function.

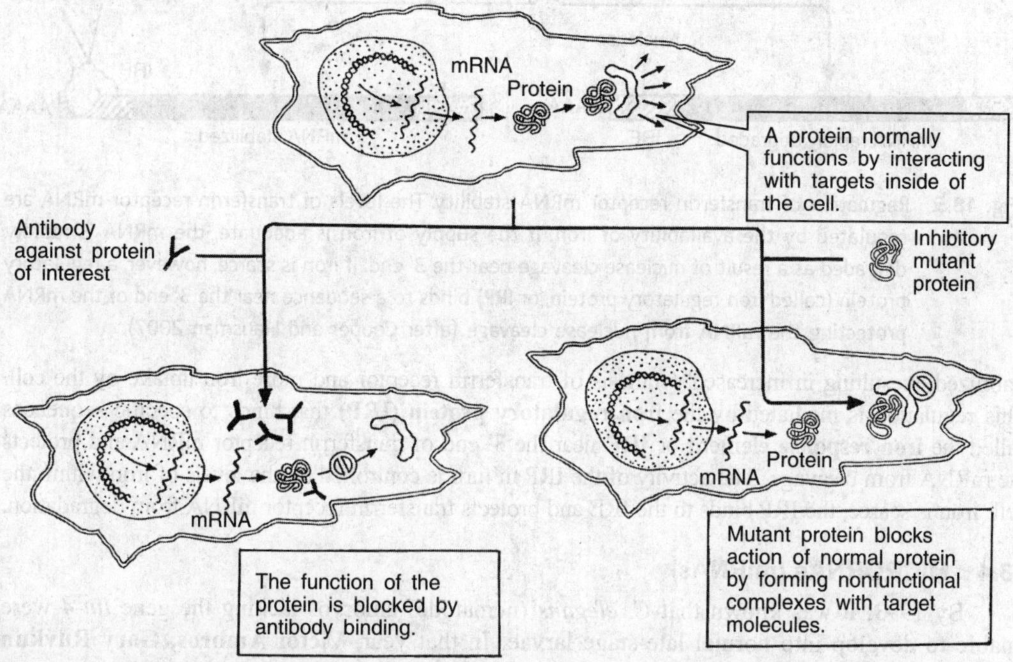

Fig. 18.4. Direct inhibition of protein function. Microinjected antibodies can bind to proteins within cells, thereby inhibiting their normal function. In addition, some mutant proteins interfere with the function of normal proteins-for example, by competing with the normal protein for interaction with target molecules (after Cooper and Hausman, 2007).

Regulation of Transferrin mRNA Stability

In addition to degrading introns, cells possess a quality-control system (called **nonsense-mediated mRNA decay**) that leads to the degradation of mRNAs that lack complete open-reading frames (*open-reading frame*) is a stretch of nucleotide sequence that does not contain stop codons and can encode a polypeptide).

The degradation of some mRNAs can also be triggered by **internal cleavage**. In many cases, cleavage is induced by siRNA or miRNA acting through the RISC complex (Fig. 18.5). Alternatively, the cleavage of some mRNAs is mediated by proteins and can be regulated in response to extracellular signals. A good example is provided by the mRNA that encodes **transferrin receptor**-a cell surface protein involved in the uptake of iron, largely as a result of modulation of the stability of its mRNA (Fig. 18.5). In the presence of adequate amounts of iron, transferrin receptor mRNA is rapidly degraded as a result of specific nuclease cleavage at a sequence near its 3′end. If an adequate supply of iron is not available, however, the mRNA is

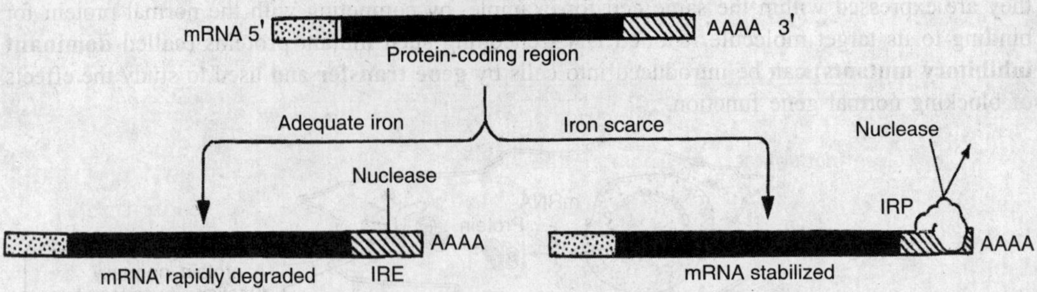

Fig. 18.5. Regulation of transferrin receptor mRNA stability. The levels of transferrin receptor mRNA are regulated by the availability of iron. If the supply of iron is adequate, the mRNA is rapidly degraded as a result of nuclease cleavage near the 3′ end. If iron is scarce, however, a regulatory protein (called iron regulatory protein, or IRP) binds to a sequence near the 3′ end of the mRNA protecting the mRNA from nuclease cleavage (after Cooper and Hausman 2007).

stabilized, rsesulting in increased synthesis of transferrin receptor and more iron uptake by the cell. This regulation is mediated by the **iron regulatory protein (IRP)** that binds to specific sequences called the **iron response element** or **IRE** near the 3′ end of transferrin receptor mRNA and protects the mRNA from cleavage. The activity of the IRP in turn is controlled by the levels of iron within the cell: iron is scarce, the IRP binds to the IRE and protects transferrin receptor mRNA from degradation.

18.4 MICRORNAs (miRNAs)

By 1993, it was known that *C. elegans* (nematode) embryos lacking the gene *lin-4* were unable to develop into normal late-stage larvae. In that year, **Victor Ambros, Gary Ruvkun** and their colleagues reported that *lin-4* gene encodes a small RNA that is complementary to segments in the 3′untranslated region (UTR) of a specific mRNA encoding the protein **LIN-14**. They proposed that, during larval development, the *lin-4* RNA binds to the complementary mRNA, blocking translaton to the next stage of development. Mutants that are unable to produce the small *lin-4* possess a abnormal high level of the LIN-14 protein and cannot transition normally to later larval stages. This was the first clear example of RNA silencing. In year 2000, it was shown that one of these tiny worm RNAs-a 21-nucleotide species called **let-7**-is highly conserved throughout evolution. Humans, for example, encode several RNAs that are either identical or nearly identical to *let-7*. This observation led to an explosion of interest in these RNAs.

It has become evident in recent years that both plants and animals produce a collection of small RNAs name **microRNAs (miRNAs)** that, because of their small size, had been overlooked for decades. As first discovered in nematodes, specific miRNAs such as *lin-4* and *let-7* are synthesized only at certain times during development, or in certain tissues of a plant or animal, and are thought to play a *regulatory role*. At this stage, we can consider an example of **selective expression** of specific miRNAs during the development of the zebra fish (Box 18.4). The size of miRNAs, at roughly 21-24 nucleotides in length, places them in the same size range as the siRNAs involved in RNAi (*i.e.*, RNA interference). Thus, miRNAs are

Box 18.4

Weinholds, **Kloosterman**, *et al*., (2005) have shown that 3 types of miRNAs are synthesized in specific tissues during embryonic development of zebra fish: 1. *miR-124a* in the nervous system, 2. *miR-206* in the skeletal tissue and 3. *miR-122* in the liver.

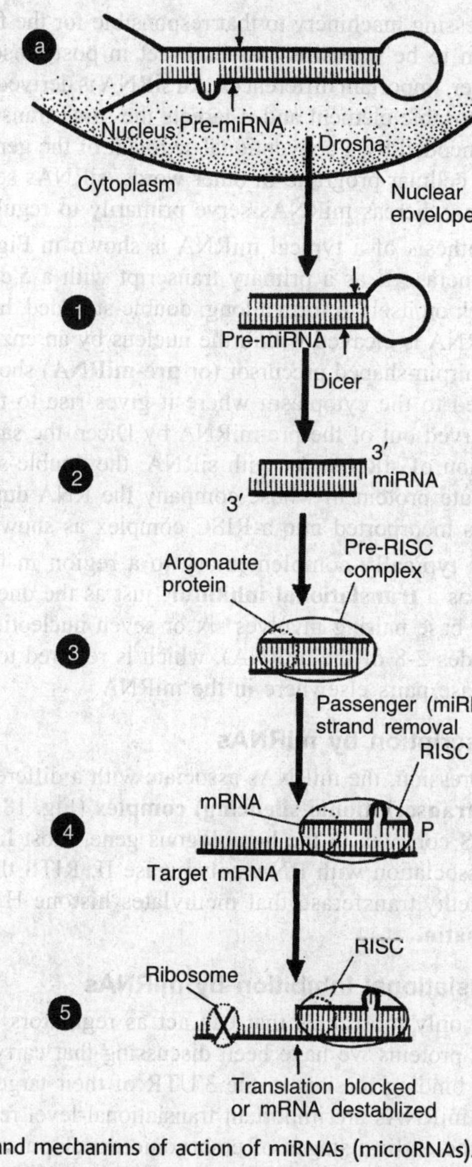

Fig. 18.6. The formation and mechanims of action of miRNAs (microRNAs). MicroRNAs are derived from single-stranded precursor RNAs that contain complementary sequences that allow them to fold back on themselves to form a double stranded RNA with a stem-loop at one end (*step 1*). This pseudo-dsRNA (or pri-miRNA) is cleaved near its terminal loop by a protein complex containing an endonuclease name Drosha to generate a pre-miRNA that has a 3′-overhang at one end. The pre-miRNA is exported to the cytoplasm where it is cleaved by Dicer into a small duplex miRNA (*step 2*) that has a 3′ overhang at both ends. The double-stranded RNA becomes associated with a protein complex containing an Argonaute protein (typically Ago1), leading to the separation of the strands and removal of the passenger strand (called miRNA*; *step 3*). The single-stranded guide miRNA then binds to a complemenatry region on an mRNA (*step 4*) and inhibits translation of the message as shown in *step 5*). Unlike siRNA, the miRNAs that inhibit translation are only partially complementary to the mRNA, hence the bulge (after Karp 2010).

produced by a similar processing machinery to that responsible for the formation of siRNAs. The miRNAs and siRNAs seem to be "**cousins**", as both act in posttranscriptional RNA silencing pathways. There are, however, important differences. An siRNA is derived from the double-stranded product of a virus or transposable element and it targets the same transcript from which it arose. In contrast, an miRNA is encoded by a conventional segment of the genome and target a specific mRNA as part of a normal cellular program. In other words, siRNAs serve primarily to maintain the integrity of the genome, whereas miRNAs serve primarily to regulate gene expression.

The pathway for synthesis of a typical miRNA is shown in Figure 18.6. The miRNA is synthesized by RNA polymerase II as a primary transcript with a 5′cap and poly(A) tail. This primary transcript folds back on itself to form a long, double-stranded, hairpin-shaped RNA called a **pri-miRNA**. The pri-miRNA is cleaved within the nucleus by an enzyme called **Drosha** into a shorter, double-stranded, hairpin-shaped precursor (or **pre-miRNA**) shown in step 1 of Fig. 18.6. The pre-miRNA is exported to the cytoplasm where it gives rise to the small double-stranded miRNA. The miRNA is carved out of the pre-miRNA by Dicer, the same **ribonuclease enzyme** responsible for the formation of siRNA. As with siRNA, the double-stranded miRNA becomes associated with an Argonaute protein in whose company the RNA duplex is disassembled, and one of the single strands is incorporated into a RISC complex as shown in Figure 18.6.

A typical miRNA is typically complementary to a region in the 3′UTR of the mRNA target (Fig. 18.6) and acts as a **translational inhibitor**, just as the ones originally discovered in nematodes. In these cases, base pairing involves six or seven nucleotides near the 5′ end of the miRNA (typically nucleotides 2-8 of the miRNA), which is referred to as the **"seed" region**, as well as a few additional base pairs elsewhere in the miRNA.

A. Regulation of Transcription by miRNAs

In transcriptional repression, the miRNAs associate with a different protein complex called the **RITS (RNA-induced transcriptional silencing) complex** (Fig. 18.7). The separated miRNA strand then guide the RITS complex to the homologous gene, most likely by base pairing with the mRNA transcript in association with RNA polymerase II. RITS then represses transcription by recruiting a histone methyltransferase that methylates histone H3 lysine-9, leading to the formation of **heterochromatin.**

B. Mechanism of Translational Inhibition by miRNAs

Proteins are not the only molecules that can act as regulators of mRNA translation and stability. Like most of the proteins we have been discussing that carry out translational control, miRNAs act primarily by binding to sites in the 3′UTR of their target mRNAs. It is becoming increasingly apparent that miRNAs are important translational-level regulators of virtually every biological process. Even the earliest stages of embryonic development require the involvement of miRNAs as evidenced by the fact that animals lacking the miRNA-producing enzyme Dicer fail to develop beyond gastrulation. Similarly, when Dicer is absent only from a particular tissue, the development of the cells of that tissue display obvious abnormalities. It is also becoming apparent that abnormalities in miRNA level play a major role in the development of many common diseases.

The role of individual miRNAs in a biological process is best studied by interfering with the production of that particular species, which can be accomplished through mutation or deletion of the gene that encodes the miRNA. Studies of this type have suggested that miRNAs govern the expression of batteries of genes that are involved in various pathways of cellular differentiation (Fig. 18.8). We can explain the importance of miRNA both in embryonic development and in adult tissues by considering examples from studies on the heart. Two members of the *miR-1-2* class of miRNA play important roles in the development of muscle tissue. The deletion of one of these miRNAs

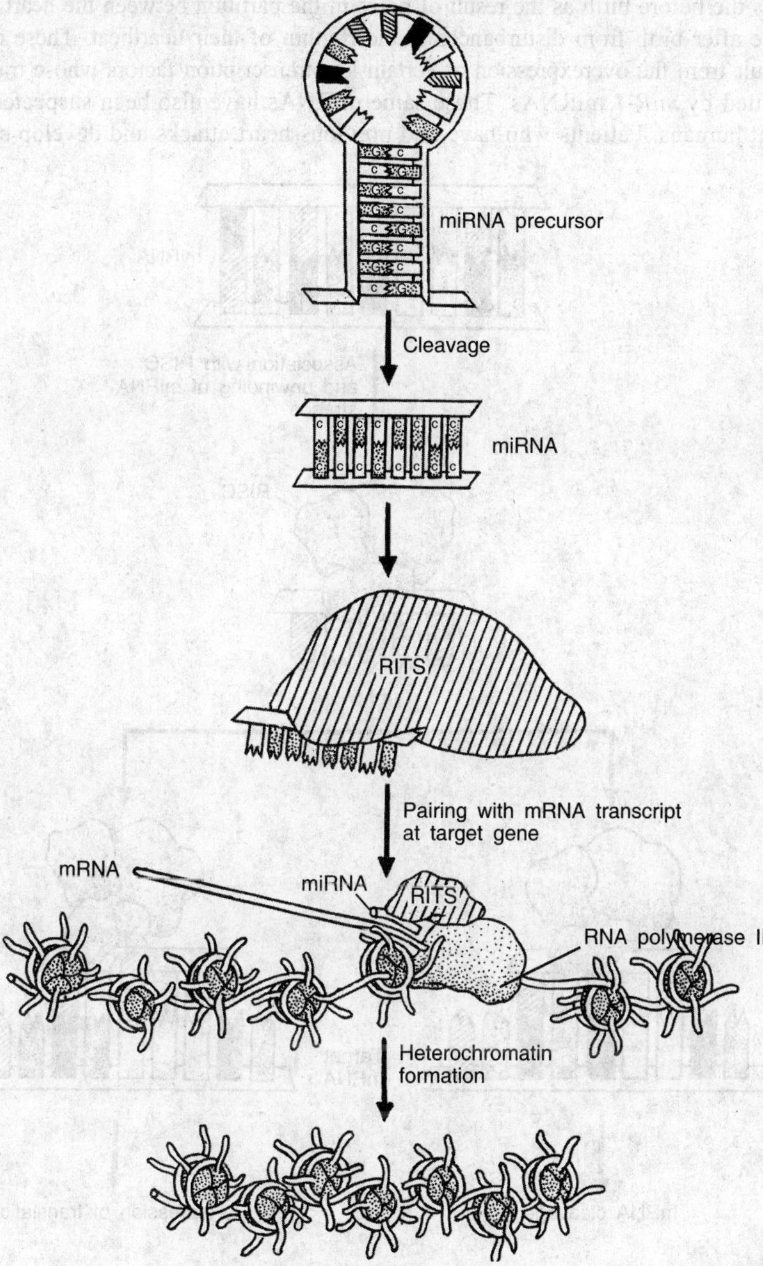

Fig. 18.7. Regulation of transcription by miRNAs. MicroRNAs are transcribed as precursors containing stem-loop structures, which are cleaved to yield short double-stranded miRNAs. Transcriptional repression is mediated by association of miRNAs with the RITS complex. The miRNA then guides RITS to the homologous target gene, most likely by base pairing with nascent mRNA transcripts associated with RNA polymerase II. RITS then recruits a histone methyltransferase that methylates lysine-9 amino acid of histone H3, leading to formaton of heterochromatin and repression of transcription (after Cooper and Hauseman 2007).

(*miR-1-2*) in mouse embryos is accompanied by dramatic **defects** in development of the heart. Many of these animals die before birth as the result of holes in the partition between the heart ventricles, while others die after birth from disturbances in the rhythm of their heartbeat. These deleterious effects may result from the overexpression of certain key transcription factors whose translation is normally inhibited by *miR-1* miRNAs. These same miRNAs have also been suspected in **heart disease** in adult humans. Patients who have had previous heart attacks and develop subsequent

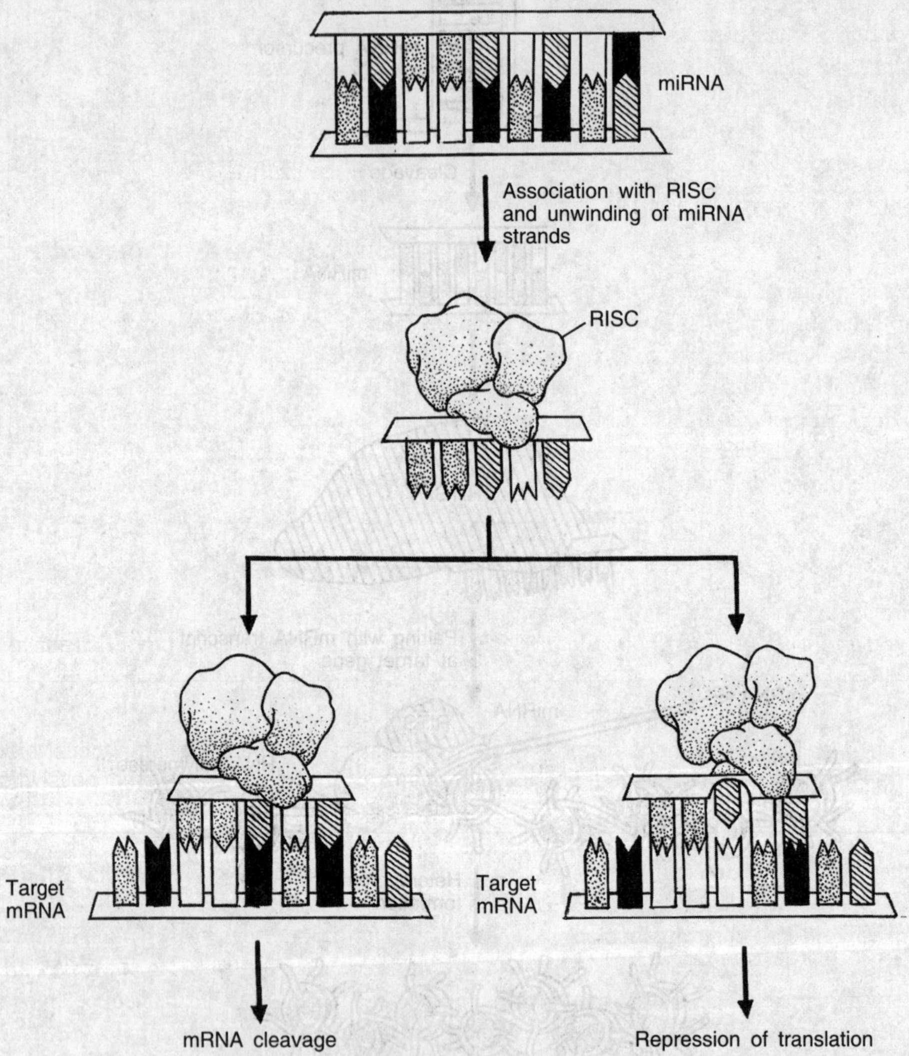

Fig. 18.8. Regulation of translation by miRNAs. MicroRNAs (miRNAs) associate with RISC complex in which the two strands of the miRNA are unwound. The miRNA then targets RISC to a homologous mRNA, leading either to mRNA cleavage or repression of translation. Cleavage generally results from miRNAs that pair perfectly with their targets, whereas translational repression results from miRNAs that are mismatched with their targets (after Cooper and Hausman 2007).

electrophysiological abnormalities tend to have elevated *miR-1* levels in their heart itssue. Two of the predicted targets of *miR-1* are mRNAs that encode (1) the **protein Cx43**, which forms the gap junctions that conduct ionic current between the muscle cells of the heart's ventricles and (2) the **protein kir2.1**, which is a potassium channel involved in repolarization of the membrane potential in heart cells. The expression of both of these proteins would be expected to be reduced in the presence of higher-than-normal levels of *miR-1*. These findings give hope that miRNAs may become a good target of therapeutic (beneficial) drugs in the treatment of heart disease.

The miRNAs likely exert their regulatory activities through several distinct mechanisms, affecting both mRNA translation and mRNA stability (Fig. 18.9).

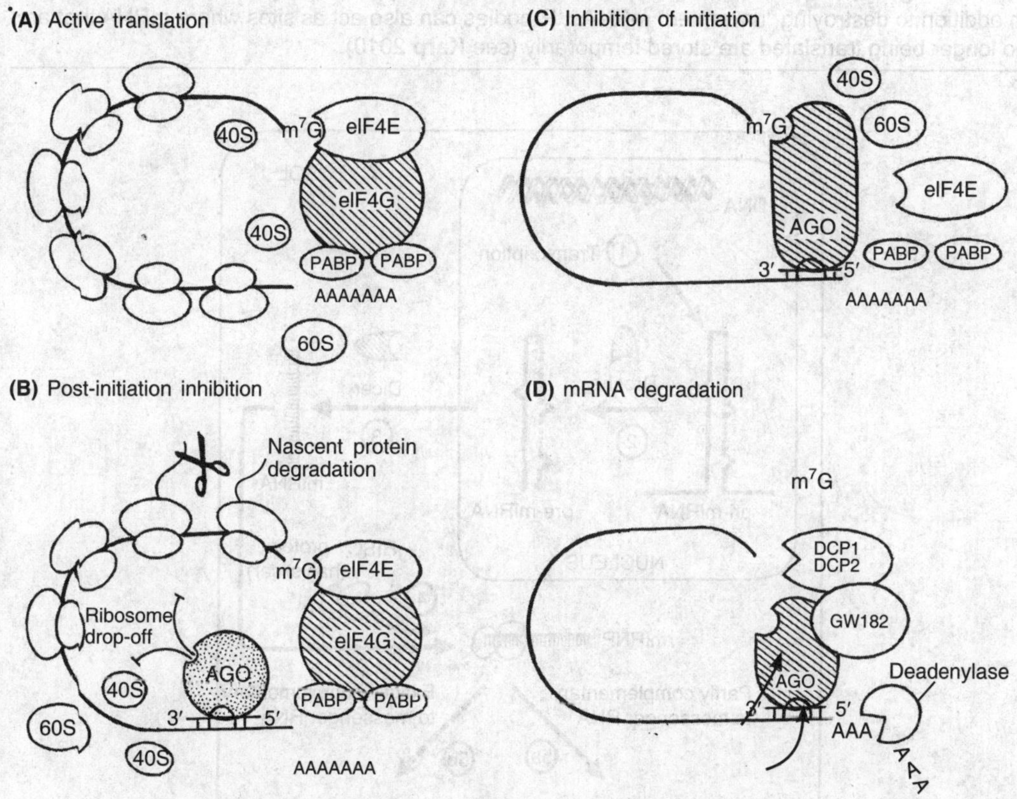

Fig. 18.9. Potential mechanisms by which miRNAs might decrease gene expression at the translational level (**G. Stefani** and **F.J. Slack** 2008). A-Active translation in the absence of miRNA regulation. B-Inhibition of gene expression at some point after translation has been initiated. In this model, the binding of RISC complex containing the Argonaute protein (AGO/Ago) leads either to degradation of the nascent protein or to the release of the ribosomes from the mRNA. C-Inhibition of gene expression at the point where the translation process is initiated. In this model, the RISC complex either inhibits the interaction between eIF4E and the 5′ end of the mRNA or prevents the assembly of the complete ribosome from its subunits. D-Inhibition of gene expression by promoting the degradation of the mRNA. In this model, the RISC complex has recruited proteins that either promote the escaping of the mRNA (by proteins DCP1/2) and/or degrade the poly(A) tail. GW182 is a component of P-bodies (after Karp 2010).

1. Early studies on nematode worms and recent studies on cultured mammalian cells indicate that miRNAs can repress the synthesis of proteins at some point after the initiation of translation has occurred (Fig. 18.9B; Box 18.5). This veiw was based initially on the observation that miRNAs bound to mRNAs are part of actively translating polysomes. One line of evidence suggests that

the bound RISC (the miRNA-Argonaute protein complex of Fig. 18.6) causes premature drop-off of ribosomes as they are engaged in translational elongation, whereas another line implicates the RISC in the degradation of the nascent polypeptide (Fig. 18.10).

Box 18.5 P-bodies

The process of mRNA degradation of mRNA in mammalian cells involves *deadenylation, decapping* and *5'-3' degradation*, all of which occur within small transient cytoplasmic granules called **P-bodies**. In addition to destroying "unwanted" mRNAs, P-bodies can also act as sites where mRNA that are no longer being translated are stored temporarily (see **Karp** 2010).

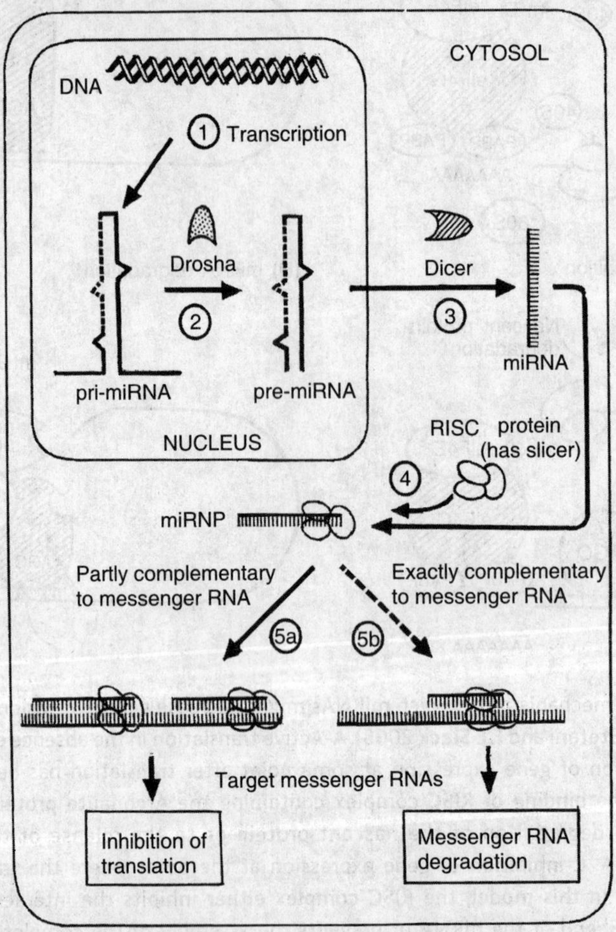

Fig. 18.10. Translational silencing by microRNAs. MicroRNAs are produced by normal cellular genes (about 200-250 genes in human beings and about 120 in nematode worms) through a multistep process. 1. A microRNA gene is transcribed into an initial primary transcript, called a pri-miRNA that folds into hairpin loop. 2. The nuclear enzyme Drosha cleaves pri-miRNA into a smaller (roughly 70 nucleotide) haripin RNA called pre-miRNA, which is exported from the nucleus. 3. Cleaved by Dicer enzyme to form a mature, single-stranded miRNA that is 21-22 nucleotides long. 4. The mature, single-stranded miRNA then joins with a group of proteins RISC containing

slicer to form a ribonucleoprotein complex (miRNP). The miRNP is targeted by its miRNA to messenger RNA (mRNA) containing sequences that are complementary to that of the miRNA. 5A. Most commonly, multiple miRNAs bind to the same messenger RNA through partly complementary sequences and together inhibit translation. 5B. Occasionally, the miRNA is exactly complemenatry to a site within a particular mRNA, resulting in messenger RNA cleavage and degradation by a mechanism similar to that observed with siRNA (after Backer *et al.*, 2006).

2. Other studies suggest that the bound RISC interferes with translation initiation (Fig. 18.9C). This view is based on numerous observations, including data suggesting that the Argonaute protein within a RISC is capable of binding to the methyl-guanine cap at the 5′ end of the mRNA and blocking the interaction of the cap with the required translation initiation factor eIF4E. Other data suggest that the RISC prevents interaction between the large and small ribosomal subunits, which is also required to initiate translation.

3. Still some studies have suggested that some miRNAs carry out tranlational-level regulation by inducing the degradation of the target mRNA (Fig. 18.9D). This is illustrated by a study of zebra fish embyos that reported a dramatic increase in the expression of one particular miRNA (miR-430) at the time when the embryo began to synthesize its own mRNAs. The appearance of this miRNA was accompained by the degradation of large numbers of maternal mRNAs that are stored in the egg at the time of fertilization and contain binding sites for the miR-430 miRNA in their 3′UTR.

Functions of microRNAs. MicroRNAs have been implicated in many processes, including the patterning of the nervous system, control of cell proliferation and cell death, leaf and flower development in plants, and differentiation of various cell types. MicroRNAs are found to affect development of cancer and have a role in mammalian development.

18.5 CLINICAL APPLICATIONS OF RNA INTERFERENCE

Medical scientists are continually searching for **"magic bullets"**, therapeutic compounds that fight a particular disease in a highly specific manner without toxic side effects. We can consider two main types of diseases-**viral infections** and **cancer**-that have become targets for a new type of **molecular "magic bullets"**. Viruses are able to plunder (*i.e.*, loot) an infected cell because they synthesize messenger RNAs that encode viral proteins that disrupt cellular activities. Most cells that become cancerous contain **mutations** in certain genes (called **oncogens**), causing them to produce mutant mRNAs that are translated into abnormal versions of cellular proteins. Let us consider what might happen if a patient with one of these diseases could be treated with a drug that destroyed or inhibited the specific mRNAs transcribed from the viral genome or mutant cancer gene while, at the same time, ignoring all the other mRNAs of the cell. Several strategies have been developed in recent years that have this goal in mind; the most recent of these strategies takes advantage of the phenomenon of **RNA interference (RNAi)**.

As already discussed in this chapter, mammalian cells can be subjected to RNAi-a process that leads to degradation of a specific mRNA-by introducing the cells to a double-stranded **siRNA** in which one of the strands is complementary to the mRNA being targeted. When cells are incubated with a 21- to 23-nucleotide synthetic siRNA, the molecules are taken up by the cells, and incorporated into an mRNA-cleaving ribonucleoprotein complex (as in Fig. 18.2), which attacks the complementary mRNA. Alternatively, RNAi can be induced within mammalian cells that have been genetically engineered to carry a gene with inverted repeats. When the gene is transcribed, the RNA product folds back on itself to form a hairpin-shaped (*i.e.*, double-stranded) siRNA precursor (similar to that in Fig. 18.6) that is processed into an active siRNA. The use of synthetic siRNA is likely to have a short-lived effect, which may or may not be beneficial depending on the condition being treated. In contrast, the use of **viral vectors** has the potential to generate a long-lasting therapeutic effect with a single application. At the same time, however, the use of viruses raises other safety issues.

As with all potential drug technologies, the first stages in development of RNAi therapeutics involved **preclinical studies** in which siRNAs were tested in cells from diseased patients or in animal models (*i.e.*, animals with induced diseases similar to those afflict or harm humans). Many of the preclinical studies suggested that RNAi held great promise in the treatment of a wide range of diseases and viral infections. We can consider just a few examples.

1. As noted above, cancer cells typically contain one or more mutated genes that encode abnormal proteins responsbile for producing the cancerous phenotype. A type of **leukemia**, for example, is caused by a gene, called *BCR-ABL*, that is formed by the fusion of two normal genes. An siRNA against the mRNA produced by the BCR-ABL fusion gene proved capable of converting the malingnant cells to a normal phenotype in culture. Similarly, an siRNA against a gene carrying an exapnded CAG tract of the type that causes Huntington's disease was shown to block the production of the abonormal protein.

2. A large number of preclinical studies probing the therapeutic value of siRNAs have focused on viral infection. Administration of siRNA complementary to **influenza**, **HIV**, or **hepatitis** viral sequences has prevented or eliminated infections by these viruses in laboratory animals. In the case of **AIDS**, it is hoped that stem cells can be isolated from a patient, transfected with a vector carrying an siRNA that targets an HIV-encoded mRNA, and then reinfused back into the patients bloodstream. Such cells have the potential to produce the siRNA more or less continuously, causing the cell and its descendants to be resistant to destruction by the virus. There are roadblocks, however, for using RNAi to treat viral infection. The hallmark of RNAi is its extraordinary sequence specificity, which can be both a virtue and a handicap. Ultimately, RNAi may prove to be ineffective against viruses, because these pathogens tend to mutate rapidly. Mutations result in changes in genome sequence, leading to the production of mRNAs that are no longer fully complementary the therapeutic siRNA.

3. The first test (**clinical trial**) of RNAi therapeutics in human patients came against a form of **macular degeneration** that is caused by the overgrowth of blood vessels behind the retina of elderly patients, causing progressive loss of vision. The excessive growth of these blood vessels is incited by the production of a growth factor, VEGF, which induces its effect by binding to the cell-surface receptor VEGFR1. When **siRNA** targeting the mRNAs encoding VEGF (or VEGFR1) were injected directly into the eyes of patients with this disease, the beneficial effect was substantial; the deterioration of vision was largely halted and the condition appeared to stabilize (Box 18.6). Given the success of these initial trials, this same siRNA (named **Bevasiranib**) is also being tested in patients experiencing **retinal deterioration** as a complication of **diabetes.**

Box 18.6

There has been a report suggesting that the beneficial effects of this agent (*i.e.*, VEGF/VEGFR1) are not the result of a direct inhibition by the siRNA of VEGF synthesis but rather an indirect effect on blood vessel growth through stimulation of a protein called **TLR3**, which is a receptor of the innate immune system (see **Karp** 2010).

4. Another series of clinical tests have begun that are directed against a respiratory virus, RSV, which is a common cause of **infant hospitalization**. In these trails, adult subjects inhale an aerosol that contains an siRNA (named **ALN-RSVO1**) directed against mRNAs encoding a key trial protein. The aerosol treatment had proven highly effective in combating these respiratory infections in laboratory animals. Clinical trials of siRNAs aimed at the treatment of high blood cholesterol, certain cancers, hepatitis, pandemic flue, and other diseases are either planned or underway.

Techniques of applications of RNAi therapeutics. Both of the diseases discussed above, age-related macular degeneration and RSV respiratory infection, can be treated with **local application** of the RNAi therapeutic. As with other types of *gene therapies*, a major obstacle in using RNAi to treat most diseases or infections is the difficulty in delivering siRNAs (or the viral vectors that encode the hairpin siRNA precursors) to the affected tissues deep within the human body. In an attempt to meet this challenge, a wide variety of **nanosized particles** have been developed that (1) either bind to siRNAs or encapsulate them and (2) target siRNAs to the appropriate tissue or organ following their injection into bloodstream.

A. Clinical trial of IBDs (=Inflammatory Bowel Diseases). In one of the clinical trial, liposome-like **nanoparticles** were constructed to contain an **antibody** that binds to the **B7 integrin** subunit on the surface of a specific subset of white blood cells known to be involved in inflammation of digestive tract. These nanoparticles were also constructed to contain an siRNA directed against the mRNA encoding **cyclin D1**, a regulatory molecule involved in the proliferation of these white blood cells. These siRNA-carrying nanoparticles were injected into the bloodstream of mice in which inflammation had been induced. The results of the experiment (**Dan Peer** *et al.*, 2008) have shown that the anti-B7 antibody has successfully targetted the therapeutic agent to the site of inflammation, and inhibition of cyclin D1 synthesis by the siRNA has led to a dramatic reversal of the disease condition. This study demonstrated the potential for siRNA therapeutics in the treatment of severe IBDs, such as **colitis** and **Crohn's disease**, which are presently treated with drugs that can have debilitating side effects.

B. Clinical trial for encephalitis. In another case, researchers have successfully targeted a virus (JEV) that infects the brains of mice, causing **encephalitis**. Of all the organs in the body, the brain is the most difficult to access, because even small molecular-weight drugs cannot penetrate the blood-brain barrier. In this study, an siRNA directed against the encephalitis virus was linked to a small fragment of a protein from the evelope of the rabies virus, this envelope protein appears to promote access to the brain by binding to acetylcholine receptors on the surfaces of nerve cells and/or cells that line the blood capillaries. In the present study, the siRNA protein complex was administered intravenously into mice that had also been infected with the encephalitis virus. The siRNA entered the brain resulting in the survivial of 80 per cent of the treated mice. None of the untreated animals survived the infection.

18.6 piRNAs (piwi-INTERACTING RNAs)

We are already aware that the movement of transposable elements (or transposons) poses a threat to the genome because it can disrupt the acitvity of gene into which element happens to insert. If this type of **genomic jumping** were to occur during the life of an adult liver or kidney cell, the consequences would likely be minimal since the transposition event affects only that particular cell and its daughters (if the somatic cell is still capable of cell division). If, however, a transposition event were to occur in a germ cell, that is, a cell that has the capability to give rise to gametes, then the event has the potential to affect every cell in an individual of the next generation. It is not surprising, therefore, that specialized mechanisms have evolved to suppress the movement of transposable elements in germ cells.

Further, it has already been discussed that one of the evident functions of siRNAs is to prevent the movement of transposable elements. Recent studies indicate that germ cells of animals express a distinct class of small RNAs, called **piwi-interacting RNAs** (or **piRNAs**) that suppress the movement of transposable elements in germ line, piRNAs get their name from the proteins with which they associate. These proteins are called **PIWIs**, and they are a subclass of the Argonaute family, the same family of proteins that associate with siRNAs and miRNAs as part of their RISC complexes. The roles of the PIWIs proteins have been best studied in fruit flies, where deletion of these proteins leads to defects in the suppression of transposon movement in germ cells and ultimate failure of gamete formation. The piRNAs and their associated PIWI proteins are also required for successful gamete formation in mammals, but their roles are poorly understood.

Differences in piRNAs and siRNAs

There are a number of important differences between piRNAs on one hand and si/miRNAs on the other:

1. The piRNAs are longer than other small RNAs, measuring about 24-32 nucleotides in length.
2. The majority of mammalian piRNAs can be mapped to a small number of huge genomic loci, some of which can encode thousands of different piRNAs.
3. The formation of piRNAs does not involve the formation of dsRNA prescursors or cleavage by the Dicer ribonuclease.

Instead, piRNA biogenesis appears to depend on the endonuclease activity of the PIWI protein acting on a long, single-stranded primary transcript. The mechanism by which piRNA act to silence target RNA expression remains unclear (**Karp** 2010).

18.7 OTHER NONCODING RNAs (ncRNAs)

Recent molecular biological findings indicate that at least two-thirds of a mouse or human genome is normally transcribed, far more than would be expected based on the number of "meaningful" DNA sequences thought to be present. In fact, most of the sequences included among this transcribed DNA are normally thought of as "junk", including large numbers of transposable elements that constitute a sizable fraction of mammalian genome. Some studies have raised the question whether there is any basic difference between a gene and an intergenic region; they report that *transcription can begin at all kinds of unexpected sites within the genome*, and that transcripts overlap extensively with one another. Other studies have revealed that cells often transcribe both strands of DNA element, producing both a **sense RNA** and an **antisense RNA**. It is also reported that antisense RNA can be synthesized by polymerases (enzymes) that had initially bound to the promoter of a protein-coding gene and then moved upstream from that site along the **opposite strand** of the DNA, that is, in the opposite direction from polymerase molecules that are transcribing the gene itself.

Just as the DNA of a cell or organism constitutes its *genome* and the proteins it produces constitutes its *proteome*, the RNAs that are synthesized by a cell or organism constitute its **transcriptome**. Why is the transcriptome of a mammalian cell is so large? In other words, why does a cell transcribe all of these various types of DNA sequences? For answering above questions following views have been forwarded by different molecular biologists:

1. According to one viewpoint, much of this pervasive transcription activity is simply the "background noise" that accompanies the complex process of gene expression. Proponents of this viewpoint cite the results of studies in mice in which blocks of the genome that lack protein-coding genes have been deleted. It was found in these studies that healthy mice can develop

even though they cannot synthesize the noncoding RNAs (ncRNAs) that would normally have been produced from the deleted DNA sequences.

2. According to a contrasting viewpoint, much of the noncoding RNA being produced is involved in diverse regulatory activities that have yet to be identified. Supporters of this viewpoint cite results suggesting that this pervasive transcription is not random; many of the noncoding transcripts display distinct and reproductive patterns of tissue specific distribution, and their origin can be traced to specific gene loci in the genome. Keep in mind that it has only been in the last decade (*i.e.*, year 2000 to 2010) or so that the existence of siRNAs, miRNAs and piRNAs have come to light and it is likely that other types of small ncRNAs remain undetected. A number of long ncRNAs (*e.g.*, **XIST** and **AIRE**) have also been identified and these are probably the tip of the iceberg. It is even possible that the vast transcriptional output of mammalian genome holds the key to explain why we have approximately the same number of genes as organisms we believe to be much less complex. Regardless of the actual explanation, one point is clear: there is a great deal that we do not understand about the many roles of RNAs in eukaryotic cells (**Karp** 2010).

QUESTIONS

Long Answer Questions

1. Give an account of RNA interferences.

2. Describe some clinical applications of RNA interference.

Short Answer Questions

1. Write shot notes on the following:

(*i*) Riboswitches;

(*ii*) Small interfering RNAs (siRNAs);

(*iii*) Regulation of transferrin mRNA stability;

(*iv*) Riboswitches;

(*v*) MicroRNAs (miRNAs);

(*vi*) Piwi-interacting RNAs (piRNAs);

(*vii*) Noncoding RNAs (ncRNAs).

Very Short Answer Questions

1. What are riboswitches?

2. Define RNA interference (RNAi).

3. What are microRNAs (miRNAs).

4. Define small interfering RNAs (siRNAs).

5. Give an idea of RNA silencing.

6. Define the Dicer.

7. What are piwiRNAs (piRNAs)?

Fill in the Blanks

1. RNA interference (RNAi) is thought to have evolved as a type of .... to protein organisms from the presence of foreign or unwanted genetic material.

2. RNA silencing is done by four RNAs, namely siRNAs, ... , ... and ncRNAs.

3. Most .... suppress gene expression by .... either termination of transcription or initiation of translation.

4. In the RNAi pathway, one of the strands of RNA duplex is called .... and other strand is called ....

5. RITS stands for ....

ANSWERS

Very Short Answer Questions

1. These are mRNAs that, once bound to a metabolite, undergo a change in their folded conformation that allows them to alter the expression of gene involved in production of that metabolite. Most riboswitches suppress gene expression by blocking either termination of transcription or initiation of translation.

2. It is a naturally occurring phenomenon in which double-stranded RNAs (dsRNAs) lead to the degradation of mRNA having identical sequences. RNAi is believed to function primarily in blocking the replication of viruses and restricting the movement of mobile elements, both of which involve the formation of dsRNA intermediates. Mammalian cells can be made to engage in RNAi by treatment of the cells with small (21 nucleotides) RNAs. These small RNAs (si RNAs) induce the degradation of mRNAs that contain the same sequence.

3. MicroRNAs (miRNAs) are small RNAs (20-23 nucleotides long) that are synthesized from many sites in the genome and involved in inhibiting translation or increasing degradation of complementary mRNAs.

4. The siRNAs are small (21-23 nucleotides) double-stranded fragments formed when double-stranded RNA initiates the response during RNA silencing. They act by either promoting the degradation of mRNAs with precisely complementary sequences or by inhibiting the transcription of genes containing precisely complementary sequences.

5. RNA silencing is a process in which small, noncoding RNAs, typically derived from longer double-stranded precursors, trigger sequence-specific inhibition of gene expression.

6. Dicer is an enzyme that cleaves double-stranded RNAs into short fragments about 23 base pairs in length.

7. The piRNAs are small RNAs (24-32 bases) that are encoded by a small number of large genomic loci that act to suppress the movement of transposable elements in germ cells. The piRNAs are derived from single-stranded precursors and do not require Dicer for processing.

Fill in the Blanks

1. Genetic immune system.
2. miRNAs, piRNAs.
3. Riboswitches, blocking.
4. Passenger strand, guide strand.
5. RNA-induced transcriptional silencing.

(CC–XII)
PRINCIPLES OF GENETICS

1

Introduction

Genetics is the study of inheritance in all of its manifestations, from the distribution of human traits in family pedigree to the biochemistry of the genetic material in our DNA (**Tamarin** 2002). The fundamental concept of genetics is that inherited traits are determined by the elements of heredity, that are transmitted from parents to offspring in reproduction; these elements of heredity are the genes.

1.1 HISTORY OF GENETICS

According to **Tamarin** (2002), the history or growth of genetics can be studied under the following four periods: 1. Before 1860; 2. 1860 to 1900; 3. 1900–1944, and 4. 1944 to the present.

1. Growth of Genetics before 1860

1665 **Robert Hooke** coined the term *cell* in his studies of cork. He saw, in fact, empty cells which were observed at a magnification of thirty power.

1674–1683 **Anton van Leeuwenhoek** discovered living organisms (protozoa and bacteria) in rainwater. Leeuwenhoek was a master lens maker and produced magnifications of several hundred power from single lenses.

1833 **Robert Brown** discovered the nuclei of cells.

1835–1839 **Hugo von Mohl** described mitosis in nuclei.

1859 **Rudolf Virchow** summed up the concept of the cell theory with his Latin aphorism *omnis cellula e cellula*—all cells come from pre-existing cells.

1859 The publication of **Charles Darwin's** *The Origin of Species* took place.

Thus, by 1959, biologists had an understanding of the continuity of cells and knew of the cell's nucleus.

2. Growth of Genetics in period from 1860 to 1900

1865 **Gregor Mendel** read his paper in the British Society for Natural History.

1866 Mendel's paper published in the proceeding of the British Society for Natural History.

1875 **O. Hertwig** described the fusion of sperm and egg to form zygote.

1879–1885 With the aid of new staining technique, **W. Flemming** described the chromosomes which were noticed by **C. Von Nageli** in 1842. Flemming studied the way chromosomes split during division and the separation of sister chromatids and their movement to opposite poles of the dividing cell during mitosis.

1880's **Theodor Boveri**, as well as **K. Rabi** and **E. van Benden**, hypothesized that chromosomes are individual structures with continuity from one generation to the next despite their "disappearance" between cell divisions.

1885 **August Weismann** stated that inheritance is based exclusively in the nucleus. He is famous for his **germplasm theory** (1883).

1887 **Weismann** predicted the occurrence of a reductional division, which we now call meiosis.

1888 **W. Waldeyer** first used the term **chromosome**.

1890 **O. Hertwig** and **T. Boveri** had described the process of meiosis in detail.

Thus, this period (*i.e.*, 1860–1900) includes the discoveries of chromosomes and their behaviour insights that shed new light on Mendel's research.

3. Growth of Genetics in period from 1900 to 1944

1900 Three biologists working independently–**Hugo de Vries, Carl Correns** and **Erich von Tschermak**– re-discovered Mendel's landmark work on the rules of inheritance.

1902 **Boveri** and **Sutton** demonstrated the presence of paired chromosomes (homologs) in diploid species.

1903 **Walter Sutton** hypothesized that the behaviour of chromosomes during meiosis explained Mendel's rules of inheritance, thus leading to the discovery that genes are located on chromosomes.

1905 **William Bateson** named the science **genetics**.

Fig. 1.1. Hugo de Vries (1848–1935). Dutch botanist who discovered Mendel's principles (or rules or laws) in 1900. He was a pioneer in plant genetics (after Gardner *et al.*, 2006).

1908 **Hardy** and **Weinberg** formulated the "Hardy-Weinberg law" relating genotypic frequencies to gene frequencies in randomly mating populations.

1909 **Johannsen** introduced the term **gene**.

1909 **Garrod's** book *Inborn Errors of Metabolism* published.

1910 **T.H. Morgan** (Nobel Prize 1933) established the *sex-linked inheritance* of white eyes in *Drosophila melanogaster* (fruit fly).

1911 **Morgan** postulated the chromosomal basis of linkage.

1913 **Alfred Sturtevant** created the first genetic map using fruit fly. He showed that genes existed in a linear order on chromosomes.

Fig. 1.2. **William Bateson** (1861–1926), English biologists who repeated Mendel's experiments and supported his principles and introduced Mendelian genetics to the English-speaking world (after Gardner *et al.*, 2006).

1927 **I. Stadler** and **H.J. Muller** showed that genes can be mutated artificially by X-rays.

1927 **Muller** (Nobel Prize 1946) reported the ClB technique to demonstrate that X-rays are mutagenic.

1928 **Griffith** discovered the phenomenon of **transformation** in *Diplococcus pneumoniae*.

1930–1932 **R.A. Fisher, S. Wright** and **J.B.S. Haldane** developed the algebraic foundations for our understanding of the process of evolution.

1931 **Creighton** and **McClintock's** (maize) and **Stern's** (*Drosophila*) papers appeared, demonstrating that *genetic recombination* is correlated with the *exchange of morphological marker on chromosomes*.

1940 **Oliver** demonstrated recombination within the lozenge functional unit in *Drosophila*.

1941 **Beadle** and **Tatum's** work (Nobel Prize 1958) on *Neurospora* was published, establishing *one gene-one enzyme concept*.

1943 **S. Luria** and **M. Delbruck** demonstrated that bacteria have normal genetic systems and thus could serve as models for studying genetic processes.

Thus, during this period (*i.e.*, 1900–1944), modern genetics flourished with the development of the chromosome theory, which showed that chromosomes are linear arrays of genes. In addition, the foundations of modern evolutionary and molecular genetics were derived.

4. Growth of Genetics in period from 1944 to present

1944 **Avery, MacLeod** and **McCarty** demonstrated that the pneumococcal "transforming principle" is DNA.

1946 **Lederberg** (Nobel Prize 1958) and **Tatum** discovered **conjugation** in bacteria.

1950 **Barbara McClintok** (Nobel Prize 1983) published her first paper on **"transposable elements"** in maize.

1952 **Hershey** (Nobel Prize 1969) and **Chase** demonstrated that the *genetic material of bacteriophage T2 is DNA.*

1952 **Zinder** and **Lederberg** discovered phage mediated **transduction** in bacteria.

1953 **Watson** and **Crick** (Nobel Prize 1962) worked out the double-helix structure of DNA using the X-ray diffraction data of **Wilkins** (Nobel Prize 1962) and base composition data of **Chargaff**.

1955 **Benzer's** first paper was published on *fine structure of the phage T4 rII* locus.

1956 **Tjio** and **Levan** established that the normal *diploid chromosome number in humans is 46.*

1957 **Fraenkal-Conrat** and **Singer** demonstrated that *the genetic information of tobacco mosaic virus is stored in RNA.*

1958 **Meselson** and **Stahl** demonstrated that *DNA replication is semiconservative.*

1958 **Kornberg** (Nobel Prize 1959) *isolated DNA polymerase I from E.coli.*

1959 **Ochoa** (Nobel Prize 1959) discovered the first RNA polymerase.

1961 **Jacob** and **Monod** (Nobel Prize 1965) proposed "*operon model*" for regulating gene expression.

1964 Colinearity between genes and polypeptide products established by the work of **Yanofsky** and colleagues and by **Brenner** and colleagues.

1964 **Temin** (Nobel Prize 1975) proposed the *DNA provirus* form of RNA tumour viruses.

1965 **Holley** (Nobel Prize 1968) worked out the first complete nucleotide sequence of a tRNA, a yeast alanine tRNA.

1966 The complete genetic code was established by the work of **Nirenberg** and **Khorana** (Nobel Prize 1968) and coworkers.

1968–1973 **W. Arber, H. Smith** and **D. Nathans**, along with their colleagues discovered and described **restriction endonucleases**, the enzymes that opened up our ability to manipulate DNA through recombinant DNA technology.

1970 **Nathans** and **Smith** (Nobel Prize 1978) isolated the first *restriction endonucleases.*

1970 *Reverse transcriptase* of RNA tumour viruses identified by **Baltimore** (Nobel Prize 1975).

1972 First *recombinant DNA* produced *in vitro* in laboratory of **Paul Berg** (Nobel Prize 1980).

1976 **Bishop** and **Varmus** (Nobel Prize 1989) demonstrated the *protooncogene to oncogene.*

1976 **Hozumi** and **Tonegawa** (Nobel Prize 1987) demonstrated *somatic rearrangements of genes encoding antibodies.*

1977 Demonstration of **introns** in eukaryotic genes by **Breathnach, Mandel** and **Chambon** and by **Jeffreys** and **Flavell**.

1977 Publication of *the DNA sequencing techniques* of **Maxam** and **Gilbert** and of **Sanger, Nicklen** and **Coulson** (Sanger and Gilbert, Nobel Prize 1980).

1977 **Sanger** and colleagues published *complete 5387 nucleotide sequence of phage* $\phi \times 174$.

1978 Discovery of "*splicing*" of adenovirus RNAs in three different laboratories.

1982 **Sanger** and colleagues published the complete 48,502-nucleotide-pair sequence of phage lambda.

1983 **Cech** and **Altman** (Nobel Prize 1989) established the existence of catalytic RNAs (*i.e.*, ribozymes).

1988 **Watson** became co-ordinator of the "*human genome project*".

1989 **Tsui, Collins** and colleagues cloned the "*cystic fibrosis gene*", the gene whose mutant alleles account for majority of the cases of this dreaded disease that afflicts about one out of every 2000 children in United States.

1997 The first mammal was cloned, a sheep named Dolly.

2000 The sequence of the entire human genome was determined.

2003 Completion of draft of the human genome.

2009 World's first cloned camel (named Inaz) in Dubai was the outcome of efforts of Indian Scientist **Nisar Ahmad Wani** and his colleagues.

Thus, this period (*i.e.*, 1944-present) is the era of molecular genetics, beginning with the demonstration that DNA is the genetic material and culminating with our current explosion of knowledge due to recombinant DNA technology. Today, genetics is a mature, but still vigorously growing science.

1.2 THREE GENERAL BRANCHES OF GENETICS

Historically, geneticists have worked in three different areas, each with its own particular problems, terminology, tools and organisms. These areas (or branches) are classical genetics, molecular genetics and evolutionary genetics.

1. Classical genetics. In this branch of genetics, geneticists are concerned with the chromosomal theory of inheritance, that is, the concept that genes are located in a linear fashion or chromosomes and that the relative positions of genes can be determined by their frequency in offspring.

2. Molecular genetics. It is the study of the genetic material: its structure, replication and expression, as well as the information revolution emanating from the discoveries of recombinant DNA techniques (genetic engineering, including the human genome project).

3. Evolutionary genetics. It is the study of the mechanisms of evolutionary changes, or changes in gene frequencies in populations. **Darwin's** concept of evolution by natural selection finds a firm genetic footing in this area of the study of inheritance (Table 1.1).

Table 1.1 The three major areas of genetics—Classical, Molecular and Evolutionary—and the topics they cover (*Source:* Tamarin 2006).

Classical Genetics	*Molecular Genetics*	*Evolutionary Genetics*
1. Mendel's principles	1. Structure of DNA	1. Quantitative genetics
2. Meiosis and mitosis	2. Chemistry of DNA	2. Hardy-Weinberg equilibrium
3. Sex determination	3. Transcription	3. Assumptions of equilibrium
4. Sex linkage	4. Translation	4. Evolution
5. Linkage	5. DNA cloning and genomics	5. Speciation
6. Crossing over and chromosome mapping	6. Control of gene expression	
7. Cytogenetics (chromosomal changes)	7. DNA mutation and repair	
	8. Extrachromosomal inheritance	

QUESTIONS

Long Answer Questions

1. Give an account of growth of genetics during year 1900 to 1944.
2. Describe the growth of genetics in period 1944 to 2009.

Short Answer Questions

1. Mention topics covered by three main branches of genetics.
2. Write short notes on the following:

(*a*) Molecular genetics; (*b*) Classical genetics.

Very Short Answer Questions

1. Define genetics.
2. What is heredity?
3. Define variation.

Fill in the Blanks

1. The term genetics was coined by ....
2. Mendel began his scientific investigations on inheritance in ....
3. Punnett square was devised by ....
4. The term allelomorph was coined by ....
5. Operon model was proposed by ....

Multiple Choice Questions

1. Genetics is the branch which deals with the study of
   (*a*) relation between plants and environment (*b*) inheritance and variation
   (*c*) cell structure (*d*) thermal structure
2. The term 'genetics' was given by
   (*a*) Morgan (*b*) Mendel
   (*c*) Johannsen (*d*) Bateson
3. Mendel was a
   (*a*) physiologist (*b*) plant breeder
   (*c*) cytologist (*d*) taxonomist
4. Hybridisation means
   (*a*) removal of stamens (*b*) mixing parental characters
   (*c*) production of large sized grains (*d*) study of pollination habit
5. Mendel carried out his investigations on garden pea from
   (*a*) 1851–1859 (*b*) 1856–1864
   (*c*) 1865–1834 (*d*) 1870–1875

ANSWERS

Very Short Answer Questions

1. Genetics is a biological science of heredity and variation.
2. Transmission of traits from parents to offspring.
3. The occurrence of differences among individuals.

Fill in the Blanks

1. Bateson 2. 1856 3. R.C. Punnett 4. Bateson
5. Jacob and Monod

Multiple Choice Questions

1. (*b*) 2. (*d*) 3. (*b*) 4. (*b*) 5. (*b*).

2

Laws of Heredity (Mendelian Laws of Inheritance)

Gregor Johann Mendel was born in Moravia in 1822 to peasant parents. In 1843 he joined an Augustinian monastery at Brno, in Czechoslovakia where he took Holy Orders. From there he went to the University of Vienna where he spent two years studying natural history and mathematics before returning to monastery in 1853, where he lived and worked till his death in 1884.

Whilst in Vienna, Mendel had become interested in the process of hybridization in plants, the different forms in which hybrid (see Box 2.1) appear and statistical relationship between them. At Brno, his breeding experiments on the garden pea (*Pisum sativum*) were designed and conducted over a period of ten years (1856–1865). He appears to be the first scientist to appreciate the probabilistic nature of inheritance and use probability reasoning in biological research. He published a paper "*Experiments on plant hybridization*" in 1866 which laid the foundations for the present-day science of genetics. He demonstrated that *characteristics do not blend but pass from parents to offspring as discrete units* (which were named *genes* by Danish botanist **Johannsen**). *These units, which appear in the offspring in pairs, remain discrete and are passed on to subsequent generations by the male and female gametes which each contain a single unit.*

Partly because his ideas were ahead of his time, and partly because he published his work in an obscure journal (*i.e.*, Proceedings of the Brunn Natural Science Society), the significance of his monumental work went unnoticed till 1900. In that year, three eminent biologists, **Hugo de Vries, Carl Correns** and **Erich von Tschermak** independently 're-discovered' Mendel's work while studying inheritance in plant hybrids. These biologists had obtained results similar to those of Mendel, only to find that Mendel had published similar findings 34 years earlier. Mendel's original paper was republished in Flora, 89, 361 (1901). **Bateson** confirmed Mendel's work by a series of hybridization experiments.

2.1 REASONS OF MENDEL'S SUCCESS

The success of Mendel's hybridization experiments depended on the following factors: 1. Luck; 2. Foresight; 3. Mathematical background, and 4. Scientific aptitude. The following factors led to Mendel's success.

I. Method of Working

1. Mendel first analysed the reasons of failure of his predecessors and then chalked out the programme of his hybridization experiments.
2. Preliminary investigations were carried out to obtain familiarity with the experimental organism.

Box 2.1

1. **Allele or allelomorph.** A variant form of a gene specifying a particular trait is called **allele**. It is one of possible mutational states of a gene, distinguished from other alleles by phenotypic effects. **Gene mutations** convert one allelic form of a gene into another. For many gene loci, there is **wild-type allele** which predominates in the population because it confers great **fitness** (*i.e.*, ability to survive and reproduce). This generally encodes the **normal, functional** product associated with the gene, and the **wild-type phenotype** reflects this normal gene activity. Other, rare alleles are designated **mutant alleles**. The quantity and/or structural properties of the encoded product of the mutant alleles may differ greatly, generating distinct **mutant phenotype**. Most genes are **diallelic** having only two alleles, but some are **multiple allelic** containing more than two alleles of a gene.
2. **Phenotypes.** They are observable properties of an organism that are genetically controlled.
3. **Hybrid.** An offspring of two animals or plants of different varieties.
4. **Transmission genetics.** It includes the study of mode of gene transmission from generation to generation. The kind of studies that Mendel performed are now included in the discipline of transmission genetics.
5. **Variation.** Divergence in one or more characteristics of an organism or biotype from those typical of or usual for its group. Variation can be discontinuous or continuous.
6. **Segregation** is the separation of alleles into different gametes and then into different offsprings, of any pair of alleles possessed by an individual. The two alleles having previously been brought together in the individual, one from each of its parents.
7. **Genotype.** Genotype is the specific allelic or genetic constitution of an organism; often the allelic composition of one or a limited number of genes under investigation.
8. **Phenocopy. Phenocopy** is an environmentally induced phenotype which is nonheritable and closely resembles the phenotype produced by a known gene. Epilepsy disease of human beings, for example, may result from either genetic or environmental causes. As a result, the definitive diagnosis of an inherited abnormality requires family pedigree data or other information indicating that the condition is inherited, in addition to the appropriate phenotypic manifestations in an individual.
9. **Genome. Genome** is the array of genes carried by an individual. It is the genetic constitution or the sum total of hereditary determinants or genes of an organism which it has received from its parents. For example, the genes for tallness in a tall pea plant is one of the allelic pair of its total genotype. Genome will include total number of gene pairs present in the chromosomes. For example, *Drosophila* has total eight chromosomes and its genotype is the total number of genes present on these chromosomes.
10. **Proteome. Proteome** represents the entire collection of proteins which are encoded by the genome. The proteome is a complex system, even though it represents only a fraction of the genome (*i.e.*, the coding sequences). This is because there are numerous ways to use single genes to generate multiple products, so the proteome contains many overlapping, structurally similar products.
11. **Phenome.** When proteome interacts with the environment, it generates the **phenome** which is the representation of the total sum of characters (phenotypes) displayed by an organism.

3. Mendel studied the inheritance of one character at a time unlike his predecessors who had considered the inheritance of an organism as a whole.
4. Mendel carried out his hybridization experiments to F_2 and F_3 generations.
5. Mendel scrupulously maintained the statistical records of all the experiments and analysed them carefully. Sufficient data were obtained to have statistical significance.
6. He experimentally produced and maintained genetically pure parent plants (*i.e.*, pure breed lines). He tested the purity by self-crossing the progeny for several generations.

7. Mendel made crossing between those genetically pure parents which have sharply visible contrasting characters.
8. He grew different pure lines in separate garden plots, preventing chances of cross pollination.

II. Selection of the Experimental Plant.

For his hybridization experiments Mendel had certain consideration in his mind about the choice of a suitable material. Mendel's considerations about the material were as follows:

1. Variation. The organisms which are to be chosen for the genetical experiments, should have a number of detectable differences and at a time only single detectable character should be considered.

2. Reproduction. The chosen organisms should be sexually reproducing (*i.e.*, by fusion of male and female sex cells) because only then the offsprings will be able to receive different characters from both the male and female parents.

3. Controlled mating. The chosen organisms should be able to mate in controlled or well-planned conditions. Because in genetical experiments sometimes we have to rear genetically pure parents by methods of controlled mating. One should maintain careful records of the offsprings of many generation.

4. Short life cycle. The chosen organisms should have very short life cycles.

5. Large number of offsprings. The organisms which have been chosen for the genetical experiments should produce large number of offsprings after each successive mating because it will help in deducing the correct conclusions.

6. Convenience in handling. The experimental species should be of a type that can be raised and maintained conveniently and inexpensively in the laboratory. For instance, the elephant will prove entirely useless material for genetical experiments than the *Drosophila*, pea plants, tomato, rats, guinea pigs, etc., which have been generally used and are still used in hybridization experiments.

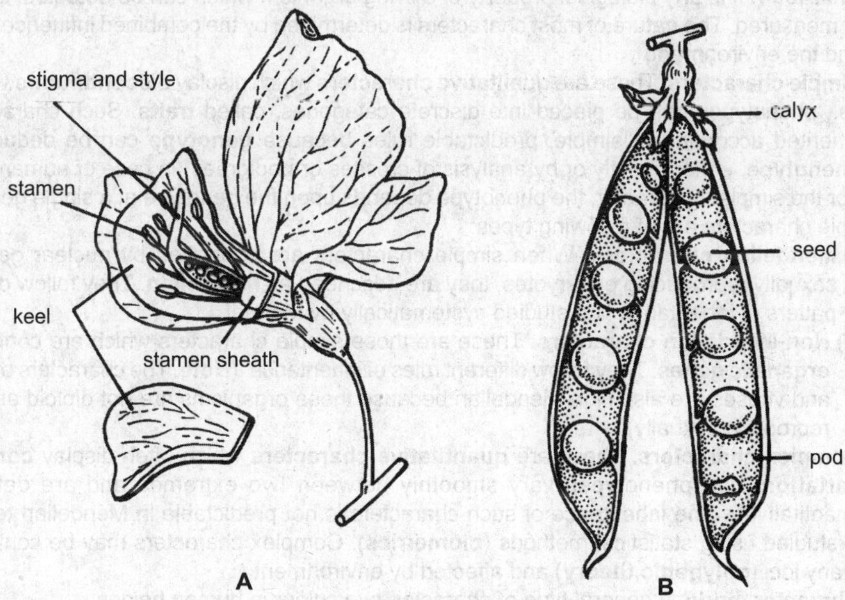

Fig. 2.1. A self-fertilizing flower of garden pea, *Pisum sativum*. A—Portion of the keel which encloses the reproductive organs has been cut open to show the stamen and stigma; B—The mature fruit (pod) containing seeds that develop from the fertilized pea flower.

Mendel selected garden pea (*Pisum sativum*) as his experimental material because it has the following advantages:

1. Garden pea plant is an annual plant. Its short life cycle make it possible to study several generations within a short period.
2. It has perfect bisexual flowers containing both male and female parts (Fig. 2.1). The flowers are predominantly self-pollinating.
3. Because of self-fertilization plants remain genetically pure. It is, therefore, easy to get pure lines for several generations.
4. Experimental cross pollination can be done easily in pea plant.

III. Selection of Traits

Mendel selected seven pairs of contrasting characters (see Box 2.2) in pea plant for his hybridization experiments (Table 2.1).

Table 2.1 Seven pairs of contrasting characters in pea plant and their dominance-recessive relationship as worked out by Mendel.

S. No.	*Characters*	*Dominant trait*	*Recessive trait*
1.	Stem length	Tall	Dwarf
2.	Flower position	Axial	Terminal
3.	Pod shape	Inflated	Constricted
4.	Pod colour	Green	Yellow
5.	Seed shape	Round	Wrinkled
6.	Cotyledon colour	Green	Yellow
7.	Seed coat colour	Coloured (Grey)	Colourless (White)

Box 2.2 Characters and Traits

1. **Character.** It is any biological property of a living organism which can be detected, described or measured. The nature of most characters is determined by the combined influence of genes and the environment.
2. **Simple characters.** These are **qualitative characters** which display **discontinuous variation**, *i.e.*, phenotypes can be placed into discrete categories, called **traits**. Such characters are inherited according to simple, predictable rules because **genotype** can be deduced from **phenotype**, either directly or by analysis of crosses or pedigrees (in case of human beings). For the simplest character, the phenotype depends upon the genotype at a single gene locus.

Simple characters are of following types:

(*i*) **Mendelian characters.** When simple characters are controlled by nuclear genome of sexually reproducing eukaryotes, they are described as **Mendelian**. They follow distinctive patters of inheritance first studied systematically by Mendel.

(*ii*) **Non-Mendelian characters.** These are those simple characters which are controlled by **organelle genes**. They follow different rules of inheritance. (**Note.** The characters of bacteria and viruses are also Non-Mendelian because these organisms are not diploid and do not reproduce sexually).

3. **Complex characters.** These are **quantitative characters** which often display **continuous variation**, *i.e.*, phenotypes vary smoothly between two extremes and are determined quantitatively. The inheritance of such characters is not predictable in Mendelian terms and is studied using statistical methods (**biometrics**). Complex characters may be controlled by many loci (**polygenic theory**) and affected by environment.
4. **Character mode.** A general type of character, eye colour in human beings.
5. **Trait.** A specific type of character, *e.g.*, blue eye colour.

IV. Mendel's Crossing Techniques

The flowers of pea plants are so constructed that the pollens of a flower normally fall on the stigma of the same flower and, thus, affects self-pollination or self-fertilization. For the required cross-pollination, the anthers have to be removed from the flower in bud stage (*i.e.*, before their maturity). This operation of removal of anthers is called **emasculation**. The stigma is protected against any foreign pollen with the help of its covering by a bag. The pollen, then at the dehiscence stage is brought from the plant to be used as a male parent and by the help of a brush is dusted on the feathery stigma of the emasculated flower. At the time of such **cross pollination**, the pollen should be mature and stigma should be receptive.

For each of seven pairs of characters listed in Table 2.1, plants with one alternative trait were used as female, and those with the other alternative as male. Reciprocal crosses were also made, *i.e.*, each of the crosses was made in two ways, depending on which phenotype is used as male or female. For example the following two crosses are reciprocal crosses:

phenotype A (*e.g.*, Tall); Male C × phenotype B (*e.g.*, Dwarf); Female
phenotype B (*e.g.*, Dwarf); Male × phenotype A (*e.g.*, Tall); Female

The population obtained as a result of crossing plants showing contrasting characters is called **F_1 generation**. The progeny of F_1 plants was then obtained by self-fertilization and it forms the **F_2 generation**. Similarly, F_3, F_4, etc., generations can also be obtained.

Further, for getting the exact results in the breeding experiments, it was necessary for Mendel to rear genetically pure variety of pea plants for a single character. Mendel adopted self-fertilization technique for it. For instance, to get pure character for tallness, he self fertilized a tall pea plant for many generations till the resulted offsprings always produced only tall plants. Likewise, he got genetically pure variety for dwarf pea plants. Mendel cross pollinated these two varieties of pea plants which were differing in a pair of contrasting characters, viz, tallness and dwarfness of the stem. When he made observations on the offsprings of first generation he found only tall plants. He allowed the self pollination in the offsprings of first generation and made further observations on the offsprings of second generation. He was astonished to note both tall and dwarf offsprings in the second generation. This showed him that the character of dwarfness disappeared in first generation but again reappeared in second generation. Further, the tall and dwarf plants of second generation were always in the ratio of 3:1 [3 (75%) tall: 1 (25%) dwarf). He self-pollinated the dwarf offsprings of second generation and found only dwarf plants in third generation. But when he self-pollinated the tall plants of second generation then he found that one-third (1/3) tall plants yield only tall plants in third generation, while rest two-third (2/3) tall plants yield tall and dwarf plants in the ratio of 3:1.

V. Mendel's Principles of Inheritance

On the basis of the results of Mendel's experiment the following seven principles have been recognised:

1. The heredity and variation of characters are controlled by factors, now called **genes**, which occur in pairs.
2. Contrasting traits are specified by different forms of each gene (different **alleles**).
3. When two dissimilar alleles are present in the same individual (*i.e.*, in a **hybrid** or **heterozygote**), one trait displays **dominance** over the other: the phenotype (see box 2.3) associated with one allele (the **dominant allele**) is expressed at the expense of that of the other (the **recessive allele**). This forms the Mendel's **phenomenon of dominance**.
4. Genes do not blend but remain discrete (**particulate**) as they are transmitted.
5. During meiosis, pairs of alleles **segregate** equally so that equivalent numbers of gametes carrying each allele are formed. This is called **principle of segregation** or **Mendel's first law**.
6. The segregation of each pair of alleles is independent from that of any other pair. This is called principle of **independent assortment** or **Mendel's second law**.

Actually Mendel himself did not postulated any genetical principle or laws, he simply gave conclusive theoretical and statistical explanations for his hybridization experiments in his research paper. However, it was **Correns**, the discoverer of Mendel's work, who thought that Mendel's discovery could be represented by these fundamental laws of heredity.

Box 2.3 Some Common Terms Used in Transmission Genetics

Allele. Broadly, a variant form of a gene specifying a particular trait. At the molecular level, a sequence variant of a gene.

Gene. Broadly a hereditary factor controlling or contributing to the control of a particular character. At the molecular level, a segment of DNA (or RNA in some viruses) which is expressed, *i.e.*, used to synthesize one or more products with particular functions in the cell.

(Gene) locus. The position of a gene on a chromosome or physical or genetic map.

Genotype. The genetic nature of an individual, often used to refer to the particular combination of alleles at a given locus.

Hemizygous. Containing one allele in a diploid cell, often used to refer to sex-linked genes.

Heredity. Passed from parent to offspring.

Heterozygous. Containing different alleles at a particular locus.

Phenotype. The outward nature of an individual, often used to refer to the nature of particular characters.

Pleiotropic gene. Affecting more than one character simultaneously.

Zygosity. The nature of alleles at a locus-homozygous, heterozygous or hemizygous.

2.2 MENDEL'S TWO LAWS

1. Law of Segregation

Mendel's 'first law—the law of segregation is also known as **law of purity of gametes**. The law states that *the hybrids or heterozygote's of F_1 generation have two contrasting characters or allelomorphs of dominant and recessive nature. These alleles though remain together for long time but do not contaminate or mix with each other and separate or segregate at the time of gametogenesis, so that each gamete receives only one allele of a character either dominant or recessive.*

For the proper understanding of the Mendel's law of segregation it will be helpful to study one of the Mendel's monohybrid cross (Fig. 2.2). Mendel crossed a homozygous red flowered pea plant with a homozygous white flowered pea plant. The F_1 heterozygotes or hybrids were found to be pink or purple flowered, thus, showing the incomplete dominance of red colour over white colour. When the F_1 hybrids were allowed to be self-fertilized, they produce both coloured (red or purple) and white flowered pea plants in F_2 generation in the ratio of three and one (3 : 1). The re-appearance of white colour in F_2 generation indicates towards the process of segregation.

Mechanism of segregation. The mechanism of segregation in above mentioned monohybrid cross between red and white flowered pea plants can be understood by assuming that the homozygous red flowered pea plant has the allele *RR* for redness and white flowered pea plant has the alleles rr for whiteness. The pea plant with *RR* alleles produces the gametes with single allele *R* and pea plant with *rr* alleles produces the gametes with the allele *r*. The gametes of both pea plants are united to form a hybrid or heterozygote having the alleles *Rr* both for redness and whiteness. Due to the phenomenon of incomplete dominance the allele *R* for red colour partially expresses itself in hybrids of F_1, while the allele *r* for white colour remains latent or recessive. Both the allele *R* and *r* remain together for long time but they do not affect each other. Neither they mix nor they contaminate each other. Because a gamete can contain only one chromosome of a homologous pair, therefore, each gamete can carry single allele *R* or *r*. At the time of gametogenesis two types of gametes are produced by F_1 hybrids in equal numbers. Half of the gametes carry the allele *R* and other half carry the allele *r*. These gametes during the process of fertilization can unite in three possible combinations, *viz.*, *RR*, *Rr* and *rr*, to produce three types of individuals in F_2 generation. Thus, in F_2 75% individuals have coloured flowers and 25% white flowers. The appearance of white colour

P.	Coloured/Red flowered pea plant	Non-coloured/white flowered pea plant
Phenotype.	Coloured (*i.e.*, Red)	Non-coloured (*i.e.*, white)
Genotype.	*RR* (Homozygous)	*rr* (Homozygous)
P. Gametes.	(*R*) (*R*)	(*r*) (*r*)

F1 Coloured (*i.e.*, Pink)
R r
Monohybrid/monoheterozygote

F_1 Cross.	*R r* F_1 Pink	×	*R r* F_1 Pink
F_1 Gametes.	(*R*) (*r*)		(*R*) (*r*)

Female gametes (eggs)

F_2

Male gametes (sperns)	*R*	*r*
R	*R R* Heterozygous coloured (*i.c.*, Red)	*r* Homozygous coloured (*i.e.*, Pink)
r	*R r* Homozygous coloured (*i.e.*, Pink)	*R r* Heterozygous non-coloured (*i.e.*, White)

Results of F_2.

F_2 Phenotypic ratio. 3 (*i.e.*, 75%) Coloured: 1 (*i.e.*, 25%) Non-coloured or 3 : 1

F_2 Genotypic ratio. 1 RR : 25 % Homozygous coloured (Red) 2 Rr : 50 % Homozygous coloured (Pink) 1 *rr* 25 % Homozygous non-coloured (White)

or 1:2:1

Fig. 2.2. Diagrammatic representation of a monohybrid cross of coloured (Red) and non-coloured (White) pea plants.

in F_2 generation indicates that in the hybrid the allele (*r*) for white colour remains along with allele (*R*) for red colour but does not mix with it or contaminated by it and it separates or segregates during gametogenesis.

Certain Other Examples of Law of Segregation

The law of segregation is universal in its application and it has been found to occur in both plants and animals. However, to understand the mechanism of segregation more clearly it will be helpful to consider an original experiment of well known geneticist **T.H. Morgan** on *Drosophila*.

Morgan crossed a homozygous long-winged (wild type) *Drosophila* with a homozygous vestigial-winged (mutant) *Drosophila*. The F_1 heterozygotes or hybrids were found to be long winged. When the F_1 hybrids were allowed to mate among themselves, they produced long-winged and vestigial-winged hybrids in three and one ratio (3 : 1) in the F_2 generation (Fig. 2.3).

In this cross the mechanism of segregation can be well understood by assuming that the homozygous long-winged *Drosophila* has a pair of alleles v^+v^+ for longness of wing and similarly homozygous vestigial-winged *Drosophila* has the alleles vv for vestigial nature of wings. The long-winged *Drosophila* fly thus produces the gametes (sperm or ova) with the single allele v^+ and the vestigial-winged fly produces the gametes with the single allele v. The gametes of both parents unite in the process of fertilization and produce a hybrid with long wings in F_1 generation. The genotype of F_1 hybrid is v^+v and in it the allele v^+ is dominant over the allele v which is recessive. At the time of gametogenesis these alleles (v^+ and v) are separated along with chromosomes to form two types of gametes, half of the gametes have the allele v^+ and other half have the allele v. These gametes unite in three possible combinations, viz., v^+v^+, v^+v and vv, to produce genotypically three types of individuals v^+v^+, v^+v and vv in 1 : 2 : 1 ratio in F_2 generation. Phenotypically here occur only three long-winged and one vestigial-winged *Drosophila*. Thus, the dominant and recessive alleles remain together for long time without contaminating or mixing with one another and segregate during gametogenesis.

2. Albinism in humans. Lack of pigment production in the human body is an abnormal recessive trait called **albinism**. Using *A* and *a* to represent the dominant (normal = melanin production) allele and the recessive (albino = no melanin production) allele respectively, the following three genotypes and two phenotypes are possible:

	Genotypes	*Phenotypes*
AA	(Homozygous dominant)	Normal (Pigment)
Aa	(Heterozygote)	Normal (Pigment)
aa	(Homozygous recessive)	Albino (No pigment)

The person with recessive *aa* genes do not produce the **tyrosinase** enzyme which is needed by melanocytes for the synthesis of (black melanin) pigment. As a result in an albino patient melanocytes are present in normal number in their skin, hairs, iris of eyes, etc., but lack in melanin pigment.

2. Law of Independent Assortment

To formulate the principle of dominance and law of segregation **Mendel** considered monohybrid crosses in which inheritance pattern of single pairs of contrasting characters were considered at a time. But he tried to find out how different characters would behave in relation to each other in their inheritance from generation to generation. For this purpose Mendel crossed two varieties of pea plants which were differing in two pairs of contrasting characters. Because such crosses yielded **dihybrids** and at a time two pairs of contrasting characters had been considered in them, therefore, these crosses were known as **dihybrid crosses**.

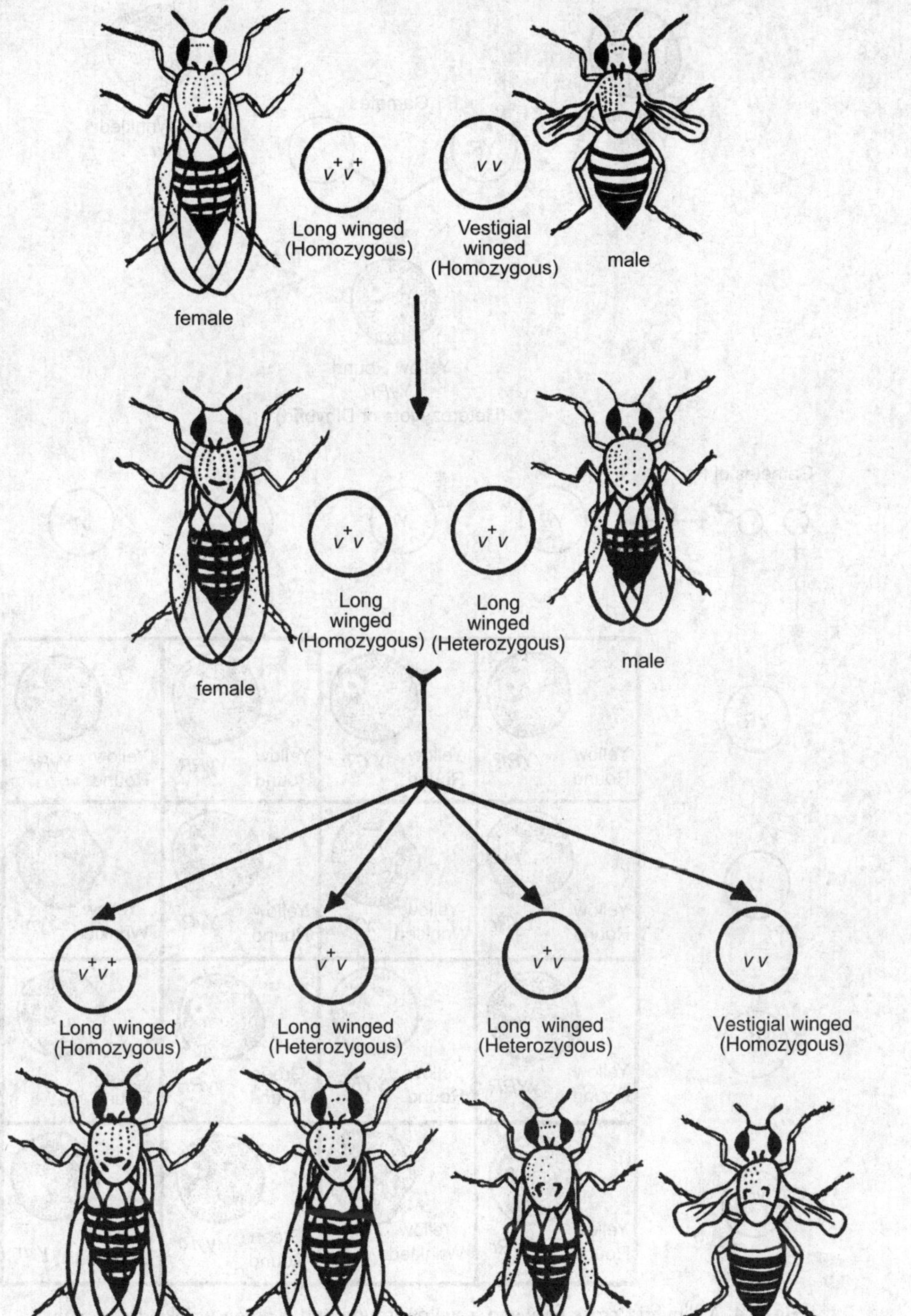

Fig. 2.3. Cross between long-winged and vestigial-winged *Drosophila*.

Mendel's dihybrid cross. In one of his hybridization experiment Mendel crossed a homozygous pea plant having **yellow round seeds** with the homozygous pea plant having **green wrinkled seeds**

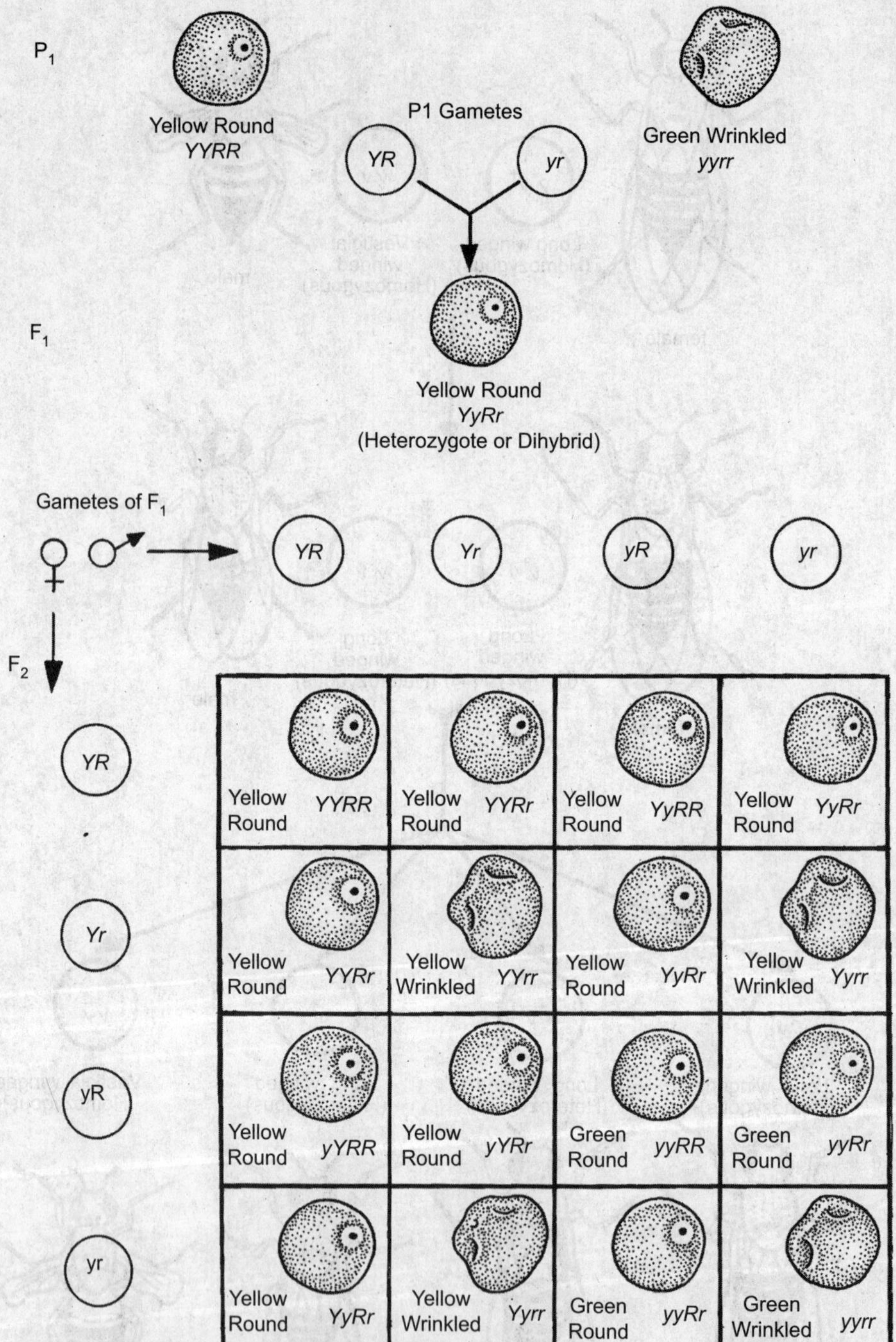

Fig. 2.4. A dihybrid cross between a yellow round and a green wrinkled pea plant.

(Fig. 2.4). The F_1 hybrids (*i.e.*, dihybrids) were found to have yellow round seeds. When the F_1 hybrids were allowed to cross among themselves they produced four types of seeds in the ratio of 9 : 3 : 3 : 1 given as follows:

1. Yellow Round – 9
2. Yellow Wrinkled – 3
3. Green Round – 3
4. Green Wrinkled – 1

Beside getting the ratio of 3 : 1 of the monohybrid crosses Mendel got the ratio of 9 : 3 : 3 : 1. This irregularity in the ratio of F_2 offspring was explained by Mendel stating that "*when the parents differ from each other in two or more pairs of contrasting characters or factors then the inheritance of one pair of factors is independent to that of the other pair of factors*". This is the Mendel's law of independent assortment.

Mechanism of independent assortment. The mechanism of independent assortment can be understood easily by assuming that the homozygous pea plant with yellow round seeds has the alleles *YY* and *RR* for the yellow colour and roundness of the seed, respectively. Similarly the homozygous pea plant with green wrinkled seeds contains the alleles *yy* and *rr* for the green colour and wrinkledness of seeds. The gametes which are produced by *YYRR* and *yyrr* plants are *YR* and *yr* types respectively. When both parents are crossed the union of both types of gametes takes place to give the F_1 hybrid (*YyRr*). The F_1 hybrid have been found to contain yellow round seeds showing the dominance of allele *Y* for yellow colour over the recessive allele *y* for green colour and the dominance of allele *R* for roundness over the recessive allele *r* for wrinkledness of seed. Now the F_1 hybrids have four types of alleles, viz., *Y* for yellow colour, *y* for green colour, *R* for round shape and *r* for rinkledness of seed.

During gametogenesis these four alleles, may combine in following four combinations:

1. The allele *Y* may be associated with the allele *R* to give rise *YR* combination.
2. The allele *Y* may be associated with the allele *r* to give rise *Yr* combination.
3. The allele *y* may be associated with the allele *R* to give rise *yR* combination.
4. The allele *y* may be associated with the allele *r* to give rise *yr* combination.

Thus, four types of alleles are assorted independently to produce four types of gametes, viz., *YR, Yr, yR* and *yr*. These four types of gametes (pollens or ovules) of F_1 hybrid unite at random in the process of fertilization and produce sixteen types of individuals in F_2 generation which are given in the Table 2.2.

Table 2.2 The phenotype and genotype of F_2 generation of a dihybrid cross between pea plants with yellow round and green wrinkled seeds.

Number of individuals	*Genotypes class*	*Phenotype class*
1	*YY RR*	Homozygous Yellow Round
2	*YY Rr, YY Rr*	
2	*Yy RR, Yy RR*	Heterozygous Yellow Round
4	*Yy Rr, Yy Rr, Yy Rr, Yy Rr*	
		Yellow Round = 9
1	*YY rr*	Homozygous Yellow Wrinkled
2	*Yy rr, Yy rr*	Heterozygous Yellow Wrinkled
		Yellow Wrinkled = 3
1	*yy RR*	Homozygous Green Round
2	*yy Rr, yy rR*	Heterozygous Green Round
		Green Round = 3
1	*yy rr*	Homozygous Green Wrinkled
		Green Wrinkled = 1
16	– –	9 : 3 : 3 : 1

Thus, the sixteen F_2 individuals have the ratio of 9 yellow round: 3 yellow wrinkled: 3 green round: 1 green wrinkled. These results have proved the law of independent assortment and showed that each pair of contrasting characters behaves independently and bears no permanent association or relation with a particular character. The allele *Y* was associated with allele *R* in parent but it does not always remain associated with it and it is also associated with the allele *r*.

Dihybrid cross in *Drosophila*. In a dihybrid cross homozygous long-winged and black-bodied *Drosophila* fly is crossed with a vestigial-winged, grey-bodied fly, they produced in F_1 long-winged grey-bodied *Drosophila*. The hybrids of F_1 generation after mating with each other produced in F_2 generation 9 long-winged grey-bodied, 3 long-winged black-bodied, 3 vestigial-winged grey-bodied and 1 vestigial-winged black-bodied *Drosophila* (Total = 16). Therefore, the 16 individuals of F_2 have the ratio of 9 : 3 : 3 : 1 (Fig. 2.5).

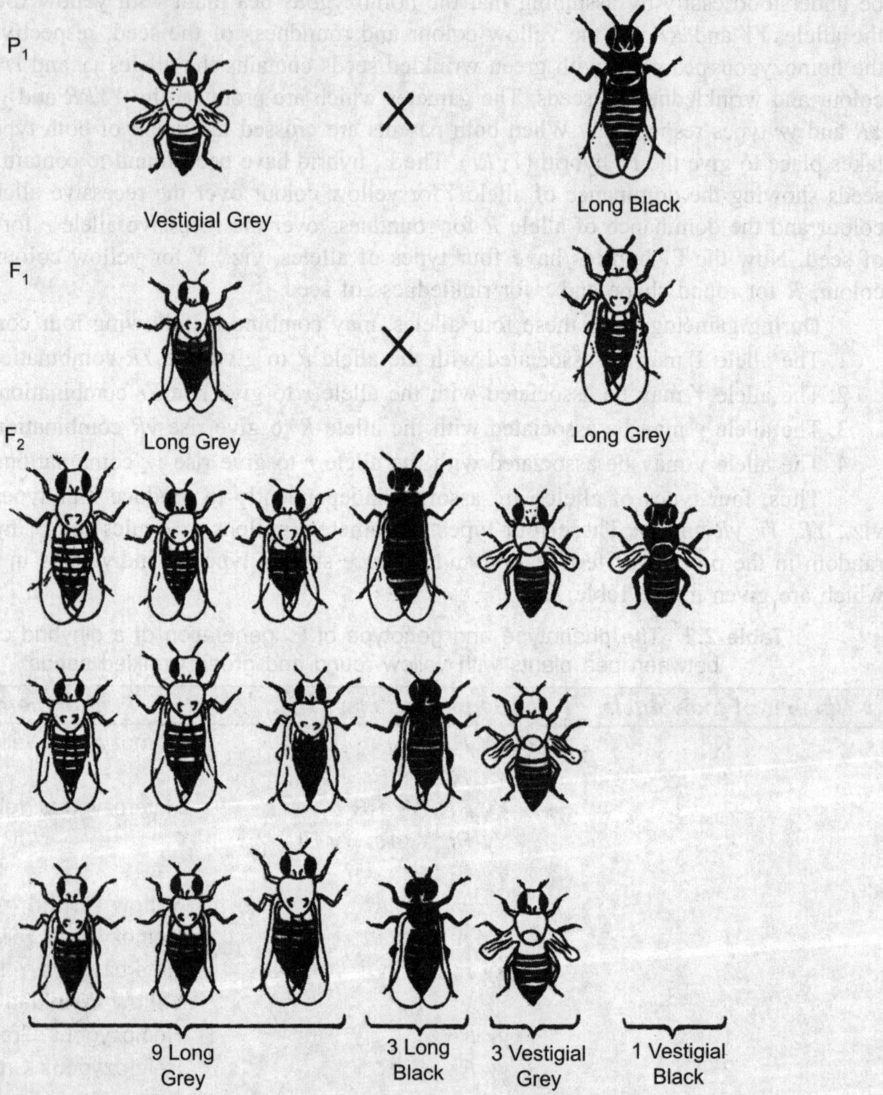

Fig. 2.5. A dihybrid cross between the vestigial-grey and long-black *Drosophila*.

Practical Application of Mendelism

Since the time of their re-discovery in 1900, Mendel's laws are being applied to various branches of breeding. In poultry, these are used for improving the varieties of fowls and their eggs. Various improved breeds of horses, dogs, sheep and cows are also obtained. A number of rust-resisting, disease-resistant and high-yielding varieties of wheat, rice, maize, etc., have been produced by cross-breeding experiments. Eugenics is mainly based on Mendelism.

2.3 CHROMOSOME THEORY OF HEREDITY (CYTOLOGICAL BASIS OF HEREDITY)

American cytologist **W. Sutton** in early 20th century noticed striking similarities between the behaviour of chromosomes during gamete formation and fertilization, and the transmission of Mendel's hereditary factors. These similarities have been summarised in Table 2.3.

Table 2.3 A summary of the similarities between events occurring during meiosis and fertilization and Mendel's hypothesis.

S. No.	*Meiosis and fertilization*	*S No.*	*Mendel's hypothesis*
1.	Deploid cells contain *pairs* of chromosomes (homologous chromosomes).	1.	Characteristics are controlled by *pairs* of factors.
2.	Homologous chromosomes *separate* during meiosis.	2.	Pairs of factors *separate* during gamete formation.
3.	*One* homologous chromosome passes into each gamete cell.	3.	Each gamete receives *one* factor.
4.	Only the *nucleus* of the male gamete fuses with the egg cell nucleus.	4.	Factors are transmitted from generation to generation as *discrete* units.
5.	Homologous pairs of chromosomes are restored at fertilization, each gamete (male and female) contributing one homologous chromosome.	5.	Each organism inherits one factor from each parent.

On the basis of the evidence suggested above, **Sutton** and **Boveri** in 1902 proposed that chromosomes were the carriers of Mendel's factors, *i.e.*, they formulated the so called **chromosome theory of heredity**. According to this theory *each pair of factors is carried by a pair of homologous chromosomes, with each chromosome carrying one of the factors*. Since the number of characteristics of any organism vastly outnumbers the chromosomes, as revealed by microscopy, each chromosome must carry many factors.

In 1909, term **factor** was replaced by the term **gene** by **Johannsen**. Whilst term gene is used to describe the unit of heredity, its variants are called **alleles** which influence phenotypic expression. Alleles occupy corresponding positions or **loci** on the homologous chromosomes.

Chromosomal Explanation of Segregation

Mendel's law of segregation of factors could now be explained in terms of the separation of homologous chromosomes which occurs during anaphase I of meiosis and the random distribution of alleles into gamete cells. These events have been summarised in Fig. 2.6.

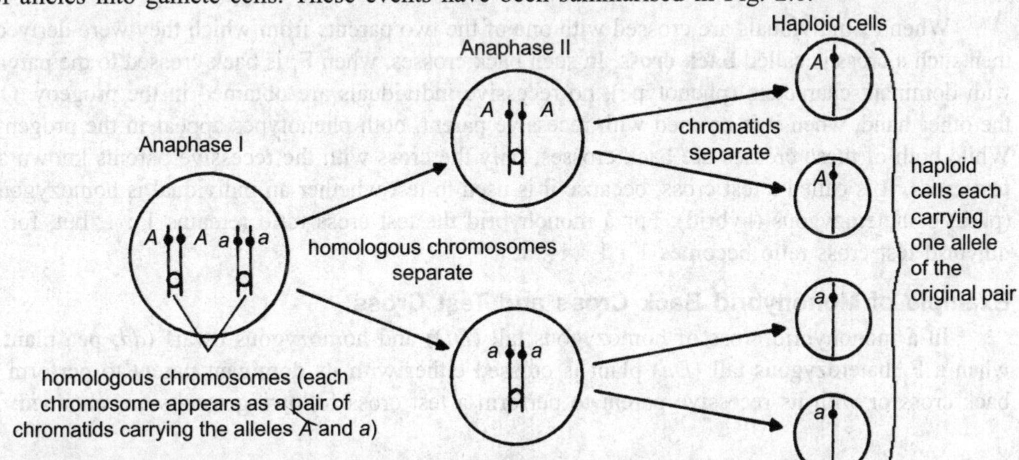

Fig. 2.6. Mendel's principle of segregation of factors (alleles) *A* and *a* described in terms of the separation of homologous chromosomes which occurs during meiosis.

Chromosomal Explanation of Independent Assortment

Mendel's law of independent assortment may also be explained in terms of the movement of chromosomes during meiosis. During gamete formation, the distribution of each allele from a pair of homologous chromosomes is entirely independent of the distribution of alleles of other pairs. This situation is described in Fig. 2.7. It is the random alignment or assortment of homologous chromosomes on the equatorial spindle during metaphase I of meiosis, and other subsequent separation during metaphase I and anaphase I that leads to the variety of alleles recombination in the gamete cells.

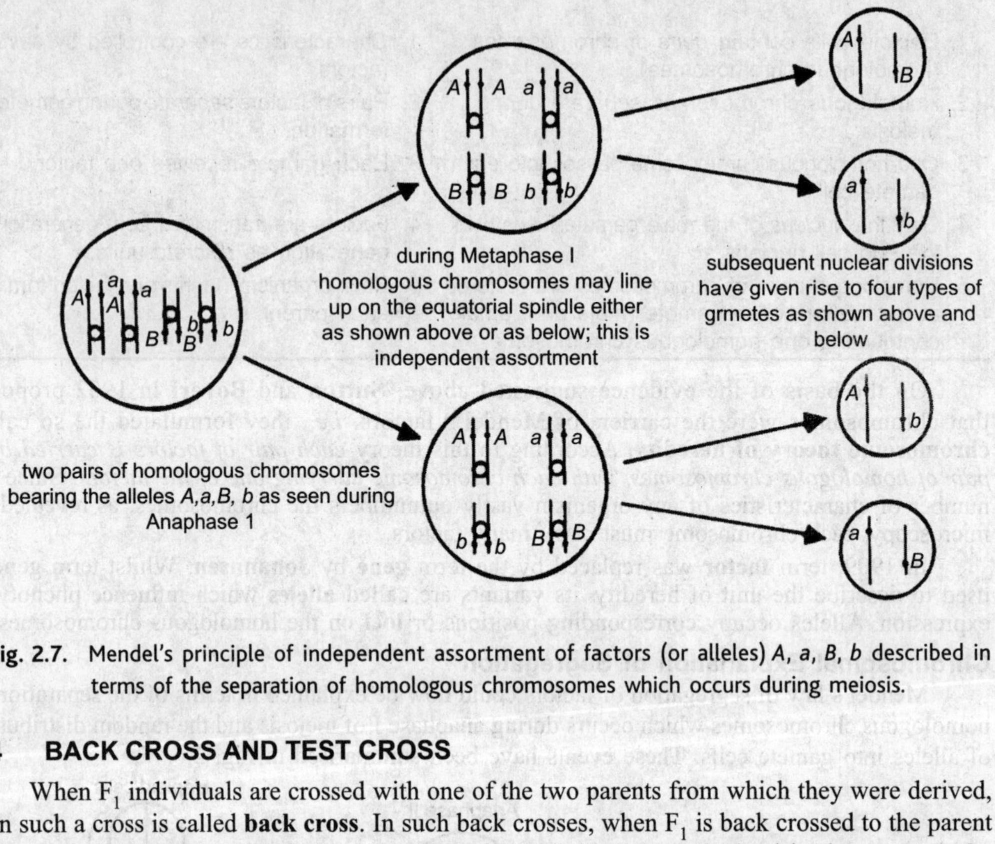

Fig. 2.7. Mendel's principle of independent assortment of factors (or alleles) *A, a, B, b* described in terms of the separation of homologous chromosomes which occurs during meiosis.

2.4 BACK CROSS AND TEST CROSS

When F_1 individuals are crossed with one of the two parents from which they were derived, then such a cross is called **back cross**. In such back crosses, when F_1 is back crossed to the parent with dominant characters (phenotype), no recessive individuals are obtained in the progeny. On the other hand, when it is crossed with recessive parent, both phenotypes appear in the progeny. While both of these crosses are back crosses, only the cross with the recessive parents known as **test cross**. It is called a test cross, because it is used to test whether an individual is homozygous (pure) or heterozygous (hybrid). For a monohybrid the test cross ratio remains 1 : 1, but, for a dihybrid test cross ratio becomes 1 : 1 : 1 : 1.

Example of Monohybrid Back Cross and Test Cross

In a monohybrid cross of homozygous tall (*DD*) and homozygous dwarf (*dd*) pea plants, when a F_1 haterozygous tall (*Dd*) plant is crossed either with its dominant parent to perform a back cross or with its recessive parent to perform a test cross following results are obtained:

P_1:	Homozygous tall *DD*	× ↓	Homozygous dwarf *dd*
		Dd	
F_1:		Heterozygous tall	
A. Back cross:	F_1 Tall *Dd*	× ↓	P_1 Tall *DD*
Back cross progeny:	½ *DD* Homozygous tall	:	½ *Dd* Heterozygous tall

(Back cross produces all tall)

B. Test Cross:	F_1 Tall *Dd*	× ↓	P_1 Dwarf *dd*
Test cross progeny:	½ *Dd* Heterozygous tall	:	½ *dd* Homozygous dwarf

(Test cross produces a ratio of 1 : 1 of tall and dwarf)

Example of Dihybrid Test Cross

The test cross of a heterozygous yellow round seeded pea plant with a double recessive parent (green wrinkled, yyrr), yields the test cross genotypic and phenotypic ratio of 1 : 1 : 1 : 1 as follows:

Parents:	F_1 Heterozygous Yellow Round *Yy Rr* ↓	×	F_1 Homozygous Green Wrinkled *yy rr* ↓
Gametes:	*YR, Yr, yR, yr*		*yr*

Test cross progeny: ¼ *Yy Rr* : ¼ *Yy rr* : ¼ *yy Rr* : ¼ *yy rr*
or ¼ Yellow Round : ¼ Yellow Wrinkled: ¼ Green Round : ¼ Green Wrinkled
or 1:1:1:1.

2.5 DISTINCTIONS

1. Phenotype and Genotype

Table 2.4 Differences between phenotype and genotype.

S. No.	Phenotype	S. No.	Genotype
1.	Phenotype is the expression of the character of an individual. Character may be morphological (such as form, sex, colour, behaviour) or biochemical (ABO blood group).	1.	Genotype is the genetic make up of the individual which determines the characters.
2.	Phenotype can be read out from the individuals by direct observation.	2.	Genotypes can be ascertained from the progeny of the individuals.
3.	Individuals having identical phenotypes may or may not have the same genotype.	3.	Individuals having identical genotypes usually have the same phenotype provided they are present in the same environment.

2. Homozygous and Heterozygous

Table 2.5 Differences between homozygous and heterozygous individuals.

S. No.	*Homozygous*	*S. No.*	*Heterozygous*
1.	These individuals possess identical alleles of a gene such as in pea plant either dominant alleles *DD* for for tallness or recessive alleles *dd* for dwarfness.	1.	These individuals possess different alleles of a gene such as a in pea plant dominant allele *D* and a recessive allele *d*.
2.	They produce only one type of gametes, e.g., only *D* by *DD* individuals and *d* by *dd* individuals.	2.	They produce two types of gametes in case of monoheterozygote (monohybrid), e.g., *D* and *d*.
3.	They breed true to the particular character, *i.e.*, when a dominant parent (or a recessive parent is self-pollinated or self-fertilized, its all the offsprings possess the same genotype and phenotype).	3.	When self-pollinated or self-fertilized, heterozygote of a monohybrid cross produces three different genotypes (*i.e.*, *DD*, *Dd* or *dd*) have two or three different phenotypes.

3. Dominant and Recessive

The allele which gains phenotypic expression in F_1/F_2 hybrid (heterozygote) is known as **dominant allele**, while the allele which is unable to express itself in the presence of the dominant allele is called the **recessive allele**. Though recessive allele fails in expressing itself in hybrids, it is transmitted to the next generation without change. It expresses only in the homozygous state.

For example, a heterozygous pea plant has one allele for tallness (*D*) and one allele for dwarfness (*d*). The allele for tallness (*D*) is dominant in nature and has it phenotypic expression both in homozygous and heterozygous states. The allele for dwarfness (*d*) is recessive and is expressed only in homozygous condition.

QUESTIONS

Long Answer Questions

1. Why did Mendel select *Pisum sativum* for his experiments?
2. Describe Mendel's method of experimentation. How was the Mendel's work re-discovered? Briefly describe the factors that led to Mendel's success.
3. State and discuss Mendel's laws. Why was Mendel successful in enunciating these laws.
4. Describe Mendel's laws of inheritance with the help of suitable examples. (*C.C.S.U. Meerut, 2009*)
5. Write an essay on Mendelism and its practical applications.
6. Give an account of the Mendel's laws of inheritance and discuss their importance in genetics. (*C.C.S.U. Meerut, 2007*)
7. Write an essay on Mendelian inheritance.
8. Describe Mendel's two laws of heredity and explain their mechanism.
9. Explain the Mendelian laws of segregation and independent assortment.
10. Explain Mendel's laws of segregation with the help of two examples.
11. With suitable examples explain Mendel's law of independent assortment.
12. Describe in detail the Mendel's law of independent assortment giving suitable example and also describe the seven characters chosen by Mendel to perform his cross-breeding experiments.
13. Give a critical appreciation of Mendel's laws of Genetics. Describe the Law of Independent Assortment and state reasons why this law is not universally applicable.

14. State Mendel's law of independent assortment in man. Brown eye colour (*B*) is dominant over blue eye colour (*b*). A brown eyed man, whose mother was blue eyed, married a brown eyed woman whose father was blue eyed. What are genotypes of man and woman.

(**Hint.** Male *Bb*; Female *bb*).

15. State and explain Mendel's law of independent assortment and explain the following terms: 1. Recessive characters; 2. Heterozygous; 3. Genotype; 4. Test cross; 5. Monohybrid cross.

Short Answer Questions

1. Give a brief life sketch of Gregor Johann Mendel.

2. Write briefly the contribution to biology by Mendel.

3. State briefly what is meant by 'dominant' and 'recessive' as applied to inherited character.

4. Describe the chromosomal basis of Mendel's crosses.

5. Explain giving examples Mendel's law of segregation. Interpret it in the light of the chromosome theory of heredity.

6. Explain "what is the heredity bridge between parents and offspring".

7. Differentiate between the following:

(*i*) Phenotype and Genotype
(*ii*) Genotype and Phenotype
(*iii*) Testcross and Backcross (*C.C.S.U. Meerut 2007, 2008*)
(*iv*) Homozygous and Heterozygous
(*v*) Dominant and Recessive
(*vi*) Dominance and Segregation
(*vii*) Monohybrid and Dihybrid crosses
(*viii*) Gene and Allele.

8. Write short notes on the following:

(*i*) Mendelism
(*ii*) Monohybrid cross
(*iii*) Dihybrid cross
(*iv*) Mendel's first law or Law of segregation
(*v*) Law of independent assortment or Independent assortment
(*vi*) Test cross
(*vii*) Allele or Allelomorph
(*viii*) Phenotype
(*ix*) Genotype
(*x*) Phenocopies
(*xi*) Genome
(*xii*) Proteome
(*xiii*) Phenome
(*xiv*) Second law of Mendel (*H.N.B. Garhwal 2009*)
(*xv*) Mendel's law of independent assortment (*H.N.B. Garhwal 2009*)
(*xvi*) Difference between test cross and back cross and their significance (*H.N.B. Garhwal 2009*)
(*xvii*) Back cross.

9. What is test cross and back cross? What is its significance?

10. What is test cross? How it helps in interpreting of the results of a monohybrid cross?

11. What is the principle of independent assortment? How did Mendel test this hypothesis with the help of test cross?

Very Short Answer Questions

1. What do the following genetic symbols mean? *AA*, *Aa*, *aa*.
2. Name the scientist who re-discovered Mendel's law.
3. What is pure line?
4. Who is considered as the 'Father of Genetics'?
5. When did the results of Mendel's experiments published?
6. What do the two forms of genes called?
7. What is hybridization?
8. Define the test cross and Back cross.
9. What is meant by heterozygote?
10. Define the allele.
11. Define F_2 generation.
12. What is dihybrid cross? *(H.N.B., Garhwal, 2009)*

Coin One Word for the Following:

1. A group of individuals which produce offsprings for several generations having same specific character.
2. A study of resemblances and differences between the parents and their offsprings.
3. An individual which does not breed true for its character.
4. A group of plants having same genes for a character.
5. A group of individuals showing same expression for a character.
6. A cross of F_1 individual with the homozygous recessive parent.

Yes or No Questions

1. The term gene was first proposed by Morgan.
2. The test cross is between F_1 hybrid and double dominant.

Fill in the Blanks

1. Mendel's work was re-discovered by three biologists ...., Carl Corens and .... in 1900.
2. The mating of genetically related individuals is called ....
3. .... confirmed Mendel's work by hybridization experiments.
4. An offspring of two animals or plants of different varieties is called .... or ....
5. .... is the specific allelic or genetic constitution of an organism.
6. A diploid or polyploid organism with different alleles at a particular locus is called ....
7. Mendel's first law, the law of segregation, is also know as ....
8. The chromosome theory of heredity was proposed by .... in 1902.
9. A diploid or polyploid organism with identical alleles at a locus is called ....
10. A cross between homozygous recessive and heterozygous organism is called .... *(H.N.B., Garhwal, 2008)*

Multiple Choice Questions

1. The various forms of a given gene are called

(*a*) genotype (*b*) phenotype

(*c*) gamete (*d*) alleles

2. The expression of more than one character by a single gene is called.

(*a*) pleiotropism (*b*) codominance

(*c*) epistasis (*d*) none

3. Mendel has chosen pea plant for his experiments to get successful results because (*M.S.U., Tamil Nadu, 2006*)

(*a*) it had a short life

(*b*) it had several pairs of contrasting characters

(*c*) it was easy to breed

(*d*) all the above

4. One of the Mendel's pure strains of pea plants has green peas. How many different kinds of ovum could such a plant produce with regard to pea colour? (*M.S.U., Tamil Nadu, 2006*)

(*a*) 1 (*b*) 2

(*c*) 4 (*d*) 8

5. Emasculation is achieved by

(*a*) removal of anthers (*b*) removal of style

(*c*) removal of petals (*d*) removal of sepals

6. Organisms phenotypically similar but genotypically different are said to be

(*a*) heterozygous (*b*) monozygous

(*c*) multizygous (*d*) homozygous

7. When a gene exists in more than one form, the different forms are called

(*a*) heterozygous (*b*) complementary genes

(*c*) genotypes (*d*) alleles

8. Pea plants

(*a*) are cross-pollinated (*b*) can be selfed

(*c*) are selfed and can be cross-pollinated (*d*) can be selfed and are cross-pollinated

9. Phenomenon of dominance and recessiveness is the result of

(*a*) back cross (*b*) incomplete dominance

(*c*) dihybrid cross (*d*) monohybrid cross

10. Law of independent assortment can be explained with the help of

(*a*) test cross (*b*) back cross

(*c*) monohybrid cross (*d*) dihybrid cross

11. A cross made between F_1 hybrid and any of its parent is

(*a*) back cross (*b*) monohybrid cross

(*c*) dihybrid cross (*d*) reciprocal cross

12. The cross between homozygous recessive and its hybrid is called

(*a*) back cross (*b*) test cross

(*c*) monohybrid cross (*d*) dihybrid cross

ANSWERS

Very Short Answer Questions

1. *AA*—homozygous dominant; *Aa*—heterozygous dominant; *aa*—homozygous recessive.

2. Hugo De Vries, Carl Correns and Eric Von Tschermak.

3. Homozygous population.

4. Gregor Johann Mendel.

5. 1866.

6. Alleles.

7. Crossing of two genetically different individuals is called hybridization.

8. Back cross. The cross of an F_1 hybrid to one of the parental types. The offspring of such a cross are referred to as back cross generation or back cross progeny.

Test cross. Back cross to the recessive parental type, or a cross between genetically unknown individuals with a fully recessive tester to determine whether an individual in question is heterozygous or homozygous for a certain allele. Test cross is also used as a test cross for linkage.

Example. Test crosses in pea plant:

Test cross 1	Homozygous Tall *DD*	×	P_1 recessive parent (or tester) Homozygous Dwarf *dd*
Gametes:	All *D*		All *d*
Back cross Progeny:		*Dd* All long but heterozygous	

Result: Individual in question is homozygous tall.

Test cross 2	F_1 Tall (*i.e.*, Heterozygous tall) *Dd*	×	P_1 recessive dwarf parent *dd*
Gametes	1/2 *D*: 1/2 *d*		All *d*
Back cross Progeny:	1/2 *Dd* Heterozygous tall	: 1/2 *dd* Homozygous dwarf	

Result: 1:1 phentypic ratio indicates that individual in question is a monohybrid or single heterozygote for dominant allele *D* for tallness and recessive allele *d* for dwarfness.

9. A heterozygote (or hybrid) possesses two contrasting alleles of a gene, but only one allele of the two is able to express itself, while the other remains hidden. The allele which gains phenotypic expression in the F_1 heterozygote is called **dominant allele**, while the other allele which fails to express itself in the presence of the dominant allele is the **recessive allele**. Recessive allele, though, is unable to express itself in F_1, is transmitted to the next generation without change. For example, a heterozygous pea plant has one allele for tallness (*D*) and one allele for dwarfness (*d*). The allele for tallness (*D*) is dominant and the allele for dwarfness (*d*) is recessive.

10. Alleles are variants of a gene controlling the same characteristic (*e.g.*, length of pea stem) but producing different effects (*e.g.*, tall or dwarf) and occupying corresponding positions (loci of gene) on homologous chromosomes. An allele may be a dominant allele, recessive or null allele, wild type allele or mutant allele.

The genes for the characters described by Mendel occurred in two forms, *i.e.*, they were **diallelic**. For many genes, however, a greater degree of **allelic variation** is apparent, *i.e.*, they are **multiple allelic**. The human ABO blood group locus, for instance, has three physiologically distinct alleles, *i.e.*, I^A, I^B, or *i*.

11. It stands for **second filial generation** which is resulted by interbreeding or selfing of F_1 offspring (which are heterozygotes). F_2 phenotypic ratio of a monohybrid cross is 3 : 1 and of a dihybrid cross is 9 : 3 : 3 : 1.

12. A cross between homozygous parents differing in two respects. Progeny of such a cross is called dihybrid; it is heterozygous for two pairs of alleles.

Coin One Word for the Following:

1. Pure line
2. Genetics
3. Heterozygous
4. Homozygous
5. Phenotype
6. Test cross.

Yes or No Questions

1. No
2. No.

Fill in the Blanks

1. Hugo de vries, Erich von Tschermak.
2. Inbreeding.
3. Bateson.
4. Hybrid, heterozygous.
5. Genotype.
6. Heterozyote.
7. Law of purity of gametes.
8. Sutton and Boveri.
9. Homozygote.
10. Test cross.

Multiple Choice Questions

1. (*d*) **2.** (*a*) **3.** (*d*) **4.** (*a*) **5.** (*a*) **6.** (*a*)
7. (*d*) **8.** (*c*) **9.** (*d*) **10.** (*d*) **11.** (*a*) **12.** (*b*).

3

Dominance, Incomplete Dominance and Codominance

3.1 PHENOMENON OF DOMINANCE

The cross between the pea plants differing in single pair of contrasting characters is known as **monohybrid cross.** As we have already noticed that when Mendel made a monohybrid cross (Fig. 3.1) between tall and dwarf pea plants then only tall pea plants appeared in the first filial generation (F_1). But when the F_1 progeny were allowed to be self-fertilized, both tall and dwarf characters appeared in the second filial generation or F_2. This shows that in F_1 hybrid the character of tallness dominates or conceals the character of dwarfness and so the character of dwarfness could not express itself in F_1 generation. The character which expresses itself (*i.e.,* tall) in F_1 generation is called by Mendel as **dominant character;** while the character which remained unexpressed or latent had been called **recessive**. According to these results Mendel described the phenomenon of dominance in following way: *in crossing between pure (homozygous) organisms for contrasting characters of a pair, only one character of the pair appears in the first filial generation.*

In pea plant Mendel found following characters to be dominant or recessive in various pairs of contrasting characters (See Table 3.1). In each of Mendel's crosses, the trait associated with one allele was fully dominant over the other, so that the phenotype of the heterozygote was identical to that of the dominant homozygote. At the biochemical level, such **complete dominance** often reflects the presence of a (recessive) **null allele** which is totally compensated by the presence of a (dominant) functional allele.

Certain Examples of Phenomenon of Dominance

After Mendel several geneticists tested the validity of the phenomenon of dominance on several plants and animals. They found its wide application in various plants and animals. A few important examples are described as follows:

1. Phenomenon of Dominance in Plants. Besides pea plant the phenomenon of dominance has also been observed in the following plants (See Table 3.2).

2. Application of Phenomenon of Dominance in Animals

A. The phenomenon of dominance is also applicable well to the animals. For instance, when homozygous black guinea pig is crossed with a homozygous white guinea pig (Fig. 3.2) then all hybrids of first filial generation (F_1) are found to be black. The black hybrids of F_1 when mated among themselves they produced black and white offsprings in 3:1 ratio. This shows that black coat colour dominates over white coat colour.

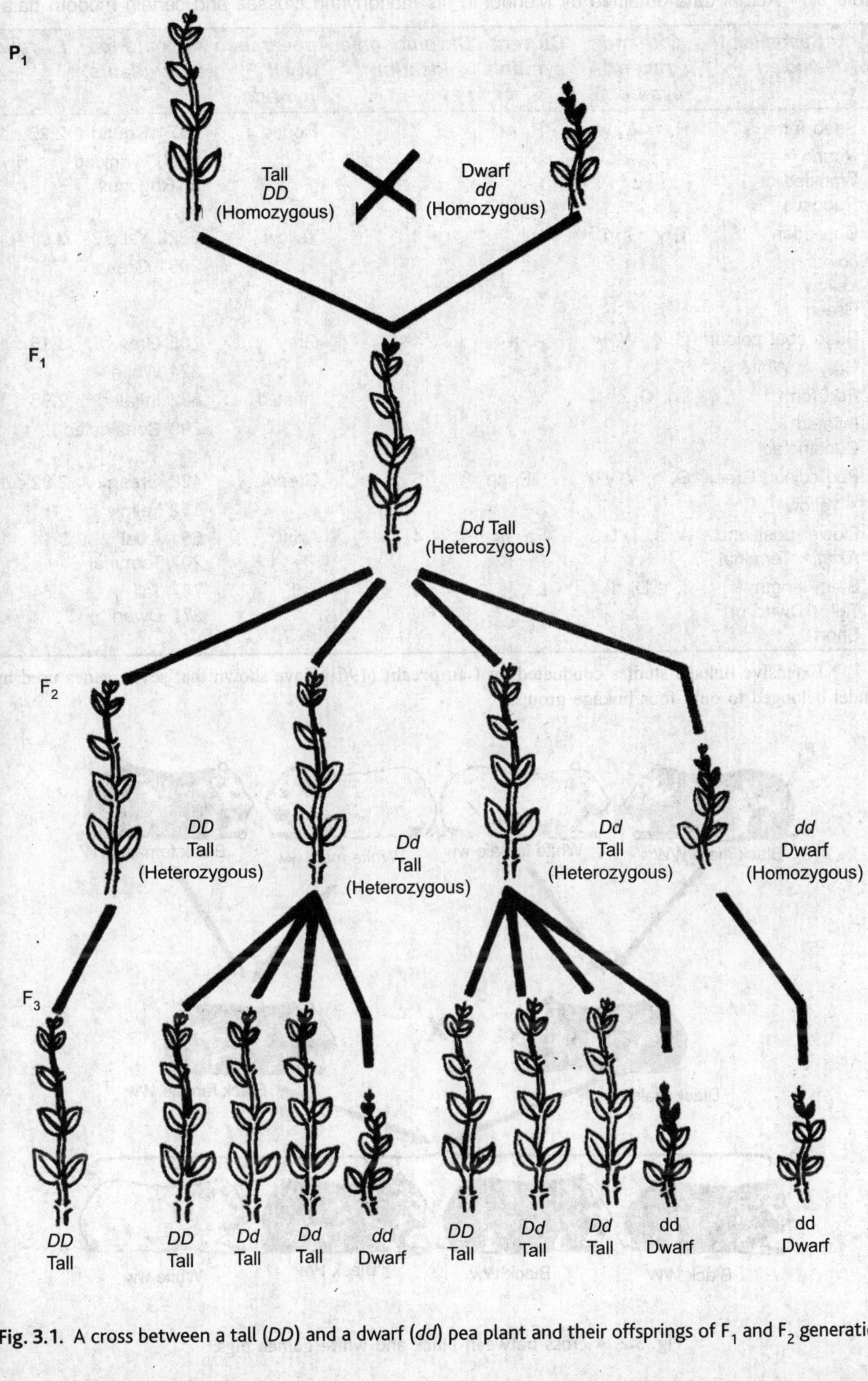

Fig. 3.1. A cross between a tall (*DD*) and a dwarf (*dd*) pea plant and their offsprings of F_1 and F_2 generations.

Table 3.1 Actual data obtained by Mendel in his monohybrid crosses and certain modern data.

S. No.	Character studied	Old and recent symbols	Current symbols	Chromosome location*	Appearance of all F_1 hybrids	Appearance of F_2 plants	F_2 ratio
1.	Seed form: Round × Wrinkled or Rugosus	R, r; W, w	R, r	7	Round	5474 Round 1850 Wrinkled or Rugosus	2.96 : 1
2.	Cotyledon colour: Yellow × Green	Y, y; G, g	I, i	1	Yellow	6022 Yellow 2001 Green	3.01 : 1
3.	Seed coat colour: Grey × White	G, g; W, w	A, a	1	Grey	705 Grey 224 White	3.15 : 1
4.	Pod form: Inflated × Constricted	I, i; C, c	V, v	4	Inflated	882 Inflated 299 Constricted	2.95 : 1
5.	Pod colour: Green × Yellow	G, g; Y, y	GP, gp	5	Green	428 Green 152 Yellow	2.82 : 1
6.	Flower position: Axial × Terminal	A, a; T, t	Fa, fa	4	Axial	651 Axial 207 Terminal	3.14 : 1
7.	Stem length: Tall × Dwarf or Short	T, t; D, d	Le, le	4	Tall	787 Tall 277 Dwarf	2.84 : 1

* Extensive linkage studies conducted by **Lamprecht** (1961) have shown that seven genes used by Mendel belonged to only four linkage groups.

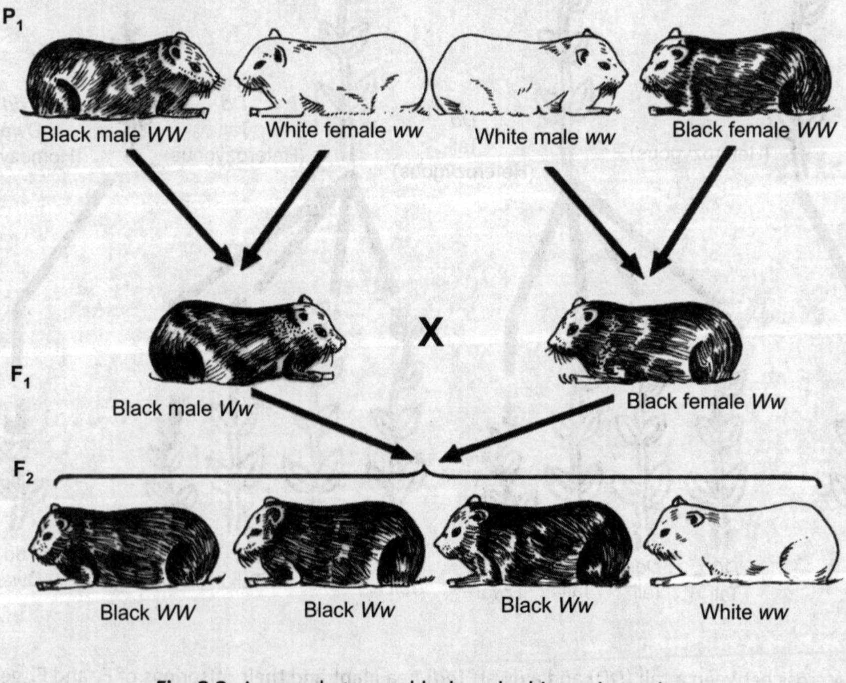

Fig. 3.2. A cross between black and white guinea pigs.

Table 3.2 The dominant and recessive characters in plants.

S. No.	Name of the plant	Dominant	Recessive
1.	Nettle	Serrated leaves	Smooth margined leaves
2.	Sunflower	Branched habit	Unbranched habit
3.	Cotton	Coloured lint	White lint
4.	Maize	Round starchy kernel	Wrinkled sugary kernel
5.	Snapdragon	Red flower	Non-red flower
6.	Barley	Beardlessness	Beardness
7.	Wheat	Susceptibility to rust	Immunity to rust
8.	Tomato	Two celled fruit	Many celled fruit

B. Certain other examples of phenomenon of dominance in animals. The dominant and recessive traits or characters of some animals can be tabulated in following table:

Table 3.3 The dominant and recessive characters in animals.

S. No.	Name of animal	Body character	Dominant	Recessive
1.	Cat	Skin colour	Tabby or Stripped	Black or blue
		Length of hair	Short hairs	Long hairs (Angora)
2.	Dog	Skin colour	Grey	Black
		Tail	Stumpy	Normal tail
3.	Cattle	Colour of face	White	Coloured
		Horn	Polled or hornless	Horned
4.	Horse	Skin colour	Black	Red
		Movement	Trotting	Pacing
5.	Sheep	Hair or wool or fleece	White	Black
6.	Swine or Pig	Skin colour	Black	Red
		Hoof	Uncleft	Normal
7.	Salamander	Body colour	Dark	Light
8.	Fruit fly (*Drosophila*)	Eye colour	Red	White
		Wings	Flat and yellow	Curled and white
		Body colour	Grey	Black
9.	Land snail	Shape of shell	Unbanded shell	Banded shell

C. Dominant and recessive characters in human beings. In human beings the following traits normally behave as the dominant and recessive traits (Table 3.4).

Table 3.4 Some normal dominant and recessive traits in human beings.

S. No.	Body part	Character	Dominant	Recessive
1.	Hair	Form	Curly	Straight
		Colour	Dark	Light
2.	Skin	Colour	Dark	Light
		Pigment (with melanin)	Normal	None (albinism, without melanin pigment)
3.	Eyes or Iris	Colour	Brown	Blue

Mechanism of dominance. Mendel carefully studied the results of his experiments and it became evident to him that there is a clear cut difference in between the actual visible character and that something which caused its production. Because the character cannot be present as such in sex cells (sperm and ova) which form the only link between a new individual and the parents.

Therefore, there must be something which represents the characters and is responsible for their production. This something of Mendel is now called the **factor** or **gene**. The gene, thus, can be considered as the unit of inheritance that is transmitted in a gamete and determines or controls the development of a character by interaction with the other genes, the cytoplasm and the enviornment.

The cytological investigations have now established that the genes are the units of deoxyribonucleic acid (DNA) which along with nucleoproteins (histones and honhistones) constitute the thread like stainable structures the **chromosomes.** The chromosomes are specific in number, shape and size to a particular species.

A diploid cell has two sets of chromosomes which come from two different parents (male and female) via gametes (sperm and ova). The chromosomes of similar size and nature often form pairs during meiotic cell division and such identical chromosomes are known as **homologous chromosomes**. Each character of a pair of contrasting characters is represented by an **allele**. (When a gene for a unit character contains two or more alternative forms, they are called **allelomorphs** or **alleles**). All alleles of a gene are produced due to mutation of a wild gene (or normal gene). Thus, homozygous tall pea plant has two identical alleles *DD* on both gene loci of the homologous chromosomes (Fig. 3.1); likewise, homozygous dwarf pea plant contains *dd* alleles.

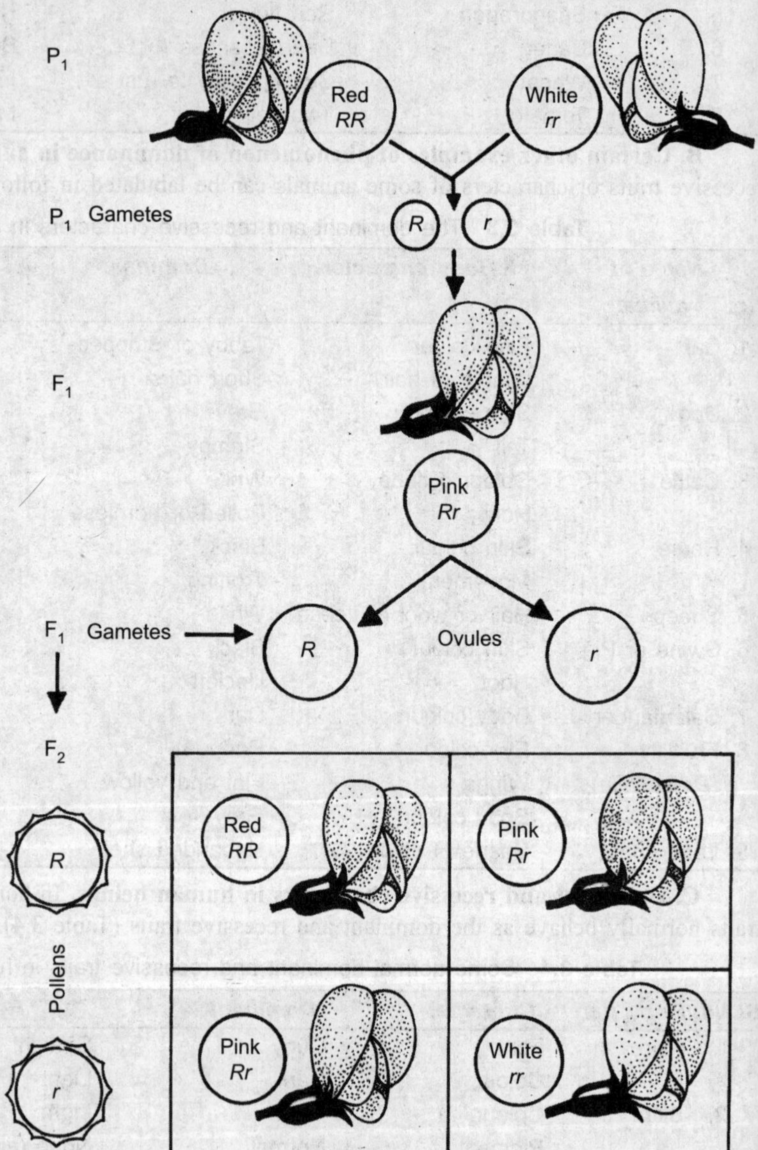

F_2 ratio: 1 Red : 2 Pink : 1 White or 1 : 2 : 1

Fig. 3.3. A cross between a red flowered and a white flowered pea plant showing incomplete dominance.

During the gametogenesis, the homologous chromosomes with *DD* or *dd* genes are separated and each chromosome with *D* or *d* gene is passed to the gamete. The gametes of both parents unite during the process of fertilization and produce a new individual containing both tall (*D*) and dwarf (*d*) characters. This new individual of first generation (F_1) contains two different genes (*i.e.,* alleles *D* and *d*) of a contrasting pair of characters, therefore, it is known as **heterozygote** or **hybrid**. Because the hybrids of F_1 have tall stems so the character of tallness (*D*) is considered as **dominant** and because the character for dwarfness (*d*) could not express itself in F_1 generation, therefore, it is considered as **recessive**.

Variation in Dominance Relations

Mendel reported full dominance and recessiveness for all the seven gene pairs (or allele pairs) he studied (see Table 3.1). He might have been selective in his choice of pea characters to study, because variations on the basic theme appear quite often in analysis. For instance, in some cases the phenotypes of heterozygotes are found to be different from either of the homozygotes (see Box 3.1).

Box 3.1 Types of Dominance

1. **Complete dominance.** The dominant allele completely masks the effect of the recessive allele. The phenotype of the heterozygote is identical to that of one of the homozygote. The monohybrid ratio is 3: 1. This is classical dominance effect described by Mendel, and often occurs where the recessive allele is null.
2. **No dominance and partial dominance (Incomplete dominance).** Neither allele is fully dominant over the other. The phenotype of the heterozygote is somewhere in between those of the homozygotes and the monohybrid ratio is 1 : 2 : 1 (Fig. 3.3).
If the heterozygote phenotype is exactly intermediate between the two homozygotes, **there is no dominance.** If the phenotype is closer to one homozygote than the other, there is **partial dominance.**
3. **Over dominance and underdominance.** The phenotype of the heterozygote lies outside the range characterized by those of the homozygote. If the heterozygous phenotype is greater than either homozygous phenotype, the locus shows **overdominance**; if less, it shows **underdominance**. The monohybrid ratio is 1 : 2 : 1.
4. **Codominance.** The phenotype associated with each allele is expressed independently of that of the other. Codominance occurs when there is no competition between alleles. The monohybrid ratio is 1 : 2 : 1.
5. **Pseudodominance.** An allele appears dominant because the locus is hemizygous. This is applicable to sex-linked loci in the heterogametic sex, *e.g.*, in male mammals (sex-linkage) and to individuals with chromosome deletion or chromosome loss.

3.2 INCOMPLETE DOMINANCE

Sometimes in a heterozygote the dominant allele does not completely mask the phenotypic expression of the recessive allele and there occurs an intermediate phenotype in the heterozygote. This is called **incomplete dominance**.

Examples. 1. When a red flowered pea plant (*RR*) is crossed with white flowered pea plant (*rr*) then the F_1 hybrid pea plants are found to have pink flowers (*Rr*). It shows that allele for red colour could not completely dominate the allele for white colour as shown in the Fig. 3.3. In such a case, F_2 phenotypic ratio and genotypic ratio are the same, (Fig. 3.2) as follows:

F_2 phenotypic ratio = 1 Red: 2 Pink : 1 White

F_2 genotypic ratio = 1 *RR*: 2 *Rr* : 1 *rr*

2. In four-o'clock plants (*Mirabilis jalapa*) or snapdragons (*Antirrhinum majus*), when a pure line or homozygous with red petals (C_1C_1) is crossed to a pure line with white petals (C_2C_2), the F_1 progeny has no red petals but pink petals (C_1 C_2). It shows incomplete dominance. If a F_2 is produced, its progeny exhibits the following results:

¼ Red petals –1C_1C_1 ; ½ Pink petals – 2C_1C_2; ¼ White petals –1C_2 C_2

3. When a homozygous Andalusian fowl with black feathers is crossed with a homozygous fowl with splashed white feathers the F_1 hybrids are found to contain blue feathers (Fig. 3.4).

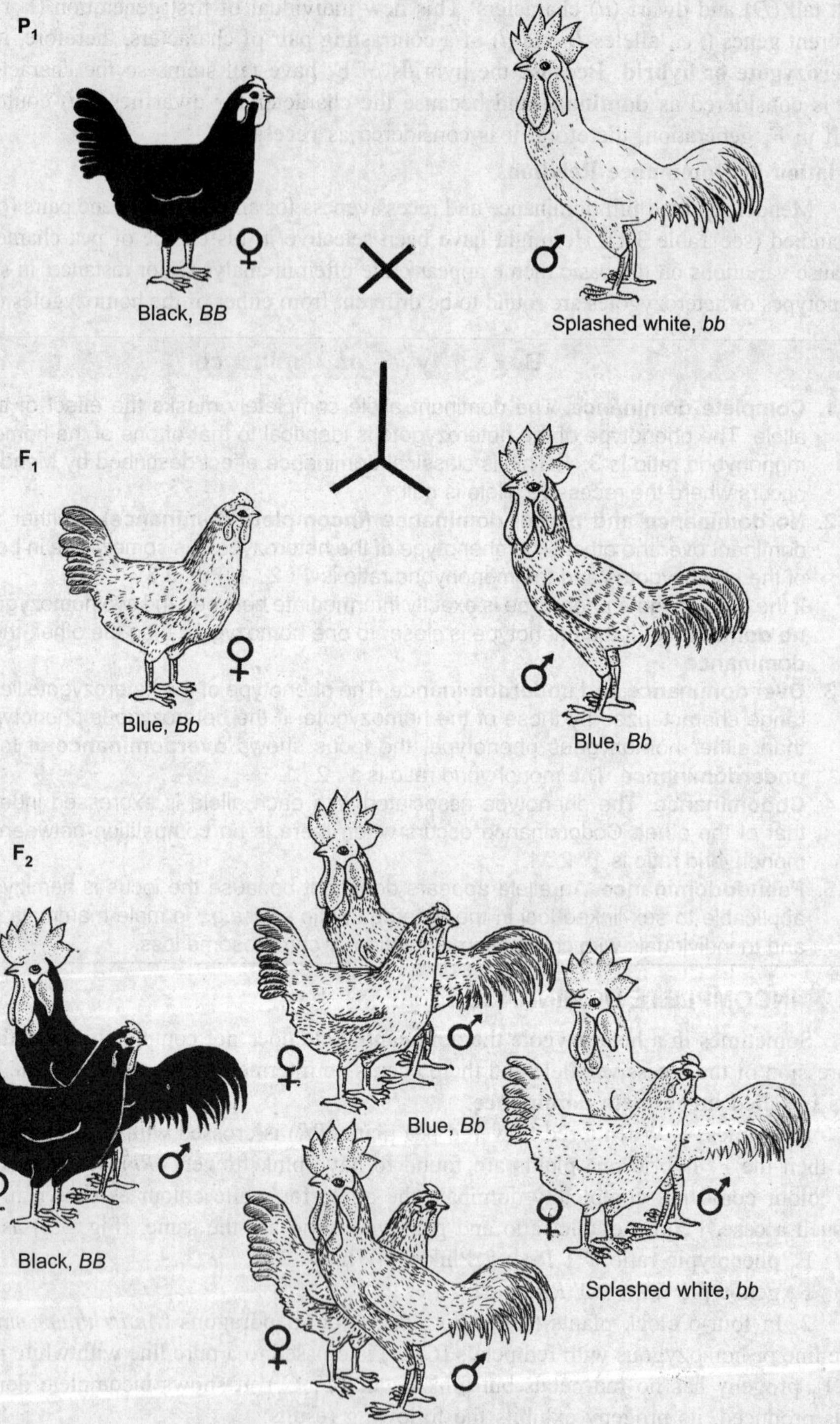

Fig. 3.4. A monohybrid cross between black and splashed white Andalusian fowl showing the incomplete dominance.

3.3 CODOMINANCE

Sometimes both alleles of a gene in a heterozygote have an equal phenotypic expression. In a sense, codominance is no dominance at all, the heterozygote showing the phenotypes of both homozygotes. Hence, heterozygote genotype gives rise to a phenotype distinctly different from either of the homozygous genotypes.

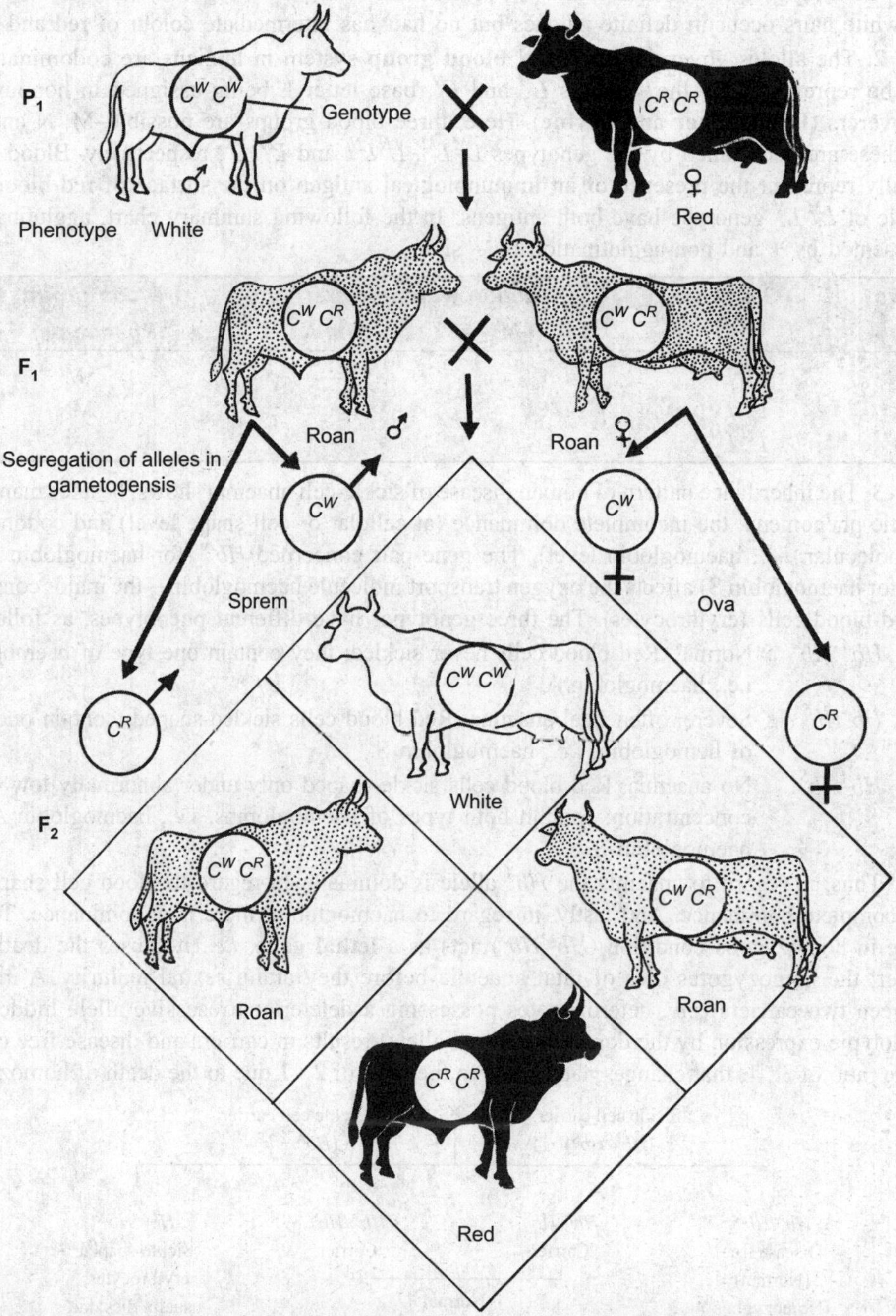

Fig. 3.5. A monohybrid cross in a white-coated and a red-coated cattle showing codominance.

Symbolism for codominant alleles. For codominant alleles, all upper case base symbol with different superscripts are used. The upper case letters indicate that each allele can expres itself to some degree even when in the presence of its alternative allele (heterozygous).

Example 1. The coat colour of the shorthorn breed of cattle represent a classical example o codominance. When a cattle of red coat (C^RC^R) is crossed with the cattle of white coat (C^WC^W) the F_1 heterozygote or hybrid is found to possess roan coat (C^RC^w; Fig. 3.5). In roan coat the re and white hairs occur in definite patches but no hair has intermediate colour of red and white.

2. The alleles governing the **M-N blood group** system in humans are codominants an may be represented by the symbols L^M and L^N, base letter L being assigned in honour of it discoverers (**Landsteiner** and **Levine**). Here, three blood groups are possible–M, N and MN and these are determined by the genotypes L^ML^M, L^NL^N, and $L^M\,L^N$, respectively. Blood group actually represent the presence of an immunological antigen on the surface of red blood cells People of $L^M\,L^N$ genotype have both antigens. In the following summary chart, agglutination i represented by + and non-agglutination by – sign:

S. No.	*Genotype*	*Reaction with antisera:* *Anti-M*	*Anti-N*	*Blood group (Phenotype)*
1.	$L^M L^M$	+	–	M
2.	$L^M L^N$	+	+	MN
3.	$L^N L^N$	–	+	N

3. The inheritance pattern of human disease of sickle-cell anaemia shows, besides many othe genetic phenomena, the incomplete dominance (at cellular or cell shape level) and codominanc (at molecular, *i.e.*, haemoglobin level). The gene pair concerned Hb^A (for haemoglobin A) an Hb^S for haemoglobin S) affects the oxygen transport molecule haemoglobin—the major constitue of red blood cells (erythrocytes). The three genotypes have different phenotypes, as follows:

- $Hb^A\,Hb^A$: Normal. Red blood cells never sickled; they contain one type of haemoglobin i.e., haemoglobin A.
- Hb^SHb^S : Severe, often fatal anaemia. Red blood cells sickled-shaped; contain one type of hemoglobin, *i.e.*, haemoglobin S.
- Hb^AHb^S : No anaemia. Red blood cells sickle-shaped only under abnormally low oxyge concentration; contain both types of haemoglobins, *i.e.*, haemoglobin A and haemoglobin S.

Thus, in regard to anaemia the Hb^A allele is dominant. In regard to blood cell shape the is incomplete dominance. And lastly, in regard to haemoglobin, there is codominance. The Hb allele in homozygous condition ($Hb^S\,Hb^S$) acts as a **lethal gene**, *i.e.*, it causes the death of i bearer; the homozygotes dies of fatal anaemia before they attain sexual maturity. A marriag between two carriers (*i.e.*, heterozygotes possessing a deleterious recessive allele hidden fro phenotypic expression by the dominant normal allele) results in carriers and disease free childre in the ratio of 3 : 1, that changes, later on, into the ratio of 2 : 1 due to the death of homozygote

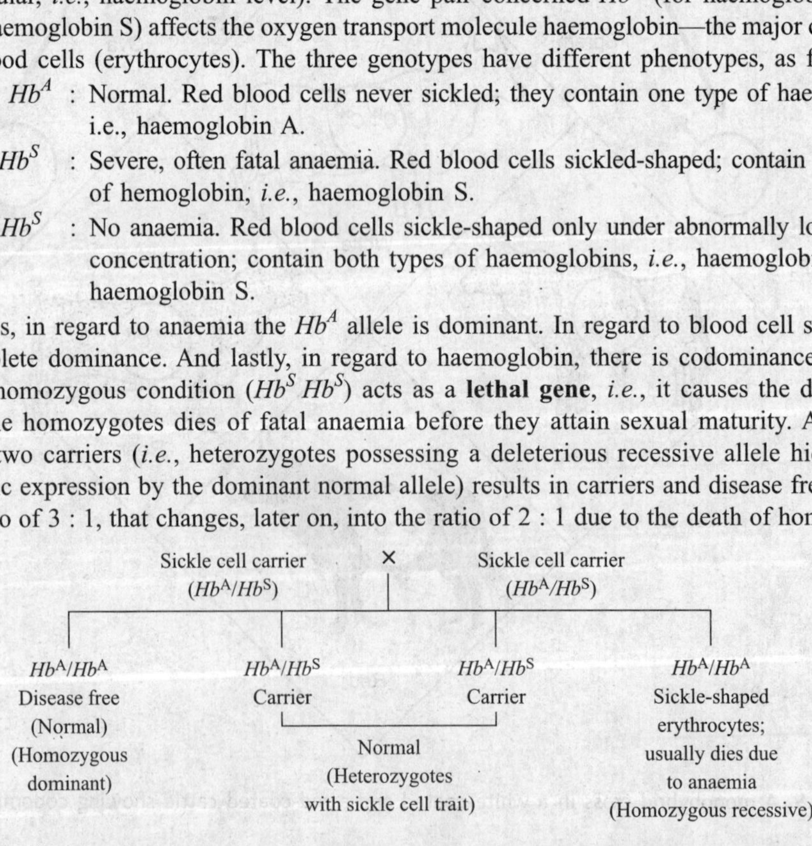

3.4 DEVIATION FROM MENDEL'S MONOHYBRID AND DIHYBRID PHENOTYPIC RATIOS

1. Deviation from Mendel's Monohybrid (3 : 1) Ratio

Mendel's monohybrid ratio of 3 : 1 becomes modified in case of incomplete dominance (*i.e.*, 1 : 2 : 1) and codominance (1 : 2 : 1) (see Sections 3.2 and 3.3).

2. Deviation from Mendelian Dihybrid (9 : 3 : 3 : 1) Ratio.

The Mendelian dihybrid phenotypic ratio of 9 : 3 : 3 : 1 is obtained only when the alleles at both gene loci display dominant and recessive relationship. If one or both gene loci have incompletely dominant alleles, or codominant alleles or lethal alleles, the dihybrid ratio becomes modified variously, such as follows:

(1) 3 : 6 : 3 : 1 : 2 : 1 Ratio

When the dihybrid parent have dominant and recessive alleles at one gene locus and codominant alleles at second gene locus, the F_2 9 : 3 : 3 : 1 phenotypic ratio becomes 3 : 6 : 3 : 1 : 2 : 1.

Example. In cattles, hornless or polled (*P*) condition is dominant to horned (*p*) condition, and trait of white (*W*) coat is codominant to the trait for red (*w*) coat colour. The mating of homozygous white polled (*PP WW*) cattle with a homozygous red horned (*pp ww*) cattle produces F_1 heterozygotes (dihybrids) with the phenotype of hornless, roan and genotype of *Pp Ww*. These hornless roan F_1 dihybrids produce a F_2 progeny in the ratio of 3 : 6 : 3 : 1 : 2 : 1 as illustrated in following diagram:

P_1 : Phenotype Hornless White × Horned Red

Genotype *PP WW* *pp ww*

↓ ↓

P_1 Gametes: *PW* *pw*

F_1: *PpWw*

Hornless Roan

F_2: 3 Hornless White: 6 Hornless Roan: 3 Hornless Red:

1 Horned White: 2 Horned Roan: 1 Horned Red (or 3:6:3:1:2:1).

(2) 1 : 2 : 1 : 2 : 4 : 2 : 1 : 2 : 1 Ratio

When each parent of a dihybrid cross has incompletely dominant alleles at both gene loci, then in F_2 large number of phenotypic classes are produced.

Example. In snapdragons, red flower colour (*R*) is incompletely dominant to white flower colour (*r*) and trait for broadness of leaf (*B*) is incompletely dominant to the trait for narrowness of leaf (*b*). The dihybrid cross between red, broad plant (*RR BB*) and white, narrow plant (*rr bb*) produces heterozygotes (dihybrids) having pink flowers and leaves of intermediate width and genotype of *Rr Bb*. These F_1 dihybrids produce the F_2 progeny in 9 phenotypic classes as follows:

P_1 : Phenotype Red Broad × White Narrow

Genotype *RR BB* *rr bb*

↓ ↓

P_1 : Gametes: *RB* *rb*

F_1 : Phenotype Pink Intermediate

Rr Bb

F_2 : 1 Red Broad; 2. Red Intermediate: 1 Red Narrow: 2 Pink Broad: 4 Pink Intermediate: 2 Pink Narrow: 1 White Broad: 2 White Intermediate: 1 White Narrow (or 1:2:1:2:4:2:1:2:1).

(3) 3 : 1 : 6 : 2 Ratio

When the F_1 dihybrids have dominant-recessive alleles at one gene locus and recessive lethal alleles at second gene locus, the F_2 offsprings manifest the phenotypic ratio of 3 : 1 : 6 :2.

(4) 1 : 2 : 1 : 3 : 4 : 2 Ratio

When the F_1 dihybrids contain codominant alleles at first gene locus and recessive-lethal alleles at second gene locus, then their F_2 progeny display the phenotypic ratio of 1 : 2 : 1 : 3 : 4 : 2.

(5) 4 : 2 : 2 : 1 Ratio

This phenotypic ratio is obtained when at both gene loci of F_1 dihybrids occur the recessive lethal alleles.

QUESTIONS

Long Answer Questions

1. Explain giving example, what do you understand by complete and incomplete dominance.
2. The classical Mendelian ratios, such as 3 : 1, 9 : 3 : 3 : 1, etc., do not by any means occur in all crosses. Explain.
3. In 4 'O' clock plants red flower colour *R* is incompletely dominant over the white *r*. The heterozygous plants are with pink flowers. If a red coloured 4 'O' clock plant is crossed with a white one, what will be flower colour of F_1, of the F_2, of the offspring of the cross F_1 with its red parent and its white parent.

Short Answer Questions

1. Write short answer/notes on the following:
   (*i*) Dominance
   (*ii*) Incomplete dominance
   (*iii*) Codominance
   (*iv*) Back cross
2. Differentiate between the following:
   (*i*) Complete and incomplete dominance
   (*ii*) Codominance and incomplete dominance
   (*iii*) Dominance and codominance

Very Short Answer Questions

1. What is dominance?
2. Define incomplete dominance?
3. What is codominance?
4. What are codominant alleles?

Fill in the Blanks

1. An allele (or phenotype) that does not express itself in the heterozygous condition is called..............
2. A gentical cross with the phenotype of each sex reversed as compared with the original cross is called .... cross.
3. The situation in which both alleles of the heterozygote influence the phenotype is called ....
4. .... is the relationship of alleles in a heterozygote that shows the individual expression of each allele in the phenotype.

True and False Questions

(*i*) *Mirabilis jalpa* hybrid for pink colour is an example of codominance. (*H.N.B., Garhwal 2008*)

Multiple Choice Questions

1. Recessive gene can express in

(*a*) homozygous condition (*b*) heterozygous condition

(*c*) both the above conditions (*c*) none of these conditions

2. A gene is said to be dominant if

(*a*) it expresses its effect both in homozygous condition

(*b*) it expresses its effect only in heterozygous condition

(*c*) it expresses its effect both in homozygous and heterozygous condition

(*d*) it never expresses its effect in any condition

3. Pink flower in the progeny of parents of *Antirrhinum majus*, having red and white flowers, are produced because of

(*a*) dominance (*b*) codominance

(*c*) epistasis (*d*) incomplete dominance

4. Alleles that produce independent effects in heterozygous condition are called

(*a*) codominant alleles (*b*) epistatic alleles

(*c*) complementary alleles (*d*) supplementary alleles

5. Mendel did not propose

(*a*) segregation (*b*) independent assortment

(*c*) dominance (*d*) incomplete dominance

6. A blue Andalusion fowl is the product of mating between black fowl and white fowl. A cross of blue fowl with another blue fowl results in

(*a*) 1 black : 2 white : 1 blue (*b*) 1 black : 2 blue : 1 white

(*c*) 2 black : 1 blue : 1 white (*d*) none of these

ANSWERS

Very Short Answer Questions

1. Condition in which one member of an allele pair is manifested to the exclusion of the other.

2. Expression of two alleles in a heterozygote that allows the heterozygote to be distinguished from either of its homozygous parents.

3. The condition in heterozygote where both members of an allelic pair contribute to phenotype, which is then a mixture of the phenotypic trait produced in either homozygous conditions.

4. Alleles that produce independent effects when heterozygous.

Fill in the Blanks

1. Recessive.

2. Reciprocal.

3. Incomplete dominance.

4. Codominance.

True and False

(*i*) False.

Multiple Choice Questions

1. (*a*) **2.** (*c*) **3.** (*d*) **4.** (*a*) **5.** (*d*) **6.** (*b*).

4

Multiple Alleles

A given phenotypic trait (character) of an individual depends on a single pair of genes, each of which occupies a specific position called the gene **locus**, on a homologous chromosome. Moreover, a particular gene has been found to occur in two alternative forms or **allelomorphs**, one being dominant and other recessive; one being wild form and the other mutant form. For example, a gene (L) for length of wings of *Drosophila* may occur in two alternative forms : a allele or gene (L^+) for normal development of wings and another allele or gene (L^{vg}) for vestigial wings. Because, most flies have normally developed wings, so, it can be. easily concluded that gene L^+ is the original form of gene or allele from which the other form of gene or allele (L^{vg}) might have originated by certain mutational event at sometime in past. The gene L^+ for normal development of wing is called the normal or **wild type allele** of the gene L and usually symbolized as L^+, while the mutated gene L^{vg} for vestigial wing is called reduced type or **mutant allele** of the gene L. A fly with normal wings, thus, has two wild type alleles (L^+L^+) and the vestigial winged fly has two mutant alleles (L^{vg} L^{vg}). Both of these allelic forms (L^+ and L^{vg}) of gene L occur at corresponding positions on genetically identical (homologous) chromosomes of same or different individual.

If the mutant allele has developed from the wild form of allele due to mutation, one may expect that the wild form of allele can mutate in more than one way. The mutant form of allele too can mutate once again to give rise to another mutant form of allele. Therefore, it is possible to have more than two allelic forms, *i.e.*, **multiple alleles**, of one kind of gene. Although only two actual alleles of a gene can exist in a diploid cell (and only one in a haploid cell), the total number of possible different allelic forms that might exist in a population of individuals is often quite large. This situation is termed as **multiple allelism**, and the set of alleles itself is called a **multiple allelic series** (Box 4.1).

4.1 CHARACTERS OF MULTIPLE ALLELES

The most important and distinguishing features of multiple alleles are summarized below:

1. Multiple alleles of a series always occupy the same locus in the chromosome.
2. Because, all the alleles of multiple series occupy same locus in chromosome, therefore, no crossing-over occurs within the alleles of a same multiple allele series.
3. Multiple alleles always influence the same character.
4. The wild type allele is nearly always dominant, while the other mutant alleles in the series may show dominance or there may be an intermediate phenotypic effect.
5. When any two of the mutant multiple alleles are crossed, the phenotype is mutant type and not the wild type.

Box 4.1 Allelic Variation

The character described by Mendel occurred in two forms, *i.e.*, they were **diallelic**. For many characters, however, a greater degree of **allelic variation** is evident. The human ABO blood group locus, for example, has three physiologically distinct alleles, and in the extreme example of the self-incompatibility loci of clover and tobacco, over 200 different alleles may be detected in a given population. The observed allelic variation also depends upon the level at which the phenotype is determined. *At the molecular level, there is often more diversity than is evident at the morphological level* because many of the alleles identified as sequence variant or **protein polymorphism** are neutral with respect to their effect on the morphological phenotype; these are termed **isoalleles.** No matter how many alleles can be distinguished for a particular locus in a population, only two are present in the same diploid individual at anyone time. Morphologically distinct alleles can be often arranged in order of dominance, as so called **allelic series**.

Symbolism for Multiple Alleles

The dominance hierarchy is defined at the beginning of each problem involving multiple alleles. A capital letter is commonly used to designate the allele which is dominant to all other alleles in the series.

The corresponding lower case letter designates the allele which is recessive to all others in the series. Other alleles which are intermediate in their degree of dominance between these two extremes, are usually assigned the lower case letter with some suitable superscript.

4.2 EXAMPLES OF MULTIPLE ALLELES

The best examples of multiple allelic system have been observed in coat colour of rabbits, wings of *Drosophila* and blood groups in human beings.

1. The *C* gene in Rabbit (Coat Colour)

The coat of rabbit may have following different colours.

***(i)* Full colour.** The coat of the ordinary (wild type) rabbit is reffered to as **"agouti"** or **full colour**, in which individuals have banded hairs, the portion nearest the skin being gray, succeeded by a yellow band, and finally a black or brown tip. The genes for full colour may be represented by capital letter C or c^+.

***(ii)* Chinchilla.** In some individuals, the coat lacking the yellow pigment and due to the optical effect of black and gray hairs, have the appearance of silver-gray. The gene for chinchilla is represented as c^{ch}.

***(iii)* Himalayan.** The Himalayan type coat is white except for black extremities (nose, ears, feet and tail). The eyes are pigmented. The gene for Himalayan coat is represented by c^h.

***(iv)* Albino.** The albino coat totally lacks in pigmentation and the eyes of an albino also remain pink due to lack of pigment in iris of eye. The gene for albino is represented by c.

Crosses of homozygous agouti (c^+c^+) and albino (cc) individuals produce a uniform agouti F_1; interbreeding of the F_1 produces an F_2 ratio of 3 agouti : 1 albino (Fig. 4.2). Two third of F_2 agouti are found to be heterozygous by testcrosses. Thus, it is a case of monohybrid inheritance, with agouti completely dominant to albino. Likewise, crosses between chinchilla and agouti (Fig. 4.3) produce all agouti individuals in the F_1 and a 3 agouti : 1 chinchilla ratio in the F_2. Such complete dominance of agouti also occurs on Himalayan (Fig. 4.4). Further crosses, reveal that c^{ch} allele for chinchilla though is recessive to c^+ allele for agouti coat or skin, is incompletely dominant over Himalayan (c^h) and albino (c) alleles (Fig. 4.5 and Fig. 4.6). Likewise, c^h allele for Himalayan coat is recessive to c^+ (agouti) and c^{ch} (chinchilla), but dominant over albino (c, Fig. 4.7). The results of all these crosses exhibit that c^+ agouti, c^h (Himalayan) and c (albino) are allelic to each other and the alleles of this multiple allelic series have following dominance hierarchy:

$$c^+ > c^{ch} > c^h > c$$

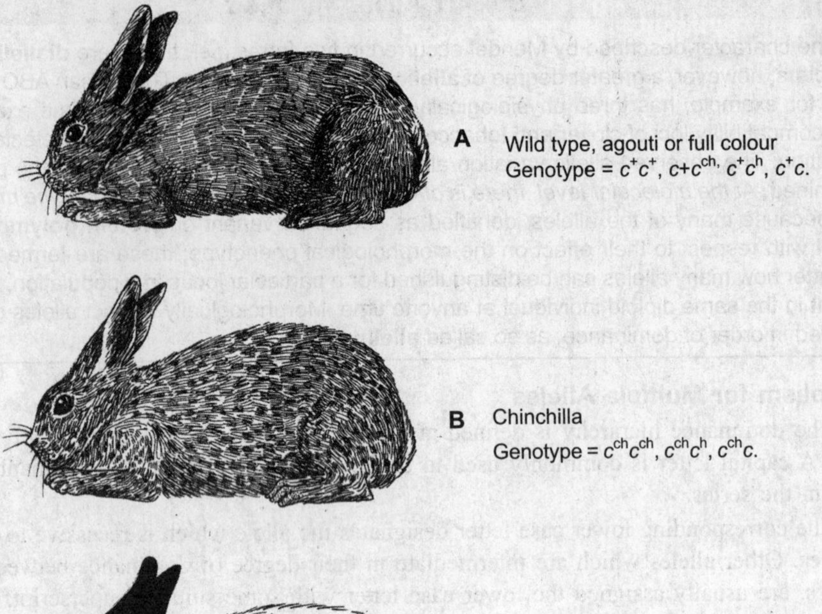

Fig. 4.1. Different coat colours in rabbits and their respective genotypes.

The possible phenotypes and their associated genotypes of this multiple allelic series has been summarized in Table 4.1.

P$_1$: Agouti × Albino
c^+c^+ ↓ cc

Agouti

F$_1$: c^+c

F$_2$: $1c^+c^+ : 2c^+c : 1cc$

3 Agouti : 1 Albino

Fig. 4.2. A monohybrid cross between agouti and albino rabbits.

P$_1$: Agouti × Chinchilla
c^+c^+ ↓ $c^{ch}c^{ch}$

Agouti

F$_1$: c^+c^{ch}

F$_2$: $1c^+c^+ : 2c^+c^{ch} : 1c^{ch}c^{ch}$

3 Agouti : 1 Chinchilla

Fig. 4.3. A monohybrid cross between agouti and chinchilla rabbits.

P_1: Agouti c^+c^+	× ↓	Himalayan c^hc^h	
F_1:	Agouti c^+c^h		
F_2:	$1c^+c^+ : 2c^+c^h : 1c^hc^h$		
	3 Agouti : 1 Himalayan		

Fig. 4.4. A monohybrid cross between agouti and Himalyan (or Russian) rabbits.

P_1: Chinchilla $c^{ch}c^{ch}$	× ↓	Himalayan c^hc^h	
F_1:	Light gray $c^{ch}c^h$		
F_2:	$1c^{ch}c^{ch} : 2c^{ch}c^h : 1c^hc^h$		
	1 Chinchilla : 2 Light gray: 1 Himalayan		

Fig. 4.5. A monohybrid cross between chinchilla and Himalayan rabbits. Showing incomplete dominance of chinchilla on Himalayan.

P_1: Chinchilla $c^{ch}c^{ch}$	× ↓	Albino $c\ c$	
F_1:	Light gray $c^{ch}c$		
F_2:	$1c^{ch}c^{ch} : 2\ c^{ch}c : 1c\ c$		
	1 Chinchilla :2 Light gray : 1 Albino		

Fig. 4.6. A monohybrid cross between chinchilla and albino rabbits showing incomplete dominance of chinchilla over albino.

P_1: Himalayan c^hc^h	× ↓	Albino $c\ c$	
F_1:	Himalayan c^hc		
F_2:	$1c^hc^h : 2c^hc : 1c\ c$		
	3 Himalayan : 1 Albino		

Fig. 4.7. A monohybrid cross between Himalayan and albino rabbits.

Table 4.1 The phenotypes and genotypes of multiple allelic series for coat colour in rabbit.

Phenotypes	*Genotypes*
Full colour (Agouti)	c^+c^+, c^+c^{ch}, c^+c^h, c^+c
Chinchilla	$c^{ch}c^{ch}$
Light gray	$c^{ch}c^h$, $c^{ch}c$
Himalayan	c^hc^h, c^hc
Albino	cc

2. Skin Colour in Rodents (Mice)

In mice, the skin or coat colour is determined by two series of multiple alleles; both allelic series occur at different chromosomes. Series number 1 is at the black locus.

In mice, the skin or coat colour is governed by a multiple allelic series. The gene *A* controlling the agouti pattern in mice has several alleles. Among the members of the series a gradation exists in dominance, similar to that of the *c* series in rabbits, but in this case two members, A^y (yellow) and A^l (agouti light belly) are dominant over the wild-type allele A^+ (agouti). The alleles of agouti (*A*) series in mice can be listed according to dominance in Table 4.2.

In this multiple allelic series, the allele A^y, when homozygous, (*i.e.*, $A^y A^y$) is lethal early in embryonic development but produces yellow colour when in heterozygous condition with other alleles.

The second series of multiple alleles for skin colour is located at the albino locus and comprises of four alleles. The normal allele in this series is for **gray coat colour** and is represented by a^+.

The other alleles in this series are **mutant alleles**—albino (a), medium light (a^m) and extreme light (a^l). All the three mutant alleles are recessive to the normal allele.

Table 4.2 Agouti (A) series of alleles in mice, listed in order of dominance.

Phenotype	*Allele*
Yellow	A^y
Agouti light belly	A^l
Agouti	A^+
Black and Tan	a^t
Nonagouti (Black)	a
Dominance hierarchy:	$A^y > A^l > A^+ > a^t > a$

3.A, B, AB and O Blood Group in Human Beings. Refer Section 4.3.

4. Rh factor. Refer Section 4.3.

5. Eye Colour in *Drosophila*

In *Drosophila*, normal red eye colour is determined by a X-linked type gene. White eyed *Drosophila* was one of the first mutants known in the fruit flies. The traits of red eye and white eye exhibited simple dominant recessive relationship. Subsequently, different shades between red and white were recovered. About a dozen different alleles are now known to occur at this locus; they are red or wild type (w^+) through coral (w^{co}), blood (b^{bl}), eosin (w^e), cherry (w^{ch}), appricot (w^a), honey (w^h), buff (w^{bf}), tinged (w^t), pearl (w^p) and ivory (w^i), to white (w). All of these were considered, on the basis of F_2 ratios, to form a multiple series, wild being dominant to all others and white recessive to all-$w^+ > w^{co} > w^{bl} > w^e > w^{ch} > w^a > w^{bf} > w^t > w^p > w^i >$ w. When any two recessive alleles were brought together, intermediate types, called **compound**, is obtained. However, some of the members of this multiple allelic series have been found to be pseudoalleles by **Lewis** (1951) (see Chapter 13: Fine structure of Gene).

6. Self-sterility Alleles.

A series of self-sterility alleles insures cross pollination in many plants, it is well known for long that some plants just will not self-pollinate in contrast to selfing in pea plant by Mendel. Thus, a single plant may produce both male and female gametes but pollen grains of this plant fail to fertilize the ovules of the same plant; so as a result no seed will ever be produced. The same plants, however, will cross with certain other plants, so evidently they are not sterile. This phenomenon is called **self-incompatibility** or **self-sterility.** **Kolreuter** (1764) described self-sterility in tobacco plant (*Nicotiana*). Later on, the phenomenon of self-sterility is found to be common in many other dicot and monacot plants such as sweet cherries, petunias and evening primroses.

4.3 INHERITANCE OF HUMAN BLOOD GROUP

1. A, B, AB and O Blood Group in Humans.

Landsteiner in 1900 and 1902 discovered two kinds of **agglutinogens** or **antigens**, called **A** and **B antigens** from the surface of red blood cells of human blood. He found that out of A and B antigens, a person may contain either one (*i.e.*, A or B antigen) or neither of them. Accordingly, he recognised three kinds of

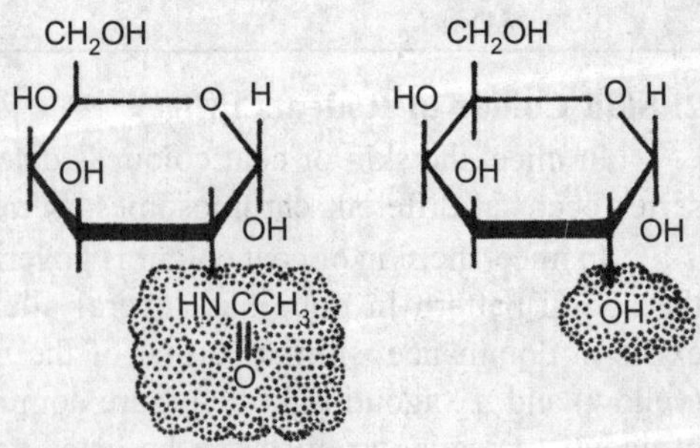

Fig. 4.8. Sugars associated with the surface of erythrocytes in blood type A and type B persons.

blood types or **blood groups: type A, type B,** and **type O.** The fourth and most rare, the **AB blood group** or type, was discovered in 1902 by two of **Landsteiner's** students, **Von Decastello** and **Sturli**. For A and B antigens, there occur two **agglutinins** or **antibodies:** anti-A (or α) and anti-B (or β). Recent, chemical investigations have shown that A and B antigens are not proteins but, are mucopolysaccharides (sugars + amino acids) of 300,000 molecular weights.

The four blood types have different agglutinizing properties. To determine the blood group types of different persons, an **agglutination test** is performed. On a glass slide is placed a drop of type A serum (containing anti-B antibodies) and a separate drop of type B-serum (containing anti-A antibodies). When a drop of type O blood is added to each drop there is no agglutination or clumping of red blood cells in either drop takes place. This shows that O blood group has neither A nor B antigen. If a drop of type B blood is added agglutination occurs with type A serum; type A red blood cells are agglutinated by type B serum and type A B red blood cells are agglutinated by both sera. The agglutination tests for four types of human blood has been illustrated in Fig. 4.9.

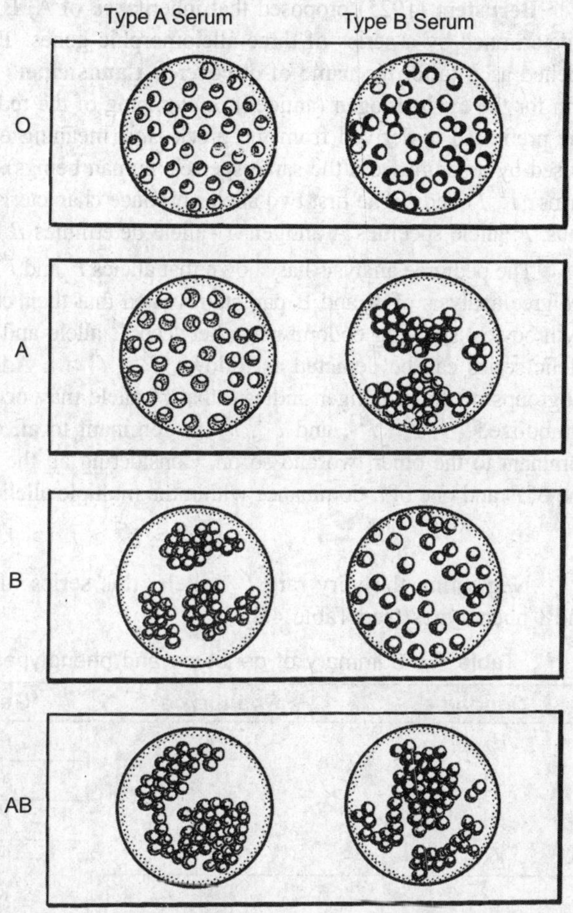

Fig. 4.9. Agglutination tests for A, B, AB and O human blood groups.

Each person, therefore, can use the blood of its own blood group in emergency, otherwise, clumping of red blood cells may take place, if blood group of different type is transfused in him. The clumped red blood cells occlude capillaries and, thus, deprive vital organs of normal blood supply and may lead to death. The characteristics of blood groups and the types of transfusions can be summarized in following Table 4.3.

Table 4.3 Human blood groups.

Blood groups (phenotype)	*Antigen in red blood cells*	*Antibodies in plasma*	*Can give blood to groups*	*Can receive blood from group*	*Genotype*
O	None	Anti-A Anti-B	O, A, B, AB	O	*ii*
A	A	Anti-B	A, AB	O, A	I^AI^A or I^Ai
B	B	Anti-A	B, AB	O, B	I^BI^B or I^Bi
AB	A and B	None	AB	O, A, B, AB	I^AI^B

Multiple Allelic Inheritance of A, B, AB and O Blood Types

Bernstein (1925) proposed that inheritance of A, B, AB and O blood types of human beings is determined by a series of three allelomorphic genes. The gene controlling blood types has been labelled as ***L*** (after the name of discoverer, **Landsteiner**) or ***I*** [from **isoagglutination**, the technical term for the **agglutinogen** (antigen) or clumping of the red blood cells by an agglutinin or antibody]. The prefix **iso** is derived from the greek *isos*, meaning equal and indicates that the agglutinations caused by a serum from the same species, human beings). The *I* gene exists in three different allelic forms : I^A, I^B and *i*. The first two alleles produce characteristic antigens on the surface of erythrocytes. Thus, I^A allele specifies A antigen, I^B allele determines *B* antigen and *i* allele specifies no antigen.

The pedigree analysis has shown that alleles I^A and I^B have dominance over *i* allele. Likewise the pedigree analysis of A and B parents revealed that their children have both A and B antigens on the erythrocytes, showing codominance between I^A allele and I^B allele. The dominance hierarchy of this allelic series can be depicted as follows: $I^A = I^B > i$. Additional studies that take into account the subgroups of the A antigen indicate that I^A allele may occur in at least four allelic forms. These are symbolized I^{A1}, I^{A2}, I^{A3}, and I^{A4}. I^{A1} is dominant to all other I^A alleles, I^{A2} is recessive to I^{A1} but dominant to the other two and so on. Considering all the six alleles of gene *I*, *i.e.*, four forms of I^A, one of I^B and one of *i*, dominance within the multiple allelic series can be shown in the following way:

$$[(I^{A1} > I^{A2} > I^{A3} > I^{A4}) = I^B] > i$$

Neglecting the very rare I^{A4} allele, this series of multiple alleles produces 15 genotypes and 8 phenotypes (see Table 4.4).

Table 4.4 Summary of genotype and phenotypes of A, B, AB and O blood group.

Genotype	*Phenotype*	*Genotype*	*Phenotype*
$I^{A1}\ I^{A1}$		$I^{A1}\ I^B$	A_1B
$I^{A1}\ I^{A2}$		$I^{A2}\ I^B$	A_2B
$I^{A1}\ I^{A3}$	A_1	$I^{A3}\ I^B$	A_3B
$I^A\ i$		$I^B\ I^B$	
$I^{A2}\ I^{A2}$		$I^B i$	B
$I^{A2}\ I^{A3}$	A_2	*ii*	O
$I^{A2}\ i$			
$I^{A3}\ I^{A3}$			
$I^{A3}\ i$	A_3		

2. The H Antigen and Bombay Phenotype

Antigens A and B of A, B and O blood phenotype are synthesized from a precursor mucopolysaccharide in the presence of the dominant allele of another pair designated as *H* and *h*. With genotypes *HH* or *Hh* the precursor is converted to an **H antigen** which, in turn, in the presence of I^A and/or I^B allele is partly converted to antigen A and/or antigen B. Gene *h* is termed **amorph** because it is producing no demonstrable product. So long as persons are of genotype *H*—(*i.e.*, *HH* or *Hh*), persons of group A produce antigens A and H, group B persons produce antigens B and H, and group AB persons produce antigens A, B and H. However, group O persons produce only antigen H if they are of the genotype *ii H*— (Fig. 4.10). On the other hand, blood of person of genotype -*hh* does not react with anti-A, anti-B, or anti-H. This is very rare **Bombay phenotype** (*i.e.*, one case in 13,000 persons; it is so named because it was first described in a family from the Bombay metropolitan. The allele *h* is found to be **epistatic** to the multiple alleles at the A-B-O locus. Erythrocytes of person having- -*hh* genotype give no reaction with anti-A or anti-B sera (even though they possess I^A or I^B genes); in fact, they contain no antigen of this multiple allelic series.

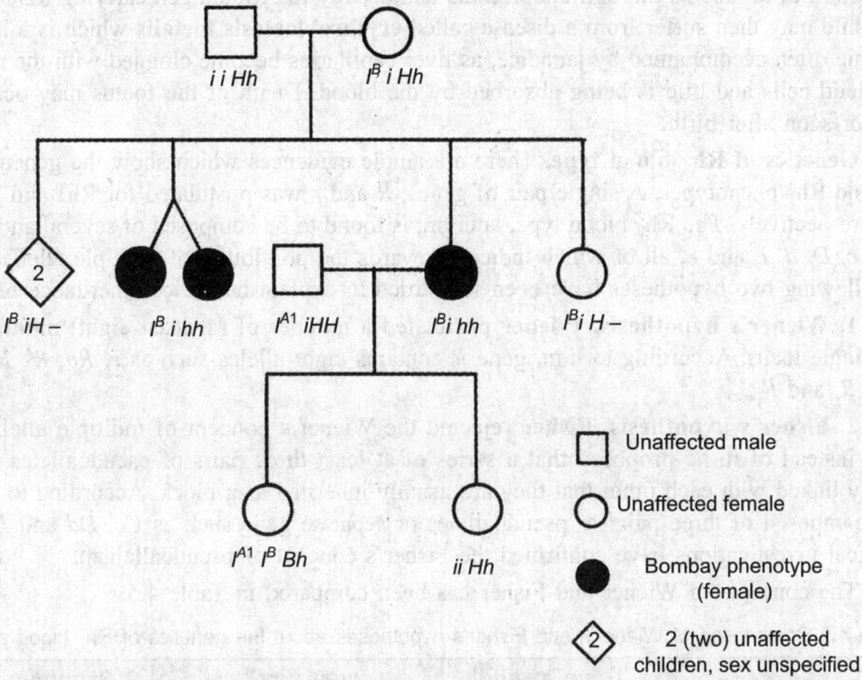

Fig. 4.10. Pathways leading to production of antigens on the red blood cells or erythrocytes showing how the Bombay phenotype is caused.

3. Rh Factor

The surface of erythrocytes (RBC) of some individuals contain one more type of antigen called **Rh factor** besides the A and B antigens. It is named after the *Macaca rhesus* monkey in which Rh factor was first discovered by **Landsteiner** and **Wiener** in 1940. Human beings are found to contain eight different types of Rh antigens.

The production of Rh antigen (Rh blood phenotype) depends on three closely set autosomal genes (pseudoallels). If any one of them is dominant, a Rh antigen is produced, but if all of them are recessive, no Rh antigen is formed. The individuals possessing the Rh antigen are called **Rh-positive** (Rh^+) and those lacking it are **Rh-negative** (Rh^-). Both of these types of persons are normal and none has natural anti-Rh antibodies in their blood plasma. However, a Rh-negative person can develop these antibodies on receiving Rh antigens through transfusion of Rh-positive blood. Such a blood transfusion will be safe only when the recipient had never been exposed to Rh-positive blood earlier. If already exposed, the previously developed anti-Rh antibodies will agglutinate the donor's RBC. In fact, "the degree of RBC agglutination depends upon the amount of anti-Rh antibodies present. The high concentration of anti-Rh causes severe agglutination of RBC which sometimes proves fatal.

Erythroblastosis foetalis. The incompatibility (mismatch) of Rh-positive and Rh-negative bloods may also be noticed in death of the child before or soon after birth. If a Rh-negative women marries a Rh-positive man and bears a Rh-positive foetus, sometime due to some placental defect, some of the foetal RBC carrying Rh antigens may pass into her own blood stream and cause the production of anti-Rh antibodies. The concentration of anti-Rh antibodies is gradually built up in the mother and she, thus, becomes **sensitized** only at or just before birth of the first Rh-positive child. In a second or subsequent pregnancy involving a Rh-positive child, these anti-Rh antibodies may return to the foetus through the placenta and destroy the Rh antigen carrying RBC of foetus. The child may then suffer from a disease called **erythroblastosis foetalis** which is a haemolytic anaemia often accompanied by jaundice, as liver capillaries become clogged with the remains of red blood cells and bile is being absorbed by the blood. Death of the foetus may occur before birth or soon after birth.

Genetics of Rh^+ blood type. There are ample evidences which show the genetic basis of Rh^+ and Rh^- phenotypes. A single pair of genes, *R* and *r* was postulated for Rh^+ and Rh^- blood types respectively. The Rh^+ blood type, later on, is found to be composed of several antigens such as *C, c, D, d, E* and *e*, all of which indicate towards the possibility of multiple allelism of gene *R*. Following two hypotheses have been forwarded to explain nature of inheritance of *R* gene:

1. Wiener's hypothesis. Wiener postulated a number of (at least eight) multiple alleles at a single locus. According to him, gene *R* contains eight alleles such as *r*, *Ro*, *R′*, *R″*, R_1, R_2, R_x or R_z and R_y.

2. Fisher's hypothesis. Fisher rejected the Wiener's concept of multiple allelism for R gene, instead of it, he proposed that a series of at least three pairs of pseudoalleles remain so closely linked with each other that they are usually inherited as a block. According to him, gene R is composed of three pairs of pseudoalleles or separate gene such as *Cc, Dd* and *Ee*. Recent genetical investigations have confirmed the Fisher's concept of pseudoallelism.

The concepts of Wiener and Fisher has been compared in Table 4.5.

Table 4.5 Comparison of Wiener's and Fisher's hypotheses about the genetics of Rh^+ blood phenotype.

	Gene symbol	*Antigen produced*	*Phenotypes*
Wiener	*Fisher*		
r	*cde*	none	Rh^-
Ro	*cDe*	Ro	Rh^+
R′	*Cde*	R′	Rh^+
R″	*cdE*	R″	Rh^+
R_1	*CDe*	Ro and R″	Rh^+
R_2	*cDE*	Ro and R″	Rh^+
Rx or *Rz*	*CDE*	R′ and R″	Rh^+
Ry	*CdE*	R′ and R″	Rh^+

QUESTIONS

Long Answer Questions

1. What are multiple alleles? Describe in brief the inheritance of anyone character determined by multiple alleles of *Drosophila*.
2. Explain the inheritance of multiple alleles with a suitable example.
3. Write an essay on multiple allelism.
4. Discuss the inheritance of skin colour in rodents.
5. Write an essay on the genetics of ABO type of blood group in human beings.
6. Write an essay on the various human blood groups.
7. Explain the inheritance of blood groups of human beings.
8. Describe different blood groups in human beings and discuss their genetic basis.
9. Describe the immunological criterion in blood grouping.

10. What do you understand by blood groups? Discuss them in detail. Discuss their significance.

11. Explain Rh factor. How can a marriage between homozygous rhesus-negative female and Rh^+ male cause erythroblastosis foetalis?

Short Answer Questions

1. Give a brief account of multiple allelism. (*M.S.U. Tamil Nadu, 2006*)

2. Write short notes on the following:
   (*i*) Multiple alleles
   (*ii*) Multiple allelism
   (*iii*) Blood groups
   (*iv*) Effect of transmission of blood
   (*v*) Inheritance of blood groups
   (*vi*) Rh^- blood groups
   (*vii*) Erythroblastosis foetalis
   (*viii*) H^+ antigen and Bombay phenotype.

Very Short Answer Questions

1. What are multiple alleles?
2. Define Rh factor.
3. Can a Rh^- person be given the blood of Rh^+ person?
4. Predict the possible blood group or groups of children whose parents have blood group 'O' and 'A'.
5. Does multiple alleles of a gene undergo crossing over?

Fill in the Blanks

1. The character described by Mendel occurred in two forms, *i.e.*, the gene for each Mendelian character is ....
2. The human ABO blood group locus has three physiologically distinct alleles. This gene is an example of ....
3. Four alleles of the gene *C* for coat colour in rabbit are agouti, ... , Himalayan and ....
4. .... eyed *Drosophila* was one of the first mutants known in fruit flies.
5. Rare AB blood group of human beings was discovered by .... and ....
6. Gene for H antigen is linked with .... phenotype.
7. The .... factor was named after the monkey *Macaca* .... It was discovered by .... and Wiener in 1940.

Multiple Choice Questions

1. Inherited Rh gene is found in
   (*a*) Rh^+ individual
   (*b*) Rh^- individual
   (*c*) AB blood group individuals
   (*d*) O blood group individuals
2. Marriage between one of the following pairs may cause child death
   (*a*) Rh^- woman and Rh^+ man
   (*b*) Rh^- woman and Rh^- man
   (*c*) Rh^+ woman and Rh^- man
   (*d*) Rh^+ woman and Rh^+ man

3. A, B, O blood group system is due to

(*a*) epistasis (*b*) multiple factor inheritance

(*c*) incomplete dominance (*d*) multiple allelism

ANSWERS

Very Short Answer Questions

1. It is a condition in which a particular gene occurs in three on more allelic forms in a population of organisms.

2. Rh factor is an antigen in the red blood corpuscles of certain people. A pregnant Rh-negative woman carrying an Rh^- positive child may develop antibodies, causing the child to develop a hemolytic disease.

3. No.

4. O and A.

5. No.

Fill in the Blanks

1. Diallelic.

2. Multiple alleles.

3. Chinchilla, albino.

4. White.

5. Von Decastello, Sturli.

6. Bombay.

7. Rh, *rhesus*, Landsteiner.

Multiple Choice Questions

1. (*a*) **2.** (*a*) **3.** (*d*).

5

Lethal Genes and Pleiotropic Genes

5.1 LETHAL GENES

Lethal genes are mutant genes and result in the death of the individual which carries them. Death of the individual occurs either in the prenatal or postnatal period prior to sexual maturity. A **fully** (completely) **dominant lethal allele** kills both in homozygous and heterozygous states. Individuals with a dominant lethal allele die before they can leave progeny. Therefore, the mutant dominant lethal is removed from the population in the same generation in which it arose. **Recessive lethal genes** kill only when they are in a homozygous state and they may be of two kinds: 1. One which has no obvious phenotypic effect in heterozygotes and 2. One which exhibits a distinctive phenotype when heterozygous.

The **completely lethal genes** usually cause death of the zygote, later in the embryonic development or even after birth or hatching. Complete lethality, thus, is the case where no individuals of a certain genotype attain the age of reproduction. However in many cases lethal genes become operative at the time the individuals become sexually mature. Such lethal genes which handicap but do not destroy their possessor are called **subvital, sublethal** or **semilethal** genes. The lethal alleles modify the 3 : 1 phenotypic ratio into 2 : 1.

Examples of Lethal Alleles

A. Lethal alleles in plants. In plants, recessive lethal alleles are known which produce **albinism**, where absence of chlorophyll is lethal (fatal) to them. Following two examples illustrate this fact:

1. In snapdragons (*Antirrhinum majus*) three types of plants occur: 1. Green plants with chlorophyll; 2. Yellowish green plants with carotenoids, usually are referred to as pale green, **golden** or **auria** plants and 3. White plants without any chlorophyll. The homozygous green plants have the genotype *CC* and the homozygous white plant has the genotype *cc*. The auria plants have the genotype *Cc* because they are heterozygotes of green and white plants. When two such auria plants are crossed, the F_1 progeny has identical phenotypic and genotypic ratio of 1 : 2 : 1 (*viz.*, 1 green (*CC*): 2 auria (*Cc*) : 1 white (*cc*). But the white plants because lack chlorophyll pigment, therefore, die to modify the ratio of 1 : 2 : 1 into 1 : 2 or 2 : 1. In this case the homozygous recessive genotype (*cc*) is lethal.

F_1 heterozygote:	Auria	×	Auria		
	Cc		*Cc*		
F_2:	1 *Cc*	:	2*Cc*	:	l*cc*
	Green		Auria		White (lethal)
or	1*CC*	:	2 *Cc*	or	1 : 2

Thus, *c* allele exhibits a lethal effect when homozygous and a distinctive phenotypic effect (*e.g.*, auria) when heterozygous.

2. In maize (*Zea mays*) the amount of chlorophyll is controlled by a recessive allele (g) which exhibits a lethal effect in homozygous (gg) and in heterozygous condition (Gg) has phenotype similar to homozygous condition for dominant gene GG. It modifies 3 : 1 phenotypic ratio into 2 : 1.

F_1 heterozygote:	Green	×	Green		
	Gg		Gg		
F_2:	1 GG	:	2 Gg	:	lgg
	Green		Green		White (lethal)
or	1GG	:	2 Gg	or	1 : 2 or 2 : 1.

B. Lethal alleles in animals. Among animals, the following three examples exhibit the role of recessive lethal alleles:

1. The inheritance of mouse body colour was studied by the French geneticist, **L. Cuenot** in 1905. The coat colour of mice is governed by a multiple allelic series (see Chapter 4) in which A allele determines **agouti** or mousy-coloured coat, A^Y allele determines **yellow** coat and a allele forms **black** coat. The dominance hierarchy is as follows: $A^Y > A > a$. The A^Y allele also acts as a **recessive lethal**, since in the homozygous state (A^YA^Y), it kills the individual in early embryonic stage (*i.e.,* during gastrulation). Thus, when two yellow coated heterozygotes ($A^Y A$) are crossed, they produce a progeny showing a ratio of 2 : 1 since homozygous yellow (A^YA^Y) individuals are never borned due to lethal effect of A^Y gene:

Parents:	Yellow	×			Yellow
	A^YA				A^Ya
	(Hybrid of yellow and agouti)	↓			(Hybrid of yellow and black)
Progeny:	1A^YA^Y	:	1$A^Y A$: 1A^Ya	:	1 Aa
	1 Homozygous yellow (die in uterus)		2. Heterozygous Yellow		1 Agouti

or 2 Yellow: 1 Agouti or 2 : 1.

2. In the chicken an incompletely dominant gene, creeper (cp) in heterozygous condition (cp/+) cause "creeper" condition. The creeper birds have much shortened and deformed legs and wings, giving them a squatty appearance and creeping gait. A cross of two creeper birds yields viable offsprings in the ratio of 2 creepers: 1 normal. The homozygous creepers having such a gross deformities that they die during incubation.

F_1:	$cp/+$	× ↓	$cp/+$		
	Creeper (Heterozygous)		Creeper (Heterozygous)		
F_2:	1 cp/cp	:	2$cp/+$	:	1+/+
	Creeper (Homozygous; dies)		Creeper (Heterozygous)		Normal (Homozygous)

3. In cattle, a recessive lethal gene in homozygous condition (aa) causes calves to born "amputated" which die soon after birth. The cross between two carriers (a+) produces the following result:

Parents:	Normal	×	Normal
	a+		a+
P gametes:	(a)(+)		(a)(+)
Progeny:	laa	:	2a + : 1 ++
	Amputated (die)		Normal

4. In *Drosophila* various sublethal and sex-linked lethal genes have been reported. The genes for vestigial wings and genes for white eyes are best examples of sublethal genes of *Drosophila*. Both of these genes reduce the viability of flies up to great extent. In *Drosophila*, certain recessive lethal genes such as curly wings (*Cy*), plum eyes (*Pm*) and stubbles (*Sb*) influence the viability of the flies when present in homozygous condition.

C. Lethal alleles in human beings. In humans several hereditary diseases have lethal effects. Few important lethal genes of human beings are following:

1. Congenital ichthyosis. One of the most typical cases of a recessive lethal gene in human beings is expressed in congenital ichthyosis. At birth children afflicted with this disease have a crusted leathery skin with deep fissures down to the subcutaneous tissue; the fissures lead to bleeding, infection and death. Congenital ichthyosis occurs only when there occurs homozygous condition for its recessive lethal genes.

2. Infantile amaurotic idiocy. A recessive allele in homozygous condition causes a fatal disease called **infantile amaurotic idiocy** in juvenile stage. Bearers of this genotype begin to lose their eye sight between the age of four to seven years. The complete blindness is followed by mental degeneration and finally death before adolescence.

3. Thalassemia. Thalassemia or **Cooley's anaemia** is a haemoglobin disease somewhat similar to sickle cell anaemia. It occurs mostly in children (in India and other countries such as Italy, Greece and Syria) and is nearly 100 per cent fatal (lethal). Thalassemia is controlled by a single gene *c* which in homozygous condition (*cc*), produces the severe Cooley's anaemia or **thalassemia major** and causes death of the patient. The heterozygous condition of this lethal gene (*Cc*) results in a mild form of the disease called **thalassemia minor** or **microcythemia**.

5.2 PENETRANCE

The ability of a given gene or gene combination to be expressed phenotypically to any degree is called **penetrance**. It is of following two kinds:

1. Complete penetrance. Most dominant and recessive genes in homozygous conditions and many completely dominant genes even in heterozygous conditions give their complete phenotypic expressions. Such genes are called to have **complete penetrance**.

Examples of Complete Penetrance

(*i*) In pea the alleles (*RR*) for red flowers and the alleles (*rr*) for white flowers have complete penetrance in homozygous conditions.

(*ii*) In *Drosophila* the recessive alleles (*vv*) for vestigial wings in homozygous conditions have complete penetrance.

(*iii*) In guinea pigs the dominant allele '*B*' for black coat has complete penetrance both in homozygous and heterozygous conditions.

2. Incomplete penetrance. Some genes in homozygous as well as in heterozygous conditions fail to provide complete (cent per cent) phenotypic expression of them. Such genes are called to have **incomplete penetrance**.

Examples of Incomplete Penetrance

(*i*) Polydactyly in human beings is thought to be produced by a dominant gene *P*. The normal condition with five digits on each limb is produced by the recessive genotype (*pp*). Some heterozygous individuals (*Pp*) are not polydactylus and, therefore, has a penetrance of less than 70%.

(*ii*) In human beings, the tendency to develop diabetes mellitus (a condition in which there is an excess of sugar in the blood) is controlled by certain genes. However, not everyone carrying the genes for diabetes actually develops the condition, for these genes have incomplete penetrance.

Effects of environment on penetrance. The environmental factors and genetical background have some definite effect on the degree of penetrance of a gene. For example, when various twins which carry genes for diabetes mellitus are studied, it is found that the disease appears only in those cases which ate more carbohydrate foods (starch and sugars).

Expressivity.

A trait though penetrant, may be quite variable in its phenotypic expressions. The degree of effect produced by a penetrant genotype is called **expressivity.**

Example of expressivity. In human beings the polydactylous condition may be penetrant in the left hand (6 fingers) and not in the right (5 fingers); or it may be penetrant in the feet and not in the hands.

Effects of environment on expressivity. The expressivity of a given gene is often influenced by environmental conditions. For instance, the expressivity of completely penetrant gene for vestigial wings in *Drosophila* is influenced by the temperature at which the fly develops, with the effect being most obvious at lower temperature.

Other examples of environmental effects on the expressivity of a gene includes such cases as the differences in the severity of symptoms of an inheritable allergy, or the differences in height of identical twins who have been raised in different home (with different diets), or who have had different medical histories (one with a serious childhood disease, the other escaping this disease).

5.3 PLEIOTROPISM

The Mendelian concept of a gene controlling a single character has also expanded to take into account genes which affect several characters simultaneously (**plciotropy**) and genes which cooperate in groups to control individual characters (*i.e.,* quantitative inheritance).

A single gene often influences more than one phenotypic trait. However, it may be that one gene may cause evidently well marked expression of some phenotypic trait (**major effect**) than the others with less evident phenotype (**secondary effect**). Most genes have their multiple effects and are called **pleiotropic genes**. The phenomenon of multiple effect (multiple phenotypic expressions) of a single gene is called **pleiotropism**.

Examples of Pleiotropism

(*i*) In *Drosophila* the recessive gene for vestigial wings cause vestigial wings in homozygous condition. However, careful observations show that other traits as well are affected: (*i*) The tiny wing-like balancer behind the wings; (*ii*) Certain bristles; (*iii*) The structure of the reproductive organs; (*iv*) Egg production is lowered, and, (*v*) Longevity is reduced.

(*ii*) In human, the gene for disease **phenylketonuria** has pleiotropic effect and produces various abnormal phenotypic traits, collectively called **syndrome**. For example, the affected individuals secrete excessive quantity of amino acid phenylalanine in their urine, cerebrospinal fluid and blood. They become short stature, mentally deficient, with widely spaced incisors, with pigmented patches on skin, with excessive sweating, and with non-pigmented hairs and eyes.

(*iii*) The A^Y allele for yellow coat in mice is also a good example of pleiotropic gene. It affects two characters: coat colour and survival. It is most probable that both effects of the A^Y allele are the result of same basic cause which promotes the yellowness of coat in a single dose and death in double dose. Genetic analysis has revealed that lethal pleiotropic A^Y allele basically affects the cartilage of mice and cause death.

Sickle cell anaemia. The inheritance pattern of the trait of sickle-cell anaemia in humans presents an example of various genetical phenomena such as codominance, incomplete dominance, lethal effect, pleiotropism and polymorphism. When an allele results in a number of related changes, then such changes are collectively called syndrome. The syndrome, called sickle cell anaemia, is caused by an autosomal mutant allele, Hb^s, that in homozygous condition causes the production of an abnormal haemoglobin, called **haemoglobin S**; this indicates **primary effect** of mutant gene. In 1949, **Pauling** and his co-workers reported that the formation of an abnormal haemoglobin (*i.e.,* Hb^S) which differed in its electrophoretic mobility from normal haemoglobin (Hb^A or adult haemoglobin) was the cause of this hereditary disease.

Further, due to **subsidiary effect** of abnormal haemoglobin (*i.e.*, Hb^S), red blood cell of some persons suffered from this fatal form of haemolytic anaemia undergo a reversible alteration in shape when oxygen tension of the plasma falls slightly and they assume elongate, filamentous and sickle cell forms. Such sickle-shaped red blood cells show a lower life span, since they clump together and often causes vascular obstruction and are rapidly destroyed. This results in severe anaemia and death of the patient due to damaged heart, kidney, spleen and brain as a result of their clogged blood vesseles.

Biochemical investigations have shown that globin protein of haemoglobin is a **heteropolymer** having two identical α polypeptide chains, each with 141 amino acid residues and two identical β polypeptide chains, each with 146 amino acid residues. In 1957 **Vernon Ingram** showed that β chain amino acid of 6th position is occupied by **glutamic acid** in haemoglobin A, when this amino acid is substituted by another amino acid, the **valine**, it forms the haemoglobin S:

	1		2		3		4		5		6		7		8	
Hb^A =	val	-	his	-	leu	-	thr	-	pro	-	glu	-	glu	-	lys	-
HbS =	val	-	his	-	leu	-	thr	-	pro	-	val	-	glu	-	lys	-

At the molecular level, it is found that the genetic code of DNA undergoes a change from GAG to GTC.

Pleiotropy of sickle cell allele. The mutant allele Hb^s causes the production of abnormal haemoglobin. As a consequence the shape of the RBC containing it becomes distorted and sickle-shaped. Homozygous individuals (Hb^S Hb^S) normally die early in life due to severe anaemia caused by premature destruction of the sickled red blood cells. But heterozygotes who have both normal (haemoglobin A) and abnormal (haemoglobin S) haemoglobins and who also have mild anaemia, are naturally protected against contracting malaria as the parasite cannot live in these distorted cells. Heterozygotes may, therefore, survive better in region where malaria is endemic. Such population have both normal individuals and individuals heterozygous for the gene. The latter act as carriers of the gene from generation to generation as shown below:

Hb^A Hb^S × Hb^A Hb^S

Sickle cell carrier ↓ Sickle cell carrier

	¼ Hb^A Hb^A	½ Hb^A Hb^S	¼ Hb^S Hb^S
	Normal; haemoglobin A; no sickling of RBC	Sickle cell trait; carriers; produce both type of haemoglobin due to codominance; sickling under reduced O_2 tension	Sickle cell anaemia; normally die early in life; produce haemoglobin S only
Heterosis:	More susceptible to malarial infection in prone areas	Better adapted for malaria in prone area	Suffer form lethal anaemia

Polymorphism of sickle cell trait. The existence of two or more genetically different classes in the same interbreeding population at the same time, in the same habitat, is known as **polymorphism**. These classes are controlled by two competing alleles in the same locus. In **balanced polymorphism**, there establishes an intermediate equilibrium between the frequencies of these alleles and two classes are maintained together in the same population. For example, in case of sickle-cell anaemia superiority retains the two types at equilibrium. Here, heterosis or superiority of heterozygotes ($Hb^A\ Hb^s$ which is resistant to malaria) over the normal homozygotes ($Hb^A\ Hb^A$) maintains the two alleles of haemoglobin (*i.e.*, normal and sickling) in an equilibrium.

QUESTIONS

Long Answer Questions

1. Give a detailed account of lethal genes.
2. Describe pleiotropism with suitable examples.

Short Answer Questions

1. Make a brief account of pleiotropism. (*M.S.U. Tamil Nadu 2006*)
2. Write short notes on the following:
   (*i*) Lethal genes; (*ii*) Sickle cell anaemia;
   (*iii*) Sickle cell anaemia and sickle cell gene; (*iv*) Pleiotropism.
3. Define and explain sickle cell anaemia. Mention its genetic significance. (*M.S.U. Tamil Nadu 2006*)

Very Short Answer Questions

1. What happens to human babies having $Hb^S\ Hb^s$ genotypes?
2. What happens when gene A^Y is present in mice in homozygous condition?
3. What is plëiotropism?

True or False Questions

1. Sickle cell anaemia as an example of lethal gene. (*H.N.B. Garhwal 2008*)

Fill in the Blanks

1. .... genes are .... genes which result in the death of the individual which carries them.
2. A lethal gene which handicaps but does not destroys its possessor is called .... gene.
3. In plants, ... lethal alleles are known which produce albinism.
4. Thalassmia or .... anaemia is a haemoglobin disease of .... and is 100 percent ....
5. Phenomenon of multiple phenotypic expression of single gene is called ....
6. Pleiotropic effect of sickle cell anaemia includes .... , incomplete dominance, ... effect and polymorphism.

Multiple Choice Questions

1. A gene which kills the bearer is termed as: (*M.S.U. Tamil Nadu, 2006*)
   (*a*) polethal (*b*) lethal
   (*c*) important (*d*) protective
2. Pleiotropic genes are those which have:
   (*a*) single phenotype (*b*) multiple phenotype
   (*c*) multiple genotype (*d*) single genotype

3. Sickle cell anaemia has not been eliminated from the African population because
   (*a*) it is controlled by dominant genes
   (*b*) it is controlled by recessive genes
   (*c*) it is not a fatal disease
   (*d*) it provides immunity against malaria
4. Sickle cell anaemia is caused due to:
   (*a*) valine at 6th position at β chain of Hb
   (*b*) glutamic acid at 6th position in β chain of Hb by valine
   (*c*) glutamic acid at 6th position in α chain of Hb
   (*d*) valine at the 6th position in α chain of Hb by glutamic acid
5. In which of the following disorders, blood has a defective haemoglobin?
   (*a*) haemophilia
   (*b*) haematuria
   (*c*) haematoma
   (*d*) sickle cell anaemia.
6. The inheritance of blood group in human beings shows
   (*a*) quantitative characters
   (*b*) qualitative characters
   (*c*) multiple characters
   (*d*) polygenic characters

ANSWERS

Very Short Answer Questions

1. Such children suffer from disease called sickle cell anaemia and die.
2. Death occurs due to lethal gene A^Y.
3. The phenomenon whereby a single mutant affects several apparently unrelated aspects of the phenotype.

Fill in the Blanks

1. Lethal, mutant.
2. Sublethal.
3. Recessive, albinism.
4. Cooley's, children, lethal.
5. Pleiotropism.
6. Codominance, lethal.

Multiple Choice Questions

1. (*b*) 2. (*b*) 3. (*d*) 4. (*a*) 5. (*d*) 6. (*b*).

6

Interactions of Genes (Factor Hypothesis)

During the discussion of Mendel's monohybrid and dihybrid crosses, we encountered with the fact that for the determination of single phenotypic trait of an organism, two alleles or allelomorphs of a single gene interacted in various ways. Such as out of two allelomorphs of a single gene, one allelomorph might show simple (complete) dominance over the action of other which was recessive; or both allelomorphs might have partial or incomplete dominant relationship or both allelomorphs might have equal expression or codominant relationship. These kinds of genic or genetic interactions occur in between the two allelomorphs of a single type of gene and are usually referred to as **intra-alleic** or **allelic genetic interactions.** These kinds of genetic interactions give the classical ratios of 3 : 1 and 9 : 3 : 3 : 1. But, in addition to intra-allelic genetic interactions, **factor hypothesis, non-allelic** or **inter-allelic genetic (genic) interactions** also occur. In inter-allelic genetic interactions, the independent (non-homologous) genes located on the same or on different chromosomes interact with one another for the expression of single phenotypic trait of an organism. The discovery of the interallelic genetic interactions has been made after Mendel. Some different type of non-allelic interactions which generate modified dihybrid ratios have been described in Table 6.1.

6.1 SUPPLEMENTARY GENES OR COLLABORATOR FACTORS (9 : 3 : 1 : 1)

In certain cases, two pairs of genes determine a same phenotype but assorted independently, produce new phenotypes by mutual non-epistatic interactions and the F_2 phenotypic ratio 9 : 3 : 3 : 1 remains unaltered. Two pairs of genes which interact to affect size and shape of comb but are independently transmitted exist in chicken.

Example

Combs in fowl (9 : 3 : 3 : 1). The classical case of genetic interaction of two genes is discovered by **Bateson** and **Punnett** (1905–1908) in fowls. There are many different breeds of domestic chicken. Each breed possesses a characteristic type of comb. The Wyandotte breed has a comb called "rose", the Brahmas breed has a comb called "pea", and the Leghorns have a comb called 'single'. Each of these types breeds true.

A cross of chicken with a rose comb to one with a single comb produces 3/4 rose and 1/4 single, showing dominance of rose over single. Another cross between pea combed and single combed chickens produces pea and single combed chickens in the ratio of 3 : 1, showing dominance of pea over single. But, when a rose combed chicken crossed with that of pea combed, the F_1 progeny was found with a different type of comb known as 'walnut' (Malay breed). When the F_1 walnut combed chickens were breed together, in F_2 all four types of combs, *i.e.*, 9/16 walnut, 3/16 rose, 3/16 pea and 1/6 single appeared (Fig. 6.1).

Table 6.1 Some different types of non-allelic interactions.

S. No.	*Interaction*	*Description*
1.	**None**	Two segregating loci which are unlinked and control distinct characters. Dihybrid ratio is 9 : 3 : 3 : 1.
2.	**Additive effect**	The segregating loci contribute to the same character in an additive fashion. Alleles are generally regarded as positive or neutral in effect and resulting phenotypes is determined by the net contribution of all addition loci. The dihybrid ratio is 9 : 3 : 3 : 1. (**Note.** For further details see Chapter 16 of Polygenic or Quantitative Inheritance).
3.	**Complementary action of genes**	A situation where specific alleles at two gene loci are required to generate a particular phenotype. The dihybrid ratio would be 9 : 7 if **dominant** alleles were required at both loci and 15 : 1 if **recessive** alleles were required at both loci.
4.	**Epistasis**	The situation where an allele at one locus prevent or masks the effect of gene expression at a second locus, which is said to be **hypostatic.** Epistasis may be **dominant** or **recessive** and dihybrid ratio would be 12 : 3 : 1 and 9 : 3 : 4, respectively.
5.	**Suppression**	An allele at one locus (the **suppressor allele**) prevents expression of a specific (and usually deleterious) allele at a second locus by compensating in some way for its effect. **The term is used to describe an interaction where a mutation compensates for a second mutation found at different locus and restore the wild-type phenotype.** Suppression may be **dominant** or **recessive** dependent upon whether one or two alleles are required at the *suppressor locus*. The dihybrid ratio would be 12 : 3 : 1 and 9 : 3 : 4, respectively. Epistasis differ from suppression in that the former is gene-specific but not allele-specific (*i.e., epistasis prevents the expression of all alleles at a particular locus*), whereas the latter is gene-specific and allele-specific (*i.e.*, suppression compensates for the effect of a particular allele at the second locus).
6.	**Synergism, Enhancement**	Similar to suppression mechanism expect that the **enhancer allele** specifically increases the effect of second mutation instead of suppressing it. As with suppressor alleles, enhancer alleles may be dominant or recessive. Synergic enhancement differs from additive effects because the enhancer locus alone does not contribute to the phenotype associated with the target (enhanced) locus, *i.e.*, if the target allele is not present, the enhancer locus has no effect on the phenotype.

These peculiar results were interpreted by **Bateson** and **Punnett** as follows: The rose comb is caused by the combination of homozygous recessive genes "*pp*" and homozygous or heterozygous dominant genes '*RR*' or '*Rr*'. The pea comb is supposed to be produced by combination of a homozygous recessive condition (*rr*) and homozygous or heterozygous dominant condition (*PP* or *Pp*). While, the single type comb is produced by the double recessive, *rrpp*, genes. Thus, *R* gene determines the shape of rose comb and *P* gene determines the shape of pea comb, but when both genes happen to come together in single individual due to cross between rose and pea combed chickens, they interact to produce a walnut comb in F_1. In the cross of two walnut chickens, two genes interact variously and produces four types of offsprings in F_2.

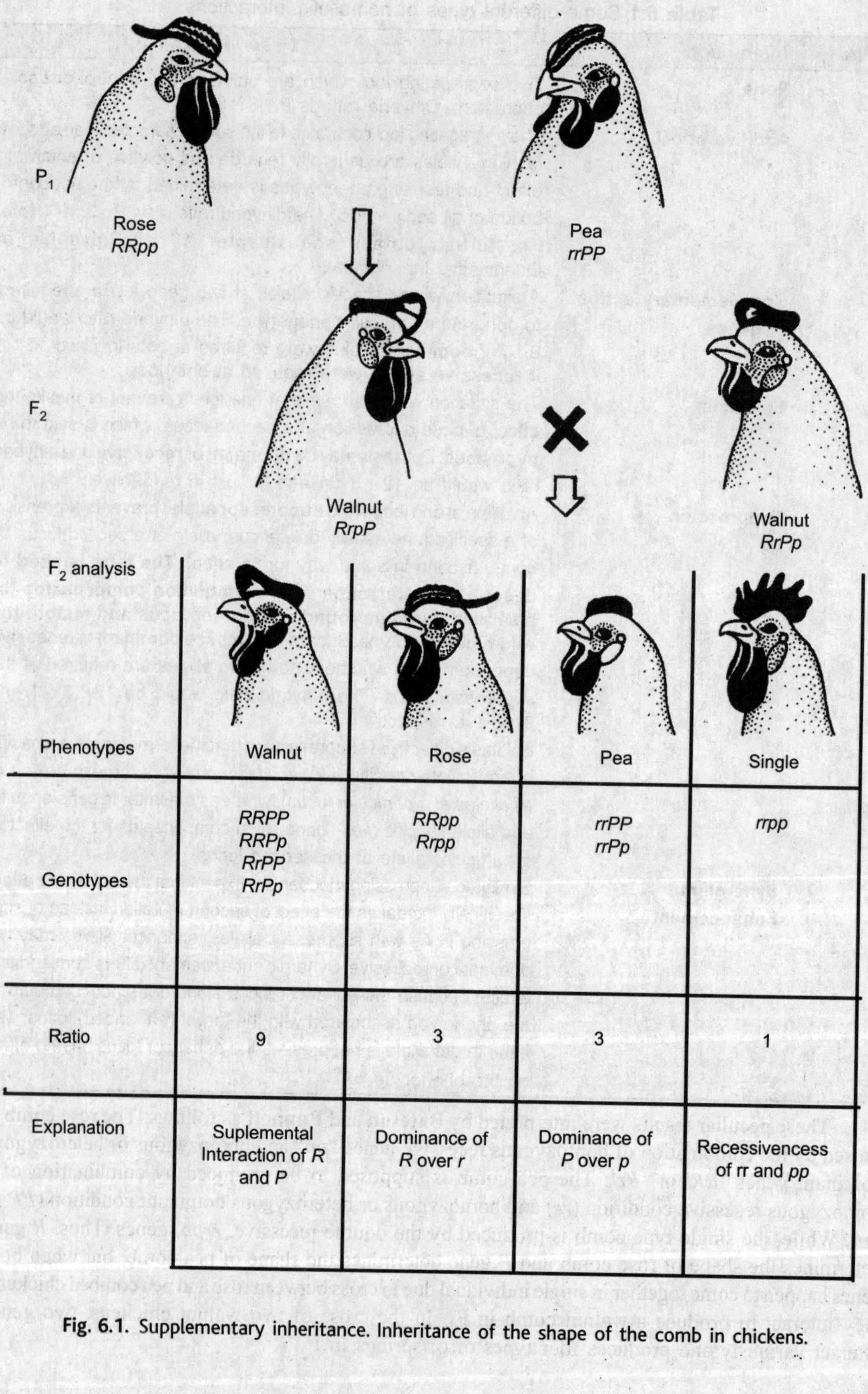

Fig. 6.1. Supplementary inheritance. Inheritance of the shape of the comb in chickens.

Thus, here two pairs of genes interact to produce comb size and shape in fowl. During the inheritance of combs in fowls, the genes themselves do not determine the development of a character (presence or absence of comb) and simply modify a character determined by a basic gene and, therefore, known as **supplementary** or **modifying genes.**

6.2 DOMINANT COMPLEMENTARY GENES OR DUPLICATE RECESSIVE GENES (9 : 7)

If both gene loci have homozygous recessive alleles and both of them produce identical phenotypes, the F_2 ratio 9 : 3 : 3 : 1 would become 9 : 7. In such case, the genotypes *aa BB, aa Bb, AA bb, Aa bb,* and *aa bb* produce one phenotype (Table 6.2). Both dominant alleles when present together, complement each other and are called **dominant complementary genes** and produce a different phenotype. A case of such complemental inheritance, resulting from the combined action of complemental genes is known in sweet peas.

Table 6.2 Mode of action of complementary genes.

S. No.	Epistatic allele	Hypostatic allele	Phenotypic expression of alleles
1.	*aa*	*BB, Bb, bb*	No phenotype production
2.	*AA, Aa, aa*	*bb*	
3.	*AA, Aa*	*BB, Bb*	Phenotypie expression due to dominant alleles on both loci (complementation of gene action)

Example

When a pure line variety of white flowered sweet pea (*Lathyrus odoratus*) was crossed with another pure line variety of white flowered sweet pea, in F_1 purple or red flowered plants were produced (Table 6.3). The F_1 plants when self-pollinated or crossed among themselves, produced the F_2 generation with the phenotypic ratio of 9 coloured and 7 white flowered plants.

Table 6.3 Mode of action of complementary alleles in the production of coloured flowers in sweet pea.

Epistatic alleles		Hypostatic alleles	Phenotypic expression of allele	F_2 Phenotypic ratio
1.	*c c*	*EE, Ee, ee*	Neither *c* nor *E* or *e*	White = 7
2.	*CC, Cc*	*e e*	Neither *C* or *c* nor *e*	
3.	*CC, Cc*	*EE, Ee*	Both *C* + *E* (complementation)*	Coloured = 9

* Chromogen production due to complementation of *C* and *E* (*i.e.*, dominant epistatic and hypostatic alleles).

These surprising results could be understood by analysing the mechanism of colour production in flowers. A given enzyme (genetically controlled as to absence or presence in a given individual) acts upon chromogen (a colourless colour base whose absence or presence is also genetically controlled) to produce the purple or red colour of flowers. The dominant allele or alleles (*CC* or *Cc*) of gene *C* are responsible for the presence of chromogen, while the homozygous recessive alleles (*cc*) of this gene are responsible for the absence of chromogen. Likewise, the dominant alleles of gene E in homozygous (*EE*) or heterozygous (*Ee*) conditions caused the production of an enzyme which is necessary for colour production from chromogen, while homozygous recessive (*ee*) condition does not produce any such enzyme.

The appearance of 9 : 7 ratio instead of 9 : 3 : 3 : 1 ratio from the cross of two white flowered sweet pea plants (*CC ee* × *cc EE*) can be illustrated in Fig. 6.2.

P: White flower *CC ee* × White flower *cc EE*

↓ ↓ ↓

P gametes: (*Ce*) (*cE*)

F_1: Purple flower *Ce Ee* × Purple flower *Cc Ee*

F_1 Female gemetes → *CE* *Ce* *cE* *ce*

F_1 Male gemetes ↓

F_2:

	CE	*Ce*	*cE*	*ce*
CE	*CC EE* Purple	*CC Ee* Purple	*Cc EE* Purple	*Cc Ee* Purple
Ce	*CC Ee* Purple	*CC ee* White	*Cc Ee* Purple	*Cc ee* White
cE	*Cc EE* Purple	*Cc Ee* Purple	*cc EE* White	*cc Ee* White
ce	*Cc Ee* Purple	*Cc ee* White	*cc Ee* White	*cc ee* White

F_2 phenotypic ratio: 9/16 Purple : 7/16 White or 9 : 7.

Fig. 6.2. Complementary genes. Production of 9 : 7 phenotypic ratio in sweet peas due to complementary genes.

6.3 RECESSIVE COMPLEMENTARY GENE OR DUPLICATE DOMINANT GENE (15 : 1).

If the dominant alleles of both gene loci produce the same phenotype without cumulative effect, the 9 : 3 : 3 : 1 ratio is modified into a 15 : 1 ratio (Table 6.4). They are also called **recessive complementary genes** since both *aa* and *bb* are necessary for the phenotype.

Table 6.4 Mode of action of duplicate dominant genes.

	Epistatic allele	*Hypostatic allele*	*Phenotypic expression of alleles*
1.	*AA, Aa*	*bb*	Same phenotype (*A, B, A + B*) (Presence of one dominant allele on both or either loci result in same phenotype)
2.	*AA, Aa*	*BB, Bb*	
3.	*a a*	*BB, Bb*	
4.	*a a*	*bb*	Another phenotype (*a* or *b*) (Absence of dominant allele on both or either loci results in different phenotype)

Example

The seed capsules of shepherd's purse (genus *Capsella*) occur in two different shapes, *i.e.*, triangular and top-shaped. When a plant with triangular seed capsule is crossed with one having top-shaped capsule, in F_1 only triangular, character appears. The F_1 offsprings by self crossing produce the F_2 generation with the triangular and top-shaped seed capsules in the ratio of 15 : 1. Two independently segregating dominant genes (*A* and *B*) have been found to influence the shape of the capsule in the same way (Table 6.5). All genotypes having dominant alleles of both of these genes (*A* and B) would produce plants with triangular-shaped capsules. Only those with the genotype *aa bb* would produce plants with top-shaped capsules. The results of this example has been shown in Fig. 6.3.

Table 6.5 Mode of action of duplicate genes in producing triangular seed capsules in *Capsella*.

Epistatic alleles	*Hypostatic alleles*	*Phenotypic expression of allele*	*Phenotypic ratio*
1. *AA, Aa*	*bb*	Phenotype of either or both dominant alleles, *i.e.*, *A* or *B* or both	Triangular = 15
2. *AA, Aa*	*BB, Bb*		
3. *a a*	*BB, Bb*		
4. *a a*	*bb*	Recessive phenotype of *a* and *b* alleles	Top-shaped = 1

P:	Triangular *AA BB*	×	Top-shaped *aa bb*
P gametes:	(*AB*)	↓	(*ab*)
F_1:	Triangular *Aa Bb*	×	Triangular *Aa Bb*

F_1 Female gemetes →
F_1 Male gemetes ↓

F_2:	*AB*	*Ab*	*aB*	*ab*
AB	*AA BB* Triangular	*AA Bb* Triangular	*Aa BB* Triangular	*Aa Bb* Triangular
Ab	*AA Bb* Triangular	*AA bb* Triangular	*Aa Bb* Triangular	*Aa bb* Triangular
aB	*Aa BB* Triangular	*Aa Bb* Triangular	*aa BB* Triangular	*aa Bb* Triangular
ab	*Aa Bb* Triangular	*Aa bb* Triangular	*aa Bb* Triangular	*aa bb* Top shaped

F_2 phenotypic ratio: 15/6 triangular : 1/16 Top-shaped or 15 : 1.

Fig. 6.3. Duplicate dominant genes: a cross between two strains of *Capsella* having triangular and top-shaped seed capsules to get 15 : 1 F_2 dihybrid ratio.

6.4 EPISTASIS

The gene is a chemical determiner. Whereas a phenotypic trait results from the combined action of many genes and their products constantly interacting with the environment. The environment includes not only ecological factors such as temperature and the amount or quality of light, but also internal factors such as hormones and enzymes. The enzymes are proteins and the specific molecular organization of protein is determined by genes. The enzymatic proteins perform catalytic function in various cellular chemical (metabolic) reactions and causing the splitting or union of various molecules. Each cellular chemical reaction involves stepwise conversion of one substance called **precursor** into another, called **end product**. Each step being mediated by a specific enzyme. All the subsequent steps of a chemical reaction constitute the **biosynthetic pathway**.

Thus, a simplest biosynthetic pathway includes various steps, each step is catalyzed by a specific enzymatic protein and each enzymatic protein in its turn depends on a specific gene for its production. For example, we may consider a simple biosynthetic pathway which transforms a precursor substance 'P' into the end product 'C' in following three subsequent steps:

$$\begin{array}{ccccccc} & g_1^+ & & g_2^+ & & g_3^+ & \\ & \downarrow & & \downarrow & & \downarrow & \\ \textbf{P}\text{ (precursor)} — & e_1 & \longrightarrow \textbf{A} — & e_2 & \longrightarrow \textbf{B} — & e_3 & \longrightarrow \textbf{C}\text{ (end product)} \end{array}$$

In this biosynthetic pathway each metabolite (A, B, C) is produced by the catalytic action of different enzymes (e_1, e_2, e_3 ... e_x, specified by different wild type genes (g_1^+, g_2^+, g_3^+, ... g_x^+). When more than two or more genes become involved in the specification of enzymes for different steps of a common biosynthetic pathway, the phenomenon of genetic interaction occurs. If substance C is essential for the production of a normal phenotype and the recessive mutant alleles g_1, g_2, g_3 produce defective enzymes, then a mutant or abnormal phenotype would result from a genotype homozygous recessive at any of the three loci. If wild gene g_3^+ becomes mutant, the conversion of metabolite B to C does not occur and substance B tends to accumulate in excessive quantity; if g_2^+ becomes mutant, substance A will accumulate. Thus, the mutant genes caused "**metabolic blocks**" in synthetic pathway.

An organism with a mutation only in gene g_2^+ could produce a normal phenotype, if it was given either substance B or C, but an organism with a mutation in gene g_3^+ has a specific requirement for substance C for the production of normal phenotype. Thus, gene g_3^+ becomes dependent upon g_2^+ for its expression as a normal phenotype. If the genotype is homozygous for the recessive g_2 allele, then the biosynthetic pathway ends with substance A. Neither g_3^+ nor its recessive allele g_3 has any effect on the phenotype. Thus, genotype (g_2g_2) can hide or mask the phenotypic expression of alleles at the g_3^+ locus. Originally a gene or locus which suppressed or masked the action of a gene at another locus was termed **epistatic gene.** The gene or locus which was suppressed by a epistatic gene was called **hypostatic gene**. Later studies revealed the fact that both loci or genes (*i.e.*, epistatic and hypostatic) could be epistatic to one another. Presently, the term epistasis (Greek, standing upon) is used for almost any type of allelic genetic interaction.

Difference Between Dominance and Epistasis

The phenomenon of dominance involves intra-allelic gene suppression, or the masking effect which one allele has upon the expression of another allele at the same locus, while the phenomenon of epistasis involves inter-allelic gene suppression or the masking effect which one gene locus has upon the expression of another. The classical phenotypic ratio of 9 : 3 : 3 : 1 observed in the progeny of dihybrid parents becomes modified by epistasis into ratios which are various combinations of the 9 : 3 : 3 : 1 groupings.

Epistasis is of following two main types:

I. Dominant Epistasis (12 : 3 : 1)

When out of two genes, the dominant allele (*e.g.*, *A*) of one gene masked the activity of alleles of another gene (*e.g.*, *B*) and expressed itself phenotypically, then *A* gene locus is said to be epistatic to the *B* gene locus (Table 6.6). Because, the dominant allele *A* can express itself only in the presence of either *B* or *b* allele, therefore, such type of epistasis is termed as **dominant epistasis**. The alleles of hypostatic locus or gene *B* will be able to express themselves phenotypically only when gene locus *A* may contain two recessive alleles (*aa*). Thus, the genotype *AA BB* or *Aa Bb* and *AA bb* or *Aa bb* produce the same phenotype whereas the genotype *aa BB* or *aa Bb* and *aa bb* produce two additional phenotypes. The dominant epistasis modify the classical ratio of 9 : 3 : 3 : 1 into 12 : 3 : 1 ratio.

Table 6.6 Mode of dominant epistasis.

S. No.	Epistatic allele	Hypostatic allele	Phenotypic expression of alleles
1.	*AA, Aa*	*BB, Bb, bb*	*A*
2.	*aa*	*BB, Bb*	*B*
3.	*aa*	*bb*	*b*

Example

Dominant epistatis in dogs. Among dogs, the colours of coats depend upon the action of two genes. One gene locus has a dominant epistatic inhibitor allele (*I*) of coat colour pigment (see Table 6.7). The allele *I* prevents the expression of colour allele at another independently assorting, hypostatic gene locus (*B* or *b*) and produces white coat colour. The alleles of hypostatic gene locus (*BB, Bb*, or *bb*) express only when two recessive alleles (*ii*) occur on the epistatic locus, *i.e.*, *ii BB* or *ii Bb* produces black and *ii bb* produces brown individuals. When two such white coat colour dogs are crossed, in F_1 the white, black and brown coat colours appear in 12 : 3 : 1 ratio as shown in Fig. 6.4.

Table 6.7 Mechanism of dominant epistasis in dogs.

Epistatic alleles		Hypostatic alleles	Phenotypic expression of allele	F_2 Phenotypic ratio
1.	*II, Ii*	*BB, Bb, bb*	*I* (no pigment)	White = 12
2.	*ii*	*BB, Bb*	*B* (Black)	Black = 3
3.	*ii*	*bb*	*b* (Brown)	Brown = 1

White (Male) × White (Female)

P: *Ii Bb*   *Ii Bb*

P Female gemetes → / **P Male gemetes ↓**	*IB*	*Ib*	*iB*	*ib*
IB	*II BB* White	*II Bb* White	*Ii BB* White	*Ii Bb* White
Ib	*II Bb* White	*II bb* White	*Ii Bb* White	*Ii bb* White
F_1: *iB*	*Ii BB* White	*Ii Bb* White	*ii BB* Black	*ii Bb* Black
ib	*Ii Bb* White	*Ii bb* White	*ii Bb* Black	*ii bb* Brown

F_1 Phenotypic ratio: 12/16 White : 3/16 Black : 1/16 Brown or 12 : 3 : 1.

Fig. 6.4. Checkerboard derived from a cross between two heterozygous white coated dogs showing 12 : 3 : 1 ratio due to dominant epistatic inhibitor genes.

2. Recessive Epistasis (9 : 3 : 4)

Sometimes the recessive alleles of one gene locus (*aa*) mask the action (phenotypic expression) of alleles of another gene locus (*BB, Bb* or *bb* alleles) (see Table 6.8). This type of epistasis is called **recessive epistasis**. The alleles of *B* locus express themselves only when epistatic locus

Table 6.8 Mode of action of recessive epistasis.

S. No.	*Epistatic allele*	*Hypostatic allele*	*Phenotypic expression of alleles*
1.	*a a*	*BB, Bb, bb*	*a*
2.	*AA, Aa*	*BB, Bb*	*B*
3.	*AA, Aa*	*bb*	*b*

has dominant alleles (*e.g.*, *AA* or *Aa*). Due to recessive epistasis the phenotypic ratio 9 : 3 : 3 : 1 becomes modified into 9 : 3 : 4 ratio.

Example

Recessive epistasis in mice. In mice various types of epistatic genetic interactions have been reported. The most interesting case is of recessive epistasis in coat colours. The common house mouse occurs in a number of coat colours, *i.e.*, agouti, black and albino. The agouti colour pattern is commonly occurred one (wild type) and is characterized by colour banded hairs in which the part nearest the skin is gray, then a yellow band and finally the distal part is either black or brown. The albino mouse lacks totally in pigments and has white hairs and pink eyes.

Table 6.9 Mode of action of recessive epistasis in alleles for coat colour in mice.

Epistatic alleles	*Hypostatic alleles*	*Phenotypic expression of allele*	*F_2 Phenotypic ratio*
1. *cc*	*AA, Aa, aa*	*c*	Abino = 4
2. *CC, Cc*	*AA, Aa*	*A*	Agouti = 9 (due to supplementary genes)
3. *CC, Cc*	*aa*	*a*	Black = 3

When a homozygous black (*CC aa*) is crossed with a homozygous albino (*cc AA*) in F_1 all agouti (*Cc Aa*) offsprings appear. When, the F_1 agouti are crossed among themselves in F_2 agouti, black and albino offsprings appear in the ratio of 9 : 3 : 4 as shown in the Fig. 6.5.

P_1: Black *CCaa* × Albino *cc AA*

P_1 gametes: (*Ca*) ↓ (*cA*)

F_1: Agouti *Cc Aa*

F_1 Male gemetes → **F_1 Female gemetes ↓**	*CA*	*Ca*	*cA*	*ca*
CA	*CC AA* Agouti	*CC Aa* Agouti	*Cc AA* Agouti	*Cc Aa* Agouti
Ca	*CC Aa* Agouti	*CC aa* Black	*Cc Aa* Agouti	*Cc aa* Black
cA	*Cc AA* Agouti	*Cc Aa* Agouti	*cc AA* Albino	*cc Aa* Albino
ca	*Cc Aa* Agouti	*Cc aa* Black	*cc Aa* Albino	*cc aa* Albino

F_2:

F_2 Phenotypic ratio: 9/16 Agouti : 3/16 Black : 4/16 Albino.

Fig. 6.5. Checkerboard derived from a cross between coloured (black) and albino mice showing 9 : 3 :4 ratio due to recessive epistatic genes.

Supplementary genes. In the cross between black (*CC aa*) and albino (*cc AA*) mice, one thing becomes apparent that two independent pairs of genes (*i.e.*, *C* – *c* and *A* – *a*) have interacted in the production of the phenotypic trait (*i.e.*, coat colour) in such a way that one dominant (*C*) produces its effect whether or not the second (*A*) is present, but the second (*A*) gene can produce its effect only in the presence of the first. These genes (*i.e.*, *C* and *A*) have been termed as **supplementary genes.**

6.5 DOMINANT AND RECESSIVE INTERACTIONS (13:3)

Sometimes, the dominant alleles of one gene locus (*A*) in homozygous (*AA*) and heterozygous (*Aa*) condition and the homozygous recessive alleles (*bb*) of another gene locus (*B*) produce the same phenotype, the F_2 phenotypic ratio becomes 13 : 3 instead of 9 : 3 : 3 : 1. In such case, the genotype *AA BB, AA Bb, AA bb, Aa BB, Aa Bb* and *Aa bb* produce same phenotype and the genotype *aa BB, aa Bb* and *aa bb* produce another but same phenotype (Table 6.10).

Table 6.10 Mode of interaction between dominant and recessive alleles.

S. No.	Epistatic allele	Hypostatic allele	Phenotypic expression of alleles
1.	*AA, Aa*	*BB, Bb, bb*	*A* inhibits *B* or *b* (No phenotype of *B* or *b*)
2.	*a a*	*BB, Bb, bb*	*a* does not inhibit *B* or *b* (phenotype of *B* or *b*)

Example

In Leghorn type of fowl the white colour of feather is caused by the dominant genotype *CC II*, similarly the white colour of feathers of Plymouth Rock breed is caused by the recessive genotype *cc ii* (Table 6.11). When both white varieties of fowl are crossed, in F_1 white coloured hybrids appear. The F_1 hybrids in F_2 produce the white and coloured offsprings in the ratio of 13 : 3, as have been illustrated in Fig. 6.6.

Table 6.11 Mode of interaction between dominant and recessive genes.

Epistatic alleles		Hypostatic alleles	Phenotypic expression of allele	F_2 Phenotypic ratio
1.	*II, Ii*	*CC, Cc, cc*	*I* (dominant inhibitor)	White = 12
2.	*ii*	*CC, Cc*	*C* (due to recessive inhibitor *i*)	Coloured = 3
3.	*ii*	*cc*	*c* (due to *c* and *i*)	White = 1

P: White Leghorn × White Plymouth Rock

CC II × *cc ii*

↓ ↓ ↓

P gametes: (*C I*) (*c i*)

F_1: White × White

Cc Ii *Cc Ii*

F_1 Female gemetes → *CI* *Ci* *cI* *ci*

F_1 Male gemetes ↓

F_2:

	CI	*Ci*	*cI*	*ci*
CI	*CC II* White	*CC Ii* White	*Cc II* White	*Cc Ii* White
Ci	*CC Ii* White	*CC ii* Coloured	*Cc Ii* White	*Cc ii* Coloured
cI	*Cc II* White	*Cc Ii* White	*cc II* White	*cc Ii* White
ci	*Cc Ii* White	*Cc ii* Coloured	*cc Ii* White	*cc ii* White

F_2 Phenotypic ratio: 13/16 White : 3/16 coloured or 13 : 3

Fig. 6.6. A cross between two white coloured breeds of fowls to get 13 : 3 F_2 dihybrid ratio.

QUESTIONS

Long Answer Questions

1. What is factor hypothesis? Write various types of interactions of genes. Explain with example any one type of interaction of gene.
2. Write an essay on the interaction of genes. (*CCS 2009*)
3. What do you understand by gene interaction? Explain with suitable examples supplementary and complementary genes.
4. What is Interaction of genes? Describe it with a suitable example.
5. What do you understand by intra-allelic and interallelic interactions? Explain the interallelic interaction by taking suitable examples.
6. Explain the inheritance of complementary genes.
7. What does gene interaction means? How does epistasis differ from dominance?
8. What is epistasis of gene? Explain with suitable examples.
9. Explain with the help of suitable examples the modification of Mendelian 9 : 3 : 3 : 1 ratio to (*i*) 9 : 7; (*ii*) 9 : 3 : 4; and (*iii*) 13.3.
10. What do you understand by 'Interaction of genes'? What is the difference between 'Dominance and Epistasis'? Describe with examples.
11. Discuss the different type of gene interactions. (*H.N.B., Garhwal 2008, 2009*)
12. Describe the kinds of epistatic interaction with suitable examples. (*M.S.U. Tamil Nadu 2006*)

Short Answer Questions

1. Write short note on the following:
   (*i*) Epistasis
   (*ii*) Interaction of genes
   (*iii*) Intra-allelic and inter-allelic interaction
   (*iv*) Complementary genes
2. Write differences between following:
   Complementary genes and supplementary genes.

Very Short Answer Questions

1. Who proposed factor hypothesis?
2. Define epistasis.
3. Define hypostatic gene.
4. Name the type of genetic interaction in which action of one non-allelic gene is completed by the other.
5. How does a walnut comb fowl produced in a generation?
6. What phenotypic ratio is found in the F_2 generation in a cross between two varieties having duplicated genes for a single trait?
7. What is the result of cross between two white flowered varieties (*CCpp* × *CcPp*) of *Lathyrus odoratus*?
8. What is complementation?

Fill in the Blanks

1. A gene or locus which tends to suppress or masked the action of a gene at another locus is termed ....
2. The gene locus which is suppressed by a epistatic gene is called ....

3. Dominant epistasis produces a F_2 phenotypic ratio of ....

4. .... epistasis produces a F_2 phenotypic ratio of 9 : 3: 4.

5. A situation where specic alleles at two gene loci are required to generate a particular phenotype is called ....

Multiple Choice Questions

1. The phenomenon in which an allele of one gene suppresses the activity of an allele of another gene, is known as

(*a*) segregation (*b*) dominance

(*c*) epistasis (*d*) suppression

2. The ratio of 9 : 7 is due to

(*a*) lethal genes (*b*) supplementary genes

(*c*) complementary genes (*d*) epistatic gene

3. In the duplicate dominant gene for shape of fruit in shepherd's purse (*Capsella*), the ratio of progeny in F_2 generation is

(*a*) 9 : 3 : 3 : 1 (*b*) 12 : 3 : 1

(*c*) 13 : 1 (*d*) 15 : 1

ANSWERS

Very Short Answer Questions

1. Bateson.

2. Epistasis is masking of the action of alleles of one gene by allelic combinations of another.

3. A gene whose expression is masked by an epistatic gene.

4. Complementary gene interaction.

5. A walnut comb of fowl is produced when the dominant alleles of rose (*R*) and pea (*P*) are present together (*R* – *P* –).

6. 15 : 1.

7. In F_1 generation purple flowered (*CcPp*) variety is produced.

8. Complementation means production of the wild-type phenotypes by a cell or an organism that contains two mutant genes. If complementation occurs, the mutants are almost certainly nonallelic.

Fill in the Blanks

1. Epistatic gene. **2.** Hypostatic gene.

3. 12 : 3 : 1. **4.** Recessive.

5. Complementation.

Multiple Choice Questions

1. (*c*) **2.** (*c*) **3.** (*d*).

7

Sex Linkage (Sex-Linked Inheritance)

In Mendel's crosses we have already observed that the progeny of a cross between two individuals of pure lines is the same regardless of which individual is female or which individual male or we can say that sex makes no difference in Mendel's crosses. But the Mendel's laws are not applicable on those genes which are exclusively located either in X or Y chromosome. It has been observed that the genes occurring only in the X chromosomes are represented twice in female (because female contains 2X chromosomes) and once in male (because male has only one X chromosome). Moreover, if the recessive type of genes occur in X chromosome of males, they express themselves phenotypically. Because in such case Y chromosome contains no dominant allelomorph or gene to overcome the recessive gene of X chromosome. The genes which occur exclusively on the X chromosome (mammals, *Drosophila*, *Melandrium*, etc.) or on the analogous Z chromosome (in birds and other species with ZO or ZW mechanism of sex determination) are called **X- or Z -linked genes**. The genes which exclusively occur in Y chromosome are called **holandric genes**. The inheritance of X- or Z-linked and holandric genes is called **sex-linked inheritance**. In XX-XY type organisms, sex-linked genes (see Fig. 7.1) can be classified into following three types:

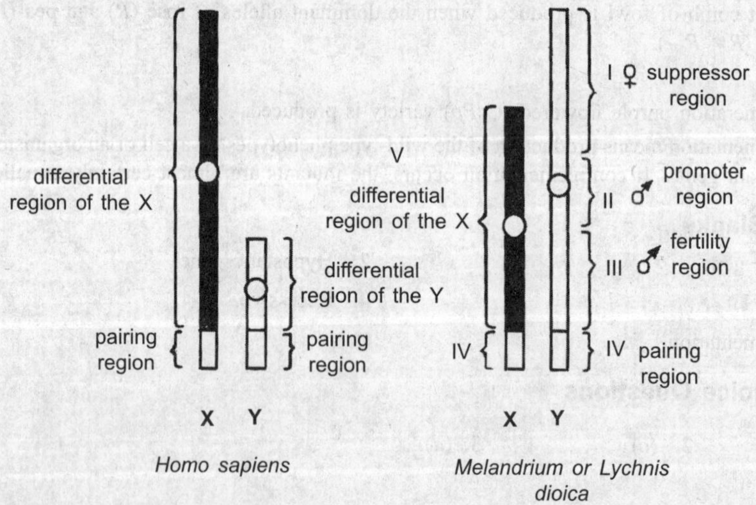

Fig. 7.1. Differential and pairing regions of sex chromosomes of humans and of the plant *Melandrium album*.

A. X-linked. The X-linked type sex-linked inheritance is performed by those genes which are localized in the non-homologous sections of X-chromosome, and that have no corresponding allele in Y chromosome. The X-linked genes are commonly known as **sex-linked genes**.

B. Y-linked. The Y-linked type sex-linked inheritance is performed by those genes which are localized in the non-homologous section of Y chromosome, and that have no alleles in X-chromosome. The Y-linked genes are commonly known as **holandric genes** (Greek, *holos* = whole, and *andros* = man).

C. XY-linked. The XY-linked type sex-linked inheritance is performed by those genes which are localized in homologous sections of X and Y chromosomes.

Box 7.1

The **sex chromosomes** control **sex determination** in animals and are asymmetrically distributed between the sexes. There is a **homogametic sex** which possesses a pair of identical sex chromosomes and thus produces one type of gamete, and **heterogametic sex**, which possesses either a pair of non-identical sex chromosomes or a single unpaired sex-chromosome and produces two types of gametes.

In mammals, females are homogametic: they carry two copies of the X-chromosome where as males carry one X-chromosome and one Y-chromosome and are therefore heterogametic. There, are two short regions of homology between the X-and Y-chromosomes, the **major** and **minor pseudoautosomal regions**, which facilitate pairing during meiosis. The major pseudoautosomal region is the site of an obligatory cross-over and genes located there are inherited in normal autosomal fashion (pseudoautosomal inheritance). Other genes are described as **sex-linked** because their expression depends upon how the sex chromosomes are distributed.

7.1 INHERITANCE OF X-LINKED (SEX-LINKED) GENES

Basic Characteristics of Inheritance of X-linked Recessive Genes

The X-linked genes exhibit following characteristic patterns of inheritance:

1. The differential region of each chromosome (*i.e.*, X) contains genes that have no counterparts on the other kind of sex chromosome. These genes, whether dominant or recessive, show their effects in the male phenotype. Genes in the differential regions are called **hemizygous** ("half-zygous") in the males.

2. The X-linked recessive genes show the following two more peculiar features: **criss-cross** pattern of inheritance (*i.e.*, in criss-cross inheritance, a X-linked recessive gene is transmitted from P_1 male parent (father) to F_2 male progeny (grandsons) through its F_1 heterozygous females (daughters), which are called **carriers** and different F_1 and F_2 results (ratios) in the reciprocal crosses.

3. The X-linked recessives can be detected in human pedigrees (also in *Drosophila*) through the following clues:

(*i*) The X-linked recessive phenotype is usually found more frequently in the male than in the female. This is because an affected female can result only when both mother and father bear the X-linked recessive allele (*e.g.*, $X^AX^a \times X^aY$), whereas an affected male can result when only the mother carries the gene. Further, if the recessive X-linked gene is very rare, almost all observed cases will occur in males.

(*ii*) Usually none of the offspring of an affected male will be affected, but all his daughters will carry the gene in masked hetero-

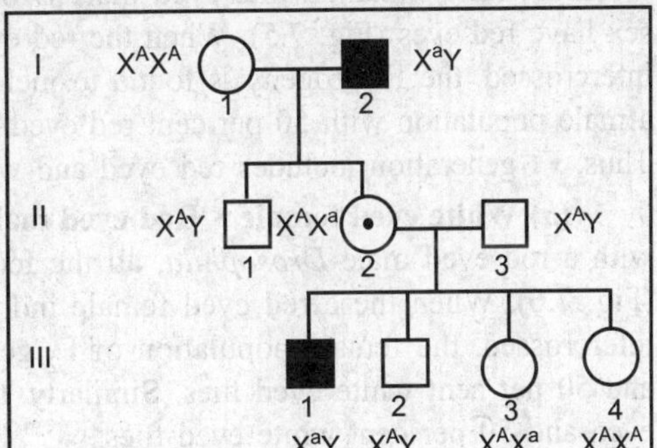

Fig. 7.2. Pedigree showing how X-linked recessive genes are expressed in males, then carried unexpressed by females in the next generation, to be expressed in their sons. III.3 and III.4 heterozygous or carrier females are not distinguished phenotypically.

zygous condition, so one half of their sons (*i.e.*, grandsons of F_1 father) will be affected (Fig. 7.2).

(*iii*) None of the sons of an affected male will inherit the X-linked recessive gene, so not only will they be free of the defective phenotype, but they will not pass the gene along to their offspring.

4. Dominant X-linked genes can be detected in human pedigrees (also in *Drosophila*) through the following clues:

(*i*) It is more frequently found in the female than in the male of the species.

(*ii*) The affected males pass the condition on to all of their daughters but to none of their sons (Fig. 7.3).

(*iii*) Females usually pass the condition (defective phenotype) on to one-half of their sons and daughters (Fig. 7.4).

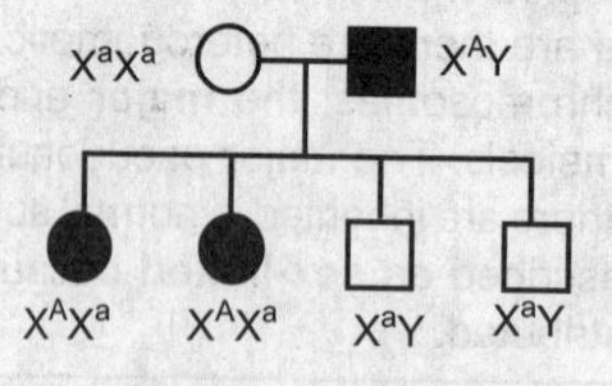

Fig. 7.3. Pedigree chart showing how X-linked dominants are expressed in all the daughters of affected male.

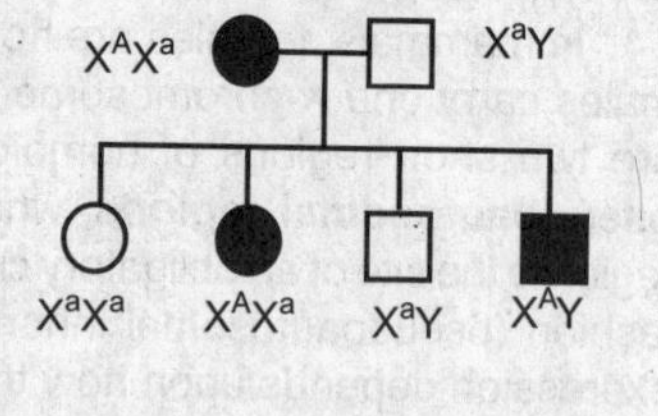

Fig. 7.4. Pedigree chart showing that females affected by a X-linked dominant condition usually are heterozygous and pass the condition to one-half of their progeny.

1. Inheritance of X-Linked Gene for Eye Colour in *Drosophila*

In *Drosophila*, the gene for white eye colour is X-linked and recessive to another X-linked, dominant gene for red-eye colour. It was discovered by **Morgan** in 1910. Following crosses between white eyed and red eyed *Drosophila* will make clear the characteristic criss-cross inheritance of gene for white eyed colour in it:

(*a*) Red eyed female × White eyed male. If a wild red eyed female *Drosophila* is crossed with a mutant white eyed male *Drosophila*, all the F_1 individuals irrespective of their sex have red eyes (Fig. 7.5). When the red eyed male and red eyed female individuals of F_1 are intercrossed, the F_2 progeny is found to include an exclusively red eyed female population and a male population with 50 per cent red eyed individuals and 50 per cent white eyed individuals. Thus, F_2 generation includes red eyed and white eyed individuals in the ratio of 3: 1.

(*b*) White eyed female × Red eyed male. When a white eyed female *Drosophila* is crossed with a red eyed male *Drosophila*, all the female individuals in the F_1 generation are red eyed (Fig. 7.6). When these red eyed female individuals and white eyed male individuals of F_1 are intercrossed, the female population of F_2 generation is found to include 50 per cent red eyed and 50 per cent white eyed flies. Similarly, the male population of F_2 includes 50 per cent red eyed and 50 per cent white eyed flies.

The inheritance of X-linked recessive gene for white eyes can be understood more properly by considering the behaviour of X and Y chromosome, such as follows:

In the cross between red-eyed female and white-eyed male *Drosophila*, the red-eyed female contains the gene ++ for red colour of eye. The white-eyed male contains a single gene *w* for white colour of eye which remains located in X chromosome. This one allelic condition of male is termed **hemizygous** condition in contrast to the homozygous or heterozygous possibilities in the female. The female being homogametic produces only one type of gametes or eggs each with the gene +

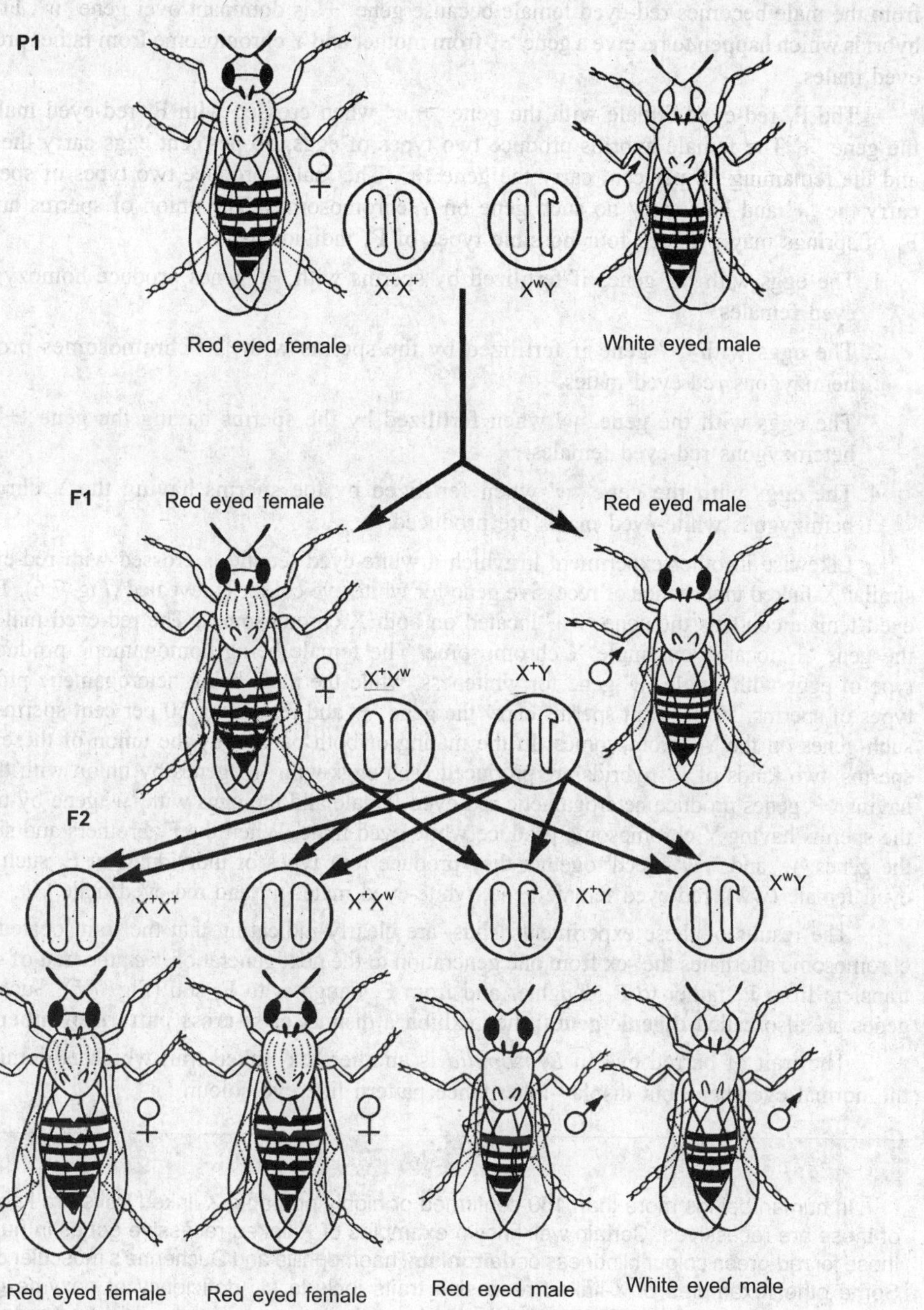

Fig. 7.5. A cross between red-eyed female and white-eyed male showing sex-linked inheritance in *Drosophila*.

for red coloured eyes. The male being heterogametic produces two types of gametes, 50 per cent sperms with X chromosome containing *w* gene, 50 per cent sperms with 'Y' chromosome without *w* or + gene. The gametes of both parents unite in fertilization to produce F_1 progeny. The F_1 hybrids which receive a X-chromosome with + gene from the female and a X-chromosome with *w* gene from the male becomes red-eyed female because gene '+' is dominant over gene '*w*'. Further the hybrids which happen to receive a gene '+' from mother and Y chromosome from father produce red eyed males.

The F_1 red-eyed female with the gene '+ *w*' when crossed with F_1 red-eyed male having the gene '+' The female hybrids produce two types of eggs, 50 per cent eggs carry the gene '+' and the remaining 50 per cent carry the gene '*w*'. The males produce two types of sperms, half carry the '+' and half carry no such gene on Y chromosomes. The union of sperms and ova of F_1 offsprings may produce four possible types of F_2 individuals:

1. The eggs with '+' genes if fertilized by sperms with '+' genes produce homozygous red-eyed females.
2. The eggs with '+' gene if fertilized by the sperms with 'Y' chromosomes produce the hemizygous red-eyed males.
3. The eggs with the gene '*w*' when fertilized by the sperms having the gene '+' produce heterozygous red-eyed females.
4. The eggs with the gene '*w*' when fertilized by the sperms having the Y chromosome hemizygous white-eyed males are produced.

Likewise in other experiment in which a white-eyed female is crossed with red-eyed male, similar X-linked inheritance of recessive gene for white eye colour is revealed (Fig. 7.6). The white-eyed female contains the gene '*ww*' located on both X chromosomes. The red-eyed male contains the gene '+' located on single X chromosome. The female being homogametic produces single type of eggs with single '*w*' gene for whiteness, while the male being heterogametic produce two types of sperms, 50 per cent sperms carry the gene '+' and remaining 50 per cent sperms carry no such genes on the Y chromosomes. In the mating of both parents by the union of these eggs and sperms, two kinds of F_1 hybrids are produced. The eggs with '*w*' genes by union with the sperms having '+' genes produce heterogametic red-eyed female and the eggs with '*w*' gene by union with the sperms having Y chromosome produce white-eyed male. When the F_1 brothers and sisters with the genes '*w*' and '+*w*' breed together they produce four types of individuals in F_2 such as white-eyed female (ww), red-eyed female (+*w*), white-eyed male (*w*) and red-eyed male (+).

The results of these experiments, thus, are clearly indicating that the trait located on a sex chromosome alternates the sex from one generation to the next generation, *i.e.*, the trait of white eyes transfers from P_1 father to F_1 daughter and from F_1 daughter to F_2 son (Fig. 7.5). Such X-linked genes are also called **digenic genes**; they exhibit a distinct **criss-cross pattern of inheritance**.

The trait of barred eye in *Drosophila* is another sex-linked trait which is dominant over the normal eye shape but display inheritance pattern like eye-colour.

Box 7.2

In human beings more than 150 confirmed or highly probable X-linked traits are known; most of these are recessives. Certain well known examples of X-linked recessive genes in humans are those for red-green colour blindness or **daltonism**, haemophilia and Duchenne's muscular dystrophy. Some other examples of X-linked recessive traits include (1) deficiency of enzyme glucose-6 phosphate dehydrogenase (G6PD deficiency) in erythrocytes causing haemolytic anaemia during allergy reaction of persons for the drugs such as sulphonamides or for the broad bean (*Vicia faba*), called **favism;** (2) two forms of diabetes insipidus; (3) one form of anhidrotic ectodermal dysplasia (absence of sweat glands and teeth); (4) absence of central incisors; (5) certain forms of deafness; (6) spastic parapelagia (*i.e.* tetanoid or partial paralysis of lower extremities with increased irritability and spasmodic contraction of the muscles); (7) uncontrollable rolling of the eye balls (nystagmus); (8) a form of cataract; (9) night blindness; (10) optic atrophy; (11) juvenile glaucoma (hardening of eye ball); (12) juvenile muscular dystrophy and (13) white frontal patch of hair.

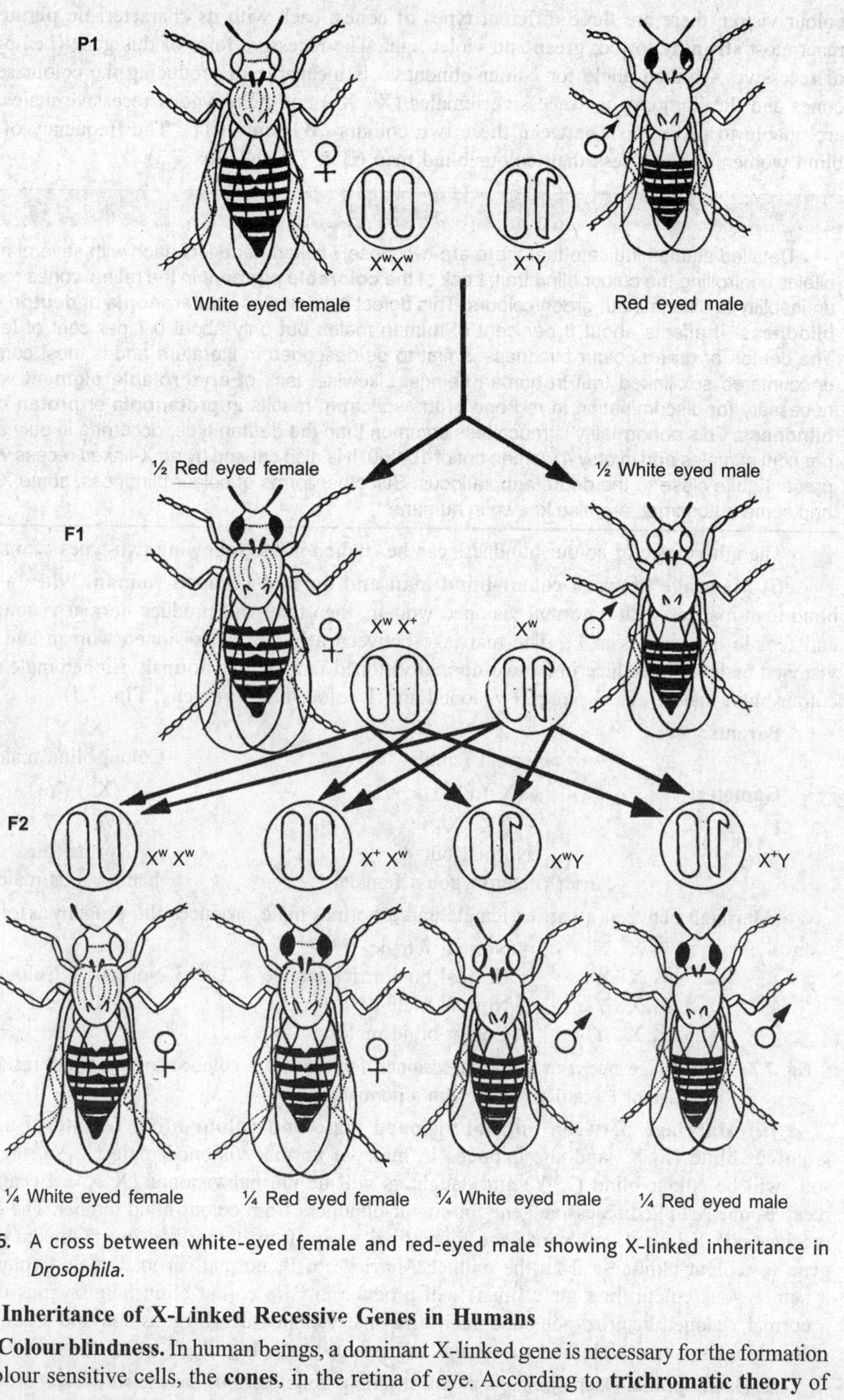

Fig. 7.6. A cross between white-eyed female and red-eyed male showing X-linked inheritance in *Drosophila*.

2. Inheritance of X-Linked Recessive Genes in Humans

1. Colour blindness. In human beings, a dominant X-linked gene is necessary for the formation of the colour sensitive cells, the **cones**, in the retina of eye. According to **trichromatic theory** of colour vision, there are three different types of cones, each with its characteristic **pigment** that react most **strongly** to red, green and violet light. The recessive form of this gene (*i.e.*, presence of recessive X-linked allele for colour blindness) is incapable of producing the colour sensitive cones and the homozygous recessive females ($X^C X^C$) and hemizygous recessive males (X^CY) are unable to distinguish between these two colours (**Wilson,** 1911). The frequency of colour blind women is much less than colour blind man (Box 7.2 and Box 7.3).

Box 7.3

Detailed studies indicate that there are two closely linked gene loci each with several multiple alleles controlling the colour blind trait. Lack of the **colorable pigment** in the retinal cones results in an inability to discriminate green colours. This defect is known as **deuteranopia** or **deuton colour blindness**. It affects about 8 per cent of human males but only about 0.7 per cent of females. The deuton or green colour blindness is first to be described in literature and is most commonly encountered sex-linked trait in human beings. Likewise, lack of **erythrolable pigment** which is necessary for discrimination in red end of the spectrum results in **protanopia** or **protan colour-blindness**. This abnormality is much less common than the deutan type, occurring in only about 2 per cent of males and in only 4 women out of 10,000. It is also caused by an X-linked recessive gene present quite close to the deuteranopia locus. Still other forms of colour blindness, some X-linked and some autosomal, are also known in humans.

The inheritance of colour blindness can be studied in the following two types of marriages:

***(i)* Marriage between colour-blind man and normal visioned woman.** When a colour-blind man marries with a normal visioned woman, then they will produce normal visioned male and female individuals in F_1. The marriage between a F_1 normal visioned woman and normal visioned male will produce in F_2 two normal visioned female, one normal visioned male and one colour-blind male (*i.e.,* 3 Normal visioned and 1 colour-blind progeny; Fig. 7.7).

Parents:	$X^+ X^+$ Normal female	×	$X^C Y$ Colour-blind male
Gametes:	(X^+) (X^+)		(X^C) (Y)
F_1:	½ $X^+ X^C$ Normal but carrier (heterozygous) female		½ $X^+ Y$ Normal but hemizygous male

(Marriage between a carrier female and a normal male produces the progeny as follows—)

F_2:	¼ $X^+ X^+$	= Normal female	3 Normal: 1 Colour-blind or 3 : 1
	¼ $X^+ X^C$	= Normal but carrier female	
	¼ $X^+ Y$	= Normal male	
	¼ $X^C Y$	= Colour-blind male	

Fig. 7.7. A marriage between a normal visioned female and a colour-blind man and result of the marriage of F_1 carrier female with a normal male.

***(ii)* Marriage between normal visioned male and colour-blind female.** If a woman is colour blind (X^CX^C)and she happens to marry a normal visioned male (X^+Y), then all F_1 sons will be colour blind (X^CY) and daughters will be normal visioned (X^+X^C). Because male receives one X-linked recessive gene for colour-blindness from colour-blind mother. The daughter receives one X-linked dominant gene for normal vision from father and one X-linked recessive gene for colour-blindness from the mother. Marriage of F_1 normal carrier female (women) with a hemizygous colourblind male (man) will produce in F_2 a colour-blind homozygous daughter, a normal visioned heterozygous daughter, a normal visioned hemizygous son and a hemizygous colour-blind son (*i.e.,* 1:1 ratio of normal and colour-blind; Fig. 7.8).

2. Haemophilia. Haemophilia is the most serious and notorious disease which is more common in men than women. This is also known as bleeder's disease. The person which contains the recessive

Parents:	$X^C X^C$ Colour-blind female	×	$X^+ Y$ Normal male
Gametes:	$(X^C)(X^C)$		$(X^+)(Y)$
F_1:	$\frac{1}{2} X^+ X^C$ Normal but carrier (heterozygous) female		$\frac{1}{2} X^C Y$ Colour-blind but hemizygous male

(Marriage of F_1 carrier female with a colour blind male produces the progeny as follows–)

F_2:	$\frac{1}{4} X^+ X^C$	= Normal but carrier female	2 Normal: 2 Colour-blind or 1 : 1
	$\frac{1}{4} X^C X^C$	= Colour-blind female	
	$\frac{1}{4} X^+ Y$	= Normal male	
	$\frac{1}{4} X^C Y$	= Colour-blind male	

Fig. 7.8. A marriage between a colour-blind female and a normal male and result of the marriage of F_1 carrier female with a colour-blind male.

gene for haemophilia lacks in normal clotting substance (**thromboplastin**) in blood so minor injuries cause continuous bleeding and ultimate death of the person due to haemorrhages. This hereditary disease was reported by **John Cotto** of Philadelphia in 1803 in man. Recently, two types of X-linked haemophilia have been recognized:

(*a*) Haemophilia A. It is characterized by lack of antihaemophilic globulin (Factor VIII). About four fifths of the cases of haemophilic are of this type.

(*b*) Haemophilia B. It is also called "**christmas disease**" after the family in which it was first described in detail. Haemophilia B results from a defect in plasma thromboplastic component (factor IX). This is milder form of haemophilia.

Woodliff and **Jackson** in 1966 have found the occurrence of both types of haemophilia in an unusual family, both segregating independently; therefore, they concluded that the two loci (*i.e.*, haemophilia A and haemophilia B) were far apart on the X chromosome (*i.e.*, having a distance of more than 40 map units between them).

Haemophilia is well known in the royal families of Europe, where it is traceable to Queen Victoria of England, who must have been heterozygous (carrier). No haemophiliac is known in her ancestry; hence, it is concluded that her haemophilia allele arose from a mutant gamete. Since it is caused by recessive X-linked gene, a lady may carry the disease (*i.e.*, she is carrier but nonsufferer and would transmit it to 50 per cent of her sons, even if the father is normal (Fig. 7.9).

Parents:	$X^+ X^h$ Normal (Carrier) mother	×	$X^+ Y$ Normal father
Gametes:	$(X^+)(X^h)$		$(X^+)(Y)$

Progeny:	$\frac{1}{4} X^+ X^+$	= Normal daughter	3 Normal: 1 Haemophilic or 3 : 1
	$\frac{1}{4} X^+ X^h$	= Normal but carrier daughter	
	$\frac{1}{4} X^+ Y$	= Normal son	
	$\frac{1}{4} X^h Y$	= Haemophilic son	

Fig. 7.9. Inheritance of haemophilia disease in human beings.

3. Inheritance for Z-linked gene for coloured (black or red) feathers in chicken. In case of birds, moths, butterflies, etc., the females are heterogametic and males are homogametic quite unlike that of *Drosophila* and man. Here also sex-linked genes follows the "crisscross" pattern but from mother through heterozygous F_1 sons to grand-daughters of F_2. The common example of ZW-ZZ sex-linkage is plymouth rock chicken.

In Plymouth rock chicken the gene for barred feathers is dominant and the gene for black or red unbarred feathers is recessive. Both the genes are Z-linked. A barred male chicken contains two genes for barring bacause it has two sex chromosomes (ZZ). When this barred male with the gene BB is crossed with unbarred female containing single recessive gene 'b' in its Z chromosome (W chromosome contains no genes), produce in F_1 only barred males and females. (see **Sinnott**, **Dunn** and **Dobzhansky**, 1958; Fig. 7.10). These F_1 barred males and females when inbred produce in F_2 a hemizygous non-barred female, a hemizygous barred female, a heterozygous barred male and a homozygous barred male (*i.e.,* a ration of 3:1).

P_1. Nonbarred female (Hen) × Barred male (cock)
(Z^bW) ($Z^B Z^B$)

P_1 gametes. (Z^b) (W) (Z^B) (Z^B)

F_1. $Z^b Z^B$: Z^B W
Barred Male Nonbarred female

F_2.

Sperms ↓ / Ova →	Z^B	W
Z^b	$Z^b Z^B$ Barred Male	Z^b W Nonbarred female
Z^B	$Z^B Z^B$ Barred Male	Z^b W Barred Female

F_2 ratio: 3 Barred : 1 Nonbarred or 3 : 1.

Fig. 7.10. A classical cross between a barred cock and a non-barred hen showing Z-linked inheritance (3 : 1 F_2 ratio).

In another cross (Fig. 7.11) when barred hens and non-barred cocks are crossed, the F_1 has half barred male and half non-barred females. These by inbreding produce in F_2 half barred males and females and half non-barred males and females (*i.e.,* a ration of 1:1).

Besides fowl, many birds, *e.g.,* pigeon, duck, canary, moths, butterflies, all reptiles, some fishes and amphibians which have XX male and XY or XO female type of sexes showing sex-linked inheritance of (ZW-ZZ) type.

P_1. Barred female (Hen) × Nonbarred (Black/Red) male (cock)
(Z^BW) ($Z^b Z^b$)

P_1 gametes. (Z^B) (W) (Z^b) (Z^b)
ova sperms

F_1. W Z^b × $Z^B Z^b$
Nonbarred female Barred male

F_2.

Sperms ↓ / Ova →	W	Z^b
Z^B	WZ^B Barred female	$Z^B Z^b$ Barred male
Z^b	W Z^b Nonbarred female	$Z^b Z^b$ Nonbarred male

F_2 ratio: 2 Barred individuals : 2 Nonbarred individuals or 1: 1

Fig. 7.11. A cross between a barred hen and a nonbarred cock showing sex-linked (Z-linked) inheritance (1 : 1 F_2 ratio).

7.2 HOLANDRIC GENES

Genes in the non-homologous region of the Y chromosome pass directly from male to male (Fig. 7.12). In man, the Y-linked or holandric genes such as **ichthyosis hystrix gravis hypertrichosis** (excessive development of hairs on pinna of ear) are transmitted directly from father to son. Recently, certain other holandric genes have been reported in humans, *e.g.*, genes for H-Y antigen, histocompatibility antigen, spermatogenesis, height (stature) and slower maturation of individual. The Y-linked gene of H-Y antigen is located on short arm of Y chromosome, while genes controlling spermatogenesis occur on the long arm of Y chromosome. Y chromosome of *Drosophila* too contain Y-linked gene for male fertility.

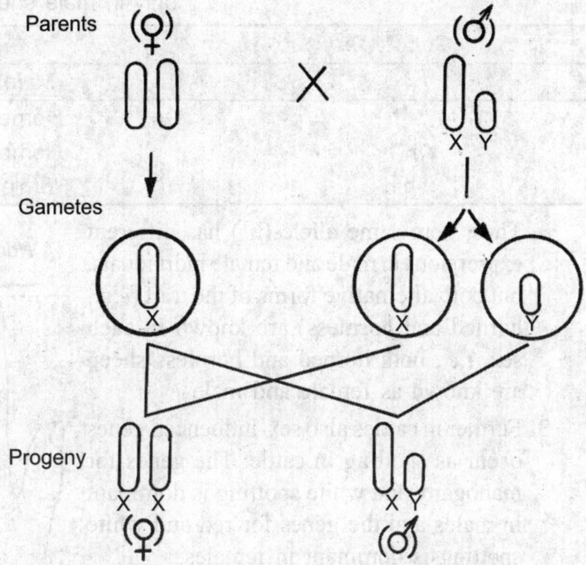

Fig. 7.12. Mode of inheritance of Y-linked genes from father to son.

7.3 SEX INFLUENCED GENES

Sex influenced genes are those whose dominance is influenced by the sex of the bearer. Thus, male and female individuals may be similar for a particular trait but give different phenotypic expressions of the same trait.

Table 7.1 In heterozygotic human females (Bb) dominant allele B for baldness fail to make its phenotypic expression.

Genotype	Phenotypes	
	Men	Women
BB	Bald	Bald
Bb	Bald	Non-bald
bb	Non-bald	Non-bald

Examples.

1. In man the baldness may occur due to disease, radiation or thyroid defects but in some families baldness is found to be inherited trait. In such inherited baldness the hairs gradually become thin on head top, leaving ultimately a fringe of hair low on the head and commonly known as **pattern baldness**. The gene B for baldness is found to be dominant in males and recessive in females (Table 7.1). In heterozygous condition it expresses itself only in the presence of male hormones (in male sex).
2. In sheep, the genes for the development of horns is dominant in males and recessive in female (Table 7.2).

Table 7.2 Phenotypic expression of dominant sex-influenced gene for horned trait occurs only in male sheep.

Genotype	***Phenotypes***	
	Males	***Females***
h^+h^+	Horned	Horned
h^+h	Horned	Hornless
h h	Hornless	Hornless

Thus, here same allele (h^+) has different expressions in male and female individuals, but both alternative forms of the trait (*e.g.*, horned and hornless) are known in each sex, *i.e.*, both horned and hornless sheep are known as female and male.

3. Further in cattles also sex-influenced genes occur as spotting in cattle. The genes for mahogany and white spotting is dominant in males and the genes for red and white spotting is dominant in females.
4. **Length of index finger in human.** A pair of sex-influenced genes control the length of index finger in humans. Index finger is mostly long in females and short in males in comparison to ring finger (Fig. 7.13).

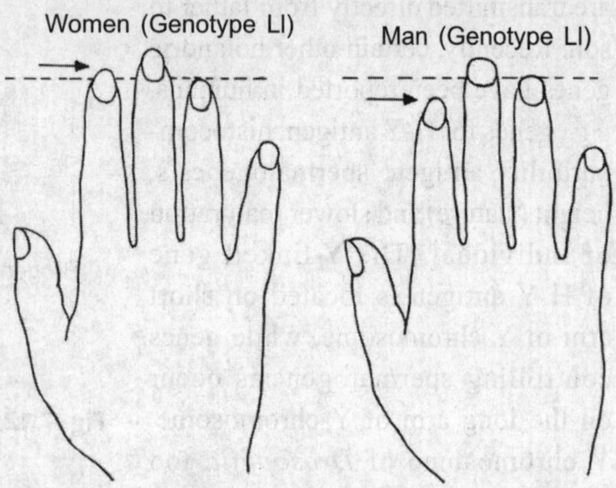

Fig. 7.13. Length of index finger in human female and human male.

7.4 SEX-LIMITED GENES

Sex-limited genes are autosomal genes whose phenotypic expression is determined by the presence or absence of one of the sex hormones. Their phenotypic effect is limited to one sex or other. In other words, the penetrance of a sex-limited gene in one sex remain zero. The sex-limited genes are mainly responsible for secondary sex characters in cattle, humans and fowl.

Examples

1. The bulls have genes for milk production which they transmit to their daughters, but they or their sons are unable to express this trait. The production of milk is, therefore, limited to variable expression only in the female sex.
2. Beard development in human beings is a sex limited trait as men normally have beards, whereas women normally do not. Likewise, the genes for male voice, body hair and physique are autosomal in human beings, but they are expressed only in the presence of androgens which are absent in females.
3. In chicken the recessive gene (*h*) for cock feathering is male sex-limited (*i.e.*, it is penetrant only in male environment).

Genotype	***Phenotypes***	
	Males	***Females***
HH	Hen-feathering	Hen-feathering
Hh	Hen-feathering	Hen-feathering
hh	Cock-feathering	Hen-feathering

7.5 NON-DISJUCTION

The failure of homologs (at meiosis) or sister chromatids (at mitosis) to separate properly to opposite poles is called **non-disjunction**. The sex chromosomes pass through the phenomenon of non-disjunction and present various interesting situations. The non-disjunction may be of following types:

1. Primary Non-disjunction.

The phenomenon of primary non-disjunction was discovered by **C.B. Bridges** (1916) during his classical matings of *Drosophila melanogaster*. Let us consider one of his classical crosses as follows:

The gene for wild type red eyes (+) is carried by the X chromosome; a recessive allele (v) produces vermilion eyes in homozygous females (X^VX^V) and in all males (X^VY). Normally vermilion-eyed females mated to red-eyed males produce only red-eyed daughters and vermilion eyed sons (Fig. 7.14).

Parents:		$X^V X^V$ Vermilion female (or mother)	×	$X^+ Y$ Red male (or father)
Gametes:	**Eggs:**	(X^V)		(X^V)
	Sperms:	(X^+)		(Y)
F_1:	½ $X^+ X^V$ Red female (or daughter)	:		½ $X^V Y$ Vermilion male (or son)

Fig. 7.14. Results of a normal cross between vermilion-eyed female and red-eyed male *Drosophila*.

Crosses of this type, however, in rare cases produce unexpected vermilion-eyed daughters and red-eyed sons with a frequency of one per 2,000 to 3,000 offspring. **Bridges** speculated that these unusual progeny are due to a failure of the X chromosomes in an XX female to disjoin during meiosis I of oogenesis. This phenomenon was termed **primary disjunction** by Bridges and this event tends to produce three types of eggs: the majority of eggs with normal single X chromosome and a small number of eggs with their two X chromosomes or no X at all. These eggs on normal fertilization produce the following results:

Parents:		$AAX^V X^V$ Vermilion female		×		$AAX^+ Y$ Red male
Gametes:	**Eggs:**	(AX^V) (numerous)	+	$(AX^V\text{-}X^V)$ (rare due to non-disjunction)	+	(AO) (rare called **nullo**)
	Sperms:	(AX^+)		+		(AY)

F_1:
1. $AA\,X^+X^v$ = Red female (numerous; normal)
2. $AA\,X^+X^V\text{-}X^V$ = Red "metafemales" (rare; die)
3. $AA\,X^+O$ = Red male (rare; sterile)
4. $AA\,X^VY$ = Vermilion male (numerous; normal)
5. $AA\,X^V\text{-}X^VY$ = Vermilion female (rare: fertile)
6. AAOY = Die (rare)

Fig. 7.15. Inheritance of eye colour in *Drosophila* due to primary non-disjuction of sex chromosomes.

The **metafemales** (AA XX-X) are weak and seldom live beyond the pupal stage; the AAOY individuals die in the egg stage. Note that the AA X-XY female indicate that the presence of a Y chromosome does not determine maleness itself, though males without it (AAXO) are sterile. From these results three facts became established: 1. In *Drosophila* sex determination occurs according to genetic balance theory of Bridges. 2. In *Drosophila*, Y chromosome contains genes for the spermatogenesis. 3. It provides strong support to chromosome theory of inheritance (*i.e.*, Mendelian factors or genes are carried on the chromosomes).

2. Secondary Non-disjunction.

In next cross, **Bridges** mated the exceptional vermilion-eyed females (AAX^V-X^VY) that arose as a result of primary non-disjunction to normal red-eyed males (AAX^+Y). The progeny of this cross included 96 per cent female with red eyes and only 4 per cent females had vermilion eyes indicating that there was 4 per cent secondary non-disjunction (Fig. 7.16).

Thus, primary non-disjunction may occur in either XX females or XY males. In the former it leads to the production of XX and O eggs. Occurrence in the first meiotic division of male produces XY and O sperms results. Secondary non-disjunction, on the other hand, occurs in XXY females, where it gives rise to XX, XY, X and Y eggs (Fig. 7.17). As the term non-disjunction implies, these aberrant gametes are produced only as a result of failure of the sex chromosomes to disjoin after synapsis; they are not physically attached.

Similar results were obtained by **Bridges** (1916) from the cross of red eyed male and white eyed female *Drosophila*.

Parents: AAX^+X^V-X^VY (Vermillion females) × $AA\,X^+Y$ (Red males)

Gametes:

Egg: $(AX^V) + (AX^VY) + (AX^VX^V)$ [Due to normal disjunction] $+ (AY) + (AX^V) + (AX^VY)$ [Due to secondary non-disjunction]

Sperms: $(AX^+) + (AY)$

F_2:

			Sperms		
			AX^+(50%)	AY (50%)	
Eggs		AX^VX^V (4%)	$AAX^+X^VX^V$ Metafemale (die) (2%)	$AAX^v\,X^vY$ Vermilion female (2%)	4% secondary exceptional progeny phenotype
	X-Y pairing (16%)	AY (4%)	AAX^+Y Red male (2%)	AAYY (die) (2%)	96% regular (*i.e.*, expected) progeny phenotypes
		AX^V (4%)	AAX^+X^V Red female (2%)	$AA\,X^VY$ Vermilion male (2%)	
		$AX^V\,Y$ (4%)	AAX^+X^VY Red female (2%)	$AAX^VY\,Y$ Vermilion male (2%)	
	X-X pairing (84%)	AX^V (42%)	AAX^+X^V Red female (21%)	AAX^VY Vermilion male (21%)	
		AX^VY (42%)	AAX^+X^VY Red female (21%)	AAX^VYY Vermilion male (21%)	

Fig. 7.16. Result obtained from secondary non-disjuction.

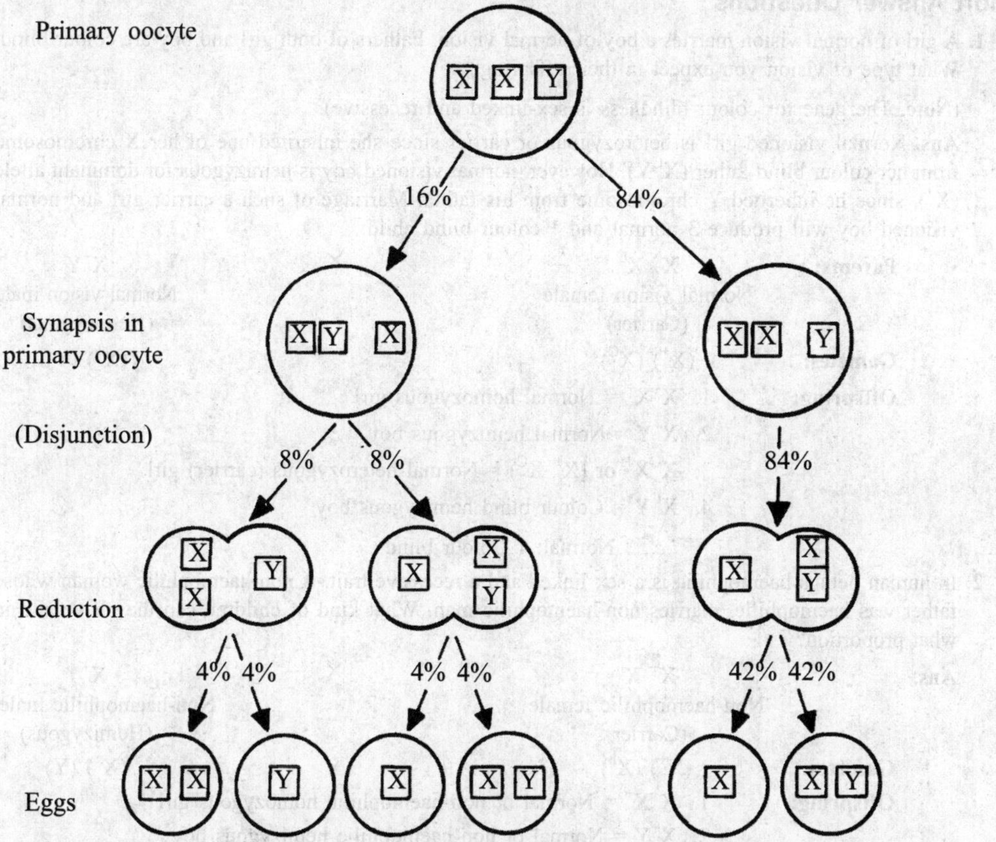

Fig. 7.17. Secondary non-disjunction in X-XY female *Drosophila*, resulting in 46 per cent X, 46 per cent XY, 4 per cent XX, and 4 per cent Y eggs.

QUESTIONS

Long Answer Questions

1. Write an essay on sex-linked inheritance.
2. What is sex-linked inheritance? Illustrate sex-linked inheritance with suitable examples.
3. What is sex-linked inheritance? Explain this phenomenon with reference to human beings or *Drosophila* giving suitable examples.
4. Describe sex-linked inheritance in the fruitfly, *Drosophila*.
5. Give an account of the chromosomal basis of sex-linked inheritance in *Drosophila* and human beings.
6. What is sex-linked inheritance? Explain the inheritance pattern of haemophilia/red green blindness in human beings.
7. What is criss-cross inheritance? Describe mechanism of inheritance of haemophilia in human beings.
8. What do you mean by sex-linked inheritance? Describe the role of genes with reference of haemophilia and colour blindness.
9. Colour blindness is found more often in man than in women. Give reasons.
10. What is sex linked inheritance? Explain with suitable examples in human beings. (*C.C.S.U., Meerut 2008*)
11. What is sex linked inheritance? Explain it with reference to man. (*H.N.B. Garhwal 2009*)

Short Answer Questions

1. A girl of normal vision marries a boy of normal vision. Fathers of both girl and boy are colour blind. What type of vision you expect in their offspring?

(**Note.** The gene for colour blindness is sex-linked and recessive).

Ans. Normal visioned girl is heterozygous or carrier since she inherited one of her X chromosome from her colour blind father (X^CY). However, normal visioned boy is hemizygous for dominant allele (X^+), since he inherited Y chromosome from his father. Marriage of such a carrier girl and normal visioned boy will produce 3 normal and 1 colour blind child:

Parents: $X^+ X^C$ Normal vision female (Carrier) × X^+Y Normal vision male (Hemizygous)

Gametes: (X^+) (X^C) — (X^+) (Y)

Offspring:
1. X^+X^+ = Normal homozygous girl
2. X^+Y = Normal hemizygous boy
3. X^CX^+ or ($X^+ X^C$) = Normal heterozygous (carrier) girl
4. X^CY = Colour blind hemizygous boy

i.e., 3 Normal: 1 Colour blind.

2. In human beings haemophilia is a sex linked and a recessive trait. A non-haemophilic woman whose father was haemophilic, marries non-haemophilic man. What kind of children can they have and in what proportion?

Ans. $X^+ X^h$ Non-haemophilic female (Carrier) × X^+Y Non-haemophilic male (Hemizygous)

Gametes: (X^+) (X^h) × (X^+) (Y)

Offspring:
1. X^+X^+ = Normal or non-haemophilic homozygous girl
2. X^+Y = Normal or non-haemophilic hemizygous boy
3. X^+X^h = Normal or non-haemophilic carrier (heterozygous) girl
4. X^hY = Haemophilic hemizygous boy

or 3 Non-haemophilic : 1 Haemophilic.

3. Can a normal son have a colour-blind mother? Explain in only five sentence.

Ans. A normal son (X^+Y) gets his X-linked dominant allele for normal vision from his mother. A colour blind mother would have two homozygous recessive alleles for the colour blindness (*i.e.*, X^CX^C), Therefore mother of a normal visioned son is always normal visioned (*i.e.*, X^+X^+ or X^+X^C) never colourblind (X^CX^C).

4. Describe basic characteristics of inheritance in sex-linked characters.

5. Discuss the inheritance of Y-linked genes.

6. What are holandric genes? How does their inheritance differ from the inheritance of digenic genes? Describe with suitable examples, the inheritance of digenic genes in man.

7. Write a short note on non-disjunction.

8. What is non-disjunction? Explain this phenomenon with suitable examples.

9. Give an account of sex-influenced characters.

10. Give a brief account of sex-limited genes.

11. How colour blindness is inherited in man? (*M.S.U. Tamil Nadu 2006*)

12. Describe the sex linked inheritance with suitable examples. (*C.C.S.U., Meerut 2009*)

13. Describe the sex-linked inheritance in *Drosophila*. (*C.C.S.U., Meerut, 2007*)

14. Describe the sex-linked inheritance and genetic disorder in human. (*H.N.B., Garwal, 2008*)

15. Write short notes on the following:

(*i*) Colour blindness

(*ii*) Haemophilia

(*iii*) Sex-linked genes

or

Sex-linked inheritance

or

Sex-linked characters

(*iv*) Carrier

(*v*) Sex-linked inheritance in human beings

(*vi*) Sex-linked and sex-limited genes

16. Differentiate between X-linked and sex-limited characters.

17. Show diagrammatically the inheritance of eye colour in *Drosophila.*

Very Short Answer Questions

1. What do you mean by sex-chromosome?

2. What do you mean by sex-linked genes?

3. Define holandric inheritance.

4. Define hologenic inheritance.

5. What is digenic sex linked character?

6. What do you mean by diandric sex-linked inheritance?

7. A son cannot receive X-linked gene directly from his father, why?

8. What is sex-limited trait?

9. Define sex-influenced traits.

Yes or No Question

1. Muller propounded the chromosome theory of heredity.

Fill in the Blanks

1. Haemophilia is an example of a human trait with .... pattern of inheritance.

2. In *Drorophila*, the recessive X-linked gene for white eyes was discovered by .... in 1910.

3. The X-linked or sex-linked genes are .... genes and exhibit .... pattern of inheritance.

4. .... deficiency in erythrocytes of human beings causes .... during allergy reaction.

5. The genes which exclusively occur in Y chromosomes of human beings are called .... genes.

6. Genes in differential region of X chromosome of male *Drosophila* are .... since they are present is 50 per cent of sperms.

Multiple Choice Questions

1. A gene present only in one dose on the X-chromosome in heterogametic males is called

(*a*) heterozygous (*b*) hemizygous

(*c*) homozygous (*d*) none of the above

(*M.S.U., Tamil Nadu 2006*)

2. The number of Barr body in XXY men is

(*a*) one (*b*) two

(*c*) none (*d*) not known

3. Which of the following is sex linked?
   (*a*) sickle-cell anaemia
   (*b*) albinism
   (*c*) haemophilia
   (*d*) phenylketonuria
4. In protanopia, a person cannot distinguish
   (*a*) green colour
   (*b*) red colour
   (*c*) blue colour
   (*d*) blue and green colour
5. According to **Gardner** *et al.,* (1981), in humans, only two Y-linked genes have been discovered till date. These are
   (*a*) *H-Y* and *SRY*
   (*b*) *H-Y* and hairy ear pinna
   (*c*) *SRY* and hairy ear pinna
   (*d*) *H-Y, SRY* and hairy ear pinna
6. Sex linked characters are always
   (*a*) lethal
   (*b*) recessive
   (*c*) dominant
   (*d*) pleiotropic
7. Colour blindness, in which all colours are perceived as gray, is termed as
   (*a*) chromasia
   (*b*) dichromasia
   (*c*) monochromasia
   (*d*) all of these
8. Moustaches and beared in human males are example of
   (*a*) sex-linked traits
   (*b*) sex-limited traits
   (*c*) sex-differentiating traits
   (*d*) sex-determining traits
9. Deuteronopia is a disease when there is no perception of ________ colour
   (*a*) red
   (*b*) blue
   (*c*) green
   (*d*) white
10. One of the following is not true of haemophilia
   (*a*) royal disease
   (*b*) bleeder's disease
   (*c*) X-linked disease
   (*d*) more common in males than in females
   (*e*) Y-linked disease

ANSWERS

Very Short Answer Questions

1. Chromosomes which differ in two sexes (male and female sexes).
2. Genes located on sex-chromosomes, X and Y are called sex-linked genes. These genes are located on non-homologous parts of two chromosomes.
3. Characters passes from father to son directly (*e.g.*, Y-linked genes in human male).
4. Character goes directly from mother to daughter (female to female, *e.g.*, W-linked genes in birds).
5. When a character passes from male to male through female (*e.g.*, X-linked recessive gene such as allele for haemophilia in humans).
6. When a character passes from female to female through male (*e.g.*, Barred plumage in poultry).
7. Son inherits only Y-chromosome from his father.
8. In case of sex-limited trait expression of a trait is only in one sex. Examples–milk production in mammals; horns in Rambouiltet sheep, egg production in chicken. Sex-limited traits may be controlled by sex-linked or autosomal loci.
9. Sex-limited traits are also called **sex-conditioned traits**. These are traits that appear more often in one sex than in another.

Yes or No Question

1. No.

Fill in the Blanks

1. Criss-cross.

2. T.H. Morgan.

3. Digenic, criss-cross.

4. G6PD, haemolytic anaemia.

5. Holandric.

6. Hemizygous.

Multiple Choice Questions

1. (*b*) **2.** (*a*) **3.** (*c*) **4.** (*b*) **5.** (*a*) **6.** (*b*)
7. (*c*) **8.** (*b*) **9.** (*c*) **10.** (*e*).

8

Linkage

The hereditary units or genes which determine the characters of an individual are carried in the chromosomes and an individual usually has many genes for the determination of various different characters. As there are more genes than the chromosomes, it can be expected that each chromosome contains more than one gene. The genes for different characters may be either situated in the same chromosome or in different chromosomes. When the genes are situated in different chromosomes, the characters they control appear in the next generation either together or apart, depending on the chance alone. They assort independently according to Mendel's law of independent assortment. But if the genes are situated in the same chromosome and are fairly close to each other, they tend to be inherited together. This type of coexistence of two or more genes in the same chromosome is known as **linkage**.

8.1 DIFFERENCE BETWEEN LINKAGE AND INDEPENDENT ASSORTMENT

The difference between independent assortment and linkage can be understood by the following two examples:

Example 1. Genes on different chromosomes assort independently giving a 1 : 1 : 1 : 1 test cross ratio which is as follows:

P_1:	*AA BB*	×	*aa bb*
P_1 gametes:	(*AB*)	↓	(*ab*)
F_1:		*Aa Bb*	
Test cross:	F_1 *Aa/Bb*	×	P_1 *aa/bb*
Gametes:	(*AB*) (*Ab*) (*aB*) (*ab*)		(*ab*)
F_2:	¼ *Aa Bb*: ¼ *Aa bb*: ¼ *aa Bb*: ¼ *aa bb* or 1 : 1 : 1 : 1.		

Example 2. Linked genes do not assort independently but tend to stay together in the same combination as they were in the parents. In the following figure, the genes to the left of the slash line (*/*) are on one chromosome and those on the right are on the homologous chromosome.

P_1:	*AB/AB*	×	*ab/ab*
P_1 gametes:	(*AB*)	↓	(*ab*)
F_1:		*AB/ab*	
Test cross:	F_1 *AB/ab*	×	P_1 *ab/ab*
Gametes:	(*AB*) (*ab*)		(*ab*)
F_2:	½ *AB/ab*	:	½ *ab/ab* or 1 : 1.

This type of large deviation of F_2 results of example 2 from a 1 : 1 : 1 : 1 ratio in the test cross progeny of dihybrid is an evidence for linkage.

Further, linked genes are separated from each other by the process of crossing over. In that case, linkage is termed as **incomplete linkage** and causes genetical variations due to recombination of genes.

8.2 THEORIES OF LINKAGE

Linkage can be defined as close physical association between two loci on a chromosome, preventing independent assortment (see Box 8.1).

Box 8.1 Linkage

The term **linkage** has three distinct meanings in genetics.

1. Firstly, linkage (or **linking**) describes the coiling of double-stranded DNA and is used to quantify topological parameters of duplex DNA.
2. Secondly, linkage indicates when a particular gene (and by extension, its associated phenotype) is found on a certain chromosome, *e.g.*, genes and the characters they control associated with the X-chromosome are described as **X-linked** (*i.e.*, **sex-linked inheritance**).
3. Finally, **genetic linkage** refers to genes which tend to be inherited as a unit rather than assorting independently. This reflects close **physical linkage**, *i.e.*, genes are physically joined together on the same chromosome.

Development of Concept of Linkage

The hypothesis that linked genes tend to remain in their original combinations because of their location in the same chromosome was advanced by **T.H. Morgan** in 1911. But prior to Morgan, **W. Sutton** and **T. Boveri** (1902), **Sutton** (1903) and **Bateson** and **Punnet** (1906) had given some hints about the phenomenon of linkage. It is strange that **Mendel** could not detect the phenomenon of linkage in pea plant, since extensive linkage studies of **Lamprecht** (1961) have demonstrated that seven genes used by Mendel belonged to only four linkage groups (chromosomes 1,4,5 and 7). The views of various geneticists about the phenomenon of linkage can be represented as follows:

1. Sutton-Boveri chromosome theory of heredity. Sutton and **Boveri** (1902) concluded that genes or Mendelian factors were located on chromosomes. This generalization is now known as **chromosome theory of inheritance (or of heredity)**. This theory provides a physical basis of heredity.

2. Sutton's view on linkage. Sutton (1903) predicted that the chromosomes are the bearers of the units of heredity or genes and since the number of hereditary units is much larger than the number of chromosomes, therefore, each chromosome must contain a number of genes. He further stated that since the chromosomes move as units during meiosis to the gametes, all the genes which are situated in the same chromosome will be linked together. As a result each species of animals and plants would have a specific number of groups of linked genes which would correspond with the number of chromosomes found in that species. Unfortunately **Sutton** could not prove his predictions experimentally.

3. Bateson and Punnet's coupling and repulsion hypothesis. Bateson and **Punnet** (1905–1908) formulated the 'hypothesis of coupling and repulsion' to explain the unexpected F_2 results of a dihybrid cross between a homozygous sweet pea (*Lathyrus odoratus*) having dominant alleles for blue or purple flowers (*RR*) and long pollen grains (*Ro Ro*) with another homozygous double recessive plant (*rr*, *roro*) with red flowers and round pollen grains. When they test crossed a heterozygous blue or purple long (*Rr*, *Ro ro*) plant with recessive parent (*rr*, *ro ro*), besides getting the 1 : 1 : 1 : 1 test cross ratio, they received phenotypic ratio of 7 : 1 : 1 : 7 as has been illustrated in following figure:

Parent:	Blue or purple long (*RR RoRo*)	×	Red Round (*rr roro*)
F_1:		All Blue or purple long (*Rr Roro*)	
Test cross:	F_1 Blue or purple long (*Rr Roro*)	×	Red Round (*rr roro*)

Test cross progeny:	Blue or purple long	=	192
	Red round	=	182
	Blue or purple round	=	23
	Red long	=	30
			427

Test cross ratio: 7 Blue or purple Long : 1 Blue or purple Round : 1 Red Long : 7 Red Round or 7 : 1 : 1 : 7.

The 7 : 1 : 1 : 7 test cross ratio clearly indicated that there was a tendency in the dominant alleles (*R Ro*) to pass together to the same gamete. Similar was the case with recessive alleles (*r ro*). This tendency of dominant or recessive alleles to inherit together was explained as **"gametic coupling"** by **Bateson** and **Punnett**.

Further, when they crossed blue or purple round (*RR ro ro*) with red long (*rr Ro Ro*), the F_1 hybrids were found to be heterozygous blue or purple long (*Rr Roro*). The F_1 hybrid when test crossed with recessive (*rr roro*) parent, the test cross ratio was 1 blue or purple long : 7 red long: 7 blue or purple round : 1 red round, as has been illustrated in following figure:

Parent:	Blue or purple Round (*RR roro*)	×	Red Long (*rr RoRo*)
F_1:	All Blue or purple Long (*Rr Roro*)		
Test cross:	F_1 Blue or purple Long (*Rr Roro*)	×	P_1 Red Round (*rr roro*)

Test cross progeny: 1/16 Blue or purple Long : 7/16 Blue or purple Round : 7/16 Red Long: 1/16 Red Round or 1 : 7 : 7 : 1.

Hence, the two dominant pairs of alleles repelled each other. The tendency of both dominant or both recessive alleles to repel each other, so that the gametes of genotypes of *R ro* and *r Ro* are formed more frequently, was termed **repulsion**.

Bateson and Punnett could not explain the exact reasons of coupling and repulsion, and it was **T.H. Morgan** who while performing experiments with *Drosophila*, in 1910, found that coupling or repulsion was not complete. He further suggested that the two genes are found in coupling phase or in repulsion phase, because they are present on the same chromosome (coupling) or on two different homologous chromosomes (repulsion). Such genes are then called **linked genes** and the phenomenon of inheritance of linked genes is called **linkage** by Morgan. The terms **cis** and **trans** were employed later (**Haldane**, 1942), the former replacing coupling and the latter, repulsion.

4. Morgan's views on linkage. Morgan stated that the pairs of genes of homozygous parent tend to enter in the same gametes and to remain together, whereas same genes from heterozygous parent tend to enter in the different gametes and remain apart from each other. He further stated that the tendency of linked genes remaining together in original combination is due to their location in the same chromosome. According to him the degree or strength of linkage depends upon the distance between the linked genes in the chromosome. Morgan's concept about the linkage developed the theory of linear arrangement of genes in the chromosomes which helped the cytogeneticists in the construction of genetic or linkage maps of chromosomes.

5. Chromosome theory of linkage. Morgan along with **Castle** formulated the chromosome theory of linkage which is as follows:

1. The genes which show the phenomenon of linkage are situated in the same chromosomes and these linked genes usually remain bounded by the chromosomal material so that they cannot be separated during the process of inheritance.

2. The distance between the linked genes determines the strength of linkage. The closely located genes show strong linkage than the widely located genes which show the weak linkage.
3. The genes are arranged in linear fashion in the chromosomes.

Why did Mendel failed to detect linkage?

Because genes for some of the traits Mendel followed, are on different chromosome pairs; linked traits with which he worked (seed colour/flower colour; pod colour/plant height, among others) are so far apart on their chromosomes that genes for these traits appear to segregate randomly. In one other case Mendel did not make the cross that would have revealed linkage.

8.3 KINDS OF LINKAGE

T.H. Morgan and his co-workers by their investigation on the *Drosophila* and other organisms have found two types of linkage, *viz.*, complete linkage and incomplete linkage.

1. Complete linkage. The complete linkage is the phenomenon in which parental combinations of characters appear together for two or more generations in a continuous and regular fashion. In this type of linkage the genes are closely associated and tend to transmit together.

Example. The genes for bent wings (*bt*) and shaven bristles (*svn*) of the fourth chromosome mutant of *Drosophila melanogaster* exhibit complete linkage.

Complete linkage in male *Drosophila*. In most of the organisms crossing over takes place both in males and females. But in male *Drosophila* and female silkworm, *Bombyx mori* (see **Swanson,** 1957) crossing-over takes place either very rarely or not at all. This becomes clear from Morgan's experimental results from *Drosophila*. In 1919, **T.H. Morgan** mated gray bodied and vestigial winged (b^+ vg/b^+ vg) fruit flies with flies having black bodies and normal wings (bvg^+/bvg^+). F_1 progeny had gray bodies and normal long wings (b^+vg/bvg^+), indicating thereby that these characters are dominant. When F_1 males (b^+ vg/bvg^+), were backcrossed (*i.e.*, test crossed) to double recessive females (*bvg/bvg* or black vestigial), only two types of progeny (one with gray bodies and vestigial wings, *b+vg/bvg* and the other with black bodies and normal wings (bvg^+/bvg) instead of four types of phenotypes were obtained.

Parents:	Gray, Vestigial b^+vg/b^+vg	×	Black, Long bvg^+/bvg^+
Gametes:	(b^+vg)		(bvg^+)
F_1:		All Gray, Long b^+vg/bvg^+	
Test cross:	F_1 male Gray, Long b^+vg/bvg^+	×	Female Black, Vestigial *bvg/bvg*
Gametes:	(b^+vg) (bvg^+) (only two types of gametes due to complete linkage and lack of crossing over in male *Drosophila*)		(*bvg*)
Test cross ratio:	½ Gray, Vestigial b^+vg/bvg	:	½ Black, Long or 1: 1. bvg^+/bvg

Here, the use of the testcross is very important. Because one parent (the tester) contributes gametes carrying only recessive alleles, the phenotypes of the offspring represent the gametic contribution of the other double heterozygote parent. So the genetical analyst can concentrate on one meiosis and forget the other. This is in contrast to the situation in an F_1 selfing where there are two sets of meiotic divisions to consider one for the F_1 male parental gametes and one for the F_1 female.

2. Incomplete linkage. The linked genes do not always stay together because homologous non-sister chromatids may exchange segments of varying length with one another during meiotic prophase. This sort of exchange of chromosomal segments in between homologous chromosomes is known as **crossing over**. The linked genes which are widely located in chromosomes and have chances of separation by crossing over are called **incompletely linked genes** and the phenomenon of their inheritance is called **incomplete linkage**.

Example. The incomplete linkage has been reported in female *Drosophila* and various other organisms such as tomato, maize, pea, mice, poultry and human beings, etc. Here, the examples of incomplete linkage have been considered only for *Drosophila* and maize.

Incomplete linkage in female *Drosophila*. When F_1 females of the Morgan's classical cross in *Drosophila* between gray, vestigial (b^+vg/b^+vg) and black, normal or long (bvg^+/bvg^+) were test-crossed to double-recessive (bvg/bvg) males, all four types of progeny were obtained in following ratio, showing occurrence of crossing over:

Parents:	Gray, Vestigial b^+vg/b^+vg	×	Black, Long bvg^+/bvg^+
Gametes:	(b^+vg)		(bvg^+)
F_1:		Gray, Long b^+vg/bvg^+	
Test cross:	F_1 Female Gray, Long b^+vg/bvg^+ ↓	×	Male Black, Vestigial bvg/bvg ↓
Gametes:	(b^+vg) (bvg^+) = Non-crossovers (b^+vg^+) (bvg) = Recombinants		(bvg)

Test cross ratio:	1. Gray, Vestigial; b^+vg/bvg	= 41.5%	83% parental combination showing linkage
	2. Black, Long; bvg^+/bvg	= 41.5%	
	3. Gray, Long; b^+vg^+/bvg	= 8.5%	17% recombinants due to crossing over
	4. Black, Vestigial; bvg/bvg	= 8.5%	

Similar different test cross ratios (showing complete linkage in males and incomplete linkage in females) were obtained for F_1 males and females of *Drosophila* by **Bridges**, one of the student of Morgan. He made a cross between fruit flies having wild dominant alleles for red eye colour and normal wings (pr^+vg^+/pr^+vg^+) and having mutant recessive alleles for purple eye colour and vestigial wings ($pr\ vg/pr\ vg$).

2. Incomplete linkage in maize. In *Zea mays* (maize) a case of incomplete linkage between the alleles for colour and shape of the seed has been observed by **Hutchinson**. When a maize plant with seeds having coloured and full endosperm (*CS/CS*) is crossed with another plant having recessive alleles for colourless, shrunken seeds (*cs/cs*), the F_1 heterozygotes are found with the phenotype of coloured full and genotype of *CS/cs*. When F_1 hybrid is test crossed with double recessive parent (*cs/cs*) four classes of descendants are obtained instead of two as shown in following figure:

Parents:	Coloured Full *CS/CS*	×	Colourless Shrunken *cs/cs*
F_1:		Coloured Full (*CS/cs*)	
Test cross:	F_1 Coloured Full *CS/cs*	×	Colourless Shrunken *cs/cs*

Test cross results:	Coloured :	Coloured :	Colourless :	Colourless :
	Full	Shrunken	Full	Shrunken
	CS/cs	*Cs/cs*	*cS/cs*	*cs/cs*
	48%	2%	2%	48%

The test cross results are clearly showing that parental combination of alleles (*e.g.*, *CS/cs* and *cs/cs*) are those expected from complete linkage and appear in 96% cases, the other two are new combinations (*e.g.*, *Cs/cs* and *cS/cs*) and appear in 4% cases. Thus, in 4% cases crossing over has occurred between linked genes.

Arrangement of Linked Genes.

In an individual, which is heterozygous for two pairs of linked genes, the linkage can be either of the following two types:

1. Cis-arrangement. The dominant alleles of both linked genes are located in one member of the homologous pair of chromosome and their recessive alleles are located in the other chromosome of the pairs. This arrangement is known as **cis-arrangement** and the heterozygotes with such arrangement (*AB/ab* or *R Ro/r ro* in Fig. 8.1A) are known as **cis-heterozygotes**.

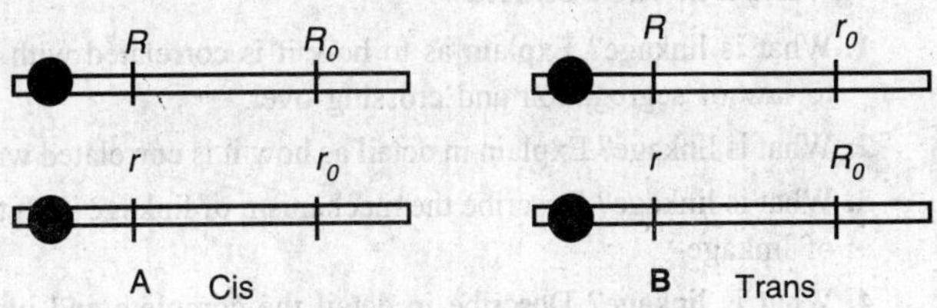

Fig. 8.1. Cis (A) and trans (B) arrangement of two pairs of linked genes of *Lathyrus odoratus*.

2. Trans-arrangement. The dominant allele of one linked gene and recessive allele of the other linked gene are located on one chromosome of the homologous pair and the recessive allele of first linked gene and dominant allele of the second linked gene are located in the second chromosome of the homologous pair (*Ab/aB* or *Rro/rRo* in Fig. 8.1B). Such an arrangement of a dominant and recessive alleles of two linked genes in the same chromosome is called **trans-arrangement** and heterozygotes with such arrangement are called **trans-heterozygotes**.

Significance of Linkage

The phenomenon of linkage has one of the great significance for the living organisms that it reduces the possibility of variability in gametes unless crossing over occurs.

8.4 LINKAGE GROUPS

All the linked genes of a chromosome form a **linkage group**. A linkage group is a collection of genes which demonstrate linkage and therefore map together. In physical terms, linkage group is equivalent to a chromosome. Because, all the genes of a chromosome have their identical genes (alleomorphs) on the homologous chromosome, therefore linkage groups of a homologous pair of chromosome is considered as one. The number of linkage group of a species, thus, corresponds with haploid chromosome number of that species.

Example.

1. *Drosophila* has 4 pairs of chromosomes and 4 linkage groups.
2. Each human being has 23 pairs of chromosomes and 23 linkage groups.
3. Corn (*Zea mays*) has 10 pairs of chromosomes and 10 linkage groups.

However, in organisms where female or male sex has dissimilar sex chromosomes (*e.g.*, human beings, *Drosophila*, fowl, etc.), one more linkage group occurs than the haploid number.

Example.

1. Female human beings: = 22 pairs of autosomes or non-sex chromosomes + 1 pair of homomorphic X chromosomes
= 22 autosomal linkage groups + 1 X chromosomal linkage group
= 23 linkage groups.

2. Male human beings = 22 pairs of autosomes + 2 heteromorphic sex chromosomes, *i.e.*, 1 X chromosome + 1Y chromosome
= 22 autosomal linkage group + 1X chromosomal linkage group + 1Y chromosomal linkage group
= 24 linkage groups.

QUESTIONS

Long Answer Questions

1. What is linkage? Explain as to how it is correlated with Mendel's law of independent assortment or to law of segregation and crossing over.

2. What is linkage? Explain in detail as how it is correlated with Mendel's laws of independent assortment.

3. What is linkage? Describe the mechanism of linkage with the help of an example. State the significance of linkage.

4. What is linkage? Describe in detail the complete and incomplete linkage.

5. Describe in detail the phenomenon of linkage.

6. Define linkage. Give an account of different theories of linkage with a note on its significance.

7. Describe the process of linkage and give its significance.

8. Give an account of linkage in *Drosophila melanogaster*.

9. Describe the independent assortment law of Mendel. How it is affected by linkage? *(C.C.S.U. Meerut, 2008)*

10. Define linkage and write about the theories of linkage. *(M.S.U., Tamil Nadu, 2006)*

Short Answer Questions

1. Mendel studied seven pairs of contrasting characters in the garden pea. Why do you suppose, did he not discover the principle of linkage?

2. What is coupling and repulsion?

3. What are the advantages of test cross method for determining linkage relation?

4. Write short notes on the following:
(*i*) Linkage group
(*ii*) Linkage
(*iii*) Complete linkage.

5. Give differences between the following:
(*i*) Linkage and law of independent assortment
(*ii*) Complete linkage and incomplete linkage
(*iii*) Linkage and crossing over
(*iv*) Cis-arrangement and trans-arrangement.

Very Short Answer Questions

1. *Drosophila melanogaster* has 8 chromosomes and human beings have 46. How many linkage groups are there in each case?

2. Define linkage.

3. Name the genes that do not follow the Mendel's law of independent assortment.

4. Write the name of an organism in which no crossing over is observed.

Fill in the Blanks

1. All the genes which occur in a given chromosome are known as ....

2. If test cross of offsprings of a dihybrid cross show a ratio of 1 : 1 : 1 : 1 it means the genes are ....

3. Mendel's law of independent assortment fails in case of ....

4. Chromosome theory of linkage is formulated by ....

Yes or No Questions

1. Muller propounded the chromosome theory of heredity.

2. The term gene was first used by Morgan.

3. Linkage reduces the frequency of hybrid. *(H.N.B. Garhwal, 2008)*

Multiple Choice Questions

1. Bateson and Punnet (1909) discovered linkage while working on

(*a*) *Pisum sativum* (*b*) *Lathyrus odoratus*

(*c*) *Drosophila melanogaster* (*d*) *Lychmis alba*

2. The genes controlling the seven pea characters studied by Mendel are now known to be located on how many different chromosomes?

(*a*) seven (*b*) six

(*c*) five (*d*) four

3. When two or more characters do assort independently in F_2 generation, it indicates that characters are not

(*a*) dominant (*b*) recessive

(*c*) linked (*d*) lethal

4. Linked genes may be separated by

(*a*) crossing over (*b*) polyploidy

(*c*) gene mutation (*d*) segregation

5. When a cluster of genes show linkage behaviour they

(*a*) do not show a chromosome map (*b*) show recombination during meiosis

(*c*) do not show independent assortment (*d*) induce cell division

ANSWERS

Very Short Answer Questions

1. *D. melanogaster* has 4 linkage groups; each human being has 23 linkage groups.

2. Tendency of the genes located on the same chromosome to remain together in the process of inheritance is called linkage.

3. Linked genes.

4. Male fruitfly (*Drosophila melanogaster*).

Fill in the Blanks

1. Linked genes **2.** Not linked

3. Linked genes **4.** Morgan and Castle

Yes or No Questions

1. No **2.** No

3. Yes

Multiple Choice Questions

1. (*b*) **2.** (*d*) **3.** (*c*) **4.** (*a*) **5.** (*c*).

9

Crossing Over

In the chapter of linkage, we have stated that the genes located in the same chromosome show linkage. These linked genes may either remain together during the process of inheritance and, thus, showing complete linkage or they may be segregated or separated during gametogenesis and, thus, displaying the incomplete linkage. The incomplete linkage takes place due to the occurrence of new combinations or recombinations of linked genes. The recombination in its turn is accomplished through a process known as **crossing over** in which the non-sister chromatids of homologous chromosomes, exchange the chromosomal parts or segments. In another words "*the crossing over is a process that produces new combinations (recombinations) of genes by interchanging of corresponding segments between non-sister chromatids of homologous chromosomes*". The chromatins resulting from such interchanges of chromosomal parts are known as **cross overs**. Thus, a cross-over is a site of chromosome breakage and strand exchange; it is called **Holliday junction** in molecular terms and leads to homologous recombination between the parental chromosomes and the production of recombinant combination of alleles. [**Note** that *crossing over* refers to the exchanges of segments of chromosomes (the process), whereas *recombination* implies the formation of new combination of genes (one possible outcomes of the process, see **Gardner** *et al.*, 2001)].

The term **crossing over** was coined by **Morgan**. No crossing over occurs in male *Drosophila* and female silk worm, *Bombyx mori*. Certain external agents such as heat shock, chemicals, radiations, etc., have profound effect on crossing over.

Characteristics of Crossing Over

1. Crossing over or recombination occurs at two levels (*i*) at gross chromosomal level, called **chromosomal crossing over** and (*ii*) at DNA level, called **genetic recombination**.
2. A reciprocal exchange of material between homologous chromosomes in heterozygotes is reflected in crossing over.
3. The crossing over results basically from an exchange of genetic material between non-sister chromatids by break-and-exchange following replication.
4. The frequency of crossing over appears to be closely related to physical distance between genes on chromosome and serves as a tool in constructing genetic maps of chromosomes.

9.1 TYPES OF CROSSING OVER

According to its occurrence in the somatic or germ cells following two types of crossing over have been recognized:

1. Somatic or mitotic crossing over. When the process of crossing over occurs in the chromosomes of body or somatic cells of an organism during the mitotic cell division it is known as **somatic** or **mitotic crossing over**. The somatic crossing over is rare in its occurrence and it has no genetical significance. The somatic or mitotic crossing over has been reported in the body or somatic cells of *Drosophila* by **Curt Stern** and in the fungus *Aspergillus nidulans by* **G. Pontecorvo** (Box 9.1).

Box 9.1 Parasexual Exchange

Any recombination occurring in the absence of meiosis (*i.e.*, in the absence of sexual reproduction) is termed **parasexual exchange**, and this is the form of genetic exchange predominantly seen in bacteria (*i.e.*, gene transfer in bacteria). Parasexual exchange in eukaryotes occurs between **organelle genome** and between nuclear chromosomes in somatic cells (**mitotic recombination**).

2. Germinal or meiotic crossing over. Usually the crossing over occurs in germinal cells during the gametogenesis in which the meiotic cell division takes place. This type of crossing over is known as **germinal** or **meiotic crossing over**. The meiotic crossing over is universal in its occurrence and is of great genetic significance.

9.2 MECHANISM OF MITOTIC RECOMBINATION

Mitotic recombination between homologous chromosomes lead to **mitotic segregation**, where heterozygous loci segregate into somatic daughter cells generating reciprocal homozygous clones identifiable by their distinct phenotypes. The first evidence for mitotic recombination came from *Drosophila*.

Example. Stern (1936) observed that *Drosophila* females which are heterozygous for sex-linked recessive genes, yellow body colour (*y*) and singed (=burned or scorched) hairs and bristles (*sn*), are phenotypically like wild types but occasionally show yellow spots on their body. Homozygous recessive individuals for these genes are yellow bodied and with bent blunted bristles. Normally, all the cells of a heterozygous should be normal and should not express either of these genes. Yellow skin spots can appear only if cells are homozygous for *y*, indicating thereby that whenever yellow spots appear in heterozygous flies, there is prior crossing over between *y* and *sn* alleles. Stern found that equal sections of singed and yellow tissues lie adjacent to each other or form **twin spots**. This indicate that such tissues arise as a result of mitotic crossing over, rather than due to mutation. Somatic or mitotic crossing over occurs at a four strand stage and during this process there is pairing of homologous chromosomes (Fig. 9.1).

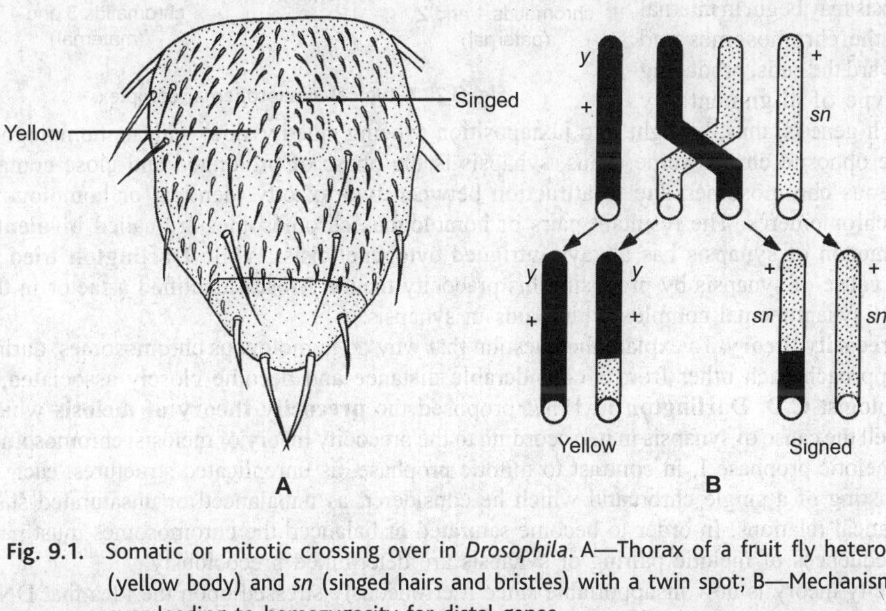

Fig. 9.1. Somatic or mitotic crossing over in *Drosophila*. A—Thorax of a fruit fly heterozygous for *y* (yellow body) and *sn* (singed hairs and bristles) with a twin spot; B—Mechanism of crossing over leading to homozygosity for distal genes.

Mitotic recombination can be stimulated by exposure to X-rays, which generate double-strand breaks in DNA. This suggests that the process may be a by-product of DNA-repair.

A second form of mitotic recombination occurs not between homologous chromosomes but between identical chromatids–**sister chromatid exchange (SCE)**. This is also stimulated by damage to DNA.

9.3 MECHANISM OF MEIOTIC CROSSING OVER

The process of crossing over includes following stages in it, *viz.*, synapsis, duplication of chromosomes, crossing over and terminalization. The chromosomes which tend to undergo recombination due to meiotic crossing over necessarily complete two functions: 1. 99.7 per cent replication of DNA and 75 per cent synthesis of histones, both of which take place prior to onset of prophase I and 2. attachment of each chromosome by its both ends (**telomeres**) to the nuclear envelope (*i.e.*, to nuclear lamina) *via* the specialized structure, called **attachment plaques** (see **Suzuki** *et al.*, 1986). This event occurs during the **leptotene** stage of prophase I and though each chromosome at this stage is visually long and thin thread, but contains material of two sister chromatids (*i.e.*, two DNA molecules plus almost duplicated amount of histones).

1. Synapsis.

Synapsis or intimate pairing between the two homologous chromosomes (one maternal and another paternal) is initiated during **zygotene** stage of prophase I of meiosis I. Synapsis often starts when the homologous ends of the two chromosomes are brought together on the nuclear envelope and it continues inward in a zipper-like manner from both ends, aligning the two homologous chromosomes side by side (*e.g.*, mammals). In other cases, synapsis may begin in internal regions of the chromosomes and proceed toward the ends, producing the same type of alignment. By synapsis each gene is, thus, brought into juxtaposition (=being side by side) with its homologous gene on the opposite chromosome. Thus, synapsis is the phase of prolonged and close contact of homologous chromosomes due to attraction between two exactly identical or homologous regions or chromomeres. The resultant pairs of homologous chromosomes are called **bivalents**. The phenomenon of synapsis has always intrigued cytogeneticists. While **Darlington** tried to explain the cause of synapsis by proposing his precocity theory, **Moses** identified a factor in the formation of synaptonemal complex which aids in synapsis.

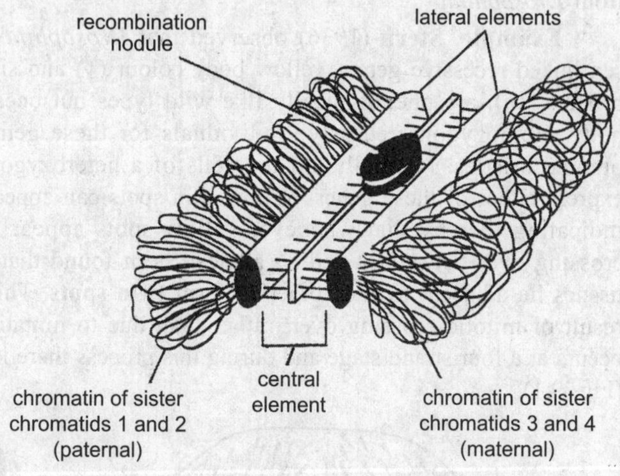

Fig. 9.2. A typical synaptonemal complex.

(*a*) Precocity theory. To explain the question that why do homologous chromosomes, during synapsis, approach each other from a considerable distance and become closely associated, a British cytologist **C.D. Darlington** in 1937, proposed the **precocity theory of meiosis** which embraces well the cause of synapsis in it. According to the precocity theory of meiosis, chromosomes enter into meiotic prophase I, in contrast to mitotic prophase, as unreplicated structures, each of which consisting of a single chromatid which he considered as unbalanced or unsaturated state in electrostatical relations. In order to become saturated or balanced the chromosomes must pair. Thus, the sequences of meiotic pairing or synapsis are determined precociously.

Precocity theory is now insupportable since it erroneously stressed upon the idea that DNA synthesis or chromosome duplication takes place later in pachytene or diplotene and then results in separation of homologous chromosomes.

(b) Synaptonemal complex. Montrose J. Moses (1956) has revealed a highly organized structure of filaments called **synaptonemal complex** in between the paired chromosomes of zygotene and pachytene stages in crayfish by electron microscopy. Synaptonemal complex has also been observed in a wide variety of species of plants and animals.

In electron-micrographs the synaptonemal complex appears as three parallel dense lines that lie equally spaced in a plane and are flanked by chromatin. The elements of two lateral lines usually appear densest, while the element of central line is of variable prominence. Some fine transverse strands also cross between lateral elements, connecting them with the central element. Though, the morphology of lateral and central elements may vary from species to species, but the basic structure and the spacing of the synaptonemal complex is constant within the species.

Cytochemical studies have shown that the lateral elements of synaptonemal complex are rich in DNA, RNA and proteins, but that the central element of it contains mainly RNA, protein and little DNA.

Functions of synaptonemal complex. The synaptonemal complex is found to contribute both chiasma formation and crossing over. For instance, the synaptonemal complex may serve crossing over by facilitating effective synapsis in one or more of the following ways—(*a*) to maintain pairing in fixed state for an extended period, (*b*) to provide a structural framework within which molecular recombination may occur, and (*c*) to segregate recombination DNA from the bulk of the chromosomal DNA.

Robert King (1970) suggested that the synaptonemal complex may orient the non-sister chromatids of homologs in a manner to facilitate enzymatically induced exchanges between their DNA molecules. In 1971, **D.A. Comings** and **T.A. Okada** have shown electron microscopically that synapsis occurs at two levels one at chromosomal level and the other at the molecular level. According to them, the synaptonemal complex pulls homologous chromosomes into approximate association with each other but plays no role in molecular pairing of DNA strand.

Later on each event of recombination due to crossing over is found to be performed by a large protein assembly of 90 nm diameter, called **recombination nodule**, which is placed on the synaptonemal complex (**Carpenter**, 1977, 1987). Each nodule marks the site of a large multienzyme 'recombination machine' which brings local regions of DNA on maternal and paternal chromatids together across the 100-nm wide synaptonemal complex.

2. Duplication of Chromosomes

The synapsis is followed by duplication of chromosomes (in **pachytene**). During this stage, each homologous chromosome of bivalents splits longitudinally and form two identical sister chromatids which remain held together by an unsplitted centromere. The longitudinal splitting of chromosomes is achieved by the separation of already duplicated DNA molecules along with certain chromosomal proteins. At this stage each bivalent contains four chromatids, so it is known as **tetrad**.

3. Crossing Over by Breakage and Union

It is well evident that crossing over occurs in the homologous chromosomes only during the four stranded or tetrad stage. Homologs continue to stay in synapsis for days during pachytene stage and chromosomal crossing over occurs due to exchange of chromosomal material between non-sister chromatids of each tetrad. In pachytene, the recombination nodules become visible between synapsed chromosomes.

During the process of crossing over, two non-sister chromatids first break at the corresponding points due to the activity of a nuclear enzyme known as **endonuclease**. Then a segment on one side of each break connects with a segment on the opposite side of the break, so that the two non-sister chromatids cross each other. At this stage 3 per cent synthesis of DNA (*i.e.*, P–DNA replication) occurs to fill the gap. The fusion of chromosomal segments with that of opposite one takes place due to the action of an enzyme known as **ligase**. The crossing of two chromatids is known as **chiasma** (Gr., *chiasma* = cross) formation. The crossing over, thus, includes the breaking of chromatid segments, their transposition and fusion.

Chiasma frequency or percentage of crossing over. The crossing over may take place at several points in one tetrad and may result in the formation of several chiasmata. The number of chiasmata depends on the length of the chromosomes because the longer the chromosome the greater the number of chiasmata. In a species each chromosome has a characteristic number of chiasmata. The frequency by which a chiasmata occurs between any two genetic loci has also a characteristic probability. The more apart two genes are located on a chromosome, the greater the opportunity for a chiasma to occur between them. The closer two genes are linked lesser the chances for a chiasma occurring between them.

Limits of recombination. If two gene loci are so far apart in the chromosome that the probability of chiasma forming between them is 100% then 50% of the gametes will be parental type (non-cross-over) and 50% recombinant (cross over) type. Such dihybrids give the test cross ratio of 1 : 1 : 1 : 1. Recombination between two linked genes cannot exceed 50% even when multiple cross overs occur between them.

Box 9.2 Recombination Proteins in Eukaryotes

RAD52 group of genes is involved in recombinational repair. **RND51** gene is inducible by DNA damage. **RNA50** gene encodes a DNA-binding protein which functions in the repair of double strand breaks in mitotic cells and induces and processes such breaks in meiotic cells. The product of **RAD52** gene is essential for both mitotic and meiotic crossing over, *i.e.*, in the repair of double-strand DNA breaks and mating-type switching in yeast cells. RAD52 interacts with RAD51. **RAD54** encodes **helicase** required for recombination and other forms of DNA repair.

4. Terminalisation.

After the occurrence of process of crossing over, the non-sister chromatids start to repel each other because the force of synapsis attraction between them decreases. During **diplotene**, desynapsis begins, synaptonemal complex dissolves and two homologous chromosomes in a bivalent are pulled away from each other. During **diakinesis**, chromosomes detaches from the nuclear envelope and each bivalent is clearly seen to contain four separate chromatids with each pair of sister chromatids linked at their centromeres, while non-sister chromatids that have crossed over are linked by chiasmata. The chromatids separate progressively from the centromere towards the chiasma and the chiasma itself moves in a zipper fashion towards the end of the tetrad (Fig. 9.3). The movement of chiasma is known as **terminalisation**. Due to the terminalisation the homologous chromosomes are separated completely.

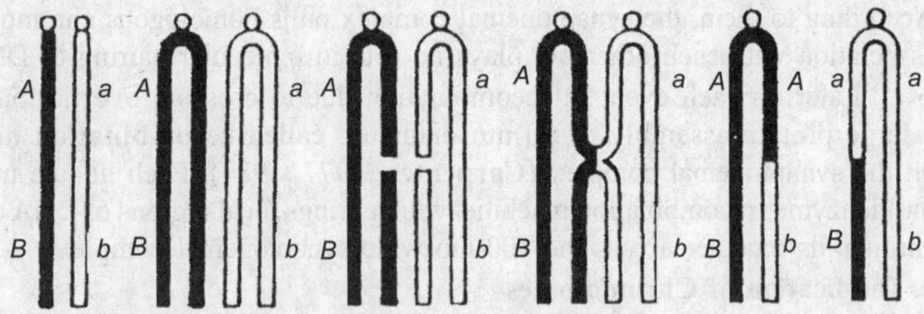

Fig. 9.3. Diagram showing the mechanism of crossing over.

9.4 KINDS OF CROSSING OVER

According to the number of chiasma following types of crossing over have been described.

1. Single crossing over. When the chiasma occurs only at one point of the chromosome pair then the crossing over is known as **single crossing over**. The single crossing over produces two cross over chromatids and two non-cross over chromatids.

2. Double crossing over. When the chiasmata occur at two points in the same chromosome, the phenomenon is known as **double crossing over**. In the double crossing over, the formation of each chiasma is independent of the other and in it four possible classes of recombination occur.

In the double crossing over following two types of chiasma may be formed:

(*i*) **Reciprocal chiasma.** In the reciprocal chiasma the same two chromatids are involved in the second chiasma as in the first. Thus, the second chiasma restores the order which was changed by the first chiasma and it produces two non-cross over chromatids. The reciprocal chiasma occurs in two strand double crossing over in which out of four chromatids only two are involved in the double crossing over (Box 9.3).

(*ii*) **Complimentary chiasma.** When both the chromatids taking part in the second chiasma are different from those chromatids involved in the first chiasma, the chiasma is known as **complimentary chiasma.** The complimentary chiasma produces four single cross overs but no non-cross over. The complimentary chiasma occurs when three or four chromatids of the tetrad undergo the crossing over.

3. Multiple crossing over. When crossing over takes place at more than two places in the same chromosome pair then such crossing over is known as **multiple crossing over**. The multiple crossing over occurs rarely.

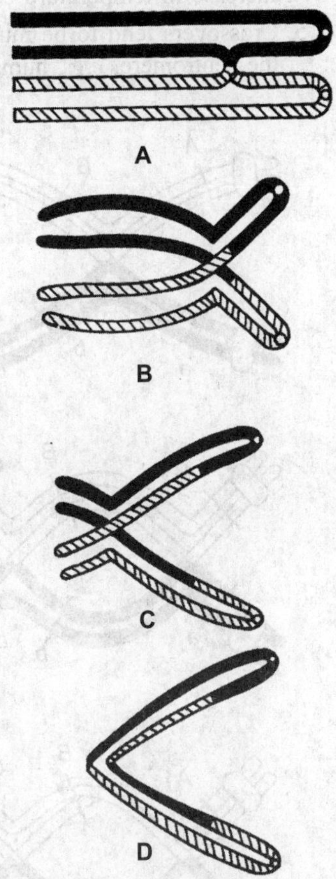

Fig. 9.4. Diagram showing the terminalisation.

Factors Affecting Crossing Over.

Many extrinsic and intrinsic factors have found to affect the rate of crossing over. Among these are the effects of age, sex, temperature, water content, ionizing radiation, certain chemicals, heterochromatic regions, chromosomal aberrations such as inversions, etc. Some examples of factors affecting the process of crossing over are as follows:

1. The frequency of crossing over decreases with increasing age in female *Drosophila*.
2. Crossing over is often depressed in one sex of certain species, *e.g.*, male *Drosophila*, female silkworm. There is higher frequency of recombination in females.
3. Low water contents can lead to reduced chiasma frequencies, in extreme cases it can result in complete failure of pairing and exchange.

Box 9.3 Reciprocal and Non-Reciprocal Recombination

1. **Reciprocal recombination.** Reciprocal recombination is both symmetrical and conservative, *i.e.*, information is *exchanged* between duplexes and is neither lost nor gained. Thus, if recombination occurs between parental chromosomes carrying alleles *AB* and *ab*, the reciprocal recombinants are *Ab* and *aB*.
2. **Non-reciprocal recombination.** It involves a unidirectional transfer of information. An example of non-reciprocal recombinants arising from above cross would be *Ab* and *AB*, where information has been transferred-unidirectionally from *A* to *a* allele and *a* allele has been lost. This type of recombination is also called **gene conversion**, because it appears that one gene has been converted to the other. Non-reciprocal recombination can be observed in fungal tetrads (*i.e.*, tetrad analysis).
3. Ionizing radiation increases the rate of mitotic as well as meiotic crossing over.
4. Certain chemicals such as **colchicine** and **selenium** have a reducing effect on crossing over.
5. The distance between two linked genes is the main factor for the occurrence of crossing over. If the two genes are distant apart, they have better chances of separation than those which are closely set together on the chromosome.

4. Increase in temperature enhances the amount of crossing over.
5. Crossovers tend to be much more frequent in telomeric chromosome regions than around the centromere, *e.g.*, human chromosomes.

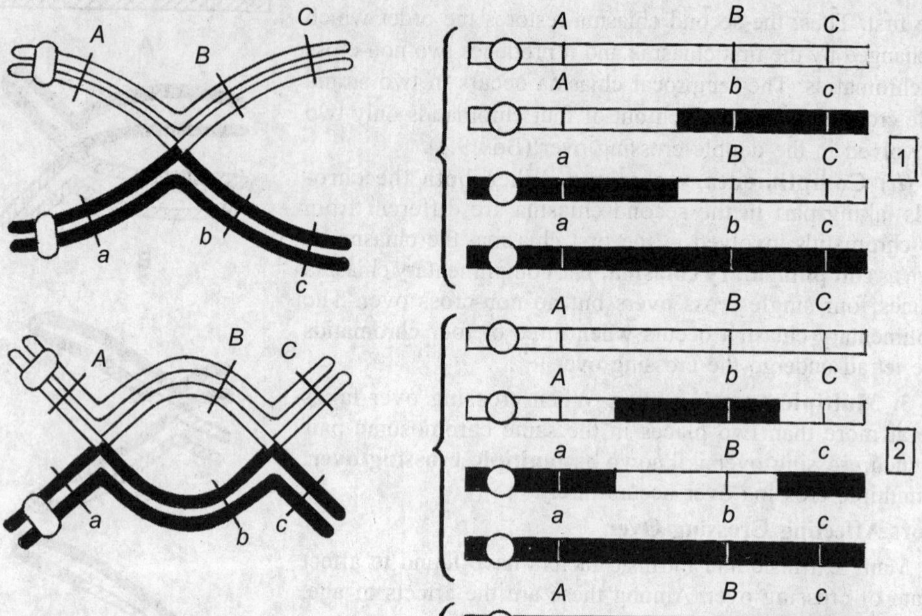

Fig. 9.5. Diagram showing the results of single and double crossing over. 1. Single chiasma. 2. Two strand double cross over. 3. Four strand double cross over. 4. Three strand double cross over.

9.5 THEORIES ABOUT THE MECHANISM OF CROSSING OVER

Following theories have been propounded to explain the mechanism of crossing over:

1. Duplication theory. This theory was proposed by **John Belling** (1928) while studying meiosis in some plant species. Belling believed that crossing over might occur during duplication of homologous chromosomes and might be brought about due to novel attachments formed between newly synthesized genes. He visualized genes as beads (described as **chromomeres**), connected by non-genic linking elements-the **interchromomeric regions**. During duplication of chromosomes, initially the chromomeres are duplicated and the newly formed chromomeres remain tightly juxtaposed to the old ones. When interchromomeric regions are synthesized to join these new genes or chromomeres, they may switch from a newly synthesized chromomere on one homologous chromosome to an adjacent chromomere of other homologue. This results in the formation of recombinants or crossovers in new sets of chromatids (Fig. 9.6).

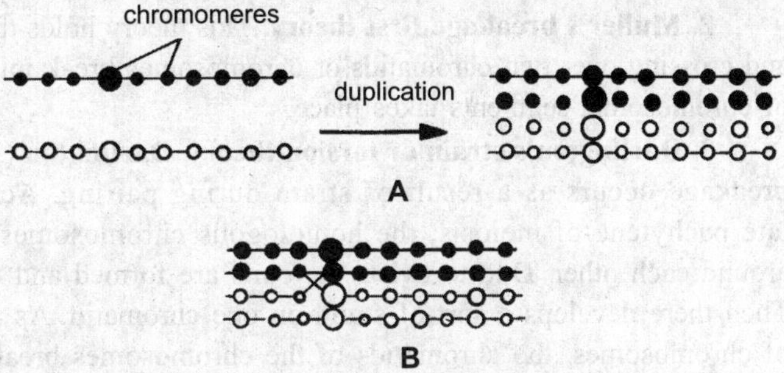

Fig. 9.6. Three steps of Belling's duplication theory.

2. Copy choice hypothesis or Switch model. To explain recombination in microorganisms, **J. Laderberg** (1955) proposed a modified version of Belling's hypothesis. This is called **copy choice mechanism** of recombination or crossing over. According to this model a daughter chromosome is formed by alternate use of recipient and donor chromosome material as a kind of model or template. The daughter chromosome is then like the recipient chromosome except for portions "copied" from the donor chromosome segment. At the end of process a recipient cell would contain (1) Original donor segment, (2) The originally recipient whole chromosome, and (3) A "hybrid" daughter chromosome. In succeeding divisions of the recipient cell, the donor segment is lost.

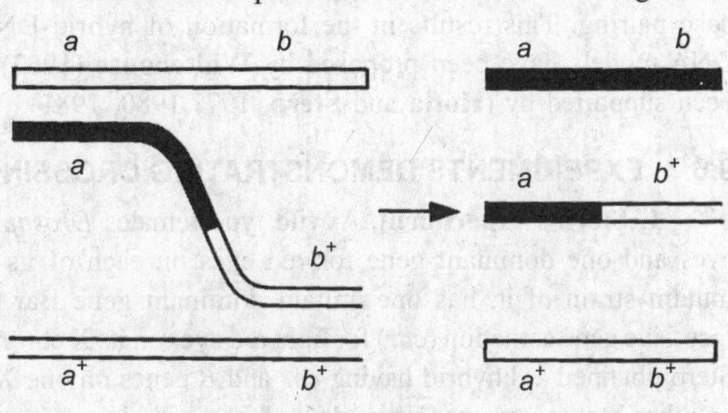

Fig. 9.7. Copy choice model for the mechanism of crossing over.

The copy choice model to explain the mechanism of recombination has been criticised for the following two reasons: (1) Copy choice model relies on conservative mode of DNA replication, but all experimental evidences suggest that DNA replication occurs in a semi-conservative manner. (2) The copy choice model proposed the involvement of only two chromatids in the process of crossing over, while cytological investigations have made it clear that crossing over occurs at four strand stage.

3. Break and exchange theory. The break and exchange theory is the most accepted theory to explain the process of crossing over. This theory states that in the crossing over breaks occur in the non-sister chromatids of the tetrad and the exchange of chromosomal segments occurs between the non-sister chromatids.

There exists a lot of controversy about the manner and time of the break of chromatids during the crossing over. Following theories have been advanced by different workers to explain the process of breaking of chromatids:

1. Serebrovsky's contact first theory. According to this theory two chromatids of a tetrad first touch and cross each other and then at the point of cross the breakage of chromatids occur. The chromosome segments then unite to form new combinations.

2. Muller's breakage first theory. This theory holds that before the occurrence of chiasma and crossing over two chromatids or chromosomes break into two fragments and then the union of chromosomal segments takes place.

3. Darlington's strain or torsion theory. Darlington (1935) propounded that chromosomal breakage occurs as a result of strain during pairing. According to this theory during the late pachytene of meiosis, the homologous chromosomes pair and become spirally twisted around each other. Due to twists the coils are formed and these are known as **relational coil**. Then there develops a sort of strain on one chromatid. As a result of the strain due to torsion of chromosomes, the chromatids of the chromosomes break at the point of contact and, thus, recombination results.

4. Hybrid DNA models. In the last three decades several hybrid DNA models have been proposed to explain the mechanism of recombination. In these models, only one strand in each of the two DNA duplexes belonging to non-sister homologous chromatids breaks. The single strands released from these breaks then pair crosswise with unbroken strands by complementary base pairing. This results in the formation of hybrid DNA segments. Some important hybrid DNA models have been proposed by **Whitehouse** (1963) and **Holliday** (1964) and these have been supported by (**Hotta** and **Stern**, 1977,1980, 1981).

9.6 EXPERIMENTS DEMONSTRATING CROSSING OVER

1. Stern's experiment. A wild type females *Drosophila* has one recessive gene for round eyes and one dominant gene for red eyes on each of its rod-shaped X chromosome, while, a mutant strain of it, has one mutant dominant gene Bar (*B*) for narrow eyes and one mutant recessive gene carnation (*car*) for light red eyes on its X chromosomes. By crossing these two strains, Stern obtained a **dihybrid** having *car* and *B* genes on one X chromosome and normal genes (+ +) on other X chromosome. He made both of the X chromosomes of this heterozygote female aberrant by treating such flies with X-rays. The X chromosome having the genes *car* and *B* was broken into two segments, one fragment having both of the genes, while, the other X chromosome having a fragment of the Y chromosome attached to it and contained normal alleles (+ +). Thus, both aberrant X chromosomes of heterozygote were cytologically detectable. A female heterozygote with the aberrant X chromosomes was mated with a normal male having X chromosome with car + alleles, four classes of eggs (*e.g.*, two types of eggs are cross overs and two types of eggs are non-cross overs) were produced which by fertilization produced following four kinds of females:

1. Carnation Bar females with broken X without any fragment of Y.
2. The red round females with unbroken X with the attached Y fragment.
3. The carnation round have the unbroken X without an attached Y fragment.
4. The red Bar have the broken X with the attached Y fragment.

Thus, flies in which crossing over was indicated phenotypically showed microscopic evidence of exchanges between homologous chromosomes. The physical or cytological basis of crossing over was thus established.

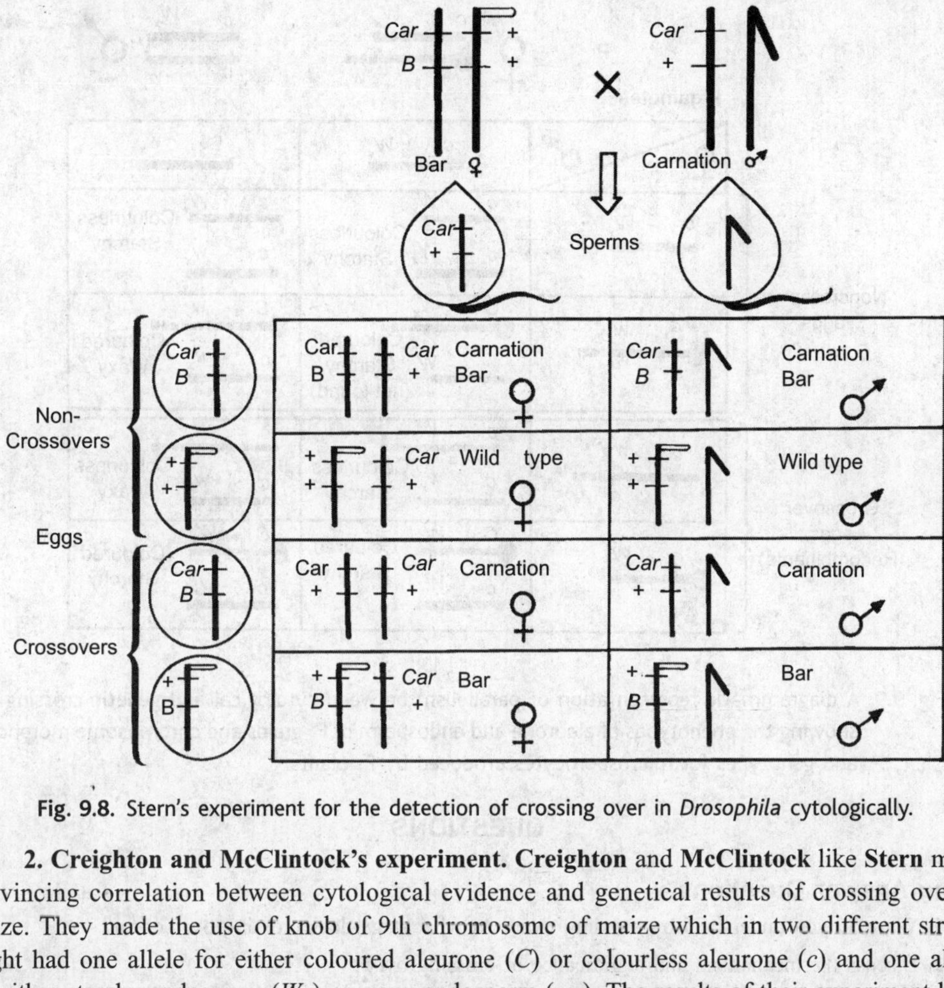

Fig. 9.8. Stern's experiment for the detection of crossing over in *Drosophila* cytologically.

2. Creighton and McClintock's experiment. Creighton and **McClintock** like **Stern** made convincing correlation between cytological evidence and genetical results of crossing over in maize. They made the use of knob of 9th chromosome of maize which in two different strains might had one allele for either coloured aleurone (*C*) or colourless aleurone (*c*) and one allele for either starchy endosperm (*Wx*) or waxy endosperm (*wx*). The results of their experiment have been illustrated in Fig. 9.9.

9.7 SIGNIFICANCE OR UTILITY OF CROSSING OVER

The study of mechanism of crossing over is significant in the following ways:

1. Crossing over furnishes a proof for the linear arrangement of linked genes on the chromosome.
2. Crossing over helps in the construction of linkage maps or genetic maps of chromosomes. Linkage maps of *Drosophila* and maize chromosomes have been constructed in details.
3. As a result of crossing over, new combinations of genes are produced which change the frequency of genes in the genetic pool of a population and therefore, play an important role in the process of evolution (micro-evolution).

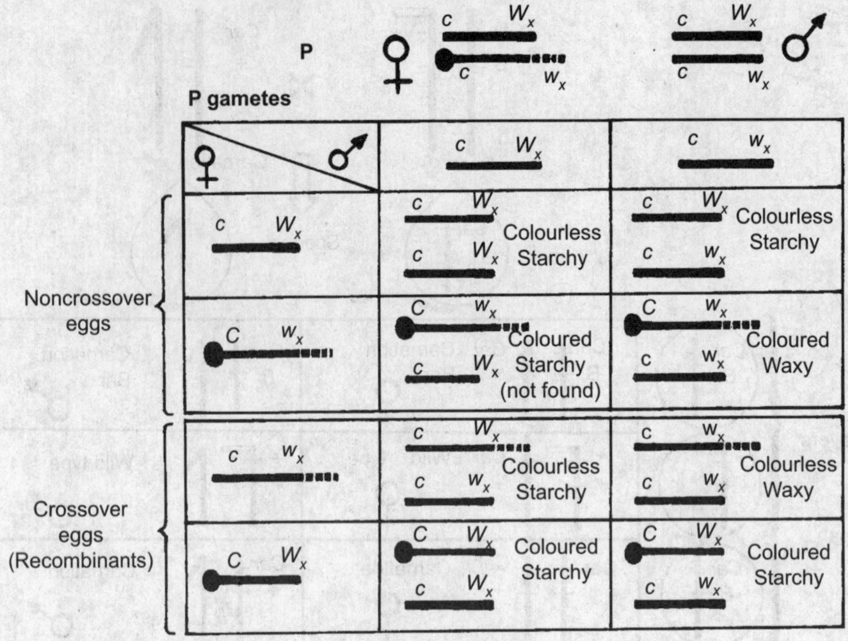

Fig. 9.9. A diagrammatic representation of parallelism between cytological and genetic crossing over showing the phenotypes of aleurone and endosperm of F_1 grains and chromosome morphology and genotypes for microsporocytes produced by F_1 plants.

QUESTIONS

Long Answer Questions

1. What do you understand by crossing over? At what stage does crossing over occur?
2. Discuss the mechanism and significance of crossing over.
3. Define crossing over. Describe various types and theories regarding crossing over and explain its mechanism.
4. Define crossing over. Explain the mechanism of crossing over and its significance.
5. Write an essay on crossing over?
6. What do you understand by crossing over? Describe the different types of crossing over.
7. Discuss the different factors influencing the crossing over.
8. Discuss the phenomenon of recombination of genes by giving suitable examples.
9. Describe the crossing over, chiasma formation and mechanism of genetic recombination.
10. What is crossing over and how does it explain recombination between linked genes?
11. What do you understand by crossing over and what part does it play in heredity?
12. Explain with example the mitotic recombination or crossing over.
13. Describe various theories to explain the mechanism of crossing over.
14. What do you understand by crossing over? Describe an experiment to show the process of crossing over.
15. Give an account of the phenomenon of linkage and crossing over. Add a note on their significance.
16. What do you mean by linkage and crossing over? Write a note on various theories proposed for explaining the mechanism of crossing over.

17. Briefly describe linkage and crossing over. What is the significance of this phenomenon?

18. Define crossing over and explain the mechanism with *Drosophila* as example.

(*M.S.U. Tamil Nadu, 2006*)

Short Answer Questions

1. Write differences between chiasma and crossing over.

2. Write short notes on the following:

(*i*) Crossing over

(*ii*) Biological significance of crossing over

(*iii*) Cytological basis of crossing over

(*iv*) Synapsis

(*v*) Theories of crossing over

(*vi*) Mechanism of crossing over

(*vii*) Precocity theory.

Very Short Answer Questions

1. What gemetes are produced by an individual whose genotype is *AB*/*ab*, if crossing over can occur between the genes?

2. Define crossing over.

3. How can genes located on a single chromosome assort independently?

4. Give one word for the following statement:

"An exchange of segments of homologous chromosomes including linked genes at meiosis".

Yes or No Question

1. Crossing over occurs in the same chromosome.

Fill in the Blanks

1. The chances of recombination between two gene loci are inversely proportional to the ....

2. Recombination of genes is caused as a result of .... between chromatids of homologous chromosomes.

3. Crossing over occurs at .... stage of meiosis.

4. Crossing over requires the formation of .... between homologous chromosomes.

5. Crossing over percentage can never exceed .... per cent. (*H.N.B. Garhwal 2009*)

6. Complete linkage is due to absence of .... between genes on same chromosome.

7. .... was discovered by M.J. Moses in 1956. It is a highly organized structure of .... occurring in between paired .... chromosomes in meiosis.

8. Events of .... due to crossing over is performed by a large protein assembly of 90 nm, is called ....

Multiple Choice Questions

1. Exchange of chromosome segments between maternal and paternal chromatids at pachytene is called

(*a*) linkage (*b*) crossing over

(*c*) disjunction (*d*) segregation

2. Crossing over occurs at

(*a*) 1 strand stage (*b*) 2 strand stage

(*c*) 3 strand stage (*d*) 4 strand stage

3. Longitudinal splitting of chromosomes and crossing over occurs in
   (*a*) leptotene (*b*) zygotene
   (*c*) pachytene (*d*) diplotene
4. The pairing of homologous chromosomes during zygotene (of meiosis) is called
   (*a*) synapse (*b*) terminalization
   (*c*) synapsis (*d*) crossing over
5. Crossing over involves
   (*a*) duplication of the chromosomes (*b*) deletion of the chromosomes
   (*c*) exchange of genetic material (*d*) addition of chromosome
6. Synaptonemal complex helps in
   (*a*) gamete formation
   (*b*) recombination during cell division
   (*c*) production of enzymes during cell division
   (*d*) chromosomal movements towards poles
7. Recombination of parental characters are due to
   (*a*) chiasma (*b*) synapsis
   (*c*) linkage (*d*) crossing over
8. Crossing over occurs in
   (*a*) leptotene stage (*b*) pachytene stage
   (*c*) anaphase stage (*d*) diakinesis stage
9. Crossing over is advantageous because it brings about
   (*a*) linkage (*b*) variation
   (*c*) inbreeding (*d*) stability

ANSWERS

Very Short Answer Questions

1. *AB*, *Ab*, *aB* and *ab*.
2. Mutual exchange of exactly equal and homologous parts of genetic material or genes between the non-sister chromatids of a pair of homologous chromosomes is called crossing over.
3. Through crossing over.
4. Crossing over.

Yes or No Question

1. No.

Fill in the Blanks

1. Difference in distance
2. Exchange
3. Pachytene
4. Synaptonemal complex
5. 50%
6. Crossing over.
7. Synaptonemal complex, filaments, homologous.
8. Recombination, recombination nodule.

Multiple Choice Questions

1. (*b*)	2. (*d*)	3. (*c*)	4. (*c*)	5. (*c*)	6. (*b*)
7. (*d*)	8. (*b*)	9. (*b*).			

10

Genetic and Physical Mapping of Chromosomes

Gene mapping is the assignment of a gene to a locus on a chromosome. By mapping the relative positions of many genes and other markers it is possible to generate a **chromosome map** or a map of entire genome. Gene maps are of following three types:

1. Genetic or linkage maps which assign genes to **linkage groups**. Genetic maps are calibrated in arbitray units, reflecting the likelihood of marker separation by recombination (*e.g.*, in meiotic mapping) or by chromosome fragmentation (*e.g.*, in **radiation hybrid mapping** and **HAPPY mapping**, Box 10.1). The principle behind genetic mapping is that the chance fragment of DNA decreases with increasing distance between them.

Genetic maps have a resolution between cytogenetic and physical maps but those based on recombination frequencies are distorted by areas of relatively high or relatively low recombination (*i.e.,* recombination hot spots and recombination cold spots; Box 10.1). Additionally, the frequency of recombinations is not the same in all species, thus, genetic map units reflect different physical distances in different species.

Box 10.1

1. **Radiation hybrid mapping.** Where chromosome breaks are introduced by irradiating somatic cell hybrids. Radiation hybrid maps are calibrated in **CentiRays** (cR_x) where *x* is the dosage of X-rays in Rads; 1 cR being equivalent to 1% frequency of separation.
2. **HAPPY mapping**. Where genomic DNA is sheared by vortexing (mixing by means of a rapid whirling or circular motion) or sonication (disrupting by treatment with high-frequency sound waves).
3. **Recombination hotspots** are often endonuclease target sites because cleavage provides single-stranded DNA for the initiation of homologous recombination.
4. **Recombination coldspots** are often sites where homologous DNA fail to synapse effectively. This usually occurs where there is steric hindrance, *e.g.*, in an individual heterozygous for a chromosome inversion or translocation where synapsis at the break points is prevented.

2. Cytogenetic map. It is created by correlating phenotypes with observable chromosomal rearrangements and deficiencies. This type of map has a low resolution and is applicable to the few species, *e.g.*, mammals and *Drosophila*, which display either natural or artificially inducible reproducible chromosome banding patterns. Cytogenetic maps of simple genomes (viruses, plasmids, etc.) are used to locate regions which differ in their physical properties (*e.g.*, denaturation mapping).

3. Physical or molecular map. It is created by ordering cloned fragments of genomic DNA and is calibrated in real units (base pairs, kilobase pairs, megabase pairs). The physical map has the highest resolution and is the ultimate aim of mapping projects.

10.1 LINKAGE MAPPING

Our discussion of linkage and crossing over has made clear so far, that 1. Because the number of genes usually exceeds the number of chromosomes in different species, therefore, many genes have to be located on same chromosomes. 2. The genes remain arranged linearly on the chromosomes and they have no option before them, except to behave according to the chromosomes during gametogenesis and inheritance, therefore, all the genes on a chromosome inherit together and are said to be linked with each other to form linkage groups. 3. The number of linkage groups corresponds to the number of homologous pairs of chromosomes or bivalents of the species. 4. The linked genes do not always remain linked, but, occasionally are departed from other members of their linkage groups by crossing over. 5. The closely linked genes have less chances of departure or frequency of crossing over than the widely located genes. 6. And each gene has definite order and location in a linkage group or chromosomes, as the crossing over frequency has been found constant for two given linked genes of a species. For example various genetic crosses have shown that in *Drosophila*, among three mutant sex-linked genes (*viz.*, genes for white eyes, yellow body and cut wings), the genes for white and yellow are always found to have 1% crossing over and similarly, white and cut genes are found to have 20% crossing over. The percentage or frequency of crossing over, thus, appears to be closely related to physical distance between genes. When one knows all the genes, linkage groups and number of linkage groups of a species, it becomes possible for him that by adopting the crossing over as a tool, he may determine the relative distances between the genes in a linkage group and also their order and may give diagrammatic representation of chromosomes showing the genes as points separated by distances proportional to the amount of crossing over. Such a diagrammatic, graphical representation of relative distances between linked genes of a chromosome is called **linkage** or **genetic map**. Because, such a linkage or genetic map is the outcome of crossing over studies, is also called **cross over map.**

Construction of a Linkage Map or Genetic Mapping

The method of construction maps of different chromosomes is called **genetic mapping**. The genetic mapping includes following processes:

1. Determination of Linkage Groups

Before starting the genetic mapping of the chromosomes of a species, one has to know the exact number of chromosomes of that species and then, he has to determine the total number of genes of that species by undergoing hybridization experiments in between wild and mutant strains. By the same hybridization techniques, it can also be easily determined that how many phenotypic traits remain always together or linked and consequently their determiners or genes during the course of inheritance. And thus, the different linkage groups of a species can be worked out.

2. Determination of Map Distance

The intergene distance on the chromosomes cannot be measured in the customary units employed in light microscopy, geneticists use an arbitary unit to measure the **map unit**, to describe distances between linked genes. A map unit is equal to 1 per cent of crossovers (recombinants); that is, it represents the linear distance along the chromosome for which a recombination frequency of 1 per cent is observed. These distances can also be expressed in **morgan units;** one morgan unit represents 100 per cent crossing over. Thus, 1 per cent crossing over can also be expressed as **1 centimorgan (lcM),** 10 per cent crossing over as **l decimorgan** and so on. The Morgan unit is named in honour of **T.H. Morgan**; however, most geneticists prefer map units.

Quite interestingly, it is now possible to calculate the size of many genes, as well as distances separating them, and to photograph genes in the electron microscope (see **Burns** and **Bottino**, 1989).

Examples

1. If a F_1 hybrid having the genotype *Ab*/*aB* produces 8% of cross over gametes *AB* and *ab*, then the distance between *A* and *B* is estimated to be 16 map units or centimorgan.
2. If the map distance between the gene loci *B* and *C* is 12 centimorgan, then 12% of gametes of genotype *BC*/*bc* should be crossover types, *i.e.*, 6% *bC*.

Because, each chiasma produces 50% crossover products, 50 per cent crossing over is equivalent to 50 map units or centimorgans. If the mean number of chiasmata is known for a chromosome pair, the total length of the map for that linkage group may be predicted:

Total length = mean number of chiasmata × 50.

Two Point Test Cross.

The percentage of crossing over between two linked genes is calculated by test crosses in which a F_1 dihybrid is crossed with a double recessive parent. Such crosses because involved crossing over at two points, so called **two point test crosses.** For example, a dihybrid having the genotype *AC*/*ac* is test crossed with a double recessive parent (*ac*/*ac*), then among F_2 test cross hybrids we may get 37% dominant genes at both gene loci (*AC*/*ac*), 37% recessive genes at both gene loci (*ac*/*ac*), 13% dominant gene at first gene locus and recessive gene at the second gene locus (*Ac*/*ac*), and 13% recessive gene at first gene locus and dominant gene at second gene locus (*aC*/*ac*). The last two groups (*i.e.*, 13% *Ac*/*ac* and 13% *aC*/*ac*) were produced by crossover gametes (13 + 13) from the dihybrid parent. Thus, 26% of all gametes (13 + 13) were of cross over types and the distance between the loci *A* and *C* is estimated to be 26 centimorgans. Because, double crossovers usually do not occur between genes less than 5 centimorgans apart, so for genes further apart, the **three point test crosses** are used.

Three Point Test Cross.

A three point test cross or trihybrid test cross (involving three genes) gives us information regarding relative distances between these genes, and also shows us the linear order in which these genes should be present on chromosome. Such a three point test cross may be carried out if three points or gene loci on a chromosome pair can be identified by marker genes. If, in addition to genes *A* and *C* indicated above, a third marker gene *B* is located in fairly close proximity in the same linkage group, all three markers may be used together in conducting a more precise analysis of the map distance and the relative position of the three points.

Suppose that we test cross trihybrid individuals of genotype *ABC*/*abc* and find in the progeny the following:

36% *ABC*/*abc*	9% *Abc*/*abc*	4% *ABc*/*abc*	1% *AbC*/*abc*
36% *abc*/*abc*	9% *aBC*/*abc*	4% *abC*/*abc*	1% *aBc*/*abc*
72% Parental type	18% Single crossovers: between *A* and *B* (region I)	8% Single crossovers: between *B* and *C* (region II)	2% Double crossover

To find the distance *A*–*B* we must count all crossovers (both singles and doubles) that occurred in region I = 18% + 2% or = 20% or 20 map units between the loci *A* and *B*. To find the distance *B*–*C* we must again count all crossovers (both singles and doubles) that occurred in region II = 8%+ 2% = 10% or 10 map units between the loci *B* and *C*. The *A* – *C* distance is, therefore, 30 map units when double crossovers are detected in a three point linkage experiment and 26 map units when double crossovers are undetected in the two-point linkage experiment above.

Without the middle marker (*B*), double crossovers would appear as parental types and hence we underestimate the true map distance (crossover percentage). In this case the 2% double crossovers would appear with the 72% parental types, making a total of 74% parental types and 26% recombinant types. Therefore, for any three linked genes whose distances are known, the amount of detectable crossovers between the two outer markers *A* and *C* when the middle marker *B* is missing is; (*A*–*B* crossover percentage) plus (*B*–*C* crossovers percentage) minus (2 × double crossover percentage).

3. Determination of Gene Order

After determining the relative distances between the genes of a linkage group, it becomes easy to place genes in there proper linear order. For example, if the linear order of three genes *ABC* is to be determined, then these three genes may be in anyone of three different orders depending upon that which gene is in the middle. For the time being we may ignore left and right end alternatives. If double crossovers do not occur, map distances may be treated as completely additive units. Now, if we suppose that the distance between the genes *A*–*B* = 12, *B*–*C* = 7 *A*–*C* = 5, we can determine the order of genes correctly in the following manner:

Case I. Let us assume that gene A is in the middle (*e.g.*, *B* – *A* – *C*):

B 12 *A*	*A* 5 *C*
B 7 *C*	

In this case because, the distances between *B*–*C* are not equitable, genes *A* cannot be in the middle.

Case II. Let us assume that gene *B* is in the middle (*e.g.*, *A*–*B*–*C*):

A 12 *B*	*B* 7 *C*
A 5 *C*	

In this case, because the distance between *A* – *C* are not equitable, therefore, gene *B* cannot be in the middle.

Case III. Let us assume that gene *C* is in the middle (*e.g.*, *A* – *C* – *B*).

A 5 *C*	*C* 7 *B*
A 12 *B*	

In this case, because the distances between *A*–*B* are equitable, therefore, gene *C* must be in the middle.

Thus, the relative distances and ordering of genes in a linkage group are determined in separate segments by two point test crosses or three point crosses, as the case may be.

4. Combining Map Segments.

Finally, the different segments of maps of a complete chromosome are combined to form a complete genetic map of 100 centimorgans long for a chromosome.

Example. For example, suppose we have to combine following three map segments:

(*i*) *a* 8 *b* 10 *c*

(*ii*) *c* 10 *b* 22 *d*

(*iii*) *c* 30 *e* 2 *d*

We can superimpose each of these segments by aligning the genes shared in common.

(*i*) *a* 8 b 10 C

(*ii*) *d* 22 *b* 10 *c*

(*iii*) *d* 2 *e* 30 *c*

Then finally we may combine the three segments into one map:

The *a* to *d* distance = (*d* to *b*) – (*a* to *b*) = 22 – 8 = 14

The *a* to *e* distance = (*a* to *d*) – (*d* to *e*) = 14 – 2 = 12

d 2 *e* 12 *a* 8 *b* 10 *c*

Model Organism for Gene Mapping

Currently, there is an intense collaborative international efforts to generate genetic and physical maps of several model organisms, *e.g.*,

1. Bacteria (*E. coli, B. sublitis*);
2. Yeast (*Sacchromyces cerevisiae* and *S. pombe*);
3. Nematode (*Caenorhabditis elegans*);
4. Fruit fly (*Drosophila melanogaster*);
5. Puffer fish (*Fugu rubripes*);
6. Zebra fish (*Danio rerio*);
7. Mouse and rats;
8. Commercial and domestic animals (cattle, pigs, sheep, cats and dogs);
9. Human beings (*Homo sapiens*);
10. Angiosperm plant (*Arabidopsis thaliana*);
11. Plants of commercial value, *e.g.*, rice, wheat and maize.

The first linkage map has been constructed for two chromosomes of *Drosophila* by **Strutevant** in 1911. The linkage maps of other chromosomes of *Drosophila* have been constructed by **C.B. Bridges.** The linkage or genetic mapping in maize has been done by **McClintock** under the leadership of **R.A. Emerson**.

Significance of Chromosome Maps

The study of chromosome maps reveals following facts:

1. Shows the number of chromosomes and the number of linkage groups of an organism.
2. Shows exact location, arrangement and exact sequence of genes on a chromosome.
3. Shows the relative distance and chances of crossing over between the genes on each of the chromosome pair.

Interference

In most higher organisms it has been found that one chiasma formation reduces the probability of another chiasma formation in an immediately adjacent region of the chromosome, probably because of physical inability of the chromatids to bend back upon themselves within certain minimum distances. The tendency of one crossover to interfere with the other crossover is called **interference**. The phenomenon of interference was discovered by **Muller** in 1911. It is found to be **inversely proportional** to the distance between the points of crossing over. It means interference is maximum over a short distance and decreases as the distance increases. In *Drosophila* interference prevents double crossing over almost together between genes ten units apart.

Types of Interference

1. Positive interference. Where one crossover inhibits another.

2. Negative interference. Where one crossover stimulates another.

In eukaryotes, it is generally observed that recombination shows **positive interference**, *i.e.*, one crossover inhibits the initiation of a second crossover nearby. Thus, when the double crossovers are altogether absent, we can say that interference is cent-per cent. If the frequency of double crossovers is equal to the expected value, it means that there is no interference. Recombination between very close markers shows evidence of **negative interference**, *i.e.*, several crossovers appear to be clustered.

Factors controlling interference. Interference varies from chromosome to chromosome and within a chromosome from one point to the other. Generally, it is maximum near the centromere and at the two ends of the chromosome (telomeres). Therefore, for accurate mapping interference in different parts of different chromosomes should be taken into consideration.

Crossing over Among three Linked Genes

A test-cross for three linked genes, *e.g.*, *C* (colourless aleurone), *sh* (shrunken grains) and *wx* (waxy endosperm) in maize/corn, yields eight types of gametes and phenotypes (Table 10.1). Remember that in a test-cross, the phenotype of a progeny is not affected by the test-cross parent as it contributes only the recessive alleles of the genes. Two of the eight types of progeny (*e.g.*, *C, Wx, Sh* and *ċ wx sh*) are the most frequent and represent the parent or non-recombinant types, while two others (*e.g.*, *C, Wx, sh* and *c wx Sh*) the least frequent and are **double crossovers**. The remaining four types (*e.g.*, *C wx sh, c Wx sh, cWx Sh* and *C wx sh*) are produced by single crossing over between the three linked genes. As a rule, the parental types are the most frequent, the double cross overs are the least frequent, while the single crossovers are intermediate between the two.

Table 10.1 The different types of progeny obtained from a test-cross for three genes *c*, *sh* and *wx* in maize (Singh, 2003).

S. No.	Genotype	Phenotype	Number	Per cent frequency*
1.	C Wx Sh/c wx sh	Coloured, nonwaxy, full	2777	39.7
2.	c wx sh/ c wx sh	Colourless, waxy, shrunken	2708	38.7
3.	C Wx sh/c wx sh	Coloured, non waxy, shrunken	116	1.7
4.	c wx Sh/c wx sh	Colourless, waxy, full	123	1.8
5.	C wx sh/ c wx sh	Coloured, waxy, shrunken	643	9.2
6.	c Wx sh/c wx sh	Colourless, nonwaxy, shrunken	626	8.9
7.	c Wx Sh/c wx sh	Colourless, nonwaxy, full	4	0.06
8.	C wx sh/c wx sh	Colourled, waxy, shrunken	3	0.04

$$\text{*Per cent frequency} = \frac{\text{Number of progeny in a phenotypic class}}{\text{Total number of progeny}} \times 100$$

Coefficient of Coincidence

Double crossovers are produced by two simultaneous crossing overs one each on either side of the gene located between two genes (*e.g.* the gene *sh* located between *c* and *wx* [Fig. 10.1]). If the occurrence of crossing over in the two regions (*e.g.*, between the genes *c* and *sh*, and between *sh* and *wx*,) were independent of each other, the frequency of double crossovers will be product of the frequencies of crossing overs in two regions. The independence of crossing over in two regions means that the occurrence of crossing over in one region does not affect the chance of its occurrence in the other region. Therefore, the expected frequency of double crossovers may be estimated from the data of Table 10.1 as follows:

The expected frequency of double crossovers may be estimated from the data of Table 10.1 as follows:

Frequency (%) of crossing over between *c* and *sh* (= α) = 3.6

Frequency (%) of crossing over between *sh* and *wx* (= β) = 18.2

The expected frequency (%) of double crossing overs = the product of the frequencies of crossing overs between *c* and *sh*, and between *sh* and *wx*.

$$= \alpha \times \beta/100$$

$$= (3.6 \times 18.2)/100$$

$$= 0.66\%$$

The product αβ (= 3.6 × 18.2) is divided by 100 since these frequencies are already in per cent. The frequency observed double crossovers is only 0.1% (= sum of the frequencies of double cross over types *C Wx sh* and *c wx Sh* = 0.06 + 0.04 per cent). The observed frequency is only 15.1 per cent (= 0.1/0.66 × 100) of the expected frequency. The ratio between the observed and the expected frequencies of double crossovers is called **coefficient of coincidence**. Thus,

$$\text{Coefficient of coincidence} = \frac{\text{Observed frequency of double crossovers}}{\text{Expected frequency of double crossovers}}$$

$$= 0.10/0.66 = 0.151$$

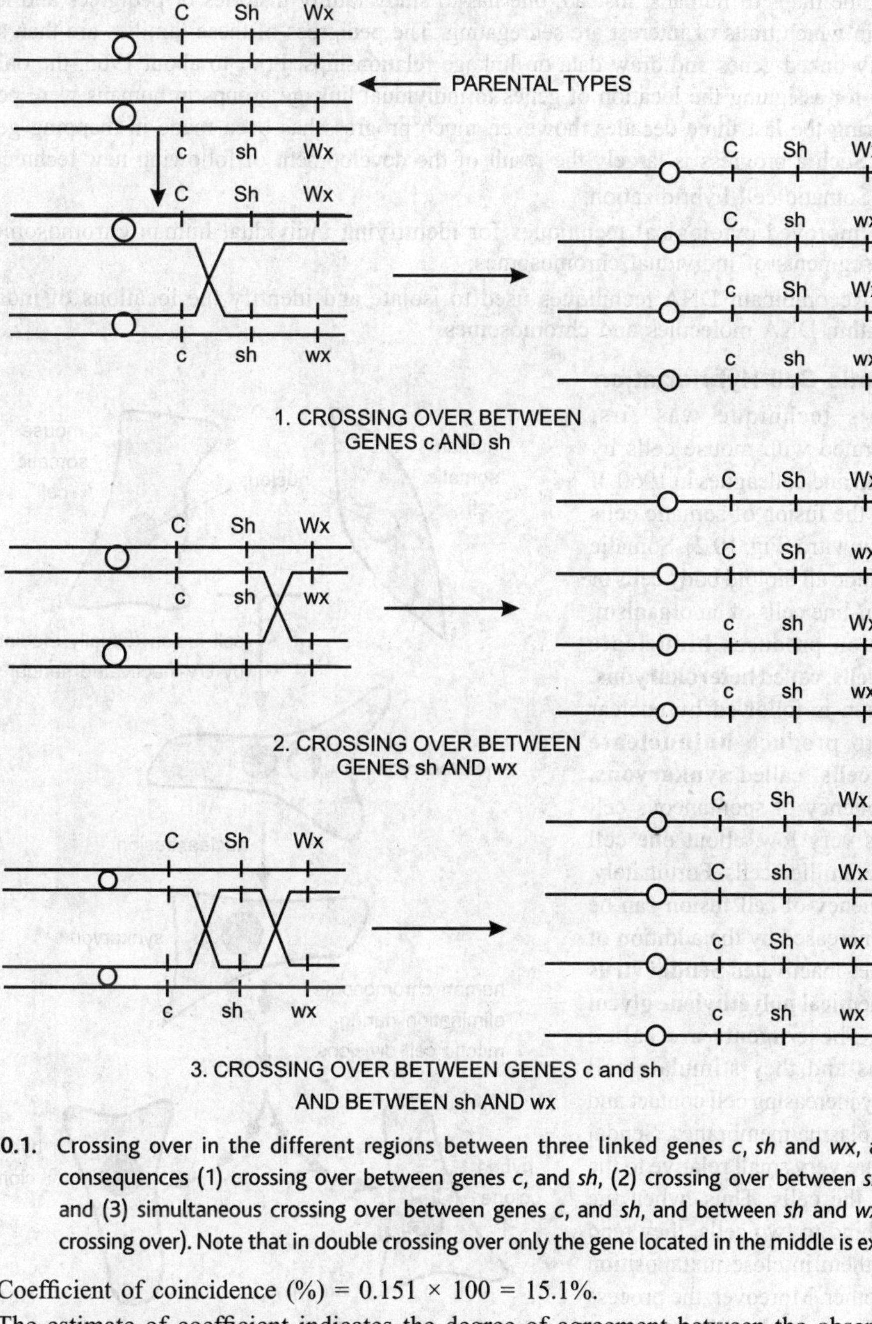

Fig. 10.1. Crossing over in the different regions between three linked genes *c*, *sh* and *wx*, and their consequences (1) crossing over between genes *c*, and *sh*, (2) crossing over between *sh* and *wx* and (3) simultaneous crossing over between genes *c*, and *sh*, and between *sh* and *wx* (double crossing over). Note that in double crossing over only the gene located in the middle is exchanged.

Coefficient of coincidence (%) = 0.151 × 100 = 15.1%.

The estimate of coefficient indicates the degree of agreement between the observed and the expected frequencies of double crossovers.

10.2 PHYSICAL MAPPING OF GENES IN HUMAN CHROMOSOMES

Apparently, human genes cannot be mapped by designed crosses of the types (*Drosophila, Zea maize, Neurospora,* yeast *etc.*) as described in earlier sections of this chapter. For preparing chromosome maps of humans, instead, one has to study family histories or pedigrees and identify families in which traits of interest are segregating. The pedigrees of these families are then studied to identify linked genes and draw data on linkage relationships. Prior to about 1965, the only data available for assigning the location of genes to individual linkage groups in humans were pedigree data. During the last three decades, however, much progress has been made in mapping genes in humans. Such a progress is largely the result of the development of following new techniques:

1. Somatic cell hybridization;

2. Improved cytological techniques for identifying individual human chromosomes and specific segments of individual chromosomes;

3. Recombinant DNA techniques used to isolate and identify the locations of individual genes within DNA molecules and chromosomes.

1. Somatic Cell Hybridization

This technique was first demonstrated with mouse cells by **G. Barski** and colleagues in 1960. It involves the fusion of somatic cells growing in vitro (Fig. 10.2). Somatic cells include all diploid body cells or non-germ-line cells of an organism. Cell fusion produces **binucleate hybrid cells,** called **heterokaryons.** Cell fusion is followed by nuclear fusion to produce **uninucleate hybrid cells,** called **synkaryons.** The frequency of spontaneous cell fusion is very low, about one cell fusion per million cells. Fortunately, the frequency of cell fusion can be greatly increased by the addition of ultraviolet-inactivated **Sendai virus** or the chemical **polyethylene glycol (PEG).** These agents are called **fusogens** and they stimulate cell fusion by increasing cell contact and altering plasma membranes. Sendai viruses are very small relative to the sizes of the cells. Thus, when the viruses bind to two cells, they tend to hold them in close juxtaposition to each other. Moreover, the process involved in normal viral penetration of cells probably facilitate fusion of the membranes of the juxtaposed cells.

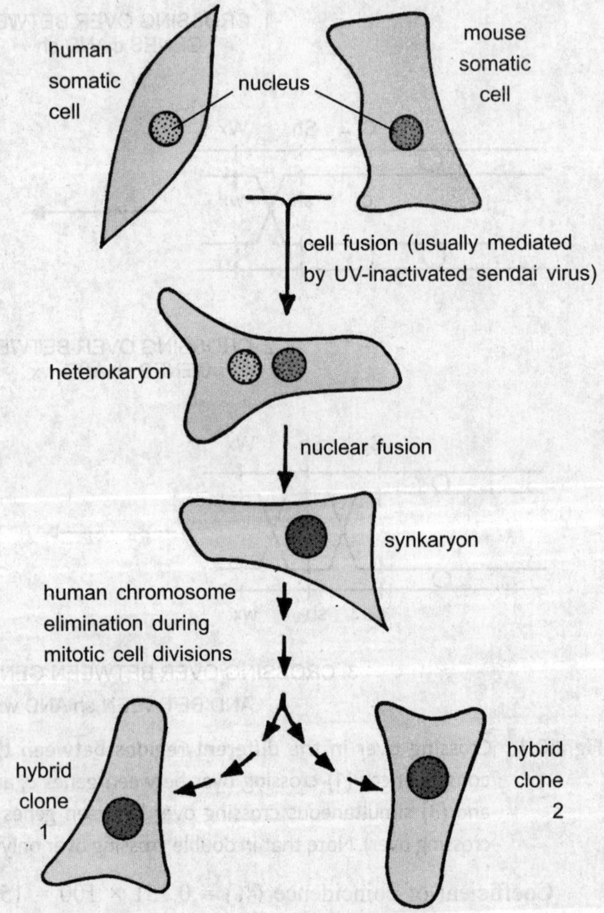

Fig. 10.2. Method of formation of stable mouse - human somatic - cell hybrids and the elimination of human chromosomes during subsequent mitotic divisions (after Gardner, Simmons and Snustad, 2001).

Somatic-cell hybrids can be produced between two cells of the same species (called **monospecific hybrids**) or two cells from different species (called **interspecific hybrids**). **Mouse-human somatic cell hybrids** have been particularly useful for mapping human genes. They are usually made using established mouse cell culture lines and human fibrocytes and leucocytes.

Mouse-human somatic-cell hybrids have following *three* important features that make them especially useful for genetic analysis:

1. **The mouse and human chromosomes are easily distinguished.** This makes the identification of human chromosomes present in hybrid cells a fairly simple task.

2. All the mouse chromosomes are usually retained in the hybrid cells, but **only 1-20 of human chromosomes are retained** in hybrid cell clones. Thus, most of the 46 human chromosomes are eliminated during subsequent mitotic divisions of the hybrid cells. Further, most importantly chromosomes are eliminated at random.

3. **Both sets of chromosomes, human and mouse are expressed in the hybrid cells.** The mouse and human genomes contain many closely related genes that code for gene-products with identical functions. Nevertheless, these gene products are usually different enough to be distinguished biochemically.

Normally, the genes are shown to be located on the specific chromosomes by correlating the phenotypes (*e.g.*, enzyme activities) controlled by these genes with the presence of individual chromosomes in hybrid cell clones. For example, suppose that human cells that make the enzyme thymidine kinase **(TK^+ cells)** were hybridized with mouse cells that cannot make this enzyme **(TK^- cells).** Assume that the hybrid cells were cultured for about 30 cell generations, and then five different cell clones were examined to determine which human chromosomes were present. If the results shown in Table 10.2 were obtained, one could deduce that the gene controlling the synthesis of thymidine kinase (gene *TK*1) is located on chromosome 17, because it is the only human chromosome present in all four TK^+ hybrid cell clones.

Table 10.2 Somatic-Cell Hybridization Data

Hybrid Clone No.	*Thymidine* Kinase Activity	*Human Chromosomes* (+ = Present, – = Absent) X	Y	1	2	4	7	9	10	15	17	18	21
1	+	+	+	–	+	–	–	+	+	+	+	–	+
2	+	+	–	+	+	–	+	+	–	+	+	+	–
3	+	–	–	+	+	+	+	–	–	–	+	–	–
4	+	–	–	+	–	–	–	–	+	–	+	–	+
5	–	+	–	–	+	–	–	+	–	+	–	–	–

Each of the 24 different human chromosomes (*i.e.*, 22 Autosomes + X chromosome + Y chromosome) now can be identified using improved staining procedures such as *G*-banding.

2. HAT Selection Procedure

Most somatic cell hybridization experiments are currently done using selection pressures that prevent the growth of the parental cells while allowing growth of hybrid cells produced by cell fusion.

Box 10.2

Somatic cell hybrids can be selected for retention of a given human chromosome or chromosome fragment if it corrects an otherwise lethal abnormality in the rodent (mouse or hamster) cell. There are following two frequently used systems:

1. **HAT selection.** Somatic cell hybrids can be forced to retain human chromosome 17 using thymidine kinase-deficient (TK^-) rodent cells and growing the hybrids in HAT (hypoxanthine-aminopterin-thymidine) medium. TK^- cells are killed in HAT medium, but are rescued by the human *TK* gene on chromosome 17.
2. **G418 selection.** Radiation hybrids can be selected for the presence of a particular human chromosome segment if it has been tagged by incorporation of a neomycin resistant (neo^R) gene. The neomycin analog *G*418 kills nonresistant cells. Neo^R is a typical example of a **dominant select marker.**

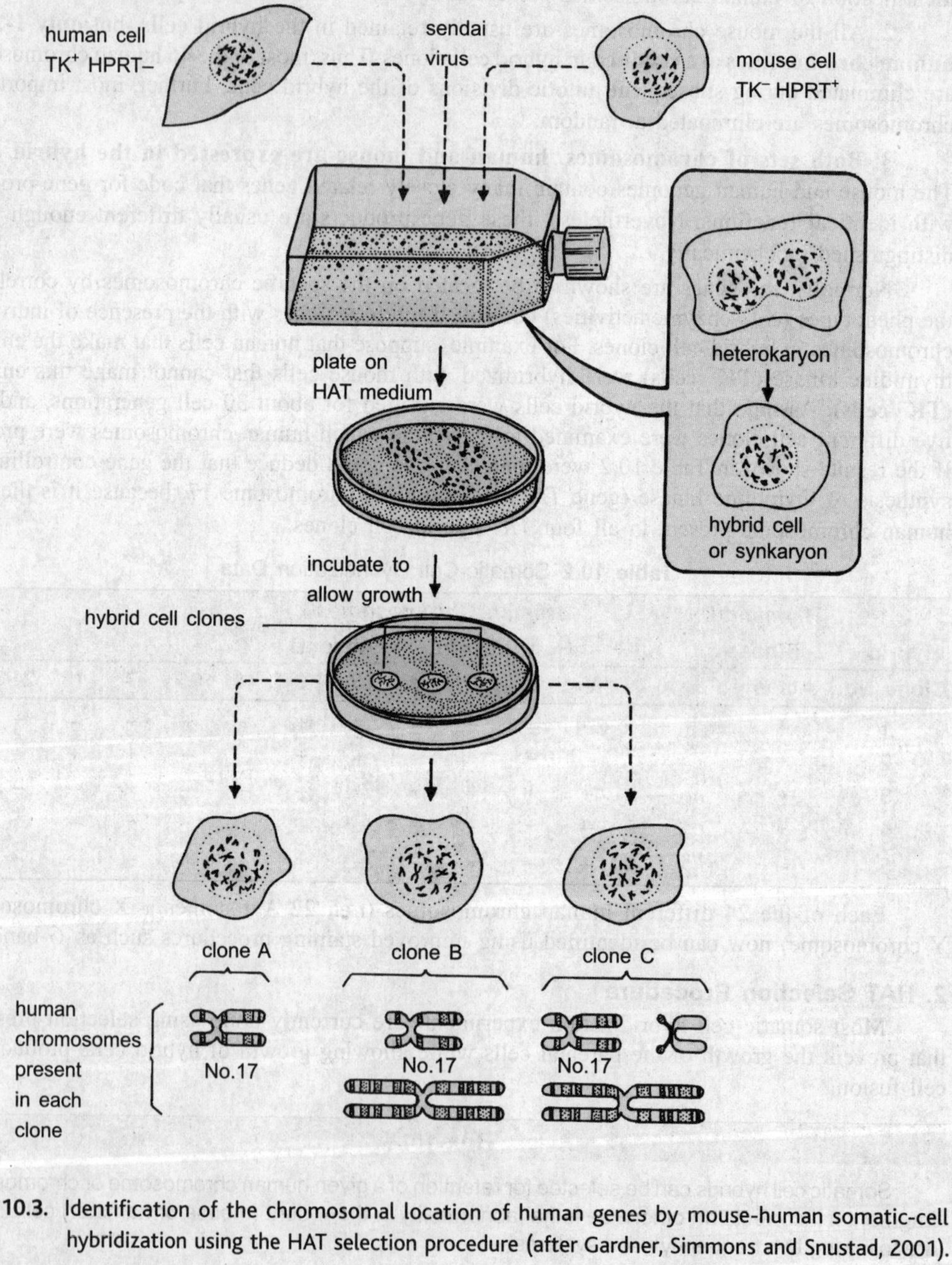

Fig. 10.3. Identification of the chromosomal location of human genes by mouse-human somatic-cell hybridization using the HAT selection procedure (after Gardner, Simmons and Snustad, 2001).

One widely used method of selection, is **HAT selection procedure** (HAT = hypoxanthine aminopterin thymidine (Box 10.2). In this technique, TK^- mouse cells are fused with human cells that cannot make the enzyme hypoxanthine-guanine phosphoribosyltransferase **(HPRT).** The HPRT enzyme is deficient in cells of individuals with Lesch-Nyhan syndrome which is caused by an X-linked recessive allele (Box 10.3).

Box. 10.3

Lesch-Nyhan syndrome is a rare and usually fatal genetic disorder of male children that is inherited as an X-linked recessive trait. It is characterised by **hyperuricemia** (= excess uric acid in the blood), mental retardation, spasticity, compulsive biting of the lips and fingers. Such patient also lacks HPRT enzyme.

In the appropriate selective medium both of these enzymes are required for growth. Thus, neither parent cell can grow, but hybrids produced by cell fusion can grow. The TK^- mouse cells are $HPRT^+$ (they make enzyme HPRT), and the $HPRT^-$ human cells are TK^+ (they make enzyme TK). Since both the mouse and the human chromosomes are expressed in the hybrid cells, they have both enzymes required for growth in the selective medium.

Biochemical basis of HAT selection method. The biochemical basis of HAT selection procedure involves two important pathways for the biosynthesis of the nucleotides required for DNA synthesis (Fig. 10.4).

1. One is a major pathway by which nucleotides are synthesized *de novo* from simple sugars and amino acids. This pathway is *blocked by the antimetabolic aminopterin*. In the presence of aminopterin, however, cells can still synthesize the nucleotides essential for growth *if certain precursors are added* to the medium and if the second pathway, the **salvage pathway** is operative.

2. Key steps of the salvage pathway are catalized by enzymes as TK and HPRT. As the name suggests, hypoxanthine-guanine phosphoribosyl-transferase catalyzes the conversion of hypoxanthine to guanine and thymidine kinase ("kinases" catalyze phosphate group transfers). Thus **TK^+ and $HPRT^+$ (= $HGPRT^+$) cells can grow in the presence of aminopterin if hypoxanthine and thymidine are added to the medium.** Medium containing these three compounds, *i.e.*, hypoxanthine, aminopterin and thymidine is called **HAT medium.** HAT medium selects for TK^+ – $HGPRT^+$ hybrid cells, since neither the TK^- $HGPRT^+$ parental mouse cells nor the TK^+ $HGPRT^-$ parental human cells can grow in such a medium.

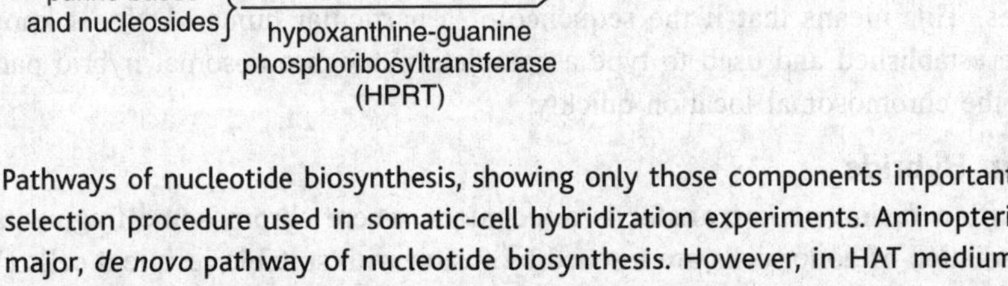

Fig. 10.4. Pathways of nucleotide biosynthesis, showing only those components important to the HAT selection procedure used in somatic-cell hybridization experiments. Aminopterin blocks the major, *de novo* pathway of nucleotide biosynthesis. However, in HAT medium, pyrimidine and purine nucleotides can be synthesized by the salvage pathway from the hypoxanthine and thymidine in this medium if the cells make thymidine kinase and hypoxanthine guanine phosphoribosyl transferase (TK^+ $HPRT^+$ cells) (after Gardner, Simmons and Snustad, 2001).

If the just described selection procedure is used, human chromosome elimination may occur at random. However those cells that have lost the human chromosome carrying the gene coding for thymidine kinase will **not** be able to grow. Only those hybrid cell clones that have retained this chromosome will grow on HAT medium. The analysis of a set of such clones should reveal that they all have one chromosome in common. In fact, clones of hybrid cells selected in this way always contain human chromosome 17, indicating that it carries the gene coding for thymidine kinase.

3. Hybrid Cell Panel

The human chromosomes in somatic cell hybrids can conveniently be identified by **PCR screening** with sets of chromosome-specific primers. By collecting hybrid cell lines with different sets of human chromosomes, it is possible to generate a **hybrid cell panel** that can be used to map any human DNA sequence to a specific chromosome. To do this, the presence of the human sequence of interest in each of the hybrid cell lines is assayed. A PCR assay can be used (with primers specific for that sequence) or the relevant DNA sequence can be labeled and used as a hybridization probe (assuming that the correct human signal can be distinguished from any similar sequences that may be present in the rodent genome). Localization of a human DNA sequence to a single chromosome can be extrapolated by deduction: the chromosome must be common to all cell lines which are positive for the test and absent from all cell lines which are negative for the test, that is, all hybrid should be concordant (**Strachan** and **Read**, 1996).

4. Use of Monochromosomal Hybrids in Human Chromosome Mapping

A disadvantage of traditional somatic cell hybrids is that the hybrid cells generally contain several human chromosomes rather than just a single human chromosome. In order to limit the amount of human genetic material transferred to a recipient rodent cell, the technique of **microcell fusion** (**Fournier** and **Ruddle**, 1977) can be applied.

Technique of microcell fusion. The first step of this technique is to subject the donor cells to prolonged mitotic arrest by continued exposure to an inhibitor of mitotic spindle formation, such as **colcemid.** As a result of this treatment the chromosome content of a cell is partitioned into discrete subnuclear packets, called **micronuclei.** The micronuclei can be physically isolated from the cells by centrifugation in the presence of **cytochalasin B** (a mitotic spindle inhibitor), resulting in the formation of **microcells,** *i.e.*, particles consisting of a single micronucleus and a thin rim of cytoplasm surrounded by an intact plasma membrane. In much the same way as for normal donor cells, the microcells can be fused with recipient cells. This is called **microcell-mediated transfer.** The resulting hybrids are known as **microcell hybrids.** Some of these hybrids contain a few donor chromosomes, but the simplest contain a single donor chromosome **(monochromosomal hybrids) (Warburton** *et al.,* 1990).

The main advantage of monochromosomal hybrids is that they provide a clearcut evidence for the presence or absence of a defined human DNA on a specified chromosome. Monochromosomal human-rodent hybrid cell lines have now been established for each of the human chromosomes. This means that if the sequence of a particular human DNA is known, a PCR assay can be established and used to type a complete monochromosomal hybrid panel, thereby establishing the chromosomal location quickly.

5. Radiation Hybrids

Irradiation fusion gene transfer is a technique where chromosome fragments generated by lethal irradiation of donor cells are rescued by fusion with suitable recipient cells (**Walter** and **Goodfellow,** 1993). In human genetics, this method has been popular for use on somatic cell hybrids containing a single human chromosome. Controlled X-ray irradiation cleaves all the chromosomes in the hybrid essentially at random into comparatively small pieces. In order to rescue fragments obtained from the human chromosome, the lethally irradiated hybrid cells is fused with rodent cells, and a selection system is applied to ensure the survival only of hybrid cells (Fig. 10.5). Cells containing human sequences can then be identified conveniently by screening for the *Alu* repeat sequence. Stable propagation of the human chromosome fragments in such cells is only possible if the fragments have inserted into the rodent chromosomes. Initially, therefore, there may be a comparatively large population of chromosome fragments before continued culture of the cells integrated into the rodent chromosomes (breakage in chromosomes are quite common, necessitating effective DNA repair systems). This results in a collection of **irradiation-induced hybrids,** often called **radiation hybrids.**

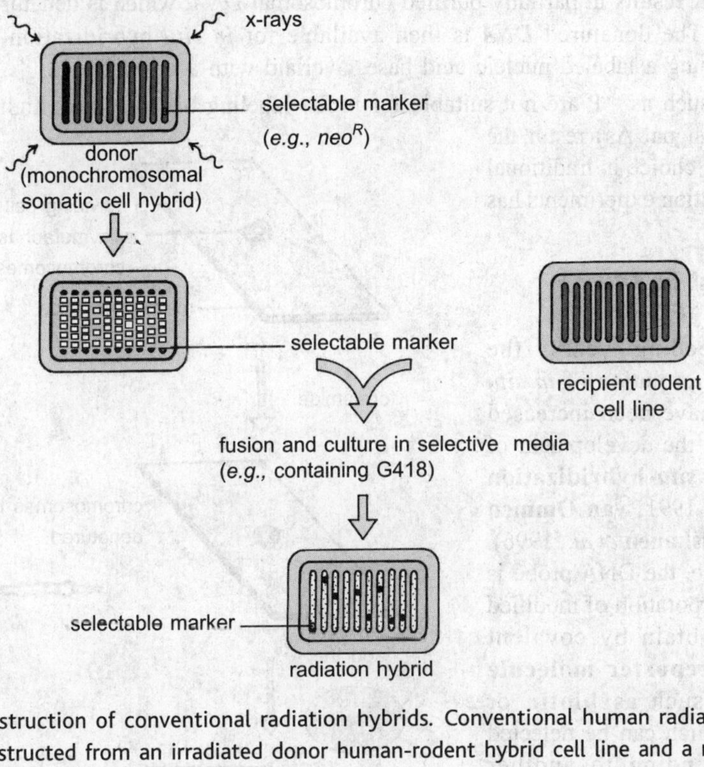

Fig. 10.5. Construction of conventional radiation hybrids. Conventional human radiation hybrids are constructed from an irradiated donor human-rodent hybrid cell line and a recipient rodent-cell line. The donor cell line contains a single human chromosome (black) and the full complement of rodent-chromosome (shaded, *i.e.*, red), one of which contains an integrated dominant selectable marker such as neo^R. *i.e.*, neomycin resistance gene. Following irradiation of the donor cell line, all the chromosomes are fragmented to give several fragments per chromosome. Culture of the fusion products in a medium containing the neomycin analog G418 selects against growth of the original recipient rodent cells. The resulting radiation hybrids will contain the selectable marker plus several fragments of the human chromosome integrated randomly into rodent chromosomes (after Strachan and Read, 1996).

When DNA samples from a panel of such radiation hybrids are screened by hybridization against a series of DNA clones, or by corresponding PCR assays, the patterns of cross-reactivity can be interpreted statistically to produce linear map order for the DNA clones (**Cox** *et al.*, 1990).

Typically about 100-200 hybrids are required to construct a map of a human chromosome. This type of approach is not practically feasible for whole gene mapping because of the huge number of hybrid cells that would be required. However, a variant form of radiation hybrid mapping involves the use of human fibroblast cells as the starting cells instead of a monochromosomal somatic cell hybrid. This has made possible of construction of reasonably high resolution maps of the entire genome with a single panel of 100 to 200 hybrids (**Walter** *et al.* 1994).

6. Chromosal *in situ* Hybridization

A single procedure for mapping genes and other DNA sequences is to hybridize a suitable labeled DNA probe against chromosomal DNA that has been denatured *in situ.* To do this, an air-dried microscope slide preparation of metaphase or prometaphase chromosomes is made, usually from peripheral blood lymphocytes or lymphoblastoid cell lines. Treatment with *RNase* and proteinase *K* results in partially purified chromosomal *DNA*, which is denatured by exposure to formamide. The denatured *DNA* is then available for *in situ* hybridization with an added solution containing a labeled nucleic acid base, overlaid with a coverslip.

Isotopes such as ^{32}P are not suitable for probe labeling because their high energy causes scattering of the signal. As a result, the radioisotopes of choice in traditional *in situ* hybridization experiments has been ^{3}H.

7. Fluorescent *in situ* Hybridization (FISH)

In succeeding years, the sensitivity and resolution of *in situ* hybridization have been increased significantly by the development of **fluorescent *in situ* hybridization (FISH) (Trask** 1991; **van Ommen** *et al.*, 1995; **Heiskanen** *et al.*, 1996). In this technique, the DNA probe is labeled by incorporation of modified nucleotides, obtain by covalent binding of a **reporter molecule** (a molecule, such as **biotin** or **digoxigenin** which can be detected by specific binding to another molecule). Following hybridization and washing to remove excess probe, the chromosome preparation is incubated in a solution containing a fluorescently labeled affinity molecule which binds to the reporter on the hybridized probe.

FISH has the advantage of providing the rapid results, which can be scored conveniently by eye using a fluorescent microscope. In metaphase spreads, positive signals show as double spots, corresponding to probe hybridized to both sister chromatids.

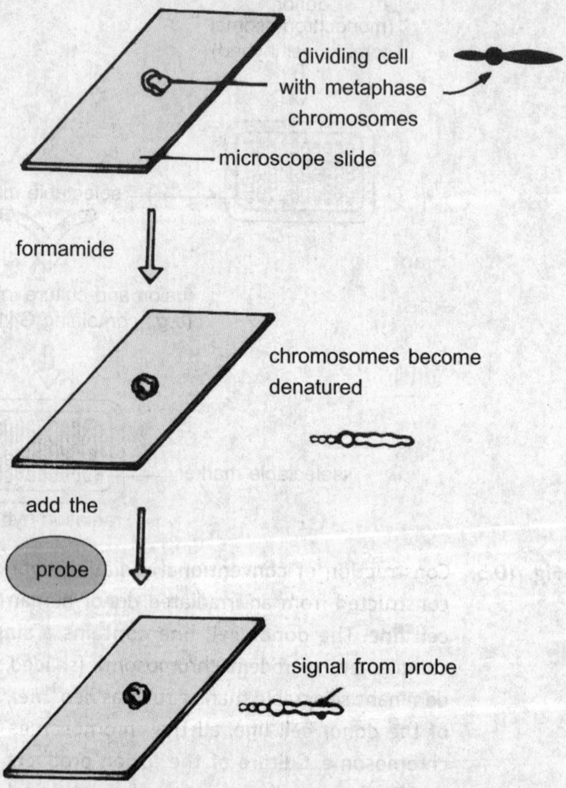

Fig. 10.6. Fluorescent *in situ* hybridization. A sample of dividing cell is dried onto a glass microscope slide and treated with formamide so that the chromosomes become denatured but do not lose their characteristic metaphase morphologies. The position at which the probe hybridizes to the chromosomal DNA is visualised by detecting the fluorescent signal emitted by the labeled DNA (after Brown, 2002).

8. Chromosome Painting

A recently developed method for visualizing each of the human chromosomes in distinct, bright colours, called **chromosome painting,** greatly simplifies differentiating chromosomes of similar size and shape. This technique is a variation of fluorescence *in situ* hybridization. It makes use of probes specific for sites scattered along the length of each chromosome. The probes are labeled with one of two dyes that fluoresce at different wavelengths. Probes specific for each chromosome are labeled with a predetermined fraction of each of the two dyes. After the probes are hybridized to chromosome and the excess removed, the sample is observed with a **fluorescent microscope** in which a detector determines the fraction of each dye present at each fluorescing position in the microscopic field. This information is conveyed to computer, and a special program assigns a false colour image to each type of chromosome (see **Lodish** *et al* 2004).

9. Sequence Tagged Site (STS) Mapping

At present the most powerful physical mapping technique, and the one that has been responsible for generation of the most detailed maps of large genomes, is STS mapping (Box 10.4). A **sequence tagged site** or **STS** is simply a short DNA sequence generally between 100-500 bp in length, that is easily recognizable and occurs only once in the chromosome or genome being studied. To map a set of STSs, a collection of overlapping DNA fragments from

Box 10.4 Three techniques of physical mapping

1. **Restriction mapping** is rapid, easy and provides detailed information, but it cannot be applied to large genomes.
2. **FISH mapping** can be applied to the large genomes and modified versions (*e.g.*, fiber-FISH) can give high-resolution data, but FISH is difficult to carry out and data accumulation is slow, map positions for no more than three or four markers being obtained in a single experiment.
3. **STS mapping** locates the positions of sequence tagged sites in a genome. It has generated most detailed maps of large genome.

a single chromosome or from the entire genome is needed. In the example shown in Fig. 10.7 a fragment collection has been prepared from a single chromosome, with each point along the

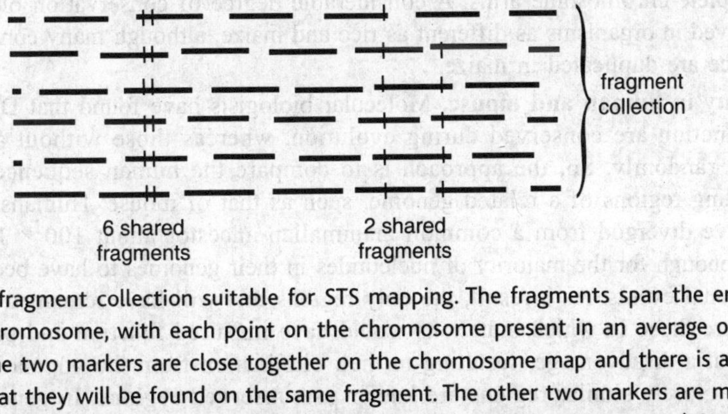

Fig. 10.7. A fragment collection suitable for STS mapping. The fragments span the entire length of a chromosome, with each point on the chromosome present in an average of five fragments. The two markers are close together on the chromosome map and there is a high probability that they will be found on the same fragment. The other two markers are more distant from one another and so are less likely to be found on the same fragment (after Brown, 2002).

chromosome represented on average five times in the collection. The data from which the map will be derived are obtained by determining which fragments contain which STSs. This can be done by hybridization analysis but PCR is generally used because it is quicker and has proven to be more responsive to automation. The chances of two STSs being present on the same fragment will, of course, depend on how close together they are in the genome. If they are very close then there is a good chance that they will always be on the same fragments; if they are further apart then sometimes they will be on the same fragment and sometimes they will not (Fig. 10.7). The data therefore can be used to calculate the distance between two markers, in a manner analogous to the way in which map distances are determined by linkage analysis. Remember that in linkage analysis a map distance is calculated from the frequency at which crossovers occur between two markers. STS mapping is essentially the same, except that each map distance is based on the frequency at which **breaks** occur between two markers.

10.3 SYNTENY

The conservation of linkage order in linkage maps of different species is known as **synteny.** In other words, synteny refers to a pair of genomes in which at least some of the genes are located at similar map positions (**Brown,** 2002).

Synteny occurs when two or more loci are on the same chromosome. Syntenic loci may or may not be linked, but linked loci are always syntenic. Syntenic loci may be far enough apart on a chromosome for crossing over to occur between them frequently (see **Friedman** *et al.,* 1996). The term syntenic has been introduced to refer to genes known to be located on the same chromosome, but for which no information is available regarding their linkage relationship (s) (see **Gardner** *et al.,* 2001).

If two organisms are sufficiently closely related then their genomes might-display synteny, the partial or complete conservation of gene order. Then it is possible to use map information from one genome to locate genes in the second genome. At one time it was thought that mapping the genomes of the mouse and other mammals, which are at least partially syntenic with the human genome, might provide valuable information that could be used in construction of the human genome map. The problem with this approach is that all the close relatives of humans have equally large genomes that are just as difficult to study, the only advantage being that a genetic map is easier to construct with an animal which, unlike humans, can be subjected to experimental breeding programs.

1. Synteny in plants. Generally there is a conservation of linkage order among related species. For example, tomato and potato linkage maps differ with respect to five inversions involving complete chromosome arms. A considerable degree of conservation of linkage order has been observed in organisms as different as rice and maize, although many conserved linkage blocks from rice are duplicated in maize.

2. Synteny in human and mouse. Molecular biologists have found that DNA sequences that have a function are conserved during evolution, whereas those without a function are free to mutate randomly. So, the approach is to compare the human sequence with that of the corresponding regions of a related genome, such as that of mouse. Humans and mice are believed to have diverged from a common mammalian ancestor about 100×10^6 years ago, which is long enough for the majority of nucleotides in their genomes to have been changed by random mutational events. As a result, the only regions that will have remained closely in the two genomes are those in which mutations would have impaired function and put the animals carrying them at a disadvantage, resulting in their elimination from the population by natural selection. Such closely similar regions are known as **conserved regions.** In general, conserved regions represent functionally important exons and regulatory sequences. In contrast, **nonconserved regions** represent DNA whose sequence is generally not critical for function.

Thus, comparative studies of this kind have revealed not only that mice and humans share most of the same genes, but also that large blocks of mouse and human genomes contain these genes in the same order, a feature called **conserved synteny.** Conserved synteny can also be revealed by chromosome painting, and this technique has been used to reconstruct the evolutionary history of our own chromosome by comparing them with those from other mammals (**Alberts** *et al.*, 2002).

Differences Between Genetic and Chromosome Maps

The frequency of crossing over usually varies in different segments of the chromosome but is a highly predictable even between any two genes loci. Therefore, the actual physical distances between linked genes bears no direct relationship to the map distance calculated on the basis of crossover percentages. However, the linear order of linked genes is identical in both maps.

Uses of Genetic Maps

The genetic maps have following uses:

(*i*) The chromosome maps display the exact location, arrangement and combination of genes in a linkage group of chromosomes.

(*ii*) They are useful in predicting results of dihybrid and trihybrid crosses.

QUESTIONS

Long Answer Questions

1. What is a chromosome map? How a chromosome map is constructed?

2. What do you understand by chromosome maps? How are these drawn? Discuss the significance of chromosome mapping.

3. Write a short essay on chromosome mapping and give its significance.

4. State how the position of central gene is determined in a three point test cross?

Short Answer Questions

1. Write short notes on the following:

(*i*) Chromosome mapping (*ii*) Linkage maps (chromosome maps)

2. Explain interference and coincidence and the correlation that exists between them.

Very Short Answer Questions

1. What is gene mapping?

2. Write other names of linkage map.

3. What is Morgan unit?

4. What is the relation between frequency of linkage with the distance between the linked genes?

5. What do you mean by recombination hotspot?

6. What is recombination coldspot?

7. Write the name of an organism which shows negative interference.

8. What is the chemical composition of HAT medium?

9. Name the inhibitors of mitotic spindle formation used in technique of microcell fusion?

True and False Questions

1. Recombination frequency is measured as centimorgan. (*H.N.B., Garawal, 2008*)

Fill in the Blanks

1. Monochromosomal hybrids contain a .... chromosome.

2. FISH stands for ....

Multiple Choice Questions

1. The unit used in indicating the distance between two genes is

(*a*) centimeters (*b*) centimorgans
(*c*) microns (*d*) angstrom

2. For the preparation of genetic maps, the recombination frequencies between genes are additive over short distances, but not over long distances due to

(*a*) multiple cross overs (*b*) lethal mutations
(*c*) epistasis (*d*) synaptonemal complex

3. Based on the relative proportion of parental and recombinant phenotypes in the test cross progeny involving two recessive genes *r* (round) *y* (yellow) controlling the seed shape and seed colour respectively, the observed percent recombination is 10. The distance between *r* and *y* is

(*a*) 10 map units (*b*) 20 map units
(*c*) 30 map units (*d*) 40 map units

4. Individuals homozygous for *cd* genes were crossed with wild type (++). The F_1 dihybrid thus produced was test crossed. It produced progeny in the following ratio:

+ +	900
c d	880
+ d	115
+ c	105

What is the distance between *c* and *d* genes?

(*a*) 5.75 units (*b*) 11 units
(*c*) 47 units (*d*) 88 units

ANSWERS

Very Short Answer Questions

1. The assignment of a gene to a locus on the chromosome is gene mapping.
2. Linkage map is also called genetic map or crossover map.
3. Morgan unit is a map distance unit.
4. Frequency of linkage is inversely proportional to the distance between the linked genes.
5. The endonuclease target sites which provide single stranded DNA for homologous recombination is recombination hotspot.
6. The sites where homologous DNA fail to synapse effectively are recombination cold spots.
7. *Aspergillus*.
8. HAT medium is required for somatic-cell hybridization experiments. It comprises of three compounds: hypoxanthine, aminopterin and thymidine.
9. Colcemid and cytochalasin *B*.

True and False Questions

1. True.

Fill in the Blanks

1. Single, donor. **2.** Fluorescent in situ hybridization.

Multiple Choice Questions

1. (*b*) **2.** (*a*) **3.** (*a*) **4.** (*b*).

11

Chromosomal Mutations

Genetics makes extensive use of deviations from the norms, and the study of chromosomes is no exception. The chromosomes of each species has a characteristic morphology (structure) and number. But, sometimes due to certain accidents or irregularities at the time of cell division, crossing over or fertilization some alterations in the morphology and number of chromosomes take place. The slightest variation in the organisation of chromosomes is manifested phenotypically and is of great genetical interest. The changes in the genome involving chromosome parts, whole chromosomes, or whole chromosome sets are called **chromosome aberrations** or **chromosome mutations**. Chromosome mutations have proved to be of great significance in applied biology—agriculture (including horticulture), animal husbandry and medicine.

Chromosome mutations are inherited once they occur and are of the following types (Box 11.1):

A. Structural changes in chromosomes:

1. Changes in number of genes
   - (*a*) Loss: deletion
   - (*b*) Addition: Duplication
2. Changes in gene arrangement:
   - (*a*) Rotation of a group of genes 180° within one chromosome: inversion
   - (*b*) Exchange of parts between chromosomes of different pairs: translocation.

B. Changes in number of chromosomes:

1. Loss, or gain, of the part of the chromosome set (aneuploidy)
2. Loss, or gain, of whole chromosome set (euploidy)
   - (*a*) Loss of an entire set of chromosomes (haploidy)
   - (*b*) Addition of one or more sets of chromosomes (polyploidy).

Both types of changes (structural and numerical) in chromosomes can be detected not only with a microscope (*i.e.*, cytologically) but also by standard genetic analysis. This gave birth to a hybrid science, called **cytogenetics** which attempts to correlate cellular events, especially those of chromosomes, with genetic phenomena.

11.1 STRUCTURAL CHANGES IN CHROMOSOMES

Structural changes in chromosome may be of the following types (Fig 11.1): 1. **deficiency** or **deletion** which involves loss of a broken part of a chromosome; 2. **duplication** involves addition of a part of chromosome (*i.e.*, broken segment becomes attached to a homolog which, thus, bears one block of genes in duplicate); 3. **inversion** in which broken segment is reattached to original chromosome in reverse order, and 4. **translocation** in which the broken segment becomes attached to a nonhomologous chromosome resulting in new linkage relations.

Further, structural, abnormalities can occur in both homologous chromosomes of a pair or in only one of them: When both homologous chromosomes: are involved, these are called **structural**

homozygotes; *e.g.*, deletion homozygate, duplication homozygote, etc. When only one homologous chromosome is involved, it is called **structural heterozygote**. In Fig. 11.1 have been shown translocation heterozygote and translocation homozygote.

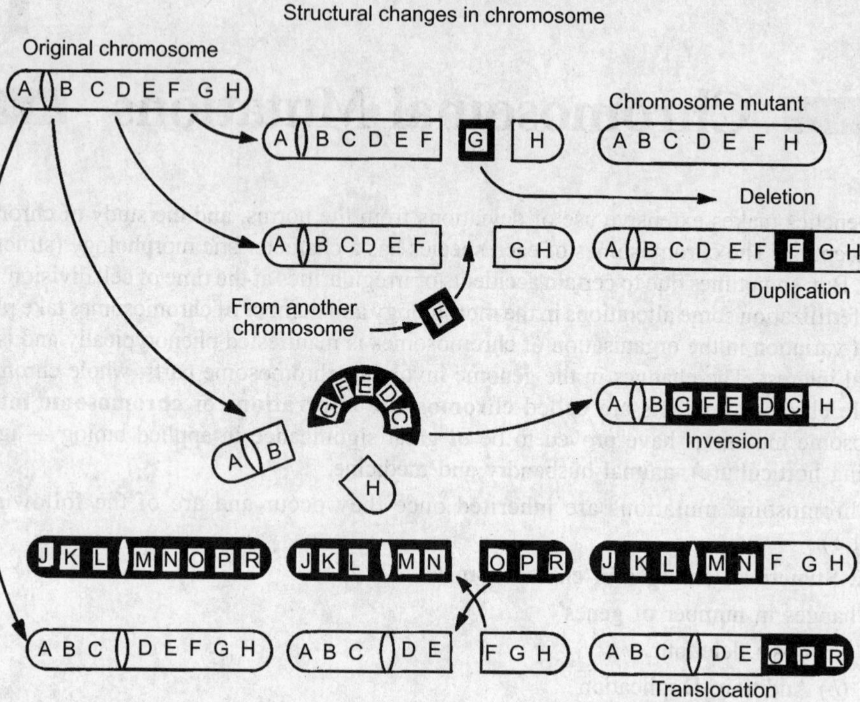

Fig. 11.1. Structural changes in chromosome.

Box 11.1

1. A **chromosome mutation** (or **chromosome aberration**) is a mutation involving a large segment of the genome. Such a mutation usually affects many genes and is often observable at the cytogenetic level, *i.e.*, it can be seen with a light microscope.
2. Chromosome mutuations are **numerical** if they involve a deviation from normal chromosome number and **structural** if they involve breakage and rearrangement of chromosome segments. Structural mutations are **balanced** if DNA is rearranged but there is no loss or gain of material, or **unbalanced** if DNA is lost or gained. All numerical mutations are unbalanced.
3. There are four consequences of chromosome mutations: disruption, fusion, position and dosage effects.
   (*i*) **Disruption and fusion effects.** These occur in structural mutations and reflect the nature of the chromosome breakpoints before and after mutation, *e.g.*, a breakpoint can interrupt a gene or separate a gene from its promoter, resulting in loss of gene function, fusion can join two genes, allowing a composite product to be synthesized perhaps with noval functions.
   (*ii*) **Position effects.** These also occur in structural mutations and reflect global influences on gene expression conferred by chromatin structure, *e.g.*, a translocation may bring a normally active gene adjacent to a region of heterochromatin causing transcriptional repression.
   (*iii*) **Dosage effect.** They occur in both structural and numerical mutations and concern the number of copies of each gene in the cell.

4. **Constitution chromosome mutation.** These arise in the germline or in the zygote and thus affects every cell in the body. In humans, these give rise to specific clinical syndromes (*e.g.*, Down's syndrome, Kleinefelter syndrome).

1. Deletion (or Deficiency)

The simplest result of breakage is the loss of a part of a chromosome. Portions of chromosomes without a centromere (called **acentric fragments**) lag in anaphase movement and are lost from reorganizing nuclei or digested by nucleases. Such loss of a portion of a chromosome (and of some genes) is called **deletion**. The chromosomes with deletions can never revert to a normal condition. If gametes arise from the cells having a deleted chromosome, this deletion is transmitted to the next generation. Further, a deletion can be terminal or intercalary (interstitial). In terminal deletion a terminal section of a chromosome is absent and it is resulted by only one break (Fig. 11.3). While in the **intercalary deletion**, an intermediate section or portion of chromosome is lost and it is caused by two breaks–one on either end of the deleted region (Fig. 11.3). Thus, in the latter case, the chromosome is broken into three pieces, the middle one of which is lost and the remaining two pieces get joined again.

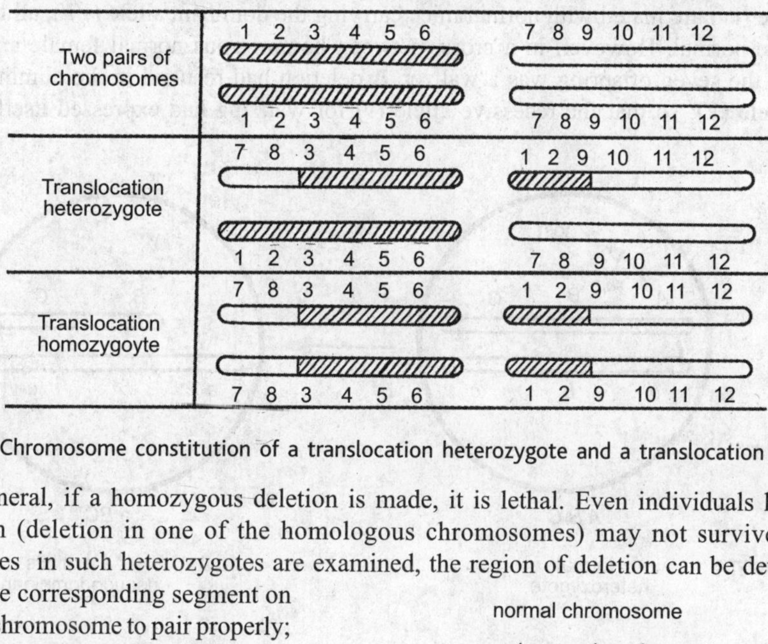

Fig. 11.2. Chromosome constitution of a translocation heterozygote and a translocation homozygote.

In general, if a homozygous deletion is made, it is lethal. Even individuals heterozygous for deletion (deletion in one of the homologous chromosomes) may not survive. If meiotic chromosomes in such heterozygotes are examined, the region of deletion can be detected by the failure of the corresponding segment on the normal chromosome to pair properly; so a "**deletion-loop**" results (Fig. 11.4). Deletion loops are also detected in polytene chromosomes of *Drosophila,* where the homologs exist in permanent state of pairing. The cytological studies of pairing between normal and deleted chromosomes have helped a lot in finding out the relative positions of genes in chromosomes.

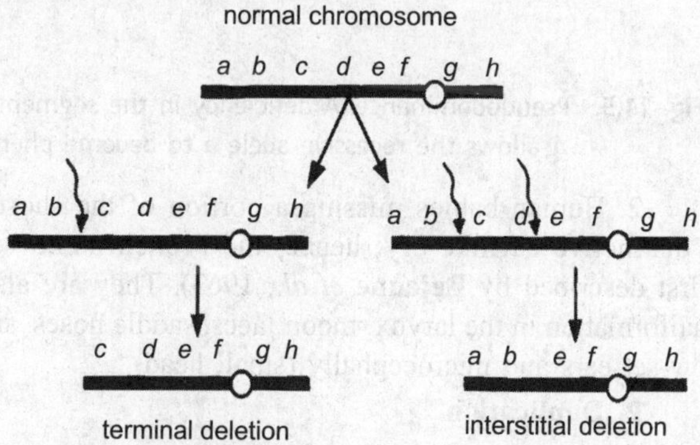

Fig. 11.3. Production of terminal and intersitial deletion. Chromosomes can be broken when struck by ionizing radiation (wavy arrows).

Genetical effects of deletion. Deletion of some chromosome regions produce their own unique phenotypes. A good example of this is a dominant **notch-**

wing mutation in *Drosophila*. In fact, this is a small deletion and acts as a recessive lethal in this regard. Further, in the presence of a deletion, a recessive allele of the normal homologous chromosome will behave like a dominant allele, *i.e.*, it will be phenotypically expressed, this phenomenon is called **pseudodominance** (Fig. 11.5).

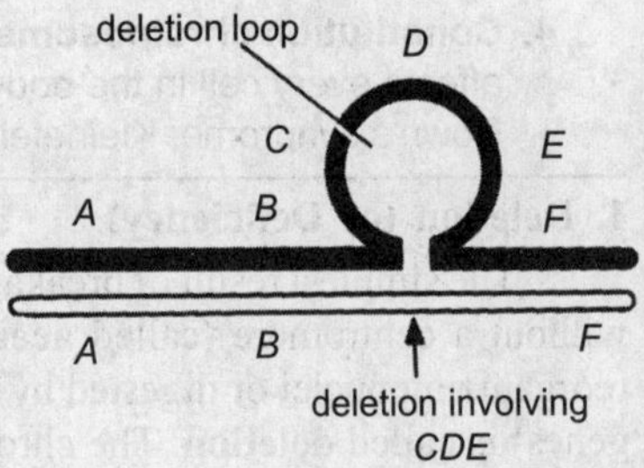

Fig. 11.4. Formation of a deletion loop during synapsis in a deletion heterozygote.

The phenomenon of pseudodominance exhibited by deficiency heterozygotes has been utilized for the location of genes on specific chromosomes and in preparing **cytological maps** in *Drosophila*, maize, bacteriophage and other organisms. Such cytological maps are often used to verify the **genetic maps** (based on linkage analysis) of these organisms.

Examples of pseudodominance (deletion). 1. **Gates** (1921) demonstrated the effects of pseudodominance. Certain mice carrying the recessive allele (v) in homozygous condition move about erratically until exhausted. Such mice are called '**waltzing mice**'. When homozygous waltzing mice (vv) are mated with normal mice carrying the dominant allele (VV), all the affecting (Vv) mice are normal. However, in a cross, between homozygous normal female and a waltzer male, one of the seven offspring was a waltzer. A deletion had resulted in the elimination of the dominant allele (V); so that the recessive allele (v) for waltzing had expressed itself.

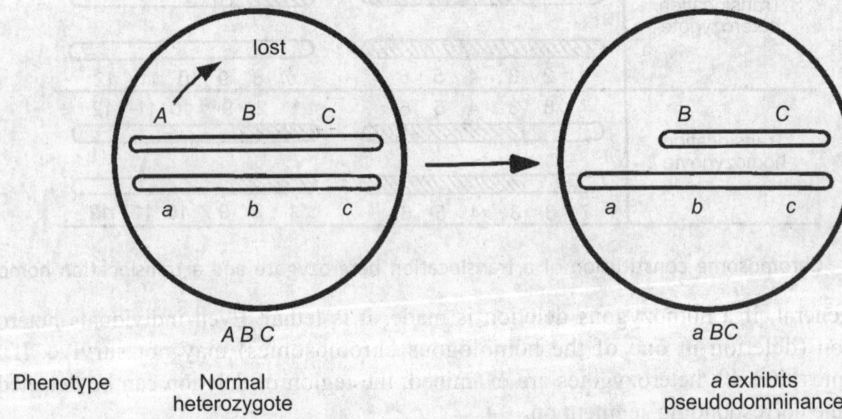

Fig. 11.5. Pseudodominance. A deficiency in the segment of chromosome bearing the dominant gene *A* allows the recessive allele *a* to become phenotypically expressed.

2. Human babies missing a portion of the short arm of chromosome 5 (autosome) have a distinctive cat-like cry; hence, the French name "**cri du chat**" (cry of the cat) **syndrome** (first described by **Lejeune** *et al.*, 1963). They are also mentally retarded (IQ below 20), have malformation in the larynx, moon faces, saddle noses, small mandibles (micrognathia), malformed low-set ears and microcephally (small head).

2. Duplication

The presence of a part of a chromosome in excess of the normal complement is known as **duplication**. Thus, due to duplication some genes are present in a cell in more than two doses. If duplication is present only on one of two homologous chromosomes, at meiosis the chromosome bearing the duplicated segment forms a **loop** to maximize the juxtaposition (during pairing) of homologous regions (Fig. 11.6).

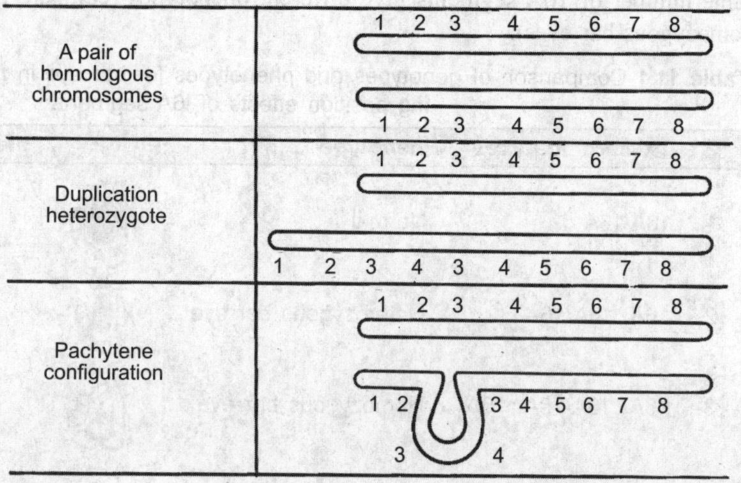

Fig. 11.6. Loop formation during chromosome pairing in a duplication heterozygote.

Extra segments in a chromosome may arise in a variety of ways such as follows:

1. Tandem duplication. In this case the duplicated region is situated just by the side of the normal corresponding section of the chromosome and the sequences of genes are the same in normal and duplicated region. For example, if the sequence of genes in a chromosome is ABC. DEFGH (The full stop depicts the centromere) and if the chromosomal segment containing the genes DEF is duplicated, the sequence of genes in tandem duplication will be ABC. DEFDEFGH.

2. Reverse tandem duplication. Here, the sequence of genes in the duplicated region of a chromosome is just the reverse of a normal sequence. In the above mentioned example, therefore, the sequence of genes due to reverse tandem duplication will be ABC. DEFFEDGH.

3. Displaced duplication. In this case the duplicated region is not situated adjacent to the normal section. Depending on whether the duplicated portion is on the same side of the centromere as the original section or on the other side, the displaced duplication can be termed either **homobranchial** or **heterobranchial**.

Example. Homobranchial duplication = ABC.DEFG DEFH

Heterobranchial duplication = ADEFB C. DEFGH

4. Transposed duplication. Here, the duplicated portion of chromosome becomes attached to a non-homologous chromosome. For example, if ABC.DEFGH and LMNOPQ. RST represent the gene sequences of two nonhomologous chromosomes, a transposed duplication will result into chromosomes with gene sequence ABC.GH and LMN DEF.OPQ. RST. Such a transposed duplication may be either interstitial (*e.g.*, LMNDEFOPQ. RST) or **terminal** (*i.e.*, LMN OPQ.RSTDEF).

5. Extra-chromosomal duplication. In the presence of centromere the duplicated part of a chromosome act as independent chromosome.

Genetical effects of duplication. Due to duplication, there occur unequal crossing over which results in deletion and reduplication which produce distinct phenotypes as shown by the following examples:

1. Bar eye in *Drosophila*. The Bar phenotype of *Drosophila* is characterized by narrower, oblong, bar-shaped eye with few facets. It is determined by a X-linked recessive allele *B*. The classical studies of **Bridges** (1936) showed that the bar trait of *Drosophila* is associated with the duplication of a segment of the X-chromosome, called **section 16A**, as observed in salivary gland chromosomes. Each added section 16A intensifies the bar phenotype (*i.e.,* duplication behaves genetically as a dominant factor). However, the narrowing effect is greater if the duplicated segments are on the same chromosome (called **position effect**) (Table 11.1). Thus, *cis* and *trans* arrangements of the same number of 16A segments give different phenotypes (compare heterozygous ultrabar and homozygous bar eyes).

Table 11.1 Comparison of genotypes and phenotypes for bar eye in *Drosophila* females showing position effects of I6A segment.

	X chromosomes	*Phenotype*		*Mean number of facets*
1.	16A/16A	Normal		779
2.	16A, 16A/16A	Heterozygous bar eye	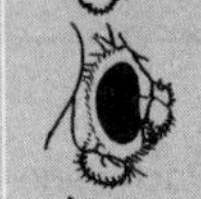	358
3.	16A, 16A/16A, 16A	Homozygous bar eye		68
4.	16A, 16A, 16A/16A	Heterozygous ultrabar (= double bar)		45
5.	16A, 16A, 16A/16A, 16A, 16A	Homozygous ultrabar		25

3. Inversion

Inversion involves a rotation of a part of a chromosome or a set of genes by 180° on its own axis. It essentially involves occurrence of *breakage* and *reunion*. The net result of inversion is neither a gain nor a loss in the genetic material but simply a rearrangement of the gene sequence. An inversion can occur in the following way: suppose that the normal order of segments within a chromosome is 1-2-3-4-5-6; breaks occur in regions 2-3 and 5-6 and broken piece is reinserted in reverse order. This results in an inverted chromosome having segments 1-2-5-4-3-6.

An **inversion heterozygote** has one chromosome in the inverted order and its homologue (homolog) in the normal order. The location of the inverted segment can be detected cytologically in the meiotic nuclei of such heterozygotes by the presence of an **inversion loop** in the paired homologs. The location of the centromere relative to inverted segment determines the genetic behaviour of the chromosomes. If the centromere is not included in the inversion it is called **paracentric inversion** and when inversion includes the centromere it is called **pericentric inversion** (Fig. 11.7). Homologous chromosomes; with identical inversions

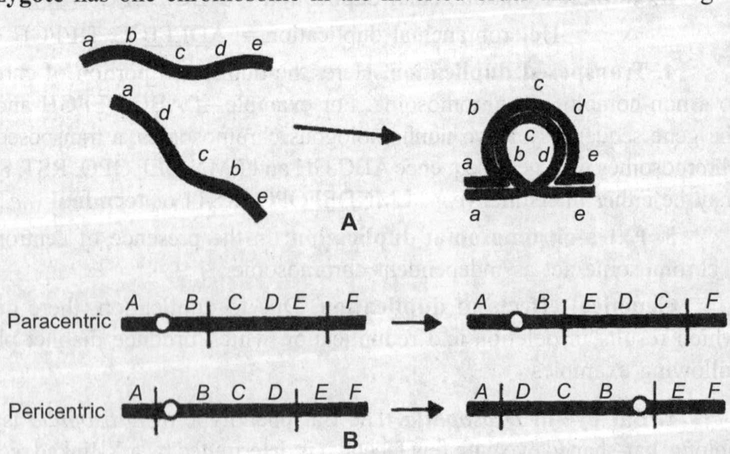

Fig. 11.7. A—An inversion loop in paired homologs of an inversion heterozygote; B—The location of the centromere relative to the inverted segment.

in each member, pair and undergo normal distribution in meiosis. However, crossing over in inversion heterozygotes produce, deletions, duplications and other curious configurations.

Advantages of inversions. Fertility of inversion homozygotes and sterility of inversion heterozygotes lead to establishment of two groups (or varieties) which are mutually fertile but do not breed well with the rest of the species. Both varities evolve in different directions and later become reproductively isolated species. There is plenty of cytological evidence to prove that such evolutionary mechanisms have and are operating in *Drosophila* and a number of other organisms.

4. Translocation

The shifting or transfer of a part of a chromosome or a set of genes to a non-homologous one, is called **translocation**. There is no addition or loss of genes during translocations, only a rearrangement (*i.e.,* change in the sequence and position of a gene). Translocations may be of following three types (Fig. 11.8).

1. Simple translocations. They involve a single break in a chromosome. The broken piece gets attached to one end of a nonhomologous chromosome.

2. Shift translocations. In this type of translocation, the broken segment of one chromosome gets inserted interstitially in a nonhomologous chromosome.

3. Reciprocal translocations. In this case, a segment from one chromosome is exchanged with a segment from another nonhomologous one, so that in reality two translocation chromosomes are simultaneously achieved.

Outcomes of reciprocal translocation. The exchange of chromosome parts between nonhomologous chromosomes creates new linkage relationships. Such translocations also drastically change the size of a chromosome as well as the position of its centromere. For example, a large metacentric chromosome is shortened by one-half in length to an acrocentric one, whereas the small chromosome becomes a large one (Fig. 11.8). Two types of translocations have been recognized: homozygous and heterozygous (Fig 11.9). The **translocation homozygotes** may have normal meiosis and in fact, are difficult to detect cytologically unless morphologically dissimilar chromosomes are involved, or banding patterns differ markedly. The **translocation heterozygotes** produce both translocated and normal chromosomes and exhibit characteristic cytological and genetical effects.

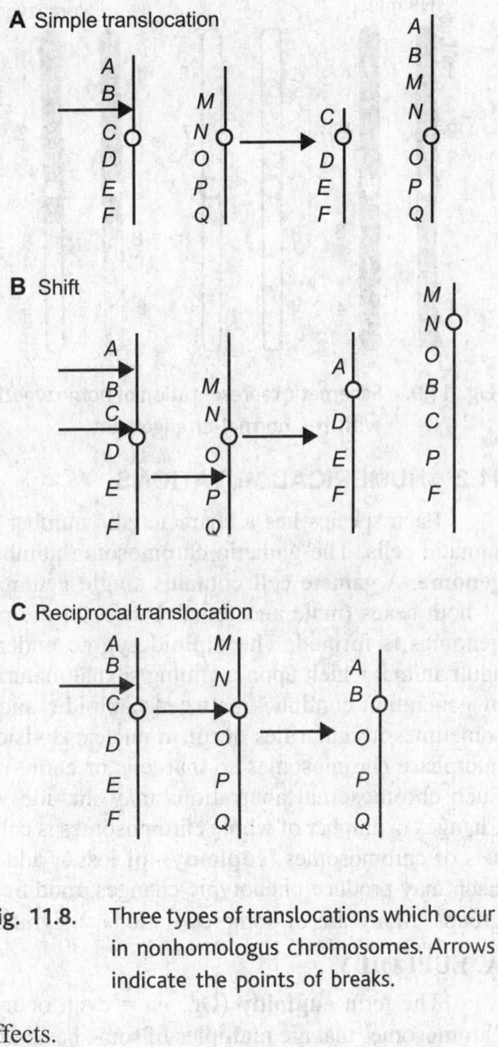

Fig. 11.8. Three types of translocations which occur in nonhomologus chromosomes. Arrows indicate the points of breaks.

The first case of translocation was studied in the evening primrose (*Oenothera*) which was originally described as a mutation by **de Vries**. In *Oenothera*, *Tradescalltia* and *Rhoeo* translocations in heterozygous condition are frequently found in nature. In many other crop plants translocations have been artificially induced by X-rays. They are well evident in many animals including humans. In *Drosophila* and other animals, translocations have also been induced by X-rays. For example, in humans, patients with **chronic myelocytic (myologenous) leukemia** (a kind of cancer) display an interesting chromosomal abnormality. In the bone marrow and in cells derived from it, is present a short chromosome, called the **Philadelphia (Ph^1) chromosome** (so named because it was discovered in that city). Detailed cytological study disclose Ph^1 to be a number 22 chromosome that has lost most of the distal part of its longer arm (22 *q*–). The deleted part of autosome 22 is translocated to one of the larger autosomes (most frequently to the distal end of chromosome 9). This translocation exhibits position effect and it is not transmitted to offspring of persons having Philadelphia chromosome (Ph^1 does not appear in gametes of the patients).

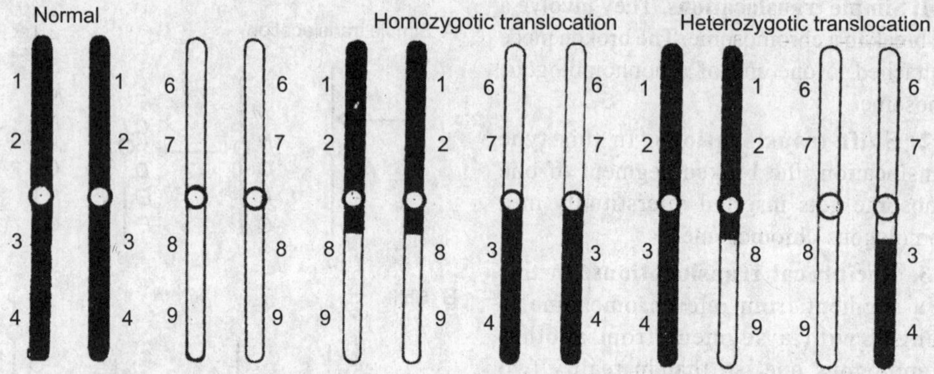

Fig. 11.9. Schematic representation of homozygotic and heterozygotic reciprocal translocations compared with the normal arrangement.

11.2 NUMERICAL MUTATIONS

Each species has a characteristic number of chromosomes in the nuclei of its gametes and somatic cells. The gametic chromosome number constitutes a basic set of chromosomes called **genome**. A gamete cell contains single genome and is called **haploid**. When haploid gametes of both sexes (male and female) unite in the process of fertilization a **diploid** zygote with two genomes is formed. The diploid zygote undergoes embryological development and forms an adult animal which upon attaining sexual maturity produces haploid gametes. And this alternation of generation continues between haploidic and diploidic generation in most species. However, sometimes irregularities occur in nuclear division or "accidents" (as from radiations) may be fall interphase chromosomes so that cells or entire organisms with aberrant genomes may be formed. Such chromosomal aberrations may include whole genomes and entire single chromosomes. Changes in number of whole chromosomes is called **heteroploidy**. Heteroploidy may involve entire sets of chromosomes (**euploidy**), of loss or addition of single whole chromosomes (**aneuploidy**). Each may produce phenotypic changes, modifications of phenotypic ratio, or alteration of linkage groups. Many are of some evolutionary significance.

A. EUPLOIDY

The term **euploidy** (Gr., *eu* = even or true; *ploid* = unit) designates genomes containing chromosomes that are multiples of some basic number (*x*). The euploids are those organisms which contain balanced set or sets of chromosomes in any number. The number of chromosomes in a basic set is called the **monoploid number**, *x*. Those euploid types whose number of sets is greater than two are called polyploid. Thus, 1*x* is **monoploid**, 2*x* is **diploid**; and the polyploid types are 3*x* (**triploid**), 4*x* (**tetraploid**), 5*x* (**pentaploid**), 6*x* (**hexaploid**) and so on: The **haploid** (**n**) refers; strictly to the number of chromosomes in gametes. In most animals and many plants the haploid number and monoploid number are the same. Hence n or x (or $2n$ or $2x$) can be used interchangeably. However, in case of polyploids the usage of n may create confusions. For example; in modern wheat x *and* n are different. Wheat has 42 chromosomes, but careful study reveals that in hexaploid there are: six rather similar but not identical sets of seven chromosomes. So, $6x = 42$, and $x = 7$. However, gametes of wheat contain 21 chromosomes, hence, $2n = 42$ and $n = 21$. A triploid hybrid of wheat, from which a hexaploid *Triticum spelta* has been obtained due to colchicine treatment; contains $2n = 3x = 21$. Such haploids, since are obtained from the polyploids (*i.e.*, cross of tetraploid emmer wheat and diploid goat grass), they are called **polyhaploids**; just to differentiate them from the normal monoploids.

The lower organisms such as bacteria and viruses are called **haploids** because they have a single set of genetic elements. However, since they do not form gametes comparable to those of higher organisms, the term monoploid would seem to be more appropriate.

1. Monoploidy

Monoploids have a single basic set of chromosomes, *e.g.*, 7 in barley and 10 in corn. Monoploidy is common in plants and rare in animals.

Monoploids in some cases are found normally and are produced due to parthenogenesis, as in male (drone) hymenopteran insects such as bees, wasps and ants. In these insects, queen and workers are diploid females. In angiosperms (flowering plants) monoploids may also originate spontaneously due to parthenogenetic development of egg. Such rare monoploids have been obtained in tomatoes and cotton under cultivation. Rarely monoploid plants may originate from the pollen tube, synergids and antipodals of the embryo sac and are called **androgenic monoploids** (or androgenic haploids).

Monoploids can be produced by artificial means by the following methods: (1) X -ray treatments, (2) delayed pollination, (3) temperature shock (cold treatment), (4) colchicine treatment. (5) distant (interspecific or intergeneric) hybridization), (6) anther or pollen culture. Among these techniques, the most important ones are distant hybridization and anther culture.

Monoploid plants have reduced size of all vegetative and floral parts. Monoploid organisms are highly sterile. For instance, a monoploid in maize ($2n = 20$) will have 10 chromosomes and the number of chromosomes in a gamete can range from 0–10. Consequently, considerable sterility is found in such monoploid maize plants.

In contrast, in monoploid male honey bees during spermatogenesis the meiosis is bypassed by mitosis. As a result, their sperms are haploid and viable.

Uses of Monoploids. In a monoploid, since there is only one copy of each chromosome and only one allele of each gene, so, in it each gene is expressed whether it is dominant or recessive. This facilitates genetic experiments and this is the reason why microorganisms have been able helpful in genetic studies. For the same reason, scientists are trying hard to develop haploid strains of the flowering plants. Success has been achieved in developing monoploid strains of *Nicotiana, Datura* and *Triticum*. From these monoploid strains have also been developed pure breeding strains which are resistant for the insecticides and also for toxic compounds normally produced by the parasites of these plants.

2. Polyploidy

Any organism with more than two genomes ($2x$) is called a **polyploid**. Many plant genera include species whose chromosome numbers constitute a euploid series. For example, the rose genus *Rosa* includes species with the somatic numbers 14, 21, 28, 35, 42 and 56. These numbers are the multiples of 7. Therefore, this is a euploid series of the basic monoploid number 7, which gives diploid, triploid, tetraploid, pentaploid, hexaploid and octaploid species. Over 10 copies of the genome are represented by numbers, *e.g.*, 12-ploid. Except diploids, rest of these belong to polyploid category. Ploidy levels higher than tetraploid are not commonly encountered in natural populations, but our most important crops and ornamental flowers are polyploids, *e.g.*, wheat (hexaploid, 6*x*), strawberries (octaploid, 8*x*), many commercial fruits and ornamental plants. Generally, polyploidy is common in plants (more common in monocots) but rare in animals.

Types of polyploidy. There are following three different kinds of polyploids: (*i*) autopolyploids, (*ii*) allopolyploids and (*iii*) autoallopolyploids. Suppose that there are four different haploid sets of chromosomes A, B_1, B_2 and C, in which B_1 and B_2 genomes are related. By using these genomes, all three types of polyploids can be derived as have been shown in Fig. 11.10.

(*a*) Autopolyploids. The autopolyploids are those polyploids, which consist of same basic set of chromosomes multiplied. For example, if a diploid species has two similar sets of chromosomes or genomes (AA), an autotriploid will have three similar genomes (AAA), and an autotetraploid, will have four such genomes (AAAA).

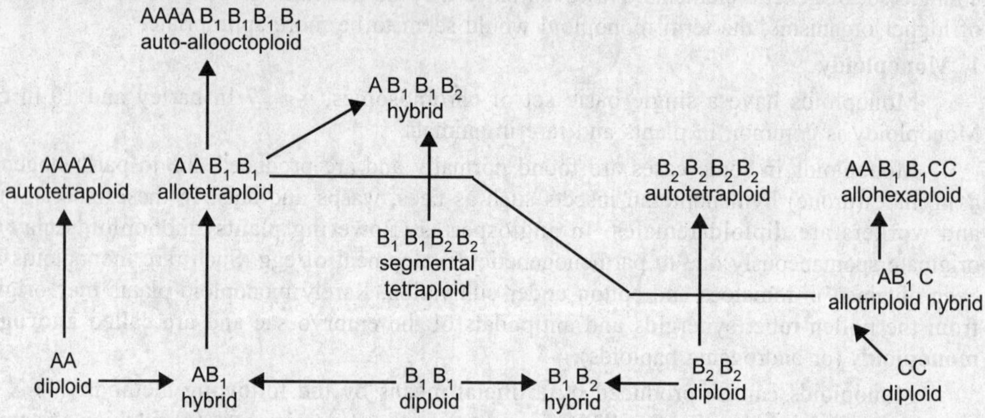

Fig. 11.10. Mode of formation of different kinds of polyploid.

(*i*) Origin and production of autopolyploids. The autopolyploids may occur in nature or may be produced artificially. When they are found in nature, their autopolyploidy nature is deduced by their meiotic behaviour. One of the common example of natural autopolyploidy is 'doob' grass (*Cynodon dactylon*) which is quite commonly cultivated in U.P. and Bihar. *Cynodon* is quite successful in the cultivation mainly due to its efficient way of vegetative propagation (since being triploid, it is sterile and setting no seed). Polyploids may arise naturally by following means: (*i*) in natural populations polyploidy may arise as a result of interference with cytokinesis, once chromosome replication has occurred; (*ii*) it may occur either in somatic tissues which give rise to tetraploid branches or during meiosis which produces unreduced gametes. All these natural inductions of polyploidy may occur due to chilling.

Some of common examples of autotriploid crop plants, which are mainly produced by artificial methods, are seedless varieties of watermelons, sugar beet, tomato, grapes and banana. Similarly, many important crop plants include **autotetraploids** such as rye (*Secale cereale*), corn (*Zea mays*), red clover (*Trifolium pratense*), berseem (*Trifolium alexandrium*), marigolds (*Tagetes*), snapdragons (*Antirrhinum*), *Phlox*, grapes, apples, *Oenothera lamarkiana* (which was recognized as mutation by **Hugo de Vries**).

Induced autopolyploidy. The autopolyploidy have been induced in many plant and animal cells by artificial means such as chemical (*e.g.*, chloral hydrate, colchicine, sulphanil amide, mercury chloride, hexachlorcyclohexane, *etc.*), radioactive substances, *e.g.*, radium and X-ray and temperature shocks. These inducers usually disturb the mitotic or meiotic spindle and cause non-segregation of already duplicated chromosomes, during cell divisions.

(*b*) Allopolyploids (Amphipolyploids). When the polyploidy results due the doubling of chromosome number in a F_1 hybrid which is derived from two distinctly different species, then, it is called **allopolyploidy** and the resultant species is called an **allopolyploid**. Let A represent a set of chromosomes (genome) in species X, and let B represent another genome in a species Y. The F_1 hybrids of these species then would have one A genome and another B genome. The doubling of chromosomes in the F_1 hybrids will give rise to allotetraploids with two A and two B genomes.

P_1: Species X (AA) (Diploid) × Species Y (BB) (Diploid)

F_1: AB
Diploid sterile hybrid

↓ Colchicine

AABB
Amphidiploid tetraploid
(Fertile)

Fig. 11.11. Formation of an amphidiploid tetraploid.

Raphanobrassica is a classical example of allopolyploidy or amphipolyploidy. In 1927, a Russian geneticist, **G.D. Karpechenko** performed a cross between radish (*Raphanus sativum*, $2n = 18$) and cabbage (*Brassica oleracea*, $2n = 18$) and in F_1 got sterile (diploid) hybrids. Among these sterile F_1 hybrids, he found certain fertile plants which were found to contain 36 chromosomes. These fertile tetraploids were called *Raphanobrassica*.

Synthesized allopolyploids. Wheat, tobacco and cotton are naturally occurring allopolyploid crops. To find out the origin of naturally occurring allopolyploids some cytogeneticists produced certain allopolyploids in laboratory by employing artificial means. Common hexaploid wheat *Triticum spelta* and tetraploid cotton *Gossypium hirsutum* furnish two such examples.

(i) Triticum spelta is a hexaploid wheat which was artificially synthesized in 1946 by **E.S. McFadden** and **E.R. Sears** and also by **H. Kihara**. They crossed an emmer wheat, *Triticum dicoccoides*, (tetraploid: 2n=28) with goat grass, *Aegilops squarrosa* (diploid; 2n=14) and doubled the chromosome number in the F_1 hybrid (Fig. 11.12). This artificially synthesized hexaploid wheat was found to be similar to the primitive wheat *T. spelta*. When the synthesized hexaploid wheat was crossed with naturally occurrng *T. spelta*, the F_1 hybrid was completely fertile. This suggested that hexaploid wheat must have originated in the past due to natural hybridization between tetraploid wheat and goat grass followed by subsequent chromosome doubling.

P_1: *Triticium dicoccoides* (Tetraploid emmer wheat) AA BB (2n = 28 ; 14 bivalents) × *Aegilops squarrosa* (Diploid goat grass) DD (2n = 14 ; 7 bivalents)

F_1: ABD
Triploid hybrid
(2n = 21; 31 univalents)

↓ Colchicine

AA BB DD
Synthesized hexaploid wheat
(*Triticum spelta*)
(2n = 42; 21 bivalents)

Fig. 11.12. Artificial synthesis of hexaploid wheat.

(ii) Gossypium hirsutum, the New world cotton plant, is another interesting example of allopolyploidy. Old world cotton, *Gossypium herbaceum*, has 13 pairs of chromosomes, while, American or "upland cotton" also contains 13 pairs of chromosomes. **J.O. Beasley** crossed the old world and American cottons and doubled the chromosome number in the F_1 hybrids. The allopolyploids, thus, produced resembled the cultivated New world cotton and when crossed with it gave fertile F_1 hybrids (Fig. 11.13). These results, thus, suggested that tetraploid *Gossypium hirsutum* originated from two diploid species, namely *G. herbaceum* (2n = 26) and *G. raimondii* (2n = 26).

***(iii) Triticale (Triticosecale* wittmack)** is the first human made cereal which has been developed in recent years and is cultivated on about one million hectares of land throughout the Globe for the commercial use. Triticale is an artificial allopolyploid which has been derived by crossing wheat

Gossypium herbaceum × *Gossypium raimondii*
(Old world cotton) (American or upland cotton)
(2n = 26; 13 bivalents) (2n = 26; 13 bivalents)
↓
F_1 hybrid
(2n = 26; 26 univalents)
↓ Colchicine
New world cotton
(*Gossypium hirsutum*)
(2n = 52; 26 bivalents)

Fig. 11.13. Artificial synthesis of new world cotton.

(*Triticum*) and rye (*Secale*). Depending upon whether *Triticum* is a tetraploid (2n = $4x$ = 28) or hexaploid (2n = $6x$ = 42), one would get hexaploid triticale (2n = $6x$ = 42; Fig. 11.14) or octaploid triticale (2n = $8x$ = 56; Fig. 11.15), respectively. In each case, only diploid rye (2n = $4x$ = 14) was used.

Triticum durum × *Secale cereale*
(Tetraploid wheat; 2n = 28) (Diploid rye; 2n = 14)
↓
F_1 hybrid (sterile)
(Triploid; 2n = 21)
↓ (Colchicine doubling)
Hexaploid triticale
(2n = 42)

Fig. 11.14. Artificial synthesis of a hexaploid triticale.

Triticum aestivum × *Secale cereale*
(Hexploid wheat; 2n = 26) (Diploid rye; 2n = 14)
↓
F_1 hybrid (sterile)
(Tetraploid; 2n = 28)
↓ Colchicine doubling)
Octoploid triticale
(2n = 56)

Fig. 11.15. Artificial synthesis of a octoploid triticale.

(c) Segmental allopolyploids. Different genomes of some allopolyploids are not quite different from each other. Consequently in these polyploids chromosomes belonging to different genomes do pair together to some extent. This indicates that segments of chromosomes and not the whole chromosomes are homologous. Therefore, such allopolyploids are called **segmental allopolyploids** (**Stebbins**, 1943, 1950). The segmental allopolyploids are intermediate between autopolyploids and allopolyploids and can be identified by their peculiar meiotic behaviour.

It is generally believed that most naturally occurring polyploids are segmental allopolyploids. Our common hexaploid bread wheat too is found to be a segmental hexaploid.

Box 11.2 Green Revolution in India

1. One of the most significant development of modern agriculture has been the production of **semi-dwarf** varieties of cereal plants (such as wheat, rice, maize, etc.).
2. **N.E. Borlaugh** was an agriculture scientist who was awarded the 1970 Nobel Peace Prize for his part in the development of improved varieties of crereal crops, important in green revolution of world.
3. **Borlaugh** and his assocaites at CIMMYT (International Centre for Wheat and Maize Improvement, Mexico) developed semi-dwarf wheat using *Norin-10* (a Japanese variety) as the source of dwarfing genes (see **M.U. Charaya**, 2008).
4. In 1963, ICAR (Indian Council for Agricultural Research) introduced many dwarf selections from CIMMYT. **Kalan Sona** and **Sonalika** were selected from these materials. For many years, these were most popular varieties of wheat in India.
5. A great majority of wheat varieties cultivated in India are semi-dwarf as these are lodging-resistant, fertilizer-responsive and high-yielding.
6. **Dr. M.S. Swaminathan** (b., 7th August 1925) is pioneer in mutation breeding (*i.e.*, induced mutation breeding) of cereal in India. He is called **"Founder of Radiation Genetics"** also in addition to **"Father of Green Revolution in India"**. While working at IARI, N. Delhi, during 1954–1972, Dr. Swaminathan developed a new variety of wheat called **Sharbati Sonora**.
7. Sharbati Sonora was produced from Sonora 64 through a mutation breeding using gamma rays.

Polyploidy in animals. Polyploidy is rare in animals but occur in flatworms, leeches and brine shrimp. In mice, also, 40 per cent liver cells are tetraploids, and about 5 per cent are octoploids. Polyploidy in humans have been found in liver cells and cancer cells. In them polyploidy is either complete or mosaic and it leads to gross abnormalities and death.

Phenotypic Effects of Polyploidy

The increase in the genome's size beyond the diploid level is often caused following detectable phenotypic characteristics in a polyploid organism:

(*i*) **Morphological effect of polyploidy.** The polyploidy is invariably related with, **gigantism**. The polyploid plants have been found to contain large-sized pollen grains, cells, leaves, stomata, xylem, etc. The polyploid plants are more vigorous than diploids.

(*ii*) **Physiological effect of polyploidy.** The ascorbic acid content has been reported to be higher in tetraploid cabbages and tomatoes than in corresponding diploids. Likewise corn meal of a tetraploid maize seed contain 40 per cent more vitamin A than corn meal from a diploid plant.

(*iii*) **Effect on fertility of polyploidy.** The most important effect of polyploidy is that it reduces the fertility of polyploid plants in variable degrees.

(*iv*) **Evolution through polyploidy.** Interspecific hybridization combined with polyploidy offers a mechanism whereby new species may arise suddenly in natural populations.

B. ANEUPLOIDY

Changes that involve parts of a chromosome set results in individuals, called **aneuploids** (Gr. *aneu* = uneven; *ploid* = unit). Aneuploidy can be either due to the loss of one or more chromosomes (**hypoploidy**) or due to addition of one or more chromosomes to the complete chromosome set (**hyperploidy**). Hypoploidy is mainly due to the substraction (or, loss) of a single chromosome, called **monosomy** ($2n$–l) or due to the loss of one pair of chromosome called **nullisomy** ($2n$–2; two, lost chromosomes are **homologs**). Likewise, hyperploidy may involve addition of either a single chromosome, called **trisomy** ($2n + 1$) or a pair of chromosomes, called **tetrasomy** ($2n + 2$). In the monoploid organisms, addition of single chromosome produces **disomy** ($n + 1$). All of these aneuploids are probably produced by non-disjunction during mitosis or meiosis.

1. Monosomy

Diploid organisms which are missing one chromosome of a single pair are monosomic with the genomic formula $2n$–1. A monosomic individual forms gametes of two types, (n) and (n–1). The n–1 gametes do not survive in plants, but, in animals that may cause genetic imbalance, which is manifested by high mortality or reduced fertility of resulted organism.

Monosomy in diploids is not tolerated, since it creates imbalance due to loss of one complete chromosome. However, in polyploids, monosomy has no apparent effect, since they have several chromosomes of same type and loss of one chromosome can be easily tolerated. The number of possible monosomics in an organism will be equal to the haploid chromosome number. For example, in common wheat, since 21 pairs of chromosomes are present, 21 possible monosomics are known. These 21 monosomics in wheat were produced by **E.R. Sears** in variety cal-led "**chinese spring**" and being used for genetic studies all over the world. Monosomics were also reported in cotton ($2n = 52$) by **J. Endrizzi** and his coworkers and in tobacco

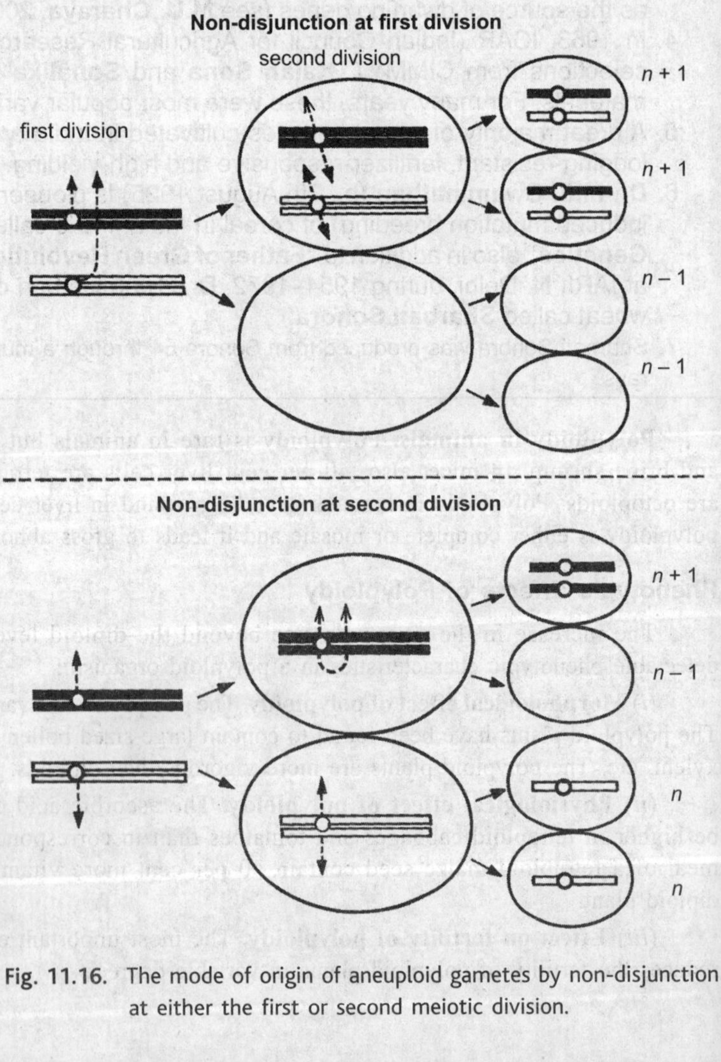

Fig. 11.16. The mode of origin of aneuploid gametes by non-disjunction at either the first or second meiotic division.

($2n = 48$) by **E.R. Clausen** and **D.R. Cameron**. Monsomics have also been produced in maize and tomato ($2n = 24$) despite their being diploids. Double mono-somics ($2n$–1–1) or triple monosomics ($2n$–1–1–1) could also be produced in polyploids such as wheat. Double monoso-mic means that the chro-mosome number is $2n$–2 like that in a nullisomic, but the missing chromo-somes are non-homolo-gous. The same explana-tion is applied to the triple monosomics.

2. Nullisomy

An organism which has lost a chromosome pair is a nullisomic. The nullisomic organism has the genomic formula ($2n$–2). A nullosomic diploid often does not survive, however, a nullosomic polyploid (*e.g.*, hexaploid wheat, $6x$–2) may survive but exhibit reduced vigour and fertility.

3. Trisomy

Trisomics are those diploid organisms which have an extra chromosome ($2n + 1$). Since the extra chromosome may belong to anyone of different chromosomes of a haploid complement, the number of possible trisomics will be equal to the haploid chromosome number. For example, haploid chromosome number of barley is 7, consequently in it seven trisomics are possible. Further, when the extra chromosome is identical to its homologs, such a trisomic is called **primary trisomic**. There are also secondary and tertiary trisomics. While the **secondary trisomic** means that the extra chromosome should be an isochromosome (*i.e.*, both chromosome arms genetically similar), a **tertiary trisomic** would mean that the extra chromosome should be the product of translocation (Fig. 11.17). Trisomics were obtained for the first time in jimson weed (*Datura stramonium*) by **A.F. Blakeslee** and **J. Belling** (1924). Since the haploid chromosome number of this species is $n = 12$, so here, 12 primary trisomics, 24 secondary trisomics and a large number of tertiary trisomics are possible. Most of the trisomics were identified by the size, shape and other morphological features of the fruit of jimson weed. In barley, such a trisomic series is produced and extensively studied by **T. Tsuchiya.**

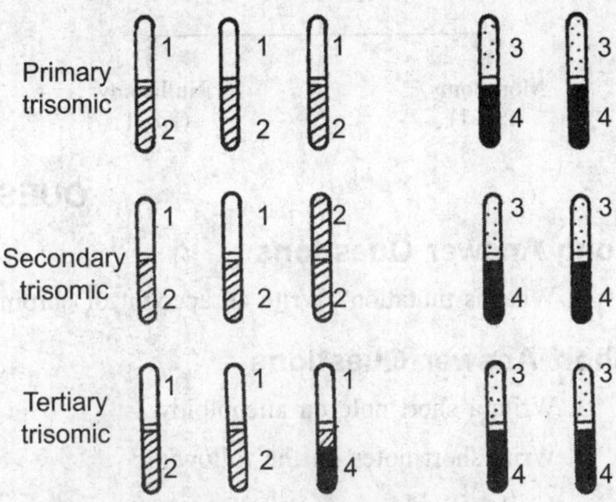

Fig. 11.17. Three kinds of trisomics.

Trisomy in humans. In human beings, the following three syndromes due to trisomy have been studied:

A. Down's syndrome (DS) or trisomy-21

B. Edward's syndrome or trisomy-18

C. Patau's syndrome or trisomy-13

(For details of these syndromes refer Chapter 87)

Trisomy in non-humans. Trisomy-22 has been reported in chimpanzees; this shows Down syndrome-like phenotypic features. **Trisomy-21** has been reported in the gorilla.

Cytology of trisomics. The trisomics have an extra chromosome which is homologous to one chromosome of the diploid complement. Therefore, it forms **trivalent** which may take a variety of shapes in primary and secondary trisomics. In a tertiary trisomic a characteristic pentavalent is observed.

4. Double Trisomy

In a diploid organism when two different chromosomes are represented in triplicate, the double trisomic is resulted. The double trisomic causes great genetic imbalance and has the genomic formula $2n + 1 + 1$.

5. Tetrasomy

The diploid organisms having two extra chromosomes are known as tetrasomic. They have the genomic formula $2n + 2$. All the 21 possible tetrasomics are available in wheat.

Different types of ploidies can be summarized in the following chart:

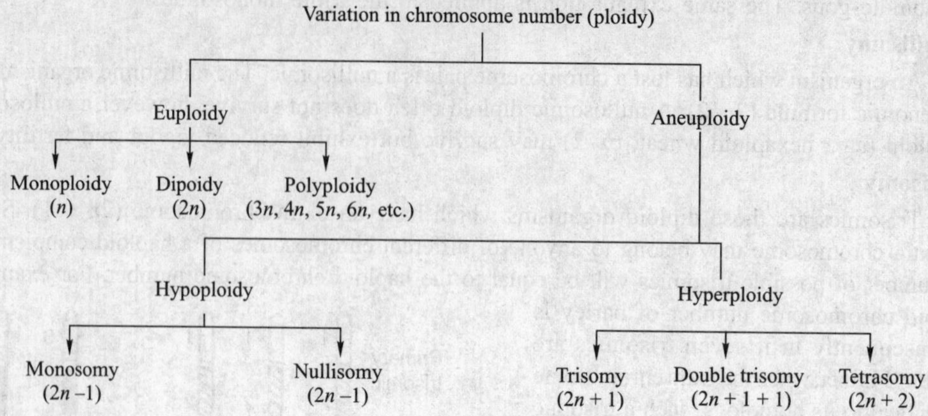

QUESTIONS

Long Answer Questions

1. What is mutation? Write an account of chromosomal mutations.

Short Answer Questions

1. Write a short note on aneuploidy.

2. Write short notes on the following:

(*i*) Polyploidy

(*ii*) Trisomy

(*iii*) Translocation.

3. Describe the chromosomal aberration. What is their significance in organic evolution? (*C.C.C.S.U. Meerut 2008*)

Very Short Answer Questions

1. Define mutation. (*H.N.B. Garhwal 2009*)

Fill in the Blanks

1. In .... , the broken segment of a chromosome becomes attached to a nonhomologous chromosome resulting in new linkage relations.

2. In intercalary deletion, an intermediate section of chromosome is .... and it is caused by .... breaks.

3. Deletion loops occur during synapsis in a ....

4. Bar eye trait of *Drosophila* is associated with the .... of a segment of the X-chromosome, called section ....

5. Dr. M.S. Swaminathan is called Founder of .... He developed a new variety of wheat called ....

Multiple Choice Questions

1. Polyploidy leads to rapid formation of new species through

(*a*) genetic recombination (*b*) isolation

(*c*) mutation (*c*) development of multiple sets of chromosomes

2. A syndrome in human in which an individuals somatic cells contain XXX set chromosomes is called (*M.S.U. Tamil Nadu 2006*)

(*a*) Klinefelter's (*b*) Turner's

(*c*) down's (*d*) superfemale

3. Genetic composition of Turner's syndrome is.....? (*H.N.B. Garhwal University 2008*)

(*a*) 44 + XXY (*b*) 44 + XO

(*c*) 44 + XXX (*d*) 44 + XYY

4. The bread wheat *Triticum aestivum* is a

(*a*) hexaploid (*b*) diploid

(*c*) haploid (*d*) heteropolyploid

5. The chromosome doubling in making polyploid plants is carried out by using

(*a*) phosphoethylene glycol (PEG) (*b*) colchicine treatment

(*c*) ethyl-methyl-sulphonate (EMS) (*d*) nicotinamide acedic acid (NAA)

6. The loss of one single chromosome creates a condition called

(*a*) trisomy (*b*) nullisomy

(*c*) monosomy (*d*) haploid

7. An autotetraploid plant bearing a genotype *AAAa* is called as

(*a*) monosomic (*b*) trisomic

(*c*) triploid (*d*) triplex

8. *Rhaphanobrassica* is a/an

(*a*) autopolyploid (*b*) allotetraploid

(*c*) hexaploid (*d*) none of these

9. Polyploidy means

(*a*) occurrence of haploid set of chromosomes

(*b*) occurrence of diploid st of chromosomes

(*c*) occurrence of three or more sets of chromosomes

(*d*) all of these

10. A polyploid derived from F_1 hybrid between two species is

(*a*) amphidiploid (*b*) hexaploid

(*c*) autodiploid (*d*) autopolyploid

11. Down's syndrome (DS) is the result of

(*a*) an extra X chromosome (*b*) an extra Y chromosome

(*c*) an additional twenty-first chromosome (*d*) an extra eighteenth chromosome

12. Colchicine interferes in

(*a*) chromosome condensation

(*b*) DNA replication

(*c*) organisation and orientation of spindle elements

(*d*) none of these

13. Who among the following merged cytology and genetics into cytogenetics?

(*a*) H.J. Muller (*b*) O.T. Avery
(*c*) G.W. Beadle (*d*) T.H. Morgan

14. Which is the human made cereal

(*a*) wheat (*b*) rice
(*c*) *Triticale* (*d*) barley

15. *Triticale* is obtained by crossing wheat with

(*a*) oat (*b*) barley
(*c*) maize (*d*) rye

16. Father of green revolution in India is

(*a*) T.S. Venkataraman (*b*) Boshi Sen
(*c*) K. Ramiah (*d*) M.S. Swaminathan

17. *Norin-10* is gene for

(*a*) dwarfness in wheat (*b*) dwarfness in rice
(*c*) flavour in rice (*d*) high yield in rice

18. Dr. M.S. Swaminathan could produce an amber-seeded high yielding variety called "Sharbati Sonora" from Mexican dwarf variety by using

(*a*) chemical mutagens (*b*) radiations of UV and gamma rays
(*c*) cross breeding techniques (*d*) tissue culture

19. The Indian dwarf wheat is

(*a*) *Triticum dicoccum* (*b*) *Triticum turgidum*
(*c*) *Triticum sphaerococcum* (*d*) *Triticum aestivum*

20. Most of our crop plants are of

(*a*) allopolyploidal origin (*b*) autopolyploidal origin
(*c*) mixed genotypic origin (*d*) structurally heterozygous origin

ANSWERS

Very Short Answer Questions

1. A change in the DNA at a particular locus in an organism. The term is used loosely to include point mutations involving a single gene change as well as a chromosomal change; the end result of this process.

Fill in the Blanks

1. Translocation. **2.** Lost, two.
3. Deletion heterozygote. **4.** Duplication, 16A.
5. Radiation genetics; sharbati sonora.

Multiple Choice Questions

1. (*d*) **2.** (*d*) **3.** (*b*) **4.** (*a*) **5.** (*b*) **6.** (*c*)
7. (*d*) **8.** (*b*) **9.** (*c*) **10.** (*a*) **11.** (*c*) **12.** (*c*)
13. (*a*) **14.** (*c*) **15.** (*d*) **16.** (*d*) **17.** (*b*) **18.** (*b*)
19. (*b*) **20.** (*a*).

12 Alteration in the Genetic Material (Point Mutation)

A gene mutation is abrupt inheritable qualitative or quantitative change in the genetic material of an organism. Since in most organisms, genes are segments of DNA molecule, so a mutation can be regarded as a change in the DNA sequence which is reflected in the change of sequence of corresponding RNA or protein molecules. Such a change may involve only one base/base pair or more than one base pair of DNA. Mutations occur in a **random** manner, *i.e.*, they are not directed according to the requirements of the organism. Most mutations occur **spontaneously** by the environmental effect, however, they can be induced in the laboratory either by radiations, physical factors or chemicals (called **mutagens**). A unicellular organism is more subjected to environmental attacks since it is at the same time a somatic or germ cell. In multicellular organisms, the germ cells are distinct cells, and are relatively protected from the environment. Mutation has a significant role to play in the origin of species or evolution.

12.1 HISTORICAL BACKGROUND

The earliest record of point mutations dates back to 1791, when **Seth Wright** noticed a lamb with exceptionally short legs in his flock of sheep. Visualising the economic significance of this short-legged sheep. *i.e.*, short-legged sheep could not cross the low stone fence and damage the crop fields in the vicinity, he produced a flock of sheep, each of which having short legs by employing artificial breeding techniques. The short-legged breed of sheep was known as **Ancon** breed. Later on, the trait of short legs was found to have resulted from a recessive mutation and the short-legged individuals were found to be homozygous recessive.

Hugo de Vries was the first hybridist who used the term "mutation" to describe the heritable phenotypic changes of the evening primrose, *Oenothera lamarckiana*. Many mutations described by de Vries in *O. lamarckiana*, are now known to be due to variation in chromosome number or ploidy and chromosomal aberrations (*viz.* **gross mutations**). The first scientific study of mutation was started in 1910, when **Morgan** started his work on fruitfly, *Drosophila melanogaster* and reported white-eyed male individuals among red-eyed male individuals. The discovery of white-eyed mutants in *Drosophila* was followed by an extensive search of other mutants of *Drosophila* by **Morgan** and his co-workers and other geneticists. Consequently, about 500 mutants of *Drosophila* have been reported by geneticists all over the world. Later on, several cases of mutations have been reported in a variety of microorganisms (*e.g.*, bacteriophages, bacteria (*Escherichia coli*), *Neurospora*, *etc.*, plants (*i.e.*, pea, snapdragon, maize, *etc.*) and animals, (*i.e.*, rodents, fowls, human beings, *etc.*).

12.2 OCCURRENCE

Mutations occur frequently in nature and have been reported in many organisms, *e.g.*, *Drosophila*, mice and other rodents, rats, rabbits, guinea pigs and human beings. In *Drosophila*, mutation causes white and pink eyes, black and yellow body colours, and vestigial wings.

In rodents the mutations are responsible for black, white and brown coats. In human beings, the mutations cause variation in hair colour, eye colour, skin pigmentation and several somatic malformations. Various genetical diseases of human beings such as haemophila, colour blindness, phenylketonuria, *etc*., form other examples of mutation in human beings.

How does a mutation act? Any change in sequence of nucleotides in the DNA will result in the corresponding change in the nucleotide sequence of mRNA. This may result in alignment of different tRNA molecules on mRNA (during protein synthesis). Thus, the amino acid sequence, and, hence, the structure and properties of the enzyme formed will be changed. This defective enzyme or structural protein may adversely affect the trait controlled by the protein. In consequence, a mutant phenotype makes its expression.

12.3 KINDS OF MUTATIONS

There exists a lot of controversy among geneticists about the possible kinds of mutations. Mutations have been classified variously according to different criteria as follows :

1. Classification of Mutation According to Type of Cells

According to their occurrence in somatic and germinal cells, following types of mutations have been classified :

A. Somatic mutations. The mutations occurring in non-reproductive body cells are known as **somatic mutations.** The genetical and evolutionary consequences of somatic mutations are insignificant, since only single cells and their daughter cells are involved. If, however, a somatic mutation occurs early during embryonic life, the mutant cells may constitute a large proportion of body cells and the animal body may be a mosaic of different types of cells. Somatic mutations have often been related with malignant (cancerous) growth. Examples of somatic mutation have been reported in *Oenothera lamarckiana* (by Hugo de Vries) and several other cases including human beings. In human beings, somatic mutation causes several fatal diseases such as *paraoxysmal nocturnal haemoglobinura*, *circumscribed neurofibroma*, *unilateral retinoblastoma* and *heterochromia of the iris*.

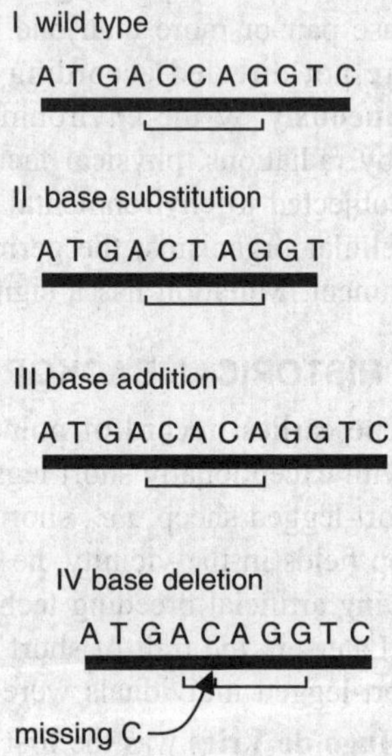

Fig. 12.1. Three types of point mutations. Only the base squence of one DNA strand is shown. Changes are shown in square; the horizontal brackets indicate the affected segment (after Freifelder, 1985).

B. Gametic mutations. The mutations occurring in gamete cells (*e.g.*, sperms and ova) are called **gametic mutations**. Such mutations are heritable and of great genetical significance. The gametic mutations only form the raw material for the natural selection.

2. Classification of Mutations According to the Size and Quality

According to size, following two types of mutations have been recognised:

A. Point mutation. When heritable alterations occur in a very small segment of DNA molecule, *i.e.*, a single nucleotide or nucleotide pair, then these types of **mutations** are called "**point mutations**". The point mutations may occur due to following types of subnucleotide change in the DNA and RNA.

1. Deletion mutations. The point mutation which is caused due to loss or deletion of some portion (single nucleotide pair) in a triplet codon of a cistron or gene is called **deletion mutation.** Deletion mutations have been frequently reported in some bacteriophages (Phage T_4).

2. Insertion or addition mutation. The point mutations which occur due to addition of one or more extra nucleotides to a gene or cistron are called **insertion mutations.** The insertion mutations can be artificially induced by certain chemical substances called **mutagens** such as acridine dye and proflavin. A proflavin molecule, it is believed, insert between two successive bases of a DNA strand, thereby stretching the strand lengthwise. At replication, this situation would allow the insertion of an extra nucleotide in the complementary chain at the position occupied by the proflavin molecule.

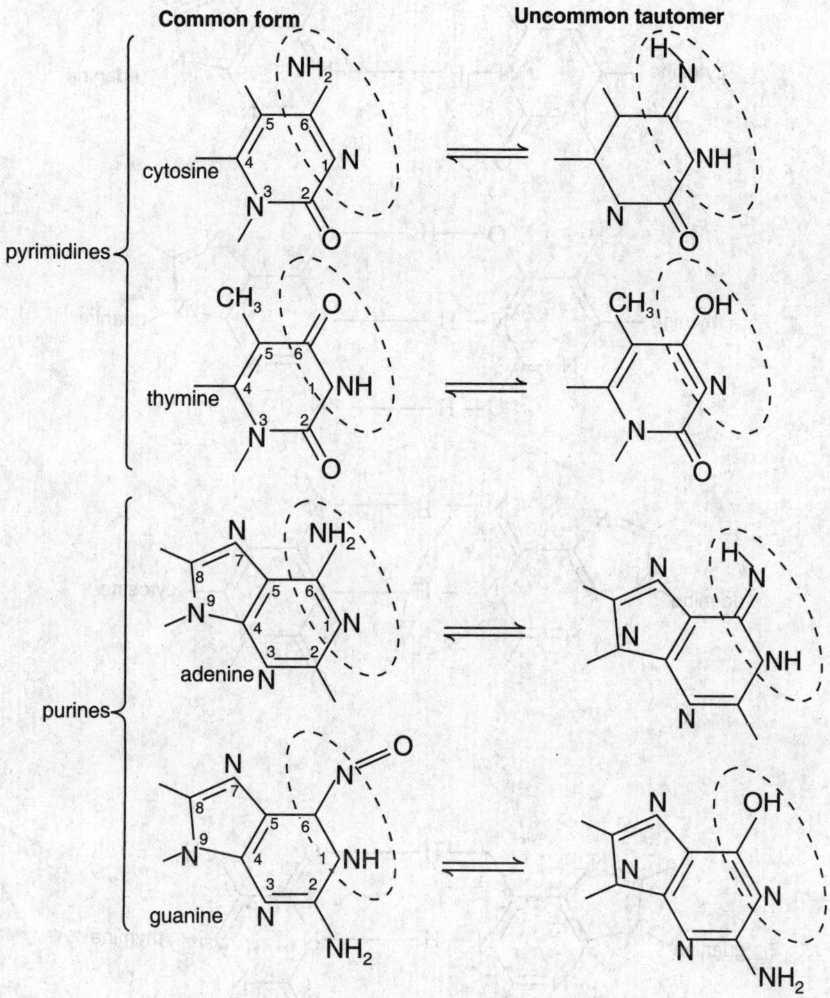

Fig. 12.2. The uncommon form of DNA bases.

The mutations which arise from the insertion or deletion of individual nucleotides and cause the rest of the message downstream of the mutation to be read out of phase, are called **frameshift mutations.** They result in the production of an incorrect, hence, inactive protein, due to which death of the cell may occur.

3. Substitution mutation. A point mutation in which a nucleotide of a triplet is replaced by another nucleotide, is called **substitution mutation.** The substitution mutation affect only a particular triplet codon. Such an altered code word (triplet codon) may designate a different amino acid and may result in the production of a protein with a single amino acid substitution. The substitution mutations alter the phenotype of an organism variously and are of great genetical significance. They may be of following types:

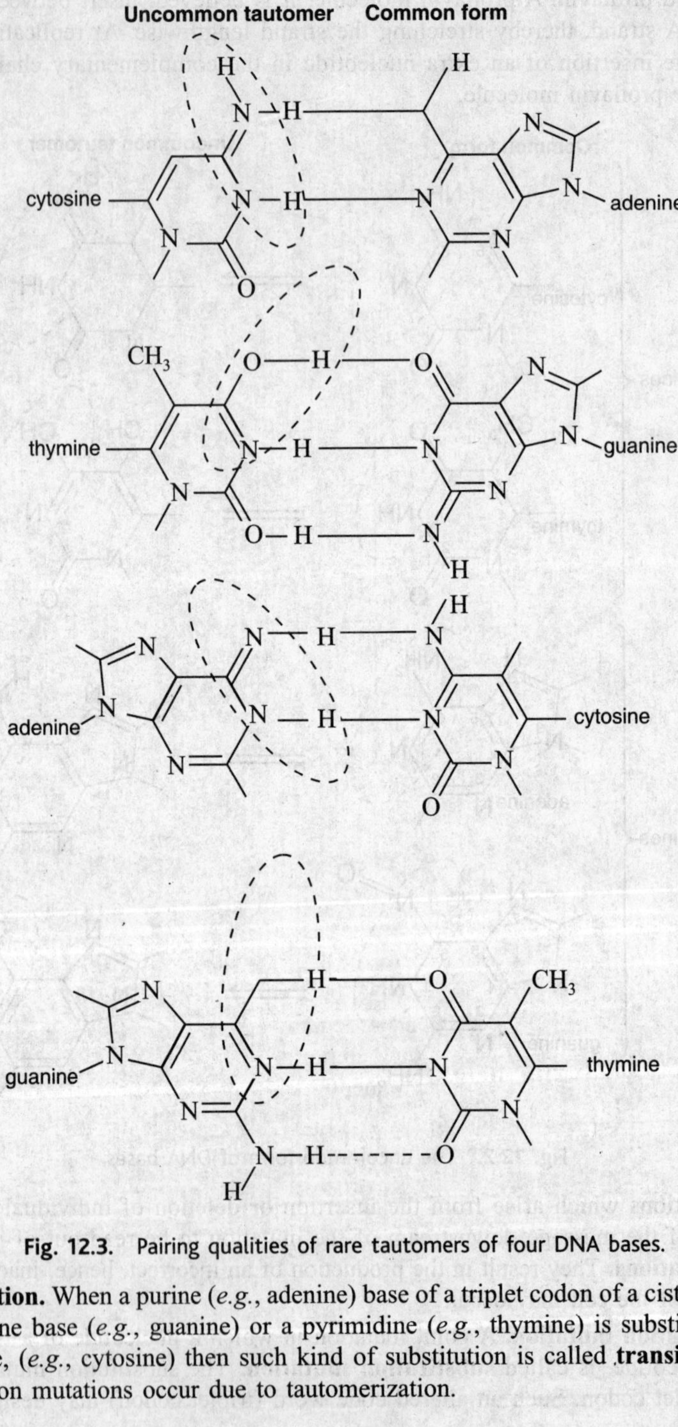

Fig. 12.3. Pairing qualities of rare tautomers of four DNA bases.

***(i)* Transition.** When a purine (*e.g.*, adenine) base of a triplet codon of a cistron is substituted by another purine base (*e.g.*, guanine) or a pyrimidine (*e.g.*, thymine) is substituted by another pyrimidine base, (*e.g.*, cytosine) then such kind of substitution is called **transition.** The transitional substitution mutations occur due to tautomerization.

Tautomerization

In a DNA molecule, normally, the purine, adenine (A) is linked to the pyrimidine, thymine (T), by two hydrogen bonds, while the purine guanine (G) is linked to the pyrimidine, cytosine (C) by three hydrogen bonds. Besides the common molecular configurations, each DNA base may have some altered uncommon molecular configuration, as has been shown in the figure 12.2.

Such uncommon forms of DNA bases are generated by single proton shifts and are called **rare states** or **tautomers**. A tautomeric shift is believed to occur when the amino (NH_2) form of adenine is changed to an imino (NH) form. Similarly, a tautomeric shift may occur in thymine changing it form the keto (C = O) form to the rare enol (COH) form. When a base occurs in its rare or tautomeric state, it cannot be linked to its normal partner. However, a purine, such as adenine in its rare state can form a bond with cytosine (besides thymine), provided the cytosine is in its normal state.

Watson and **Crick** (1953) hypothesised that the occurrence of the bases in their rare states provides a mechanism for mutation during DNA replication. If, for example, adenine in an old chain is in its rare state at the moment that the complementary new chain reaches it, cytosine can pair with it (adenine) and be added to the growing end of the new chain. The result of this type of pairing is the formation of a DNA molecule that contains an exceptional base pair. This situation is not stable and at the next replication, adenine is expected to return to its common state and to pair with thymine. Cytosine introduced into the complementary strand due to tautomeric shift in adenine, would then pair with guanine. Thus, there would be formed two kinds of DNA molecules, one that is identical to the original DNA and another that has undergone a base pair substitution of G-C for A-T. This transitionally substituted DNA molecule has altered coding at a point and results in recognisable mutation. Such mutations which formed during DNA replication are called **copy error mutations.** Such copy error mutations have been shown by Figure 12.4.

The abnormal pairing due to transitional substitution may also occur due to **ionisation** of a base at the time of DNA replication. Ionisation involves the loss of the hydrogen from the number one nitrogen of a base. For example, in its ionised state, thymine can pair with guanine, if the guanine is in its common form. In a similar fashion, guanine in its ionised state can pair with thymine in its common form. From any such unstable base pair, a transition will result following the steps outlined in Fig. 12.4 for A-T to G-C and G-C to A-T.

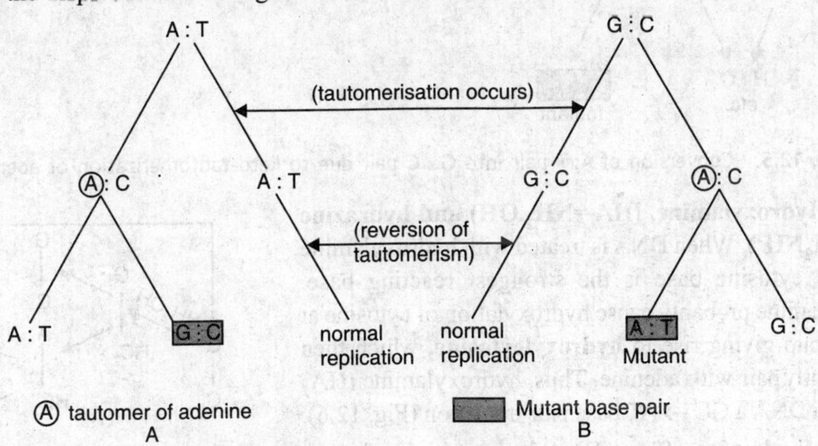

Fig. 12.4. Copy error mutation due to tautomerization of adenine.

Effect of Chemical Mutagens on Nucleotide Sequence

(*a*) Alteration in Resting Nucleic Acid

1. Deamination. Some chemical substance such as nitrous acid causes transitional mutation due to oxidative deamination of DNA bases. In the process of oxidative deamination, the amino group (NH_2) of a DNA base is replaced by hydroxyl (OH) group by the chemical mutagen. Thus, adenine is deaminated into **hypoxanthine** by nitrous acid as shown in the following figure:

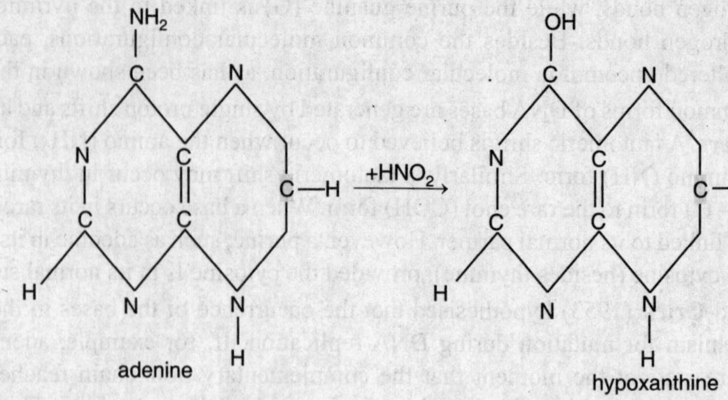

By tautomeric shift the hypoxanthine (HX) is converted into more common keto-tautomer which pairs with cytosine. The A:T pair, thus, can be converted to a G : C pair.

Similarly, deamination converts cytosine to uracil, which has pairing properties similar to thymine and in such a case G : C pair would be changed into A : T pair.

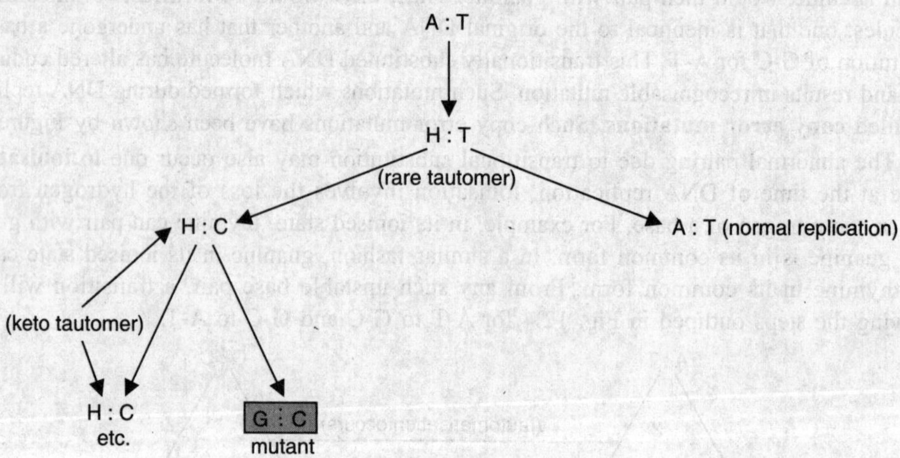

Fig. 12.5. Conversion of A: T pair into G : C pair due to keto-tautomerization of adenine.

2. Hydroxylamine, (HA =NH_2.OH) and hydrazine (HZ =NH_2NH). When DNA is treated with hydroxylamine (HA), its cytosine base is the strongest reacting base. Hydroxylamine probably cause hydroxylation of cytosine at amino group giving rise to **hydroxylcytosine,** which then subsequently pair with adenine. Thus, hydroxylamine (HA) induces in DNA a GC → AT base pair transition (Fig. 12.6).

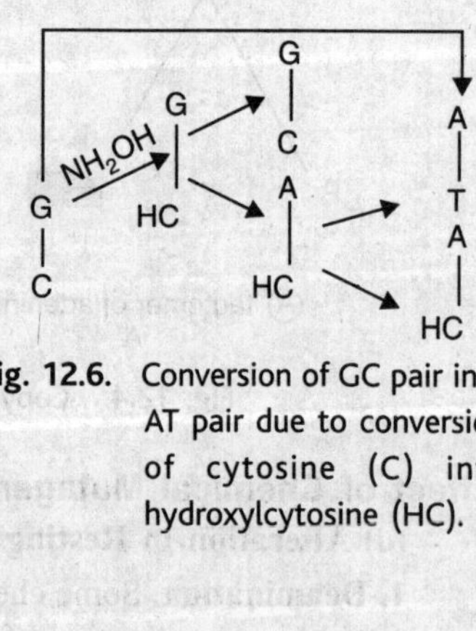

Fig. 12.6. Conversion of GC pair into AT pair due to conversion of cytosine (C) into hydroxylcytosine (HC).

The hydrazine affects DNA by breaking of rings of uracil and cytosine giving rise to **pyrazolone** and **3-aminopyrasole**, respectively. The treatment of RNA or DNA with **anhydrous hydrazine** results in the destruction of their pyrimidines.

3. Alkylating agents. Some alkylating agents carry one, two, or more alkyl groups in a reactive form and act as strong mutagens. Examples of some most extensively studied alkylating agents include **diethyl sulphate (DES)**, **dimethylsulphate (DMS), methyl methane sulphonate (MMS), ethyl ethane sulphonate (EES)** and **ethyl methane sulphonate (EMS).** These mutagens produce mutations in the following ways:

(1) They add ethyl or methyl groups to guanine. This makes guanine the base analogue to adenine.

(2) They remove the alkylated guanine. This is known as **depurination.** The loss of the base produces gaps in the DNA chain which may be filled with a wrong base, thus, producing mutation.

(3) The gap may also produce a deletion, causing mutation.

***(b)* Alteration during Replication of Nucleic Acid**

1. Base analogues. Certain chemical substances have molecular structure similar to the usual DNA bases that, if they are available, such **analogues** may be incorporated into a replicating DNA strand. For example, **5-bromouracil** (5BU) or its nucleoside **5-bromodeoxyuridine** (5-BUdR) in its usual (keto) form is a structural analogue of thymine (5-methyluracil) and it will substitute for thymine. Thus, an A-T pair becomes and remains A-BU. There is some *in vitro* evidence to indicate the BU immediately adjacent to an adenine in one of DNA strands causes the latter to pair with guanine. But, in its rare (enol) state, 5BU behaves similar to the tautomer of thymine and pairs with guanine. This converts A : T to G : C as shown in Fig. 12.7.

2-Aminopurine (2-AP) is another base analogue which is a relatively undifferentiated purine that apparently can pair with cytosine and thymine. It is thought that 2-AP acts by "switching" pyrimidines : for example, it may be incorporated opposite thymine during one round of replication and then pair with a cytosine at the next round to produce an AT → GC transition (see **Goodenough** and **Levine**,1974).

2. Inhibition of precursors of nucleic acids. There are some mutagens which interfere with the synthesis of nitrogen bases of nucleic acids such as purines or pyrimidines. Often lack of one base either causes breaks or causes pairing mistakes. For example, **azaserine** (a potent alkylating agent) inhibits purine synthesis and **urethane** (a mild alkylating agent) is an inhibitor of pyrimidine synthesis. However, urethane induced chromosome breaks are inhibited by thymine.

3. Transversion. The substitution mutation when involves the substitution or replacement of a purine with a pyrimidine or *vice versa* then that type of substitution mutation is called **transversion mutation.** The existence of transversion mutation was first of all postulated by **E. Freese** in 1959. We still have poor information about the mechanism of induction, identification and characterisation of transversion mutations. Moreover, it is extremely difficult to recognise transversion mutations genetically. However, they can be recognized only by analysis of amino acid substitutions in proteins.

Effects of Physical Conditions on Nucleotide Sequence

High temperature and low pH value are known to affect depurination or loss of purine bases. The removal of a purine from a strand of DNA leaves a gap at that point. At the time of replication, it would be possible for any of the four bases to insert in the complementary newly formed strand. If the inserted nucleotide contained a purine, the complementary strand would contain a transversion.

B. Multiple mutations or gross mutations. When changes involving more than one nucleotide pair, or entire gene, then such mutations are called **gross mutations.** The gross mutations occur due to rearrangements of genes within the genome and may be of the following types:

1. The rearrangement of genes may occur within a gene. Two mutations within the same functional gene can produce different effects depending on gene whether they occur in the *cis* or *trans* position.

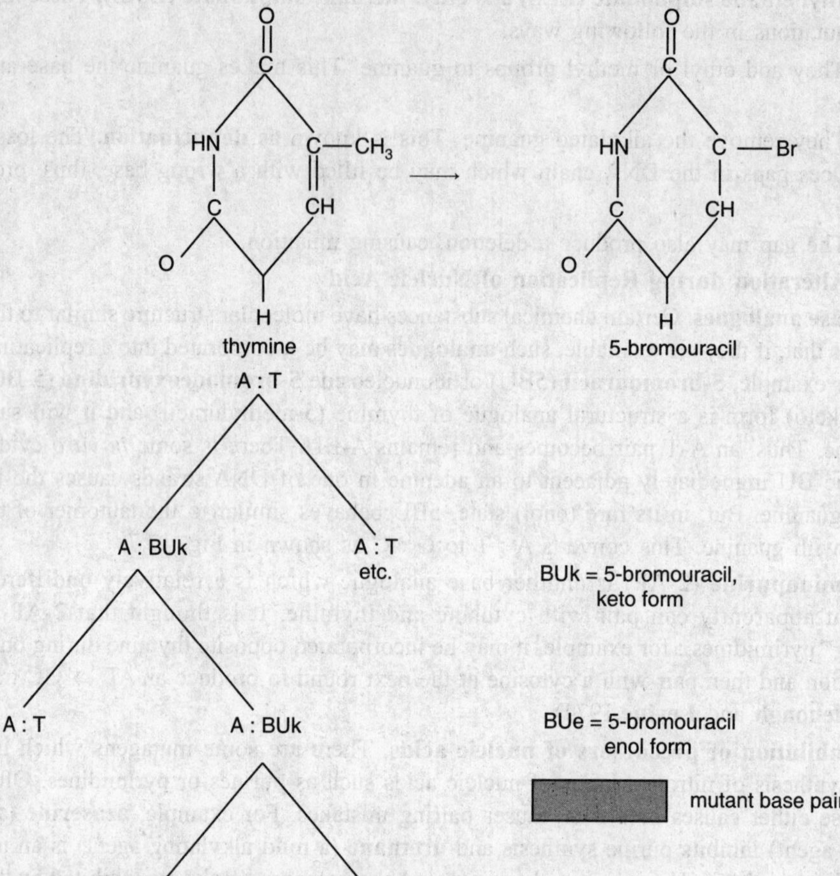

Fig. 12.7. Conversion of A : T base pair into G : C base pair due to keto and enol forms of bromouracil.

2. The rearrangement of gene may occur in number of genes per chromosome. If the number of gene replicas are non-equivalent on the homologous chromosomes, they may cause different types of phenotypic effects over the organisms.

3. Due to movement of a gene locus, new type of phenotypes may be created, especially when the gene is relocated near heterochromatin. The movement of gene loci may take place due to following method :

(*i*) Translocation. Movement of a gene may take place to a non-homologous chromosome and this is known as **translocation.**

(*ii*) Inversion. The movement of a gene within the same chromosome is called **inversion.**

3. Classification of Mutation According to the Origin

According to the mode of origin, following two kinds of mutations have been recognised:

(1) Spontaneous mutations. The spontaneous mutations occur suddenly in the nature and their origin is unknown. They are also called "**background mutation**" and have been reported in many organisms such as, *Oenothera,* maize, bread molds, microorganisms (bacteria and viruses), *Drosophila*, mice, human beings, etc.

(2) Induced mutations. Besides naturally occurring spontaneous mutations, the mutations can be induced artificially in the living organisms by exposing them to abnormal environment such as radiation, certain physical conditions (*i.e.*, temperature) and chemicals. The substances or agents which induce artificial mutations are called **mutagens** or **mutagenic agents.**

Mutagenic agents. The mutagenic agents are of the following kinds:

A. Radiations. The radiations which are important in mutagenesis are of two categories: one type is **ionising radiations** such is X-rays and gamma rays; alpha and beta rays; electrons, neutrons, protons and other fast moving particles. The second type is **non-ionising radiations** such as ultraviolet and visible light. Both types of radiations induce mutations by following methods:

(*i*) Ionising radiations as mutagens. Relatively little is known about the mechanism by which ionising radiations cause mutation. As, we are already familiar that matter composed of atoms and atoms, in turn, are made up of a positively charged atomic nucleus (with neutrons, protons) and a surrounding cons-tellation of negatively charged electrons. The charges of atomic particles remain so balanced that normal atoms are electri-cally neutral. When ionising radiations pass through matter, they dissipate their energy in part through the ejection of electrons from the outer shell of atoms and the loss of these balancing, negati-vely-charged particles (electrons) leaves atoms which are no longer neutral but are positively charged. The positively-charged atom is called **ion.** The ejected electrons move at high speed; knock other electrons free from their respective atoms and when their energy is dissipiated, become attach to other atoms and convert the atoms into negatively charged ions. To achieve their stable configuration

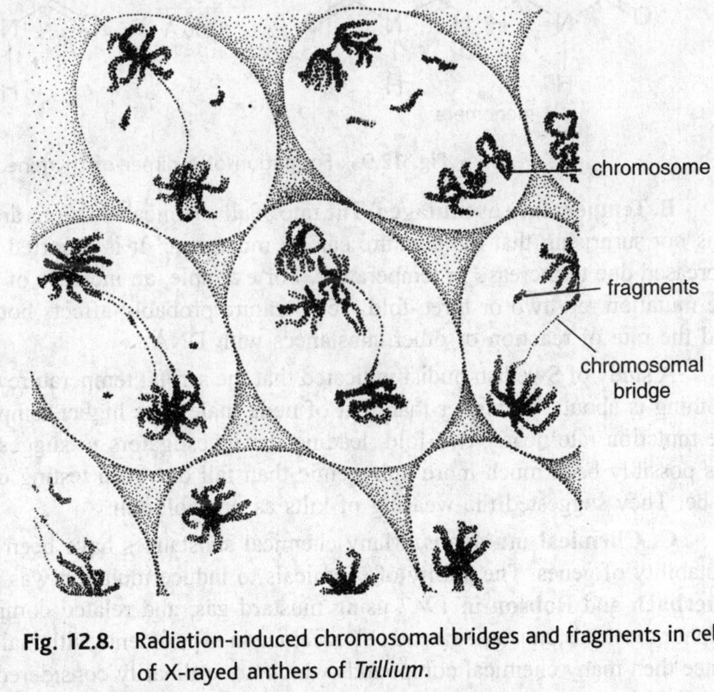

Fig. 12.8. Radiation-induced chromosomal bridges and fragments in cells of X-rayed anthers of *Trillium*.

(*i.e.*, neutral charge), ions undergo many chemical reactions and during these chemical reactions ionising radiation is thought to cause mutation.

Further, ionizing radiations cause breaks in poly-sugar phosphate backbone of DNA and, thus, causing chromosomal mutations such as break, deletion, addition, inversion and translocation. During breakage of DNA molecule due to ionising radiation, the active role of oxygen is predicted. Because, oxygen is important in the formation of H_2O_2 and H_2O in irradiated water and these products may induce breaks in DNA molecule.

***(ii)* Non-ionising radiations as mutagens.** The ultraviolet (UV) light is a non-ionizing radiation which may cause mutation. The most effective wavelength of ultraviolet for inducing mutations is about 2,600 A°. This is a wavelength that is best absorbed by DNA and a wavelength at which proteins absorb little energy. When a substance absorbs sufficient energy from the ultraviolet light, some of their electrons are raised to higher energy levels, a state called **excitation.** The excited molecule becomes reactive and mutated and is called **photoproduct.**

Dimerization. The ultraviolet radiation produces several effects on DNA, one being the formation of chemical bonds between two adjacent pyrimidine molecules in a polynucleotide and particularly, between adjacent thymine residues as shown in Fig. 12.9. As the two thymine residues associate, or **dimerize** to form a **dimer**, their position in the DNA helix becomes so displaced that they can no longer form hydrogen bonds with the opposing purines and thus regularity of the helix becomes distorted. Thus, dimerization interferes with the proper base pairing of thymine with adenine, and result in thymine's pairing with guanine. This will produce a T-A to C-G transition.

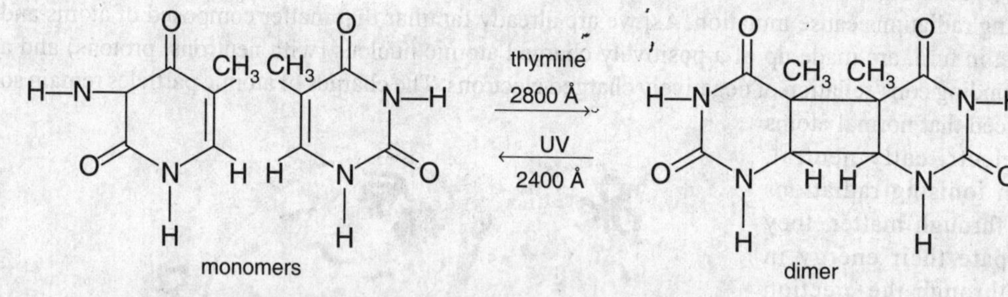

Fig. 12.9. Formation of a dimer of thymine.

B. Temperature as mutagen. The rate of all chemical reactions are influenced by temperature. It is not surprising that temperature can be mutagenic. It is reported that the rate of mutation is increased due to increase in temperature. For example, an increase of 10°C temperature increases the mutation rate two or three-fold. Temperature probably affects both thermal stability of DNA and the rate of reaction of other substances with DNA.

A study of Swedish nudist indicated that the scrotal temperature of human males in ordinary clothing is about 3°C higher than that of nude male. The higher temperature could well increase the mutation rate nearly two-fold, leading the investigators to suggest that the wearing of pants has possibly been much more unhygeinic than fall out from testing of nuclear devices threatens to be. They suggested the wearing of kılts as one solution.

C. Chemical mutagens. Many chemical substances have been responsible to increase the mutability of genes. The ability of chemicals to induce mutation was first of all demonstrated by **Auerbach** and **Robson** in 1947 using mustard gas, and related compounds as the nitrogen and sulphur mustards, mustard oil and chloracetone in experiments with male *Drosophila melanogaster*. Since then many chemical compounds which are ordinarily considered to be non-toxic have been found to be mutagenic in certain specific situations. Any chemical substance that affects the chemical environment of chromosomes is likely to influence, at least indirectly, the stability of DNA and its ability to replicate without error. A chemical mutagen can cause mutation only when it enters in the nucleus of the cell. It can affect the chromosomal DNA by following two ways :

(1) Direct gene change. Certain chemical mutagens affect DNA directly. They affect the constituents of DNA only when DNA is not replicating. For example, nitrous acid converts adenine into hypoxanthine and cytosine to uracil by deamination. Like the nitrous acid, nitrogen mustard, formaldehyde, epoxides, dimethyl and diethyl sulphonate, methyl and ethyl methanesulphonate (MMS and EMS) and nitrosoguanidine (NG) also have direct mutagenic effect on the DNA molecule.

(2) Copy error. Certain chemical compounds, called base analogues (*e.g.*, 5-bromouracil, 2-aminopurine, etc.) closely resemble with certain DNA bases and therefore, act as mutagens. During DNA replication, they are incorporated by DNA in place of the normal DNA bases. Certain other base analogues such as urethane triazine, caffeine (in coffee, tea and soft drinks), phenol and carcinogens, acridines (proflavin, etc.), have mutagenic effects. Certain inorganic substances such as manganese chloride is mutagenic for many organisms, as, they are the compounds which bind calcium and, thus, interfere with the integrity of the chromosome structure.

4. Classification of Mutation According to the Direction

According to their mode of direction, following types of mutations have been recognised :

(A) Forward mutations. In an organism when mutations create a change from wild-type to abnormal phenotype, then that type of mutations are known as **forward mutations.** Most mutations are forward type.

(B) Reverse or back mutations. The forward mutations are often corrected by error correcting mechanism so that an abnormal phenotype changes into wild-type phenotype. They may be of the following types :

(*i*) Single site mutations. Some reverse mutations change only one nucleotide in the gene and are called **single site mutations.** For example, due to forward mutation the adenine is changed into guanine and backward mutation change guanine into adenine :

$$\text{Adenine} \xrightarrow{\text{forward}} \text{Guanine} \xrightarrow{\text{reverse}} \text{Adenine}$$

(*ii*) Mutation suppressor. When a mutation occurs at a different site from the site where already primary mutation occurred and that mutated gene reverse the effects of primarily mutated gene, then such (secondary) mutations are called **mutation suppressors.** They may be of following types :

(*a*) Extragenic suppressor. The extragenic suppressor mutation occurs in a different gene from that of the mutant gene. In *E. coli*, a gene mutation suppressor gene called **rec A** (*rec* for recombination) is known which is necessary for recombination and is found to repair ultraviolet-induced thymine dimers of a gene by a process called **postreplication recombinational repair** (see **Goodenough** and **Levine,** 1974).

(*b*) Intragenic suppressor. The intragenic suppressor mutation occurs in a different nucleotide within the same gene and shift the reading frame back into register.

(*c*) Photoreactivation. In photoreactivation type reverse mutation, reversal of ultraviolet induced thymine dimers takes place by specific enzymes in the presence of visible light waves. During ultraviolet radiation, a particular enzyme is selectively bound to the bacterial DNA. During photoreactivation the enzyme is activated by visible light and that cleaves the pyrimidine or purine dimers into monomers and restores their original forms.

(*d*) **Excision repair or Dark reactivation.** In an ultraviolet (UV) induced mutation, the reverse mutation may also occur in the absence of light. According to **Howard Flanders** and **Boyce** (1964), dark reactivation includes following stages: (*i*) An enzyme possibly endonuclease makes a cut in the polynucleotide strand on either side of the dimer which may be formed due to ultraviolet radiation and excises a short, single-strand segment of the DNA. (*ii*) Another enzyme, possibly exonuclease widens the gap produced by the action of the endonuclease. (*iii*) DNA polymerase resynthesises the missing segment, using the remaining opposite strand as a template; and (*iv*) the final gap is closed by some enzymatic rejoining process, (*i.e.*, DNA ligase).

5. Classification of Mutation According to Magnitude of Phenotypic Effect

According to their phenotypic effects following kinds of mutations may occur :

1. Dominant mutations. The mutations which have dominant phenotypic expression are called dominant mutations. For example, in human beings the mutation disease **aniridia** (absence of iris of eyes) occurs due to a dominant mutant gene.

2. Recessive mutations. Most types of mutations are recessive in nature and so they are not expressed phenotypically immediately. The phenotypic effects of mutations of a recessive gene is seen only after one or more generations, when the mutant gene is able to recombine with another similar recessive gene.

3. Isoalleles. Some mutations alter the phenotype of an organism so slightly that they can be detected only by special techniques. Mutant genes that give slightly modified phenotypes are called **isoalleles.** They produce identical phenotypes in homozygous or heterozygous combinations.

4. Lethal mutations. According to their effects on the phenotype, mutations may be classified as lethals, subvitals and supervitals. Lethal mutations result in the death of the cells or organisms in which they occur. **Subvital mutations** reduces the chances of survival of the organism in which they occur. **Supervital mutations,** in contrast, cause the improvement of biological fitness under certain conditions:

6. Classification of Mutation According to Consequent Change in Amino Acid Sequence

1. Missense mutations. They change the meaning of a codon, changing one amino acid into another.

2. Temperature sensitive mutations or T_s mutations. If the substitution produces a protein that is active at one temperature (typically 30°C) and inactive at a higher temperature (usually 40 – 42°C).

3. Nonsense or chain termination mutations. They arise when a codon for an amino acid is mutated into a termination codon (UAG, UAA or UGA), resulting in the production of a shorter protein.

Since, temperature-sensitive and chain termination mutations exhibit the mutant phenotype only under certain conditions, they are called **conditional mutations**; they are the most versatile and useful mutations.

4. Silent mutations. They change a nucleotide but not the amino acid sequence because they affect the *third position* of the codon, which is usually less important in coding. This is a silent mutation because it leaves the protein sequence unchanged.

7. Classification of Mutation According to the Types of Chromosomes

According to the types of chromosomes, the mutations may be of following two kinds:

1. Autosomal mutations. This type of mutation occurs in autosomal chromosomes.

2. Sex chromosomal mutations. This type of mutation occurs in sex chromosomes.

12.4 MUTATION RATE

The frequency with which genes mutate spontaneously is called **mutation rate.** Most genes are relatively stable and mutation is a rare event. The great majority of genes have mutation rate of 1×10^{-5}, *viz.*, one gamete in 100,000 to one gamete in million would contain a mutation at a given locus. Mutations occur much more frequently in certain regions of the gene than in others. The favoured regions are called **hot spots.** The mutation rate is influenced by various factors which are as follows:

1. Genetic control of mutation rate. There are ample evidences which show that mutation rate is under genetic control, *viz.*, certain genes called **mutator genes** may increase the mutation rate in *Drosophila* (**Demere**, 1937), maize (**Rhoades**, 1938) and *E. coli* (**Goldstein**, 1955). However, certain suppressor genes may decrease the rate of mutation.

In bacteria, as well as in eukaryotes, spontaneous mutations most frequently are caused by transoposons which are segments of DNA that have a tendency to jump around the genome.

2. Viral control of mutation rate. Virus reportedly affect the mutability of host's genes. **Sprague** (1963) experimented with maize suggested that virus may cause mutation. **Baumiliar** (1967) reported that viruses increase the mutation rate in *Drosophila melanogaster*. But, still we do not know how viruses increase the mutability of host genes.

3. Environmental control of mutation rate. There are three major environmental factors that affect mutation rates, *viz*, temperature, certain radiations and chemicals.

12.5 METHOD OF DETECTION OF SEX-LINKED LETHAL MUTATION

H.J. Muller devised an easy method for detecting lethal mutations in the sex chromosomes of *Drosophila*. This is called a CLB method in which a special type of female fly is employed which carries a normal X chromosome and an abnormal X chromosome. The abnormal X chromosome contains an inversion mutation C (which prevents the chromosome to do crossing over with the normal X chromosome, therefore, called **crossover suppressor**), a recessive lethal mutational gene, I and a dominant gene B for bar-eye. In Muller's CLB technique, these CLB female flies are matcd with males which are previously treated with some mutagenic agent (such as X-rays) to cause mutation in some of their sperms. The resulting zygotes are of four types and one of these, the CIB male, fail to survive because such embryos contain a recessive lethal which expresses itself when hemizygous. Thus, only one class of male (with CLB X chromosome) male remains to fertilize the F_1 females. Each heterozygous CLB female results from the fertilization of a CLB egg and an irradiated X-bearing sperm and some of these sperms will contain mutated X-chromosomes. Mated F_1 heterozygous CLB females are distributed individually into culture tubes in which each lays fertile eggs and so produces a single F_2 culture. The culture produced by females bearing an induced lethal mutation contain only females; whereas female bearing irradiated X-chromosomes in which no recessive lethal has been induced yield cultures containing some wild-type males. Thus, if in a population of 1000 cultures, 990 contained some males and 10 contained only females, the induced rate of sex-linked, recessive lethal mutations would be 1 per cent.

Besides the CIB method, there are many more methods such as **Muller-5 method** (for the detection of sex-linked lethal mutations); **attached X- method** (for the detection of sex-linked visible mutations); **balanced lethal systems** (for the detection of autosomal mutations); and **Stadler's method** and **Singleton's method** (for the detection of specific loci in plants).

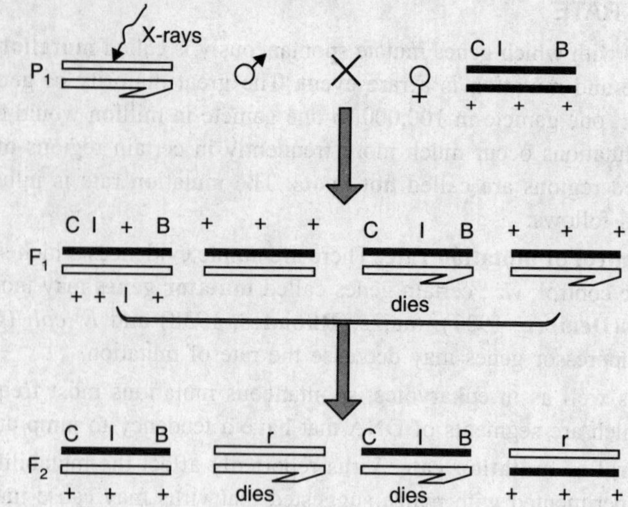

Fig 12.10. Diagram of Muller's CIB technique for detecting sex-linked (X-linked), recessive lethal mutations.

12.6 PRACTICAL APPLICATIONS OF MUTATIONS

Mutations are generally deleterious and recessive for the organisms, therefore, majority of them are of no practical value. **A. Gustafsson** has estimated that less than one in 1000 mutants produced may be useful in plant breeding. In India, several useful mutations of various cereals and other crop plants have been developed (see Table 12.1).

M.S. Swaminathan: Father of green revolution of India.

1. Wheat. In bread wheat, many useful mutations have been obtained and utilised in plant breeding, *e.g.*, branched ears, lodging resistance, high protein and lysine content, amber seed colour and awned spikelets. **Dr. M.S. Swaminathan**, one of the most distinguished and legendary figure in the field of cytogenetics and plant breeding in Indian subcontinent, had utilised amber mutation of Mexican wheat variety to develop a new variety of wheat, called **Sharbati Sonora** while working at Indian Agriculture Research Institute (IARI), New Delhi. According to **Dr. N.E. Borlaug** (Nobel Laureate), this variety of wheat paved the way for the Green Revolution in India (see **Gupta**, 1994).

2. Rice. In rice, one of the high yielding varieties **Reimei** was developed through mutations isolated after gamma irradiation. Certain developed mutants of rice are found to contain increased contents of proteins and lysine. In certain other mutant rice, the duration of crop was reduced by as many as 60 days.

3. Barley. In barley, mutations called **erectoides** and **eceriferum** have been induced. These mutants had high yields including several useful characters.

12.7 SIGNIFICANCE OF MUTATION

The vast majority of mutations are deleterious to the organism and are kept at low frequency in the population by the action of natural selection. Mutant types are generally unable to compete equally with wild-type individuals. Even under optimal environmental conditions many mutants appear less frequently than expected.

Table 12.1 List of varieties of crop plants released by the use of induced mutations (Source: **P.K.Gupta**, 1994).

S. No.	*Type of crop*	*Number of released variety up to year 1991*	*Varieties released in India*
I.	Cereals		
	1. Bread wheat	113	NP836; Sharbati Sonora, Pusa Lerma; N/I-5643.
	2. Durum wheat	25	–
	3. Rice	278	Jagannath, IT48, IT 60, Hybrid Mutant 95.
	4. Barley	229	RBD-I, DL-253.
II.	Legumes	15	Pusa Parvati (French bean).
		136	Hans (pea); Ranjan (lentil); Trombay Vishakhi (pigeon pea); Pant moong 2; TAP -7; MUM-2 (mung bean)
III.	Other Crops	223	Pusa Lal Meeruti, S-12 (tomato); Aruna, Sowbhagya (157-B), RC8 (castor bean); MCU-7, Rasmi, Pusa Ageti (cotton); RLM 198, RLM 514 (mustard); New Hybrid Bajra 5 (NH BS); Pusa 46 (pearl millet); Co 997, Co 6608 (sugar cane); MDUI (chili); JRC-7447 (jute).

QUESTIONS

Long Answer Questions

1. What are missense, nonsense, and frameshift mutations and what are the consequences of each ?
2. What do you understand by spontaneous mutations and induced mutations? Discuss variation in mutation rates and frequencies at different loci within an organism.
3. (*a*) What critical evidence is needed to distinguish a gene mutation from a minute deletion, *i.e.*, one too small to be cytologically detectable?
   (*b*) Show how you would determine the rate of mutation from a recessive to a dominant allele in human beings.
   (*c*) Describe what is meant by a tautomeric shift?
   (*d*) Are the mutational consequences of tautomeric shifts base-pair transitions or base pair transversions? Explain, using a specific example.
4. Explain :
   (*i*) Why geneticists find most mutations to be deleterious?
   (*ii*) Why, nevertheless, the mutation process is considered to be the basis of evolutionary progress?
5. Answer each of the following questions as briefly as possible:
   (*a*) Which type of mutation, one induced by a base analog or one induced by proflavin, would you expect to be more deleterious to an organism and why?
   (*b*) What evidence is there that ionisation caused by X-rays need not occur in the gene itself to cause mutation?
   (*c*) What are the possible mechanisms by which a gene may change to many different allelic forms?
   (*d*) Why are sex-linked lethal mutations easier to detect than autosomal lethals?

6. Three repair mechanisms are known in *E. coli* for the repair of DNA damage (pyrimidine dimer formation) after exposure to ultraviolet light: (1) photoreactivation; (2) excision (dark) repair; and (3) postreplication repair. Compare and contrast these mechanisms, indicating how each achives repair and how the events occurring in each may lead to gene mutations.
7. Describe different kinds of radiations and chemical mutagens utilised for induction of mutations.
8. Discuss the procedure used in the detection of sex-linked lethal mutations by CIB method.
9. Explain the difference beetween a transition and a transversion and give an example of each.
10. What is the difference beetween intragenic and intergenic suppression? Give an illustration of each.
11. Compare the effects of nitrous acid, hydroxylamine and 5-bromouracil on DNA.

Short Answer Questions

1. Write short notes on the following:
   (*a*) Base substitution; (*b*) Ionising radiation;
   (*c*) Balanced lethal; (*d*) Somatic mutation;
   (*e*) Acridine dye; (*f*) Lethal mutation.
2. Give an account of 'Frame-shift Mutation'. (*C.C.S.U. Meerut 2007*)

Very Short Answer Questions

1. Why are most mutations in structural genes recessive to their wild-type alleles?
2. Which of the following would be likely to suffer the greatest, and which the least genetic damage from radiation exposure: (*a*) a haploid, (*b*) a diploid, (*c*) a polyploid?
3. If the mutation rate of a certain gene is directly proportional to the radiation dosage and the mutation rate of *Drosophila* is observed to increase from 3% at 1000 R to 6% at 2000 R. What percentage of mutations would be expected at 3500 R?
4. The X-linked recessive mutations are more easily studied in appropriate organisms than are autosomal ones. Why?
5. Some individuals have a patch of blonde hair in a head of brown hair. What types of mutation would this be?
6. If a drastic alteration occurred in the structure of one of the genes for 28S rRNA, do you think that the translation of mRNA into protein would cease? If not, why not?
7. What possible explanations can you offer for the reversion of a mutant to the wild-type phenotype?
8. How many base pairs would have to be deleted in a mutational event to eliminate a single amino acid from a protein and not change the rest of the protein?
9. The "dotted" gene in maize (Dt) is a "mutator" gene influencing the rate at which the gene for colourless aleurone (*a*) mutates to its dominant allele (A) for coloured aleurone. An average of 7.2 coloured dots (mutations) per kernel was observed when the seed parent was dt/dt, a/a and the pollen parent was Dt/Dt, a/a. An average of 22.2 dots per kernel was observed in the reciprocal cross, How can these results be explained?
10. What is the difference between substrate and a template transition mutation?
11. 5-bromouracil, 2-aminopurine, proflavin, ethyl ethane sulphonate and nitrous acid are chemical mutagens. What does each do?

Fill in the Blanks

1. In Muller's ClB technique C stands for crossover suppressor, l for recessive .... and B for ....
2. .... mutations occur due to tautomerization of DNA bases.
3. Nitrous acid causes transitional mutation due to .... of DNA bases.
4. About .... mutants of *Drosophila* are known from all over world.

5. Heterochromia of iris in human beings is due to a fatal ....

6. Acridine dye is a .... which causes addition mutation in DNA.

Multiple Choice Questions

1. Mutation is
(*a*) change that is inherited
(*b*) change in parent not inherited
(*c*) plant growth controlling factor
(*d*) change which affects the offsprings of F_2 generation

2. Gene mutation is
(*a*) mutation in the genes of DNA
(*b*) mutation in phosphodiester linkage
(*c*) mutation in chromosomes
(*d*) change in the sequence of nitrogenous bases

3. Mutations which normally happen randomly are considered one of the raw materials for evolution because they
(*a*) are stable
(*b*) contribute new variation in organism
(*c*) cause death of organism
(*d*) none of these

4. Proflavin and acridine orange induce
(*a*) transitions
(*b*) transversions
(*c*) inversions
(*d*) frameshift mutations

5. Induction of mutation by X-rays was discovered by
(*a*) Morgan
(*b*) Hugu devires
(*c*) Muller
(*d*) Luria

6. Who is associated with "Green revolution" in India?
(*a*) B.P. Pal
(*b*) M.S. Swaminathan
(*c*) R.S. Paroda
(*d*) E.J. Butler

7. Low temperature is mutagenic in
(*a*) wheat
(*b*) maize
(*c*) rice
(*d*) mustard

8. Frequency of mutation
(*a*) varies with characters and organisms
(*b*) can be increased by X-rays
(*c*) is greatly affected by environmental factors
(*d*) all of the above

9. Which of the following mutagens can be best used in inducing mutation in microorganisms?
(*a*) X-rays
(*b*) β-rays
(*c*) UV-rays
(*d*) γ-rays

10. Why are haploids are superior to diploids in study of mutations?
(*a*) they have shorter life time
(*b*) smaller number of chromosomes
(*c*) they allow expression of recessive mutation immediately
(*d*) obtained in large numbers

ANSWERS

Very Short Answer Questions

1. Wild-type alleles usually code for complete, functional enzymes or other proteins. One active wild-type allele can often cause enough enzyme to be produced so that normal or nearly normal phenotype result (dominance). Mutations of normally functioning genes are more likely to destroy the biological activities of proteins. Only in the complete absence of the wild-type gene product would the mutant phenotype be expressed recessiveness.
2. Haploid greatest, polyploid least. Most mutations are recessive, and recessives have only a very low probability of being expressed in polyploids with their multiple sets of chromosomes bearing normal (dominant) alleles. Both dominant and recessive mutations are expressed at once in haploids.
3. 10.5%.
4. Recessive mutations are more easily detected in hemizygous males.
5. Somatic mutation.
6. Translation would not cease since numerous genes for rRNA are present in the genome. The mutation of one of these would probably not interfere with protein synthesis.
7. *Intragenic mutation within the same codon,* either restoring the original amino acid or resulting in the presence of a compatible amino acid; *intragenic mutation within the same cistron,* such as one that restores the normal reading frame; *intergenic direct suppression,* such as alteration in some component directly involved in protein synthesis, for example, tRNA; *intergenic indirect suppression* by an alteration in the cellular milieu.
8. Three, as any other number of deletion (or addition) would cause a frameshift and other amino acid changes.
9. Seed parent contributes two sets of chromosomes to triploid endosperm; one Dt gene gives 7. mutations/kernel, two Dt genes increase mutations to 22.2/kernel.
10. In replicating DNA, a transition mutation can occur by tautomerization of a base in the template strand (template transition) or entering the progeny strand (substrate transition).
11. *5-bromouracil* (pyrimidine analogue) and *2-aminopurine* (purine analogue) are incorporated into DNA as thymine and adenine, respectively. However, each undergoes tantomeric shifts more frequently than the normal base. Both cause transitions. *Nitrous acid* also promotes transitions by converting cytosine into uracil, which acts like thymine, and adenine into hypoxanthine, which acts like guanine. *Proflavin* induces insertions and deletions by intercalating and buckling DNA. *Ethyl ethane sulphonate* removes purine rings and thus promotes transitions and transversions.

Fill in the Blanks

1. Lethal gene, bar-eye.
2. Copy error.
3. Oxidative deamination.
4. 500.
5. Somatic mutation.
6. Chemical mutagen.

Multiple Choice Questions

1. (*a*) **2.** (*d*) **3.** (*b*) **4.** (*d*) **5.** (*c*) **6.** (*b*)
7. (*c*) **8.** (*d*) **9.** (*c*) **10.** (*c*).

13

The Gene (Fine Structure of Gene)

13.1 EVOLUTION OF GENE CONCEPT

The hereditary units which are transmitted from one generation to the next generation are called **genes**. A gene is the fundamental biologic unit, like the atom which is the fundamental physical unit. **Mendel** while explaining the results of his monohybrid and dihybrid crosses, first of all conceived of the genes as particulate units and referred them by various names such as **hereditary factors** or **hereditary elements** (or **Merkmal**). The term **gene** was coined by **Wilhelm Johansen** in 1909 to describe a heritable factor responsible for the transmission and expression of a given biological character but without reference to any particular theory of inheritance. Thus, in its original context, the gene had no specific material basis and could be treated purely in abstract terms, as it is in the study of transmission genetics (Mendelian inheritance).

A more precise idea of the physical and functional basis of the gene arose from many sources in the first half of the twentieth century. In 1909, **Archibald Garrod** showed that the human metabolic disorder alkaptonurea resulted from the failure of some specific enzyme and could be transmitted in an autosomal recessive fashion. He called this disorder as an **'inborn error of metabolism'**. Garrod was unaware about Mendelian inheritance and the significance of his discovery was not appreciated for 30 years, until **George Beadle** and **Edward Tatum** (1940) found that X-ray induced mutations in fungi (*e.g., Neurospora*) often caused specific biochemical defects. This led to the **one gene one enzyme** model of gene function. In 1911, **Thomas Hunt Morgan** showed that genes were located on chromosomes and were physically linked together (chromosome theory of inheritance). Thus, genes were considered as beads and chromosomes as strings of beads. Mutation was supposed to alter the bead structure and recombination (or crossing over) was regarded to involve a breakage between two beads followed by their exchange between paired chromosomes. Each bead was thought of to control one character by controlling some biochemical step. Thus, a gene was considered to be a unit of mutation, recombination and function **(Benzer)**.

Muller (1932) have compared the mutant genes with wild type genes (or standard genes) and classified them into the following types: **(1) Hypomorphs.** These are the genes which have the same effect as the standard gene but their phenotypic expression is less effective. **(2) Amorphs.** These are genes which have very less effect than the standard gene, *i.e.*, they do not cause any significant change in phenotype. *e.g.*, h antigen of Bombay phenotype. **(3) Hypermorphs.** These are genes which have greater effect than the standard gene, *i.e.*, they produce much phenotypic change. **(4) Antimorphs.** These are genes which result in opposite phenotypic effect than the standard gene, *i.e.*, they produce abnormal phenotype. **(5) Neomorphs.** These are genes which produce a phenotypic effect that is qualitatively different from that of standard gene, *e.g.*, Bar locus of *Drosophila*.

In 1944, **Oswald Avery** and colleagues showed by elimination that DNA was the genetic material. Thus, evolved a simple picture of gene: *a length of DNA in a chromosome which encoded the information for an enzyme.*

This definition of gene has to be expanded to include what is currently known about gene function: **not all genes encode enzymes** (some encode polypeptides with different functions; and some encode functional RNA molecules such as rRNA and tRNA); also in viruses, genes may be RNA not DNA. Furthermore, the information in the gene may be used selectively to generate more than one product. The Mendelian concept of a gene controlling a single character has also been expanded to include genes which affect several characters simultaneously (**pleiotropy**, *e.g.*, sickle cell anaemia) and genes which cooperate in groups to control individual characters (**quantitative inheritance**). With the advent of molecular biology, it is now possible to define the gene exactly in terms of its structure and function.

13.2 UNIT OF GENETIC STRUCTURE AND FUNCTION

(*i*) Indivisible units of genetic structure. After having established that genes were carried on chromosomes and could be mapped on them by recombination (**genetic mapping** or **linkage mapping**), individual genes came to be regarded as **fundamental** and **indivisible units** of genetic information, in terms of both structure and function, linked together in a linear fashion (the **beads on a string theory**). This view was challenged by recombination studies in T_4 bacteriophage which showed that the gene could be subdivided into smaller units. **Seymour Benzer** (1950) introduced the terms **muton, recon** and **cistron** to define indivisible units of mutation, recombination and function, respectively. By crossing independently derived mutants (*i.e.*, rII) of the same gene in a phage infection, it was shown that wild type phage (r +) could be produced. This could happen in **intragenic recombination** (recombination within a gene) if distinct sub-elements of the gene were mutated. This showed that the gene was divided into smaller units. The muton and recon are equivalent to a single nucleotide pair (of DNA molecule).

(*ii*) The gene as an indivisible unit of genetic function. Benzer's term cistron means an indivisible unit of genetic function. This can be determined by **complementation analysis** where the gene (or its product) is tested for its ability to compensate for a mutation in a homologous gene in the same cell. Successful complementation restores the wild type phenotypes.

Cis-trans test. The basis of complementation analysis is the cis-trans test (from which the term cistron arises), where pairs of independently derived mutations are considered in the *cis* and *trans* configuration. The *cis* test is a control, because of it both mutations (mutant alleles) are present in one genome, the other should be wild type at both loci and produce the normal gene product(s), thus granting the wild type phenotype upon the cell. The *trans* test is the complementation test and defines the boundaries of the unit of function. If both mutations are within the same gene, when they are present in the *trans*-configuration, each genome carries a mutant copy of the gene and no functional product would be generated in the cell—there would be no complementation (**negative complementation**). If the mutations lie in different genes, when they are present in the *trans*-configuration, each genome can supply the gene product the other lack. With all necessary gene products, the cell is wild type-there is **positive complementation** (Fig. 13.1).

Complementation analysis in bacteria and yeast established that the gene is a cistron, *i.e.* the gene could be defined as a unit of function.

13.3 GENE-CISTRON RELATIONSHIP IN PROKARYOTES AND EUKARYOTES

(*i*) Equivalence of gene and cistron in simple genomes. In prokaryotes and lower eukaryotes there is usually a simple relationship between the gene and its product. In most cases, there is one gene-one product correspondence and the gene is usually colinear with its product. In these organisms therefore, *the gene and cistron are equivalent, i.e.*, the gene is a unit of genetic function; a unit of expressed genetic information. In bacteria, the gene is synonymous with the coding region (open reading frame), whereas in eukaryotes it is synonymous with the transcription unit. This is because bacterial genes are often arranged as an **operon**, so that many products are translated from a common **polycistronic mRNA**. In eukaryotes, conversely, most genes are transcribed as **monocistronic mRNA**.

***(ii)* Nonequivalence of gene and cistron in complex genomes.** In higher eukaryotic genomes, there is often a complex relationship between the gene and its product. Most higher eukaryotic genes contain **introns**, which are intervening sequences not represented in the final product and are not therefore part of its function. While the gene represents the entire transcription unit, the cistron would be interrupted by the introns. The *cistron is therefore* equivalent to the **exons** of the eukaryotic gene.

13.4 GENE SIZE

Bacterial genes are characteristically small (average 1 kbp) and show little diversity in size, whereas genes of higher eukaryotes are large (average 16 kbp) and show great size diversity. Although higher eukaryotic genes are generally much longer than bacterial genes, the same is not true of the mRNAs derived from them. Such a difference is caused by **introns** which are intervening sequences which interrupt the transcription unit and must be spliced out at the RNA level (*i.e.*, during RNA processing). The remaining parts of the transcription unit, which become spliced together and expressed, are termed **exons**. *Gene size is inversely proportional to the percentage of exon material in the gene*. Bacterial genes generally lack introns and are therefore 100 per cent exon material. Introns are also rare in many microbial eukaryotes, such as yeast *Saccharomyces cerevisiae*, whose average, size, 1–2 kbp, is similar to that of bacteria. In humans, the smallest genes have the fewest and smallest sized introns (*e.g.*, the 500 bp histone H4 gene has no introns). Conversely, the largest human gene, the **dystrophin gene** which is 2500 kbp in length has > 95 per cent intron material. The dystrophin gene has 75 introns with an average size of 30 kbp; only 0.5 per cent of the gene is exon material.

13.5 GENE NOMENCLATURE

Genes are named according to species convention (Table 13.1). It is normal to present the names of genes, alleles, genotypes, and where necessary, the mRNA transcribed from a gene in italic and the protein products and phenotypes in Roman. In mot species, genes are designated by a symbol comprising several letters and numbers. Some species conventions (*e.g.*, *Drosophila*, *E. coli*) dictate the use of lower case letters for genes identified as recessive mutations and initial capital letters for those identified as dominant mutations. In other species, including humans, genes are designated by all capital letters.

Table 13.1 A brief summary of conventional genetic nomenclature (*Source:* Twyman 1998).

S. No.	*Species*	*Convention i.e.,*
1.	*E. coli and other bacteria*	Three lower case letters to designate an operon followed by upper case letters to denote different loci, *e.g.*, *lac* operon; lac Z, lac Y, lac A. Proteins Lac Z, Lac Y, Lac A.
2.	Yeast	Three letters identify gene function followed by a number to specify different loci, *e.g.*, *Saccharomyces cerevisiae* gene GAL 4, CDC28; proteins GAL4, CDC28.
3.	*Caenorhabditis elegans*	Three lower case letters indicative of the mutant phenotype hyphenated to a number if more than one locus involved, *e.g.*, genes unc-86, ced-9; proteins UNC-86, CED-9.
4.	Plants	Although no convention covers all plants, most named with 1–3 letter lower case symbols. *Arabidopsis* genes named in style of *Drosophila* but using capital letters, *e.g.*, gene *AGAMOUS* (*AG*), protein AGAMOUS.

S. No.	Species	Convention
5.	*Drosophila*	Name descriptive of mutant phenotype, may be represented by symbol of 1–4 letters, *e.g.*, genes *white* (*w*), *tailless* (*tll*), *hedgehog* (*hh*); proteins white, Tailless, Hedgehog.
6.	Vertebrates	Generally a 1–4 letter lower case symbol with letters and numbers, representative of gene function, *e.g.*, genes *sey, myc,* proteins Sey, Myc.
7.	Humans	As above but in capitals, *e.g.*, genes *MYC, ENO1,* proteins MYC, ENO1.

13.6 PSEUDOALLELES OR LEWIS EFFECT

Apricot eye colour in *Drosophila*. In 1951, **E.W. Lewis** reported results of one of his experiments on *Drosophila* where from a cross of apricot (w^{apr}) and white (*w*) eye he obtained F_1 having intermediate eye colour. In F_1 generation he expected segregation for apricot and white, but to his surprise, he recovered from F_2 wild type (red eyes) also, showing that the two presumed alleles were non-allelic, *i.e.*, both occupying different sites, **Lewis**, therefore, preferred to called them **pseudoalleles** and the phenomenon as **pseudoallelism**. Thus, pseudoalleles are those genes which are located almost at the same place on the linkage map but show recombination, indicating thereby that although they are functionally allelic, structurally they are non-allelic. **Pontecorvo** later preferred to use the term **Lewis effect** for the pseudoalleles. With accumulated cross over data, the distance between the two sites was placed at about 0.01.

Pseudoalleles show complementation, thereby, indicating their functional individuality. Thus, if *a* and *b* are non-allelic the parents should be represented as *aa* ++ and ++ *bb* and in F_2, wild type (++) would be recovered as expected. Thus, the appearance of wild type in F_2 generation could be explained. However, it was difficult to explain why F_1 progeny *a*+/+*b* was not wild type. For explaining this, another criterion, called **complementation**, was employed. Complementation means that in allelic forms there will be lack of complementation due to which F_1 individuals will not be wild type. But since intragenic interalleic crossing over (recombination) is possible, in the F_2 generation wild type individual appear. In F_1 individual *a*+/+*b*, the **position effect** does not allow the wild type to be expressed. As a result of crossing over, ++ and *ab* gametes are produced from the F_1 individuals and in the F_2 generation ++/++ or ++/*ab* will express the wild phenotype. Thus, it has been demonstrated that within the same gene ++/*ab* and *a*+/+*b* exhibit different phenotypes. This is known as the **cis-trans effect** (Fig. 13.1).

$$\frac{++}{ab} \text{ or } \frac{apr/w}{+/+} \qquad\qquad \frac{a+}{+b} \text{ or } \frac{apr/+}{+/w}$$

cis (wild) red trans (mutant) apricot

Fig. 13.1. Cis and trans arrangements of alleles a and b or apr and w alleles of *Drosophila* in heterozygous condition.

Why do the *w* and *apr* genes behave differently in different positions? The production of red pigment is a two-step chemical process. One gene may control one step, while the other gene in close proximity may control another step in the same reaction. When the two genes w^+ and apr^+ (++) occur next to each other on the same chromosome (cis position) they produce the normal red eye colour. When they are opposite to each other; however, chemical blocks occur that stop the reaction at one or more points. So cis-trans effect is a type of position effect of the genes with respect to each other rather than their presence or absence determines the end result.

13.7 COMPLEX GENE LOCI

The Russian geneticist **N.P. Dubinin** in 1939 recognized **step allelism** (that is, the graded effect of different alleles) in *Drosophila*. This discovery foreshadowed the idea that "genes" could be subdivided, *i.e.*, certain genes are **compound** or **complex genes**. The complex loci were discovered when techniques became available in particular organisms for exploring the fine structure of the gene. The first evidence came from *Drosophila* studies in which presumed, alleles were found to be pseudoalleles, that is, *closely linked genetic units that behaved ordinarily as if they were alleles but were not in exactly corresponding position*. They could be separated by crossing over. Virus studies then showed that many sites were present in a particular locus (*e.g.*, *rII* locus in T_4 phase). Around 1970's, studies of gene's fine structure were extended to bacteria (*Salmonella*), yeast, mold (*Neurospora*), wasp (*Mormoniella*), flies, cotton and maize. It appears that complex gene structure is basic to all organisms.

Examples of complex loci. The lozenge gene of *Drosophila* provides a good example of complex locus. In *Drosophila* several other complex loci have been demonstrated which include the following loci: *bx* (bithorax), *sb* (stubble bristles), *N* (notch wing), *g* (garnet eye colour), *sn* (singed bristles), *dp* (dumpy wing) and s (star).

Complementary interaction between alleles indicates complex loci even when recombination between them has not been observed. By this technique, the *Td* region in *Neurospora crassa* which controls the final step in tryptophan synthesis, has been shown to be a complex of several sites in linear order.

In maize, complex loci have been reported at the *A* locus, which controls pigment production; the *R* locus, which controls pigmentation; and the *wx* locus, which influence the nutrient content of the seed.

QUESTIONS

Long Answer Type Questions

1. Write an essay on gene.
2. Explain the fine structure of gene according to Benzer. (*H.N.B. Garhwal 2009*)

Short Answer Questions

1. Describe the fine structure of gene. (*H.N.B. Garwal 2008*)
2. Write short notes on the following:
   (*a*) Modern concept of gene
   (*b*) Structure of genes
   (*c*) Cistron
   (*d*) Gene
   (*e*) Pseudoalleles
   (*f*) Lewis effect
   (*g*) Complex gene loci.

Very Short Answer Questions

1. Define the gene.
2. What is cistron?
3. What is *trans*?
4. Define *cis*.
5. What is position effect?
6. What is *cis-trans* complementation test?

Fill in the Blanks

1. The term gene was coined by .... in 1909.
2. A gene was considered to be a unit of .... , ... and function by ....
3. In eukaryotes, most genes are transcribed as .... mRNA.
4. The dystrophin gene of human beings contains .... introns.

5. .... used the term Lewis effect for the pseudoalleles.

6. The lozenge gene of *Drosophila* is a ....

Multiple Choice Questions

1. The carriers of hereditary material are
   (*a*) chromosomes (*b*) genes
   (*c*) gametes (*d*) gametocytes
2. The terms cistron, muton and recon were coined by
   (*a*) Jenner (*b*) Benzer
   (*c*) Carlson (*d*) Bateson
3. A gene that specifies the amino acid sequence of a polypeptide chain is termed as
   (*a*) structural gene (*b*) regulator gene
   (*c*) operator gene (*d*) split gene
4. Smallest part of DNA which takes part in crossing over is
   (*a*) recon (*b*) muton
   (*c*) replicon (*d*) exon
5. The smallest segment of genetic material affected by mutation is
   (*a*) recon (*b*) cistron
   (*c*) muton (*d*) exon
6. A broad definition of gene would be that it is
   (*a*) a unit of heredity (*b*) a part of DNA
   (*c*) a unit of mutation (*d*) all the three

ANSWERS

Very Short Answer Questions

1. Gene is a hereditary determinant of a specific biological function; a unit of inheritance (DNA) located in a fixed place on the chromosome.
2. 1. Cistron is a unit of function in DNA. One DNA cistron specifies one polypeptide chain in protein synthesis. 2. Cistron is a term coined by **Benzer** for the smallest genetic unit that exhibits the *cis-trans* position effect; this term is synonymous with gene.
3. *Trans* means "across" and refers usually to the geometric configuration of mutant alleles across from each other on a homologous pair of chromosomes.
4. *cis* mean "on the near side of", refers to geometric configurations of atoms or mutants on the same chromosome.
5. An alteration of phenotype caused by a change in the relative arrangement of the genetic material.
6. It is a mating test to determine whether two mutants on opposite chromosomes complement each other; a test for allelism.

Fill in the Blanks

1. Wilhelm Johansen.
2. Mutation, recombination, Seymour Benzer.
3. Monocistronic.
4. 75.
5. Pontecorvo.
6. Complex locus.

Multiple Choice Questions

1. (*b*) **2.** (*b*) **3.** (*a*) **4.** (*a*) **5.** (*c*) **6.** (*d*).

14

Sex Chromosomes, Sex Determination and Sex Differentiation

14.1 SEX DETERMINATION

Members of almost all species are often divided into two sections according to the kind of gamete or sex cell produced by them, *i.e.*, male sex and female sex. The word sex has been derived from Latin word *sexus* meaning section or separation. However, some of the lowest forms of plant and animal life are found to have several sexes. For example, in one variety of the ciliated protozoan *Paramecium bursaria* there are eight sexes or "mating types" all morphologically identical. Each mating type is physiologically incapable of conjugating with its own type, but may exchange genetic material with any of the seven other types within the same variety. Further, in organisms in which the number of sex reduced to just two, sexes may reside in different individuals or within the same individual. An animal possessing both male and female reproductive organs is usually referred to as a **hermaphrodite**. The sex cells and reproductive organs form the primary sexual characters of male and female sexes. Besides these **primary sexual characters**, the male and female sexes differ from each other in many somatic characters known as **secondary sexual characters**. The phenomenon of molecular, morphological, physiological or behavioral differentiation between male and female sexes is called **sexual dimorphism**.

The phenomenon of sexual dimorphism has been a biological riddle for the thinkers and biologists of all time. People always tried to know those factors which determine the male and female sexes of a species. Literally hundreds of mistaken hypotheses and wild guesses were proposed before 1900 in vain attempts to find out a solution to the problem of determination of sex. Modern geneticists have reported many different mechanisms of determination of sex in living organisms. Some important and common mechanisms of sex determination are following:

14.2 GENETICALLY CONTROLLED SEX DETERMINING MECHANISMS

Most of the mechanisms of the determination of the sex are under genetic control and they are of following types:

1. Sex chromosome mechanism or Heterogamesis;
2. Genic balance mechanism;
3. Male haploidy or haplodiploidy mechanism;
4. Single gene effects.

A. Sex Chromosomal Mechanisms (Heterogamesis)

Sex chromosomal mechanism or chromosome theory of sex determination was proposed by **McClung** (1902). The chromosomes of an individual are of two types:

(i) **Autosomes.** The chromosomes which have no relation with the sex and contain the genes which determine the somatic characters of the individuals are known as **autosomes (A)**.

(ii) **Sex chromosomes.** The chromosomes which are responsible for the determination of sex are known as **sex chromosomes**, *e.g.,* X and Y chromosomes.

Structure of sex chromosomes. The X and Y sex chromosomes exhibit structural differences. The cytological studies have shown that the X chromosomes of most organisms are straight, rod-like and comparatively larger than Y chromosomes. The Y chromosome is smaller in size with one end slightly curved or bent to one side in Drosophila (Fig 14.1); in human beings and *Melandrium* no such curvature of Y chromosome occurs. The X chromosomes have large amount of euchromatin and small amount of heterochromatin. The euchromatin has large amount of DNA material, hence, much genetic information. The Y chromosome contains small amount of euchromatin and large amount of heterochromatin. The Y chromosome has little genetic information, therefore, sometimes it is referred to as genetically **inert** or **inactive**.

Types of Sex Chromosomal Mechanism of Sex Determination.

In dioecious diploidic organisms following two systems of sex chromosomal determination of sex have been recognized:

(*a*) Heterogametic males; (*b*) Heterogametic females.

Heterogametic males. In this type of sex chromosomal determination of sex, the female sex has two X chromosomes, while the male sex has only one X chromosome. Because, male lacks a X chromosome, therefore, during gametogenesis produces two types of gametes, 50 per cent gametes carry the X chromosomes, while the rest 50 per cent gametes lack in X chromosomes. Such a sex which produces two different types of gametes in terms of sex chromosomes, is called **heterogametic sex**. The female sex, because, produces similar type of gametes, is called, **homogametic sex**. The heterogametic males may be of following two types:

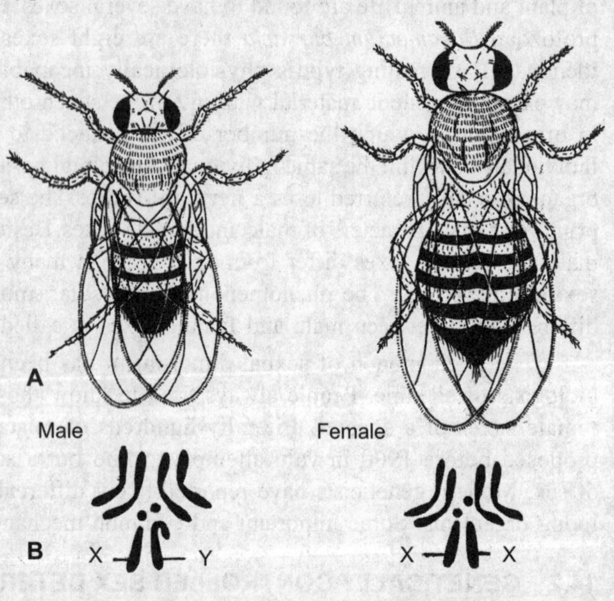

Fig. 14.1. Male and female *Drosphila melanogaster* and their chromosomes.

(i) **XX-XO type.** In certain plants (*e.g., Vallisneria spiralis, Dioscorea sinuata, etc.,*) and insects specially those of the orders Hemiptera (true bugs) and Orthoptera (grasshoppers and roaches), the female having two X chromosomes (hence, referred to as XX) and are, thus, homogametic, while the male having only one X chromosome (hence, referred to as XO). The presence of an unpaired X chromosome determines the masculine sex. The male lacking in one X chromosome produces two types of sperms: half with X chromosome and half without X chromosome. Hence the male is a heterogametic sex, producing two type of sperms. The sex of the offspring depends upon the type of sperm that fertilizes the egg (each of which carries a single X chromosome).

(ii) **XX-XY type.** In human beings, other mammals, certain insects including *Drosophila* and *Lygqeus turicicus* and in certain angiospermic plants such as *Melandrium album* (*Lychis*), *Humulus lupulus*, *Coccinia indica*, the female possesses two homomorphic X chromosomes in their body

cells (hence, referred to as XX) and they being homogametic, produce one kind of eggs, each with one X chromosome. The males of these organisms possess one X chromosome and one Y chromosome (hence, referred to as XY). The males being heterogametic and having two heteromorphic sex chromosomes produce two kinds of sperms: half with X chromosome and half with Y chromosome. The sex of embryo depends on the kind of sperm. An egg fertilized by a X-bearing sperm, produces a female, but, if fertilized by a Y-bearing sperm, a male is produced.

Sex determination in human beings. Sex has the genetic basis. Genes determining the sex in human beings are located on a special pair of chromosomes. They are called **sex chromosomes**. Accordingly, the 23 pairs of human chromosomes can be divided into two groups:

1. **Autosomes.** The 22 pairs of chromosomes carry genes for body characters and general physiology.
2. **Sex chromosomes.** One pair of chromosomes carries genes for the determination of the sex. These are represented by X- and Y-chromosomes. X-chromosome contains genes that determine female sex; Y-chromosome contains genes for male sex. In humans, sex determination is of XX-XY type (Fig. 14.2). Female is homogametic producing only one type of gametes (ova), each containing one X chromosome; male is heterogametic, producing two types of sperms, half carry X and other half carry Y chromosome.

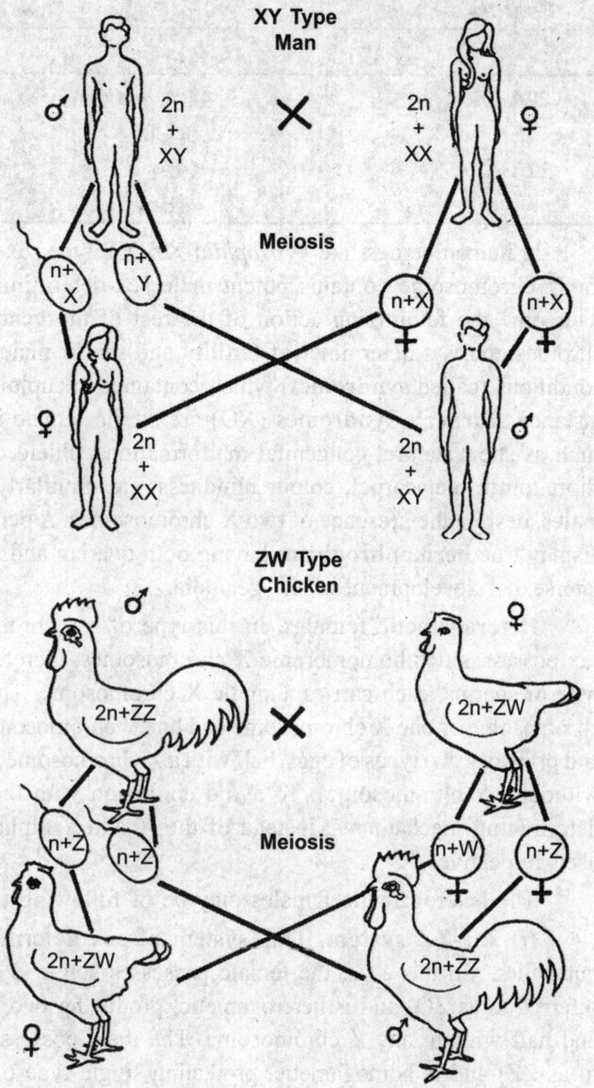

Fig. 14.2. The XX-XY and ZW-ZZ type determination of sex in human beings and fowl.

	Man (Heterogametic)		**Woman (Homogametic)**
	44A + XY		44A+XX
Sperms:	(22 A + X)	**Ova:**	(22A + X)
	(22 A + Y)		(22 A + X)

Sperms	*Ova*	
	22 A + X	22 A + X
22A+X	44 A + XX Girl	44A + XX Girl
22A+ Y	44A+XY Boy	44A+XY Boy

In human beings like *Drosophila* XX-XY type sex determining mechanism occurs but here the Y chromosome contains potent male sex-determining genes which can almost completely overcome the feminizing action of the rest of the genotype. The conclusive evidences that Y chromosome is a determiner of fertility and sex of male individual came from certain abnormal conditions (called **syndromes**) which contained aneuploidic sex-chromosomal abnormalities. For instance, **Turner's syndromes** (XO) are sterile female individuals having certain abnormalities such as short stature, congenital malformations, shield chest, pronounced webbing of the neck, short fourth metacarpel, colour blindness, etc. Similarly, **Klinefelter's syndromes (XX Y)** are males, despite the presence of two X chromosomes. A person with extra one X and Y chromosome display true **hermaphroditism** having both ovarian and testicular tissues and variable degrees of intersexual development of the genitalia.

Heterogametic females. In this type of sex chromosomal determination of sex, the male sex possesses two homomorphic X chromosomes, therefore, is homogametic and produces single type of sperms, each carries a single X chromosome. The female sex either consists of single X chromosome or one X chromosome and one Y chromosome. The female sex is, thus, heterogametic and produces two types of eggs, half with a X chromosome and half without a X chromosome (with or without a Y chromosome). To avoid confusion with that of XX-XO and XX-XY types of sex determining mechanisms, instead of the X and Y alphabets, Z and W alphabets are generally used respectively.

The heterogametic females may be of following two types:

(*i*) ZO-ZZ system. This system of sex determination is found in certain moths and butterflies. In this case, the female possesses single Z chromosome in its body cells (hence, is referred to as ZO) and is heterogametic, producing two kinds of *ova*, half with a Z chromosome and half without any Z chromosome. The male possesses two Z chromosomes (hence, referred to as ZZ) and is homogametic, producing single type of sperms, each of which carries a single Z chromosome. The sex of the offspring depends on the kind of *ovum* as shown below:

Parent: Homogametic female × Homogametic male

2A + ZO   2A + ZZ

Gametes: (A + Z) (A + O)   (A + Z) (A + Z)

Ova   Sperms

F₁: 2A + ZZ   2A + ZO

Male   Female

(*ii*) ZW-ZZ system. This system of sex determination occurs in certain insects (gypsy moth) and vertebrates such as fishes, reptiles and birds and plants such as *Fragaris elatior*. Here the female sex has one Z chromosome and one W chromosome. It is heterogametic and produces two types of ova, 50 per cent ova carry the Z chromosomes, while rest 50 per cent ova carry W chromosomes. The male sex has two homomorphic Z chromosomes and is homogametic producing single type of sperms, each carries a Z chromosome. The sex of the offspring depends on the kind of egg, the Z bearing eggs produces males but the W bearing eggs produce females (Fig. 14.2).

B. Genic Balance Mechanism

By studying sex chromosomal mechanism of sex determination, it may appear at first glance that some genes carried by the sex chromosomes (X and Y) were entirely responsible for sex. But this is not the case. Extensive experimentation of different workers (**Wilson**, 1909 ; **Bridges**, 1921 and **Goldschmidt**, 1934) on different organisms have revealed the fact that most organisms generally have inherent potentialities for both sexes and each individual is found to be more or less intermediate between male and female sexes (Hence may be referred to as **intersex**). There seems to exist a delicate **balance** of masculine and feminine tendencies in the hereditary compliment of an individual and mechanisms like the XY, ordinarily serve to trip the balance in one direction or another. Such **genic balance** mechanism of determination of sex was first of all studied in *Drosophila* by **C.B. Bridges** in 1921.

Sex determination in *Drosophila*. In *Drosophila*, the presence of Y chromosome has been found essential for the fertility of male sex but that has nothing to do with the determination of male sex. In this fly, the sex is determined **polygenically**. The sex determining genes were so distributed that the net effect results in the autosomes determining maleness and the X chromosomes femaleness. The sex of an individual then depends upon the ratio of X chromosomes to autosomes. If each haploid set of autosomes carries factors with a male-determining value equal to one (1), then each X chromosome carries factors with a female determining value of one and half (1½). Let A represent a haploid set of autosomes. In a normal male (AAXY) the male and female determinants are in the ratio of 2 : 1½ and, therefore, the genic balance is in the favour of maleness. A normal female (AAXX) has a male and female ratio of 2 : 3 and, therefore, the balance is in the favour of femaleness (See Fig. 14.3).

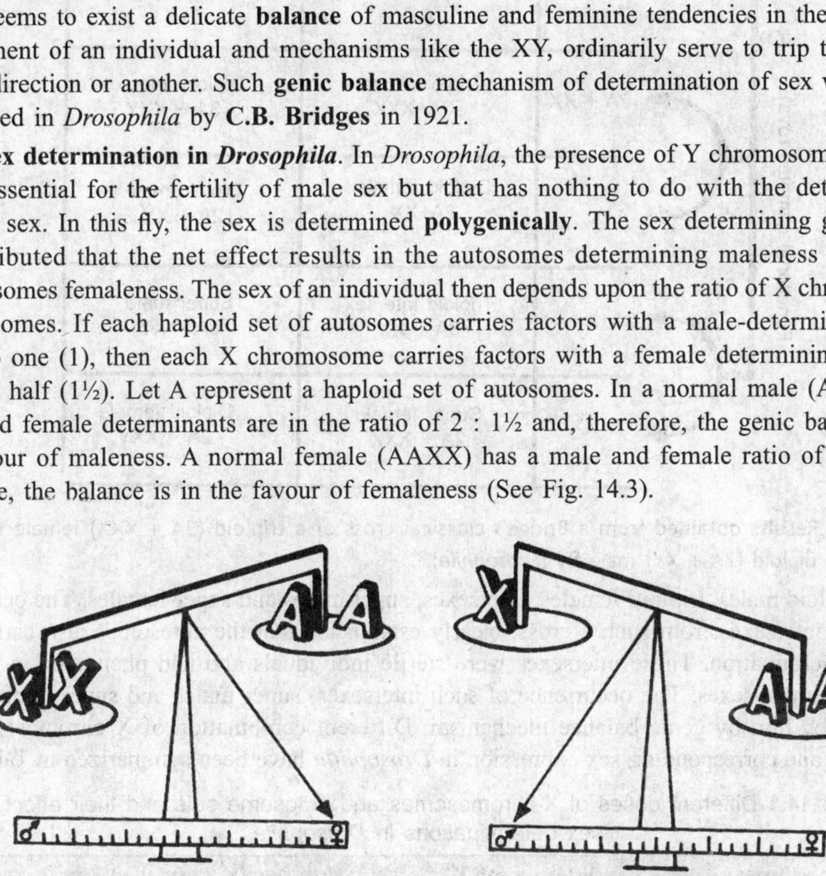

Fig. 14.3. A diagrammatic representation of genic balance theory in *Drosophila* ♀ = female sex; ♂ = male sex.

Recent studies indicate that the sex determination in chickens and probably in birds in general, is similar to that of *Drosophila*, *i.e.*, it is dependent upon the ratio between the Z chromosomes and the number of autosomal sets of chromosomes. The W chromosome is not a strong female sex determining element. Same system of genic balance in which the X/A ratio is critical, has been reported for the flowering plant (angiosperm) *Rumex acetosa*.

The genic balance theory of Bridges has been supported by the following findings:

1. Polyploid flies. Bridges crossed the experimentally produced triploid (3n individual having three whole sets of chromosomes) female *Drosophila* (3A : 3X) to diploid males (2A : XY). The results obtained from such a cross are shown in Fig. 14.4. From this cross he obtained

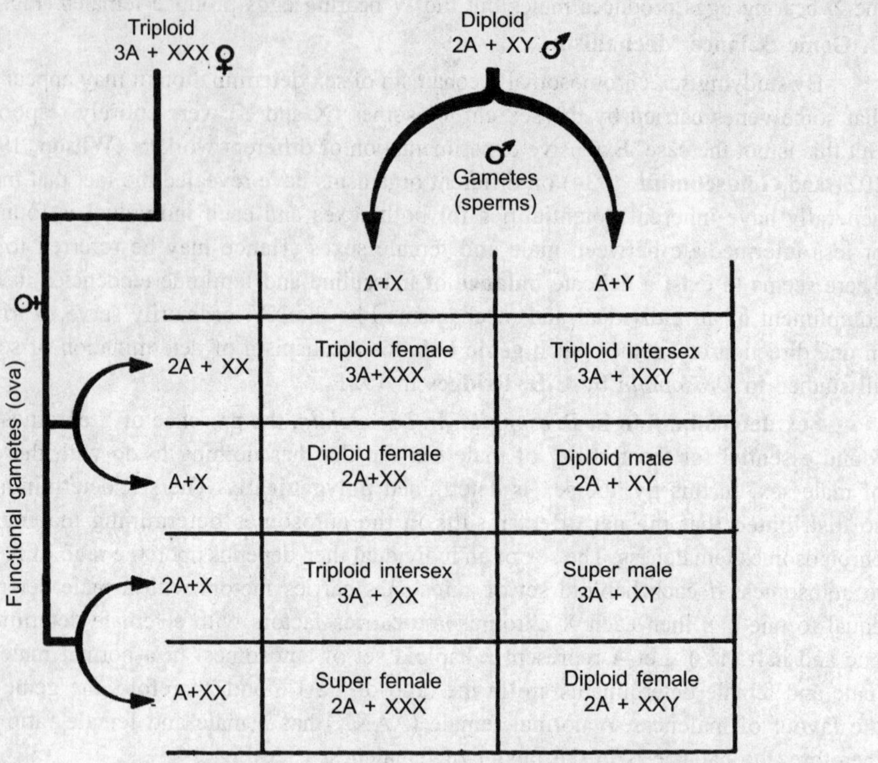

Fig. 14.4. Results obtained from a Bridge's classical cross of a triploid (3A + XXX) female fly and a diploid (2A + XY) male fly (*Drosophila*).

normal diploid males, triploid females, intersexes, super males and super females. The occurrence of triploid intersexes from such a cross, clearly established that the autosomes also carry genes for sex determination. These intersexes were sterile individuals and had phenotype in between male and female sexes. The occurrence of such intersexes, super males and super females were explained by him by genic balance mechanism. Different combination of X chromosomes and autosomes and corresponding sex expression in *Drosophila* have been summarized in Table 14.1.

Table 14.1 Different doses of X-chromosomes and autosome sets and their effect on sex determinations in *Drosophila*.

Phenotypes	*Number of X chromosomes*	*Number of autosomes (A sets)*	*Sex index No. X's/No. A sets*
Super female	3	2	1.5
Normal female — tetraploid	4	4	1.0
Normal female — triploid	3	3	1.0
Normal female — diploid	2	2	1.0
Normal female — haploid	1	1	1.0
Intersex	2	3	0.67
Normal male	1	2	0.50
Super male	1	3	0.33

As shown in the Table 4.1, when the X/A ratio is 1.0, the individual will be female and if it is 0.50, it would be male. When this balance is disturbed, the sex of individual deviate from normal male or normal female. For example, when the X/A ratio falls between 1.0 and 0.50, it would be **intersex**; when it is below 0.50 it would be **supermale** and when above 1.0, it would be **metafemale** or **super female**.

2. Gynandromorphs. Concepts of sex determination as developed for *Drosophila* are verified by the occasional occurrence of **gynandromorphs** which are individuals in which part of the body expressed male characters, whereas other parts express female characters. In a way, gynandromorphs represent one kind of **mosaic** or an organism made up of tissues of male and female genotypes. For example, a **bilateral gynandromorph** of *Drosophila* is male on one side (right or left) and female on the other (Fig. 14.5). It results due to the loss of an X-chromosome in a particular cell during development, *i.e.*, when the laggered X chromosome fails to be incorporated in a daughter nucleus and is lost forever (Fig. 14.6). If this event happens during first cleavage (or mitotic division) of the zygote, then one of the two blastomeres will have AAXX chromosomal complement and the other will have AAXO. The portion of the body developing from AAXX blastomere will be normal female and the portion developing from the AAXO blastomere will be male. The cytological examination of gynandromorphs suggested that Y chromosome does not play any role in the determination of sex in *Drosophila*.

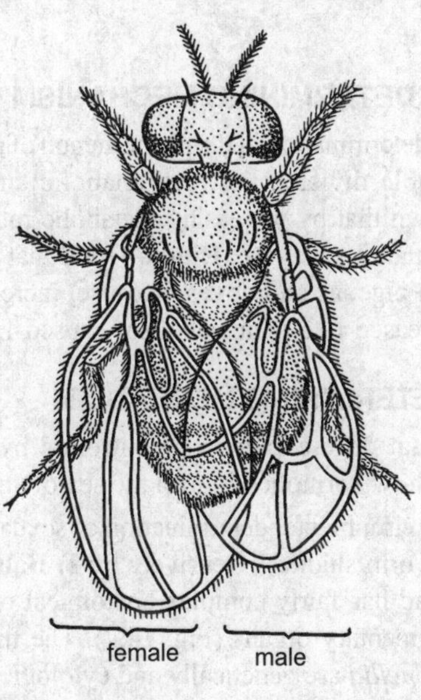

Fig. 14.5. A bilateral gynandromorph of *Drosophila*.

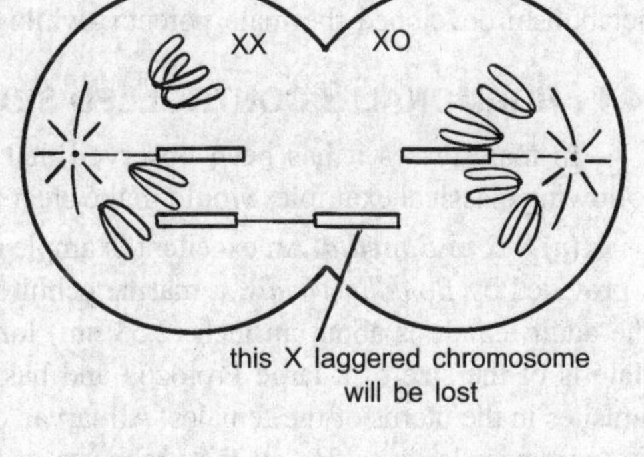

Fig. 14.6. The loss of an X chromosome during mitosis in a 2A + XX cell and formation of two types of cells—XX and XO.

C. Male Haploidy or Haplodiploidy Mechanism

Male haploidy or haplodiploidy or arrhenotokous parthenogenesis is particularly common in the hymenopterous insects such as ants, bees, sawflies and wasps (*e.g.*, *Bracon hebetor*). In these insects, since, fertilized eggs develop into diploid females and unfertilized ones into haploid males; so arrhenotoky is both a form of reproduction and a means of sex determination. Meiosis is normal in females, but crossing over and reduction division in chromosome number fail to occur during spermatogenesis in males due to their haploidy.

For example, a honeybee queen (whose diploid number is 32) can lay two types of eggs. By controlling the sphincter of her sperm receptacle (which holds sperms previously obtained in matings with males during nupital flight), she produces a fertilized egg (a diploid zygote having 32 chromosomes and developing into a female) or an unfertilized egg (a haploid zygote having 16 chromosomes and developing into a male). The diploid female zygotes can differentiate into either workers (sterile) or queens (fertile) depending on the diet they consume during their development.

D. Single Gene Control of Sex

In certain organisms, for example *Chlamydomonas*, *Neurospora*, yeast, *Asparagus*, maize, *Drosophila*, etc., individual single genes are found to be responsible for the determination or expression of sex. Following cases exemplified the single gene control of sex:

(*a*) Monogenic sex determination in *Drosophila*. In *Drosophila*, a **transformer gene (*tra*)** has been recognized which when present in homozygous condition (*tra*/*tra*) transforms a female fly into a sterile male, but, it does not act upon normal male individuals. Thus, a XX female with *tra*/*tra* genotype will be a sterile male, but, a XY male with *tra*/*tra* genotype will still be a normal male fly.

(*b*) Sex reversal gene (Sxr) in mammals. Recently a **sex reversal gene (*Sxr*)** has been discovered in human beings, so that in the presence of this gene XX female individuals may become male. Such cases of sex reversal are also reported in goat and mice. Mice also contain two other genes *Tdy* and *Tda*-1 which interact to cause sex reversal in XY male individuals to transform them into females.

14.3 METABOLICALLY CONTROLLED SEX DETERMINING MECHANISM

Certain workers have seen the possibility of sex determination in the phenomenon of metabolism. **Crew** suggested that sex is a physiological equitable division between anabolic and catabolic individuals. **A.F. Shull** and **D.D. Whitney** have shown that by increasing metabolic rate in rotifers the occurrence of male individuals increases than females. Likewise, **Riddle** found that metabolism had some definite role in the determination of sex in pigeons and doves, because, increased rate of metabolism developed the male potency, while decreased rate of metabolism caused femaleness.

14.4 HORMONALLY CONTROLLED SEX DETERMINING MECHANISM

In many cases it has been observed that sexual differentiation is controlled by hormones. Following classical examples would make clear that how hormones control the sex of an individual:

(*a*) Sex in *Bonellia*. An excellent example of environmental determination of sexual phenotype is provided by *Bonellia viridis*, a marine echiuroid worm studied extensively by **F. Baltzer** (1935). The adult female is about an inch (2.53 cm) long and has fairly complex anatomical organization. Male is of the size of a large Protozoa and has rudimentary organs (Fig. 14.7). The males live as parasites in the uterus of the females. All larvae of *Bonellia* are genetically and cytologically similar. If a particular larva settles on the proboscis of an adult female, it becomes a male individual. On the other hand, if a larva develops in isolation (*i.e.,* in water) it develops into a female. Further, if a incompletely developed male is detached from the proboscis of female, it becomes an intersex. Thus, it becomes evident that proboscis of adult female secretes some hormone-like substance which suppresses femaleness and initiating maleness in the larvae which are attached with it.

(*b*) The Crew's hen. In birds only one gonad of a normal female develops into a functional ovary. The other gonad remains rudimentary. If the functional ovary of a hen is destroyed, the rudimentary gonad develops into a testis (**Crew**, 1923). Thus, the female sex is reversed into male sex due to the phenomenon called **sex reversal**. Such a sex reversed hen can even father chicks which will be expected to show a sex ratio of two females to one male, since in birds the female is the heterogametic sex. The case of sex reversal of hen can be interpreted as follows:

During embryonic development, the XY (ZW) genotype stimulates the pituitary gland to produce female hormones that cause the gonad of the hen to develop into an ovary. After the development of ovary, the pituitary ceases to produce female hormones, due to inhibition of the pituitary by hormones produced by the ovary, thus, acting as **developmental feedback system**. The high level of female hormones secreted sequentially by the pituitary and the ovary is sufficient to suppress the action of male hormone producing cells of the body such as the steroid producing cells of the adrenal glands. When the ovary of a hen is removed, the steroid cells of adrenal become active and provoked the development of rudimentary gonad into testis which itself is an endocrine gland. Both endocrine glands (*e.g*, adrenal and testis) produce large amount of male hormones which sufficiently suppress the action of the female hormone cells of pituitary. Such cases of sex reversals have also been observed in amphibians (*e.g.*, in salamander by **E. Witschi**).

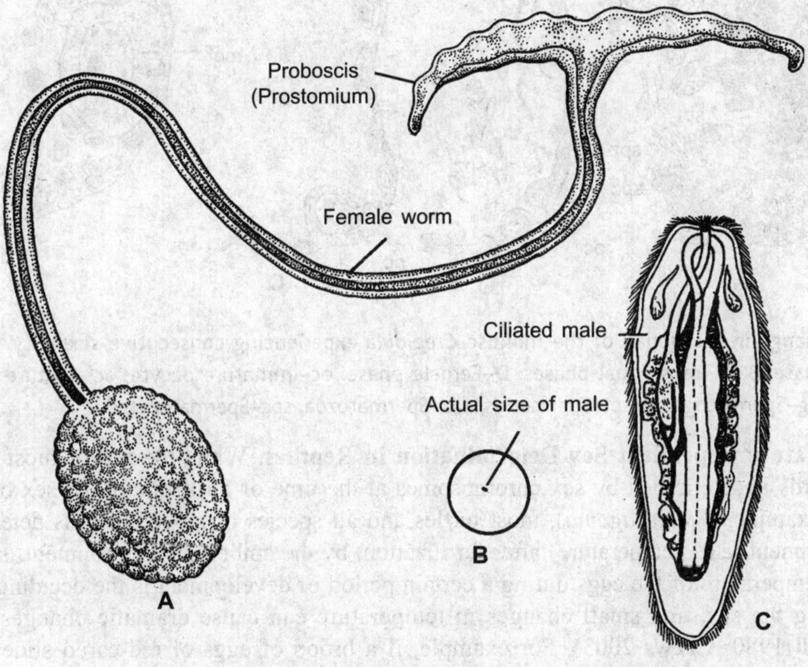

Fig. 14.7. The female (A) and male (B, C) of the echiuroid *Bonellia viridis.*

(*c*) **Hormones and freemartin.** Another classical example of hormonal control of sex determination has been found in cattles. In cattle, when twin calves of different sexes occur, the female member is usually a sterile intersex called a **freemartin**. The freemartin has external female genitalia but internal sex organs are more or less like those of male. The male twin is usually normal. **F.R. Lillie** (1917) has suggested that the formation of a freemartin was due to a fusion of the foetal membranes of the twin calves, while they were in uterus of the mother. The fusion of the foetal membranes permitted the blood of each twin to circulate in the blood vessel of the other. The male hormones produced by the male twin are presumed to suppress the differentiation of the female internal sex organs of the co-twin.

14.5 ENVIRONMENTALLY CONTROLLED SEX DETERMINING MECHANISM

In some organisms, the environment determines the sexual phenotype of the individuals. Thus, some environmental factors as size of parent body or of egg, age of parent and temperature are found to determine the sex in many cases.

Coe (1943) demonstrated in mollusc genus *Crepidula* that its every zygote contains all the genes which are necessary for the development of male and female reproductive systems. But, in this species it has been found that young *Crepidula* remains functional male and produce only sperms but no ova. As it matures the male phase is changed by transitional phase to a functional female phase. The female phase continues for the remainder of the animal's life. It is suggested that environment somehow regulates the activity of certain genes to produce only male phenotype during young stages, while inhibits them during adult stages and simultaneously initiating other genes to produce female phenotype.

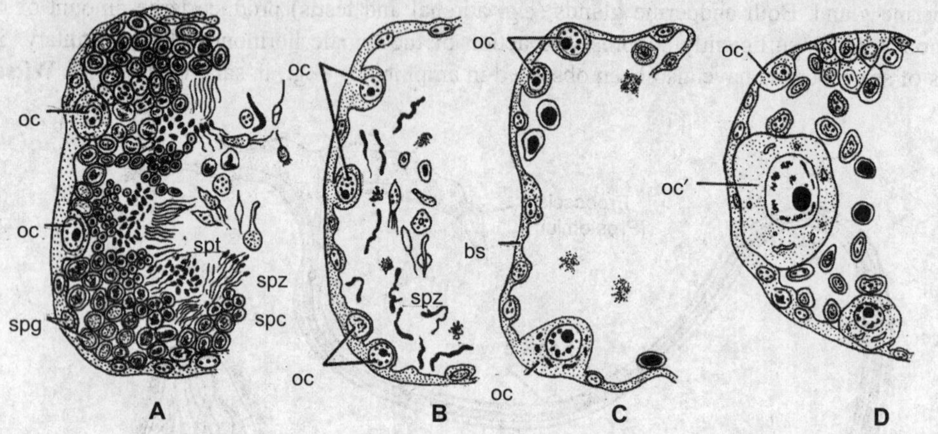

Fig. 14.8. Change in the gonad of the mollusc *Crepidula* experiencing consecutive sexuality. A–Male phase; B, C–Transitional phases: D–Female phase. oc–Immature occyte, oc′–Mature occyte, spg–Spermatogonia, spt–Spermatids, spz–Spermatozoa, spc–Spermatocytes

Temperature Dependent Sex Determination in Reptiles. While the sex of most snakes and most lizards is determined by sex chromosomes at the time of fertilization, the sex of some lizards (For example, *Agama agama*), most turtles and all species of crocodilians is determined by the environment (*e.g.*, temperature) after fertilization, by the embryonic environment. In these reptiles, the temperature of the eggs during a certain period of development is the deciding factor in determining the sex, and small changes in temperature can cause dramatic changes in the sex ratio (**Bull** 1980; **Crews** 2003). For example, if a brood of eggs of red-eared slider turtle (*Trachemys scripta*) is incubated at a temperature below 28°C, all the turtles hatching from the eggs will be male. Above 31°C, egg gives rise to a female. At temperature in between, the brood will give rise to individuals of both sexes (Box 14.1).

Box 14.1

The major disadvatage of temperature–dependent sex determination may be narrowing of the temperature limits within which a species can persist. Thus thermal pollution (either locally or due to global warming) could conceivably eliminate a species in a given area (**Janzen** and **Paukstin** 1991). Researchers (**Ferguson** and **Joanen** 1982; **Miller et.al.**, 2004) have speculated that dinosaurs may have had temperature–dependent sex determination and that their sudden demise may have been caused by a slight change in temperature creating conditions where in only males or only females hatched (Gilbert 2010).

14.6 SEX DIFFERENTIATION

Although the primary determination of sex is made at fertilization, but the embryo acquires its definite sex characteristics by a more complex mechanism called **sex differentiation**. Thus, sex determination is not a single, once-and-for all decision. But rather it is a gradual *awakening* that spreads from the ovary to testis to the reproductive tract, to the general body tissues and even to the brain itself; finally almost the whole body is overtaken by the process of sexual differentiation. In human beings and other animals sex differentiation occurs in the following steps:

1. Genetic Sex

Normal females ordinarily have two X chromosomes; normal males have one X and one Y. The genes on these sex chromosomes determine femaleness or maleness. Thus, one can speak of females having the genetic sex designation XX and males having the genetic sex designation XY.

Further, since the X-chromosome carries much more genetic information in striking contrast to Y chromosome, one might wonder how it is that the female can carry a double dose of many vital X-linked genes, whereas the male has only a single dose of these X-linked genes. Such inequality in fact cannot be tolerated and so female mammals and male *Drosophila* seem to have developed their own types of dosage compensation mechanisms.

Dosage Compensation of Genes

The term dosage compensation was coined by **Muller** (1932) and this phenomenon was originally discovered in *Drosophila*. Dosage compensation of the genes is done either by **hypoproduction due** to inactivation of one X-chromosome in homogametic female sex, as observed in mammals, or by the **hyperproduction** due to hyperactivity of X-chromosome in the heterogametic male sex as observed in *Drosophila*.

(*i*) X-chromosome inactivation in mammals. It has been demonstrated that in homogametic XX female individuals, one X-chromosome gets characteristically condensed and inactivated. Such chromatin material is called **facultative heterochromatin**, since it becomes inactive in certain part of the life cycle and resumes activity before entering the germ line. The phenomenon of inactivation of chromosome was confirmed by the observation of the Barr body (Box 14.2).

Barr body. Barr and **Bertram** (1949) reported a deeply stained chromatin body (*i.e.*, a chromocentre) in the nerve cells of female cat which was absent in the male. This chromatin body is called **sex chromatin** or **Barr body** after the name of its discoverer. Such Barr body has also been observed in most of the body cells (*e.g.*, skin, oral epithelium and blood cells) of human beings and other mammals. Human females have the Barr body in the nuclei of their body cells in higher proportion than males and are, therefore, referred to as **sex chromatin positive**. The human males are called **sex chromatin negative**.

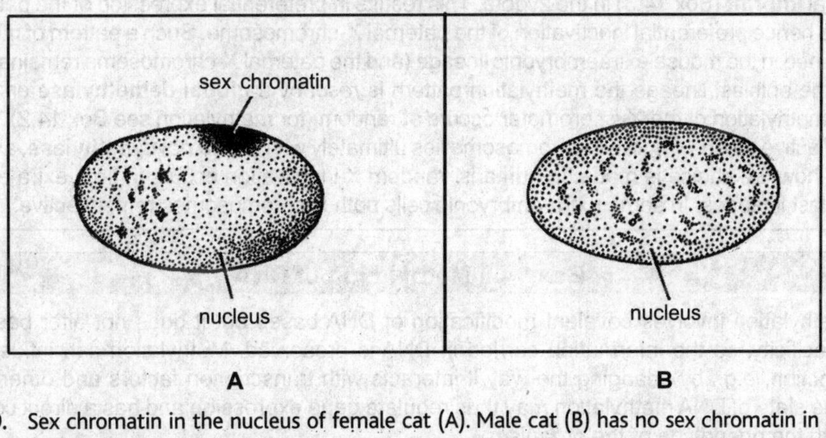

Fig. 14.9. Sex chromatin in the nucleus of female cat (A). Male cat (B) has no sex chromatin in its nucleus.

The sex chromatin appears in the interphase nucleus as a small chromocentre, heavily stained with basic dyes (Fig. 14.9). It gives a positive Feulgen reaction and has a relatively constant position in each tissue and species. It can be found in four positions: (*i*) attached to the nucleus as in nerve cells of certain species; (*ii*) attached to the nuclear membrane as in cells of epidermis or of the oral mucosa; (*iii*) free in the nucleoplasm as in neurons after electric stimulation, (Fig. 14.10) and (*iv*) as a nuclear expansion (Fig. 14.11). The best known example of nuclear expansion is that of the neutrophil leukocyte of female in which the sex chromatin (Barr body) appears as a small rod called the **drumstick**.

Box 14.2 Molecular Basis of X-chromosome Inactivation

In the somatic cells of female mammals, **dosage compensation** for the X-chromosome is promoted by **X-chromosome inactivation**. One of the X-chromosomes become condensed into **heterochromatin** and so become transcriptionally inert. The **inactive X** (called **Xi**) remains condensed throughout the cell cycle as a densely staining **Barr body**, while the **active X** (called **Xa**) acts normally.

In the female zygote, both X-chromosomes are active. The inactivation process initiates at the late blastocyst stage in the extraembryonic lineage, and slightly later in the epiblast lineage which gives rise to the somatic cells of the embryo. In embryonic tissues of placental mammals, the choice of active versus inactive X chromosome is random. Once choice is made, the process is irreversible: the active and inactive chromosomes are clonally propagated, so that an individual heterozygous for a visible X-linked marker demonstrates patchy variegation for the marker phenotype (*e.g.,* tortoise shell coat colours in cats). In extraembryonic tissue of many mammals (but not humans), the parental X-chromosome is preferentially inactivated (it is called **parental imprinting**; Box 14.2) and paternal preference occurs in both embryonic and extraembryonic tissue of marsupials.

The molecular basis of X-inactivation can be divided into four stages: **counting, initiation, propagation (spreading)** and **maintenance**. A X-linked *cis*-acting site, apparently involved in all four stages, is the **X-inactivation centre** (*Xic*), which has been mapped by studying deletions and translocations of the X-chromosome resulting in the disruption of inactivation. If the region containing *Xic* is deleted, the mutant X chromosome remains active; if the X chromosome is involved in a translocation, the fragment containing *Xic* is inactivated (as is part of the autosome to which it is attached) while other fragment remains active.

Micromapping of *Xic* originally revealed a single locus, called *X-ist*, which is expressed specifically on Xi (= inactive X chromosome). *X-ist* encodes a large **untranslated RNA** which associates with Xi in the nucleus and is thought to induce heterochromatinization by coating the X chromosome. The overall regulation of X-inactivation exhibits the balance between X-chromosome and autosomes. **Polysomy X** is tolerated because supernumerary X-chromosomes are inactivated, generating multiple Barr bodies, whereas **polyploidy** causes more X-chromosomes remain active. The nature of X chromosome counting mechanism is unknown.

During gametogenesis the promoter of *Xist* allele is subjected to sex-specific **methylation**, causing a maternal **imprint** (Box 14.3) in the zygote. This results in preferential expression of the paternal *Xist* allele and hence preferential inactivation of the paternal X-chromosome. Such a pattern of methylation is maintained in the mouse extraembryonic lineage (and the paternal X-chromosome remains inactive), while in the epiblast lineage the methylation pattern is *reset* by a **global demethylase** enzyme and *de novo* methylation of the *X-ist* promoter occurs at random (for methylation see Box 14.2). Thus, the choice of active versus inactive X chromosome lies ultimately with the *de novo* **methylase**, and it is still not clear how the choice is made. In humans, random X-inactivation occurs in both extra-embryonic and epiblast lineages. In some extra-embryonic cells both X-chromosomes remain active.

Box 14.3 Methylation of DNA

1. Methylation involves covalent modification of DNA bases but it does not alter base-pairing specificity, so the information carried in DNA is preserved. Methylation can influence DNA function, *e.g.*, by changing the way it interacts with transcription factors and other proteins. The state of DNA methylation may thus regulate gene expression and has a direct connection with the phenotype of the organism.
2. Enzymes which add methyl group to DNA, **DNA methyltransferases** (or **DNA methylases**) uses S–adenosyl methionine as methyl donor. In eukaryotes, the only modified base commonly found in DNA is 5-methyl cytosine (5-mec).

Two types of enzymes can be recognized: ***de novo* methylases** add methyl groups to unmethylated DNA at specific sites and can therefore initiate a pattern of methylation, whereas **maintenance methylases** add methyl groups to DNA which is already methylated on one strand **(hemimethylated DNA)** and thus perpetuates patterns of methylation through successive rounds of replication.

3. DNA carries two forms of information: **genetic information** in its nucleotide sequence and epigenetic information in its structure. *Epigenetic information is any heritable property of DNA which influences its activity* (this ultimately contributes to the phenotype of the organism) but which lies outside the nucleotide sequence itself). **DNA methylation is one source of epigenetic information: it is heritable** (due to maintenance methylation) **and able to control gene expression**. A methylation state is thus termed as an **epigenotype** and a change in methylation state as an **epimutation**.
4. **Parental imprinting.** An **imprint** is a form of epigenetic information, and **parental imprinting** is thus an epigenetic mechanism by which chromosome region can retain an epigenetic memory of its parental origin. In somatic cells, most genes are expressed on both maternal and paternal chromosomes but **imprinted genes** are expressed in a parent chromosome-specific manner. Conventionally, an allele is said to be imprinted if it is repressed (*e.g.*, at maternally imprinted loci, only the paternal allele is expressed). At the molecular level, *the imprint is DNA methylation* but in some loci methylation corresponds to gene activity and in others it corresponds to inactivity.

The relationship between sex chromatin and sex chromosomes has been elucidated by **Ohno, Kaplan** and **Kinosita**, 1959. It has been demonstrated that the sex chromatin is derived only from one of the two X-chromosomes. The other X chromosome behaves like a autosome and is not heteropycnotic at interphase. Later, **Lyon** (1972) confirmed the existence of Barr body in normal females, metafemales or superfemales (XXX) and in Klinefelter males (XXY).

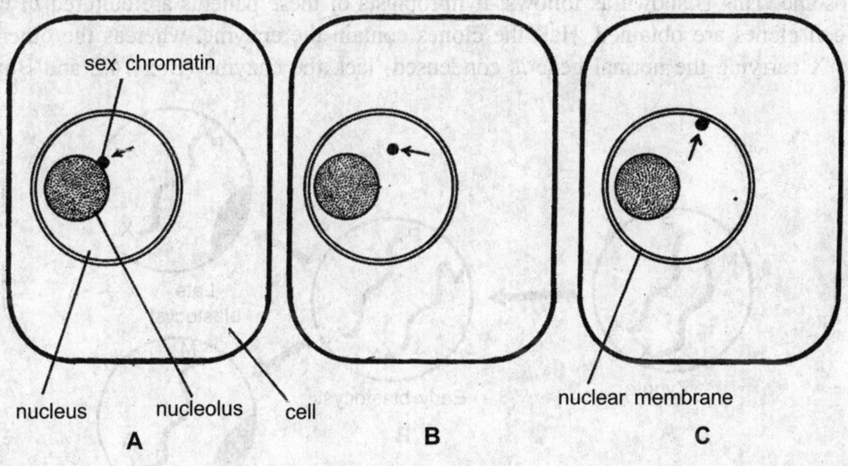

Fig. 14.10. A–Position of sex chromatin near nucleolus; B–In nucleoplasm and C–Attached to the nuclear membrane.

Lyon's hypothesis. It was demonstrated by **M.F. Lyon** of U.K. that whenever the number of X chromosomes was two or more than two, the number of Barr bodies was one less than the number of X chromosomes (nX-1; *i.e.*, one Barr body in XX females and XXY males; two Barr bodies in XXXY males and XXX metafemales). Thus, in normal female only one active X chromosome is present.

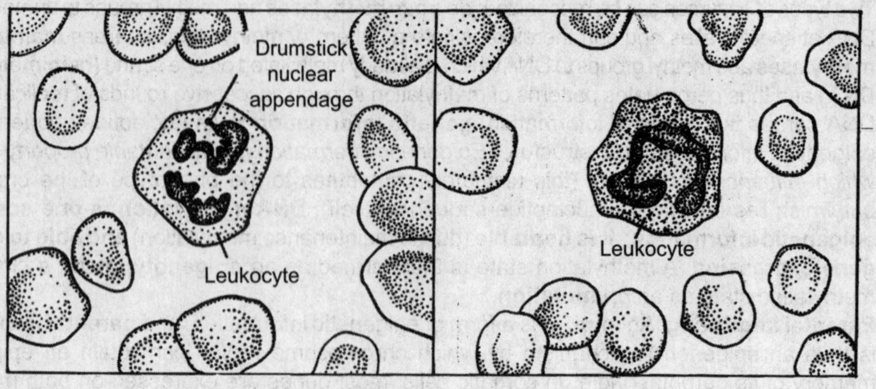

Fig. 14.11. A-Leukocyte of human female with drumstick-like sex chromatin; B-Leukocyte of human male without any drumstick.

Which of the two X chromosomes remains active in female individuals, is determined at the early stages of development. It was observed by Lyon in 1961 that each of the paternal (P) and maternal (M) X chromosomes has a chance to become inactive (*i.e.*, Barr body). In other words, the inactivation of X chromosome is a random phenomenon. This fact has been demonstrated in human diseases linked to X chromosome. The **Lesch-Nyhan syndrome**, in which a deficiency of one enzyme of the purine metabolism (*i.e.*, hypoxanthineguanine phosphoribosyl transferase) produces mental retardation and increased uric acid levels results, from a recessive mutation in the X chromosome. This is shown as follows: if fibroblasts of these patients are cultured *in vitro*, two types of cell clones are obtained. Half the clones contain the enzyme, whereas the other half (in which the X carrying the normal gene is condensed) lack the enzyme (Box 14.2 and Box 14.3).

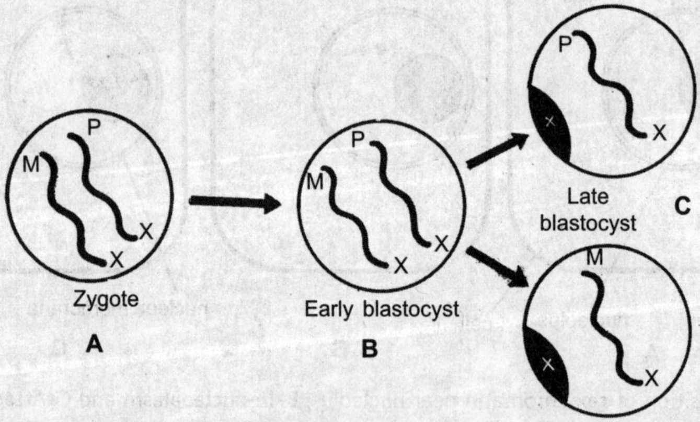

Fig. 14.12. Diagram of the evolution of XX chromosome. A–In the zygote both paternal (P) and maternal (M) X-chromosomes are euchromatic; B–In the early blastocyst the same is true as in A; C–In late blastocyst 50% of the cells have a maternal heterochromatic X chromosome and the other 50% have the paternal heterochromatic X chromosome. In diagram C formation of sex chromatin by one of the X chromosome has been shown.

In the human embryo X chromosome inactivation starts in the late blastocyst about the 16th day of life. Once the inactivation is established, it is irreversibly maintained in somatic cells, however, in germ cell line reactivation occurs at a specific stage of germ cell development (*i.e.*, entry into meiotic prophase; **Martin**, 1982).

The good illustration of X chromosome inactivation is seen with tortoise-shell cats, where the coat is a mosaic patchwork of black and yellow hair. Black hair is produced by the dominant allele *B*, and yellow by its recessive allele *b*. This gene is X-linked, so if one X chromosome contains the dominant allele *B* and other X chromosome the recessive allele *b*, random inactivation will allow both coat colours to be expressed. Male tortoise-shell cats are understandably rare, since it has only one X chromosome.

***(ii)* Hyperactivity of X chromosome in male *Drosophila*.** The phenomenon of dosage compensation in *Drosophila* has been shown to be due to hyperactivity of one X chromosome in male *Drosophila*. This fact was discovered by **Dr. A.S. Mukherjee** of Calcutta University. In mosaic individuals with XX and XO cells, it could be shown that X chromosome in XO cells was always hyperactive. Due to this fact, mutant and wild type flies showed same intensity of eye colour in male and female flies. Similarly, enzyme levels for several enzymes including **6PGD** (6 phospho-gluconate dehydrogenase), **G6PD** (glucose-6 phosphate dehydrogenase), **tryptophan pyrrolase** and **fumarase** were found to be similar in female and male flies.

2. Gonadal Sex

In the human embryos until the six weeks the gonads and primordia of the urinogenital tract are identical in males and females. At this stage (time) the gonad has already been invaded by the primary XX or XY cells. At this point, a gene or set of genes, called **testis-determining factor** or **TDF**, present in the Y chromosome causes the undifferentiated gonad to differentiate into a testis and the absence of this gene allows the gonad to become an ovary. The development into a testis starts as soon as the gonocytes (*i.e.*, primordial germ cells) from the yolk sac have finished their migration into the gonadal ridge. Gonocytes of the male (XY) migrate deeper into the gonadal blastema forming the medulla and female gonocytes (XX) remain at the periphery, forming a thick cortical layer. Hence, the XX genetic sex is ordinarily associated with **ovarian gonadal sex**, and XY is associate with **testicular gonadal sex**.

Further, identical TDF region of Y chromosome has been identified in gorillas, monkeys, dogs, cattle, horses and goats, further suggesting that this is a common sex-determining mechanism in mammals. The current theory is that the TDF gene is the **master switch** that when turned on, activate an entire series of genes whose function is sex differentiation.

H-Y antigen. It is known that the Y chromosome in mammals is a strong male inducer and directs the organogenesis of the testis. The short arm of the Y chromosome of human beings, near the centromere contains a gene responsible for the production of a protein, called **H-Y antigen**. This antigen occurs in the plasma membrane of all male tissues and possibly tends to regulate H-Y receptor cells. If these are defective, the undifferentiated gonadal cells fail to respond to H-Y antigen and testes do not form, leading the individual to mature as a female.

Cortico-medullary antagonism. Experiments in which male and female amphibians larvae (*e.g.*, toad, frog, salamander) have been surgically united to one another so that their circulation fused, gave rise to the concept of "**cortico-medullary antagonism**". If the medulla was stimulated to develop, it was thought to produce a substance that could act locally or via the circulation to inhibit the ingrowth of ovarian cortex and vice versa. For example, in case of **freemartin** which occurs in cows, sheep, pigs and goats, a female foetus that is twin to a male and shares a conjoined placental circulation undergoes a partial sex reversal of her ovaries, which interferes with the subsequent sexual differentiation of the reproductive tract.

3. Hormonal or Genital Sex

The embryonic gonads produce hormones that, in turn, determine the morphology of the external genitalia and the genital ducts. XX embryos normally develop ovaries, female external genitalia and Mullerian ducts. XY embryos, on the other hand, ordinarily develop testes, male external genitalia, and Wolffian ducts. In XX embryos the Wolffian ducts are suppressed; in XY embryos the Mullerian duct remain under-developed. Thus, there is a distinction between male and female genital sex.

For example, the human gonad differentiates into a definite testicle at the seventh week, whereas the female gonad begins to develop into an ovary between the eight and nine weeks of development. At the time of differentiation an important epigenetic factor is the production of androgens (testosterone) by somatic cells in the embryonic male gonad. In mammals, administration of testosterone hormone to the mother produces in the foetus a shift in the differentiation of XX genitalia to a male type, producing **masculine pseudohermaphroditism**. This hormone locally accelerates the development of the testis, whereas in the female the absence of the hormone permits the slower development characteristic of the ovary.

4. Somatic Sex

Production of gonadal hormone continues to increase until secondary sex characters appear at puberty. These include amount and distribution of hair (*e.g.*, facial, body, axillary, pubic), pelvis dimensions, general body proportions, subcutaneous fats over hips and thighs and breast development in human females, as well as, increased larynx size and deepening of the voice in the human male.

5. Socio-psychological Sex

In most individuals, genetic sex, gonadal sex, genital sex and somatic sex are consistent; XX persons, for example, develop ovaries, female genitalia and female secondary sexual characteristics. Ordinarily these persons are raised as females and adopt the faminine gender role under whatever culture pattern has been established in the society of which they are members. A similar consistency from genetic sex to socio-psychological sex is seen for XY individuals. Intersexuality occurs when the chromosomal contribution is inconsistent with the gonadal or other secondary sexual characters.

Sex Mosaics or Gynandromorphs in Humans

***(i)* Sex mosaics.** These are those persons whose body contains two or more populations of cells, each with a different sex chromosome complement. For example, **Murray Barr** found a girl in which cells of buccal and vaginal smears showed two Barr bodies thus indicating XXX chromosomes but her blood cells showed XO chromosomal complement. The girl showed Turner syndrome indicating XO condition. Such sex mosaics are produced due to the phenomenon of non-disjunction during mitosis.

***(ii)* Pseudohermaphroditism.** This is an intersexual condition in human beings. In pseudohermaphroditism, an individual exhibits characters of both sexes in one body. It is either male or female but owing to deficiency of sex hormones, he fails to show all the characters of his sex in full.

***(iii)* Sex reversal in human beings.** In human beings the sex is determined by XX-XY mechanism, however, sex is differentiated by the hormones. If testes of male is removed before puberty, the male secondary sexual characters fail to develop. In him female secondary sexual characters, such as body form, voice, hair pattern, *etc.*, appear. If such sex reversed male is injected with male hormones, the male sex characters reappear.

In fact, in human embryo, the gonads are quite similar regardless of the chromosomal sex of the embryo. These gonads are called **ovotestes**. Each gonad is formed of inner **testicular tissue**, and outer **ovarian tissue**. In the presence of Y chromosome, the testicular tissue develops first and secretes male hormone. These male hormones have following functions. 1. Limit the development of ovarian tissue of the gonads. 2. Accelerate development of testicular tissue. 3. Control differentiation of male external genitalia. These changes result in the formation of male foetus.

QUESTIONS

Long Answer Questions

1. What is sex determination? Name various mechanisms of sex-determination. Give an account of genetically controlled sex-determination mechanism.

2. Describe the mechanism of sex-determination in animals.
3. Write an essay on 'sex determination'.
4. Write an account of modern theories of sex determination.
5. Describe sex determination in *Drosophila*. Explain how environmental factors and internal secretions influence sex differentiation in *Bonellia* and cattle.
6. Describe sex-determination in *Drosophila*, bugs, birds, mammals. Explain how environmental factors and internal secretions influence sex determination in *Bonellia* and cattle.
7. Write an essay on chromosomal sex determination.
8. Give an account of chromosomal basis of sex determination in *Drosophila*.
9. What is sex chromosome? Describe the phenomenon of sex determination with special reference to balance theory.
10. Explain genetic (genic) balance theory of Bridge with an example.
11. Describe the mechanism of sex-determination in *Drosophila*.
12. Give an account of the chromosomal mechanism of sex determination in human beings. How does it differ from the avian type of sex determination.
13. Discuss the mechanism of sex determination in human beings. Does the Y-chromosome bear any gene for sex determination.
14. Give an account of sex differentiation.
15. What do you understand by sex determination? How does it differ from sex-differentiation? Describe the chromosome theory of sex determination.
16. What do you understand by sex-determination? How does it differ from sex-differentiation? Describe in detail the genic balance theory of sex determination and various theories of sex differentiation in different animals.
17. Explain the chromosomal mechanism of sex determination. (*M.S.U., Tamil Nadu, 2006*)
18. Discuss the role of X and Y chromosomes in the determination of sex in human beings.
19. What do you understand by sex chromosomes? Explain chromosomal theory of sex determination in animals. (*H.N.B. Garhwal 2009*)

Short Answer Questions

1. Given an account of Lyon's hypothesis.
2. Give an account of the chromosomal basis of sex determination in human beings. What are Barr bodies?
3. What is Barr body? Discuss the relationship between the Barr body and X-chromosome.
4. Describe the phenomenon of sex determination in animals. (*H.N.B. Garwal, 2008*)
5. Write short notes on the following:
   (*i*) Sex chromosomes
   (*ii*) Y Chromosome
   (*iii*) Sex determination
   (*iv*) Sex determination in *Drosophila*.
   (*v*) Environmental sex determination
   (*vi*) XY female-XX male type sex determination
   (*vii*) Dosage compensation
   (*viii*) Barr bodies
   (*ix*) Lyon's hypothesis
   (*x*) Gyandromorph

(*xi*) Sex mosaics

(*xii*) Pseudohermaphroditism

6. Write differences between the following:

(*i*) Sex determination in *Drosophila* and *Apis.*

(*ii*) Sex chromosomes and autosomes

(*iii*) Sex determination in birds and human beings.

(*iv*) Autosome and allosome

Very Short Answer Questions

1. How many Barr bodies will be found in the cells of XXXY human beings?

2. Explain how sex-chromatin helps in the identification of sex?

3. The nucleus from buccal epithelium cells of a person has a Barr body but he is a male. How do you account for this?

4. Define sex determination. (*C.C.S.U., Meerut, 2009*)

5. What are the sex chromosomes?

6. Define sex-determining Y (SRY).

Give one word for the following sentence:

1. A special condition is *Dropsophila*, where one half on the body exhibits male sex characters and the other half female sex-character.

Fill in the Blanks

1. Sex chromosomal mechanism or .... of sex determination was proposed by .... in 1902.

2. The chromosomes having genes for somatic characters of a human being are called ....

3. The XX-XY type sex determination occurs in .... and .... (fruit fly).

4. The XO human female has .... syndrome which is .... and has .... stature, webbing of neck, etc.

5. The XXY human male has .... syndrome which is characterized by .... having both ovarian and testicular tissues.

6. Moths have .... type sex determination.

7. .... and gypsy moth have .... type sex determination.

8. .... mechanism of sex determination in *Drosophila* was discovered by .... in 1921.

9. In *Drosophila*, a .... transforms a female fly into a sterile male.

10. Barr body of body calls of human female is .... heterochromatin.

11. Absence of .... gene of Y chromosome allows a undifferintiated gonad to become an ovary.

12. A .... is a female foetus of cows that is twin to a male and shares a ....

Multiple Choice Questions

1. In *Drosophila*, sex is determined by the genic balance between

(*a*) X and Y

(*b*) X and sets of autosomes

(*c*) Y and sets of autosomes

(*d*) none of the above

2. In sex determination of human beings the key factor is

(*a*) Y chromosome (*b*) X chromosome

(*c*) both (a) and (b) (*d*) none of these

3. Barr body is

(*a*) nucleolus (*b*) facultative heterochromatin

(*c*) constitutive heterochromatin (*d*) euchromatin

4. Foetal sex can be determined by examining cells from the amniotic fluid by looking for:

(*a*) chiasmata (*b*) kinetochore

(*c*) Barr bodies (*d*) autosomes

5. Human gene determining sex at seventh week of pregnancy called TDF is present

(*a*) X chromosome (*b*) Y chromosome

(*c*) 21st pair of autosomes (*d*) none

6. In *Melandrium* the sex determination type is

(*a*) XX–XY type (*b*) XX–XO type

(*c*) ZZ–ZW type (*d*) ZZ–ZO type

7. Where are Barr bodies found?

(*a*) sperm (*b*) ova

(*c*) body/somatic cells of human male (*d*) body/somatic cell of human female

8. AMH (= anti-Mullerian duct hormone) is secreted by

(*a*) Sertoli cells (*b*) Leydig cells

(*c*) by both types of cells (*d*) by none of these cells

9. Who discovered sex chromosomes

(*a*) Robert Brown (*b*) M.J.D. White

(*c*) Nettle Stevens (*d*) G.J. Mendel

10. Genetic identity of a human male is determined by

(*a*) sex chromosomes (*b*) cell organelles

(*c*) nucleolus (*d*) autosomes

11. Chromosome theory of sex determination was propounded by

(*a*) McClung (*b*) Balbiani

(*c*) Goldschmidt (*d*) Mendel

12. Lyon's hypothesis deals with

(*a*) centromere position (*b*) genetic compatibility

(*c*) genetic incompatibility (*d*) number of Barr bodies

13. Genetic balance theory was proposed by

(*a*) Bridges (*b*) Morgan

(*c*) Boveri (*d*) Lillie

14. A sex mosaic is known as

(*a*) hemizygous (*b*) hermaphrodite

(*c*) gynandromorph (*d*) heterozygous

15. A medical technician while observing a human blood smear under the microscope notes the presence of Barr body close to the nuclear membrane in the WBC. This indicates that person under investigation is

(*a*) colour blind (*b*) haemophilic

(*c*) normal female (*d*) normal male

16. Webbed neck in a human being is characteristic of

(*a*) XXY male (*b*) YY male

(*c*) XO female (*d*) XXX female

ANSWERS

Very Short Answer Questions

1. Two Barr bodies.

2. The sex chromatin or Barr body is condensed heterochromatic X-chromosome which is found in the nuclei of somatic cells of female mammals.

3. A man with Barr body will have XXY chromosomes. It means he is suffering from Kleinefelter syndrome.

4. Sex is a hereditary character and is determined by chromosomes and genes in various organisms.

5. Sex chromosomes are heteromorphic chromosomes whose distribution in a zygote determines the sex of the organism.

6. The sex switch, or testes-determining factor (TDF), in human being, located on the Y chromosome (It is celled *Sry* in mice).

Give one word for the following sentence

1. Gyandromorph.

Fill in the Blanks

1. Chromosome theory, McClung. **2.** Autosomes.

3. Human beings, *Drosophila*. **4.** Turner's, short.

5. Klinefelter's, hermaphroditism. **6.** ZO-ZZ.

7. Fowl, ZW-ZZ. **8.** Genic balance, C.B. Bridges.

9. Transformer gene. **10.** Facultative.

11. TDF. **12.** Freemartin, conjoined placental circulation.

Multiple Choice Questions

1. (*b*)	**2.** (*a*)	**3.** (*b*)	**4.** (*c*)	**5.** (*b*)	**6.** (*a*)
7. (*d*)	**8.** (*a*)	**9.** (*c*)	**10.** (*a*)	**11.** (*a*)	**12.** (*d*)
13. (*a*)	**14.** (*c*)	**15.** (*c*)	**16.** (*c*).		

15

Non-Chromosomal Inheritance (Cytoplasmic Inheritance)

In preceding and forthcoming chapters, we have discussed different roles of genes of the nuclear chromosomes in inheritance, cellular metabolism, development and mutation of the organisms in which they occur. Though, the genes of nuclear chromosomes have a significant and key role in the inheritance of almost all traits from generations to generations, but they altogether cannot be considered as the sole vehicles of inheritance, because certain experimental evidences suggest the occurrence of certain extranuclear genes or DNA molecules in the cytoplasm of many prokaryotic and eukaryotic cells. For example, bacterial cells such as *E. coli* possesses a single main chromosome in the nucloid and often extra DNA elements called **plasmids** in the cytoplasm; the eukaryotic cells possess a main complement of chromosomes in the nucleus and extra-DNA molecules or chromosomes in their mitochondria and chloroplasts (plastids). Qualifying also as extra hereditary elements are certain viruses, bacteria, and algae (which live as endosymbionts inside the cukaryotic cells). These cytoplasmic extra-nuclear genes or DNA molecules of plasmids, mitochondria, chloroplasts, endosymbionts and cellular surfaces have a characteristic pattern of inheritance which does not resemble with that of genes of nuclear chromosomes and is known by different terms such as **non-Mendelian, non-chromosomal, uniparcntal, maternal, extra-chromosomal, cytoplasmic** and **extra-nuclear inheritance**.

15.1 EVIDENCES FOR CYTOPLASMIC FACTORS (GENES)

Traits with extranuclear basis are identified by the accumulated evidence from a number of diagnostic criteria such as follows:

1. Because the female gamete contributes almost all of the cytoplasm to the zygote and male gamete (sperm or pollen) contributes only a nucleus, an inheritance pattern that differs between reciprocal crosses suggests a cytoplasmic involvement. This is clearly the basis for **uniparental** or **maternal inheritance** where the progeny always resemble one parent, most commonly the female parent (*e.g.*, shell coiling in *Limnaea peregra*).
2. Differences in reciprocal crosses which cannot be attributed to sex-linkage or some other chromosomal basis tend to associate extranuclear factors (*e.g.*, chloroplast inheritance in *Mirabilis jalapa*).
3. The uniparental inheritance of a trait which cannot be attributed to unequal cytoplasmic contributions from parental gametes may, however, involve cytoplasmic factors (*e.g.*, streptomycin resistance in *Chlamydomonas*).

4. Whenever trait fails to demonstrate the classical segregation patterns and deviates from standard ratios, the conclusion is again a cytoplasm based type of inheritance (*e.g.*, mitochondrial inheritance in yeast).
5. When the trait fails to show linkage to any known nuclear linkage groups and assort independently from nuclear genes, a cytoplasmic mode of inheritance is suggested.
6. Many types of mutants that fit the above criteria will show segregation during mitotic division. This is very common in variegated plants that carry more than one type of plastid (chloroplast) per cell.

The cytoplasmic inheritance, therefore, will be understood to be based on cytoplasmically located, independent, self-replicating nucleic acids, which differ from chromosomal genes by their location within the cell, and have their own nucleotide sequences. The smallest heritable extra chromosomal unit is called a **plasmagene** or **cytoplasmic gene**. All of the plasmagenes of a cell constitute the **plasmon**.

15.2 NON-CHROMOSOMAL INHERITANCE IN EUKARYOTES

Many geneticists have studied various cases of extra-nuclear inheritance in different eukaryotes. Certain most important examples of extra-nuclear inheritance in eukaryotes are the following:

1. Maternal Inheritance

In certain cases, it has been observed that certain characteristic phenotypic traits of F_1, F_2 or F_3 progeny are not the expression of their own genes, but rather those of the maternal parents. Such phenotypic expressions of maternal genes (genotype) may be short-lived or may persist throughout the life-span of the individual. The substances which produce the maternal effects in the progeny are found to be transcriptional products (*i.e.*, mRNA, rRNA and tRNA) of maternal genes which have been manufactured during oogenesis and which exist in the ooplasm of unfertilized eggs in the form of inactive protein coated and late translating mRNA molecules (**informosomes; masked messenger** hypothesis of **Spirin**) or inactivated rRNA and tRNA. These transcriptional products of maternal genes produce their phenotypic effects during early cleavage and blastulation when there occur little or no transcription since, maternal and paternal genes of zygote remain engaged in mitotic replication or duplication of DNA. There may be other reasons of maternal effect which are still little understood.

The maternal inheritance has been studied in some of the following cases:

(*a*) Shell coiling in *Limnaea*. In the snails (gastropods), the shell is spirally coiled. In most cases the direction of coiling of the shell is clockwise, if viewed from apex of the shell. This type of coiling is called **dextral**. However, in some snails the coiling of shell may be counter clockwise or **sinistral**. Both types of coilings are produced by two different types of genetically controlled cleavages, one being **dextral cleavage**, another being **sinistral cleavage** (Fig. 15.1).

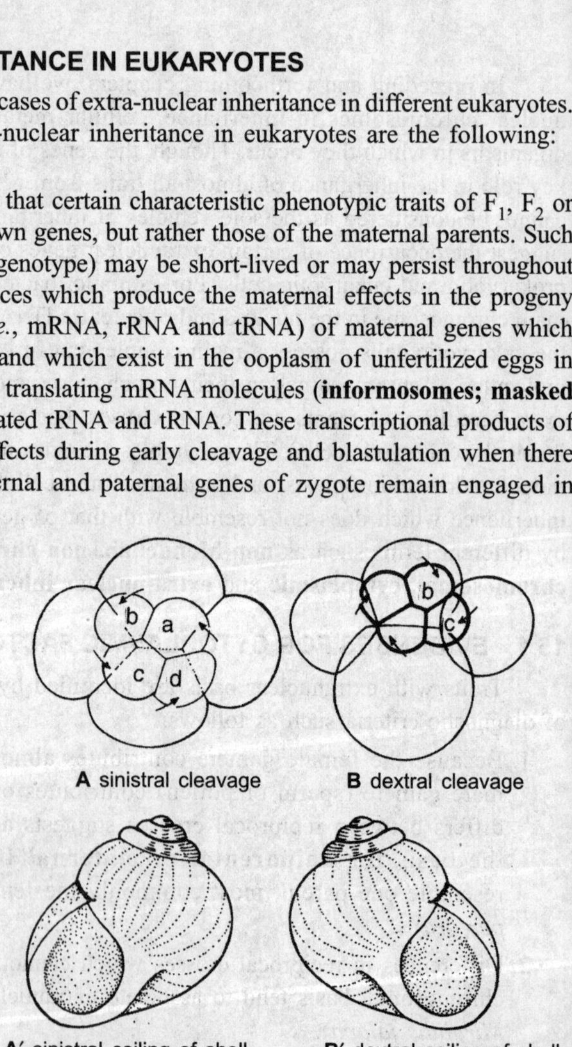

Fig. 15.1. Correlation of sinistral (A) and dextral (B) cleavage with sinistral (A′) and dextral (B′) coiling of the shell in gastropods.

There are some species of gastropods in which all the individuals are sinistral but the main interest attaches to a species in which sinistral individuals occur as a mutation among a population of normal dextral animals. Such a mutant was discovered in the freshwater snail *Limnaea peregra* (**A. Sturtevant,** 1923).

Breeding and cross breeding of dextral and sinistral snails showed that the difference between the two forms is dependent on a pair of allelomorphic genes, the gene for sinistrality being recessive (*S*), and the gene for the normal dextral coiling being dominant (S^+). The two genes are inherited according to Mendelian laws, but the action of any genic combination is visible only in the next generation after the one in which a given genotype is found.

The eggs of a homozygous sinistral individual (*SS*) are fertilized by the sperm of a dextral individual (S^+S^+), the eggs cleave sinistrally and all the snails of this F_1 generation show a sinistral coiling of the shell. Thus, the gene of sperm (S^+) do not manifest themselve, although the genotype of the F_1 generation is S^+S. If a second generation (F_2) is bred from such F_1 sinistral individuals, it is all dextral, instead of showing segregation as would be expected in normal Mendelian inheritance. In fact, segregation does take place in the F_2 generation so far as the genes are concerned, but the new genic combinations fails to manifest themselves, since the coiling is determined by the genotype of the mother. The genotype of F_1 mother being S^+S, the gene for dextrality dominates and is responsible for the exclusively dextral coiling of the second generation. Only in the F_3 generation does segregation in the ratio of 3 : 1 becomes apparent, since the individuals of the F_2 generation had the genotypes —1 S^+S^+; 2 S^+S, 1*SS*, 1/4 of them, on the average, produce eggs developing into sinistral individuals (Fig. 15.2).

It is easy to understand the results of a reciprocal cross that is, of the fertilization of the eggs of a homozygous dextral individual (S^+S^+) by the sperm of a sinistral individual (*SS*)—will lead to a somewhat different type of pedigree: the F_1 generation will be dextral (with genotype S^+S) and the F_2 generation again all dextral (with genotypic ratio of $1S^+S^+ : 2S^+S : 1SS$). The F_3 generation will show segregation among broods, just as observed in the first cross.

The whole case becomes clear if it is realized that the type of cleavage (sinistral or dextral) depends on the organization of the egg which is established before the maturation division of the oocyte nucleus. The type of cleavage is, therefore, under the influence of the genotype of the maternal parent. The sperm enters the egg after this organization is already established. Lastly, the direction of coiling of shell depends upon the orientation of the mitotic spindle of first cleavage of the zygote. If the spindle is tipped toward the left of the median line of the egg cell, the sinistral pattern will develop; conversely if the mitotic spindle is tipped toward the right of the median line of the cell, the dextral pattern will develop. The spindle orientation is, thus, controlled by the organization of ooplasm which becomes established during oogenesis and before fertilization.

2. Non-Chromosomal Inheritance by Cellular Organelles

Chloroplasts and mitochondria are organelles that contain their own DNA and protein-synthesizing apparatus. A widely held theory concerning their origin proposes that they were once infectious endosymbiotic prokaryotes that evolved such a dependence on the gene products of the host that they are no longer able to function autonomously.

This theory has been supported by the fact that the genetic components of these organelles are often similar to those found in prokaryotes. For example, the chloroplasts of certain algae and *Euglena* contain 70S type small ribosomes and "naked" chromosomes or DNA which is circular. Their protein synthesis begins with the amino acid N-formylmethionine, as does prokaryotic protein synthesis, and their DNA-dependent RNA polymerase is sensitive to the inhibitor **rifampicin**. The genetic materials of chloroplasts and mitochondria will be transmitted to offspring almost exclusively via the egg. Maternal inheritance due to chloroplast and mitochondria is well illustrated by the following examples:

(*a*) Chloroplast inheritance in variegated four-o' clock plant. The cytoplasmic or extranuclear inheritance of colour in plant by plastids was first of all described by **C. Correns** in 1908 in the

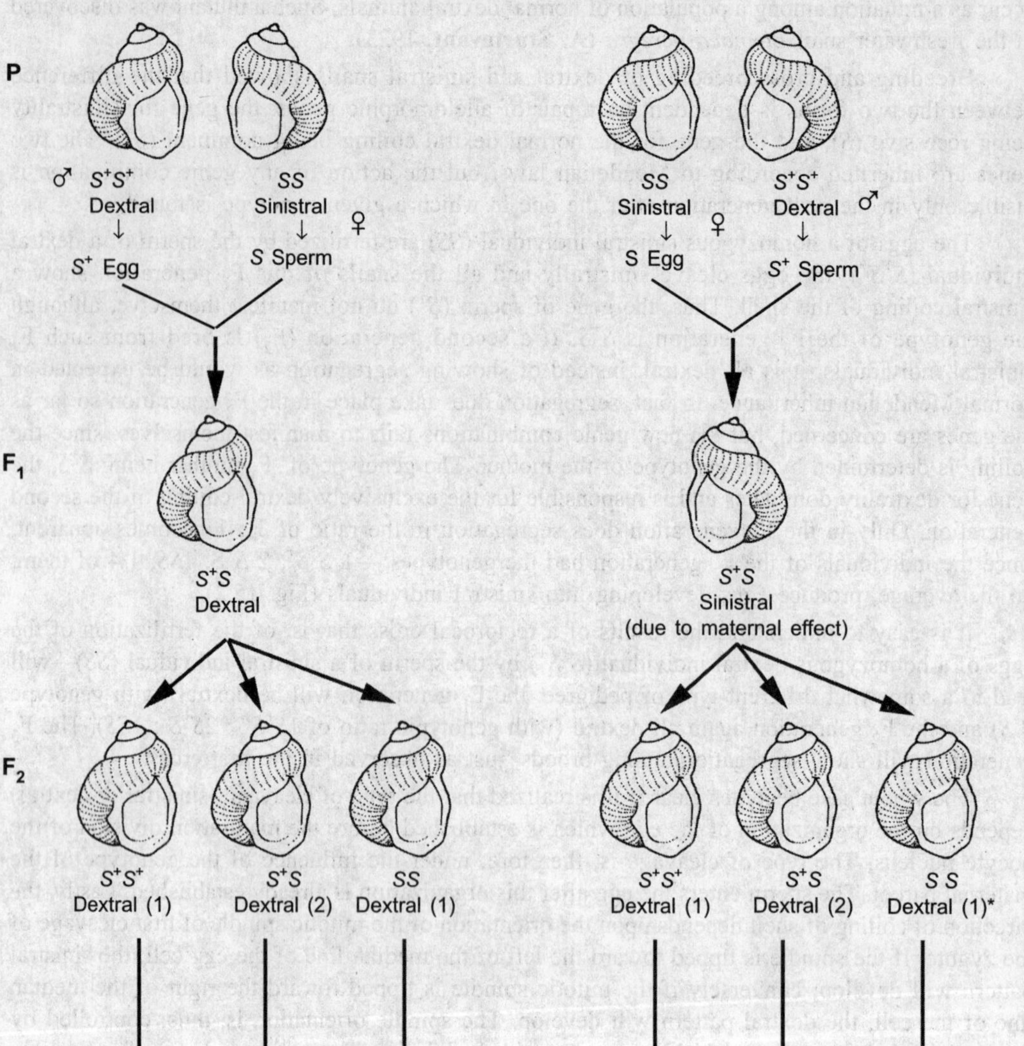

Fig. 15.2. Maternal effect in direction of coiling of the shell in *Limnaea*. (*Note*. **Dextral due to S^+ of sinistral mother at the time of gametogenesis; ®sinistral due to SS mother*).

four-o' clock plant, *Mirabilis jalapa*. In contrast to other higher plants, *Mirabilis* contains three types of leaves and parts: (1) Full green leaves or branches having chloroplast, (2) White (pale) leaves and branches having no chloroplast, (3) Variegated branches having leucoplast in white (pale) areas and chloroplast in green patches (Fig. 15.3). Because, the chlorophyll pigment of chloroplast is related with photosynthesis of food and leucoplasts are incapable to perform photosynthesis, so the white or pale parts of plant survive by receiving nourishment from green parts. **Correns** reported that flowers on green branches produced only green offsprings, regardless of the genotype and phenotype of pollen parent and likewise, flowers from the white or pale branches produced only white or pale seedlings regardless of genotype and phenotype of pollen parent. The plants developing from the white or pale seedlings die because they lack chlorophyll and cannot carry on photosynthesis. **Correns** further reported that flowers from the variegated branches yielded mixed progeny of green, white (pale) and variegated plants in widely varying ratios (Fig. 15.4). These results are summarized in Table 15.1.

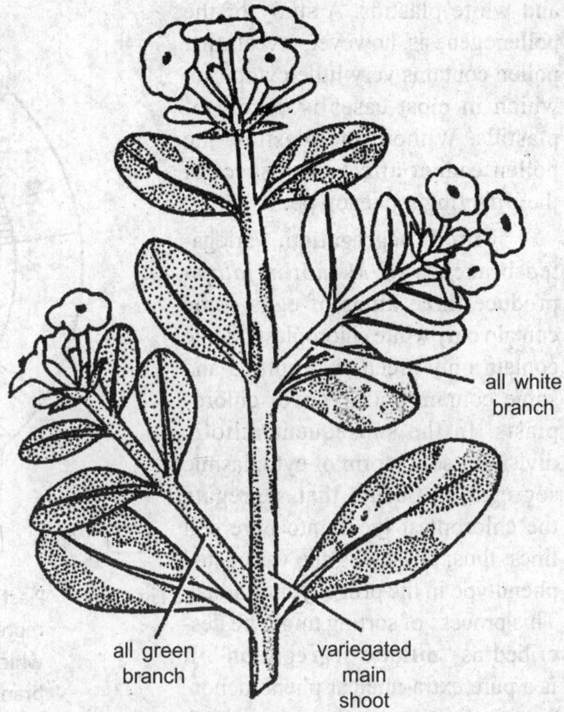

Fig. 15.3. Leaf variegation in *Mirabilis jalapa*, the four-o' clock plant. Flowers may form on any branch (variegated, green, or white), and these flowers may be used in crosses.

Table 15.1 Chloroplast inheritance in variegated four-o'clock plants.

Branch of origin of the male parent	***Branch of origin of the female parent***	***Progeny***
Green	Green	Green
	Pale or white	Pale or white
	Variegated	Green, pale or white, variegated
Pale or white	Green	Green
	Pale or white	Pale or white
	Variegated	Green, pale or white, variegated
Variegated	Green	Green
	Pale or white	Green, pale or white,
	Variegated	Variegated

The irregularity of transmission from variegated branches could be understood by considering cytoplasmic genes (plasmagenes) of plastids. A study of the egg during oogenesis in *Mirabilis* reveals that the ooplasm contains plastids like cytoplasm of other plant cells. If the egg cell is derived from green plant tissues, its ooplasm will contain coloured plastids; if derived from white plant tissues, its ooplasm will contain white plastids; if derived from variegated tissues, its cytoplasm may contain coloured plastids only, white plastids only or a mixture of coloured and white plastids. A study of the pollenogenesis, however, reveals that pollen contains very little cytoplasm which in most cases is devoid of plastids. Without the plastids, the pollen cannot affect this aspect of the offspring's phenotype.

Mitotic segregation. Variegated branches of *Mirabilis jalapa* produce three kinds of eggs: some contain only white chloroplasts, some contain only green chloroplasts and some contain both types of chloroplasts. In the subsequent mitotic divisions, some form of cytoplasmic segregation occurs that segregate the chloroplast types into pure cell lines, thus, producing the variegated phenotype in the progeny individual. This process of sorting might be described as "**mitotic segregation**". It is a pure extra-nuclear phenomenon. In mitotic segregation since both segregation and recombination of organelle genotype takes place, so it is called **cytoplasmic segregation and recombination** (its acronym is **CSAR**).

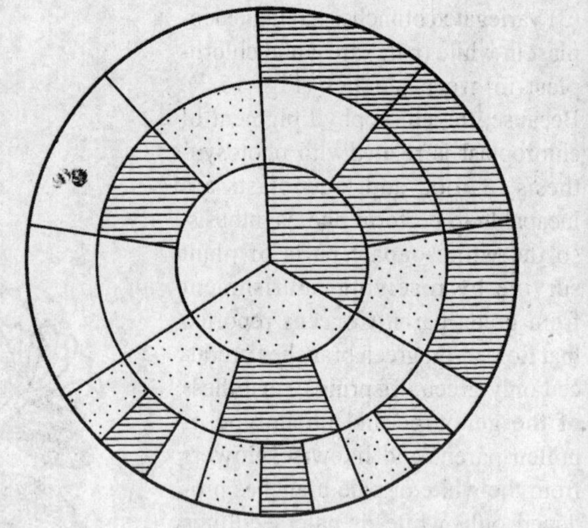

Fig. 15.4. Plastid inheritance of *Mirabilis jalapa*. the central circle represents the type of branch that produces flowers which are pollinated. Intermediate circle represents branch from which pollen is used and outer circle shows the progeny.

(*b*) **Non-chromosomal inheritance by mitochondria.** The most important work on the genetics of mitochondria done in yeast which was initiated by the discovery of petite mutants by **B. Ephrussi** (1953). Subsequently mtDNA was studied in several organisms including plants and animals.

1. Poky strain of *Neurospora*. In fungi, *Neurospora crassa* a number of mutations of mitochondria are inherited via the female parent. The best studied of these is the **poky strain** of *N. crassa*, first isolated by **Mitchell** and **Mitchell** (1952). A poky mutant differs from wild type strain of *Neurospora* in the following aspects: (1) it is slow growing; (2) it shows maternal inheritance, and (3) it has abnormal cytochromes. Of the three cytochromes—cyt a, b and c found in wild type, cyt a and cyt b are absent, and cyt c is in excess in poky mutant.

In reciprocal crosses, poky character shows maternal inheritance:

poky (female) × wild type (male) → all poky

wild type (female) × poky (male) → all wild type

However, there are other marker nuclear genes (ad^+/ad^-) which show. 1 : 1 mendelian segregation. The following evidences suggested that poky trait may be located in mitochondrial DNA: (*i*) slow growth may be due to lack of ATP energy and source of this energy is mitochondria; (*ii*) cytochromes in poky strain differ from those in wild type in quality and quantity and these cytochromes are found in mitochondria.

2. Non-chromosomal inheritance through mitochondria in human beings. In 1988, **Douglas Wallace** and his colleagues showed that **Leber's Hereditary Optic Neuropathy (LHON)** is a cytoplasmically inherited disease in human beings. The disease causes blindness due to degeneration of the optic nerve and it occurs at the age of fifteen to thirty five years. It is certain that the LHON disease is transmitted only maternally. The disease is caused by the mutation in mitochondrial DNA [*e.g.*, mutation at base pair 11778 affects one of the subunits of complex I of the electron transport chain (NADH hydrogenase), resulting in the substitution of a histidine for an arginine; see **Cooper** and **Hausman**, 2007].

3. Non-Chromosomal Inheritance by Endosymbionts

Certain intracellular parasites such as bacteria and virus particles maintain symbiotic relationship with host cells. They are self-reproducing and look like the cytoplasmic inclusions. Sometimes they exhibit an infection like transmission with a hereditary continuity of their own. Generally such symbionts are coined by letters of the Greek alphabets (sigma, kappa, mμ, etc.). The various types of infective symbionts are as follows:

***(i)* Sigma virus in *Drosophila*.** A certain strain of *Drosophila melanogaster* shows a high degree of sensitivity to carbon dioxide, whereas the wild type strain can be exposed for long periods to pure CO_2 without permanent damage. The sensitive strain quickly becomes uncoordinated in even brief exposure to low concentrations of CO_2. This trait (extra-sensitivity) is transmitted primarily, but not exclusively, through the maternal parent. Tests have disclosed that CO_2 sensitivity is dependent upon an infectious DNA virus called **sigma**, found in the cytoplasm of **CO_2 sensitives** *Drosophila*. These infective particles are transmitted normally via the egg's larger amount of cytoplasm but occasionally through the sperms as well. Carbon dioxide sensitivity may even he induced in normal flies by injections of cell free extracts of sigma particles from CO_2 sensitive flies.

***(ii)* Spirochaetes and maternal sex ratio in *Drosophila*.** Females of many *Drosophila* species can harbour a population of **spirochaete bacteria** known generally as **SR**. When SR spirochaetes infect the eggs of the host and when these eggs are fertilized, virtually all XY zygotes are killed early in embroyonic development and XX zygotes survive. Thus, the spirochaete can be considered as an endosymbiont of female but not of male *Drosophila*, and its presence in the female gives rise to the condition called **maternal sex ratio**, in which the progeny are exclusively or almost entirely female.

The SR spirochaete is infectious, for when isolated from the haemolymph of female carriers and introduced into normal females the latter become carriers. Why the female genotype permits their retention, and conversely, why XY cells are sensitive to their presence is not yet known. **K. Oishi** and **D. Poulson** (1970) have reported DNA-containing viruses in these endosymbiont spirochaetes of female *Drosophila*.

***(iii)* Kappa particles.** In 1938, **T.M. Sonneborn** reported that some races (known as "**killers**" or **killer strain**) of the common ciliate protozoan, *Paramecium aurelia* produce a poisonous substance, called **paramecin** which is lethal to other individuals called "**sensitives**". The paramecin is water soluble, diffusible and depends for its production upon cytoplasmically located particles called **kappa**. Electron microscopic observations have shown that kappa particles are about 0.4μ long symbiotic bacteria, *Caedobacter taeniospiralis*; 20 per cent of kappa bacteria of the killer strain contain a refractile protein containing "**R body**" and are called "**brights**". They are infected with a virus that controls the synthesis of toxic viral protein, the paramecin. A killer *Paramecium* may contain hundreds (*e.g.*, 400) of kappa particles. The presence of kappa particles in the killer *Paramecium* is dependent for their maintenance and replication on the chromosomal dominant gene *K*. Paramecia with nuclear genotype *kk* are unable to harbour kappa particles.

When a *Paramecium* of killer strain having the genotype *KK* or ($K^+ K^+$) conjugates with the *Paramecium* of non-killer strain having the genotype *kk*, the exconjugants are all heterozygous for Kk genes (Fig. 15.5). The *Kk* genotype suggests that both exconjugant should be killers. But this is not the case. If conjugation is normal, *i.e.*, lasts only for a short time, and no exchange of cytoplasm takes place between the two, both killers and non-killers (sensitive) are produced. However, rare or prolonged conjugation (*i.e.*, lasting for long time) permits mixing of cytoplasm of both conjugants and results killers only. The killer trait is stable only in killer strain with *KK* genotype and is unstable in sensitive strain with *kk* genotype.

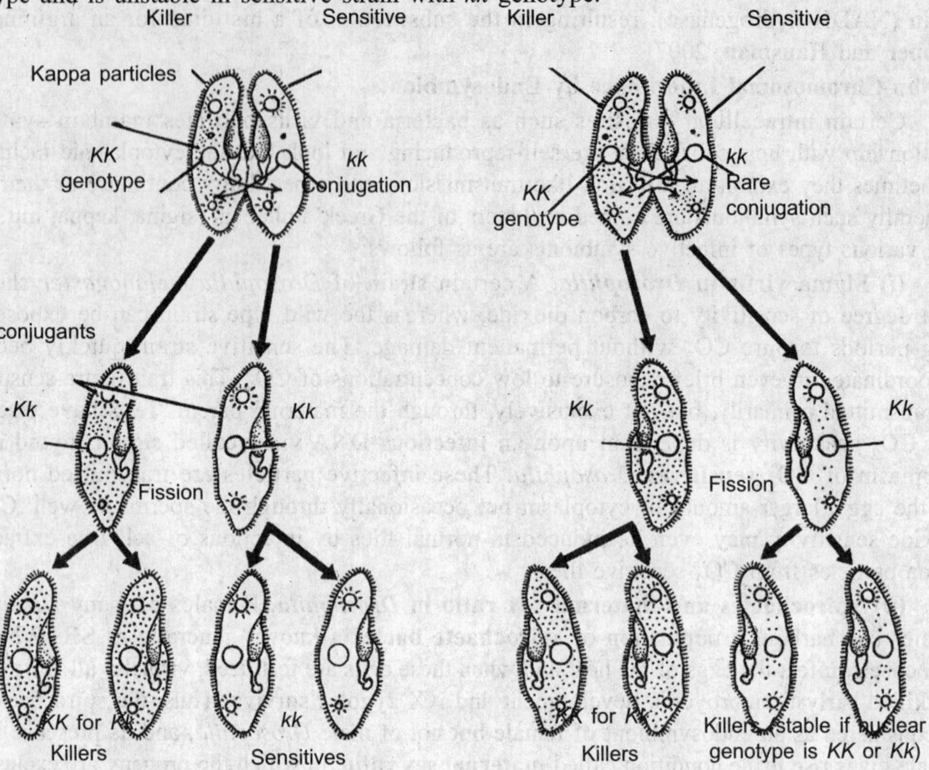

Fig. 15.5. Conjugation in *Paramecium* and the killer trait. Progeny of sensitives are killers only in rare situations where conjugation persists for a longer period so that kappa-containing cytoplasm is introduced into the conjugating sensitive. Kappa particles, however, are maintained only in the presence of *K*—nuclear genotype.

***(iv)* mμ particle.** Another type of killer trait known as **mate killer** has been reported in *Paramecium* by **R.W. Siegel** in 1952. The mate killer trait is contributed by a cytoplasmic mμ particle and a *Paramecium* with a mμ particle is called **mate killer** because when it conjugates with a *Paramecium* without any mμ particle (called **mate sensitive**) then it kills the latter. The mμ particles exist only in those cells whose micronucleus contains at least one dominant gene of either of two pairs of unlinked chromosomal genes (M_1 and M_2). The mμ particles are composed of DNA, RNA and other substances and are symbionts.

(*v*) Milk factor or Breast cancer in mice. Bittner found that females of certain lines of mice are highly susceptible to mammary cancer and this trait was found to be maternally transmitted trait. Results of reciprocal crosses between these and animals of low-cancer-incidence strain depend on the characteristic of the female parent. When the young mice of a low-incidence strain are allowed to be nursed by susceptible foster mothers produces a high rate of cancer in them. Apparently this is a case of infective agent transmitted in the milk. This so called milk factor resembles in many respects with a virus and has been discovered to be transmissible also by saliva and semen. The presence of milk factor also depends on nuclear genes.

QUESTIONS

Long Answer Questions

1. What is meant by cytoplasmic inheritance? Explain with suitable examples.
2. Give an account of cytoplasmic inheritance.
3. Write an account of non-chromosomal inheritance.
4. How will you distinguish between maternal effect and cytoplasmic inheritance? Illustrate with suitable examples.
5. How will you determine whether a particular trait is inherited via nucleus or as a cytoplasmic particle?
6. Describe the non-chromosomal inheritance. Justify with suitable examples. (*CCSU Meerut 2005*)

Short Answer Questions

1. Write short notes on the following:
   (*i*) Cytoplasmic genes
   (*ii*) Plasma genes
   (*iii*) Plastid inheritance
   (*iv*) Cytoplasmic inheritance
2. Describe cytoplasmic inheritance in *Paramecium*. (*C.C.S.U., Meerut, 2009*)
3. Differentiate between the following:
   (*i*) Nuclear inheritance and cytoplasmic inheritance
4. What is cytoplasmic inheritance? Explain with suitable examples? (*C.C.S.U., Meerut 2008*)

Very Short Answer Questions

1. Dextral and sinistral coiling in snail. (*C.C.S.U., Meerut 2008*)
2. What is cytoplasmic inheritance?
3. Define non-Mendelian genes.
4. How much cytoplasm is contributed to the zygote by the male gamete?
5. What are the kappa particles? (*C.C.S.U. Meerut, 2006*)
6. Inheritance of plasma genes found in the cell?
7. Where are plasma genes found in the cell.
8. Who studied inheritance of plasma genes found in the cell?
9. Who investigated inheritance of kappa particles in *Paramecium*?

Fill in the Blanks

1. Inheritance of sinistral and dextral shell coiling in the snail *Limnaea* is an example of .... or .... inheritance.
2. Poky trait of fungus *Neurospora crassa* occurs due to DNA of ....
3. Variegated branches of four-o' clock plant (*Mirabilis jalapa*) produce .... kinds of eggs.
4. The acronym CSAR stands for ....

Multiple Choice Questions

1. Cytoplasmic inheritance differ from nuclear inheritance in the absence of
   (*a*) similarity of reciprocal crosses (*b*) biparental contribution
   (*c*) effect of backcrossing (*d*) uniparental contribution

2. The extrachromosomal autosomes unit within bacteria is
   (*a*) phasmid (*b*) plasmid (*c*) pyramid (*d*) nucleosome
3. The cytoplasmic units of inheritance are called
   (*a*) genome (*b*) lethal gene (*c*) plasmagene (*d*) all the above
4. Extra chromosomal genes can be traced in
   (*a*) colour blind person (*b*) *Paramecium*
   (*c*) sex-linked diseased person (*d*) sex-influenced person
5. Extranuclear genes are present in
   (*a*) mitochondria (*b*) mitochondria and chloroplasts
   (*c*) mitochondria, plastids and cytoplasm (*d*) ER and cytoplasm
6. Cytoplasmic male sterility is passed down
   (*a*) biparentally (*b*) maternally
   (*c*) paternally (*d*) through bacteriophage
7. Genes for cytoplasmic male sterility in plants are located in
   (*a*) nuclear genome (*b*) cytosol
   (*c*) chloroplast genome (*d*) mitochondrial genome
8. A plasmid is
   (*a*) genetic material of a virus (*b*) extrachromosomal DNA in a bacterial cell
   (*c*) smallest bacterium (*d*) starch granule
9. When a certain character is inherited only through the female parent, it probably represents case of
   (*a*) incomplete dominance (*b*) cytoplasmic inheritance
   (*c*) medelian nuclear inheritance (*d*) multiple plastid inheritance
10. When an albino plant of maize female is crossed with normal green male plant, all plants in the progeny are albino because
    (*a*) albinism is dominant over green characters
    (*b*) plastids are inherited through maternal plants
    (*c*) green plastids of male parent becomes maternal
    (*d*) crossing result in structural change in green plants

ANSWERS

Very Short Answer Questions

1. When the shells of the snails coil to right, it is called **dextral coiling**. In contrast, when the shells of the snails coil to left, it is called **sinistral coiling.**
2. Extrachromosomal inheritance controlled by nonnuclear genomes.
3. The non-Mendelian genes are plasma genes which are located in cytoplasm: DNA of mitochondria and plastids, mRNAs of unfertilized eggs, endosymbionts, etc.
4. Negligible.
5. Kappa particles are DNA-containing, self-reproducing cytoplasmic particles in certain strains of *Paramecium aurelia*. They control a toxic substance, *Paramecin,* that is released into the culture medium and kills sensitive paramecia. Nuclear gene *K* is required for maintenance of kappa in the cytoplasm of killer strain.
6. Cytoplasm of egg.

7. In mitochondria (DNA), plastids (DNA), mRNA of ooplasm (which are transcribed by maternal genome during oogenesis), endosymbionts (*e.g.*, kappa particles).

8. L. Heritier and **Teissier**.

9. T. Sonneborn.

Fill in the Blanks

1. Maternal, cytoplasmic.

2. Mitochondria.

3. Three.

4. Cytoplasmic segregation and recombination.

Multiple Choice Questions

1. (*a*)	**2.** (*b*)	**3.** (*c*)	**4.** (*b*)	**5.** (*c*)	**6.** (*b*)
7. (*d*)	**8.** (*b*)	**9.** (*b*)	**10.** (*b*).		

16

Polygenic Inheritance

16.1 QUALITATIVE AND QUANTITATIVE TRAITS

The phenotypic traits of the different organisms may be of two kinds, *viz.*, qualitative and quantitative. The **qualitative traits** are the classical Mendelian traits of **kinds** such as form (*e.g.*, round or wrinkled seeds of pea); structure (*e.g.*, horned or hornless condition in cattles); pigments (*e.g.*, black or white coat of guinea pigs); and antigens and antibodies (*e.g.*, blood group types of human beings) and so on. Each qualitative trait may be under genetic control of two or many alleles of a single gene with little or no environmental modifications to obscure the gene effects. The organisms possessing qualitative traits have distinct (separate) phenotypic classes and are said to demonstrate **discontinuous variations**. The **quantitative traits**, however, are economically important measurable phenotypic traits of **degree** such as height, weight, shape, skin pigmentation, metabolic activity, reproductive rate, behaviour, eye-facet or bristle number in *Drosophila*, susceptibility to pathological diseases or intelligence in human beings; amount of flowers, fruits, seeds, milk, meat or egg produced by plants or animals, etc. Economically important traits such as body weight gains, mature plant heights, egg or milk production records, yield of grain per acre, etc., are also quantitative traits. The quantitative traits are also called **metric traits**. They do not show clear cut differences between individuals and form a spectrum of phenotypes which blend imperceptively from one type to another to cause **continuous variations**.

In contrast to qualitative traits, the quantitative traits may be modified variously by the environmental conditions and are usually governed by many factors or genes (perhaps 10 or 100 or more), each contributing such a small amount of phenotype that their individual effects cannot be detected by Mendelian methods but by only statistical methods. Such genes which are non-allelic and affect the phenotype of a single quantitative trait, are called **polygenes** or **cumulative genes**. The inheritance of polygenes or quantitative traits is called **quantitative inheritance**, **multiple factor inheritance**, **multiple gene inheritance** or **polygenic inheritance** (Box 16.1).

16.2 MULTIPLE FACTOR HYPOTHESIS

Two or more different pairs of alleles, with presumed cumulative effects, govern the quantitative traits. Those alleles which contribute to the trait involved are called **contributing, effective** or **active alleles**; those alleles which do not appear to do so are referred to as **non-contributing, non-effective** or **null alleles**. A gene, individually exerting a slight effect on the phenotype but along with a few or many other genes, controls a quantitative trait is called a **polygene** (a term coined by **K. Mather**). Since there are usually many genes of this kind for one quantitative trait, they are named **multiple factors**, each factor having too small an effect to be traced. This **multiple factor hypothesis** (*i.e.,* large number of genes, each with a small effect, are segregating to produce quantitative variation) has long been the basic model of quantitative genetics.

Box 16.1 Some Definitions of Quantitative Inheritance

I. **Types of complex characters.** Many characters show continuous variation, *i.e.*, phenotypes are measured in quantitative terms and cannot be placed into discrete traits. The phenotype often show a normal distribution about a mean value. Such quantitative characters are inherited in a complex manner: genotype cannot be deduced from phenotype and no simple rules of heredity can be used to predict the outcome of a cross. The inheritance of such characters is studied using statistical methods (**biometrics**).
The discipline of quantitative inheritance embraces following three types of characters:
1. **Continuous characters.** They demonstrate true continuous phenotypic variation (*i.e.*, there exist no boundaries between different phenotypes).
2. **Meristic characters.** They vary in a similar manner to continuous characters but the intrinsic nature of the character itself demands that phenotypes are placed into discrete categories, usually because the value of the phenotype is determined by counting hence such characters may also be termed **countable characters**.
3. **Threshold or dichotomous characters.** They have two phenotypes—a certain condition can be either present or absent. Such characters thus appear very much like Mendelian diallelic traits but in this case an underlying quantitative mechanism controls liability to display the phenotype, which is manifested once some triggering level has been exceeded.

II. **Factorial hypothesis.** Since quantitative characters are unsuitable for Mendelian analysis, they were originally considered to be controlled by **factors** *which were fundamentally different from the genes that control Mendelian characters*.

III. **Major and minor genes.** The relative contribution of each gene to the quantitative phenotype would be different-some loci would have strong effects and others only weak effects. These are termed major genes and minor genes, respectively.

Differences between Quantitative and Qualitative Genetics

The genetical studies of qualitative and quantitative traits are called **qualitative genetics** and **quantitative genetics**, respectively. The major differences between the two are following:

Table 16.1 Differences between qualitative and quantitative traits or genetics.

S. No.	*Qualitative genetics*	*S. No.*	*Quantitative genetics*
1.	Characters of kind.	1.	Characters of degree.
2.	Discontinuous variation; distinct phenotypic classes.	2.	Continuous variations; phenotypic-measure ments form a spectrum.
3.	Single gene effects. too slight to be detected.	3.	Polygenic control; effects of single genes
4.	Concerned with individual matings and their progeny.	4.	Concerned with population of organisms consisting of all possible kinds of matings.
5.	Analyzed by making counts and ratios.	5.	Statistical analysis give estimates of population parameters such as the standard mean and deviation.

16.3 CHARACTERISTICS OF MULTIPLE GENES

Multiple genes for quantitative traits have following characteristics:

1. Each contributing allele in the series of multiple genes produces an equal effect.
2. Effects of each contributing allele are cumulative or additive.
3. There is no dominance, rather, there exist pairs of contributing and non-contributing alleles.
4. There is no epistasis (masking of the phenotypes) among genes at different loci.

5. There is no linkage involved.
6. The environmental conditions have considerable effect on the phenotypic expression of poly-genes for the quantitative traits. The genotype determines the **range** an individual will occupy with regard to a given quantitative character; environment determines the **point** within the genetically determined range at which an individual's measurement will fall.

16.4 EXAMPLES OF QUANTITATIVE INHERITANCE

1. Kernel colour in wheat. A whole grain or seed of a cereal plant such as corn, wheat, barley, etc., is called **kernel**. Kernel colour in wheat is a quantitative trait and its inheritance was studied by Swedish geneticist **H. Nilsson–Ehle** for the first time in 1908. When he crossed a certain red strain to a white strain, he observed that the F_1 was all light red and that approximately 1/16 of the F_2 was as extreme as the parents, *i.e.,* 1/16 was white and 1/16 was red. He interpreted these results in terms of two genes, each with a pair of alleles exhibiting cumulative effect (Fig. 16.1).

P : Red kernel $R_1 R_1 R_2 R_2$ × White kernel $r_1 r_1 r_2 r_2$

F_1 : Light red $R_1 r_1 R_2 r_2$

F_2 : Summary of checker board derived results, *i.e.,* F_2 genotypic and phenotypic ratios:

Genotype	*Genotypic ratio*	*Number of contributing alleles*	*Phenotype*	*Phenotypic ratio*	
$R_1 R_1 R_2 R_2$	1	4	Red	1	Coloured (15/16)
$R_1 R_1 R_2 r_2$	2	3	Medium red	4	
$R_1 r_1 R_2 R_2$	2				
$R_1 r_1 R_2 r_2$	4	2	Light red	6	
$R_1 R_1 r_2 r_2$	1				
$r_1 r_1 R_2 R_2$	1				
$R_1 r_1 r_2 r_2$	2	1	Very light red	4	
$r_1 r_1 R_2 r_2$	2				
$r_1 r_1 r_2 r_2$	1	0	White	1	Colourless (1/16)

Fig. 16.1. Results of a cross between two varieties of wheat having red kernel and white kernel showing cumulative effect of alleles.

Each of the contributing alleles R_1 or R_2 adds some red to the phenotype of kernel colour, so that the genotypes of whites contain neither of these alleles however a red genotype contains only R_1 and R_2 alleles. These results are plotted as histograms in Fig. 16.2. Here five phenotypic classes are obtained in F_2; each 'dose' of a contributing allele for pigment production increases depth of colour. At this stage one point should be clear that in case there were two genes involved, there would be obtained 15 : 1 ratio (15 coloured: 1; see Fig. 16.1).

2. Skin colour in human beings. Another classical example of polygenic inheritance was given by **Davenport** (1913) in Jamaica. He found that two pairs of genes, *A-a* and *B-b* cause the difference in skin pigmentation between Negro and Caucasian people. These genes were found to affect the character in additive fashion. Thus, a true Negro has four dominant genes, *AABB*, and a white has four recessive genes *aabb*. The F_1 offspring of mating of *aabb* with *AABB*, are all *AaBb* and have an intermediate skin colour termed **mulatto**. A mating of two such mulattoes produced a wide variety of skin colour in the offspring, ranging from skins as dark as the original black (negro) parent to as white as the original white parent. The result of this cross have been shown in Fig. 16.3. The F_2 phenotypic and genotypic ratios have been tabulated in Table 16.2.

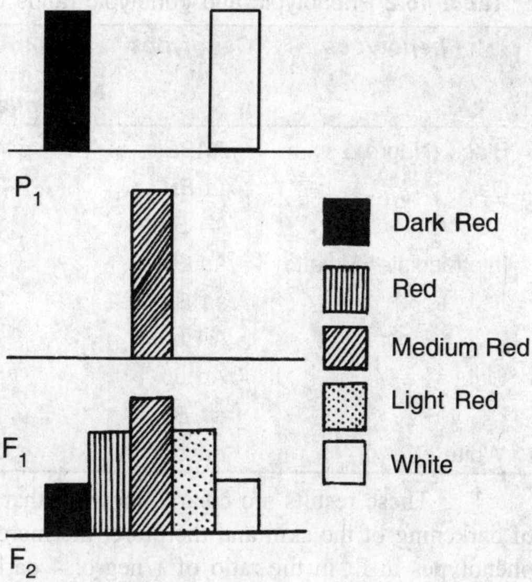

Fig. 16.2. Wheat colour as an example of a quantitative trait.

P_1: *AABB* (Negro) × *abb* (White)

P_1 gametes: (*AB*) ↓ (*ab*)

F_1: *AaBb* Mulattoes — Intermediate skin colour

Intercross: *Aa Bb* (Mulattoes) × *AaBb* (Mulattoes)

♀ / ♂	*AB*	*Ab*	*aB*	*ab*
AB	*AABB* Like negro	*AABb* Darker than mulattoes	*AaBB* Darker than mulattoes	*AaBb* Like mulattoes
Ab	*AABb* Darker than mulattoes	*AAbb* Like mulattoes	*AaBb* Like mulattoes	*Aabb* Lighter than mulattoes
aB	*AaBB* Darker than mulattoes	*AaBb* Like mulattoes	*aaBB* Like mulattoes	*aaBb* Lighter than mulattoes
ab	*AaBb* Like mulattoes	*Aabb* Lighter than mulattoes	*aaBb* Lighter than mulattoes	*aabb* Like white

F_2:

Negro colour	Colour between mulattoes and negro	Colour of mulattoes	Colour between mulattoes and white	White skin colour
(1/16)	(4/16)	(6/16)	(4/16)	(1/16)

Fig. 16.3. Checkerboard exhibiting segregation of skin colour in F_2 generation from a marriage between a negro and a white person.

Table 16.2 Phenotypic and genotypic ratios of F_2 generation of cross shown in Fig. 16.3.

Phenotypes	*Genotypes*	*Number of contributing alleles*	*Genotypic frequency*	*Ratio*
Black (Negro)	*AABB*	4	1	1
Dark	*Aa BB,*	3	2	4
	AA Bb	3	2	
Intermediate (Mulatto)	*Aa Bb*	2	4	6
	aa BB	2	1	
	AA bb	2	1	
Light	*Aa bb*	1	2	4
	aa Bb	1	2	
White	*aa bb*	0	1	1

These results are clearly showing that A and B genes produce about the same amount of darkening of the skin and therefore, the increase or decrease of *A* and *B* genes cause variable phenotypes in F_2 in the ratio of 1 negro: 4 dark: 6 intermediate: 4 light : 1 white.

Significance of Quantitative Genetics

Quantitative genetics has great agricultural importance and has helped in the increase of yield of verious economically important crops.

16.5 ENVIRONMENTAL INFLUENCE AND THE NORM OF REACTION

In fact, a large number of gene loci is not necessary for continuous variation because all biological characters are influenced to some degree by the **environment** as well as by genotype. The environment in which an organism lives will interact with the genotype to produce the phenotype. Thus, if a single genotype is exposed to a range of environments, a range of phenotypes is observed which is described as the **norm of reaction**. This explains much of the phenotypic variance in the **isogenic populations** (*i.e.*, populations where each individual has the same genotype) as all individuals are not exposed to identical environments. The degree to which a phenotype can be shaped by the environment is described as its **phenotypic plasticity**.

Simple characters (*i.e.*, monogenic Mendelian trait) thus exist because the differences between the mean phenotypic values of each genotype are larger than the norm of reaction for each genotype (In other words, the variance between genotypes is greater than the variance within genotypes). For continuous characters the opposite is true: the differences between the mean phenotypic value of each genotype is smaller than the norm of reaction of each genotype, so that the latter overlap. Such an overalp means that the genotype cannot be predicted from phenotype and Mendelian analysis is impossible – the character is quantitative. By controlling the environment, so that, the norms of reaction are small, continuous character controlled by few loci can be resolved into discrete traits and their transmission can be dissected in terms of Mendelian inheritance.

Phenotypic Plasticity

Phenotypic plasticity is the ability of an organism to react to an environment input with a change in form, state, movement, or rate of activity (**West-Eberhard** 2003). When seen in embryonic or larval stages of animals or plants this ability to change phenotype is often called **developmental plasticity**. Examples of developmental plasticity include environmental sex determination in turtles, fish and echiuroid worms (in them, the sexual phenotype was being instrucuted not by genome but by the environment, see Chapter 14). Likewise, the ability of sheer stress to activate gene expression in capillary, heart and bone tissue in human beings is example of developmental plasticity, where one can observe the effect of environmental agent on phenotype (see **Gilbert** 2010). While studies of phenotypic plasticity play a central role in plant developmental biology, the mechanisms of plasticity have only recently been studied in animals. These studies are showing that intergration of animals into ecological communities is accomplished largely through developmental interactions. Indeed, these studies show that we do not develop as autonomous entities, but rather that we are co-constructed by other organisms and that we respond to abiotic agents within our local communities (*e.g.*, temperature in case of animals).

Types of phenotypic plasticity: Phenotypic plasticity are of two main types: reaction norms and polyphenisms (**Woltereck** 1909; **Schmalhansen** 1949; **Stearn** *et al.,* 1991).

1. In a **reaction norm**, the genome encodes the potential for a **continuous range** of **potential phenotypes** and the environment the individual encounters, determines the phenotype (usually the most adaptive one). For instance, our muscle phenotype is determined by the amount of exercise our body is exposed to (even though there is a genetically defined limit to how much muscular hypertrophy is possible). Similarly, the length of male's horn in some dung beetle species *(Onthophagus)* is determined by the quantity and quality of food (*i.e.*, the dung) the larva eats before metamorphosis (**Moczek** and **Emlen** 2000). The upper and lower limits of a reaction norm are also a property of the genome that can be selected. The different phenotypes produced by environmental conditions are called **morphs** (or occasionally, *ecomorphs*).

2. **Polyphenism** refers to **discontinuous** ("either/or") **phenotypes** elicited by the environment. One obvious example is sex determination in turtles, where one range of temperatures will induce female development in the embryo, and another set of temperatures will induce male development. Between these sets of temperatures is a small band of temperatures that will produce different proportions of males and females, but they do not induce intersexual animals. Another important example of polyphenism, called **density-included polyphenism**, is the migratory desert locust *Schistocerca gregaria*. These grasshoppers exist either as a short-winged, green, solitary (low-density) morph or as a long-winged, brown gregarious morph. Cues in the environment determines this dimorphism, *i.e.,* which morphology a larva will develop upon molting (**Simpson** and **Sword** 2008).

Diet-induced polyphenisms. The effects of diet in development can be seen in the caterpillar of *Nemoria arizonaria*. When it hatches on oak trees in the spring, it has a form that blends remarkably with the young oak flowers (called **catkins**). But those larvae hatching from their eggs in the summer resemble newly formed twigs. Here, it is diet (young versus old oak leaves) that determined the phenotype (**Greene** 1989).

Diets having different amounts of proteins or differnt concentrations of methyl donors have also been found to cause different genes to be expressed in mammalian embryos. Different diets can lead to remarkably distinct phenotypes in genetically identical organisms. Diet is also largely responsible for the formation of fertile "queens" in ant (*Pheidologeton*), wasp, and bee colonies. Among these insects, each larva has the genetic potential to become either a worker or a queen; only those larvae fed adequately become queens (*e.g.,* ants). In honeybees, the extra nutrition results in demethylation of particular genes associated with ovary growth and general metabolic rate (**Foret** *et al.,* 2009). (For more details of diet-dependent and other types of polyphenisms, see **Gilbert** 2010).

QUESTIONS

Long Answer Questions

1. What are polygenes? Explain the inheritance of skin colour in man.
2. Explain multiple factors with suitable examples.
3. What are the differences between the inheritance of quantitative and qualitative characters? Give an example of each.
4. Define polygenic inheritance. Describe skin colour in man.
5. Give and account of phenotypic plasticity.

Short Answer Questions

1. What is interaction of gene? Describe the quantitative interaction of gene. *(C.C.S.U. Meerut 2008)*
2. With suitable examples discuss the influence of environment over gene action.
3. Write short note on the polygenic inheritance.
4. Write short notes on reaction norm and polyphenism.

Very Short Answer Questions

1. Name the type of genic interaction in which two or more non-allelic gene pairs affect the same character in the same way?
2. Who proposed that multiple genes are responsible for the inheritance of human skin colour?
3. What is quantitative inheritance?
4. Define Quantitative trait loci.
5. Define the phenotypic plasticity.

Fill in the Blanks

1. Quantitative traits are also called .... traits. They demonstrate .... variations.
2. Quantitative traits are governed by many genes which are called .... or .... genes.
3. Kernal colour of wheat is a .... trait and its inheritance was studied by .... in 1908.
4. Skin colour in human beings is a classical example of .... inheritance.
5. When a single genotype is exposed to a range of environments, a range of phenotypes can be observed which is described as the ....

Multiple Choice Questions

1. The polygenic genes show
   (*a*) different karyotypes (*b*) different genotypes
   (*c*) different phenotypes (*d*) none of these
2. Which one of the following is an example of polygenic inheritance?
   (*a*) skin colour in humans (*b*) flower colour in *Mirabilis Jalapa*
   (*c*) production of male honeybee (*d*) pod shape in garden pea
3. Kernel colour in wheat shows an F_2 ratio of 1 : 4 : 6 : 4 : 1. It is due to
   (*a*) different number of dominant genes (*b*) different number of recessive genes
   (*c*) two polygenes (*d*) supplementary genes

ANSWERS

Very Short Answer Questions

1. Polygenic, cumulative, additive, or quantitative gene interaction.
2. Davenport and Davenport.
3. The mechanism of genetic control of traits showing continuous variation.
4. Chromosomal regions contributing to inheritance of a quantitative trait. These regions may contain one or more polygenes that contribute to the phenotype.
5. Phenotypic plasticity is the ability of an organism to react to an environmental input with a change in form, state, movement, or rate of activity.

Fill in the Blanks.

1. Metric, continuous.
2. Polygenes, cumulative.
3. Quantitative, H.Nilsson-Ehle.
4. Polygenic.
5. Norms of reaction.

Multiple Choice Questions

1. (*c*) **2.** (*a*) **3.** (*c*).

17

Sexuality and Recombination in Bacteria and Viruses (Microbial Genetics)

In preceding chapters, at several places, we have cited certain examples of different genetic aspect of microorganisms such as viruses, bacteria, fungi, algae and protozoans. But they only are inadequate in projecting a clear and understandable image of microbial genetics. Since microorganisms have remain very advantageous and meritorious organisms for indoor genetic experimentation and since several landmark genetic discoveries (e.g., identification of DNA and RNA as the genetic material, protein synthesis, cracking of genetic code, gene regulation, etc.), some of which could fetch Nobel Prizes for their discoverers, could be made only working with these organisms, therefore, a somewhat more detailed discussion of certain genetically well-known microorganisms becomes necessary here. In this chapter, certain very important genetic principles of viruses and bacteria will be discussed.

17.1 MERITS AND DEMERITS OF VIRUSES AND BACTERIA AS THE MATERIALS FOR GENETIC EXPERIMENTATION

There are several advantages in the genetic study of microorganisms such as viruses and bacteria. Most microorganisms have only one set of chromosomes or one set of genetic instruction, as is the characteristic of the haploid gamete of a diploid organism. The genes of a haploid organism are neither dominant nor recessive, since these terms are without meaning where only one allele is present in an individual. In this respect, the study of microorganisms is relatively simple because, barring interaction, every gene which the individual possesses can be expressed in its phenotype. Moreover, the study of rare events such as mutations is facilitated by the case with which very large populations of microorganisms (*e.g.,* 10^6 to 10^9 individuals) can be raised within the confines of a petridish or a test tube. Use of selective media can easily isolate one mutant individual cell in a million. The relative simplicity of microorganisms such as viruses offers the best opportunity for studying gene structure and function at the molecular level. Most microorganisms have short generation period. For example, some bacteria divide about once every 30 minutes. Beginning with a single cell, thirty generations can go by within fifteen hours, producing approximately a billion progeny cell. All of the asexual descendants of a single bacterium are genetically identical (barring mutation) and are referred to as a **clone** or colony visible to the naked eye when grown on solid medium such as an agar plate.

One of the disadvantages of using some microorganisms for genetic studies in the relative dearth of regular recombination events. Most frequent mode of reproduction in most of the microorganisms is asexual (vegetative) reproduction by **amitosis**, a direct nuclear division without the formation of condensed chromosomes or spindle fibres. Some sexual mechanisms are also known to exist in them, and these will be described in this chapter. Another disadvantage is the relative lack of morphological variation of microorganisms. Study of the gross phenotypic characteristics of an individual is generally unfruitful because of its very small size. The morphology of a clone grown on a agar plate, however, may exhibit variation in size per unit time, shape of perimeter or growth habit, texture, colour, etc. Generally different clones may respond differentially to nutrients, dyes, drugs or pathogens in the culture medium. Wild type strains are able to grow on minimal medium (**prototrophic**), whereas some mutant strains require supplementation (**auxotrophic**). Further, though they are called haploid organisms, but most bacteria are actually multinucleate. A delay in phenotypic expression of some mutants is to be anticipated until nuclear and bacterial divisions can seggregate out the mutated nuclei from the others to form a pure clone. The more nuclei per cell the longer the delay is expected to be.

Even viruses, too small to be seen with the light microscope, can show variation in the type of cell they will parasitize (**host range**), in the rapidity with which they cause rupturing of its host cell (**lysis**), and in the shape of the "hole" (**plaque**) which appears in a confluent growth of host cells on solid media. These variations may become apparent to the investigator only after several recombinational events or "rounds of matings" have occurred, so that population dynamics may become a source of confusion in genetic analysis. Moreover, bacteria and viruses do not possess easily visibly chromosomes comparable to those of higher organisms. But many modern microtechniques and electron microscopy have greatly aided the geneticists to understand the different genetical aspects of viruses and bacteria with great precision.

17.2 SOME TECHNIQUES OF MICROBIAL GENETICS

Microbial genetics involves the following specialized techniques for understanding genotypes and phenotypes in bacteria and viruses:

1. Techniques of Cultivation of Bacteria

All organisms require an energy source, a carbon source, nitrogen, sulphur, phoshorus, several metallic ions and water. Those organisms which need an organic form of carbon are called **heterotrophs.** Those that can utilise carbon as carbon dioxide are termed **autotrophs.** All bacteria obtain their energy either by photosynthesis or chemical oxidation. Bacteria

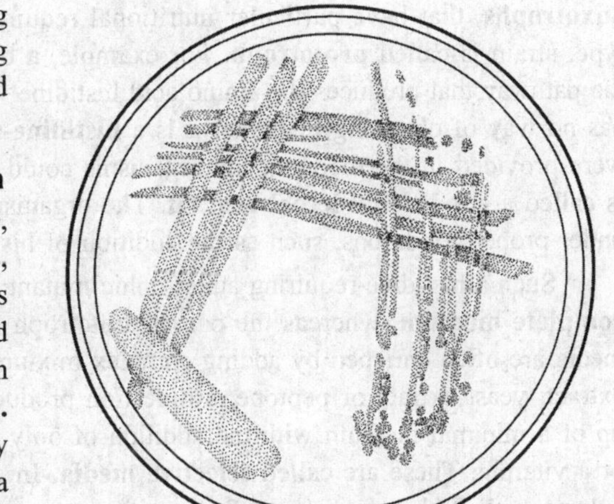

Fig. 17.1. Confluent growth (= continuous cover) of bacterial colonies on a petri dish. Bacteria were streaked on the petri dish with an inoculation loop—a metal wire with a looped end—covered with a bacteria. Streaks begin at the upper right and continued around clockwise. With a heavy inoculation on the loop, bacterial growth is confluent. Eventually, only a few bacteria are left; they form single colonies at the upper left of culture petri dish (after Tamarin, 2002).

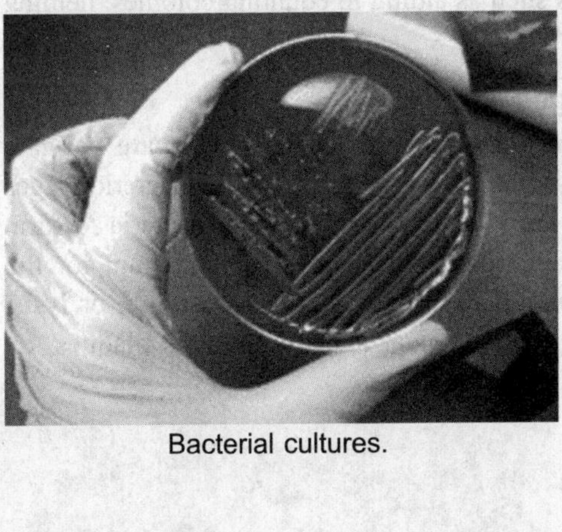

Bacterial cultures.

are usually grown in or on a chemically defined **synthetic medium,** either in liquid in flasks or test tubes, or on petri dishes or petri plates using an agar base to supply rigidity. When one cell of bacteria is placed on the medium in the petri dish, it will begin to divide. After incubation, often overnight, a colony, or clone, will exist where previously was only one cell. Overlapping colonies form a confluent growth (Fig. 17.1). (*Confluent* means flowing together so as to form one).

A culture medium that has only minimal necessities required by the bacterial species is called **minimal medium** (Table 17.1).

Table 17.1. Minimal synthetic culture medium for growing *E. coli*, a heterotrophic bacterium (*Source:* **Tamarin**, 2002).

S. No.	*Component*	*Quantity*
1.	$NH_2H_2PO_4$	1 g
2.	Glucose	5 g
3.	NaCl	5 g
4.	$MgSO_4.7H_20$	0.3 gm
5.	K_2HPO_4	1 g
6.	H_2O	1000 ml

Bacteria can also grow on a medium that supplies, in addition to their minimal requirements, the more complex substances that the bacteria normally synthesize, including amino acids, vitamins and so on. A culture medium of this kind allows the growth of strains of bacteria, called **auxotrophs,** that have particular nutritional requirements. In such a case, the parent or wild-type, strain is called **prototroph**. For example, a bacterial strain that has an enzyme defect in the pathway that produces the amino acid histidine will not grow on a minimal medium because has no way of obtaining histidine; it is a **histidine-requiring auxotroph.** If, however, histidine were provided in the medium, the organisms could grow. This type of mutant (bacterial strain) is called a **conditional-lethal mutant**. The organism (*i.e.,* bacterium) would normally die, but under proper conditions, such as the addition of histidine, the organism can survive.

Such a histidine-requiring auxotrophic mutant bacterium can grow only on an **enriched** or **complete medium**, whereas the parent prototroph could grow on a minimal medium. Culture media are often enriched by adding complex mixtures of organic substances such as blood, beef extract, yeast extract or peptone, a digestion product. Many culture media, however, are made up of a minimal medium with the addition of only one other substance, such as an amino acid or a vitamin. These are called **selective media.** In addition to minimal, complete and selective culture media, other media exist for specific purposes such as aiding in counting colonies, helping maintain cells in a nongrowth phase and so on.

2. Techniques of Cultivation of Bacteriophages

The experimental cultivation of viruses is somewhat different. Since viruses are obligate cellular parasites, they can grow only in living cells. Thus for the cultivation of bacteriophages or phages, petri dishes of appropriate culture media are inoculated with enough bacteria to form a continuous cover, called **bacterial lawn**. This, bacterial culture serves as a medium for the growth of viruses added to petri dish. Since the virus attack usually results in rupture, or **lysis**, of the bacterial cell, addition of the virus usually produces clear spots, known as plaques, on the petri dishes (Fig. 17.2 and Box 17.1).

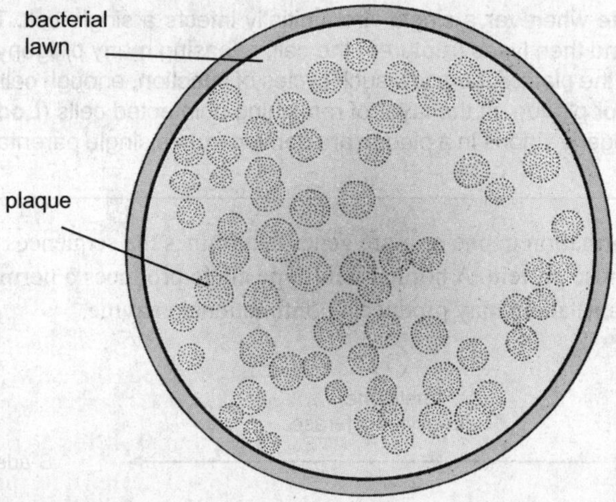

Fig. 17.2. Virus plaques (phage T1) on a bacterial lawn of *E. coli* (after Tamarin, 2002).

3. Recognition of Bacterial Phenotypes

Phenotypes of the bacteria involve following three types of traits:

***(i)* Colony morphology.** This relates simply to the form, colour and size of the colony that grows from a single cell. A bacterial cell growing on a petri dish in an incubator at 37° C divides frequently (once every twenty minutes). Each cell gives rise to a colony, or **clone**, at its original position. In a relatively short amount of time (*e.g.,* overnight), the colonies will consists of enough cells to be seen with the unaided eye. The different morphologies observed among the colonies are usually under genetic control.

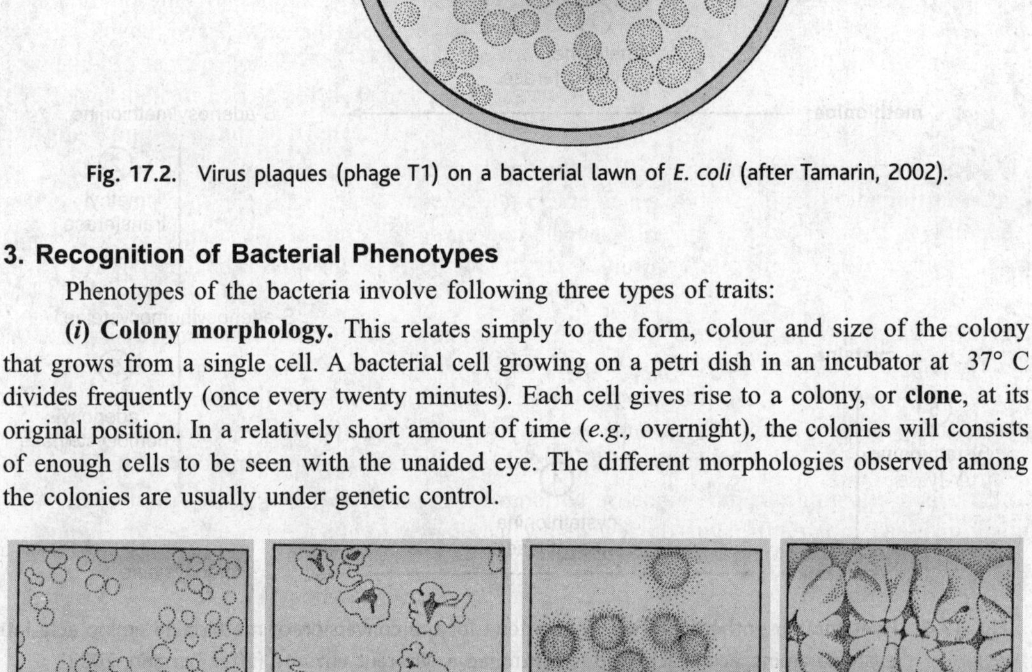

Fig. 17.3. Various types of bacterial colonies.

***(ii)* Nutritional requirements.** The second basis for classifying – bacteria by their nutritional requirements – reflects the failure of one or more enzymes in the bateria's biosynthetic pathways. For instance, if an auxotrophic bacterium has a requirement for the amino acid cysteine that the parent (wild type) strain (prototroph) does not have, then that auxotroph most likely has a nonfunctional enzyme in the pathway for the synthesis of cystenine amino acid. Figure 17.4 shows five steps in cysteine synthesis; a different enzyme controls each step. All enzymes are

Box 17.1 Plaque Assay

The number of infectious viral particles in a sample can be quantified by a **plaque assay**. This assay is performed by culturing a dilute sample of viral particles on a plate (petri dish) covered with host cells and then counting the number of local lesions, called **plaques** that develop. A plaque develops on the plate wherever a single viron initially infects a single cell. The virus replicates in this initial host cell and then lyses (ruptures) the cell, releasing many progeny virions that infect the neighboring cells on the plate. After a few such cycles of infection, enough cells are lysed to produce a visible clear area, or plaque, in the layer of remaining uninfected cells (**Lodish** *et al,* 2004).

Since all the progeny virions in a plaque are derived from a single parental virus, they constitute a **virus clone**.

---

proteins, and the information in one or more genes determines the sequences in the strings of amino acids that make up those protein. A normal wild-type allele produces a normal, functional enzyme. The alternative (mutant) allele may produce a nonfunctional enzyme.

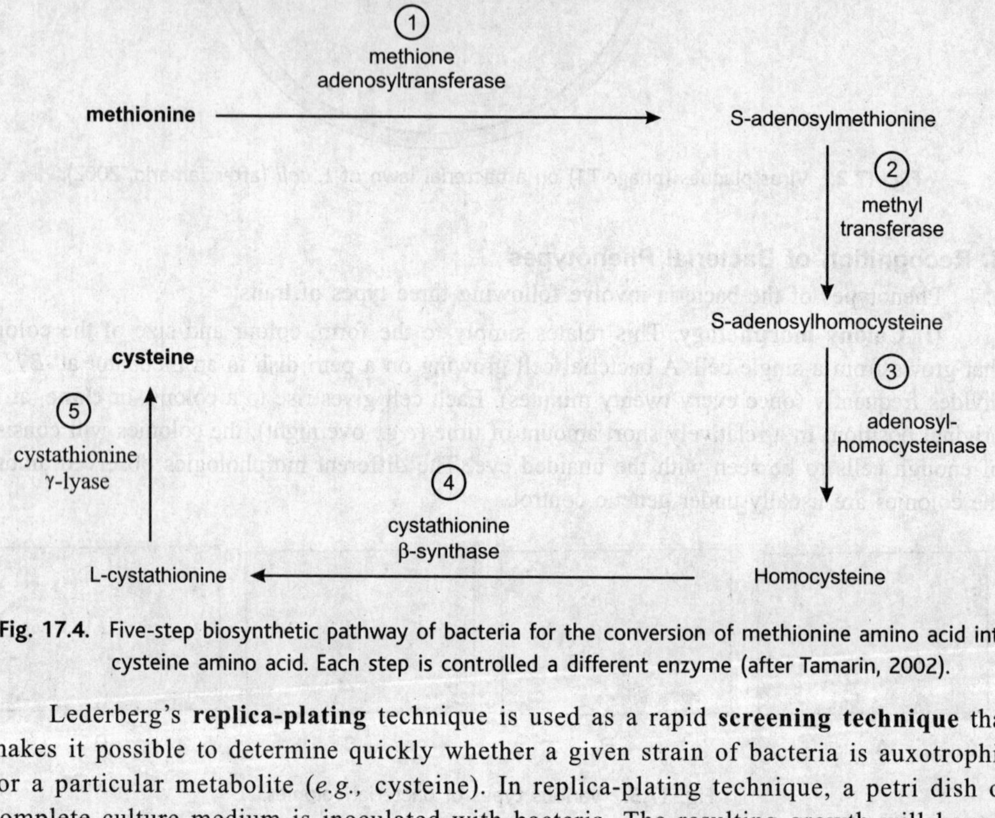

Fig. 17.4. Five-step biosynthetic pathway of bacteria for the conversion of methionine amino acid into cysteine amino acid. Each step is controlled a different enzyme (after Tamarin, 2002).

Lederberg's **replica-plating** technique is used as a rapid **screening technique** that makes it possible to determine quickly whether a given strain of bacteria is auxotrophic for a particular metabolite (*e.g.,* cysteine). In replica-plating technique, a petri dish of complete culture medium is inoculated with bacteria. The resulting growth will have a certain configuration of colonies. This petri dish of colonies is pressed onto a piece of sterilised velvet. Then any number of petri dishes, each containing a medium that lacks some specific metabolite, can be pressed onto this velvet to pick up inocula in the same pattern as the growth on the original petri dish (Fig. 17.5). If a bacterial colony grows on the complete medium but does not grow on a petri dish containing a culture medium missing a metabolite, the inference is that the colony is made of auxotrophic cells that require the absent metabolite. Samples of this bacterial (mutant strain) can be procured from the colony growing on complete medium for further stydy. The nutritional requirement of this strain is its phenotype. The methionine-requireing auxotroph of Figure 17.5 would be designated as Met$^-$, *i.e.,* methionine-minus or Met-minus.

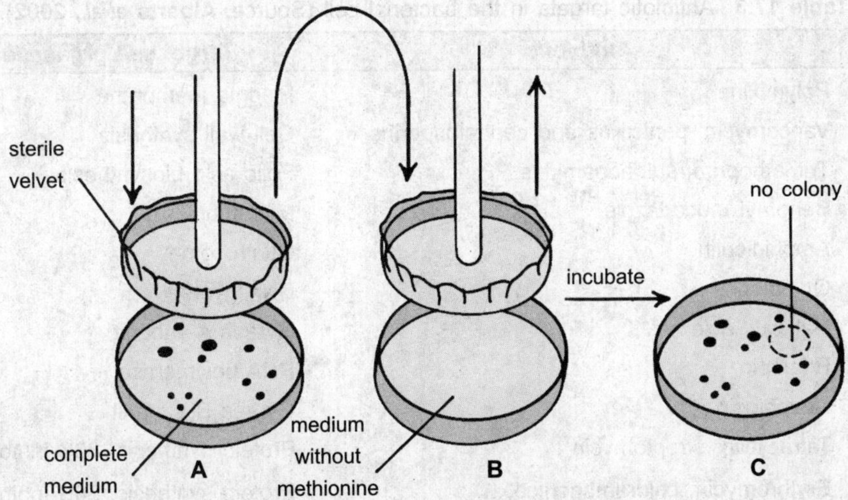

Fig. 17.5. Replica-plating technique. A-A pattern of bacterial colonies from a plate/petri dish of complete medium is transferred to a second plate of medium that lacks methionine amino acid (B). C - In the locations where colonies fail to grow on the second plate, we can conclude that the original colony was a methionine-requiring auxotroph (after Tamarin, 2002).

In context of energy sources, the plus or minus notation has a different meaning. For example, a strain of bacteria that can utilize the sugar galactose as an energy source would be Gal^+. If it could not utilise galactose, it would be called Gal^-. The latter strain will not grow if galactose is its sole carbon source. It will grow if a sugar other than galatose in present. One should note that a Met^- strain requires methionine to grow, whereas a Gal^- strain needs a carbon source other than galactose; it cannot use galactose.

***(iii)* Resistance and sensitivity.** The third common classification of bacterial phenotypes involves resistance and sensitivity to drugs, phages and other environmental insults. For example, penicillin, an antibiotic that prevents the final stage of cell wall construction in bacteria, will kill growing bacterial cells. Nevertheless, we frequently find a number of cells that do grow in the presence of penicillin. These colonies are **resistant** to the drug, and this resistance is under simple genetic control. The phenotype is penicillin resistant (Pen^r) as compared with penicillin sensitive (Pen^s), the normal condition or wild type. Numerous antibiotics are used in bacterial studies (Table 17.2 and Table 17.3).

Table 17.2 Some antibiotics and their antibacterial mechanisms.

S. No.	*Antibiotic*	*Microbial origin*	*Mode of action*
1.	Penicillin G	*Penicillium chrysogenum*	Blocks cell-wall synthesis.
2.	Tetracycline	*Streptomyces aureofaciens*	Blocks protein synthesis.
3.	Streptomycin	*Streptomyces griseus*	Interferes with protein synthesis.
4.	Terramycin	*Streptomyces rimosus*	Blocks protein synthesis.
5.	Erythromycin	*Streptomyces erythraeus*	Blocks protein synthesis.
6.	Bacitracin	*Bacillus subtilis*	—

Table 17.3 Antibiotic targets in the bacterial cell (Source, **Alberts *et al.*,** 2002).

S. No.	*Antibiotic*	*Target cell organelle*
1.	Polymixine	Plasma membrane
2.	Vancomycin, penicillins and cephalosporins	Cell wall synthesis
3.	Trimethoprim, sulphonamides	Folic acid biosynthesis
4.	Benomyl, nocodazole	Microtubules
5.	Amphidicolin	Centrosome
6.	Quinolones	DNA gyrase
7.	Nalidixic acid	Gyrase A subunit
8.	Rifampin	RNA polymerase
9.	Novobiocin	Gyrase B subunit
10.	Tetracyclin, streptomycin	Protein synthesis, 30S inhibitors
11.	Erythromycin, chloramphenicol	Protein synthesis, 50S inhibitors

Box 17.2

Various organisms produce substances that interfere with protein synthesis in other organisms. When the target organisms are bacteria, these substances act as antibiotics, (**Matthew** 2001). For example

1. **Penicillins** inhibit bacterial cell wall synthesis.
2. **Gramicidin** and **valinomycin** interefere with ionic balance across plasma membrane.
3. **Rifampicin** and **streptolydigin** blocks transcription in prokaryotes.
4. **Tetracyclin** inhibits the binding of aminoacyl tRNAs to the ribosome and thereby blocks continued translation.
5. **Streptomycin** interferes with normal pairing between aminoacyl tRNA and massage codons, causing misreading and thereby producing aberrant proteins.
6. **Erythromycin** binds to a specific site on the 17. RNA and blocks elongation by interfering with the translation step.
7. **Chloramphenicol.** Blocks elongation, apparently by acting as competitive inhibitors for the peptidyl transferase.
8. **Puromycin.** Causes premature chain elongation. Its some portion resemble the 3′ end of aminoacylated tRNA, causing premature chain release.

Drug sensitivity provides another screening technique for isolating nutritional mutations. For example, if we were looking for mutants that lacked the ability to synthesize a particular amino acid (*e.g.,* methionine), we could grow large quantities of bacteria (prototrophs) and then place them on a medium that lacked methionine but contained penicillin. Here, any growing cells would be killed. But methionine auxotrophs would not grow, and therefore, they would not be killed. The penicillin could then be washed out and the cells reinoculated onto a complete medium. The only colonies that form should be composed of methionine auxotrophs (Met^-).

Further, screening for resistane to phages is similar to screening for drug resistance. When bacteria are placed in a medium containing phages, only those bactreria that are resistant to the phages will grow and produce colonies. They can thus be easily isolated and studied.

Recognition of Viral Phenotypes

The phenotypes of bacteriophages are of two types: plaque morphology and growth characteristics on different bacterial strains. For example, T_2 an *E. coli* phage produces small plaques with fuzzy edges (genotype r^+). Rapid-lysis mutants (genotype *r*) produce large, smooth edge plaqes (Fig. 17.6). Similarly, T_4, another *E. coli* phage, has rapid lysis mutants that produce large, smooth edge plaques on *E. coli* B but will not grow at all on *E. coli* K_{12} a different strain. Here, rapid-lysis mutants illustrate both the colony morphology phenotypes and growth restriction phenotypes of phages.

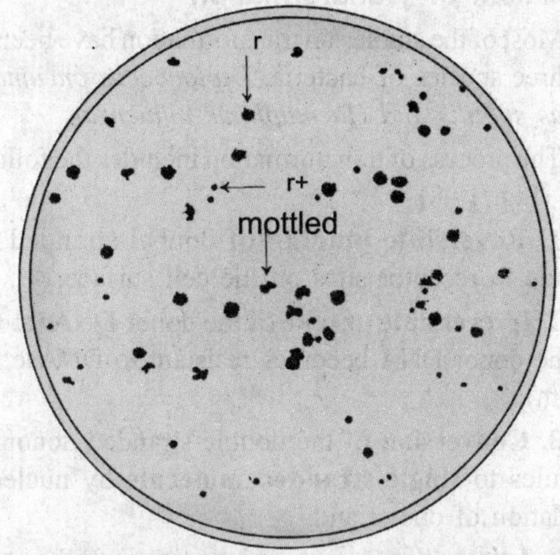

Fig. 17.6. Normal (r^+) and rapid lysis (r) mutants of phage T_2. Mottled plaques occur when r and r^+ phages grow together (after Tamarin, 2002).

17.3 SEXUAL PROCESSES IN BACTERIA AND BACTERIOPHAGES

Although bacteria and viruses are ideal subjects for biochemical analysis, *they would not be useful for genetic study if they did not have sexual processes*. If we define a sexual process as combining genetic material from two individuals, then the life cycles of bacteria and viruses include sexual processes. Although they do not undergo sexual reproduction by the fusion of haploid gametes, bacetria and viruses do undergo processes that incorporate genetic material from one cell or virus into another cell or virus, forming **recombinants**. In fact, bacteria have **three** differcnt methods to gain access to foreign genetic material : **transformation, conjugation** and **transduction.**

Phages can exchange genetic material when a bacterium is infected by more than one virus particle (virion). During the process of viral infection, the genetic material of different phages can exchange parts (or recombine) (Fig. 17.7). We will first examine the exchange processes in bacteria and then in bacteriophages and then proceed to the use of these methods for mapping bacterial and viral chromosomes.

1. Transformation. The phenomenon of transformation in bacteria was first observed in 1928 by **E. Griffith** and was examined at the molecular level in 1944 by **O. Avery Macleod** and **McCarty,** who used the process to demonstrate that DNA was the genetic material of bacteria. In transformation, a cell takes up extraneous DNA found in the environment and incorporates it into its genome (genetic material) through recombination. Not all bacteria are competent to be transformed, and not all extra-cellular DNA is competent to transform. To be competent to transform, the extracellular DNA must be double-stranded and a cell must have the surface protein, called **competence factor**, which binds to the extracellular DNA in an energy-requiring reaction. However, the bacteria that are not naturally competent can be treated to make them competent, usually by treatment with calcium chloride, which makes them more permeable.

Mechanism of Transformation

Most of the studies on transformation have been done with three species of bacteria. *Diplococcus pneumoniae, Bacillus subtilis and Haemophilus influenzae.*

The process of transformation includes the following stages (Fig. 17.8).

1. Reversible binding of doubel-stranded DNA molecule to receptor sites on the cell surface.

2. Irreversible uptake of the donor DNA (at which time the donor DNA becomes resistant to DNAse in the medium).

3. Conversion of the double-stranded donor DNA molecules to **single-stranded molecule** by nucleolytic degradation of one strand.

4. Integration (covalent insertion) of all or part of the single strand of donor DNA in to the chromosome of the recipient.

5. The segregation and **phenotypic expression** of donor gene or gens in the recombinant (transformed) cells.

Steps 2 and 3 may well be coincident effects of a single process. One attractive model (proposed by **S. Lacks** in 1962), for which there is supporting evidence in the case of pneumococcus, proposes that a specific exonuclease (or DNA "translocase") pulls one strand of donor DNA into the cell using energy derived from the degradation of the complementary strand (Fig. 17.9). Whether degradation of complementary strand of DNA actually occurs during uptake or immediately after uptake is uncertain.

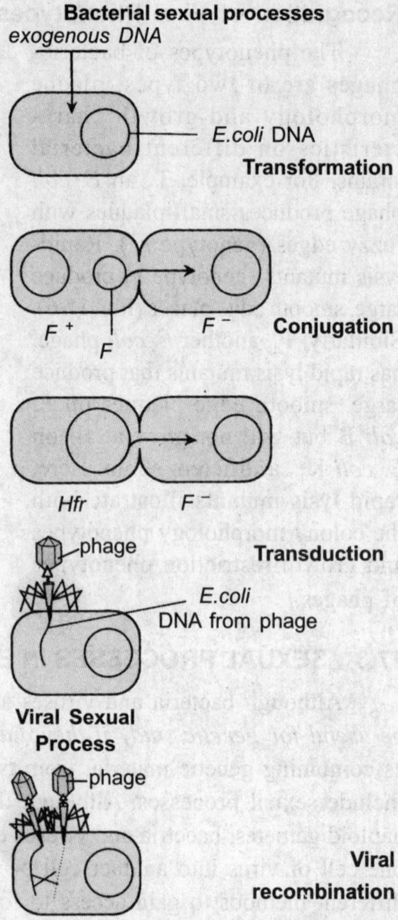

Fig. 17.7. Summary of various sexual processes of bacteria and viruses (after Tamarin, 2002).

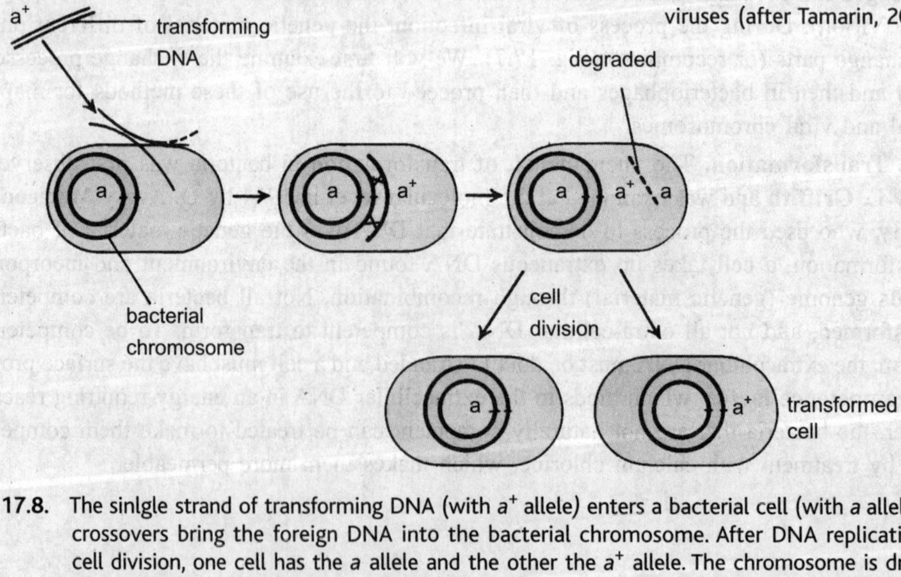

Fig. 17.8. The sinlgle strand of transforming DNA (with a^+ allele) enters a bacterial cell (with *a* allele). Two crossovers bring the foreign DNA into the bacterial chromosome. After DNA replication and cell division, one cell has the *a* allele and the other the a^+ allele. The chromosome is drawn as a double circle symbolising the double-stranded structure of DNA (after Tamarin, 2002).

The first three steps in transformation – binding, uptake and degradation of one strand of the double-stranded DNA-are not specific for homologous DNA. In fact, competent bacteria will carry out these three processes equally well given calf thymus DNA or other foreign DNAs. However, the **integration** or **DNA recombination** step, is specific for homologous DNA. (One should note here that unlike eukaryotic crossing over, this is not a reciprocal process). This is not to say that the integration of segments of heterologous (foreign) DNA never occurs. If it does occur, however, it does so at frequencies very much lower than the frequencies observed using homologous DNA. Further, although very small fragments of DNA are taken up by competent cells, a minimum length of about 500 nucleotide-pairs appears to be required for integration to occur. During integration, a single strand of donor DNA is physically inserted into the recipient chromosome, replacing a segment of one strand of the recipient chromosome (Fig. 17.9). In most transformation experiments, donor DNA fragments are about 20,000 nucleotide-pairs (or about 1/200 of the total chromosome) in length. This means that mapping experiments can be done using transformation only if the genetic markers employed are located close together on the host chromosome.

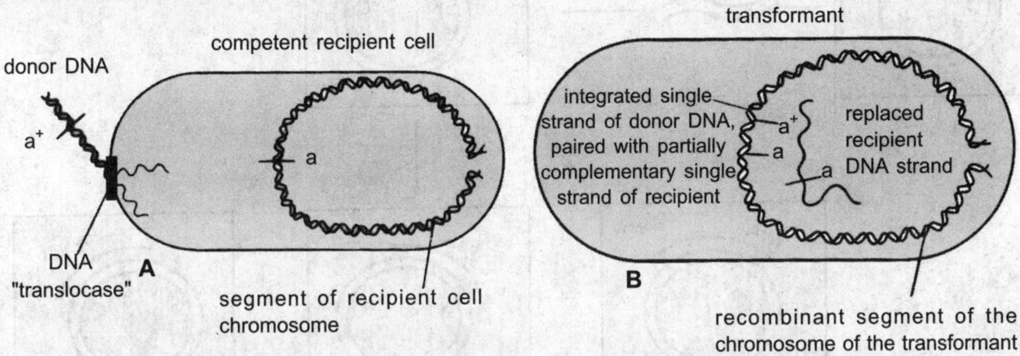

Fig. 17.9. Diagrammatic representation of the two key steps, uptake and integration, in transformation of *Pneumococcus*. A-A membrane bound exonuclease or DNA translocase "pulls one strand of donor DNA into the cell, using energy derived from degradation of the complementary strand of DNA" (Lacks model). The donor DNA fragment carries the genetic marker a^+, the recipient chromosome carries the allele marker *a*. B- The single strand of donor DNA is shown covalently integrated into the chromosome of the recipient (now called a transformant). The donor strand base-pairs most of the length with the complementary (except at *a* site) single strand of the recipient chromosome. A "mismatch" occurs at the a^+/a site where the donor strand carries the a^+ allele, whereas the recipient strand carries the *a* allele. Such DNA molecule, carrying different alleles in the complementary (partially) strands, are called **DNA heteroduplex.** DNA heteroduplexes are important intermediates in mutation, recombination and DNA repair processes. They segregate during subsequent semi-conservative replication (after Gardner *et al.*, 2001).

Transformation Mapping

The general idea of transformation mapping is to add DNA from a bacterial strain with known genotype, but with different alleles at two or more loci. We then look for incorporation of the donor alleles into the recipient strain of bacteria. More often alleles from two loci are incorporated together into the host, the closer together these loci must be to each other. Thus, we can use an index of co-occurrence that is in inverse relationship to map distance: the larger the co-occurrence of alleles of two loci, the closer together the loci must be. This is another way looking at the mapping concepts, we have discussed for eukaryotes where we discovered that the closer two loci are, the fewer the recombination between them and thus the higher the co-occurrence.

For the mapping of bacterial chromosome one has to consider another concept that is selecting for recombinant cells. In fruit flies, every offspring of a mated pair represents a sampling of the meiotic tetrad, and thus a part of the total, whether or not recombination took place. Here, however, many cells are present that do not take part in the transformation process. In a bacterial culture, for example, only one cell in a thousand might be transformed. We must thus always be sure when working with bacterial gene transfer that we count only those cells that have taken part in the process. Following example will clarify this point:

A recipient strain of the soil bacterium, *Bacillus subtilis* is auxotrophic for the amino acids tyrosine (tyr A^-) and cysteine (Cys C^-). One wishes to know that how close these loci are on the bacterial chromosome. One thus isolates DNA from a prototrophic (= wild) strain of bacteria (tyr A^+ cys C^+). This donor is added to the auxotrophic strain allow time for transformation to take place (Fig. 17.10).

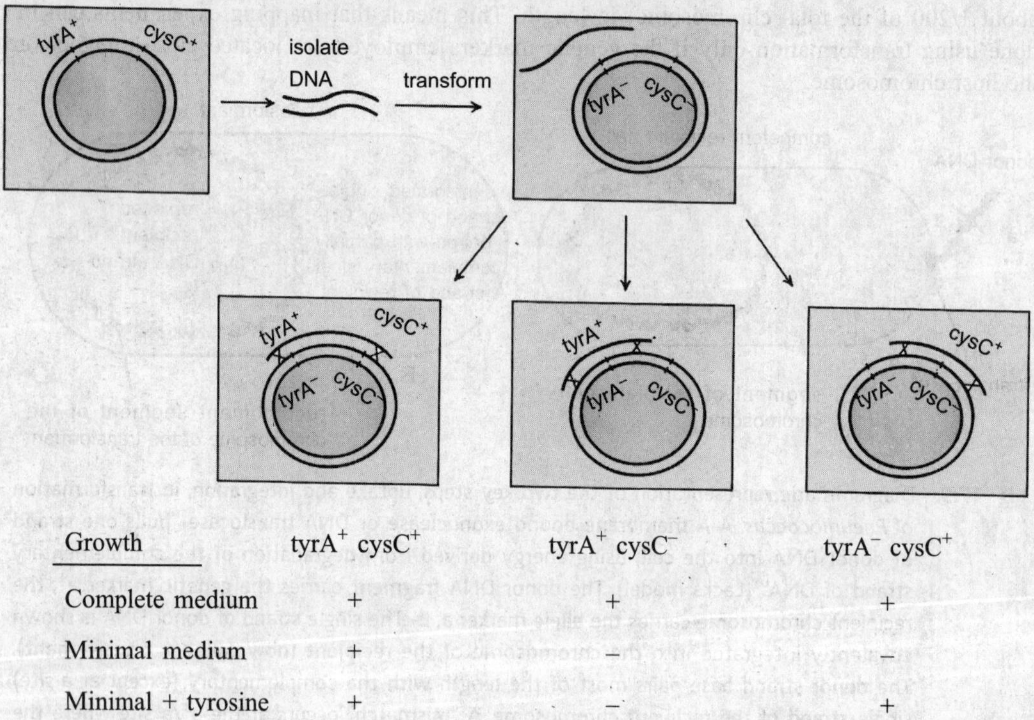

Growth	tyrA$^+$ cysC$^+$	tyrA$^+$ cysC$^-$	tyrA$^-$ cysC$^+$
Complete medium	+	+	+
Minimal medium	+	–	–
Minimal + tyrosine	+	–	+
Minimal + cysteine	+	+	–

Fig. 17.10. Transformation experiment with *Bacillus subtilis*. *A tyr* A^- *Cys* C^- strain is transformed with DNA from a *tyr* A^+ *Cys* C^+ strain. Nontransformants as well as three types of transformants (two single and one double) result. Genotypes are determined by growth characteristics on four different types of petri dishes (Fig. 17.11), (after Tamarin, 2002).

If the experiment is successful, and the loci are close enough together, then some of the recipient bacteria may incorporate donor DNA that has either both donor alleles or one or the other donor allele. Thus, some of the recipient cells will now have the *tyr* A^+ and *cys* C^+ alleles, some will now have the *tyr* A^+ allele, some will have just the donor *cys* C^+ allele and the vast majority will be of the untransformed auxotrophic genotype, *tyr* A^- *cys* C^-. One thus, needs to count the transformed cells.

This is done by removing any extraneous transforming DNA and then pouring the cells out onto a complete medium so that all cells can grow. These cells are then replica-plated on to three plates/petri dishes – a minimal medium petri dish, a minimal medium plus tyrosine petri dish, and a minimal medim plus cysteine petri dish – and allowed to grow overnight in an incubator at 37° C. In next step, colonies are counted (Fig. 17.11). Those growing on minimal medium are of genotype *tyr* A^+ $cysC^+$; those growing on minimal medium plus tyrosine but not growing on minimal medium are *tyr* A^+ *cys* C^-; and those growing or minimal medium plus cysteine but not growing on minimal medium are *tyr* A^+ *cys* C^-. The vast majority of the cells will grow on complete medium, but not on minimal medim or minimal media with just tyrosine or cysteine amino acids added. This majority is made up of the **nontransformants**, that is, auxotrophs that were not involved in a transformation event-they took up no foreign DNA.

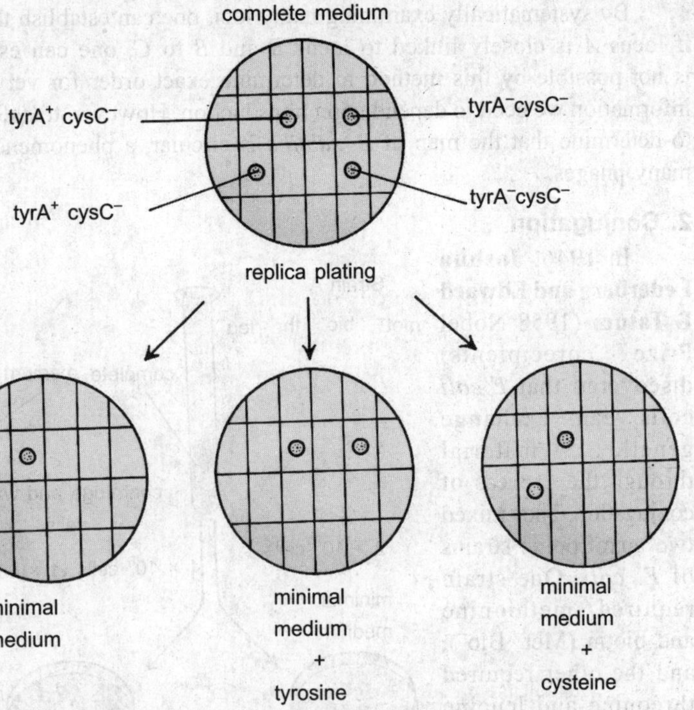

Fig. 17.11. Four patterns of growth on different media reveal the genotypes of transformed and untransformed bacterial cells. Only four colonies are shown, and a grid is added for each of identification. After transformation (Fig. 17.10) cells are plated on complete culture medium and then replica-plated on to minimal medium with either tyrosine amino acid or cysteine amino acid added (after Tamarin, 2002).

Further, as a control against reversion, the normal mutation of *tyr* A^- *to tyr* A^+ or *cys* C^- *to cys* C^+, one has to grow several petri dishes of auxotrophs in minimal medium and minimal medium with tyrosine or cysteine added. These are auxotrophs that were not exposed to prototrophic donor DNA. In next step, number of natural revertants are counted and corrections are made in experiemental numbers by the natural reversion rate. This is done to ensure that what one measures is the actual transformation rate rather than just a mutation rate that one tends to mistook for transformation. According to **Tamarin** (2002) this control should always be carried out.

From this experiment (Fig. 17.10 and 17.11), following data were obtained : twelve double transformants (*tyr* A^+ *cys* C^+), thirty one *tyr* A^+ *cys* C^- and twenty seven *tyr* A^- *cys* C^+. From these data, one can calculate the occurrence, or cotransfer index (*r*) as:

$$\text{Cotransfer index } (r) = \frac{\text{number of double transformants}}{\text{number of double transformants + number of single transformants}}$$

From above data:

$$r = 12/(12 + 31 + 27) = 0.17$$

This is a relative number indicating the co-occurrence of the two loci and thus there relative distance apart on the bacterial chromosome. One should remember that as this number increases for different pairs of loci, the loci are closer and closer together.

By systematically examining many loci, one can establish their relative order. For example, if locus *A* is closely linked to locus *B* and *B* to *C*, one can establish the gene order *ABC*. It is not possible by this method to determine exact order for very closely linked genes. For this information we need to depend upon transduction. However, transformation has allowed geneticists to determine that the map of *B. subtilis* is circular, a phenomenon found in all prokaryotes and many phages.

2. Conjugation

In 1946, **Joshua Lederberg and Edward L Tatum** (1958 Nobel Prize corecipients) discovered that *E.coli* cells can exchange genetic material through the process of conjugation. They mixed two auxotrophic strains of *E. coli*. One strain required methionine and biotin (Met^- Bio^-), and the other required threonine and leucine (Thr^- Leu^-). This cross is shown in Figure 17.12. One should remember that if a strain is Met^- Bio^-, it is without saying, wild-type for all other loci. Thus, a cell with the *Met^- Bio^-* phenotype actually has the genotype of *met^- bio^- thr^+ leu^+*. Similarly, the *Thr^- Leu^-* strain is actually *met^+ bio^+ thr^- leu^-* (note that symbols such as "*Thr*" represent phenotypes; symbols such as "*thr*" represent genotypes").

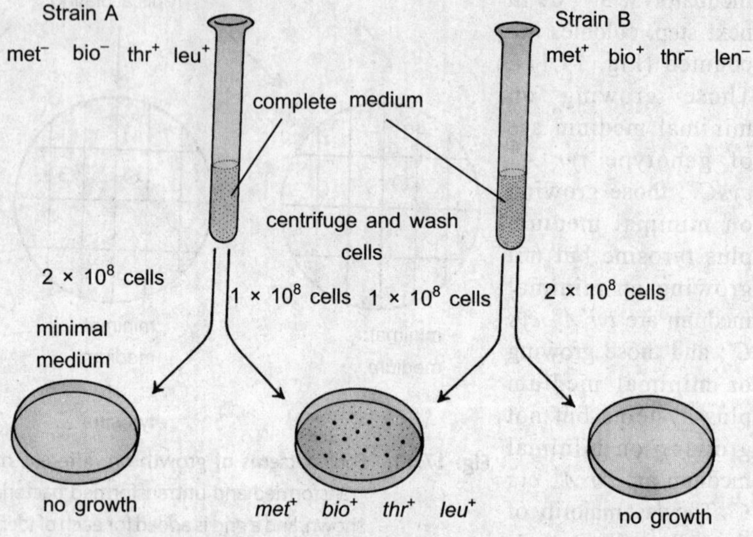

Fig. 17.12. Lederberg and Tatum's cross showing that *E.coli* undergoes genetic recombination (after Tamarin, 2002).

Lederberg and Tatum used multiple auxotrophs in order to rule out spontaneous reversion (mutation). About one in 10^6 *Met^-* cells will spontaneously become prototrophic (Met^+) every generation. However, with multiple auxotrophs, the probability that several loci will simultaneously and spontaneously revert (*e.g.*, *$met^- \rightarrow met^+$*) becomes very small. (In fact, the control petri dishes in the experiment illustrated in Figure 17.12 showed no growth for parental double mutants). After mixing the strains, Lederberg and Tatum found that about one cell in 10^7 was prototrophic (*met^+ bio^+ thr^+ leu^+*).

To rule out transformation, one strain was put in each arm of Davis U-tube (Fig. 17.12). The liquid and large molecules, including DNA, were mixed by alternate application of pressure and suction to one arm of the tube; whole cells did not pass through the filter. The result was that the fluids surrounding the cells, as well as any large molecules (*e.g.*, DNA), could be freely mixed while the cells were kept separate. After cell growth stopped in the two arms (in complete medium), the contents were plated out on minimal medium. There were no prototrophs in either arms. Therefore, cell-to-cell contact was required for the genetic material of the two cells to recombine.

F Factor and Conjugation

During conjugation, DNA is transferred from a **donor cell to a recipient cell** through a specialized intercellular connection, or **conjugation tube**, that form between them. The donor and recipient cells are sometimes referred to as male and female cells, respectively. The transfer of

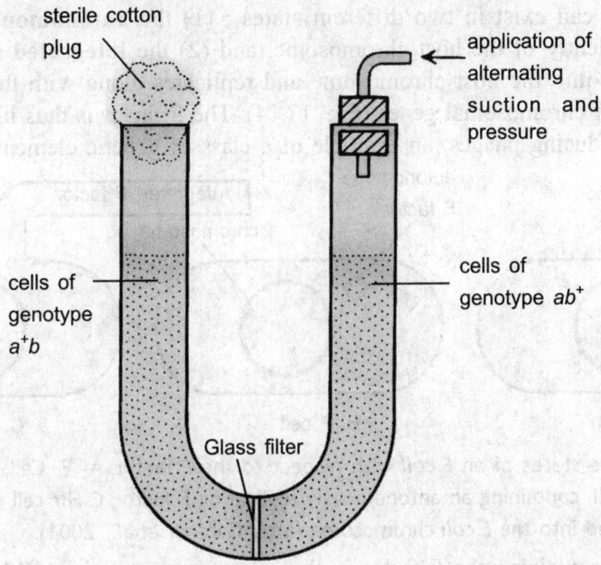

Fig. 17.13. Diagram of the U-tube experiment used to determine whether cell contact is required for recombination in bacteria. The two arms of the Davis U-tube are separated by a glass filter containing pores of size which allow the passage of DNA molecules and viruses, but not cells. Bacteria of one genotype (*e.g.*, $a^+ b$) are placed in one arm of the Davis U-tube; those of another genotype (*e.g.*, $a\ b^+$) are placed in the other arm. Alternating suction and pressure are applied to one arm of the tube during incubation to mix the medium and any small particles (such as phage) suspended in it. However, the glass filter prevents cells from passing from one arm of the tube to the other. If recombinants (*e.g.*, $a^+\ b^+$) are formed during the U-tube experiment, cell contact is not required for the recombination process to occur; thus this excludes conjugation as a possible mechanisms (after Gardner *et al.*, 2001).

genetic information is, thus, **one-way transfer** during conjugation just as in transformation and tranduction (to be described later on in this chapter), rather than a reciprocal exchange of genetic material (as in case of crossing over or recombination). Cells that have the capacity to serve as donors during conjugation are differentiated by the presence of specialized cell-surface appendages called **F pili** (Box 17.3). The synthesis of these F pili is controlled by several genes (*i.e.*, nine; see **Gardner** *et al.*, 2001) that are carried by a small circular molecule of DNA or "minichromosome" (about 94,500 nucleotide-pairs long) called a **F factor** (for fertility factor; also called "**sex factor**" or "**F plasmid**", Box 17.3). Cells carrying a F factor (*i.e.*, donor cells or male cells) form conjugation tubes and initiates DNA transfer after making contact with **cells not carrying an F factor**, called **F⁻ cells** (recipient cells or female cells).

Box 17.3

Escerichia coli cells are normally coated with hair-like **pili** (fimbriae). F^+ and Hfr cells have one to three additional pili (singular pilus) called **F-pili** or **sex pili**. During conjugation, these sex pili form a connecting bridge between F^+ (or Hfr) and F^- cells. Once a connection is made, the sex pilus then contracts to bring the two cells into contacts.

Box 17.4

The F factor is a **plasmid**, a term orinigally coined by **Lederberg** to refer to independent, self-replicating genetic particles. Plasmids are usually circles of double-stranded DNA. They form important tools of recombinant DNA technology. Plasmids are auxilliary (additional) cirlces of DNA that many bacteria carry. They are usually much smaller than the bacterial chromosome.

The F factor can exist in two different states : (1) the **autonomous state,** in which it replicates independently of the host chromosome, and (2) the **integrated state,** in which it is covalently inserted into the host chromosome and replicates along with the host chromosome like any other set of chromosomal genes (Fig. 17.14). The F factor is thus like the chromosomes of specialized transducing phages, an example of a class of genetic elements called **episomes.**

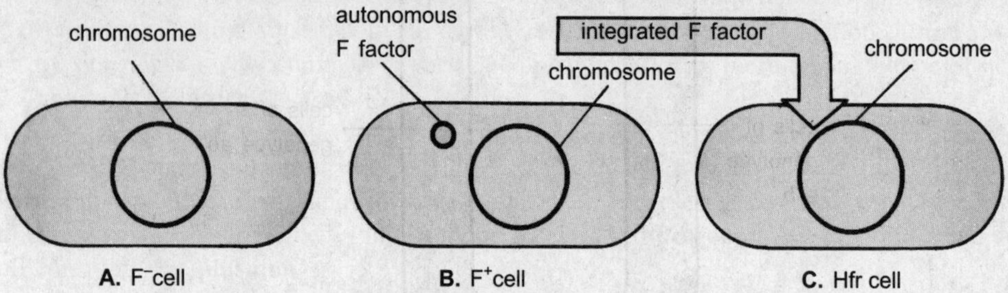

Fig. 17.14. The three states of an *E.coli* with respect to the F factor. A- F^- Cell, no F factor presents; B- F^+ cell, containing an antonomously replicating F factor; C-Hfr cell containing an F factor integrated into the *E.coli* chromosome (after Gardner *et al.,* 2001).

A donor cell containing the F factor in the autonomous state is called **F^+ cell.** When a F^+ donor cell conjugates with a F^- recipient cell, only the autonomous F factor in transferred. Both exconjugants (cells that have been involved in conjugation) become F^+ because the F factor replicates during transfer. Thus, mixing a population of F^+ cells with a population of F^- cells results in virtually all the cells in the new population becoming F^+.

The F factor can integrate into the host chromosome at any one of many sites by a mechanism that appears analogous to the integration of the chromosome of a specialized tranducing phage. This integration of the F factor is believed to be mediated by short DNA sequences called **IS elements** (Fig. 17.17). A cell carrying an integrated F factor is called a **Hfr** (for **high-frequency recombinations**). In the integrated state, the F factor mediates the transfer of a chromosome of the Hfr cell to a receipient (F^-) cell. Usually, only a portion of the Hfr chromosome is transferred before the cell separate, thus breaking chromosome (Fig. 17.16). An entire Hfr chromosome is seldom transferred.

The mechanism of transfer of DNA from a donor cell to recipient cell during conjugation appears to be the same whether just the F factor is being transferred, as in F^+ by f^- matings, or the chromosome is being transferred as in Hfr by F^- matings. Transfer is believed to be initiated by an endonucleolytic nick in one strand at a specific site (the "origin" of transfer) on the F factor. The 5′ end of the nicked strand is even transferred through the conjugation tube into the recipient cell. Transfer is believed to be coupled to **rolling circle replication**, with the intact circular strand being replicated in the donor cell and the displaced strand being replicated in the recipient cell as it is transferred (Fig. 17.16). Because the origin of transfer exists within the integrated F factor, one portion of the F factor is transferred from a Hfr cell to an F^- cell prior to the sequential transfer of chromosomal genes. The remaining part of F factor, however, is the last segment of DNA to be transferred. Thus, in Hfr by F^- matings, the recipient F^- cell acquires a complete F factor and becomes an Hfr donor. But this occurs in those rare cases when an entire Hfr chromosome, with its integrated F factor, is transferred.

Many of the details of bacterial conjugation were worked out using one particular Hfr strain called **HfrH** or **Hfr Hayes** (for **W. Hayes** who isolated it). In this strain F factor in integrated near the *thr* (threonine) and *leu* (leucine) loci (Fig. 17.16).

Box 17.5

The F factor itself has the genes for sex pillus formation and DNA transfer to a conjugating F^- cell. At least twenty two genes are involved in the transfer process, including genes for the pilus protein, nicking the DNA, and regulation of the process (**Tamarin,** 2002).

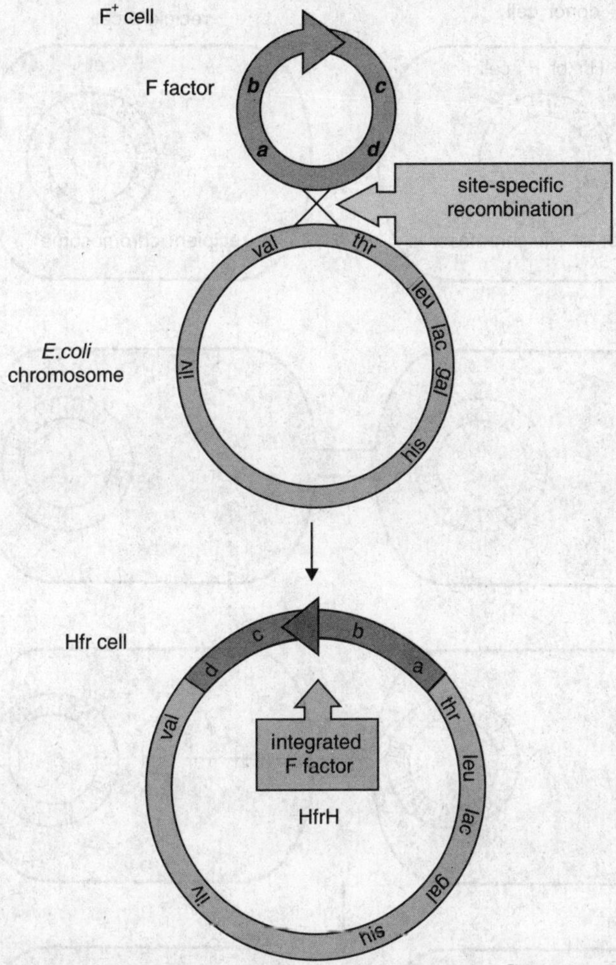

Fig. 17.15. Converson of an F^+ cell to an Hfr cell by the integration of the autonomous F factor into the host chromosome. The F factor is covalently inserted into the host chromosome by a site-specific recombination event. Integration of the F factor appears to be mediated by IS elements. The markers *a, b, c* and *d* of F factor are hypothetical. The arrowhead indicates the site of origin and direction of the sequential transfer that occurs during conjugation. The site of integration illustrated is that occurring during the formation of classic Hfr strain, Hfr H (or Hfr Hayes). A few *E.coli* genes are shown for the sake of orientation. Genes *thr, leu, his, ilv* and *val* are involved in the biosynthesis of the amino acids threonine, leucine, histidine, isoleucine and valine, respectively. Genes *lac* and *gal* are required for the utilisation of lactose and galatose, respectively, as energy soures (after Gardner *et al.,* 2001).

Role of IS Elements in Conjugation

Currently many of the properties of plasmids and episomes are known to depend on the presence of short DNA sequences called **IS elements** (IS = insertion sequences). The IS elements are also present in the main host chromosomes. These short sequences (from about 800 to about 1400 nucleotide-pairs in length) are **transposable,** that is they can move from one position to another within a chromosome, or move from one chromosome to a different chromosome. In addition, IS elements mediate the integration of episomes into host chromosomes. This is particularly evident in the case of the integration of the *E.coli* K12 F Plasmid (F factor) during the formation of Hfr's (Fig. 17.17).

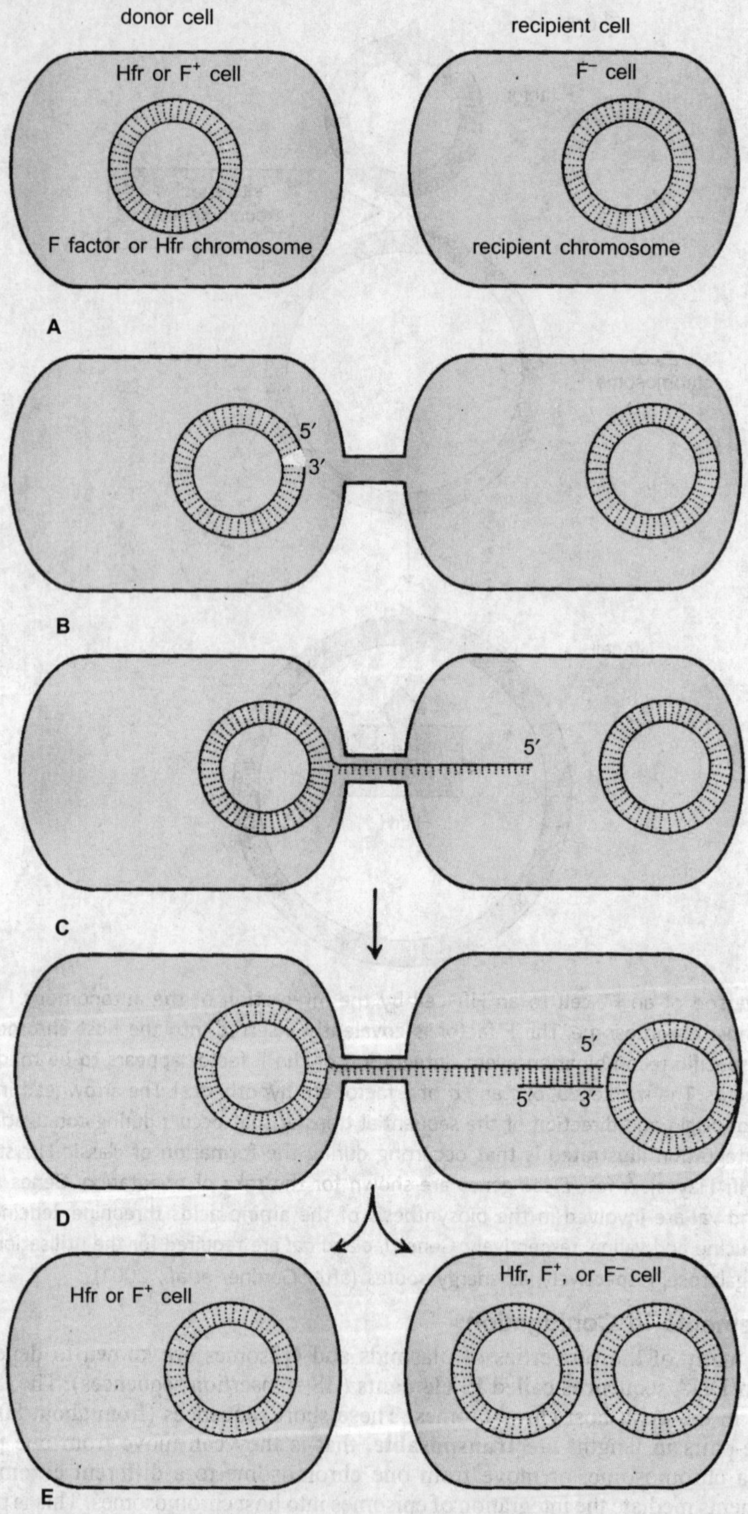

Fig. 17.16. Mechanism of DNA transfer during conjugation of bacteria (*e.g., E. coli*) (after Gardner *et al.,* 2001).

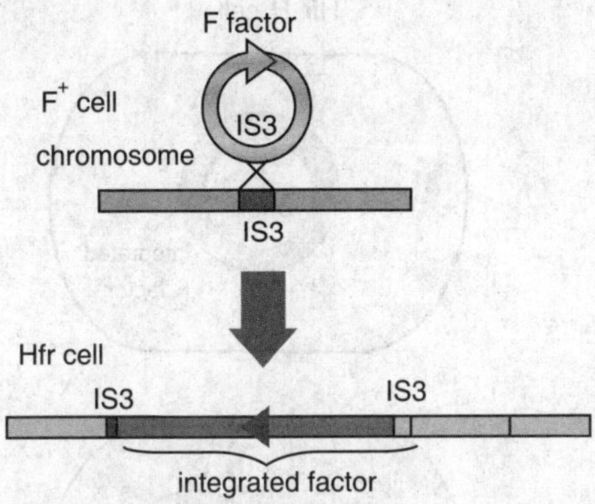

Fig. 17.17. Proposed mechanism of integration of the F factor mediated by the homology of an IS element in the chromosome and in the F factor. Such integration results in the F factor being flanked by identical IS elements when covalently inserted into the chromosome of an Hfr cell (model is based on data of Shapiro *et al.*, 1977; after Gardner *et al.*, 2001).

The first four IS elements _ IS1, IS2, IS3 and IS4 – are sequenced as 768, 1327, 1300 and 1426 nucleotide pairs in length, respectively. Several other IS elements have subsequently been identified, characterized and in some cases, sequenced.

The *E. coli* K12 chromosome apparently contains eight copies of IS1 and five copies of IS2, plus one or more copies of IS3 and IS4. The *E.coli* K12 F facctor contains one copy IS2 and two copies of IS3 elements. The positions of IS elements in the various *E.coli* strains are believed to determine the sites of integration of the F factor during the formation of Hfr strain (Fig. 17.17).

Interrupted Mating

F. Jacob and **E Wollman** (1961) devised the technique of **interrupted mating** for demonstrating that the transfer of genetic material from the donor to the recipient bacterial cell during conjugation is a linear event (Fig. 17.18). In this technique, F$^-$ and Hfr strains were mixed together in a food blender. After waiting a specific amount of time, they turned the blender on. The spinning motion separated conjugating bacterial cells and thereby interrupted their mating (*i.e.,* conjugation). Then these researchers tested the F$^-$ cells for various alleles originally in the Hfr cell. In an experiment like this, the Hfr strain is usually sensitive to an antibiotic such as streptomycin. After *conjugation is interrupted*, the bacterial cells are plated on to a culture medium containing the antibiotic, which kills all the Hfr cells. Then the genotypes of F$^-$ cells can be determined by replica-plating without fear of contamination by Hfr cells.

The matings outlined in Table 17.4 and Fig. 17.19 was carried out. In food bleander, a Hfr strain which was **sensitive** to streptomycin (str^s) but resistant to azide (azi^r) and phage T1 ($ton\ A^r$ or $T1^r$) and **prototrophic** for the amino acid leuncine (leu^+) and sugars galactose (gal^+ or $gal\ B^+$) and lactose (lac^+), was added to an F$^-$ strain that was resistant to streptomycin (str^r), sensitive to azide (azi^s), sensitive to (*phage* T_1 $T1^s$ $tonA^s$) and auxotrophic for leuncine, galactose and lactose (leu^-, $gal\ B^-$ and lac^-). After a specific number of minutes (ranging from zero to sixty), the food blender was turned on. For killing all the Hfr cells, the cell suspension as plated on a medium containing streptomycin. The remaining cells were then plated on a culture medium without leucine. The only colonies that resulted were F$^-$ recombinants. They must have received the leu^+ allele from the Hfr in order to grow on a medium lacking the leucine amino acid. Hence, all colonies had been selected to be F$^-$ recombinants. By replica plating onto specific media, investigators were able to determine the *azi, tonA, lac* and *gal B* alleles and the percentage of recombinant colonies that had the original Hfr allele (leu^+). [Note that by trial and error, it was determined that leucine should be the locus to use to select for recombinants. As one will see that during such experiments leucine locus (leu^+) entered first].

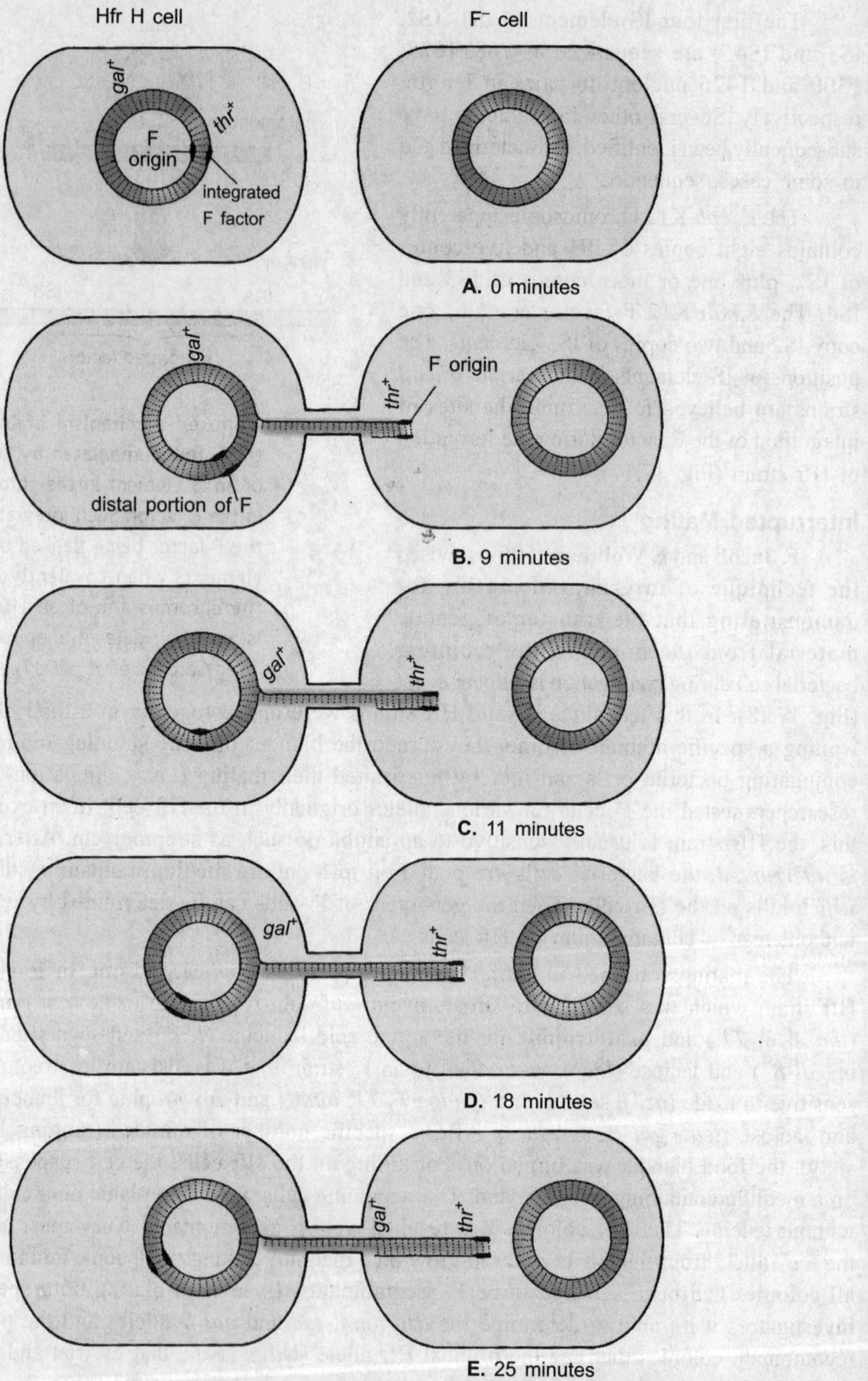

Fig. 17.18. Intepretation of the interrupted mating data of Jacob and Wollman in terms of the linear transfer of Hfr H chromosomes to the F⁻ cells. The genetic markers carried by the Hfr H strains are all dominant or codominant to the alleles carried by the F⁻ strain. The times given indicate the number of minutes since conjugation was initiated (after Gardner *et al.*, 2001).

Table 17.4 Genotypes of Hfr and F⁻ cells used in an interrupted mating experiment (Source: **Tamarin**, 2002).

S. No.	*Hfr*	*F⁻*
1.	str^s	str^r
2.	azi^r	azi^s
3.	$tonA^r$ ($T1^r$)	$TonA^s$ ($T1^s$)
4.	leu^+	leu^-
5.	gal^+ ($gal\ B^+$)	gal^- ($gal\ B^-$)
6.	lac^+	lac^-

As time of mating increases, two things happen (Fig. 17.20). First new alleles enter the F⁻ cells from the Hfr cells. The *ton A^r* allele first appears among recombinants after about ten minutes of mating, where as *gal B⁺* first enters the F⁻ cells after about twenty five minuts. This indicates a sequential entry of loci into the F⁻ cells from the Hfr (Fig. 17.20). Second, as time proceeds, the percentage or recombinants with a given allele from the Hfr increases. After fifteen minutes, about 40% of recombinants have the *tonA^r* allele from the Hfr; and after about twenty-five minutes, about 80% of the recombinants have the *tonA^r* allele. This limited percentage docs not increase with additional time. The limiting percentage is lower for loci that enter later a fact explained by the assumption that even without the food blender, mating is equally interrupted before completion by normal agitation alone.

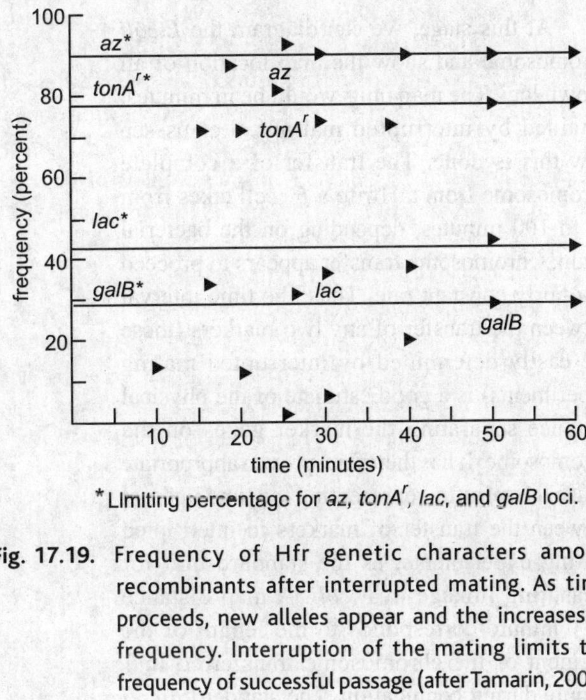

* Limiting percentage for *az*, *tonA^r*, *lac*, and *galB* loci.

Fig. 17.19. Frequency of Hfr genetic characters among recombinants after interrupted mating. As time proceeds, new alleles appear and the increases in frequency. Interruption of the mating limits the frequency of successful passage (after Tamarin, 2002).

Chromosome Mapping by Conjugation

Jacob and **Wollman**, worked with several different Hfr strains and collected data that indicated that the bacterial chromosome was circular. In this chromosome, the relative order of the loci is always the same. What differs is the point of origin and the direction of transfer. Jacob and Wollman proposed that normally the F factor is an independent circular DNA entity in the F⁺ cell, and thus during conjugation only the F factor is passed to the F⁻ cell. Since it is a small fragment of DNA, it can be passed entirely in a high proportion of conjugation before the cells separate. Sometimes, however, the F factor becomes integrated into the chromosome of the host, which then becomes an Hfr cell. Once the F factor is integrated, it determines the initiation point of transfer for the *E.coli* chromosome, as well as the direction of transfer.

During conjugation, the F factor is the last part of the *E.coli* chromosome to be passed from the Hfr cell. This explains why a Hfr cell, in contrast to a F⁺ cell, rarely passes the F factor itself.

In the original work of **Lederberg** and **Tatum**, the one recombinant in 10^7 cells most likely came from a conjugation between F^- cell and Hfr that had formed spontaneously from a F^+ cell. Integration of the F factor is diagrammed in Figure 17.21. The F factor can also reverse this process and loop out of the *E.coli* chromosome. Sometimes the F factor loops out incorrectly (Fig. 17.22), forming a *F′* **(F-prime) factor**. The passage of this *F′* factor to a F^- cell is called, **F-duction** or **sexduction.** Sexduction is not really useful in mapping, however, the process has proved very useful in studies of gene expression.

At this stage, we can diagram the *E.coli* chromosome and show the map location of all known loci. The map units would be in minutes, obtained by interrupted matings. Let us see how this is done. The transfer of a complete chromosome from a Hfr to a F^- cell takes from 90 to 100 minutes, depending on the bacterial strain. Chromosome transfer appears to proceed at a fairly constant rate. Thus, the time interval between the transfer of any two markers (these are easily determined by interrupted mating experiments) is a good estimate of the physical distance separating the marker genes on the chromosome. It has therefore, proven appropriate to use the **minute**, representing the time interval between the transfer of markers in interrupted mating experiments, as the standard unit for measuring linkage in *E.coli*. A map distance of 1 minute corresponds to the length of the segment of the chromosome transferred in 1 minute during conjugation. The standard *E.coli* **linkage map** (Fig. 17.24) is thus divided into minute intervals from O (this is arbitrarily set at the *thrA* gene) to 100 minutes on the basis of interrupted mapping. The linkage map of *E.coli* strain K12 is circular (Fig. 17.23. but it is still partial one, since it is based on data obtained on transformation and interrupted conjugation. Many corrections have been made in this map by data obtained from **transduction** (another sexual process in bacteria which will be described very shortly).

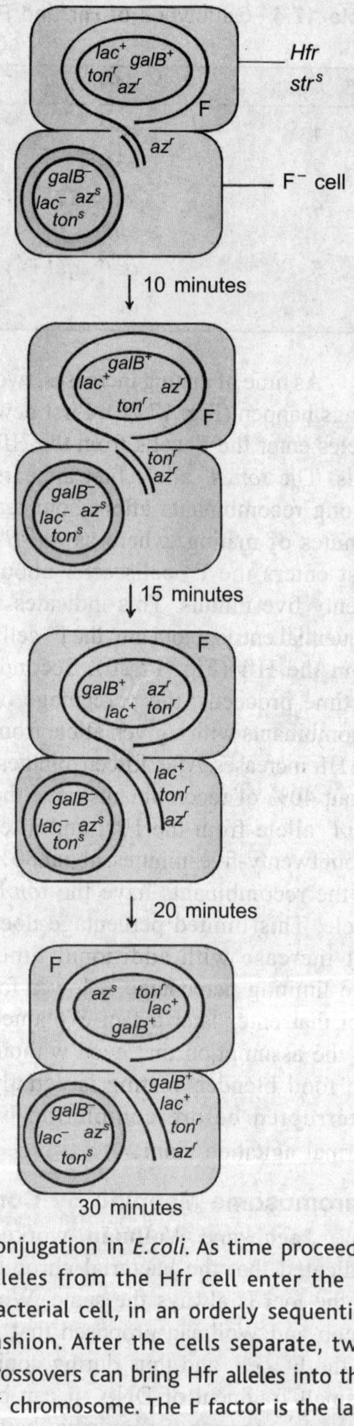

Fig. 17.20. Conjugation in *E.coli*. As time proceeds, alleles from the Hfr cell enter the F^- bacterial cell, in an orderly sequential fashion. After the cells separate, two crossovers can bring Hfr alleles into the F^- chromosome. The F factor is the last part of the Hfr chromosome to enter the F^- cell (after Tamarin, 2002).

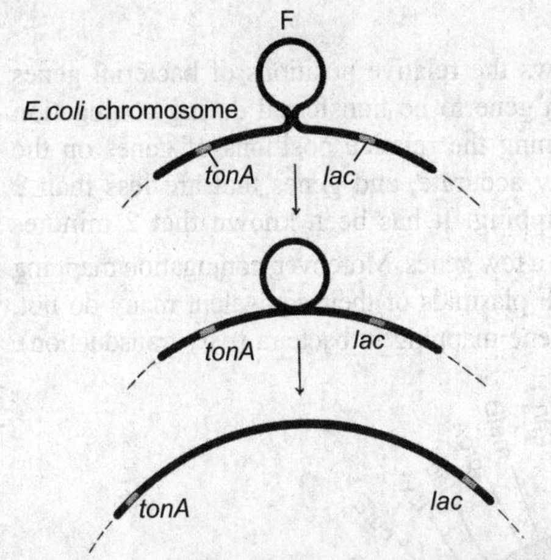

Fig. 17.21. Integration of the F- factor by a single crossover. After a simultaneous breakage in both the F factor and the *E.coli* chromosome, the two broken circles reunite to make one large circle, the Hfr chromosome. In this case, the integration occurs between the *ton A* and *lac* loci (genes) (after Tamarin, 2002).

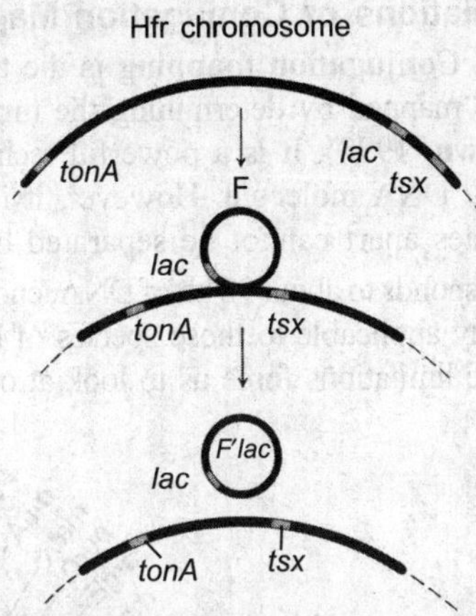

Fig. 17.22. F′ duction or sexduction. Occassionally, the F factor loops out incorrectly, taking part of the cell's genome in the loop. The circular F factor is freed by a single recombination (crossover) at the loop point (after Tamarin, 2002).

Definitions of loci have been included in Table 17.5.

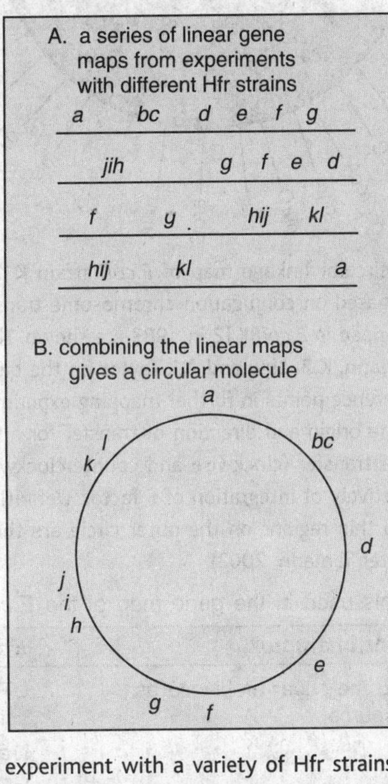

Fig. 17.23. Interrupted mating experiment with a variety of Hfr strains provided a series of linear maps (A) which when joined together, proved that the *E.coli* DNA molecule is circular (B) (after Brown, 1998).

Limitations of Conjugation Mapping

Conjugation mapping is the technique that allows the relative positions of bacterial genes to be mapped by determining the time it takes for each gene to be transferred during conjugation (**Brown,** 1998). It is a powerful technique for determining the relative positions of genes on the *E.coli* DNA molecule. However, it is not tremendously accurate, and genes that are less than 2 minutes apart cannot be separated by conjugation mapping. It has been known that 2 minutes corresponds to about 92 kb of DNA, enough space for quite a few genes. Moreover, conjugation mapping is only applicable to those species of bacteria that have F plasmids or their equivalent many do not. These limitations force us to look at other methods for gene mapping in bacteira (*e.g.,* transduction).

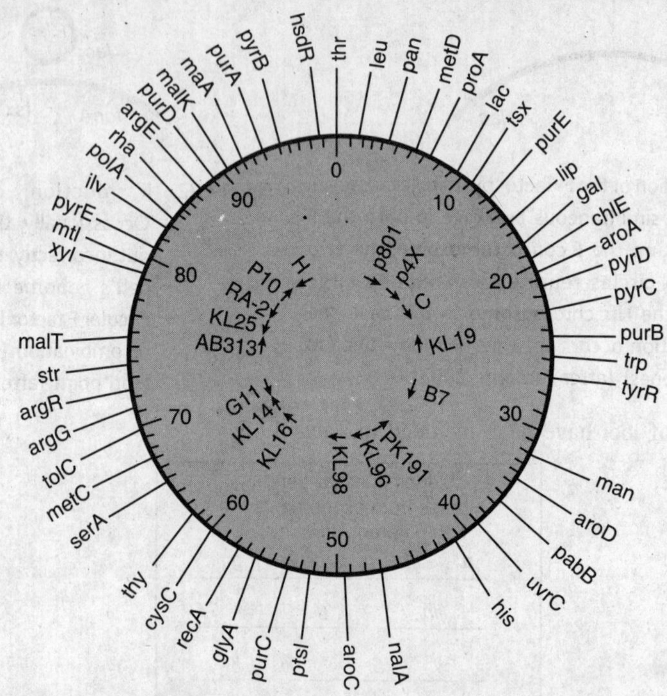

Fig 17.24. Selected loci on the circular linkage map of *E.coli* strain K12. The map is divided into 100 one-minute intervals based on conjugation chromosome transfer times. Only 52 of the 1027 loci that had been mapped in *E.coli* K12 in 1983 are shown. The loci are shown were selected in 1976 by **B.J. Bachmann, K.B. Low** and **A.I. Taylor** on the basis of greatest accuracy of map position, utility as reference points in further mapping experiments. The arrows on the dashed inner circle indicate the origin and direction of transfer for a few well known Hfr strains. The origin and direction of transfer (clockwise and counterclockwise) are determined by the site and orientation respectively of integration of F factor. Definitions of loci have been explained in Table 17.5. The two thin regions on the outer circle are the only areas not covered by P1 transducing phage (after Tamarin, 2002).

Table 17.5 Symbols used in the gene map of the *E.coli* chromosome

Genetic symbols	*Mutant character*	*Enzyme or Reaction affected*
araD	Cannot use the sugar arabinose as a carbon source	L-Ribulose-5-phosphate-4-epimerase
araA		L-arabinose isomerase
araB		L-Ribulokinase
araC		
	Requires the amino acid arginine for growth	N-Acetylglutamate synthetase

Genetic symbols	*Mutant character*	*Enzyme or Reaction affected*
argB		N-Acetyl-γ-glutamokinase
argC		N-Acetylglutamic
argH		γ-semiachdelyde dehydorgenase
argG		Acetylornithine-d-transaminase
argA		Acetylornithinase
argD		Ornithine transcarbamylase
argE		Ornithine transcarbamylase
		Argininosuccinic acid synthetase
argF		Argininosuccinase
argR	Arginine operon regulator	
aroA, B, C	Requires several aromatic amino acids and vitamins for growth	Shikimic acid to 3-Enolpyruvyl-shikimate-5-phosphate
aroD		Biosynthesis of shikimic acid
azi	Resistant to sodium azide	
bio	Requies the vitamin biotin for growth	
carA	Requires uracil and arginine	Carbamate kinase
*car*B		
cblA-E	Cannot reduce chlorate	Nitrate-chlorate reductase and hydrogen lysase
cysA	Requires the amino acid cysteine for growth	3-Phosphoadenosine-5-phosphosulphate to sulphide
cysB		Sulphate to sulphide; four known enzymes
cysC		
dapA	Requires the cell-wall compoent diaminopimelic acid	Dihydrodipicolinic acid synthetase
dapB		N-Succinyl-diaminopimelic acid deacylase
dap + bom	Requires the amino acid precursor homoserine and the cell-wall component diaminopimelic acid for growth	Aspartic semialdehyde dehydrogenase
dnaA-Z	Mutation, DNA replication	DNA biosynthesls
Dsd	Cannot use the amino acid D-serine as a nitrogen source	D-serine deaminase
fla	Flagella are absent	
galA	Cannot use the sugar galactose as a carbon source	Galactokinase
*ga*l*B*		Galactose-1-phosphate uridyl transferase
galD		Uridine-diphosphogalactose-4-epimerase
		Serine hydroxymethyl transferase
glyA	Requires glycine	
gua	Requires the purine guanine for growth	
H	The H antigen is present	
his	Requires the amino acid histidine for growth	Ten known enzymes*
bsdR	Host restriction	Endonuclease R
ile	Require the amino acid isoleucine for growth	Threonine deaminase
ilvA	Requires the amino acids isoleucine and valine for growth	α-Hydroxy-β-keto acid rectoisomerase
ilvB		α, β-Dihydorxyisovaleric dehydrase*
ilvC		Transaminase B
ind (indole)	Cannot grow on tryptophan as a carbon source	Tryptophanase
λ *(att* λ)	Chromosomal location where prophage λ is normally inserted	

Genetic symbols	Mutant character	Enzyme or Reaction affected
lacI	*Lac* operon regulator	
lacY	Unable to concentrate β-galactosides	Galactoside permease
lacZ	Cannot use the sugar lactose as a carbon source	β-Galatosidase
lacO	Constitutive synthesis of lactose operon proteins	Defective operator
leu	Requires the amino acid leucine for growth	Three known enzymes*
lip	Requires lipoate	
lon (long form)	Filament formation and radiation sensitivity are affected	
lys	Requires the amino acid lysine for growth	Diaminopimelic acid decarboxylase
lys + met	Requires the amino acids lysine and methionine for growth	
λ rec, malT	Resistant to phage λ and cannot use the sugar maltose	Regulator for two operons
malK	Cannot use the sugar maltose as a carbon source	Maltose permease
man	Cannot use mannose sugar	Phosphomannose isomerase
melA	Cannot use melibiose sugar	Alpha-galactosidase
met A-M	Requires the amino acid methionine for growth	Ten or more genes
mtl	Cannot use the sugar mannitol as a carbon source	Two enzymes
muc	Forms mucoid colonies	Regulation of capsular polysaccharide synthesis
nalA	Resistant to nalidixic acid	
O	The O antigen is present	
pan	Requires the vitamin pantothenic acid for growth	
pabB	Requires *p*-aminobenzoate	
phe A, B	Requires the amino acid phyenylalanine for growth	
pho	Cannot use phosphate esters	Alkaline phosphatase
pil	Has filaments (pili) attached to the cell wall	
plsB	Deficient phospholipid synthesis	Glycerol 3-phosphate acyltransferase
polA	Repairs deficiencies	DNA polymerase I
proA }	Requires the amino acid proline for growth	
proB }		
proC }		
ptsI	Defective phosphotransferase system	Pts-system enzyme I
purA }	Requires certain purines for growth	Adenylossuccinate synthetase
purB }		Adenylosuccinase
purC, E }		5-Aminoimidazole ribotide (AIR) to 5-aminoimidazole-4-(N-succino carboximide) ribotide
purD }		Biosynthesis of AIR
pyrB }	Requires the pyrimidine uracil for growth	Aspartate transcarbamylase
pyrC }		Dihydroorotase
pyrD }		Dihydroorotic acid dehydrogenase
pyrE }		Orotidylic acid pyrophosphorylase
pyrF }		Orotidylic acid decarboxylase
R gal	Constitutive production of galactose	Repressor for enzymes involved in galactose production
R1 pho, R2 pho	Constitutive synthesis of phosphatase	Alkaline phosphatase repressor
R try	Constitutive synthesis of tryptophan	Repressor for enzymes involved in tryptophan synthesis
RC (RNA control)	Uncontrolled synthesis of RNA	
rec A	Cannot repair DNA radiation damage or recombine	

* Denotes enzymes controlled by the homologous gene loci of *Salmonella typhimurium*.

Genetic Symbols	*Mutant character*	*Enzyme or Reaction affected*
rbaA-D	Cannot use the sugar rhamnose as a carbon source	Isomerase, kinase, aldolase, and regulator
rpoA-D	Problems of transcription	Subunits of RNA polymerase
serA	Requires the amino acid serine for growth	3-Phosphoglycerate dehydrogenase
serB		Phosphoserine phosphatase
str	Resistant to or dependent on streptomycin	
suc	Requires succinic acid	
supB	Suppressed ochre mutations	t-RNA
tonA	Resistant to phages T1 and T5 (mutants called B/1, 5)	T1, T5 receptor sites absent
tonB	Resistant to phages T1 (mutant called B/1)	T1 receptor site absent
T6, colK rec absent	Resistant to phage T6 and colicine K	T6 and colicine receptor sites
T4 rec	Resistant to phage T4 (mutants called B/4)	T4 receptor site absent
tsx	T6 resistance	
thi	Requires the vitamin thiamine for growth	
thlC	Tolerance to colicine E1	
thr	Requires the amino acid threonine for growth	
thy	Requires the pyrimidine thymine for growth	Thymidylate synthetase
trpA	Requires the amino acid tryptophan for growth	Tryptophan synthetase, A protein
trpB		Tryptophan synthetase, B protein
trpC		Indole-3-glycerolphosphate synthetase
trpD		Phosphoribosyl anthranilate transferase
trpE		Anthranilate synthetase
tyrA	Requires the amino acid tyrosine for growth	Chorismate mutase T-prephenate dehydrogenase
tyrR		Regulates three genes
uvrA-E	Resistant to ultraviolet radiation	Ultraviolet-induced lesions in DNA are reactivated
valS	Cannot charge Valyl-tRNA	Valyl-tRNA synthetase
xyl	Cannot use the sugar xylose as carbon source	

17.4 VIRAL GENETICS

A virus possesses all the genetic information necessary for the construction of new viruses but little or none of the apparatus necessary for such construction. It therefore, must infect a living cell and divert the host's biosynthetic apparatus into the business of synthesizing virus progeny. The chromosome of DNA viruses may be either in the form of **single stranded DNA** (*e.g.,* ϕ X 174 bacteriophage) ; or **circular, duplex DNA** (*e.g.,* polyoma virus); or of **rod-shaped double-stranded** DNA when outside its host cell, but soon after infection the linear chromosome closes on itself to form a covalently linked ring with superhelical twists (*e.g.,* lambda (λ) bacteriophage); or of rod-shaped Watson-Crick double helixes (*e.g.,* T-even phages).

Geneticists have taken advantage of the relatively simple constitutions of viruses, their short generation span, various life cycles, and the numbers of progeny available for analysis to study many basic aspects of genetics. Mutation rate, the fine structure (molecular detail) of genes in viruses, and the mapping of genes on the viral nucleic acids all have been studied in great detail. However, the bulk of our knowledge about viruses has come from the study of those which infect bacterial hosts, the bacteriophage.

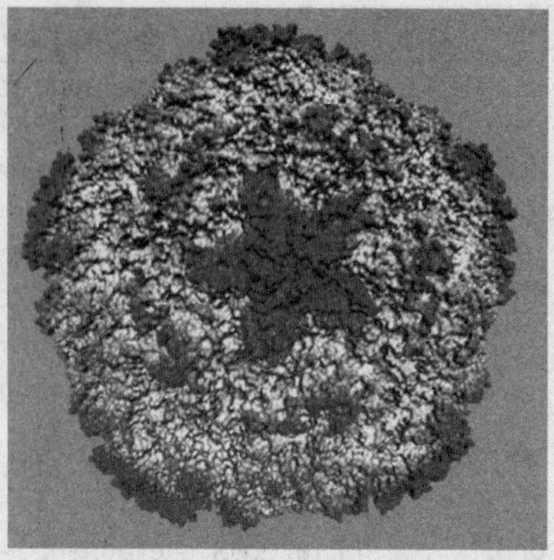

Avian flu virus.

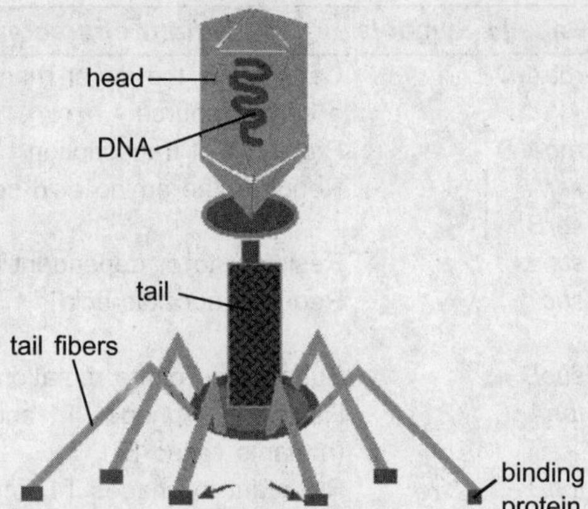

Structure of bacteriophage.

Bacteriophages are choosen for genetic mapping, because, each of them, like other viruses, possesses only one chromosome so that all of its genes appear within a single linkage group. Further, the molecular nature of chromosomes of bacteriophage is well known and each of their chromosome possesses a small number of genes so that the ordering of all or most of the genes on the chromosome can actually be visualized.

Life Cycles of Bacteriophages

Phages are obligate intra-cellular parasites. Phage genetic material enters the bacterial cell after the phage has adsorbed to the cell surface. Once inside, the viral genetic material takes over metabolism of the host cell. During the infection process, the cell's, genetic material is destroyed, while the viral genetic material is replicated many times. The viral genetic material then controls the mass production of various protein components of the virus. New virus particles are assembled within the host cell, which bursts open (is lysed), releasing a **lysate** of hundreds of viral particles to infect other bacteria (Fig. 17.25).

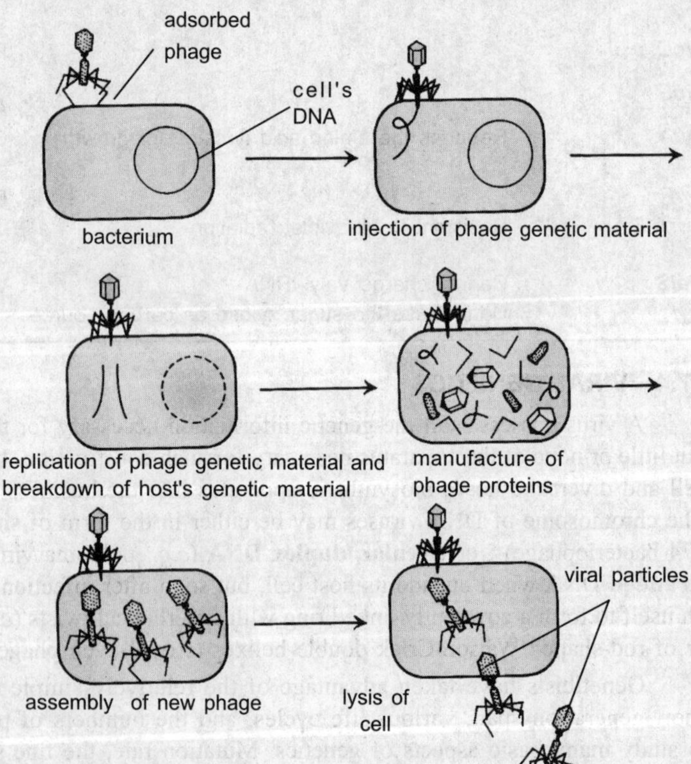

Fig.17.25. The life cycle of T4 phage which infects cell of *E. coli* (after Tamarin, 2002).

Recombination

Much genetic work on the phages has been done with a group of seven bacteriophages of *E. coli,* called the *T series* (T-odd: T1, T3, T5, T7; T-even: T2, T4 and T6) and several others including phage λ (lambda). Phages can undergo recombination processes when a cell is infected with two genetically distinct virons.

mapped by recombination. As an example, we can consider the host-range (*h*) and rapid-lysis (*r*) loci of T2 phage (Box 17.7). Rapid-lysis mutants (*r*) of the T-even phages produce large sharp-edged plaques. The wild-type produces a smaller, more fuzzy-edged plaques (see Fig. 17.6).

Box 17.6

For understanding recombination in phages, **Hershey** and **Rotman** in 1949 infected an *E.coli* culture with two different genetically marked T2 phages $r\ h^+$ and $r^+\ h$ – each at a multiplicity of infection of about five. Roughtly 98% of the progeny had the parental genotypes, but 2% consisted of about equal numbers of $r\ h$ and $r^+\ h^+$ phage. On replating, the $r\ h$ and $r^+\ h^+$ phages bred true, indicating that these genotypes were due to recombination. Furthermore, if *E.coli* was mixedly infected with the recombinants, genetic recombination occurred, again producing 98% $r\ h$ and $r^+\ h^+$ and 2% $r\ h^+$ and $r^+\ h$ phage (the original parental genotype) (see *Maloy et al.*, 1994).

Alternatively alleles are known also for host-range loci, phage loci that determine the strains of bacteria the phage can infect. For example, T2 can infect *E.coli* cells. These phages can be designated as $T2h^+$ for the normal host range. The *E.coli* is then called Tto^s, referring to their sensitivity to the T2 phage. In the course of evolution, an *E.coli* mutant arose that is resistant to the normal phage. This mutant strain is named Tto^r for T2 resistance. In the further course of evolution, the phages have produced mutant forms, that can grow on the Tto^r strain of *E.coli.* These phage mutants are designated as T2h for host-range mutants. Remember, host-range signifies a mutation in the phage genome, whereas phage-resistance indicates a mutation in the baceterial genome.

Max Delbruck (a 1969 Nobel Laureate) in 1945 developed mixed indicators which can be used to demonstrate four phage phenotypes on the same petri dish. A bacterial lawn of mixed Tro^r and Tto^s is grown in the petri dish. On this lawn, the rapid-lysis phage mutant (*r*) produce large plaques, whereas the wild-type (r^+) produce smaller plaques. Phages with host-range mutation (*h*) lyse both Tto^r and Tto^s bacteria. They produce the plaques that are clear (but appear dark). Since phages with the wild-type host-range allele (h^+) can only infect the Tto^s bacteria, they produce turbid plaques. The Tto^s bacteria growing within these plaques produce the turbidity (Fig. 17.26).

From the wild stock of phages, we can isolate host-range mutants by looking for plaques on a Tro^r bacterial lawn. Only *h* mutants will grow. These phages can then be tested for the *r* phenotype and the double mutants isolated. Once the two strains (double mutants and wild type) are available, they can be added in large numbers to sensitive bacteria (Fig. 17.26). Large numbers of phages are used to ensure that each bacterium is infected by at least one of each phage type, creating the possibility of recombination within the host bacterium. After a round of phage multiplication the phages are isolated and plated out on Delbruck's mixed-indicator stock. From this growth, the phenotype (and, hence the genotype) of each phage can be recorded. The percentage of recombinants can be read directly from the petri dish/plate. For example, on a given petri plate there might be:

$$h\ r = 46\ ;\ h^+\ r^+ = 52$$

$$h^+\ r = 34\ ;\ h\ r^+ = 26$$

The first two, $h\ r$ and $h^+\ r^+$, are the original, or parental, phage genotypes. The second two categories result from recombination between the *h* and *r* loci on the phage chromosome. A single crossover in this region produces the recombinants. Note that with phage recombination, parental phages are counted, since every opportunity was provided for recombination within each bacterium. Thus, every progeny phage arises from a situation which recombination could have taken place. The proportion of recombinants is:

$$\text{Recombination frequency} = \frac{\text{Number of recombinant phage}}{\text{Total number of phage}} \times 100\%$$

$$= (34 + 26) / (46 + 52 + 34 + 26)$$

$$= 60 / 158 = 0.38 \text{ or } 38\% \text{ map units}$$

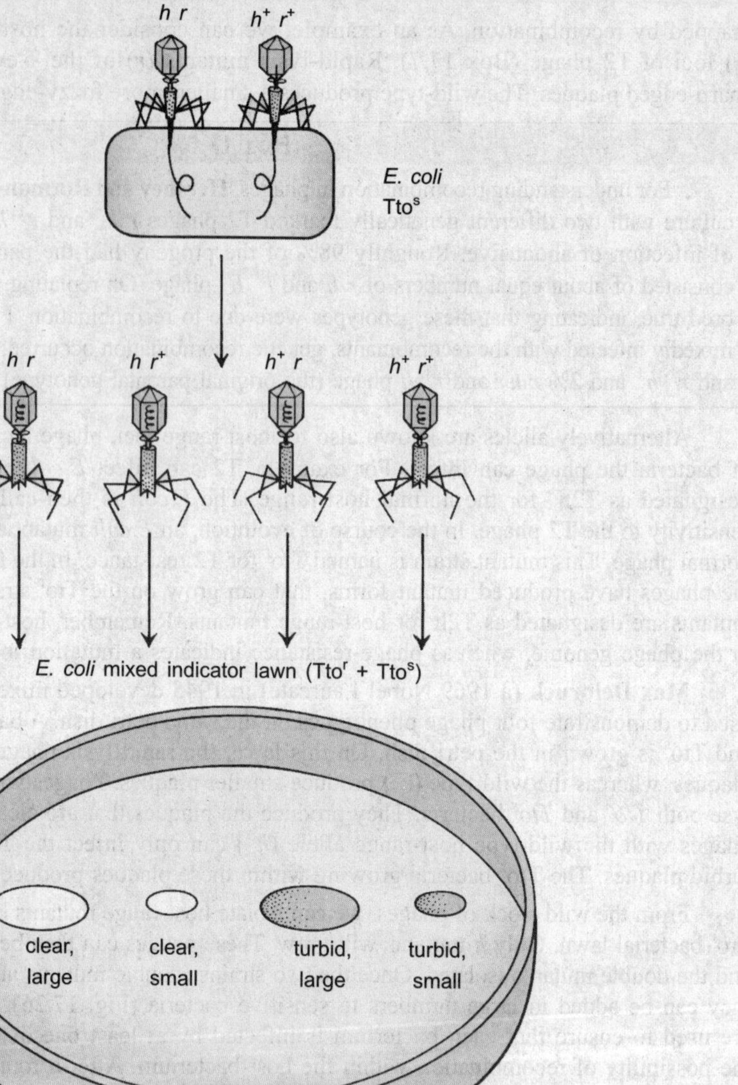

Fig. 17.26. Crossing *h r* and $h^+ r^+$ phages. Enough of both types of phages are added to sensitive bacterial cells (Tto^s) to ensure multiple infections. The lysate, consisting of four genotypes, is grown on a mixed-indicator bacterial lawn (Tto^s and Tto^r). Plaques of four types appear indicating the genotypes of the parental and recombinant phages (after Tamarin, 2002).

This percentage recombination is the map distance, which is a relative index of distance between loci : the greater the physical distance, the greater the amount of recombination, and thus the larger, the map distance. One map unit (one centimorgan) is equal to 1% recombinat offspring.

Lysogeny

Certain phages are capable of two different life-cycle stages (Fig. 17.27). Some of the time, they replicate in the host cytoplasm and destroy the host cell. At other time, these phages are capable of surviving in the host cell. The host is then referred to as **lysogenic** and the phage as **temperate.** (The term lysogeny means giving birth of lysis).

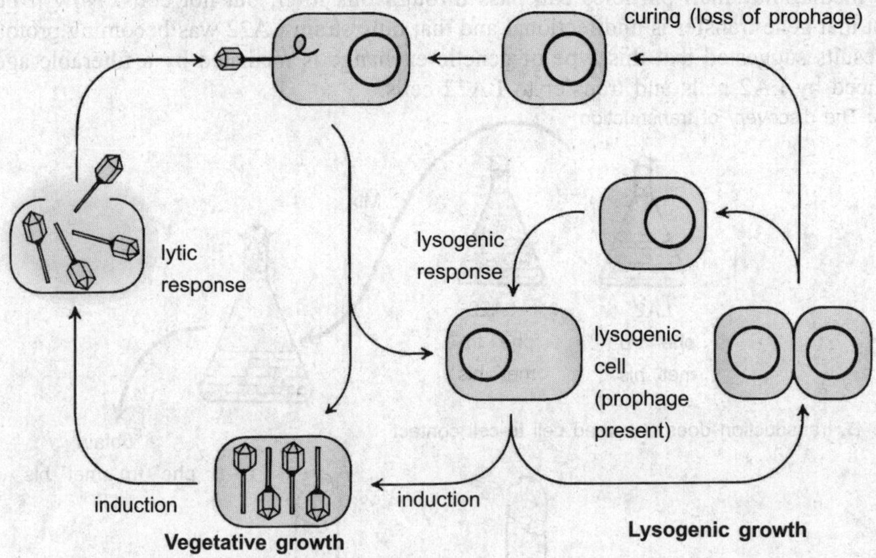

Fig. 17.27. Alternative life-cycle stages of a temperate phage (lysogenic and vegetative growth (after Tamarin, 2002).

Most of the research on lysogeny has been done on phage lambda (λ). The λ prophage integrates into the host chromosome; other prophages such as P1, exist as independent plasmids. Phage λ, unlike the F factor, attaches at a specific point, termed *att* λ. This locus can be mapped on the *E.coli* chromosome; it lies between the galactose (*gal*) and biotin (*bio*) loci. When the phage is integrated, it protects the host from further infection (*i.e.,* superinfection) by other λ phages. The integrated phage is now termed a **prophage.** Presumably it becomes integrated by a single crossover between itself and the host after apposition at the *att* λ *site*. This step resembles the F-factor integration.

A prophage can enter the lytic cycle of vegetative growth by a process of **induction,** which involves the excision of the prophage followed by the virulent or lytic stage of the viral life cycle. Induction can take place through a variety of mechanisms, including UV-irradiation and passage of the integrated prophage during conjugation (*i.e.,* zygotic induction).

Transduction

Transduction is a phenomenon in which bacterial DNA is transferred from one bacterial cell to another by a phage particle. Phage particles that contain bacterial DNA are called **transducting particles** (**Maloy** *et al.,* 1994).

Conjugation and transduction are based on totally different mechanism for gene transfer from donor bacterium to recipient bacterium. Instead of physical contact between cells, with transduction gene transfer is effected by a bacteriophage.

The discovery of transduction. The process of transduction was discovered by **Joshua Lederberg** and **Norton Zinder** in 1952, not with *E.coli* but with related enterobacterium *Salmonella typhimurium.* They mixed together two auxotrophic strains of this organism, one called **LA22** (having genotype phe^- trp^- met^+ his^+), and one called **LA2** (phe^+ trp^+ met^- his^-), about one 1 lac (*i.e.,* 100,000) cells. We are already aware that prototrophs can arise only by way of gene transfer between bacteria (Fig. 17.28).

Initially it was thought that these results must be due to a conjugative process similar to that characterized in *E.coli*, but further examination suggested that this was not the case. In particular a modification of the basic experiment proved that cell to cell contact is not needed for gene transfer between LA2 and LA22 : prototrops were still obtained when individual cultures were placed in opposite arms of a U-tube, separated from each other by a sinistered glass filter (Fig. 17.28B).

Culture media and small particles can pass through this filter, but not cells. Now it becomes apparent that gene transfer is undirectional and that only strain LA22 was becoming prototrophic. These results suggested that this type of genetic exchange is mediated by a filterable agent that is produced by LA2 cells and transfer to LA22 cells.

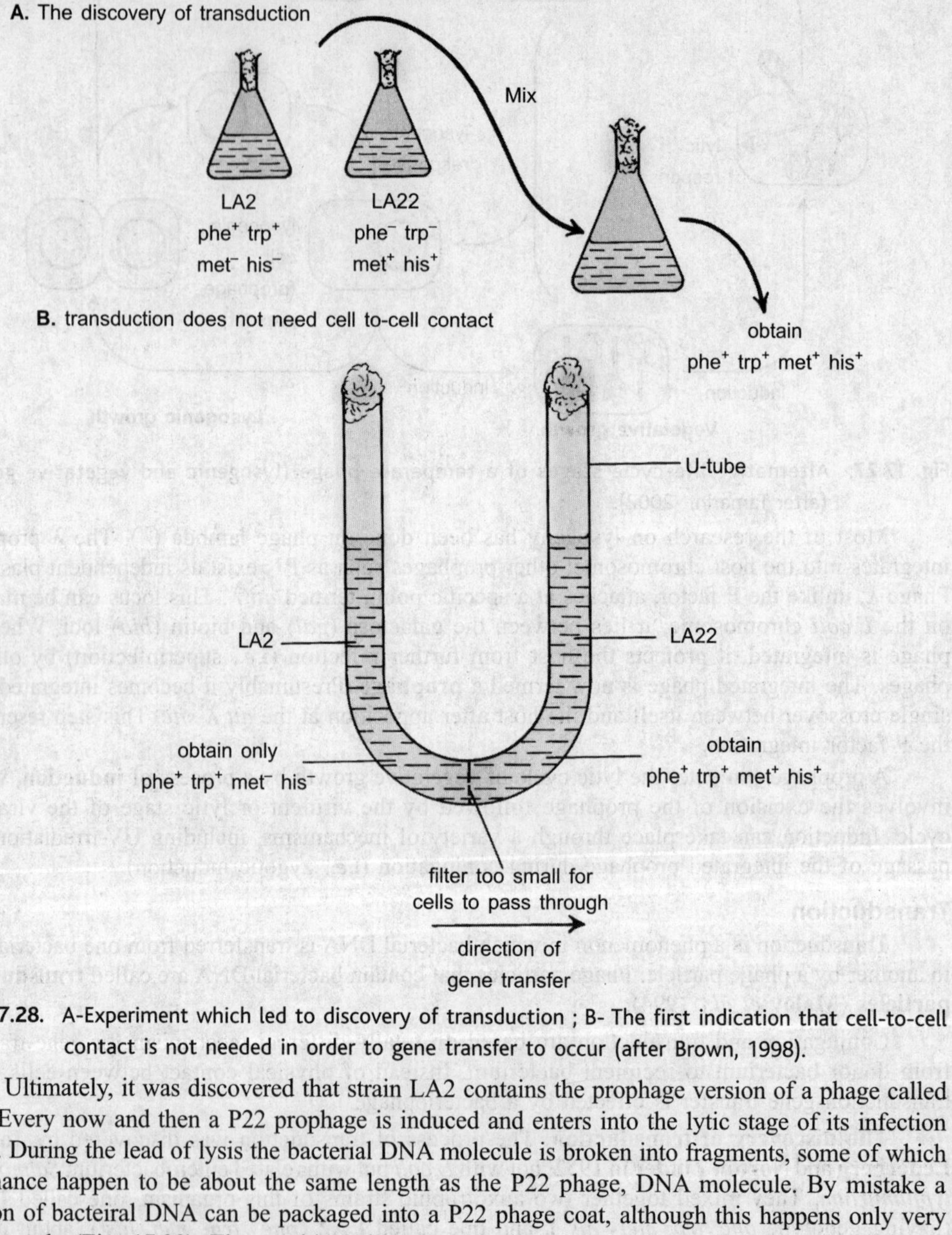

Fig. 17.28. A-Experiment which led to discovery of transduction ; B- The first indication that cell-to-cell contact is not needed in order to gene transfer to occur (after Brown, 1998).

Ultimately, it was discovered that strain LA2 contains the prophage version of a phage called P22. Every now and then a P22 prophage is induced and enters into the lytic stage of its infection cycle. During the lead of lysis the bacterial DNA molecule is broken into fragments, some of which by chance happen to be about the same length as the P22 phage, DNA molecule. By mistake a portion of bacteiral DNA can be packaged into a P22 phage coat, although this happens only very infrequently (Fig. 17.28). The resulting phage particle is still infective, as infection is a function purely of the protein components of the phage coat but the infection is *abortive* because of course the bacterial DNA fragment injected into a new cell cannot itself direct synthesis of new phages.

The bacterial DNA fragments transferred into a new cell can, however, recombine with the new cell's DNA. This means that gene transfer can occur : the donor cell is the one that was lysed by the phage, and the recipient cell is the one that is infected by the transducing phage particles.

Types of transduction. Transduction occurs when a bacteriophage particle carries a segment of the chromosome from one bacterium (the donor) to another bacterium (the recipient), facilitating subsequent recombination of the genetic markers of the two cells. Transduction is of following two types.

1. Generalised transduction. In this case, a random segment of bacterial DNA is "wrapped up" during phage maturation in place of, or along with, the phage chromosome in a few "progeny" particles, called **transducing particles.** Thus, *generalized transducing phages* can transports any gene of donor cell to the recipient cell. Since all the genes of the donor are represented in a population of these transducing particles, this type of transduction was named, '*generalized*' transduction. In some cases, generalized transducing particles contain only bacterial DNA. In other cases, they contain both phage and bacterial DNA. Any one transducing phage tends to contain only one segment of host DNA, representing 1/100 to 1/50 of the total donor chromosome.

2. Restricted transduction or specialized transduction. In this type of transduction, a recombination event involving the host chromosome and the phage chromosome occurs, producing a phage \chromosome containing a segment of bactereial DNA. Bacteriophage lambda, the best-known specialized transducing phage, for example, usually mediates transduction of only the *gal* (required for the utilization of galactose as an energy source) and *bio* genes (genes essential for biosynthesis biotin) of *E.coli.*

Mechanism of Generalized Transduction

Generalized transduction is mediated by some virulent bacteriophages and by certain temperate bacteriophages whose chromosomes are not integrated at specified attachment sites on the host chromosome. Generalized transducing particles are produced during the lytic cycles of these phages.

Further, not all virulent bacteriophages mediate transduction. The T-even pahges such as T2, T4 and T6, for example, degrade the host DNA and reutilize the mononucleotides produced for the synthesis of phage DNA. The host DNA is thus, not available for packaging in progeny particles. Other viruses may not degrade the host DNA at all, and, since the host chromosome is too large to be packaged intact, they would not be able to form transducing particles. In still other viruses, the maturation process may be highly specific for phage DNA, excluding the packaging the fragments of host DNA. In any case, only a limited number of the virulent phages known to mediate transduction. Of the generalized transducing phages. *E.coli* phage P1, *Salmonella* phage P22 and *Bacillus subtilis* phages PBS1 and SP10 have been extensively used for genetic fine structure mapping (*i.e.,* mapping mutant sites within individual genes or short segments of bacterial chromosome).

After a transducing phage injects a fragment of host DNA into a recipient cell, that DNA may either (1) be integrated into host chromosome in a manner similar to the integration of transforming DNA, except that the integrated segment is double stranded, or (2) remain free in the cytoplasm. If it is not integrated, it will not replicate and will be transmitted to only one progeny cell during each cell division. The genes located on the zduced chromosome fragments may be expressed even if they are not integrated. Bacterial cells carrying nonintegrated transducing fragments are called **abortive transductants.** They are prtially diploids and and can be used to carry out complementation tests. The **complementary test** provides the operational definition of the gene. It is used to determine whether different mutations are in the same gene or in different genes.

For P22 phage, the rate of transduction is about once for every 10^5 infecting phages. Since a transducing phage can carry only 2 to 2.5% of the host chromosome, only genes very close to each other can he transduced together (this is called **cotransduction**). Cotransduction can, thus, help to fill in the details of gene order over short distances after interrupted mating or transformation is used to ascertain the general pattern. Thus, transduction (*i.e.,* cotransduction) like the transformation (*i.e.,* co-occurence) is a relative indicator of map distance.

Mechanism of Specialized Transduction

Specialized transduction is mediated by temperate bacteriophages whose chromosomes are able to integrate at once or a few specified attachment sites on the host chromosome.

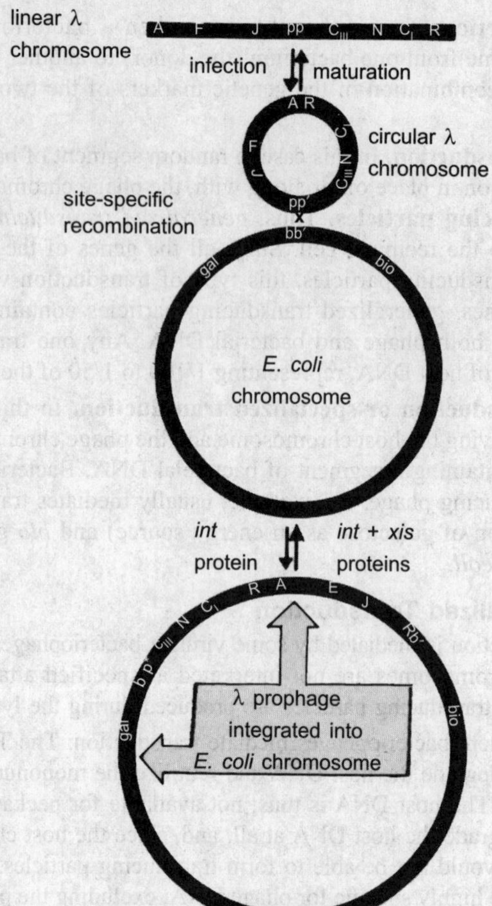

Fig. 17.29. Integration and excision of the phage lambda chromosome. When the lambda chromosome is packaged inside the phage head, it exists as a linear molecule with gene sequence *A* through *R,* as shown at the top. After infection, it is converted to its circular intracellular form. When lambda enters the lysogenic pathway, it undergoes a site- specific recombination event between the *pp'* site on the phage chromosome and the lambda attachment site, *bb'*, on the *E. coli* chromosome. This integrative recombination step is catalyzed by the lambda *int* gene-product. In the prophage state, the lambda gene sequence is *p'-C III-R-A-J-p.* This sequence is a circular trasformation of the gene sequence in the linear form packaged in the lambda head, which results from a specific endonuclease cut between genes *A* and *R* during phage maturation. During spontaneous or UV induced excision, essentially the reverse of integration occurs. We can visualize this as the lambda prophage "looping out" to form a "figure 8" structure; this step is followed by a site specific recombination event between the *bp'* sequence and *pb'* sequence. The site-specific excision step requires the product of the *int* gene plus the product of another lambda gene called *xis.* Normal excision products include an intact circular lambda chromosome plus an intact circular *E. coli* chromosome. **Landy** and **Ross** have sequenced both the lambda chromosome attachment site (*pp'*) and the *E. coli* chromosome lambda attachment site (*bb'*), as well as the lambda prophage *E. coli* DNA junctions (*bp'* and *pb'*) and found that all four contain the identical 15 nucleotide-pair sequence

–G C T T T T T T A T A C T A A–

–C G A A A A A A T A T G A T T–

For this to be true, the site–specific recombination events involved in the integration and excision must occur within or at one end of these 15 nucleotide- pair sequences (after Gardner *et al.,* 2001).

The chromosomes of temperate phages of this type are thus capable of both (1) **autonomous replication** (*i.e.*, replication independent of the replication of the host chromosome) and (2) **integrated replication** (*i.e.*, replication as a segment of host chromosome).

Integration of the chromosome of a speciatized transducing phage, such as the coliphage lambda, involves a recombination event between the circular intracellular form of the phage chromosome and the circular bacterial chromosome at **specific attachment sites** on the two chromosomes. Such *site-specific* recombination event results in the covalent linear insertion of the phage chromosome into the chromosome of the bacterium (Fig. 17.29). In its integrated state, the phage chromosome is called a **prophage.** The lytic genes of virus, those involved in viral reproduction and lysis of the host, are **repressed** (turned off) when the chromosome is in the prophage state. A bacterium harbouring a prophage is said to be **lysogenic;** the prophage–host relationship is called **lysogeny.** A lysogenic cell is **immune** to secondary infections by the same virus (homologous to prophage), because the lytic genes of the infecting virus will be repressed just as those of the prophage are repressed.

Temperate phages undergo rare (about one in 10^5 cell divisions) spontaneous transitions from the lysogenic or prophage state to the lytic state. Such transitions can also be induced, for example, by irradiation with ultraviolet light. During the switch from the lytic growth (Box 17.8), the prophage is **excised** from the host chromosome (Fig. 17.29) and starts replicating autonomously. The **excision** process is **site-specific** integration and excision processes are catalyzed by enzymes encoded by phage genes. Since, only loci adjacent to the phage attachment site can be transduced in this process, specialized transduction has not proven very useful for mapping the host chromosome (see **Tamarin,** 2002).

Box 17.7

The "genetic switch" from the lysogenic to the lytic cycle (and vice versa) is very useful to the phage, whose purpose is to reproduce as many offspring as possible. When the host cell of infecting bacteriophage is growing slowly and is low on energy, the phage becomes lysogenic Then, when the host's health is restored to a level that allows maximal phage reproduction, then the lytic cycle proceeds. In the laboratory, ultraviolet radiation is a potent inducer of the switch to the lytic cycle (**Purves** *et al.*, 1998).

Mapping by Transduction

Transduction is used to establish **gene order** and **map distance** in bacterial chromosomes. Gene order is established by **two-factor transduction.** For example if gene *A* is cotransduced with gene *B* and gene *B* with gene *C,* but *A* is never cotransduced with gene *C,* we have established the gene order *A B C* (Table 17.6). **Three-factor transduction** is found to be more valuable than two-factor transduction. For example, if genes *A, B* and *C* are usually contransduced, we can find the order and relative distances by taking advantage of the rarity of multiple crossovers. Let us use the prototroph (A^+ B^+ C^+) strain of bacteria to make transducing phages that then infect the A^- B^- C^- strain of bacteria.

Table 17.6 Gene order established by two-factor contransduction. An $A^+B^+C^+$ strain of bacteria is infected with phage. The lysate is used to infect an $A^-B^-C^-$ strain. The transductants are scored for the wild-type alleles they contain. These data include only those bacteria transduced for two or more of the loci. Since *AB* cotransductants and *BC* cotransductants occur but no *AC* types, we can infer the *ABC* order.

S. No.	*Transductants*	*Number*
1.	A^+ B^+	30
2.	A^+ C^+	0
3.	B^+ C^+	25
4.	A^+ B^+ C^+	0

For detecting bacterial cells that have been transduced for one, two or three of the loci, we need to eliminate the nontransduced cells. In other words, after transduction, there will be $A^- B^- C^-$ cells in which no transduction event has taken place. There will also be seven classes of bacteria that have been transduced for one, two, or all three loci:

1. $A^+ B^+ C^+$
2. $A^+ B^+ C^-$
3. $A^+ B^- C^-$
4. $A^+ B^- C^-$
5. $A^- B^+ C^+$
6. $A^- B^+ C^-$
7. $A^- B^- C^+$

The simplest way to select for transduced bacteria is to select bacteria in which the wild-type has replaced at least one of the loci. For example, *if*, after transduction, we grow the bacteria in minimal media with the requirements of B^- and C^- added, all the bacteria that are A^+ will grow. Remember that without the requirement of A^- bacteria, no A^- bacteria will grow. Hence, although we lose the $A^- B^+ C^+$, $A^- B^+ C^-$ and $A^- B^- C^+$ categories, we also lose the $A^- B^- C^-$, untransduced bacteria. In this example, the *A* locus is the selected locus : we must keep in mind that we have an incomplete, although informative data set. Replica-plating allows us to determine genotypes at the *B* and *C* loci for A^+ bacteria.

In this example, colonies that grow on complete medium without the requirement of the *A* mutant are replica-plated onto complete medium without the requirement of the *B* mutant and then on to complete medium without the requirement of the *C* mutant. In this way, each transductant can be scored for the other two loci (Table 17.7). Now let us take all these selected transductants in which the A^+ allele was incorporated. These can be of four categories: $A^+ B^+ C^+$, $A^+ B^+ C^-$, $A^+ B^- C^+$ *and* $A^+ B^- C^-$.

Table 17.7 Method of scoring three-factor transductants (Source: **Tamarin**, 2002).

		Minimal Medium		
Colony Number	***Colony A Requirements***	***Without B requirements***	***Without C requirements***	***Genotype***
1.	+	+	–	$A^+ B^+ C^-$
2.	+	–	–	$A^+ B^- C^-$
3.	+	–	–	$A^+ B^- C^-$
4.	+	+	+	$A^+ B^+ C^+$
5.	+	–	+	$A^+ B^- C^+$
6.	...	...	...	...
7.	...	...	...	...
8.	...	...	...	...

We can now compare the relative numbers of each of these four categories. The rarest category will be caused by the event that brings in the outer two markers, but not the central one, because this event requires four crossover (Fig. 17.30). Thus, by looking at the number of transductants in various categories, we can determine that the gene order is *A B C* (Table 17.7) since the $A^+ B^- C^+$ category is rarest.

The Table 17.8 also includes calcutations of the relative cotransduction frequencies. Remember that in all organisms and viruses, the higher the frequency of co-occurrence of transduction between the alleles of two loci, the closer those loci are on the chromosome. We usually measure the separation of loci by corssing over between them. The closer together, the lower those crossing over values are and, hence, the smaller the measure of map units apart. Here, as with transformation, we are measuring the co-occurrence directly; therefore, the measure-cotransductance- is the inverse of map distance. In other words the greater the cotransduction rate, the closer the two loci are; the more frequently two loci are transduced together, the closer they are and the higher the cotransduction value will be.

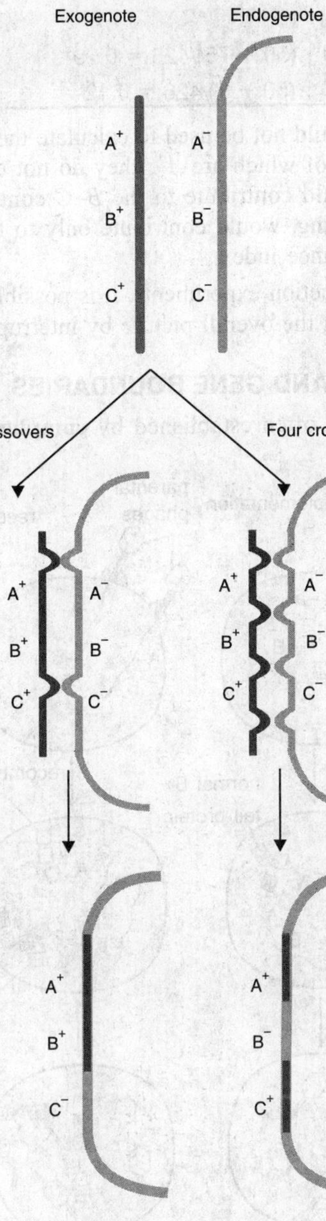

Fig. 17.30. The rarest transduction requires four crossovers.

Table 17.8. Number of transductant and relative cotransduction frequencies in the experiment used to determine the *A B C* gene order (see Table 17.7).

S. No.	*Class*	*Number*
1.	$A^+ B^+ C^+$	50
2.	$A^+ B^+ C^-$	75
3.	$A^+ B^- C^+$	1
4.	$A^+ B^- C^-$	300
	Total	426

Relative contransductance

$$A - B : (50 + 75)/426 = 0.29$$
$$A - C : (50 + 1)/426 = 0.12$$

The data in Table 17.8 should not be used to calculate the *B–C* cotransduction rate because the data are selected values; all of which are A^+; they do not encompass the total data. Missing is the $A^- B^+ C^+$ group that would contribute to the *B–C* contransductance rate. The $A^- B^+ C^-$ and $A^- B^- C^+$ groups, also missing, would contribute only to the totals in the denominator, not in numerator of the cotransductance index.

From these sorts of transduction experiments, it is possible to round out the details of map relations in *E.coli* after obtaining the overall picture by interrupted mating (see **Tamarin,** 2002).

17.5 COMPLEMENTATION AND GENE BOUNDARIES

Gene boundaries are most often established by **complementation test.** The principle of complementation is as follows:

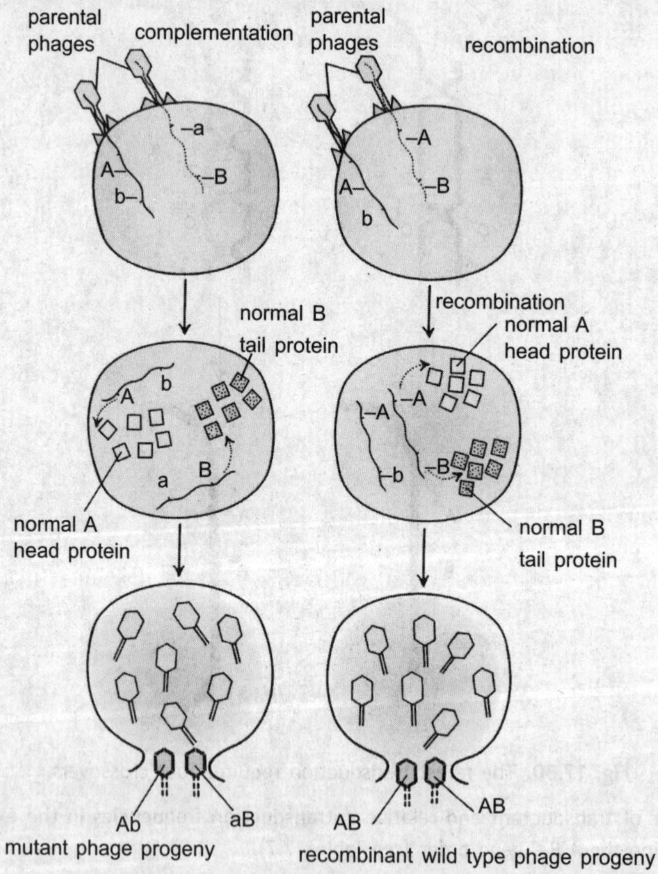

Fig. 17.31. Salient differences between complementation and recombination as illustrated for two mutant phage strains. One strain, *A, b*, carriers a normal gene for the *A* head proteins but a defective tail protein gene *b*. The other strain, *aB*, carries a normal *B* tail protein gene but an abnormal head protein gene *a*. During *complementation*, normal *A* and *B* proteins are synthesized under the direction of the complementing chromosomes, and normally constructed phages emerge that are mutant (*Ab* and *aB*) in genotype. During recombination, normal proteins also synthesized and normally constructed phages emerge, this time carrying normal (*AB*) genotypes. It should be clear that the phenomenon of complementation will operate in a cell where recombinant chromosomes have also arisen (after Goodenough and Levine, 1974).

If two homologous chromosomes carry mutations with different genes–let one chromosome is mutant in gene P and the other in gene Q–and the two chromosomes come to reside in the same cell, one chromosome will direct the synthesis of a mutant gene product but a normal Q product and the other will direct the synthesis of a normal P but a mutant Q product. That is, the cell will come to contain both normal P and normal Q products and thus normal function can occur. In such a case two mutant chromosomes are said to **complement** each other, and it is concluded that the *two mutations lie within the boundaries of two different genes,* say P, then neither chromosome will be able to direct the synthesis of a normal P product and even when both chromosomes come to reside in the same cell normal P function will not be observed. The two mutant chromosomes in this case do not complement each other, and it can be inferred that *both mutations lie within the boundaries of the same gene.*

Comparison of complementation and recombination – Complementation and recombination are two quite different processes, as summarized below.

1. Complementation involves the interaction of gene products (usually proteins) to produce a normal phenotype, whereas wild-type recom-binants produce a normal phenotype because a normal sequence of genes has actually been created along a chromosome via genetic exchange.

2. The ability to generate a normal phenotype is genetically transmitted from parent to offspring when the chromosomes arise by recombination, but progeny produced as a result of complementation remain individually defective in genotype.

3. Recombination can occur within a gene, whereas complementation can ordinarily occur only between one gene and another.

Complementation test. The actual design of a complementation test depends on organism and on the kinds of mutations being studied. For *amber* mutant strains of a λ bacteriophage, one approach is to mixedly infect a suspension of **nonpermissive** bacteria (bacterial strains that cannot suppress nonsense mutations) with equal multiplicities of two mutant phage strains, allow sufficient time for a single lytic cycle to occur, and count the total number of phages produced by plating the lysate onto a lawn of **permissive** bacteria (bacterial strains which can suppress *amber* nonsense codons via mutant forms of tRNA). When such experiments are performed with *sus* mutant strains of λ it is found that if the two strains can complement each other, 100 to 250 phages (a normal burst) will emerge per infected cell, whereas if the two strains are noncomplementary then fewer that 0.1 phage will emerge per infected cell. In other words, for every cell that produces one phage nine produce none at all, or more realistically, for every cell that release a normal burst of at least 100 phage progeny, 99 cells produce no phage progeny.

QUESTIONS

Long Answer Questions

1. What are the differences between a heterotroph and an auxotroph? A minimal and a complete medium? An enriched and a selective medium?
2. Describe the experiment, which originally demonstrated sexuality in bacteria. Add a brief note on the necessity of conjugation.
3. Describe the role of sex factor in bacterial conjugation with particular reference to F^+ and Hfr strains. How do the different Hfr strains differ?
4. Describe the life cycle of a bacteriophage following it from infection to lysis. Discuss differences between temperate and virulent phages in this connection.
5. Describe the U-tube experiment of Zinder and Lederberg to demonstrate transfer of genetic material due to transduction.

6. Describe the different technique used for preparation of linkage maps in bacteria and viruses. How do these techniques differ from those used in eukaryotic system.
7. An *E.coli* lawn is formed on a petri dish/plate containing complete medium. Replica-plating is used to transfer material to plates containing minimal medium and combinations of the amino acids arginine and histidine (Fig. 17.32). Give the genotype of the original strain of bacteria as well as genotypes of the odd colonies found growing on the plates.
8. What genotypic notation indicates alleles that make a bacterium
   (*i*) Resistance to penicillin?
   (*ii*) Sensitive to azide?
   (*iii*) Require histidine for growth?
   (*iv*) Unable to grow on galactose?
   (*v*) Susceptible to phage T1 infection?

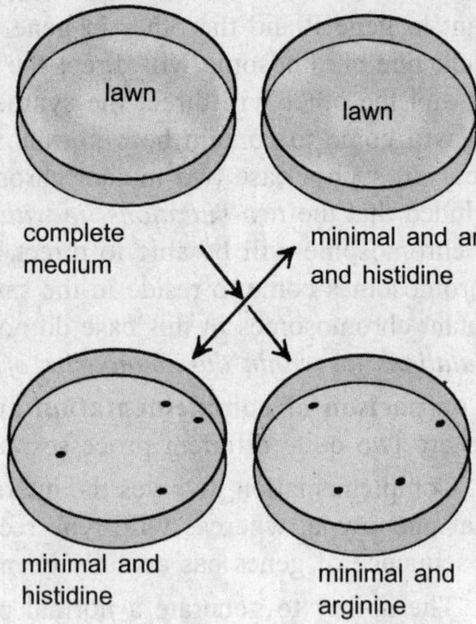

Fig. 17.32.

Short Answer Questions

1. Describe the technique of bacterial chromosome mapping by conjugation.
2. Write short notes on the following:
   (*i*) Transformation in bacteria.
   (*ii*) Conjugation in bacteria.
   (*iii*) Lysogeny.
   (*iv*) Transduction.

Very Short Answer Questions

1. What are differences between a plaque and a colony?
2. An *E.coli* cell is placed on a petri dish/plates containing lambda (λ) phages. It produces a colony overnight. By what mechanism might it have survived?
3. (*a*) What are genotypic differences between F^- cells, F^+ cells, and Hfr cells?
   (*b*) What are the phenotypic differences ?
   (*c*) By what mechanism are F^- cells converted to F^+ cells?
4. (*a*) Of what use are F′ factor in genetic analysis?
   (*b*) How are F′ factors formed?
   (*c*) By what mechanism does sex duction occur?
5. What are the basic differences between generalized transduction and specialized transduction?
6. (*a*) What are IS elements?
   (*b*) What role(s) do IS elements play in recombination?
   (*c*) What role(s) do they play in the evolution of genetic elements.
7. How can genes be mapped by interrupted mating conjugation experiments?
8. What does the term cotransduction means? How can cotransduction frequencies be used to map genetic markers?

Fill in the Blanks

1. Wild type strains of bacteria are able to grow on minimal culture medium. They are called ....

2. Mutant strains of bacteria which require supplementation in their culture media for their growth are called ....

3. The antibiotic .... inhibits the binding of aminoacyl tRNA to the ribosome and thereby blocks continued translation.

4. The phenomenon of transformation in bacteria was examined at molecular level by .... and .... in 1944.

5. Bacteriophages are .... intracellular parasites of ....

6. .... is a phenomenon in which bacterial DNA is transferred from one bacterial cell to the another by a ....

Multiple Choice Questions

1. Viruses are no more "alive" than isolated chromosomes because

(*a*) they both require RNA and DNA

(*b*) they both need food molecules

(*c*) they both require oxygen for respiration

(*d*) they both require the environment of a cell to replicate

2. In bacteriophages genetic material is

(*a*) RNA (*b*) DNA

(*c*) both RNA and DNA (*d*) none of these

3. T_2 has

(*a*) ds DNA (*b*) ss DNA

(*c*) ss RNA (*d*) ds RNA

4. Circular single stranded (ss) DNA occur in bacteriophage

(*a*) T2, T4 (*b*) $\phi \times 174$, M13

(*c*) T3, $\phi 6$ (*d*) δ,T5

5. Conjugation in bacteiral was discovered by

(*a*) Watson and Crick (*b*) Lederberg and Tatum

(*c*) Jacob and Monod (*d*) Zinder and Lederberg

6. The process of conjugation was discovered by Lederberg and Tatum in

(*a*) *Diplococcus pneumoniae* (*b*) *Escherichia coli*

(*c*) *Salmonella* (*d*) *Pneumococcus*

7. Bacteria reproduce sexually by

(*a*) endospores (*b*) transformation

(*c*) conidia (*d*) exopores

8. Transformation experiment was performed on which bacteria

(*a*) *E.coli*

(*b*) *Diplococcus pneumoniae*

(*c*) *Salmonella*

(*d*) *Pasteurella pestis*

9. The process of genetic recombination involves direct contact between donor and recipient bacteria. It is known as

(*a*) transformation (*b*) transduction

(*c*) conjugation (*d*) none of these

10. In male (F^+) cells, plasmid/episome lies free in the cytoplasm or may be integrated with the chromosome of the bacterial cell. The latter condition is known as

(*a*) F factor
(*b*) Hfr (= high frequency of recombination)
(*c*) both of these
(*d*) none of these

11. In bacterial conjugation, the transfer of genetic material is undirectional. The donor cell transfer part of its genome through the passage formed by

(*a*) flagella (*b*) dissolution of cell wall
(*c*) sex pili (*d*) none of these

12. Bacterial conjugation between F^+ and F^- results in the formation of merozygotes which are partial diploids. The use of F^+ elements to create partial diploid is called

(*a*) meromixis (*b*) amphimixis
(*c*) sexduction (*d*) sexuality

13. The bacterial process in which genetic material is carried by some external agency is called

(*a*) translation (*b*) transformation
(*c*) transduction (*d*) conjugation

14. Genetic recombination by transduction in bacteria was discovered first in

(*a*) *Salmonella typhimurium* (*b*) *Escherichia coli*
(*c*) *Streptococcus pneumoniae* (*d*) *Agrobacterium tumifaciens*

15. Clear areas of agar phates containing phage particles as well as bacteria are called

(*a*) holes (*b*) plaques
(*c*) lawn (*d*) bacteriophages

16. Bacterial transduction involves

(*a*) direct uptake of genetic material DNA from another source
(*b*) transfer of genetic material from one cell to another through conjugation
(*c*) mutation of gene
(*d*) transfer of genetic material DNA from one bacterium to another through a virus.

ANSWERS

Very Short Answer Questions

1. A *colony* is a visible mass of cells derived usually from a single progenitor. A *plaque* is the equivalent growth of phages on a bacterial lawn, producing a cleared area lacking intact bacteria.

2. The bacterium could have survived and produced a colony if it was an λ–free area, it became lysogenic (and thus resistant to further phage attack), or it may genetically resistant to phage lambda.

3. In F^- cells, no factor is present; in F+ cells, autonomous F factor is present; in Hfr cells, integrated F factor is present. (*b*) F^+ and Hfr cells have F pili; F^- cells do not. (*c*) F^- cells are converted to F^+ cells by the conjugative transfer of F factor from F^+ cells; Hfr cells are formed when F factors in F+ cells become integrated into the chromosomes of these cells.

4. (*a*) F′ factors are useful for genetic mapping and, particularly, for *cis-trans* or complementation tests in bacteria. (*b*) F′ factors are formed by abnormal excision of F factors from Hfr chromosomes. (*c*) By the conjugative transfer of an F′ factor from a donor cell to a recipient (F^-) cell.

5. *Generalised transduction.* (*i*) transducing particles often contain only host DNA; (2) transducing particles may carry any segment of the chromosome. Thus all host genes are transduced. *Specialised transduction.* (1) transducing particles carry a recombinant chromosome, which contains both phage DNA and host DNA; (2) only host genes that are adjacent to the prophage integration site and transduced.

6. IS (or insertion sequences) are short (800-1400 nucleotide pairs) DNA sequences that are transposable, that is, capable of moving from one position in a chromosome to another position or from one chromosome to another chromosome. (*b*) IS elements mediate recombination between non-homologous DNA molecules or sequences (so called "illegimate recombination"). (*c*) IS elements are implicated in the evolution of episomes and plasmids by mediating recombination between otherwise nonhomologous genetic elements.
7. By interrupting conjugation at various times after the donor and recipient cells are mixed (using a blender or other form of agitation), one can determine the length of time required to transfer a given genetic marker from an Hfr cell to an F^- cell. Since the chromosome is transferred in a linear sequence, the positions of genetic markers can be ordered relative to each other.
8. Cotransduction refers to the simultaneous transduction of two different genetic markers to a single recipient cell. Since bacteriophage particles can package only 1/100 to 1/50 of the total bacterial chromosome, only markers that are relatively closely linked can be cotransduced. The frequency of cotransduction of any two markers will be an inverse function of the distance between them on the chromosome. As such, this frequency can be used as an estimated of the linkage distance. Specific cotransduction-linkage functions must be prepaid for each phage-host system studied.

Fill in the Blanks

1. Prototrophic.
2. Axotrophic.
3. Tetracyclin.
4. O. Avery Macleod, McCarthy.
5. Obligate, bacteria.
6. Transduction, phage particle.

Multiple Choice Questions

1. (*d*) **2.** (*b*) **3.** (*a*) **4.** (*b*) **5.** (*b*) **6.** (*b*)
7. (*b*) **8.** (*b*) **9.** (*c*) **10.** (*b*) **11.** (*c*) **12.** (*c*)
13. (*c*) **14.** (*a*) **15.** (*b*) **16.** (*d*).

18 Plasmids, IS Elements, Transposons and Retroelements

18.1 PLASMIDS

Plasmids are autonomous self-replicating double-stranded circular DNA molecules that are maintained in the cell in a stable and characteristic number of copies. A plasmid is a replicon that is stably inherited in an extrachromosomal state. Most, but not all, plasmids are dispensable as they are not essential for the survival of bacterial cells, except under certain environmental conditions.

The word **plasmid** was originally used by **J. Lederberg** (1952) for any extrachromosomal hereditary determinat. This term is now restricted for those accessory DNA circles, which are found in bacteria (or in cell organelles such as mitochondria in eukaryotes), in addition to the main chromosome.

The nucleoid DNA molecule is not the only repository for genes in the bacterial cell. Bacteria often also harbour small independent DNA molecules, the plasmids, which carry genes not found on the main DNA molecule. Plasmids are circular molecules that lead an independent existence in the bacterial cell (Fig. 18.1).

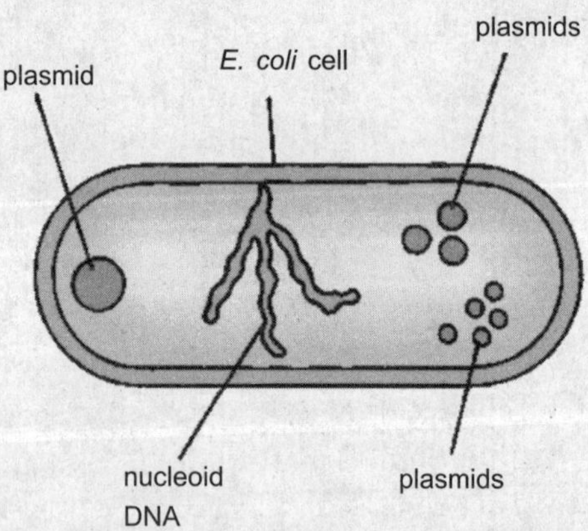

Fig. 18.1. Plasmids in a bacterial cell.

Plasmids carry one or more genes which are often responsible for a useful characteristic. For example, the ability of bacteria to survive in normally toxic concentrations of antibiotics such as chloramphenicol or amplicillin is often due to the presence in bacterium of a plasmid that carries antibiotic – resistance genes. Genes carried by plasmids also provide to bacteria the ability to utilise certain complex compounds such as toulene as a carbon source.

All plasmids possess at least one DNA sequence that can act as an origin of replication (replicon), so they are able to multiply in the cell independently of the main DNA molecule (Fig. 18.2A). The smaller plasmids make use of the cells own DNA replicative enzymes to make copies of themselves, whereas some of the larger ones carry genes that code for special enzymes that are specific for plasmid replication. A few types of plasmids are also able to replicate by inserting them into the main DNA molecule (Fig. 18.2B). These integrative plasmids are called **episomes** and they may be stably maintained in this form through numerous cell divisions, but always at some stage exist as independent elements.

Characteristics of Plasmids

Plasmids share the following general properties:

1. They are genetic elements made up of DNA; 2 they are smaller than and separate from the main chromosome, and 3. They are capable of replication. Some plasmids such as sex factors have unique properties such as 1. They control bacterial conjugation, 2. They are reversibly inserted in the bacterial chromosome and 3. They are transferred from one bacterium to another.

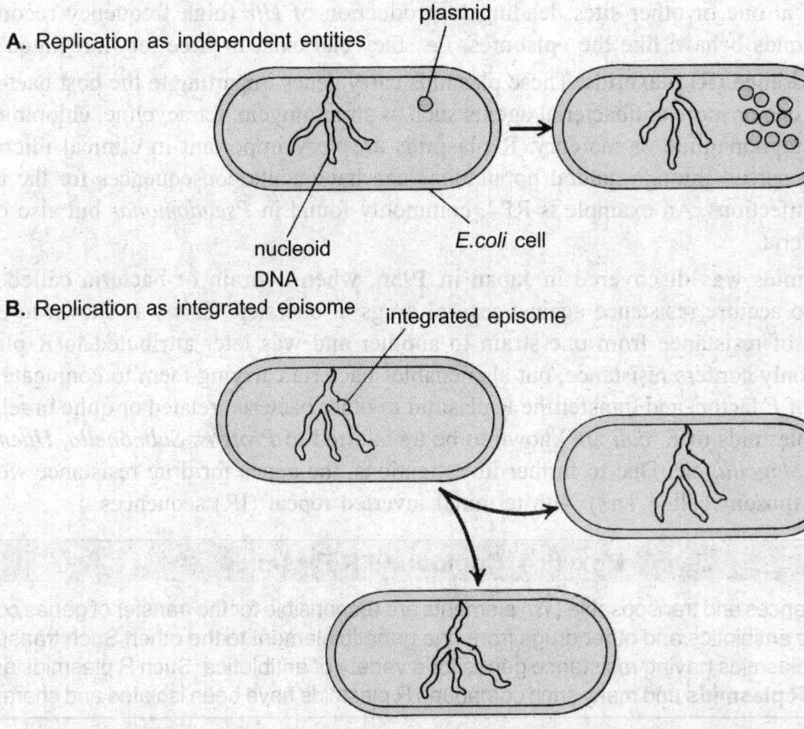

Fig. 18.2. Two methods of replication of the plasmids (after Brown, 1998).

Further, bacteriophages which are included in the category of episomes are sometimes excluded from the category of plasmids, since they show unrestricted replication, leading to lysis. But some phages such as λ (lambda) phage share most of the properties of plasmids, are grouped with them by some workers.

Classification of Plasmids

Almost all species of bacteria harbour plasmids and a large number of different types are known (Table 18.1). They are classified according to the genes they carry and the characteristics that these genes confer on the host bacterium. According to this classification there are following *five* main types of plasmids:

Table 18.1 Types of special feature of some bacterial plasmids.

S. No.	*Type of plasmid*	*Special properties*
1.	*F, R1, Col*	Fertility
2.	*Col E1*	Bacteriocin production
3.	*R6*	Heavy metal resistance
4.	*Ent*	Entrotoxin production
5.	*Com*	Metabolism of camphor
6.	*Ti* (in *Agrobacterium tumefaciens*)	Tumorigenicity in plants

1. **Fertility (F) plasmids.** These plasmids direct conjugation between different bacteria. They are also called **conjugative** or **transmissible plasmids.** An example is the *F* plasmid of *E. coli.* During conjugation of *E. coli,* when F^+ strain (carrying *F* factor) comes in contact with F^- strain, two processes may follow:

One involving transfer of *F* factor, so that F^- becomes F^+ and second involving transfer of a segment of bacterial chromosome, which is facilitated by insertion of *F* factor within the bacterial chromosome at one or other sites, leading to production of *Hfr* (high frequency recombinant). Some F plasmids behave like the **episomes,** *i.e.,* they can exist in 'free' or 'integrated' state.

2. **Resistance (R) plasmids.** These plasmids carry genes imparting to the host bacterium the resistance to one or more antibacterial agents such as streptomycin, tetracycline, chloramphenicol, ampicillin, sulphonamide or mercury. R plasmids are very important in clinical microbiology because their spread through natural populations can have acute consequences for the treatment of bacterial infections. An example is RP4, commonly found in *Pseudomonas* but also occurring in other bacteria.

R plasmids was discovered in Japan in 1956, when a strain of bacteria called *Shigella* was shown to acquire resistance against several drugs in one step. This was shown to be due to transmission of resistance from one strain to another and was later attributed to R plasmid. R plasmid not only confers resistance, but also enables bacteria carrying them to conjugate (even in the absence of *F* factor) and transfer the R plasmid to other bacteria, related or quite unrelated. For example, R plasmids of *E. coli* are known to be transferred to *Proteus, Salmonella, Haemophilus, Pasteurella, Shigella,* etc. Due to further investigations, the genes for drug resistance were found to be a **transposon** (called Tn3) with terminal inverted repeat (IR) sequences.

Box 18.1 Compound R Plasmids

IS sequences and transposable (*Tn*) elements are responsible for the transfer of genes controlling resistance to antibiotics and other drugs from one genetic element to the other. Such transpositions generate R plasmids having resistance genes for a variety of antibiotics. Such R plasmids are called **compound R plasmids** and many such compound R plasmids have been isolated and characterized.

Sometimes the R plasmids are found to contain two distinct parts: one being genes for the resistance (called R factor) and the other responsible for transfer of plasmid (called **resistance transfer factor** or **RTF**). But in some other cases such as penicillin resistance in *Staphylococcus,* the resistance may not be associated with ability to transfer (*RTF*). In such a case transfer of drug resistance may be brought about by **transduction.**

3. **Col plasmids.** These plasmids carry genes coding for colicins, the toxic proteins that kill other bacteria. An example is plasmid *Col E1* of *E. coli.* Colicin is produced by *Escherichia coli, Shigella* and *Salmonella* due to their *Col* plasmids. *Col* plasmids resemble *F* factor or *R* factor. Like transfer of *R* factor, transfer of *Col* factor does not depend on the *F* factor. Since transfer of *Col* factor can occur between two F^- strains of bacteria, one carrying *Col* and other lacking it. There are, however, some *col* plasmids (*e.g., Col* E1) which require the help of sex factor for its transfer. Due to this property, *R* plasmids and *Col* plasmids are also called **nonconjugative** or **nontransmissible plasmids.**

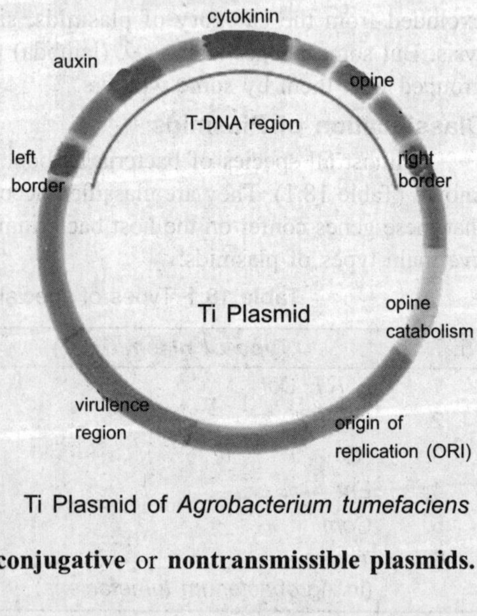

Ti Plasmid of *Agrobacterium tumefaciens*

4. Degradative plasmids. These plasmids allow the host bacterium to metabolize unusual molecules such as toluene and solicylic acid. Several examples occur in the *Pseudomonas* genus of bacteria, for instance TOL of *Pseudomonas putida.*

5. Virulence plasmids. These plasmids confer pathogenicity on the host bacterium. The best known example is the *Ti* plasmid of *Agrobacterium tumefaciens,* which induces crown gall tumors in dicotyledonous plants.

Replication, Transfer and Recombination in Plasmids

All plasmids discovered so far are small circular DNA molecules. The DNA circles may be single or multiple structures which are linked together in chains. The size of these circular DNA molecules varies from 30μm (*e.g.*, F^+) in plasmid of *E. coli,* making 20% of the main chromosome or can be less than 1μm as in P16C of *E. coli* (Table 18.2). Colicin plasmids such as *Col E1, E2* and *E3* are usually having a length of 2.3 μm. The number of plasmids in a bacterial cell varies from a few in case of *F* and *R* plasmids to as many as 20 – 30 in case of *Staphylococcus* plasmids (Box 18.2).

Table 18.2 Properties of various plasmids

S. No.	*Plasmid*	*Type*	*Size (kb)*	*Copy number*	*Bacterium*
1.	*F*	Fertility	95	1 – 2	*E. coli*
2.	*Rbk*	Resistance	37	12	*E. coli*
3.	*Col E1*	Colicin	6.36	10 – 15	*E. coli*
4.	*TOL*	Degradative	117	1 – 2	*Pseudomonas putida*
5.	*Ti*	Virulence	213	1 – 2	*Agrobacterium tumefaciens*

Box 18.2 Copy Number of Plasmids

Each plasmid is maintained in the bacterial cell at a characteristic **copy number** mainly due to its replication control system. There are following methods for the regulation of plasmid's copy number in a cell:

1. **Single Copy plasmids.** They are maintained in one copy per bacterial cell. Their replication control is the same as that of the bacterial host chromosome so that they replicate and segregate with the bacterial chromosome. Such a replication control is called **stringent control.**
2. **Multicopy plasmids.** They exist in several, but a characteristic number of copies (typically 10 to 20 copies) per cell. Their replication control permits more than one replications to occur while the host cell chromosome replicates once. This type of replication control is called **relaxed control;** it allows a pool of origins to exist in each cell.

The DNA of plasmids replicates in **semi-conservative** manner. While initiation of replication is controlled by plasmid genes, other stages of replication such as elongation (polymeriz-ation) and termination are controlled by bacterial genes. Further, when a plasmid is inserted in the bacterial chromosome, bacterial control on plasmid replication is absolute, but when it is free, the control is less than absolute.

The property of transfer of plasmid from one bacterium to another is facilitated by the formation of a hair-like extracellular growth, called **sex pilus.** Sex pilus is formed of outer membrane of the donor bacterium. The *F* sex pilus is up to 20 μm in length and 8 nm wide. In *E. coli,* transfer is controlled by plasmid genes *tra-A, tra-B, tra-C,* etc., and mutations in these genes lead to failure of transfer. But, when two plasmids containing mutations for different *tra-* genes are introduced in the same cell, pilus may be formed and transfer may occur due to complementation between mutants.

In the process of transfer, one of the two strands of plasmid DNA is nicked (broken) at a site called **origin of transfer** (*Ori-T*) and the linear strand thus formed moves to the recipient bacteria. The single strands in the donor as well as in recipient cell then synthesize complementary strands, so that duplex DNA plasmid are reconstituted in both cells (Fig. 18.3).

Lastly, plasmid DNA may undergo **recombination** either with another plasmid DNA or with bacterial DNA. In the bacterium chromosome, there are **insertion sequences (IS),** which are homologous in recombination. There are atleast 17 such sites in *E. coli* genome. **Mapping** of genes in plasmid DNA has also been achieved. Methods employed in the plasmid mapping involved **recombination** and **complementation** between strains have overlapping deletions.

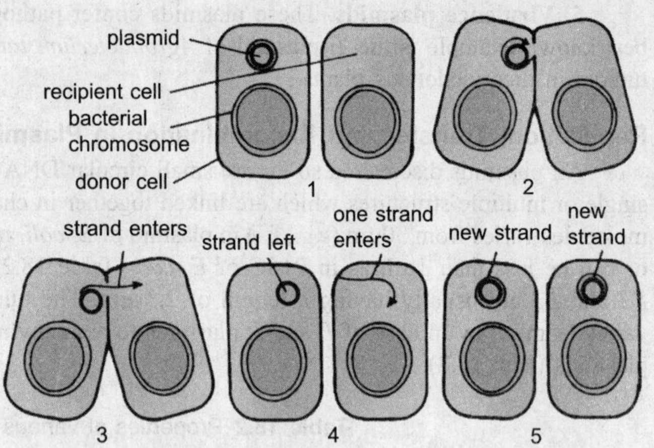

Fig. 18.3. Mechanism of transfer of plasmid (*e.g., F, R*) from one bacterium to another.

18.2 IS ELEMENTS AND TRANSPOSABLE ELEMENTS

A great part of classical genetic analysis has been devoted to the localization of genes on chromosomes. Accordingly, if we map a gene today, we except to see it in the same place tomorrow. However, our discovery of mobile genetic elements has modified that view to some extent. We now know that some segments of the genome can move quite readily from one place to another. The movement of those elements has effects on the phenotype of the organism, primarily at the transcriptional level.

Transposable element, mobile gene, jumping gene or **transposon** is a segment of DNA that can move from one position in a genome to another (see **Alberts** *et. al.,* 2004). **Retrotransposon** is a type of transposable element that moves by being first transcribed into RNA copy that is then reconverted to DNA by reverse transcriptase and inserted elsewhere in the chromosome.

Thus, a transposable element is a DNA segment that translocates to other sites in the genome, essentially independent of sequence homology. Usually, such elements are flanked by short, inverted repeats of 20 to 40 base pairs at each end. Insertion repeats into a structural gene can produce a mutant phenotype. Insertion and excision of transposable elements depend on two enzymes, **transposase** and **resolvase.** Such transposable elements have been identified in both prokaryotes and eukaryotes (**Klug** and **Cummings,** 2003).

Discovery and History

In the 1940s, in a study of the genetics of kernel mottling in maize, **Barbara McClintock** discovered an element that not only regulated the mottling but also caused breakage of the chromosome carrying the genes for colour and consistency of the kernel. The element was called **dissociation** (*Ds*).

Mapping data showed that the chromosome breakage always occurs at or very near the location of *Ds.* McClintock's critical observation was that *Ds* does not have a constant location but occasionally moves to a new position (*transposition*), causing chromosome breakage at the new site. Furthermore, *Ds* moves only if a second element, called *activator* (*Ac*) is also present. *Ac* itself moves within the genome and can cause in the expression of genes at or near its insertion site, alternations similar to the modifications that result from the presence of *Ds.* McClintock (1956) referred to *Ac* and *Ds* as **controlling elements.** Both of these form a two - factor system of transposable elements of maize. The term **transposable elements** was coined by another greater geneticist **Alexander Brink.**

In the mid - 1960s **Susum Mitsuhashi** first suggested that the gene responsible for resistance to several antibiotics were mobile and could move between bacterial plasmids and chromosomes. The first observation of jumping genes at the molecular level came in the early 1970s. A number of independent researchers, including **Peter Starlinger** and **James Shapiro,** visualised a unique class of mutations affecting different genes in various bacterial strains. The term **transposon** was coined by **R.W. Hedges** and **A.E. Jacob** in 1974 for a DNA segment which could move from one DNA molecule (or chromosome) to other and carried resistance for antibiotic ampicillin. In 1975, **David Hogness** and his colleagues **David Finnegan, Gerald Rubin** and **Michael Young** identified a class of mobile genes in *Drosophila melanogaster* that they designate as *Copia.*

Transposition and Site-specific Recombination

In general type of recombination, DNA rearrangements occur between DNA segments that are very similar in sequence. Although these rearrangements can result in the exchange of alleles between chromosomes, the order of the genes on the interacting chromosome typically remain the same. A second type of recombination, called **site - specific recombination,** moves specialised nucleotide sequences, the *mobile genetic elements,* between nonhomologous sites within a genome. Due to this reason, this type of recombination is also called **illegimate recombination (Brown,** 1998). Here, movement can occur between two different positions in a single chromosome, as well as between two different chromosomes.

Unlike general recombination, site - specific recombination is guided by recombination enzymes that recognise short, specific nucleotide sequences present on one or both ends of the recombining DNA molecules. Each type of mobile element generally encode the enzyme that mediates its own movement and contains special sites upon which the enzyme acts. Many elements also carry other genes. For example, virus encode coat proteins that enable them to exist outside cells, as well as essential viral enzymes. The spread of mobile elements that carry antibiotic resistance genes is a major factor underlying the widespread distribution of antibiotic resistance in bacterial population (e.g., spread of resistance for antibiotics such as ampicillin and tetracycline). Site - specific recombination mainly occurs by **transposition** or **transpositional site - specific recombination** which involves breakage reactions at the ends of the mobile DNA segments embedded in chromosomes and the attachments of those ends at one of many different nonhomologous target DNA sites. It does not involve the formation of heteroduplex DNA.

In *transposition,* a specific enzyme, usually encoded by the transposon and called a **transposase,** acts on a specific DNA sequence at each end of the transport — first disconnecting it from the flanking DNA and then inserting it into a new target DNA site. There is no need for homology between the ends of the element and the insertion site. Most transposons move only very rarely (once in 10^5 cell generations for many elements in bacteria).

Types of Transposons

On the basis of their structure and transposition mechanisms, transposons can be grouped into following *three* large classes (Table 18.3):

1. DNA only transposon. The mobile element exists as DNA throughout its life cycle — the translocating DNA segment is directly cut out of the donor DNA and joined to the target site by a transposase enzyme.

2. Retroviral - like retrotransposons. They move by a less direct mechanism. An **RNA polymerase enzyme** first transcribes the DNA sequence of the mobile element into RNA. The enzyme **reverse transcriptase** then transcribes this RNA molecule back into DNA using the RNA as a template, and it is this DNA copy that is finally inserted into a new site in the genome. For historical reasons, the transposase - like enzyme that catalyzes this insertion reaction is called an **integrase** rather than a transposase.

3. Non retroviral retrotransposons. Each of these transposons moves by making a DNA copy of an RNA molecule that is transcribed from it. However, the mechanism involved for these nonretroviral retrotransposons is distinct from that in which the RNA molecule is directly involved in the transposition reaction.

All these three types of transposable elements range in length from 1000 to about 12000 nucleotide pairs; each family contains many members. In addition to transposable elements, there are selected viruses that can move in and out of host cell chromosomes; these viruses are related to the first two classes of transposons.

Table 18.3 Some features of three major classes of transposable elements (Source **Alberts** *et al.*, 2004).

Class description and structure	*Genes in complete element*	*Mode of movement*	*Examples*
1. DNA only transposons  Short inverted repeats at each end of transposable unit	Encodes transposase enzyme	Moves as DNA, either excising or following a replicative pathway	1. Pelement (*Drosophila*) 2. *Ac - Ds* (maize) 3. *Tn3* and *IS1* (*E. coli*) 4. *Tam3* (snapdragon)
2. Retroviral - like retrotransposons  Directed repeated long terminal repeats (LTRs) at ends	Encodes reverse transcriptase and resemble retrovirus	Moves via an RNA intermediate produced by promoter in LTR	1. *Copia* (*Drosophila*) 2. *Ty1* (yeast) 3. THE - 1 (human) 4. Bs1 (maize)
3. Nonretroviral retrotransposons 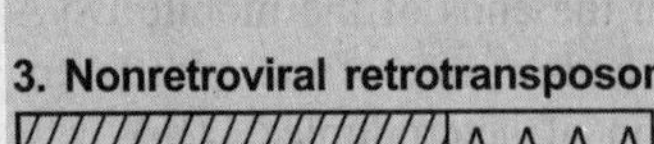 Poly A at 3′ end of RNA transcript; 5′ end is often truncated	Encodes reverse transcriptase	Moves via an RNA intermediate that is often produced from a neighboring promotor	1. Felement (*Drosophila*) 2. *L1* (human) 3. *Cin4* (maize)

Mode of Transposition of DNA Transposons

DNA transposons may have the following *two* methods of transposition (Fig. 18.4):

1. Replicative transposition. This transposition mechanism involves direct interaction between the donor transposon and the target site resulting in the copying of the donor element, *e.g.,* Shapiro's model.

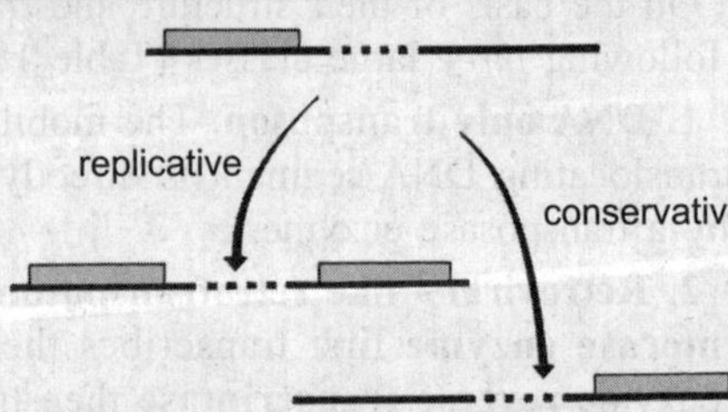

Fig. 18.4. Replicative and conservative transposition. Retroelements transpose replicatively via an RNA intermediate (after Brown, 2002).

2. Conservative transposition. This transposition mechanism involves excision of the element and re-integration at a new site, *e.g.*, cut-and-paste model.

Both mechanisms require enzymes which are usually coded by genes within the transposons. DNA transposons use either the replicative or conservative pathway (some can use both).

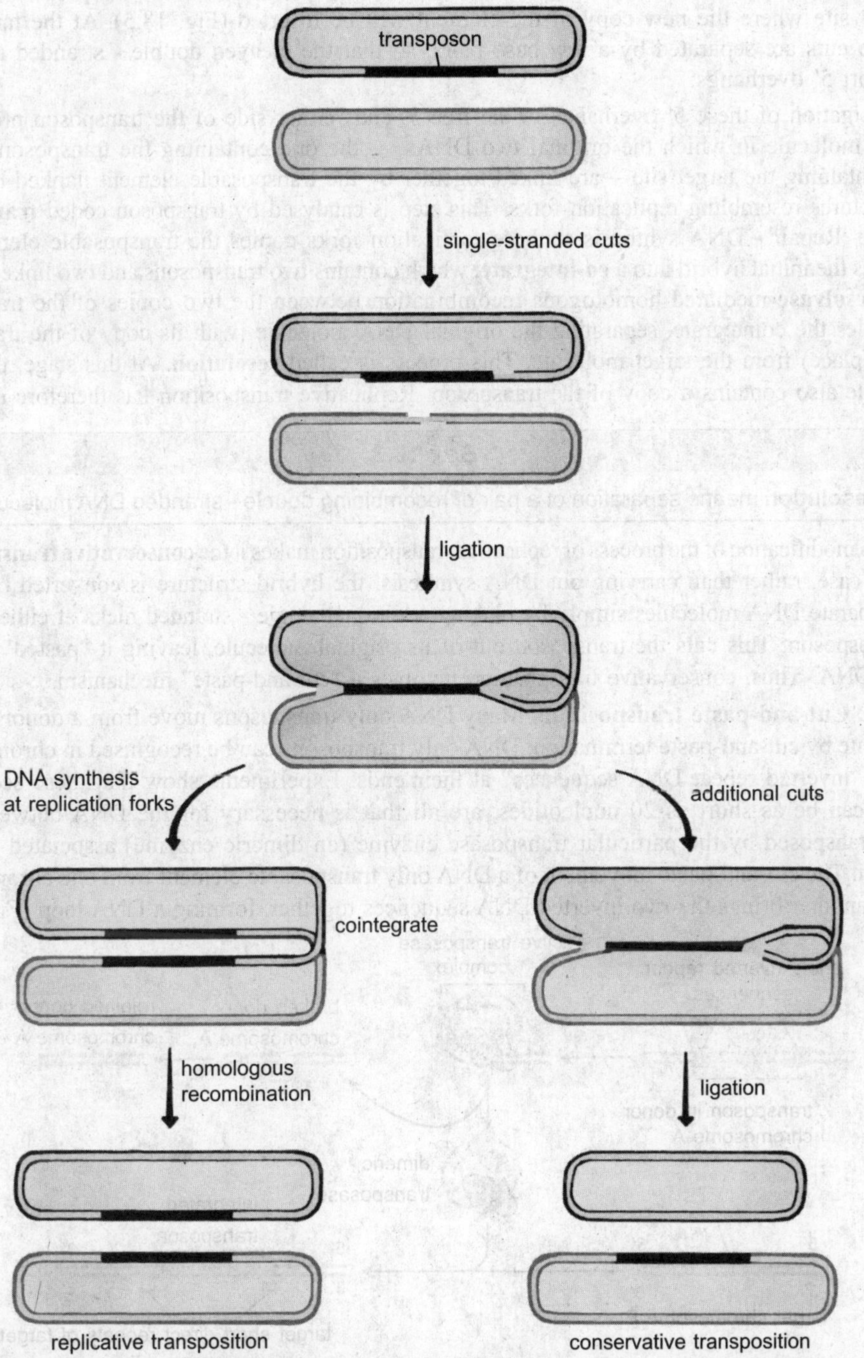

Fig. 18.5. A model for the process resulting in replicative and conservative transposition.

1. Shapiro's model of transposition: A number of models for replicative and conservative transposition have been proposed over the years but most are modifications of a scheme originally outlined by **J.A. Shapiro** (1979). According to Shapiro's model, the *replicative transposition* of a bacterial element such as Tn3 - type transposon or a transposable phage is initiated by one or more **endonucleases** that make single stranded cuts on either side of the transposon and in the target site where the new copy of the element will be inserted (Fig. 18.5). At the target site, the two cuts are separated by a few base pairs, so that the cleaved double - stranded molecule has short 5′ overhangs.

Ligation of these 5′ overhangs to the free 3′ ends either side of the transposon produces a hybrid molecule in which the original two DNAs — the one containing the transposon and the one containing the target site – are linked together by the transposable element flanked by a pair of structures resembling replication forks. This step is catalysed by transposon coded **transposase** enzyme. Repair - DNA synthesis at these replication forks copies the transposable element and converts the initial hybrid into a **co-integrate,** which contains two transposons and two linked DNAs. Now **resolvase** mediated homologous recombination between the two copies of the transposon uncouples the cointegrate, separating the original DNA molecule (with its copy of the transposon still in place) from the target molecule. This process is called **resolution.** At this stage, the target molecule also contains a copy of the transposon. Replicative transposition has therefore occurred.

Box 18.3

Resolution means separation of a pair of recombining double - stranded DNA molecules.

A modification of the process of replicative transposition makes it the **conservative transposition.** In this case, rather than carrying out DNA synthesis, the hybrid structure is converted back into two separate DNA molecules simply by making additional single - stranded nicks at either side of the transposon. This cuts the transposon out of its original molecule, leaving it "pasted" into the target DNA. Thus, conservative transposition involves a "cut-and-paste" mechanism.

2. Cut-and-paste transposition. Many DNA only transposons move from a donor site to a target site by cut-and-paste termination. DNA only transposons can be recognised in chromosomes by the "inverted repeat DNA sequences" at their ends. Experiments show that these sequences, which can be as short as 20 nucleotides, are all that is necessary for the DNA between them to be transposed by the particular transposase enzyme (an dimeric enzyme) associated with the element. The cut-and-paste movement of a DNA only transposable element from one chromosomal site to another brings the two inverted DNA sequences together, forming a DNA loop (Fig. 18.6).

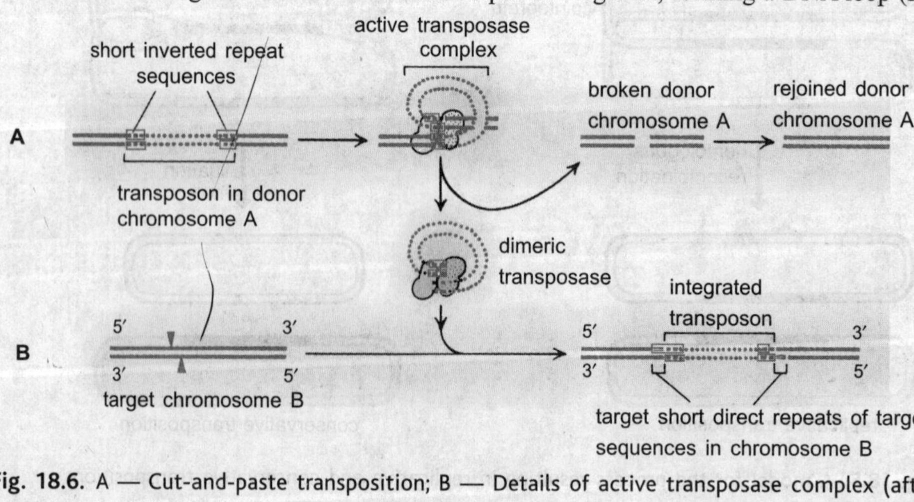

Fig. 18.6. A — Cut-and-paste transposition; B — Details of active transposase complex (after Alberts *et al.*, 2004).

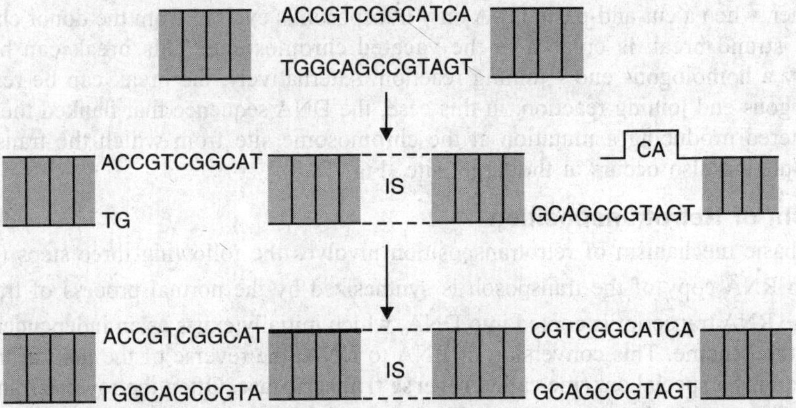

Fig. 18.7. Production of target site duplication by the insertion of IS element. The element causes breaks in the target DNA at the sites indicated by the arrows. After insertion, DNA synthesis fills in the gaps left by the staggered gaps, producing direct repeats on each side of the element (after Gardner *et al.*, 2001).

Insertion into the target chromosome, catalysed by the transposase enzyme, occurs at random site through the creation of staggered breaks in the target chromosome. As a result, the insertion site is marked by a short direct repeat of the target DNA sequence (Fig 18.6).

Since, the breaks made in two target DNA strands are staggered (Fig. 18.7), two short, single - stranded gaps are initially formed in the product DNA molecule, one at each end of the inserted transposon. These gaps are filled in by a host cell **DNA polymerase** and **DNA ligase** to complete the recombination process, producing a short duplication of the adjacent target DNA sequence. These flanking direct repeat sequences serve as convenient markers of a prior transpositional site - specific recombination event.

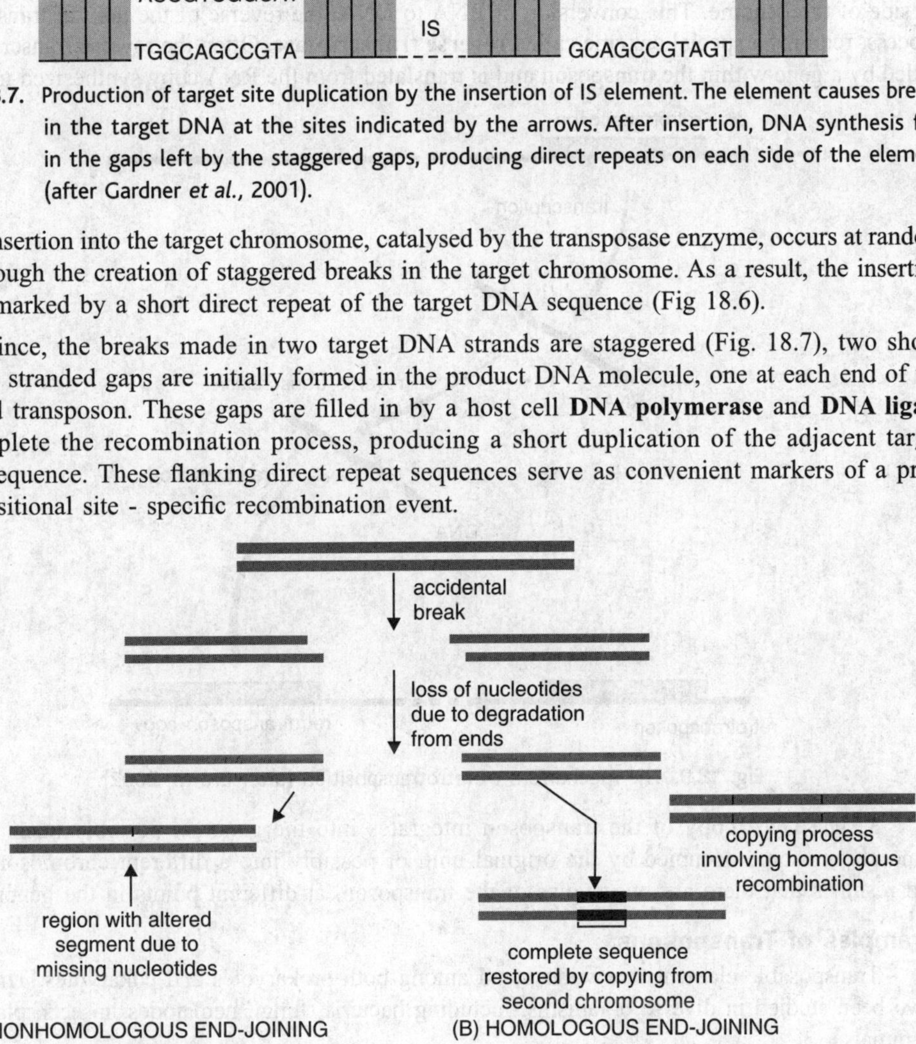

Fig. 18.8. Two different types of end - joining for repairing double - strand breaks. A nonhomologous end - joining alters the original DNA sequence when repairing broken chromosomes. B - Homologous end - joining. Here undamaged chromosome act as the template for transferring genetic information to the broken chromosome, repairing it with no change in DNA sequence (after Alberts *et al.*, 2004).

Further, when a cut-and-paste DNA only transposon is excised from the donor chromosome, a double - strand break is created in the vacated chromosome. This break can be perfectly "healed" by a homologous end - joining reaction. Alternatively, the break can be resealed by a nonhomologous end joining reaction; in this case, the DNA sequence that flanked the transposon is often altered producing a mutation at the chromosome site from which the transposon was excised. Mutation also occurs at the target site (Fig. 18.8).

Mechanism of Retrotransposition

The basic mechanism of retrotransposition involves the following three steps (Fig. 18.9):

1. An RNA copy of the transposon is synthesized by the normal process of transcription.

2. The RNA transcript is copied into DNA, which initially exists as an independent molecule outside of the genome. This conversion of RNA to DNA, the reverse of the normal transcription process, requires a special enzyme called **reverse transcriptase.** Often the reverse transcriptase is coded by a gene within the transposon and is translated from the RNA copy synthesized in step 1.

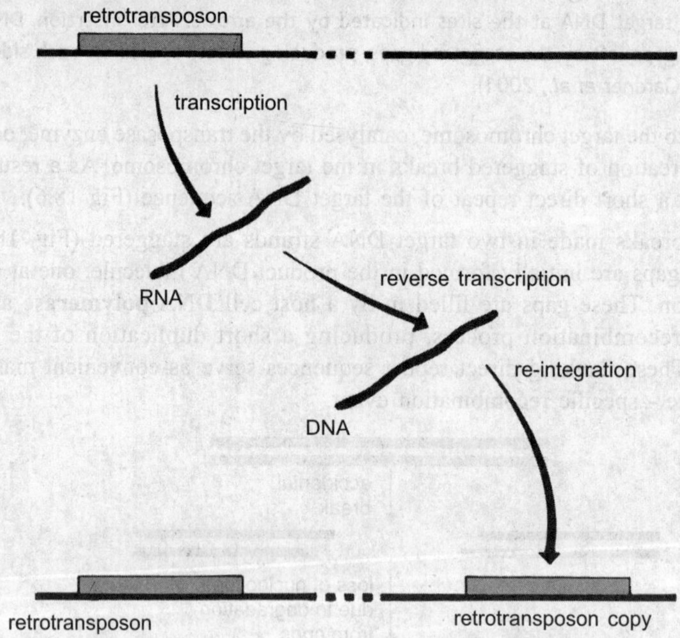

Fig. 18.9. The mechanism of retrotransposition (after Brown, 2002).

3. The DNA copy of the transposon integrates into the genome, possibly back into the same chromosome occupied by the original unit, or possibly into a different chromosome. The end result is that there are two copies of the transposon, at different points in the genome.

Examples of Transposons

Transposable elements are widespread among both prokaryotes and eukaryotes. Thus, they have been studied in diverse organisms, including bacteria, fungi, nematodes, insects, plants and mammals.

1. Bacterial Transposons

***(i)* Insertion sequence (IS).** During the 1960s geneticists were puzzled by a set of unusual mutations which were occasionally found in operons of *E.coli* and other bacteria. These are called **polar mutations** because they affect not only the gene in which the mutation occurs but also other genes downstream of the mutation (Fig. 18.10). Several types of polar mutations are known and often they are ones that *introduce a transcription termination signal into the operon.*

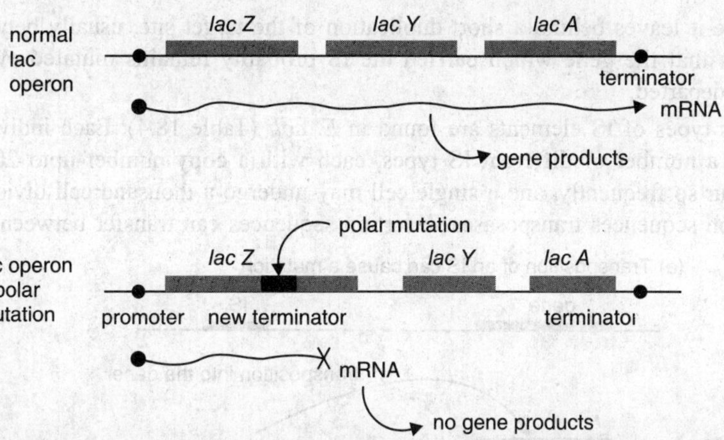

Fig. 18.10. The effect of a polar mutation in the *lac* operon. The polar mutation in *lacZ* prevents expression of *lacY* and *lacA*, even though the last two genes are not themselves altered (after Brown, 1998).

In contrast to other known mutations these polar mutations are always **irreversible.** Further, if a plasmid is introduced into a bacterium that carries a polar mutation of this type, then similar irreversible polar mutations suddenly appear in genes carried by the plasmid (Fig. 18.11). It happens as though the mutation is able to jump from one DNA molecule to another.

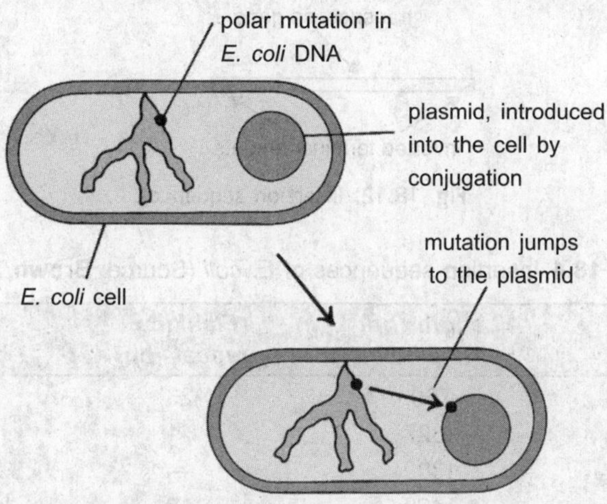

Fig. 18.11. The unusual type of polar mutation that is able to jump from one position to another on *E. coli* DNA molecule (after Brown, 1998).

The first hint of an explanation of this strange phenomenon appeared when the plasmids that had become mutated were studied. It was found that these plasmids had increased in size, generally a kilobase or more. When the DNA sequences of the mutated plasmids were examined it was discovered that the mutations were caused by the insertion into the target genes of new pieces of DNA, which were subsequently called **insertion sequences** or **ISs** (Fig. 18.12). These insertion sequences possess the remarkable capability of being able to move around the genome from site to site by a process dependent on recombination and transposition. Moreover, when an IS transposases

Box 18.4

IS elements are the simplest forms of transposons. They are so called because they were discovered as spontaneous insertions within bacterial operons that prevented transcription or translation of concerned gene into which insertion has occurred.

to a new site it leaves behind a short duplication of the target site, usually between 5 to 15bp, which means that the gene which carried the IS probably remains mutated even after the IS element has departed.

Various types of IS elements are found in *E. coli* (Table 18.4). Each individual bacterium may contain a number of different IS types, each with a copy number upto 20. Transposition does not occur so frequently, and a single cell may undergo a thousand cell divisions before one of its insertion sequences transposases. Insertion sequences can transfer between bacteria during

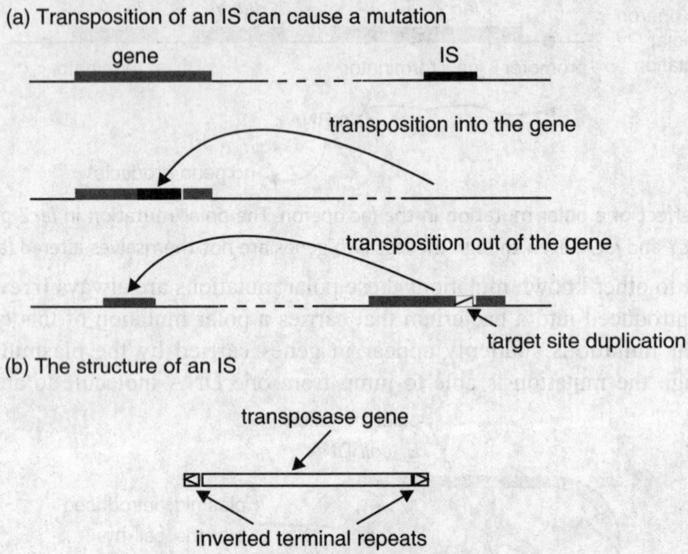

Fig. 18.12. Insertion sequences.

Table 18.4 Insertion sequences of *E. coli* (Source: **Brown,** 1998).

S. No.	*Insertion sequence*	*Length (bp)*	*Flanking repeat (bp)*	*Target site duplication (bp)*
1.	IS1	769	23	9
2.	IS2	1327	41	5
3.	IS3	1300	—	1
4.	IS4	1428	18	11
5.	IS5	1195	16	4
6.	IS10	1329	22	9
7.	IS50	1531	9	9

conjugation, and can also transfer between bacteria during conjugation and can also transfer between related species. The most part of each IS element is taken up by a single gene (except IS1 which has two) coding for a transposase enzyme that catalyses the transposition event. The IS element has at its ends a pair of **inverted repeats** (Box 18.5) between 9 and 41 bp in length, depending on the IS type (Fig. 18.12).

Box 18.5 Inverted Repeats

Consider the following squence of a DNA molecule:

5′ — A G G C A T — 3′
3′ — T C C G T A — 5′

Inverted repeat of above DNA sequence will be:

5′ — A G G C A T A T G C C T — 3′
3′ — T C C G T A T A C G G A — 5′

Repeats may occur adjacent to one other, as shown here, or may flank a different DNA sequence, as with the repeats at the ends of insertion sequences and Tn - 3 type transposons.

***(ii)* Composite transposons.** They consist two insertion sequences flanking an internal region which usually includes at least one gene, often coding for resistance to an antibiotic (Fig. 18.13A). Transposition occurs because of the transposase enzyme coded by IS elements. There are several different composite transposons (Table 18.5), some over 10kb in size. Often these types of element are carried by the plasmids.

Table 18.5 Composite and Tn3 - type transposons of *E. coli*

	Transposons	*Size (bp)*	*Flanking element*	*Gene carried*
(*a*)	Composite transposons			
	(*i*) Tn 5	5700	IS50	Kanamycin resistance
	(*ii*) Tn 9	2500	IS1	Chloramphenicol resistance
	(*iii*) Tn10	9300	IS10	Tetracycline resistance
(*b*)	Tn3 - type transposons			
	(*i*) Tn 3	~ 5000	38 bp inverted repeats	Ampicillin resistance
	(*ii*) Tn 501	~ 5000	38 bp inverted repeats	Mercury resistance

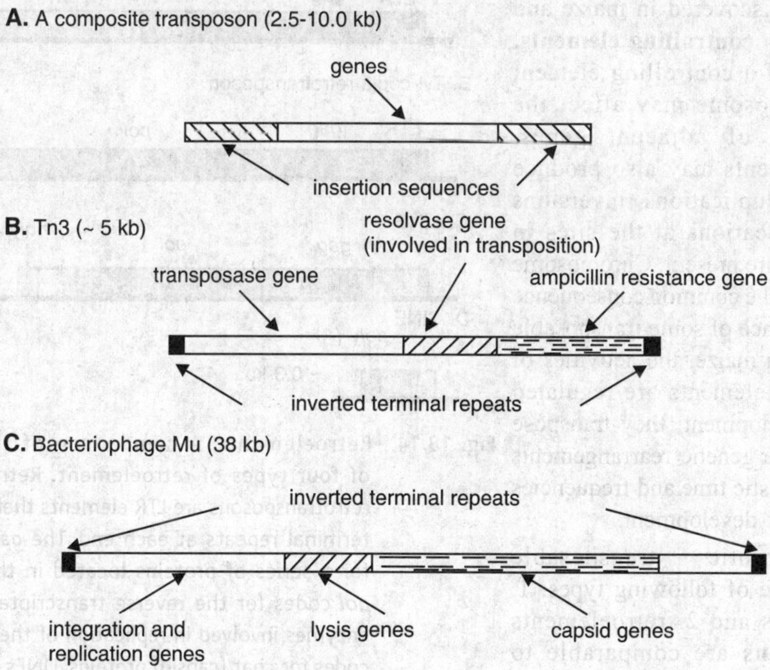

Fig. 18.13. Three types of bacterial transposable elements.

***(iii)* Tn3 - type transposons.** They are not flanked by IS elements but are still able to transpose because they carry a different transposase enzyme (Fig. 18.13, Table 18.5).

***(iv)* Transposable phages.** Certain phages such as Mu and D108 transpose as part of their normal infection cycle (Fig 18.13C). This means that besides inserting themselves into the host DNA (like phage λ), the genomes of these phages are able to move about from place to place in the host DNA (unlike phase λ). Transposable phages may be transposons that have evolved genes for protective proteins, enabling them to exist outside the cell, or they may be ordinary phages that have learned to transpose (see **Brown,** 1998).

Significance of bacterial transposable element. Transposons have no apparent function. Many evolutionary geneticists suggested that transposons are an evolutionary accident that once created, are self - maintaining. Since they may exist without an noticeable benefit to the host's phenotype, transposons have been referred to as **selfish DNA,** *i.e.*, they exist only to further their own existence (propagation), contributing nothing to the cell as a whole. However, it has been established that bacteria which carry a large number of transposons have a **selective advantage** over those with few transposons, possibly because transposons increase the mutation rate of the bacterium to adapt to changing environmental conditions.

In higher organisms, the retroviruses, which include some viruses implicated in cancer as well as the AIDS virus, resemble transposable elements. In the laboratory, transposable elements are used as **mutagens** because they inactivate genes that they transpose into, or possibly modulate the expression of a gene downstream from an insertion site (*e.g.*, in lac operon of *E. coli*) in both cases, a phenotypic change indistinguishable from a mutation.

2. Eukaryotic Transposons and Retroelements

Eukaryotes contain a variety of transposable elements that were originally discovered in maize and described as **controlling elements.** Insertion of a controlling element in a chromosome may affect the expression of adjacent genes. These elements may also produce deletions, duplications, inversions and translocations at the sites in which they are present. Chromosome breakage is the common consequence of the presence of some transposable elements. In maize, the activities of controlling elements are regulated during development; they transpose and promote genetic rearrangements at characteristic time and frequencies during plant development.

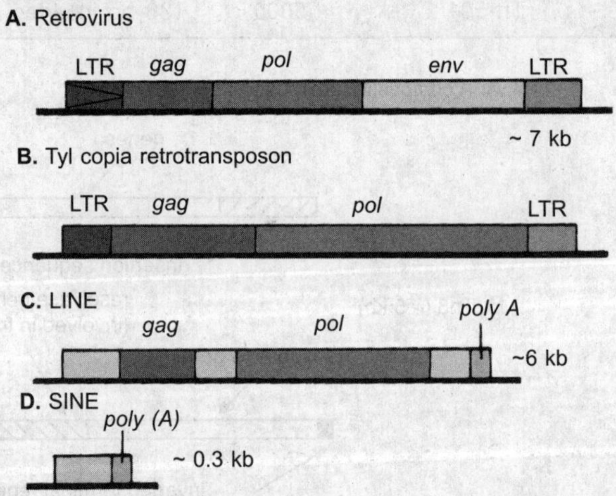

Fig. 18.14. Retroelements. A comparison of the structure of four types of retroelement. Retroviruses and retrotransposons are LTR elements that possess long terminal repeats at each end. The *gag* gene codes for a series of proteins located in the virus core; *pol* codes for the reverse transcriptase and other enzymes involved in replication of the element. *env* codes for coat (capsid) proteins. LINEs and SINEs are non - LTR retroelements or retroposons. Both have a poly (A) region (a long series of A nucleotide) at one end (after Brown, 2002).

Eukaryotic transposable elements are of following types: 1. transposons and 2. retroelements **Transposons** are comparable to the bacterial transposons, and have no life outside the genome, *e.g.*, controlling elements of maize. **RNA transposons** or **retroelements** are features of eukaryotic genomes but have not so far been discovered in prokaryotes (see **Brown** 2002). They have attracted great attention because of their close similarity with free - living viruses, called **retroviruses.** These viruses include both benign forms and also virulent types such as the human immunodeficiency viruses (HIV) that cause AIDS. These structural relationships are illustrated in Figure 18.14 and are of following types:

1. Retroviruses. These are exogenous viruses whose genomes are made of RNA. Their genetic organisation have been shown in (Fig. 18.14). They infect many types of vertebrates. Once inside a cell, the RNA genome is copied into DNA by the reverse transcriptase enzyme specified by the viral *pol* gene, and the DNA copy integrates into the host genome. New viruses can be produced by copying the integrated DNA into RNA and packaging this into virus coat proteins, the latter coded by the *env* genes on the virus genome.

2. Endogenous retroviruses (ERVs). These are retroviral genomes integrated into vertebrate chromosomes. Some are still active and might, at some stage in a cell's lifetime, direct the production of exogenous viruses, but most are decayed relics that have lost the capacity to form viruses (**Patience** *et al.,* 1997). These inactive sequences are interspersed repeats of the genome and they are not capable of additional proliferation.

3. Retrotransposons. These elements have sequences similar to ERVs but are features of the non - vertebrate eukaryotic genomes (*i.e.*, plants, fungi, invertebrate and microbial eukaryotes) rather than vertebrates. Retrotransposons are of many types and each has very high copy numbers in some genomes. Most of the interspersed repeats in maize are retrotransposons; in maize these elements form almost half of the genome (**San Mignel** *et al.,* 1996). There are two types of retrotransposons:

***(i) Ty3 / gypsy* family.** The members of this family have the same set of genes as an ERV, *e.g.*, Ty3 genes of yeast, *gypsy* gene of *Drosophila.*

***(ii) Ty1 copia* family.** The members of this family lack the *env* gene.

Both of these types of retrotransposons are able to transpose but the absence of the *env* gene means that the *Ty1/copia* group cannot form infectious virus particles. In fact, despite the presence of *env* in the *Ty3/gypsy* genome, some of these elements can form viruses and hence should be regarded as invertebrate retroviruses (**Song** *et al.,* 1994; **Peterson - Burch** *et al.,* 2000). Although technically they are interspersed elements, retrotransposons are sometimes found in clusters in a genome sequence as a result of the presence of preferred integration sites for transposing elements.

The three types of retroelements described so far are **LTR elements,** as they have long terminal repeats at either end which play a role in the transposition process (Fig. 18.14). Other retroelements do not have LTRs. These are called **retroposons** and in mammals include the following (Table 18.6):

1. LINEs (Long Interspersed Nuclear Elements)

They contain a reverse - transcriptase - like gene probably involved in the retrotransposition process (Fig 18.14C). An example is the human mobile element LINE - 1 which is 6.1 kb long and has a copy number of 516000 in the human genome (Table 18.7).

2. SINEs (Short Interspersed Nuclear Elements)

They do not have a reversed transcriptase gene but can still transpose, probably by "borrowing" reverse transcriptase enzymes that have been synthesized by other retroelements (Fig. 18.14D).

The commonest SINE in the human genome is *Alu*, which has a copy number of over one million (Table 18.8). *Alu* seems to be derived from the gene for the 7SL RNA, a non - conding ScRNA (= small cytoplasmic RNA) involved in transport of proteins around the cell. The first *Alu* element may have arisen by the accidental reverse transcription of a 7SL RNA molecule and integration of the DNA copy into the human genome. Some *Alu* elements are copied into RNA, providing the opportunity for proliferation of the element.

Examples of Eukaryotic Transposons

A. Transposons in Maize

In maize at least 14 different transposable systems are known, *e.g.*, *En, As-Ds, mu*, etc. Each of these elements is present in multiple copies in maize genome. These have the following general characteristics:

1. Autonomous and nonautonomous controlling elements. The controlling elements of maize may be grouped into two classes: 1. autonomous controlling elements; and 2. nonautonomous controlling elements. **Autonomous controlling elements** have the ability to excise and transpose themselves and corresponding nonautonomous elements. **Nonautonomous controlling elements** do not have the ability to transpose; they transpose only when an autonomous member of same family is present elsewhere in the genome. The nonautonomous controlling elements may be derived from autonomous controlling elements by the loss of *trans*-acting functions needed for transposition. The autonomous controlling elements of maize are *Ac* ('activator'; V9 copies/ genome), *MuI* (mutator; > 10 copies/genome), *Spm* ('suppressor mutator') and *Dd* ('dotted'). The nonautonomous controlling elements are *Ds* (dissociator, released to *Ac*), *dspm* (defective *Spm*; related to *Spm*) and *I* ('inhibitor'; related to *En*). The nonautonomous controlling elements possess the sites such as terminii, on which transpoases act. They also have the defective *cis*-acting functions which make the elements permanently stable (since they can no longer be activated to transpose).

2. Change of phase. Autonomously controlling elements may be reversibly inactivated during certain phases of plant development; this is called **'change of phase'**. The phenomenon of change of phase is heritable but relatively unstable so that a controlling element cycles between active and inactive states during development. The inactivation of both *Ac* and *Mu* seems to be due to cytosine **methylation** in the target sequence CAG/GTC. Both *Ac* and *Mu* elements have several such target sites in each element. There may be self - regulative controls of transposition; an increase in the frequency of its transposition.

Example of transposons of maize are the following:

1. En controlling element. The *En* element (*En* = enhancer) codes for a single enzyme called **transposase,** which is believed to induce its transposition. A deletion of the *En* element during coding for this enzyme results in the element *I* ('inhibitor'); *I* element is present in several copies in each maize line but it is nontransposable in the absence of *En* which acts as the signalling for *I*. A similar situation exists for the *Ac - Ds* system maize chromosome, it may generate three - base duplication, *e.g.*, ATA, ATA, etc., in the gene from which it moves out; produce an inversion for a segment of the gene; or even cause duplication of entire gene. In view of these genetical effects, transposable elements are often referred to as **genome reorganizers.**

2. *Ac-Ds* controlling elements. *Ac* (activator) controlling elements are 4,563 bp long, encode a single mRNA of 807 codons (which encode transposase enzyme) from a gene having

5 exons. They have 11 bp inverted repeats at their ends. *Ac* elements generate repeats of 8bp at the sites of their insertion. The nonautonomous counterparts, of *Ac* elements occur in a variety of forms, *e.g.*, *Ds1, Ds2d1, Ds2d2, Ds6, Ds9*, etc., all of which contain the same 11 bp inverted repeats at their ends. *Ds* (= dissociation) elements are shorter than *Ac*, and the length of deletion varies from merely 194 bp in *Ds9* to over 2.5 kb in *Ds6* (Fig. 18.15). Evidently, non-autonomous controlling elements (*Ds* elements) are derived from autonomous elements by deletions or other changes that inactivates their transposase enzyme coding regions, but the sites on which this enzyme acts remain intact. Members of *Ds1* family, however, are different their only resemblance to *Ac* is that they possess the same 11 bp terminal inverted repeats. *Ds1* controlling elements are short and were possibly not derived directly from *Ac* elements.

Like the bacterial composite transposon such as *Tn10*, transposition of *Ds* elements occur by a nonreplicative mechanism. The *recipient site* is frequently on the same chromosome as the *donor site* and often quite close to it. In them, transposition almost always occurs after DNA replication, and usually only of the two copies produced by replication undergoes transposition. The basis for this may be similar to that for *Tn10* and could involve asymmetrical **methylation sites** as C* GTG / G* CAC. After replication, the daughter DNA duplexes will have different methylated sequences, *viz.*, C* GTG / GCAC and CGTG / G* CAC; one of these sequences may be recognised by the transposase enzyme more efficiently than the other.

The donor site suffers a break, which could lead to a simple loss of the chromosome fragment distal to the centromere (hence lacking a centromere) or produce a breakage - fusion - bridge cycle. A loss of acentric fragment allows the expression of recessive allele present on the corresponding region of the homologous chromosome. The breakage - fusion - bridge cycle, however, continues through the future generations causing changes (deletion, duplication) in the affected chromosome.

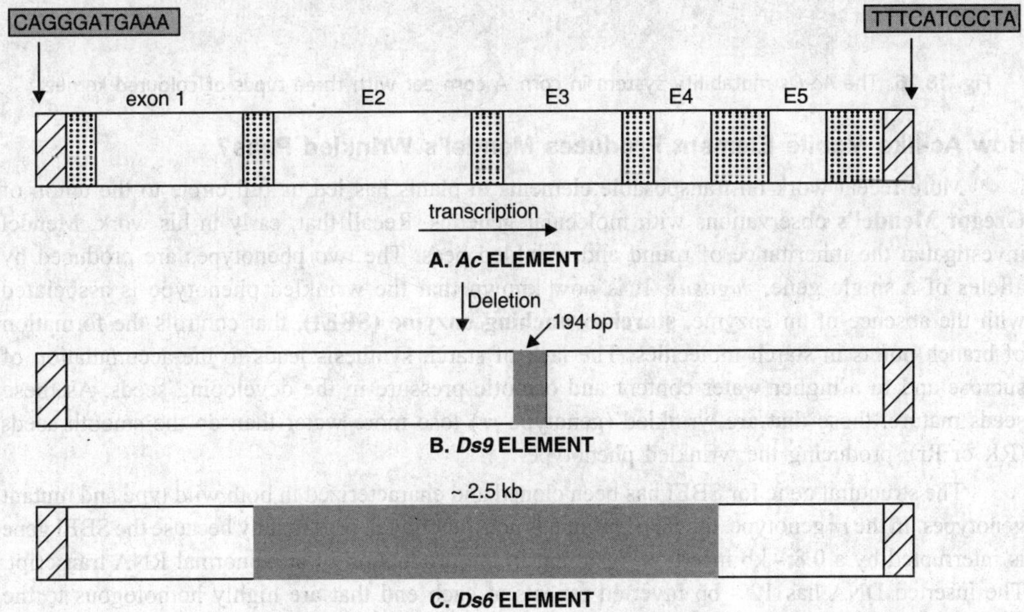

Fig. 18.15. Organization of maize *Ac* and *Ds* elements. A — an *Ac* element produces a single mRNA from a gene having 5 exons (depicted as Exon 1, E2, E3, E4 and E5) and ends in 11 bp inverted repeats. B— and C— Ds elements are usually derived by deletion of internal sequences from the *Ac* element; the region of deletion is depicted as solid black area).

How purple spots are produced in maize kernels?

Ds element affects the phenotype (in maize) by blocking expression of the genes it transposes into as well as causing the loss of alleles in acentric chromosome fragments lost when *Ds* breaks its chromosome. In some maizes or corns we can see *three* kinds of kernels in the ear of corn - *purple, bronze without purple spots* and *bronze with purple spots*. The purple kernels result from the presence of dominant functioning alleles providing enzymes in the pathway for purple pigment. In those kernels that are bronze without spots, *Ds* elements have transposed into both copies of the B_z^2 locus, disrupting the pigment pathway. Without the *Ac* element present, the *Ds* elements remain in place and the kernels are a uniform bronze colour. In those bronze kernels with purple spots, the *Ac* element has entered the genome in the genetic cross. In the presence of *Ac*, *Ds* leaves its site in some of the cells, restoring the activity to the B_z^2 locus. This restored activity results in purple spots in those cells and their progeny with the functioning B_z^2 allele (**Tamarin**, 1996). Purple spots difer in size based on time of excision of the *Ds* element during the development of the kernel: early yields large spots; late yields small spots.

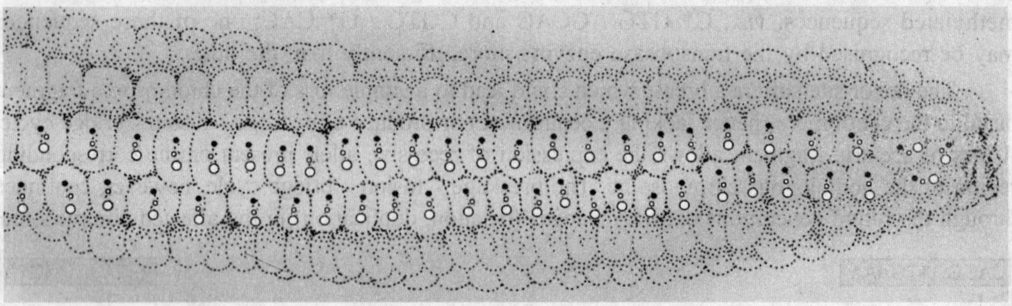

Fig. 18.16. The *Ac-Ds* mutability system in corn. A corn ear with three types of coloured kernels.

How Ac-like Mobile Element Produces Mendel's Wrinkled Peas?

More recent work on transposable elements in plants has led us full circle to the union of **Gregor Mendel's** observations with molecular genetics. Recall that, early in his work, Mendel investigated the inheritance of round and wrinkled peas. The two phenotypes are produced by alleles of a single gene, *rugosus*. It is now known that the wrinkled phenotype is associated with the absence of an enzyme, **starch-branching enzyme (SBEI),** that controls the formation of branch points in starch molecules. The lack of starch synthesis leads to the accumulation of sucrose and to a higher water content and osmotic pressure in the developing seeds. As these seeds mature, those that are wrinkled (genotype *rr*) lose more water than do the smooth seeds (RR or Rr), producing the wrinkled phenotype.

The structural gene for SBEI has been cloned and characterized in both wild type and mutant genotypes. In the *rr* genotype, the SBEI protein is non-functional, presumably because the SBEI gene is interrupted by a 0.8 - kb insertion, resulting in the production of an abnormal RNA transcript. The inserted DNA has 12 - bp inverted repeats at each end that are highly homologous to the terminal sequences in the transposable element *Ac* from maize and to other *Ac* - like elements from snapdragons and parsley. Terminal repeated sequences and the genetic information encoding a transposase enzyme appear to be universal components of transposons in all organisms studied.

3. *Spm-dSpm* controlling elements. The *Spm* (suppressor mutator) elements have 13bp inverted repeats at their termini and have two genes. Gene *tnpA* has 11 exons that are spliced to produce a 2,500 base mRNA, which encodes TnpA protein. Protein TnpA binds to a 12bp consensus regions of *Spm.* The action of protein TnpA is required for excision, but the excision of itself is effected by a hypothetical protein called TnpB. The protein TnpB is presumably coded by a 6000 base mRNA produce by alternative splicing of *tnpA* transcript; this mRNA contains the two open - reading frames, ORF - 1 and ORF - 2, both are present in the first large intron of gene *tnpA* (Fig. 18.17). ORF - 1 and ORF - 2 are present in an alternatively spliced RNA of 6,000 bases; this RNA is present at ~ 1% of *tnpA* mRNA. Protein TnB protein presumably binds the 13 bp terminal inverted repeats, and cleaves the termini for transcription.

A group of *Spm* derivatives are weakly autonomous; they are denoted as *Spm-w.* These controlling elements may show deletion in ORF - 1 and ORF - 2 regions; this indicates that *TnpB* function during transposition may be bypassed or substituted. The *dSpm* controlling elements show deletion of *tnpA* exon sequences and the length of deleted sequences varies greatly.

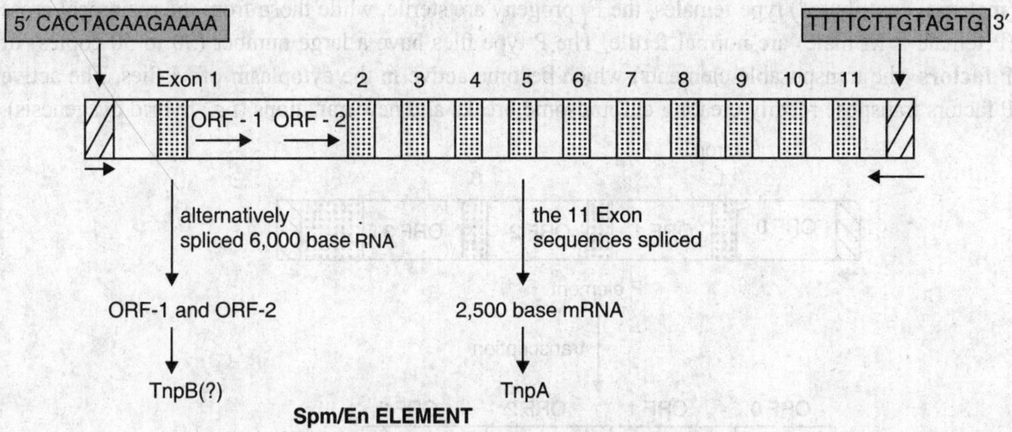

Fig. 18.17. The organization and function of *spm/En* controlling elements. The *Spm* and *En* elements differ at < 10 positions.

4. Dotted (*dt*) element. Dotted (*Dt*) element was discovered by **Marcus Rhoades** in 1938, when from a maize's ear of a pure breeding pigmented genotype of maize among kernels a dihybrid segregation ratio of 13: 3: 1 (*i.e.*, 13 pigmented: 3 dotted: 1 colourless) was obtained. This was surprising because no segregation was expected. The genetic analysis suggested that the original phenotype (full colour) was controlled by two genes with the genotype A_1, A_1, *dt dt* and that two mutations ($A_1 \rightarrow a_1$ and $dt \rightarrow Dt$) gave the genotype A_1 a_1 Dt *dt* which gave the following phenotypic and abbreviated genotypic ratio (Fig. 18.18): 12/16 pigmented (A_1Dt = 9/16; A_1dt = 3/16): 3/16 dotted (a_1Dt): 1/16 colourless (a_1 dt).

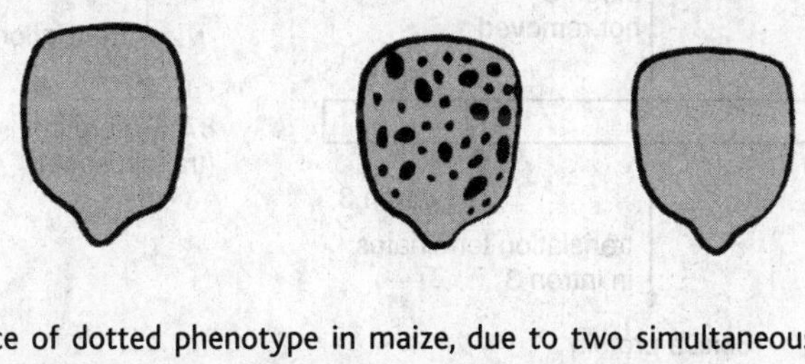

Fig. 18.18. Appearance of dotted phenotype in maize, due to two simultaneous mutations ($A_1 \rightarrow a_1$) and dt → *Dt*).

The dotted phenotype (appears in kernels and leaves of maize) in a_1Dt, would probably result from the reverse mutation ($a_1 \rightarrow A_1$) in some of the somatic cells, and the frequency of this mutation should be unusual high. This led to discovery of a_1 as an unstable mutant allele (its instability is dependent on *Dt*) showing high rate of reverse mutation. It was also shown that once reverse mutations have occurred, they are stable, reverted A_1 being transmitted to the progeny in a stable form. It means that *Dt* causes mutations in a_1 to A_1, which remain stable.

B. Transposons in *Drosophila*

The genome of *Drosophila melanogaster* has several transposable elements, *e.g.*, P factor.

P factor element. In *D. melanogaster,* F_1 hybrids derived from certain strains show a series of defects, including high mutation rate, high frequency of chromosomal aberrations, non - disjunctions and sterility. The phenomenon by which such abnormal hybrids are produced is called **hybrid dysgenesis.** For example when P (paternal contributing) type of males are mated to M (maternal contributed) type females, the F_1 progeny are sterile, while those from the reciprocal cross (P female × M male) are normal fertile. The P type flies have a large number (30 to 50 copies) of **P factors** (the transposable elements) which become active in the cytoplasm of M flies. The active P factors transpose readily creating chromosome breaks and new mutations (*i.e.,* hybrid dysgenesis).

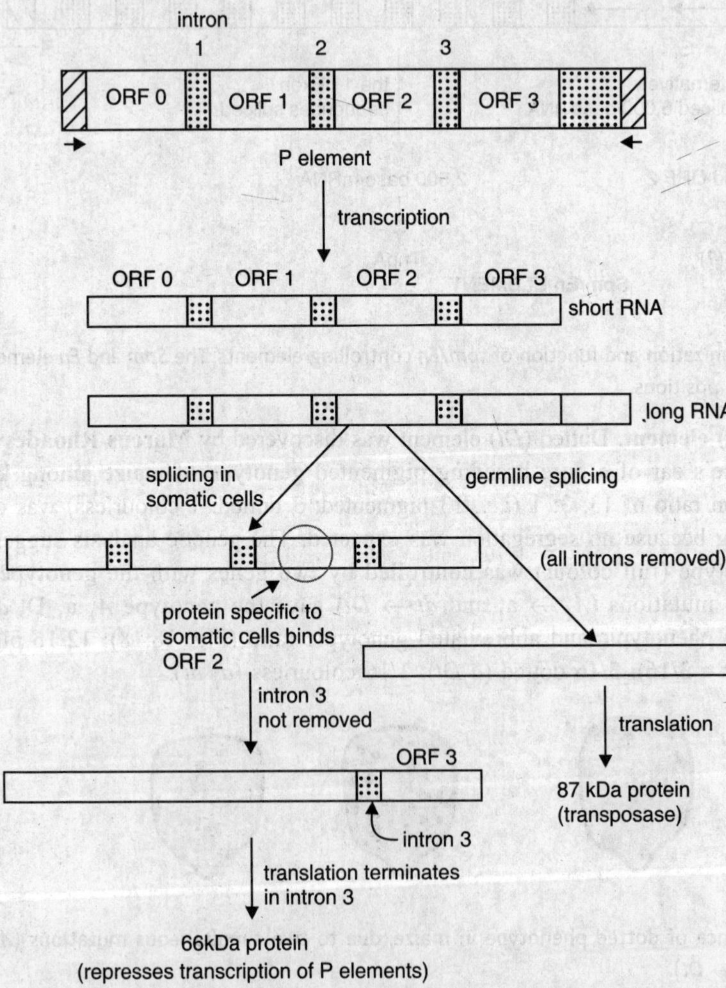

Fig. 18.19. Tissue - specific splicing of P element transcript leads to transposition in germline cells of F_1 individuals from crosses between P males and M females.

The P elements vary in length, the longest being ~ 2.9 kb. They all have inverted terminal repeats of 31 bp (Fig. 18.19) and generate direct repeats of 8 bp in target DNA. The full length P factors have 4 ORFs (= open reading frames), while their shorter types arise by internal deletions in them. The shorter P elements arise more frequently so that about 2/3 of the elements present in P males are of this type. Transposition of P elements is nonreplicative like that of *Tn10* (of *E. coli*), but the breaks in donor chromosomes are often repaired by using the corresponding sequence from homologous chromosomes. Transposition requires ~ 150 bp of terminal DNA and transposase enzyme binds to 10bp sequences that are adjacent to the 31bp inverted terminal repeats. Further, transposition of P elements is tissue - specific in that it occurs only in the germline cells; it begins at the stage when rapid cell division starts in the germ line. This tissue specificity occurs due to the tissue specific alternative splicing of P elements transcripts as follows:

1. Germline splicing. It is a defective splicing pattern in which the 3 introns separating the 4 ORFs are spliced out generating an mRNA that encodes 87 kDa transposase enzyme. This enzyme induces P element transposition (Fig. 18.19).

2. Somatic splicing. It is induced by a protein, present in somatic cells of *Drosophila.* This protein binds to sequences in ORF-2 and prevents the removal of 3rd intron, which has a termination codon. The mRNA generated by this splicing produces a 66 kDa protein that does not contain information from ORF-3. This protein represses transcription of P elements, thus, it stops transposition and forms the basis for reciprocal differences seen in P × M crosses of *Drosophila* (Fig. 18.19).

P cytotype and M cytotype. *Drosophila* of P strain contain P elements in their genome (*P chromosomes*) and their cytoplasm has the 66 kDa protein; the cytoplasm of such flies is called **P cytoplasm** and they are said to have **P cytotype.** In contrast, flies belonging to M strain do not have P elements in their chromosomes (*M chromosomes*) and lack the 66 kDa protein in their cytoplasm. These flies are said to have **M cytoplasm** that gives rise to the **M cytotype.** The 66 kDa protein is a highly stable protein and is evidently present in very large quantities in the eggs so that the P cytotype effect on transposition of P elements may continue for more than one generation.

Thus, when P males are mated to P females, or M males are mated to P females, the eggs contain the 66 kDa repressor protein. The transcription of P elements, therefore, remains suppressed and there is no transposition either in somatic or germline cells. But when M females are crossed to P males, the eggs lack the 66 kDa protein. P elements are, therefore, transcribed to produce the 66 kDa protein in somatic cells which suppresses further transcription of P elements. But in germline cells, defective splicing of P element transcripts leads to the production of transposase; this helps in P element transposition and hybrid dysgenesis.

The P element is highly invasive when placed into a new population. The source of P element seems to be another species of *Drosophila.* Since hybrid dysgenesis reduces inbreeding, it helps in specification by creating breeding barriers. Thus, multiple hybrid dysgenesis system may develop in different population; this would lead to inability to intermate and therefore to specification.

18.3 EXAMPLES OF RETROPOSONS (RETROELEMENTS)

1. Copia elements in *Drosophila*. The *copia* elements are 5000 bp long, have direct terminal repeats of 276 bp, they are always intact, sometimes they occur in circular form and they have a single long (4,227 bp) reading frame. *Copia* is present in 20 – 60 copies per genome depending on the strain of the fly and is represented by a large number of closely related sequences (the copia family; Table 18.6). These sequences code for about abundant poly $(A)^+$ mRNAs. The name *Copia* is based on the presence of this sequence in copius amounts in cellular RNA of *Drosophila.*

In cultured cells, the number of *copia* elements per genome increases upto 2 to 3 times the number found *in vivo.* In cultured cells, the rate of transposition increases to 10^{-3} to 10^{-4} events per generation. A direct repeat of 5bp of target DNA is generated by insertion of a *Copia* element. Each of the terminal direct repeats of *Copia* elements are identical, indicating their relationship with retroposons. The single *Copia* ORF has homologies with *gag* and *pol* sequences of retroviruses.

Table 18.6 Properties of some families of Copia like elements

S. No.	Family	Number/ genome	Length (kb)	Length of LTR (bp)	Length of inverted terminal repeat (bp)
1.	Copia	60	5.0	276	13/17
2.	412	40	7.5	481	8/10
3.	297	30	7.0	412/415	3
4.	*mdg*1	25	7.2	442/444	13/16
5.	*mdg*3	15	5.5	283/268	15/18
6.	B 104	80	8.7	429	3
7.	17.6	40	6.9	512	0
8.	*mdg*4/gypsy	10	7.3	479	415

The mRNAs from *Copia* sequences are translated into a variety of proteins *in vitro,* ranging from 18 to 51 kDa, whose functions remain unknown. The *Copia* elements must have originated from retroviruses and they are able to transpose (but the mode of transposition is unknown). Integration of *Copia* element produces mutations, a classical example being white-apricot (w^a) mutation for eye colour caused by insertion of a copia-like element.

2. FB elements of *Drosophila*. The fold - back or FB elements range from 500 to 5,000 bp (base pair) and occur in about 30 copies per genome. They carry long (250 to 1,250 bp) inverted repeats at the termini. The repeats may be adjoining (FB1, FB5) or may be separated (FB3, FB4, FB7, FB8, FB9), but can always fold back to form stem and loop structures. FB elements cause mutations by insertion or by their effect on gene expression (due to their presence near a control region). FB elements are also found flanking another class of transposon called *TE* (*e.g., TE1, TE28, TE77, TE98*). TE1 was first identified due to transposition of white eye locus from X-chromosome to second chromosome.

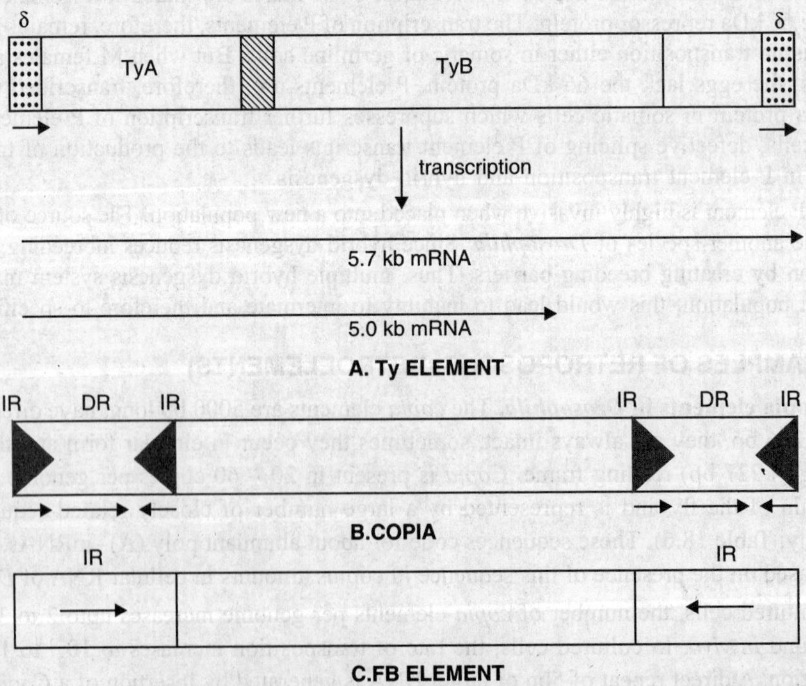

Fig. 18.20. A — T_y element of yeast; B — *Copia* element of *D. melanogaster* and C — *FB* element of *D. melanogaster.* IR, inverted repeat; DR, direct repeat.

3. T_y elements in yeast. Yeast T_y (= transposon *y*east) elements are an example of retroposons. T_y elements are very divergent. There are two main types of T_y elements, *viz.*, *T_y1* and *T_y917*. They transpose at a frequency of ~ 10^{-7} to 10^{-8} through an RNA intermediate. This RNA is reversed transcribed to yield a DNA duplex that becomes inserted into the yeast genome generating a 5bp long direct repeats in yeast DNA. Such a transposition causes mutation in target gene due to insertion. Transpositions of T_y elements is controlled by genes within the T_y elements; they form noninfective virus - like particles, which accumulate within such cells in which transposition has been induced. Only some of the yeast T_y elements are active, while most of them have lost the ability to transpose.

Each T_y element is 6.3 kb long and has 330 bp direct repeats at its ends (no inverted repeats or IRs are found in bacteria); these repeats are called **delta (δ) repeats.** All T_y elements share δ repeats, a long region at the left end, another region in the centre and short region adjacent to the right δ repeat (Fig. 18.20). In a typical yeast genome, there are ~ 30 copies of T_y1 type and ~ 6 copies of the T_y 917 type element; in addition, there are ~ 100 copies of independent δ elements; these are called **solo deltas.** The sequences of δ elements very greatly, but the two δ elements of a single T_y element are identical. Thus, δ sequences seem to be similar to the LTR sequences of retroviruses. Although, the two δ sequences of a T_y element are identical, yet a *promoter* is active in the δ repeat at the right end; this feature is also found in the retroviruses. The T_y elements can be excised by homologous recombination between the δ sequences, leaving behind a solo δ sequence in the host DNA.

The T_y elements are transcribed into two species of poly $(A)^+$ mRNAs of which one is 5 kb long and the other is 5.7 kb long. These mRNAs form > 75% of the total mRNA of a haploid yeast cell. Transcription for both mRNA species begins ~ 95 bp from the left end of T_y element. Further, the T_y elements have two open - reading frames (ORFs), which are read in different phases and overlapped by 13 codons. One ORF is *T_yA* which encodes a DNA - binding protein. The other ORF is *T_yB* which contains regions that have homologies with reverse transcriptase, protease and integrase sequences of retroviruses. The organization and functions of *T_yA* and *T_yB* are analogous to the behaviour of retroviral *gag* and *pol* functions. Thus, *T_yB* is expressed only when a frameshift event allows the termination codon at the end of *T_yA* to be bypassed; as a result, *T_yB* product is fused with that of *T_yA*.

Moreover, the T_y elements comprise regions of homology within and between chromosomes. Recombination between T_y elements usually damages the chromosome either by deletion, inversion or other more effective structural changes in chromosomes, *e.g.*, dicentric chromosomes. Recombination between T_y elements seem to occur in bursts and gene conversion occurs between T_y elements present at different locations. Quite interestingly, although both the δ repeats of an active T_y element have th same sequence, a promoter is active in one δ elements, while a terminator is functional in the other. The T_y element behaves like a retrovirus that has lost its *env* gene so that it is unable to properly package its genome into virus particles.

4. Retrotransposons in plants. Retrotransposons are ubiquitous in plants. They make upto 50% to 80% of the genomic DNA (*e.g.*, in maize), thus, playing a major role in gene and genome evolution in plants. For example, LTR – retrotransposons are 100 bp to 13.5 kbp in size (Table 18.7) and possess the following coding sequences: 1. *gag,* encoding proteins involved in maturation and packaging of RNA, and proteins necessary for integration of retrotransposons into the genome; (*ii*) *pol,* encoding **reverse transcriptase (RT)** and **Rnase H** activities, needed for replication/transposition to a new position on a chromosome. The *T_y1-copia* and *T_y3-gypsy* types of LTR-retrotransposons differ in order of the above coding sequences, particularly the *int* and *Rnase* sequences. Among non-LTR retrotransposons, LINEs, though simpler and smaller in size (1000 bp to 6 kbp), still encode many of the proteins encoded by the LTR retrotransposons. For instance, they have *gag* and *pol,* but not *int* and instead have endonuclease (*en*). SINEs do not encode any of these genes and are derived from RNA polymerase III products such as tRNA. In plant's chromosome, the retrotransposons are either dispersed on all chromosomes or can be clustered in some regions such as telomeres, centromeres, pericentromeric regions and also in the interstitial heterochromatic regions. An individual retrotransposon can be present as a single copy or it may be represented by upto as many as 50,000 copies per genome. There can be dozens of different types of retrotransposons within a species (*e.g.*, maize). Some of known plant retrotransposons have been listed in Table 18.7.

Table 18.7 Some selected retrotransposons in plants (Source: **Gupta**, 2004)

Types of retrotransposons	*Size (in kb)*	*Number of copies*	*Plant / Crop*
I. LTR retrotransposons			
1. T_y–1 Copia group			
(*i*) BARE - 1	12.1	> 50,000	Barley
(*ii*) Opie - 1	8.7	> 30,000	Maize
(*iii*) PREM - 2	9.5	> 10,000	Maize
(*iv*) SIRE - 1	11.0	> 7,200	Maize
(*v*) Tnt - 1	5.3	> 100	Maize
2. T_y– gypsy group			
(1) Cerebra	~ 10.0	1,500	Barley
(2) Cinful - 1	8.6	20,000	Maize
(3) Grand - 1	13.5	> 1,300	Maize
(4) Zeon - 1	7.3	> 1,250	Maize
(5) Cyclops - 2	12.5	1,000	Pea
(6) RIRE3	~ 10.5	10	Rice
II. Non-LTR retrotransposons			
1. *LINES*			
(*i*) Cin 4	1 – 6.8	50 – 100	Maize
(*ii*) Del 2	18.5	25,000	*Lilium*
2. *SINEs*			
(*i*) p - SINE1	0.125	> 100	Rice
(*ii*) $S1_{Bn}$	0.170	500	*Brassica napus*

5. Tam 1 element in Snapdragon (*Antirrhinum majus*)

In snapdragon, Tam1 element is responsible for pattern of colour seen on petals of some variegated strains and is 17 kb long. *Tam1* element is found in an unstable mutant, *Nivea recurrens,* where *Tam1* is present in the region of *chalcone synthase* gene and causes 3bp target site duplication. *Tam1* element also has 13bp inverted repeats at its termini, which are identical to those in Spm18 of maize. Other transposable elements of this series are numbered *Tam2* to *Tam10.* Putative transposable elements are also suggested to be present in soya bean.

6. Interspersed Repeats and Transposable Elements of Human Genome

The draft sequences have shown that approximately 62% of the human genome comprises **intergenic regions,** the parts of the genome that lie between genes and which have no known function. These sequences used to be called **junk DNA** but the term is falling out of favour of modern molecular biologists. Recently, bulk of the intergenic DNA is found to be made up of repeated sequences of one type or another: 1. **genome wide** or **interspersed repeats,** whose individual repeat units are distributed around the genome in an apparently random fashion; 2. **tandemly repeated DNA,** whose repeat units are placed next to each other in an array (Fig. 18.21).

Interspersed repeats include four types of mobile elements: SINEs, LINEs, LTR elements and DNA transposons. An interesting feature of these interspersed repeats is that each type appears to be derived from a **transposable element**, a mobile segment of DNA which is able to move around the genome from one place to another. Many of these mobile elements leave copies of themselves when they move, which explains how they propagate and become common throughout the genome. There are two main classes of transposable elements: those that transpose via an RNA intermediate and those that do not. LINEs, SINEs and LTR elements are examples of the first class and DNA transposons are examples of the second class. Characteristic features, types and subtypes of different genome - wide repeats have been tabulated in Table 18.8.

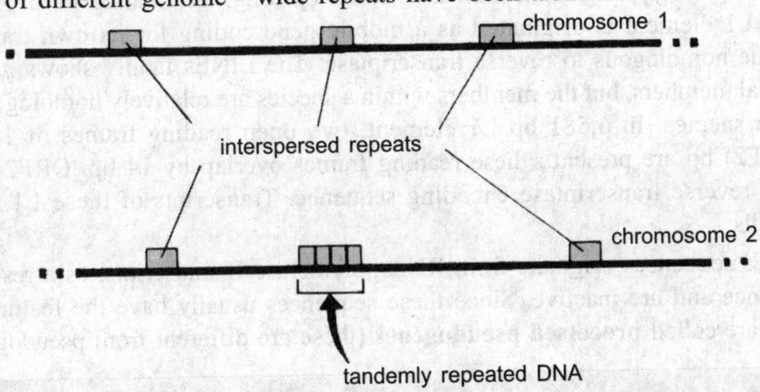

Fig. 18.21. The two types of repetitive DNA: interspersed repeats and tandemly repeated DNA.

Just for example, SINEs form most numerous genome - wide repeats of human genome. They comprise three subtypes: *Alu* element (approximately 1,090,000 copies in the genome), *MIR* (393,000 copies) and *MIR3* (75,000). Altogether genome wide repeats make up 44% of the draft genome sequences, some 1400 Mb of DNA (**Brown**, 2002).

Table 18.8 The types of genome - wide repeats in the human genome (Source: **Brown**, 2002).

S. No.	*Type of repeat*	*Subtype*	*Approximate number of copies in the human genome*
1.	SINEs		**1,558,000**
		1. Alu	1,090,000
		2. MIR	393,000
		3. MIR3	75,000
2.	LINEs		**868,000**
		1. LINE-1	516,000
		2. LINE-2	315,000
		3. LINE-3	37,000
3.	LTR elements		**443,000**
		1. ERV class I	112,000
		2. ERV (k) class II	8,000
		3. ERV (L) class III	83,000
		4. MaLR	240,000
4.	DNA transposons		**294,000**
		1. hAT	195,000
		2. Tc-1	75,000
		3. PiggyBac	2000
		4. Unclassified	22,000

7. Mammalian Retroposons

Mammals contain a variety of retroposons; most of these belong to the following two families:

1. LINEs (Long Interspersed Elements). They are derived from transcripts of RNA polymerase II.

2. SINEs (Short Interspersed Elements). They are derived from transcripts of RNA polymerase III.

L1 retroposon is a good example of LINEs sequence. It is found in 20 to 50 thousand copies per genome. A typical L1 element is ~ 6,500 bp long, terminating in an A-rich region. It appears that L1 element is originated as a mobile gene coding for its own transposition as its reading frame homologous to reverse transcriptase. The LINEs family shows great variation among individual members, but the members within a species are relatively homologous compared to those within species. In 6,581 bp L1 element, two open reading frames of 1,137 (ORF1) and 3,900 (ORF2) bp are present.; these reading frames overlap by 14 bp. ORF2 has a region homologous to reverse transcriptase encoding sequence. Transcripts of these L1 sequence are found in the cells.

The LINEs sequences originate from RNA polymerase II transcripts; as a result, they lack promoter sequence and are inactive. Since these sequences usually have the features of mature transcript, they are called **processed pseudogenes** (these are different from pseudogenes). These

Box 18.6

A **processed pseudogenes** is a pseudogene that results from integration into the genome of a reverse transcribed copy of mRNA (**Brown**, 2002).

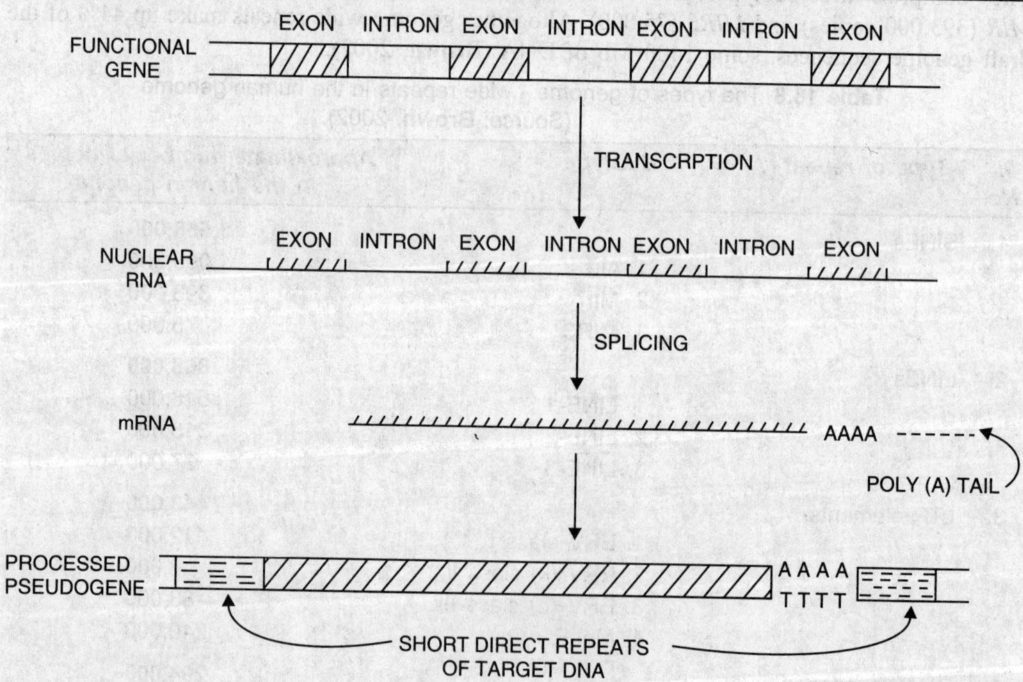

Fig. 18.22. Mode of origin and features of processed pseudogenes (*i.e.*, LINEs sequences). They have 3′ AT sequence, their 5′-end corresponds exactly with mRNA and they are flanked by short direct repeats.

pseudogenes have their 5′ ends having exact similarity with mRNA, many of them have exactly joined exon sequences and they usually end in a short 3′ stretch of AT base pairs. This 3′ stretch is probably derived from the poly (A) tail of the mRNA (Fig. 18.22). Presumably these pseudogenes have been originated from a mobile gene ecncoding its own transposition. It indicates that L1 elements belong to the viral superfamily of retroposons. The processed pseudogenes are flanked on either side by a short direct repeat, each of which is presumed to have been generated by a transposon - like insertion event.

The ***Alu* family** is an good example of SINEs sequence of human genome. These sequences are so named because they all are cleaved at a single site located at 170bp by the restriction enzyme *Alul*. The sequence of *Alu family* are ~ 300 bp long, occur in ~ 300,000 copies per human genome and are widely dispersed in her genome. The human *Alu* family seems to have originated by a 130bp tandom duplication with an unrelated sequence of 31bp inserted into the right half of the dimer.

The *Alu* sequence is related to 7SL RNA, which is a component of the signal recognition particle. The 5′ terminal 90 bases of 7SL RNA are homologous to the left end of the left half (130 bp) of an *Alu* element, while the 40 bases at the 3′-end of 7 SL RNA are homologous to the right end of the left half of *Alu* sequence; the central 160 bases of 7 SL RNA are absent from the *Alu* sequence. The gens for 7 SL RNA are transcribed by RNA polymerase III, and the SINEs are derived from transcripts of this RNA polymerase III. It is likely that the *Alu* sequences originated from the transcripts of 7 SL RNA genes by a deletion of the central 160 bases of this RNA.

The presence of multiple copies of *Alu* sequence in a single RNA molecule can produce secondary structure in the RNA. The presence of *Alu* sequences in the form of inverted repeats seems to be responsible for most of the secondary structure found in mammalian nuclear RNAs.

Members of *Alu* family resemble transposons in being flanked by short repeats of target DNA; the lengths of these DNA repeats are different for individual members of the family. *Alu* sequences have a 14 bp region that is almost identical with a sequence present at the origin of replication of papova virus such as SV40 and hepatitis B virus.

Other examples of SINEs sequence include ***B1 family*** which occurs in mice and ***Alu - equivalent family*** of chinese hamster.

QUESTIONS

Long Answer Questions

1. What a plasmid is and provide examples of the different types of plasmid that are known.
2. Contrast the various transposable genetic elements in bacteria, maize, *Drosophila,* and humans. What properties do they share?

Short Answer Questions

1. Write short notes on the following:
   (*i*) Starch-branching enzyme of pea.
   (*ii*) Transposons of *Drosophila.*
   (*iii*) Copia elements.
   (*iv*) Ty elements of yeast.
   (*v*) Transposable elements in human genome.

Very Short Answer Questions

1. The initial discovery of IS units in bacteria involved their presence upstream (5′) to three genes controlling galactose metabolism. All three genes were affected despite not being contiguous with the IS unit. Offer an explanation as to why this occurs.

2. In maize, the recessive allele *bz* produces a lighter colour in the aleurone than does the dominant allele, B_z. Ears on a homozygous b_z/b_z plant were fertilised by pollen from a homozygous B_z/B_z plant. The resulting cobs contained kernels that were uniformly dark except for a few on which light spots occurred. Suggest an explanation.
3. Sometimes solo copies of the LTR of a retrotransposon, such as *gypsy,* are found in *Drosophila* chromosomes. How might these solo LTRs originate?
4. *Escherichia coli* cells that are resistant to kanamycin are infected with bacteriophage lambda. Phage particles that were purified from the lysed cells were used to infect a kanamycin sensitive strain called C107 and cells that were lysogenised were picked from turbid plaques. These cells were subsequently plated on a medium containing kanamycin and several kanamycin resistant colonies appeared. Suggest an explanation for the transfer of kanamycin resistance to the C107 strain.

Fill in the Blanks

1. The term plasmid was originally used by .... in 1952.
2. In bacterial cells, the antibiotic-resistance genes are often located on ....
3. Fertility plasmids of *E.coli* are also called .... or .... plasmids.
4. The term .... was coined by R.W. Hedges and .... in 1974.
5. Transoposon is a .... which could move from one DNA molecule or chromosome to another DNA.
6. .... are transposable genetic elements found in eukaryotic DNA that move through the reverse transcription of an RNA intermediate.
7. .... is a good example of LINEs sequences of mammals.
8. .... enzyme acts at the site of inverted repeats for recognizing and mobilising the IS (.................. ....) elements.
9. In LTR-transposon, LTR stands for ....

Multiple Choice Questions

1. Certain genetic elements that regularly "jump" to new locations, often inactivating the gene into which they become inserted, are called

   (*a*) episomes (*b*) heterochromatin
   (*c*) transposons (*d*) overlapping genes
2. The jumping genes in maize were discovered by

   (*a*) Beadle and Tatum (*b*) Barbara McClintok
   (*c*) H. G. Khorana (*d*) T. H. Morgan
3. Genes which confer antibiotic resistance on bacteria are located on

   (*a*) chromosomal DNA (*b*) plasmid
   (*c*) RNA (*d*) polysome
4. Plasmids found in cells of bacteria are molecules of

   (*a*) DNA (*b*) RNA
   (*c*) proteins (*d*) none of these
5. Which of the following is called resistance transfer factor?

   (*a*) Col - factor (*b*) R - factor
   (*c*) F - factor (*d*) none of these
6. Plasmid is a

   (*a*) double - stranded circular DNA (*b*) single - stranded DNA
   (*c*) extrachromosomal linear DNA (*d*) none of the above

ANSWERS

Very Short Answer Questions

1. It is probable that the IS occupied or interrupted normal function of a controlling regions related to the galactose genes.
2. Transposon - induced mutation or loss of a B_z allele in the endosperm.
3. Through crossing - over between the LTRs of a gypsy element.
4. A transposon carrying the kanamycin resistance gene transposed from the original cells to some of the infected phage and subsequently conferred resistance on some of the lysogenised C107 cell, produced by the second infection.

Fill in the Blanks

1. J. Lederberg.
2. Plasmids.
3. Cojugative, transmissible.
4. Transposon, A.E. Jacob.
5. DNA segment.
6. Retrotransposons.
7. L1 retroposon.
8. Transposase, insertion sequence.
9. Long, terminal repeats.

Multiple Choice Questions

1. (*c*) **2.** (*b*) **3.** (*b*) **4.** (*a*) **5.** (*b*) **6.** (*a*).

INDEX

D

E

N

O

P

Q

R

S

T

U

V

W

X

Y

Z